Climate Change and the Law

AF699084

IUS GENTIUM

COMPARATIVE PERSPECTIVES ON LAW AND JUSTICE

VOLUME 21

Series Editors

Mortimer N.S. Sellers
University of Baltimore

James Maxeiner
University of Baltimore

Board of Editors

Myroslava Antonovych, *Kyiv-Mohyla Academy*
Nadia de Araújo, *Pontifical Catholic University of Rio de Janeiro*
Jasna Bakšic-Muftic, *University of Sarajevo*
David L. Carey Miller, *University of Aberdeen*
Loussia P. Musse Félix, *University of Brasilia*
Emanuel Gross, *University of Haifa*
James E. Hickey Jr., *Hofstra University*
Jan Klabbers, *University of Helsinki*
Cláudia Lima Marques, *Federal University of Rio Grande do Sul*
Aniceto Masferrer, *Univeristy of Valencia*
Eric Millard, *West Paris University*
Gabriël Moens, *Murdoch University*
Raul C. Pangalangan, *University of the Philippines*
Ricardo Leite Pinto, *Lusíada University of Lisbon*
Mizanur Rahman, *University of Dhaka*
Keita Sato, *Chuo University*
Poonam Saxena, *University of Delhi*
Gerry Simpson, *London School of Economics*
Eduard Somers, *University of Ghent*
Xinqiang Sun, *Shandong University*
Tadeusz Tomaszewski, *Warsaw University*
Jaap de Zwaan, *Erasmus University Rotterdam*

For further volumes:
http://www.springer.com/series/7888

Erkki J. Hollo • Kati Kulovesi • Michael Mehling
Editors

Climate Change and the Law

LAW LIBRARY
JUN 0 5 2013
GOLDEN GATE UNIVERSITY

Springer

Editors
Erkki J. Hollo
Faculty of Law
University of Helsinki
Helsinki, Finland

Kati Kulovesi
Department of Law
University of Eastern Finland
Joensuu, Finland

Michael Mehling
Ecologic Institute
Washington, DC, USA

ISBN 978-94-007-5439-3 ISBN 978-94-007-5440-9 (eBook)
DOI 10.1007/978-94-007-5440-9
Springer Dordrecht Heidelberg New York London

Library of Congress Control Number: 2012953930

© Springer Science+Business Media Dordrecht 2013
This work is subject to copyright. All rights are reserved by the Publisher, whether the whole or part of the material is concerned, specifically the rights of translation, reprinting, reuse of illustrations, recitation, broadcasting, reproduction on microfilms or in any other physical way, and transmission or information storage and retrieval, electronic adaptation, computer software, or by similar or dissimilar methodology now known or hereafter developed. Exempted from this legal reservation are brief excerpts in connection with reviews or scholarly analysis or material supplied specifically for the purpose of being entered and executed on a computer system, for exclusive use by the purchaser of the work. Duplication of this publication or parts thereof is permitted only under the provisions of the Copyright Law of the Publisher's location, in its current version, and permission for use must always be obtained from Springer. Permissions for use may be obtained through RightsLink at the Copyright Clearance Center. Violations are liable to prosecution under the respective Copyright Law.
The use of general descriptive names, registered names, trademarks, service marks, etc. in this publication does not imply, even in the absence of a specific statement, that such names are exempt from the relevant protective laws and regulations and therefore free for general use.
While the advice and information in this book are believed to be true and accurate at the date of publication, neither the authors nor the editors nor the publisher can accept any legal responsibility for any errors or omissions that may be made. The publisher makes no warranty, express or implied, with respect to the material contained herein.

Printed on acid-free paper

Springer is part of Springer Science+Business Media (www.springer.com)

Foreword

In a remarkably short span of time, climate change has become deeply embedded in important areas of the legal system. Greenhouse gas mitigation is now a major focus of environmental and energy legislation in a number of jurisdictions around the world. Meanwhile, adaptation to the negative impacts of climate change is having a growing effect on the normative systems governing land use and agriculture, water resources management, health policy, and other fields.

As a global challenge calling for collective solutions, climate change has elicited substantial rule-making at the international level, which in turn percolates through the broader legal system to the regional, domestic and local planes. But addressing climate change through law is not only a question of multiple governance levels. Because of its cross-cutting nature, climate change can lead to spillover effects; some of the some of the issues recently discussed by legal scholars include overlaps between rules addressing climate change and other subject matters, such as international trade or biodiversity conservation.

Both domestically and internationally, moreover, the governance of climate change has seen a lively exploration of new regulatory philosophies, harnessing innovative and flexible instruments. It has also witnessed the active involvement of non-state actors, and in many ways stretches the conceptual boundaries of traditional jurisprudence. Given this proliferation of relevant norms and institutions, is it premature to assume the emergence of a new area of law, 'climate change law'? If its existence can indeed be affirmed, what common principles, objectives and other shared categories define it?

Climate Change and the Law is the first monograph to systematically and comprehensively address these doctrinal questions. It assembles several of the most recognized experts in the field to identify relevant trends and common themes from a comparative perspective. The editors would like to thank Kaisa Huhta and Gordon Bradshaw for their valuable research assistance during the editorial process.

June 2012

Erkki J. Hollo
Kati Kulovesi
Michael Mehling

Preface

Humans everywhere put their trust in certainty, consistency, a sense of justice and fairness. Once they trust, they gain a greater confidence to act in every walk of life. Broader, faster action at all levels of government, business and society is required to respond to the long-term challenge of climate change. In this respect, law and the legal process are critical elements to any policy response to climate change and its adverse effects. *Climate Change and the Law* provides a helpful review of the emergence of a new discipline, its core principles and legal techniques, and its relationship and potential interaction with other disciplines. It is particularly timely because, in Durban, South Africa, at the 2011 UN Climate Conference, governments of the world agreed to craft a universal climate agreement with legal force, to be adopted in 2015 and to come into force no later than 2020.

As an instrument for policy implementation, law provides a normative and institutional framework for managing and responding to climate change. It translates policy precepts into binding legal norms. A legal regime may provide for the establishment of legally binding targets to limit and reduce greenhouse gas emissions and sanction non-compliance. It can thereby channel human behaviour along pathways consistent with a low emissions economy. Institutional processes may be established and vested with power and authority to carry-out assessments, monitor trends and compliance, incentivize action, and enforce legal requirements. Legal processes are critical in addressing social crises and disputes arising from the impacts of climate change. At the global level, international law provides a framework for cooperation amongst states in responding to a global problem that requires a multilateral response. This book critically examines the current international legal framework and undertakes a comparative legal survey of national climate law, bringing together views from a broad array of perspectives, and a diverse group of authors. This work will undoubtedly be a valuable reference for the on-going global efforts to construct the post-2020 international climate change regime.

Christiana Figueres

Executive Secretary
United Nations Framework
Convention on Climate Change

Contents

1 **Introduction: Climate Change and the Law** 1
Kati Kulovesi, Michael Mehling, and Erkki J. Hollo

Part I Climate Law as an Emerging Discipline

2 **Implementing Climate Governance: Instrument Choice and Interaction** 11
Michael Mehling

3 **Exploring the Landscape of Climate Law and Scholarship: Two Emerging Trends** 31
Kati Kulovesi

4 **Climate Change and Justice: Perspectives of Legal Theory** 63
Felix Ekardt

Part II International Climate Law – Architecture and Institutions

5 **Foundations of International Climate Law: Objectives, Principles and Methods** 83
Rowena Maguire

6 **Alternative Venues of Climate Cooperation: An Institutional Perspective** 111
Camilla Bausch and Michael Mehling

7 **Analyzing Soft Law and Hard Law in Climate Change** 143
Antto Vihma

8 **Compliance and Enforcement in the Climate Change Regime** 165
Meinhard Doelle

Part III International Climate Law – Cross-Cutting Issues

9 **The New Framework for Climate Finance Under the United Nations Framework Convention on Climate Change: A Breakthrough or an Empty Promise?** ... 191
Yulia Yamineva and Kati Kulovesi

10 **Climate Justice: The Clean Development Mechanism as a Case Study** ... 225
Tomilola Eni-ibukun

11 **Legal Aspects of Climate Change Adaptation** ... 257
Jonathan Verschuuren

12 **Climate Change and Human Rights** ... 287
Timo Koivurova, Sébastien Duyck, and Leena Heinämäki

Part IV International Climate Law – Sectoral Issues

13 **Managing the Fragmentation of International Climate Law** ... 329
Harro van Asselt

14 **No Need to Reinvent the Wheel for a Human Rights-Based Approach to Tackling Climate Change: The Contribution of International Biodiversity Law** ... 359
Elisa Morgera

15 **The Role of REDD in the Harmonisation of Overlapping International Obligations** ... 391
Annalisa Savaresi

16 **Climate Change and Trade: At the Intersection of Two International Legal Regimes** ... 419
Kati Kulovesi

17 **Climate Law and Geoengineering** ... 447
Ralph Bodle

Part V Comparative Climate Law

18 **Climate Law in the United States: Facing Structural and Procedural Limitations** ... 473
Michael Mehling and David John Frenkil

19 **Canada and the Kyoto Protocol: An Aesop Fable** ... 489
Jane Matthews Glenn and José Otero

20 Climate Law and Policy in the European Union: Accidental Success or Deliberate Leadership? 509
Michael Mehling, Kati Kulovesi, and Javier de Cendra

21 Climate Law in Germany 523
Felix Ekardt

22 Climate Law in the United Kingdom 537
Colin T. Reid

23 Climate Law and Policy in Russia: A Peasant Needs Thunder to Cross Himself and Wonder 551
Yulia Yamineva

24 Australia: From "No Regrets" to a Clean Energy Future? 567
Sharon Mascher and David Hodgkinson

25 Climate Law and Policy in Japan 585
Hitomi Kimura

26 Sustainable Development and Climate Policy and Law in China 597
Christopher Tung

27 India's Evolving Climate Change Strategy 605
Patodia Rastogi

28 Climate Change Responses in South Africa 619
Michael Kidd and Ed Couzens

29 Climate Change Policy and Legislation in Brazil 639
Haroldo Machado-Filho

30 Climate Law in Latin American Countries 653
Soledad Aguilar and Eugenia Recio

Index 679

Contributors

Soledad Aguilar holds an LL.M from the London School of Economics and Political Science. She specializes in international environmental law and policy and was a former negotiator for the Government of Argentina on climate change and biodiversity. She currently works for international organizations and academic institutions in the assessment and design of national and international climate change regulations.

Camilla Bausch works in the Senior Management and as Senior Fellow at the Ecologic Institute, an internationally recognized think tank. She has been part of the German delegation to the UN climate change negotiations for many years. She works on projects dealing with the future of the UN climate regime, emissions trading, international cooperation in climate protection and the improvement of the legal framework for the energy sector. Dr. Bausch is also Associate Editor of the journal Carbon & Climate Law Review, a quarterly journal on climate regulation and the carbon market.

Ralph Bodle is Senior Fellow and Coordinator of Ecologic Legal at the Ecologic Institute, Berlin, where he works in and writes on international and European environmental law and policy. He is a member of the German delegation to the UN climate negotiations, working on the post-2012 climate regime and legal aspects of climate finance. He holds a Ph.D. in law from Humboldt University, Berlin, and an LL.M. from the University College London, University of London. Before joining Ecologic Institute, he was a practising lawyer at law firms advising and representing the public sector on environmental and public law.

Ed Couzens holds the degrees of PhD from the University of KwaZulu-Natal; LLM Environmental Law from the Universities of Natal and Nottingham; and BA Hons LLB from the University of the Witwatersrand. He is currently an Associate Professor in the School of Law, University of KwaZulu-Natal, Durban. He is an admitted attorney of the High Court, RSA. He has been since 2006 a co-editor of the University of Eastern Finland/United Nationa Environment Programme Review of

International Environmental Law-making and Diplomacy; and has been since 2001 an assistant editor on the South African Journal of Environmental Law and Policy.

Meinhard Doelle, B.Sc., LL.B, LL.M, JSD, is Professor and Associate Dean, Research, at the Schulich School of Law, Dalhousie University, in Halifax, Canada, where he specializes in environmental law. From 2000 to 2006, he served as a non-governmental member of the Canadian delegation to the UN climate change negotiations. He continues to follow the negotiations as an official observer. Professor Doelle has written on a variety of environmental law topics, including climate change, energy, compliance, and public participation in environmental decision-making. His recent books include: *Promoting Compliance in an Evolving Climate Regime* (2012); *Environmental Law: Cases and Materials* (2009); and *From Hot Air to Action: Climate Change, Compliance and the Future of International Environmental Law*.

Sébastien Duyck is a researcher at the Northern Institute for Environmental and Minority Law (Arctic Center/University of Lapland). His main research areas include public participation in intergovernmental processes, international climate change law, and the interplay between human rights and climate change. Sébastien is currently writing his doctoral dissertation on the procedural rights of stakeholders in the international climate change regime.

Felix Ekardt is Professor for environmental law and legal philosophy at the University of Rostock, board member of the Baltic Sea Institute for Environmental Law, Director of the Research Unit Sustainability and Climate Policy in Leipzig, and a visiting professor at various universities, e.g. University of Leipzig (Faculty of Philosophy). He is working as a political advisor on international and national level in several projects and advisory commissions, for instance as a member of the German soil protection commission of the German Federal Government. His research interests include: environmental/climate law and policy, interdisciplinary theory of sustainability, theory of justice, theory of human rights, and governance.

Tomilola Eni-ibukun is an independent climate change consultant with a Ph.D. in law from the University of Dundee. She also teaches foundation courses in law at the University of St Andrews. Her expertise is in the areas of climate change law, particularly the Clean Development Mechanism and carbon markets, and sustainable development law. She has worked extensively in these areas, both independently and for the *Earth Negotiations Bulletin*/International Institute for Sustainable Development.

Leena Heinämäki is a Senior Researcher in the Northern Institute for Environmental and Minority Law, Arctic Centre, University of Lapland. Her special expertise is on indigenous peoples' rights in international law in general, and in the Arctic context in particular. She also has a more general research interest in environmental rights, connections between human rights and the environment, as well as the law related to the sustainable development, with an attempt to create an integral or holistic approach for international law. She is a vice-leader in the research Thematic Network on Arctic Law and has participated in many international research projects relating to the legal issues of environmental protection and indigenous peoples.

David Hodgkinson is an Associate Professor at the University of Western Australia Law School and Special Counsel at international law firm Clayton Utz. He is the coauthor of the book *Global Climate Change: Australian Law and Policy* (LexisNexis) and the editor of the looseleaf service *Australian Climate Change Law and Policy*. He also leads a global project the aim of which is to draft a treaty to address the climate change displacement problem.

Erkki J. Hollo is Professor Emeritus in Environmental Law at the University of Helsinki in Finland. He is Director of the Finnish Environmental Law Institute (FELI), and holds leading positions in numerous academic societies in Finland and throughout Europe (in environment, legal sciences, agriculture, development and humanities). He is also the President of Honour of the European Council for Rural Law (CEDR); Board Member of the Finnish Academy of Sciences and Humanities; and Board Member of the European Association of Environmental Law (EELA).

David John Frenkil is an energy attorney and publisher of EfficiencyLaw.com, an online journal that provides coverage of electric power and energy development. He provides counsel on transactional, regulatory and policy matters for electric power companies, natural gas pipelines and renewable energy developers. Prior to joining the firm, he worked with the U.S. Department of Energy and SunEdison on issues related to power purchase agreements, transmission development and federal energy financing. He has advised current and prospective state legislators in Maryland and California on energy policy, while also consulting a US biofuels company and South American landowner on carbon mitigation strategies.

Michael Kidd is Professor of Law at the University of KwaZulu-Natal, Pietermaritzburg, South Africa. He holds a B.Com., LL.B, LL.M and Ph.D degrees, all from the University of Natal.

Hitomi Kimura is Assistant Professor of International and Japanese Environmental Law at Otsuma Women's University. She holds an LL.M. from Temple University and an M.A. from Nagoya University. She also taught English classes at the University of Tokyo, Graduate School of Arts and Sciences as part time. After working as corporate consultant and as researcher at UFJ Institute, currently Mitsubishi UFJ Research & Consulting, she has worked on climate law and policy as a researcher and fellow at the Institute for Global Environmental Strategies.

Timo Koivurova is a Research Professor and the Director of the Northern Institute for Environmental and Minority Law, Arctic Centre, University of Lapland. His expertise lies in the areas of environmental law, indigenous law as well as law of the sea, in particular how these bodies of law apply in the Arctic. He has led several international research projects and has been involved as an expert in several international processes. Professor Koivurova is also an editor-in-chief of the *Yearbook of Polar Law* – together with Gudmundur Alfredsson – and an editor in the *Nordic Journal of International Law* and the *Nordic Journal of Environmental Law*.

Kati Kulovesi is Adjunct Professor in Climate Law at the Department of Law, University of Eastern Finland. She holds LL.M degrees from the University of Helsinki and LSE, and a Ph.D. from the London School of Economics and Political Science where she has also taught international and European environmental law to LL.M. students. She is managing partner of CC Law – Climate Change Law Consulting and has, since 2003, worked on climate law and carbon markets also as a practicing practicing lawyer and consultant, advising governments, international organizations, carbon funds and various other organizations. She has followed the UN climate negotiations since 2001, mostly as Team Leader/Writer for the *Earth Negotiations Bulletin*/International Institute for Sustainable Development.

Haroldo Machado-Filho holds a Ph.D. in International Law from the Graduate Institute of International Studies, Geneva. He is a special adviser on climate change and sustainable development issues, currently working at at the UN Development Programme, Brazil. He has been a negotiator of the Brazilian Delegation to the Conference of the Parties and to the Subsidiary Bodies under the United Nations Framework on Climate Change and the Kyoto Protocol since 1998, and is also a Lead Author of the Fifth Assessment Report of the Intergovernmental Panel on Climate Change, Working Group III. In addition, he is a Member of the Global Climate Change and International Relations Network at the University of Brasília.

Rowena Maguire, Ph.D., is a lecturer within the law school at Queensland University of Technology in Australia and a research affiliate at the Cambridge Climate Change Centre for Mitigation Research. Rowena teaches undergraduate and postgraduate courses in international law, environmental law and property law at the Queensland University of Technology. Rowena's doctoral study examined the international regulation of sustainable forest management and she is currently working with the Forest Carbon and Law Program at Cambridge to look at the practical translation of international forest policy (reducing emissions from deforestation and forest degradation). Other research interests include governance of the international climate regime, climate justice and the intersection between the environment and indigenous and human rights.

Sharon Mascher is a Professor in the Faculty of Law at the University of Calgary where she teaches and researches in the areas of natural resources, energy and environmental law. Sharon was previously an Associate Professor in the Faculty of Law at the University of Western Australia, during which time she was the Deputy Director, Climate Change, in the University of Western Australia's Centre for Mining, Energy and Resources Law.

Jane Matthews Glenn is Professor Emeritus in the Faculty of Law and the School of Urban Planning and Associate Member of the McGill School of Environment at McGill University, Montreal, Canada. Her research interests are in the general area of property law and include urban planning, housing, environmental and agricultural law.

Michael Mehling is President of the Ecologic Institute in Washington D.C., an environmental think tank with partner offices in Berlin, Brussels and Vienna, and serves as Adjunct Professor at Georgetown University. Michael has published widely on climate change law and policy, led advisory projects for a range of government agencies and intergovernmental organizations, and taught at universities in Europe, North America, and the developing world. In 2007, he launched the Carbon & Climate Law Review (CCLR), the first academic journal focused on climate regulation and the carbon market. He is registered with the Bars of Berlin and the District of Columbia.

Elisa Morgera holds an LL.M. degree from the University College London and a Ph.D. from the European University Institute, Florence. She is a Lecturer in European Environmental Law and Director of the LL.M. Programme in Global Environment and Climate Change Law, School of Law, University of Edinburgh, UK.

José Otero is a Ph.D. candidate in the School of Urban Planning at McGill University, Montreal, Canada. His doctoral research is in the area of climate change adaptation, examining the role of planning institutions and land use laws as determinants of adaptive capacity in Canadian coastal communities. He is a graduate of George Washington University Law School and practiced as an attorney in the US for many years, specializing in government contracts law.

Patodia Rastogi is an independent consultant and was formerly International Fellow at the Center for Climate and Energy Solutions (previously the Pew Center on Global Climate Change), Virginia, United States.

Eugenia Recio holds an M.Phil. from the Complutense University, Madrid, Spain. She specializes in international and Latin American environmental law and policy, and works for international organizations in the assessment and design of national and international climate change regulations. She is an expert on international climate policy and has written extensively on India and climate change, and various issues related to the international climate negotiations.

Colin T. Reid is Professor of Environmental Law at the University of Dundee. He is a graduate of Oxford and Cambridge Universities and worked at Aberdeen University before moving to Dundee in 1991. He has taught Environmental Law and many other courses over the years and has written on many topics, especially biodiversity law. He is the founding Convener of the Environmental Law Section of the Society of Legal Scholars, an active member of the UK Environmental Law Association and a member of the editorial board of several environmental journals.

Annalisa Savaresi is visiting researcher at the International Center for Climate Governance, Venice, Italy, and Ph.D. candidate at the Faculty of Law, University of Copenhagen. Her research deals with the law-making process associated with the establishment of the REDD+ mechanism under the UNFCCC. She is member of the IUCN Commission on Environmental Law and recently joined the International Institute for Sustainable Development team of writers for the *Earth Negotiation Bulletin*. In the fall of 2012 she will join the School of Law, University of Edinburgh, as a teaching and research fellow.

Christopher Tung is a partner in the Hong Kong office of global law firm K&L Gates. He has been based in Hong Kong since 1992 and practices dispute resolution, construction and environmental law. He specializes in climate change and clean energy matters and is a leading advocate of sustainable development law and practice in Asia. Chris is a senior research fellow of the Centre for International Sustainable Development Law based at McGill University and is currently a member of the Zero Carbon Building Task Force of the Hong Kong Construction Industry Council.

Harro van Asselt is Postdoctoral Researcher at the Stockholm Environment Institute, visiting research associate with the Environmental Change Institute, University of Oxford and visiting researcher with the Institute for Environmental Studies, VU University Amsterdam.

Jonathan Verschuuren is a Professor of International and European Environmental Law at Tilburg University (the Netherlands) and faculty member of the Tilburg Sustainability Center. His research mainly focuses on the role of international and EU environmental law in legal practice, on such issues as biodiversity conservation and climate change. Currently his research focuses on climate change adaptation law. Verschuuren has written more than 200 publications in the field of environmental law, including several books and many articles in outstanding refereed journals throughout the world. He is or has been a visiting professor at the Universities of Connecticut (USA), Leuven (Belgium), and Sydney (Australia), as well as at North West University (South Africa).

Antto Vihma recently completed his Ph.D. at the University of Helsinki, Faculty of Social Sciences, analyzing effectiveness and legitimacy of international agreements. He works as a research fellow in the Global Security Research Programme at the Finnish Institute of International Affairs (FIIA), conducting academic research and consulting governments and NGOs on UN climate talks.

Yulia Yamineva holds a Ph.D. in International Studies from the University of Cambridge and is currently Postdoctoral Researcher at the Department of Law, University of Eastern Finland. She followed the UNFCCC negotiations in 2007–2009 for the *Earth Negotiations Bulletin* and was a consultant on climate finance based at the UNFCCC Secretariat in 2010–2011.

Abbreviations

AAU	Assigned Amount Unit
AC	Adaptation Committee
ACIA	Arctic Climate Impact Assessment
ADP	Ad Hoc Working Group on the Durban Platform for Enhanced Action
AFB	Adaptation Fund Board
AOSIS	Alliance of Small Island States
AWG-KP	Ad Hoc Working Group on Annex I Parties' Further Commitments under the Kyoto Protocol
AWG-LCA	Ad Hoc Working Group on Long-term Cooperative Action under The Convention
CAA	Clean Air Act
CARB	California Air Resources Board
CBD	Convention on Biological Diversity
CCPR	International Covenant on Civil and Political Rights
CCS	Carbon Capture and Storage
CESCER	International Covenant on Economic, Social and Cultural Rights
CDM	Clean Development Mechanism
CER	Certified Emission Reduction
CITES	Convention on International Trade in Endangered Species
CMS	Convention on the Conservation of Migratory Species of Wild Animals
CO_2	Carbon dioxide
COP	Conference of the Parties
COP/MOP	Conference of the Parties serving as meeting of the Parties to the Kyoto Protocol
DOE	Designated Operational Entity
EB	Enforcement Branch of the Kyoto Protocol's Compliance Committee
EPA	Environmental Protection Agency
ER	emission reduction
ERT	expert review team
ERU	Emission Reduction Unit
EITs	countries with economies in transition to a market economy

ETS	Emissions Trading Scheme
EU	European Union
EU ETS	European Union Emissions Trading Scheme
FB	Facilitative Branch of the Kyoto Protocol's Compliance Committee
FCPF	Forest Carbon Partnership Facility
GATT	General Agreement on Tariffs and Trade (WTO)
GATS	General Agreement on Trade in Services (WTO)
GCF	Green Climate Fund
GDP	Gross Domestic Product
GEF	Global Environment Facility
GHG	greenhouse gas
GIS	Green Investment Scheme
IAR	international analysis and review
ICA	international consultation and analysis
ICC	Inuit Circumpolar Council
ICJ	International Court of Justice
ICZM	integrated coastal zone management
IPCC	Intergovernmental Panel on Climate Change
ITLOS	International Tribunal for the Law of the Sea
JVETS	Japanese Voluntary Emissions Trading Scheme
JI	Joint Implementation
LDCs	Least Developed Countries
LDCF	Least Developed Country Fund
LULUCF	land use, land-use change and forestry
MEAs	multilateral environmental agreements
MRV	measurement, reporting and verification
MoU	memorandum of understanding
NAAQS	national ambient air quality standards
NAMA	nationally appropriate mitigation action
NAPA	national adaptation plan of action
NERC	North American Electric Reliability Corporation
NSPS	new source performance standard
ODA	official development assistance
OECD	Organization for Economic Development and Cooperation
OHCHR	Office of the High Commissioner for Human Rights
PSD	Prevention of significant deterioration
REC	renewable energy certificate
REDD	reducing emissions from deforestation and forest degradation in developing countries
REDD+	reducing emissions from deforestation and forest degradation in developing countries, and forest conservation, sustainable management of forests and enhancement of forest carbon stocks
RPS	renewable portfolio standard
SBs	UNFCCC Subsidiary Bodies
SBI	Subsidiary Body for Implementation

SBSTA	Subsidiary Body for Scientific and Technological Advice
SCCF	Special Climate Change Fund
SCM	Agreement on Subsidies and Countervailing Measures (WTO)
SD	sustainable development
SIDS	small island developing States
STAR	System for Transparent Allocation of Resources (GEF)
UN	United Nations
UNCCD	United Nations Convention to Combat Desertification
UNCLOS	United Nations Convention on the Law of the Sea
UNDP	United Nations Development Programme
UNEP	United Nations Environment Programme
UNFCCC	United Nations Framework Convention on Climate Change
US	United States
VCLT	Vienna Convention on the Law of Treaties
VCS	voluntary carbon standard
WTO	World Trade Organization

Chapter 1
Introduction: Climate Change and the Law

Kati Kulovesi, Michael Mehling, and Erkki J. Hollo

Abstract The United Nations Framework Convention on Climate Change (UNFCCC) celebrates its twentieth anniversary in 2012. The two decades following its adoption have witnessed a significant evolution in legal responses to climate change. The international climate regime itself has grown considerably and evolved into a highly specialized area of international law and legal expertise. Its evolution has inspired the expansion of climate law also at the regional, national, subnational and transnational levels. The emphasis of climate law has traditionally been on mitigation, but adaptation, finance, technology and capacity building have recently asserted their place as key elements of climate change law and policy.

1.1 Exploring the Relationship Between Climate Change and the Law

The United Nations Framework Convention on Climate Change (UNFCCC) celebrates its twentieth anniversary in 2012. The two decades following its adoption have witnessed a significant evolution in legal responses to climate change.

K. Kulovesi (✉)
Department of Law, University of Eastern Finland,
PO Box 111, 80101 Joensuu, Finland
e-mail: kati.kulovesi@uef.fi

M. Mehling
Ecologic Institute, 1630 Connecticut Ave. NW, Suite 300,
20009 Washington, DC, USA
e-mail: Michael.Mehling@Ecologic-Institute.us

E.J. Hollo
Faculty of Law, University of Helsinki, Yliopistokatu 3,
00014 Helsinki, Finland
e-mail: erkki.hollo@helsinki.fi

E.J. Hollo et al. (eds.), *Climate Change and the Law*, Ius Gentium: Comparative Perspectives on Law and Justice 21, DOI 10.1007/978-94-007-5440-9_1,
© Springer Science+Business Media Dordrecht 2013

The international climate regime itself has grown considerably and evolved into a highly specialized area of international law and legal expertise. Its evolution has inspired the expansion of climate law also at the regional, national, subnational and transnational levels. The emphasis of climate law has traditionally been on mitigation, but adaptation, finance, technology and capacity building have recently asserted their place as key elements of climate change law and policy.

Given the enormous complexity of the underlying challenge, legal activity around climate change has witnessed the active exploration of innovative regulatory instruments and approaches. The range of legal approaches to mitigate greenhouse gas emissions include regulations, standards, environmental permitting, taxes, emissions trading, offsetting mechanisms, financing schemes and other inventive instruments. Meanwhile, the global average temperature is increasing and the impacts of warming are already observable on all continents. A failure to halt the continuing growth of global greenhouse gas emissions and turn them towards a declining trend is projected to have dramatic consequences for a number of countries, populations and ecosystems. Adaptation to the harmful impacts of climate change is therefore having a growing effect on normative systems governing, for instance, land use and agriculture, water resources, coastal management and health policy. management and health policy. The adaptation challenge is also increasingly drawing attention to links between climate change, biodiversity and human rights.

Against the backdrop of lively regulatory activity around climate change, this books sets out to explore the relationship between climate change and the law. Is climate change law emerging as a new legal discipline? If so, what common principles, objectives and other shared categories define it? How does climate change law relate to other areas of law? The book approaches these questions by exploring the rich diversity of international, regional, national, sub-national and transnational legal responses to climate change. While the book seeks to place the emphasis on doctrinal questions, its 30 chapters also address a range of substantive and institutional issues. They illustrate that normative activity around climate change has seen a lively exploration of new regulatory philosophies, harnessing innovative and flexible instruments. Furthermore, the active involvement of multiple governance levels and various non-state actors has stretched in many ways the conceptual boundaries of traditional jurisprudence. Given the global nature of the climate change challenge, countries sometimes seek to use climate law to influence the behavior of actors located outside their geographical boundaries. The European Union (EU), with its global climate leadership aspirations, is an obvious example. Others could, however, soon follow the suit. Such developments highlight doctrinal questions concerning the migration of legal norms and regulatory innovations on the one hand, and the legal implications of climate change unilateralism on the other. Because of its cross-cutting nature, climate change law can also easily affect neighboring areas of law. The relevant doctrinal questions in this regard relate to fragmentation and ways of making the relevant international legal regimes, such as those dealing with trade, biodiversity, human rights and climate change, mutually supportive. These are some of the broad themes explored in this book.

1.2 Structure and Organization

The book consists of five parts. Part I is dedicated to exploring climate law as an emerging legal discipline. In Chap. 2, Mehling therefore highlights the unique nature of the objectives, principles and instruments that form the normative acquis devoted to addressing climate change, which allows for a better definition of its boundaries vis-à-vis other areas of law and justifies its conceptual understanding as a distinct body of law. In Chap. 3, Kulovesi analyzes emergent trends in climate law and scholarship. She draws attention to questions concerning multi-level governance, migration of legal norms and relevance of the private sector and voluntary intiatives for climate law. In Chap. 4, Ekardt addresses the relationship between climate change, legal theory and justice, drawing a normative vision of a justice-based framework for global climate governance. His conclusion is that a governance framework based on the concept of "one human, one emission right" and a global system of transferable emission rights is the approach most likely to maximize enjoyment of freedom across generations and regions and, by extension, to achieve justice in climate policy.

Part II focuses on international climate law. It addresses questions concerning architecture and institutions. Chapter 5 by Maguire analyzes objectives, principles and methods of climate law. It draws attention to the foundations of the international climate regime, including the Convention's ultimate objective in Article 2, its key principles in Article 3, and the methods of the regime in Article 4. In addition, Maguire examines state sovereignty and responsibility, preventative action, cooperation, sustainable development, precaution, polluter pays and common but differentiated responsibilities, evaluating the incorporation of these concepts into the international climate regime. In Chap. 6, Bausch and Mehling survey alternative venues of climate change cooperation from an institutional perspective, assessing the past performance of different regimes and fora active in climate change mitigation, and inferring necessary conditions for their ability to contribute to meaningful progress in international climate cooperation. In Chap. 7, Vihma discusses the role of hard and soft law in the international climate change regime. His argument is that a notable turn toward soft law is taking place in terms of developed country mitigation commitments. He contends, however, that the UN regime is at the same time becoming harder by enhancing transparency of actions by all major economies. Chapter 8 by Doelle tracks the work of the Compliance Committee under Kyoto Protocol. He reviews the basic features of the Committee, including its Enforcement and Facilitative Branches, and provides an overview of the key issues brought before the Committee in 2006–2012. Doelle then assesses the effectiveness of the Enforcement Branch in light of the first seven issues of implementation brought before it.

As indicated above, the scope of international climate law has expanded from its original focus on mitigation and is now increasingly addressing questions related to adaptation, technology and finance. At the same time, questions concerning justice, equity and human rights are also crucial in the context of legal responses to climate change. Part III is dedicated to such cross-cutting issues. In Chap. 9, Yamineva and Kulovesi describe and analyze the reformed framework for climate finance under the UNFCCC. They argue that the recent establishment of the Green Climate Fund constitutes an important milestone and progress has also been made in other respects. They conclude, however, that long-standing divides and mistrust between developed and developing countries have shaped the negotiations and continue to be reflected in their outcomes. In their view this, together with the lack of clarity over long-term sources of finance, casts shadows over the future effectiveness of the new framework. In Chap. 10, Eni-ibukun focuses questions concerning climate justice in light of experiences from the Kyoto Protocol's Clean Development Mechanism (CDM). She proceeds from the argument that justice considerations are intimately linked to the climate change, and that the climate change regime contains a range of provisions, tools and measures that seek to promote justice. One of such tools is the CDM. Eni-ibukun analyzes the CDM from the point of view of distributive justice, defining what distributive justice in the CDM means, examining what it should look like, and identifying the main causes for the lack of distributive justice in the CDM. In Chap. 11, Verschuuren discusses legal aspects of climate change adaptation. He provides an overview of the adaptation-related international legal obligations under the UNFCCC. He also demonstrates the impact of adaptation considerations on the relevant fields of law, highlighting marine and coastal adaptation, water management, biodiversity conservation, planning and land use, buildings and infrastructure, energy and telecommunications, and migration. Verschuuren argues that there are hardly any areas of law that are not impacted by climate change, and considers the need for changing the law to facilitate adaptation measures. Chapter 12 by Koivurova, Duyck and Heinämäki explores the relationship between climate change and human rights. The authors argue that while this linkage has been given little attention, its importance is likely to grow in the coming years. They demonstrate how climate change affects the enjoyment of human rights and has already led to a human rights petition against the United States. They also explore implications of human rights for the functioning of the climate change regime, such as the emerging rights to participate in environmental decision-making. They also consider whether human rights can, or even should, influence the future design of the climate change regime will be examined.

Given its complexity, climate change affects most sectors of the economy and society. For this reason, the international climate regime is closely related to several other international legal regimes in the form of overlapping subject matters and legal rules. These themes are explored in Part IV, which focuses on sectoral issues. In Chap. 13, van Asselt discusses the fragmentation of international climate change law and describes interactions between the relevant international legal regimes. He also examines various management strategies with a view to enhancing synergies and mitigating conflicts between climate-related international legal regimes. The conclusion by van Asselt is that the application of various strategies is necessary to manage the fragmentation of international climate law. Chapter 14 by Morgera draws attention to linkages between the international biodiversity and climate change regimes, and highlights ways in which international biodiversity law contributes to the fight against climate change by addressing negative impacts on biodiversity and community livelihoods of measures to address climate change and adopting the ecosystem approach to climate change mitigation and adaptation. Morgeera argues that positive interaction between the international biodiversity and climate change regimes can promote a human rights-based approach to the development of the international climate change regime and its implementation at the national level. In Chap. 15, Savaresi focuses on the role of REDD+ (reducing emissions from deforestation and forest degradation in developing countries; and the role of conservation, sustainable management of forests and enhancement of forest carbon stocks in developing countries) in harmonazing overlapping international obligations relating to climate change, biodiversity and human rights. She highlights links between biodiversity and human rights law and the subject matter of REDD+, also demonstrating that some steps have already been taken to address potential overlaps. The question of overlaps is therefore not a merely theoretical one. Savaresi offers an account of the ongoing debate, providing a snapshot of its evolution, as well as some predictions on its outcome. In Chap. 16, Kulovesi focuses on the intersection between the international climate change and trade regimes. She identifies potentially sensitive areas in the relationship between the UNFCCC and the World Trade Organization (WTO), including sustainability requirements targeting processes and production methods, as well as measures targeting carbon leakage and competitiveness concerns. Kulovesi also highlights institutional and doctrinal challenges related to fragmentation of international law, identifying problems that could arise if a climate change related dispute was considered by the WTO dispute settlement system. She concludes that the trade and climate regimes are increasingly relevant for each other and that they are not necessarily rivals – both could benefit from identifying and promoting unexploited synergies between the two regimes. However, closer cooperation and institutional coordination may be needed in the future in order to avoid mutually unhelpful institutional and legal clashes. Chapter 17 by Bodle discusses the question of geoengineering. Bodle provides an overview of geoengineering techniques and the existing international law applicable to them. He indicates that geoengineering techniques are not prohibited as such, and are hardly addressed by international law. He therefore argues that they pose fundamental challenges to international climate law. According to Bodle, the main challenge for policy makers is deciding whether and how to get involved in geoengineering without providing an incentive or excuse for stepping away from reducing emissions. He also stresses the need to clearly separate scientific input and political decision-making.

Part V entitled "Comparative Climate Law" traces the evolution of climate change law and policy in a number of countries and regions. This part serves to demonstrate that developments at the international level have often given important impetus to the evolution of climate law and policy in several national jurisdictions and regionally. Still, many of the detailed solutions have been driven by domestic considerations. Part IV begins with a focus on North America. In Chap. 18, Mehling and Frenkil assess the landscape of climate and energy law in the United States, and identify procedural and structural obstacles to greater domestic policy ambition.

Glenn and Otero proceed to describe the evolution of the Canadian legal framework for climate policy in Chap. 19, explaining the controversial position of Canada with respect to the Kyoto Protocol as a consequence of inconsistent federal leadership. Following these two chapters on North America, the focus then shifts to the European Union (EU) and some of its Member States. In Chap. 20, Mehling, Kulovesi and de Cendra explore the development of climate law in the EU and suggest explanations for the leadership role that the EU has assumed on climate change in the international community – a role that goes so far as to include highly controversial unilateral measures. In Chap.s 21 and 22 Ekardt and Reid focus on national-level developments in Germany and the United Kingdom, respectively. Ekardt elaborates his interpretation of German climate law in Chap. 21, drawing a critical picture of the body of rules adopted in that country to counter climate change, and highlighting the unintended consequences of several measures, such as rebound effects from energy efficiency regulation and displacement effects from rules on renewable energy promotion. Chapter 22 by Reid indicates that there is no single legislative source for the United Kingdom's legal response to climate change. He explains that initial measures to tax large energy users, enable participation in the EU Emissions Trading Scheme and encourage renewable electricity generation have subsequently been joined by the Climate Change Acts operating at UK and Scottish levels. These Acts set demanding targets for reductions in greenhouse gas emissions and break new and uncertain legal ground in making these legally binding on Ministers. Reid also demonstrates that separate legislation in Scotland highlights the difficulties that dealing with pervasive issues, especially those with an EU and international dimension, pose for sub-national governments with distinct political ambitions but limited jurisdiction.

In Chap. 23, Yamineva discusses the evolution of climate law and policy in Russia. She argues that for a long time, Russia's climate policy remained underdeveloped and lagged behind other countries. The presidential term of Dmitry Medvedev and his modernisation agenda brought about the necessary transformation as the Climate Doctrine adopted in 2009 acknowledged the anthropogenic nature of climate change, setting principles and goals for mitigation and adaptation policies. She also discusses the development of a comprehensive framework for energy efficiency and energy conservation arguing that, if fully implemented, it will lead to significant reductions in greenhouse gas emissions. In Chap. 24, Mascher and Hodgkinson describe developments in Australia. They disucss the Clean Energy Act 2011, which will, for the first time, introduce a carbon price into the Australian economy. They argue that the passage of the Act marks a momentous step forward for Australia, a country that until now has been dominated by a domestic climate change policy of 'no-regrets.' Chapter 25 by Kimura focuses on the evolution of climate change law and policy in Japan. She explains that the Japanese regulatory approach combines a framework law, the Law Concerning the Promotion of the Measures to Cope with Global Warming, and specific laws, as well as the proactive use of voluntary approaches such as Keidanren's Voluntary Action Plan. Unique policies have also been introduced, including the Japanese Voluntary Emissions Trading Scheme and bilateral offset mechanisms. According to Kimura, the, Japanese decision not to participate in the second commitment period under the Kyoto Protocol spells out a gloomy future for a quick passage of the Basic Bill to Cope with Global Warming Bill.

The next four chapters describe national developments in the four largest emerging economies, namely China, India, South Africa and Brazil. Chapter 26 by Tung is dedicated to climate law in China. He argues that China has taken significant steps to advance sustainable development and transition to a low carbon economy. Since 1994, a national sustainable development strategy has underpinned the creation of policies and law that directly and indirectly impact the environment and climate change. While indicating that the policy and legislative process has been broadly successful, Tung also draws attention to problems. In conclusion, he offers recommendations on how problems could be avoided and how sustainable development objectives and principles could be strengthened in the implementation of Chinese laws and policies directly or indirectly impacting climate change and low carbon economy objectives. In Chap. 27, Patodia Rastogi analyzes the development of India's national climate change strategy. She argues that India, along with most other developing countries, has viewed climate change as an environmental concern that first and foremost must be addressed by the industrialized west. According to her, development challenges are India's priority and domestic action on climate change has been minimal and in so far that it existed, it has primarily been viewed as a 'co-benefit' of another policy. Patodia Rastogi argues that in 2008, a dramatic shift was seen in India's approach to addressing climate change due to the release of the National Action Plan on Climate Change, a comprehensive framework policy where climate change is the central focus. In Chap. 28, Kidd and Couzens describe climate change responses in South Africa. They argue that despite a long history of climate policy development, there is insufficient legislation addressing climate change. They discuss relevant policy documents, including the 2011 White Paper on the National Climate Change Response and also consider South Africa's energy policy. They conclude that continuing on the current path means that it will be extremely difficult to reconcile the goals of strong economic growth and poverty alleviation with environmental protection generally, and South Africa's international commitments in the climate change issue-area specifically. In Chap. 29, Machado-Filho describes climate change policy and legislation in Brazil. He highlights common obligations for all Parties established under Article 4.1 of the UNFCCC. He then focuses on recent policies and legislation on climate change adopted in Brazil, which are fundamental for the implementation of commitments under the UNFCCC. Machado-Filho argues that new developments, including the voluntary quantified target for reducing emissions announced in 2009 and encapsulated in the National Policy on Climate Change, demonstrates that Brazil has moved from "due diligence" measures, with a view to respecting the obligations formulated under international law, towards the goal for real contribution to the combat against climate change.

Finally, in Chap. 30, Aguilar and Recio provide an overview of the evolution of climate law in Latin American countries, arguing that climate change law in Latin America is in its infancy, although advancing at a steady pace. They explain that most countries in the region have adopted soft law instruments, including climate change strategies, and, in some cases, climate change plans of action or sectoral action plans for adaptation or forestry. Brazil, Mexico, Colombia and Ecuador have more coherent legal frameworks for climate change, although only Brazil had adopted a substantive climate change law at the time of writing. Aguilar and Recio conclude that frameworks related to climate change mitigation are more advanced than those dealing with adaptation, even though several Latin American countries identify adaptation as a key priority for their future development. They also argue that policy implementation remains challenging, with mainstreaming across sectors, allocation of budget resources and presidential support being identified as crucial elements and recurring challenges. The chapter also finds that sub-national entities are increasingly involved in the development and implementation of climate change policy tools at the local level.

Part I
Climate Law as an Emerging Discipline

Chapter 2
Implementing Climate Governance: Instrument Choice and Interaction

Michael Mehling

Abstract At all levels of regulation, the legal response to both causes and impacts of climate change has shifted away from a segmented array of isolated measures and initiatives on specific aspects of global warming, such as policies to manage energy demand or promote research on sustainable alternatives, to an increasingly sophisticated network of regulatory standards, market mechanisms, and other innovative approaches. While the first elements of a new area of law are arguably emerging in the shape of common principles and objectives for sustainable energy use, the countless rules devoted to climate change are still but loosely related and far from becoming a coherent normative framework.

2.1 Introduction

With energy production and consumption accounting for a vast majority of anthropogenic greenhouse gas emissions, climate policy invariably affects larger and also more sensitive areas of society, compelling change in nearly all domains of social behaviour and, notably, constraining economic activity at a much broader scale than any other area of environmental governance. As a result, decision makers have openly embraced alternative policy approaches based on flexible markets and price incentives, in the hope of limiting harmful effects on the economy and competitive distortions in the global marketplace. While the reasoning behind this changed orientation is understandable, the rapid growth and evolution of new

Michael Mehling is President of the Ecologic Institute, Washington DC, and an Adjunct Professor at Georgetown University.

M. Mehling (✉)
Ecological Institute, 1630 Connecticut Ave. NW, Suite 300,
20009 Washington, DC, USA
e-mail: mehling@ecologic-institute.us

E.J. Hollo et al. (eds.), *Climate Change and the Law*, Ius Gentium: Comparative Perspectives on Law and Justice 21, DOI 10.1007/978-94-007-5440-9_2,
© Springer Science+Business Media Dordrecht 2013

mechanisms has also brought along new shortcomings, giving rise to conflicts at the level of individual rules and principles, all the way to systemic tensions within the overall configuration of the legal system. Partly, this can be ascribed to a dramatic change in the conception and focus of environmental regulation: as economic considerations acquire greater weight in decision making, increased preoccupation with the cost and efficiency of policies has resulted in a variety of flexible market incentives joining or supplanting more conventional performance and quality standards.[1]

Such difficulties have also overshadowed the design and implementation of many domestic climate policy portfolios. Looking back on the early stages of domestic climate regulation process, one might often garner the impression of an incremental, barely coordinated strategy, resulting in a coincidental rather than intended assortment of regulatory devices, not seldom based on overly rushed legislative schedules,[2] substantive disagreement between rival government agencies, and the challenge of balancing international commitments with domestic legal and political realities. Faced with changing demands in a politically exposed issue area, legislators and administrators have been mandated with elaborating an operational regime for activities which, previously, had been subject to no form of regulation. Confused by the unfolding disarray and widespread misinformation, affected stakeholders have often voiced their irritation at the lack of coherence and systematisation in climate law and policy.

And yet, as this area of law matures, one can already perceive efforts to streamline the current diversity of rules through shared definitions, common objectives, and dynamic referencing between different acts of legislation. Against the backdrop of efforts in several national jurisdictions to systematise the diversity of environmental statutes, ordinances, decrees, and other relevant sources of law in a uniform code, it should hardly surprise that suggestions have also been made to harmonise climate policy under a single domestic legal act, marking a departure from piecemeal regulation to an integrated system for the management of our atmosphere. Several countries have indeed gone down that path, illustrating the growing systemic coherence of a distinct area of law.

At the international level, nations seeking to cooperate on climate change have always been forced to navigate a fine line between substance and process, general principles and specific rules, formal obligations and political commitments. Many of the core issues have been so divisive that progress has only been possible at the expense of specific and binding normative outcomes. As the negotiations on a future climate regime unfold, it is becoming increasingly evident that international cooperation itself is undergoing fundamental change.

[1] See generally Tom Tietenberg, "Economic Instruments for Environmental Regulation" 6.1 *Oxford Review of Economic Policy* (1990), at 17.

[2] One might also draw attention to the current approach to political representation, which favours short-term measures over long-term strategic policies by exerting pressure on elected politicians to provide demonstrable results in time for the next popular vote, see generally Anthony Downs, *An Economic Theory of Democracy* (New York, N.Y.: Harper & Row, 1957); Joseph A. Schumpeter, *Capitalism, Socialism and Democracy* (New York, N.Y.: Harper & Row, 1942).

High levels of normative and analytical uncertainty, the complex nature of interrelated issues, and substantial costs associated with any meaningful policy efforts have all strengthened the role of actors beyond the nation state, and also prompted the exploration of innovative approaches to climate governance, for instance by harnessing market instruments.[3] Likewise, the traditional model of intergovernmental cooperation centred on a binding treaty is starting to give way to a more fragmented topography of regional and bilateral networks and partnerships, where informal consultations take the place of legally enshrined rights and obligations, allowing states prepared to cooperate to do so "without unduly restricting their freedom of action."[4]

In many ways, this evolution also has far reaching implications for the legal nature of climate cooperation. If current trends are any indication, the global response to climate change beyond 2012 will see a shift in emphasis from binding obligations to more loosely organised coordination and facilitation in a system based on voluntary pledges, where national policy developments displace negotiated arrangements as the new benchmark of climate efforts.[5] As one observer has remarked about the outcome of recent negotiations, rather than adopting "a detailed, binding framework for furthering global climate cooperation", the international community has instead embraced "a general political statement that privileges the voluntary actions of states and devalues the role of international law and global climate governance."[6]

Should the crucial feature of enforcement also soften as it evolves towards responses more 'in harmony with the cooperative spirit'[7] required for climate cooperation, it could raise questions about the very role and limitations of international law.[8]

After all, it would imply that climate cooperation is ultimately determined only by the interests, at any given time, of the regime participants. Whether commitments are

---

[3] Frank Biermann, "Beyond the Intergovernmental Regime: Recent Trends in Global Carbon Governance", 2 *Current Opinion in Environmental Sustainability* (2010), 284.

[4] Patricia Birnie, "International Environmental Law: Its Adequacy for Present and Future Needs", in Alexander Hurrell and Benedict Kingsbury, *The International Politics of the Environment* (Oxford: Oxford University Press, 1992), 51–84, at 54.

[5] Jacob Werksman and Kirk Herbertson, "The Aftermath of Copenhagen: Does International Law have a Role to Play in a Global Response to Climate Change?", 25 *Maryland Journal of International Law* (2010), 109–142; see also, more broadly, Lavanya Rajamani, "Addressing the 'Post-Kyoto' Stress Disorder: Reflections on the Emerging Legal Architecture of the Climate Regime", 58 *International and Comparative Law Quarterly* (2009), 803–834.

[6] David Hunter, "Implications of the Copenhagen Accord for Global Climate Governance", 10 *Sustainable Development Law & Policy* (2010), 4–15, at 4, referring to the "Copenhagen Accord" adopted at the 15th Session of the Conference of the Parties to the United Nations Framework Convention on Climate Change (COP).

[7] Critically Martti Koskenniemi, "Breach of Treaty or Non-Compliance: Reflections on the Enforcement of the Montreal Protocol", 3 *Yearbook of International Environmental Law* (1992), 123–162, at 147.

[8] For a polemic, yet relevant argument about the limitations of international law in affecting state behaviour, see Jack L. Goldsmith and Eric A. Posner, *The Limits of International Law* (Oxford: Oxford University Press, 2005); for an impassioned counterargument, see Mary E. O'Connell, *The Power and Purpose of International Law* (Oxford University Press, 2008).

enshrined in law would then become largely irrelevant, displacing binding norms to an anachronistic realm of burdensome procedures, an obstacle, some might even argue, in the formulation of effective cooperation strategies. In such a system, a "country that deliberately fails to abide by … legally binding commitments under the Kyoto Protocol is also likely to resist the application of punitive consequences, regardless of whether these consequences are made legally binding or not."[9] But that must surely beg the question: what normative force is then left to international climate law?

For international lawyers, this question will resonate with a latent anxiety about the changing role and perception of their discipline, a departure from the application of objective rules in a coherent and enforceable system of norms to the politically guided management of technical, fragmented regimes.[10] Indeed, climate cooperation and its study appear particularly amenable to new vocabularies of governance, legitimacy and compliance, where preoccupation with the seemingly archaic language of formal international law and its binary focus on the observance or violation of rights and obligations may seem entirely outdated.[11] Aside from revising our understanding of climate cooperation, therefore, do we also need to leave behind the tools of international jurisprudence and reconceptualise the climate regime and its commitments through the lenses of more novel ways of thinking about international cooperation, such as transnational governance and global administrative law?[12]

2.2 Exploring the Boundaries of Domestic Climate Law

2.2.1 Instrument Choice at the Domestic Level

Decision makers seeking to address the causes and effects of climate change can take recourse to a portfolio of policy instruments, including pricing controls and quantity rationing,[13] performance standards, subsidies, agreements, and

---

[9] Anita M. Halvorssen and Jon Hovi, "The Nature, Origins and Impact of Legally Binding Consequences: The Case of the Climate Regime", 6 *International Environmental Agreements: Politics, Law and Economics* (2006), 157–171, at 158.

[10] Martti Koskenniemi, "The Fate of International Law: Between Technique and Politics", 70 *The Modern Law Review* (2007), 1–32.

[11] For a critique of the ongoing turn to political science vocabularies, see Martti Koskenniemi, "Legitimacy, Rights and Ideology: Notes towards a Critique of the New Moral Internationalism", 7 *Associations: Journal for Legal and Social Theory* (2003), 349–374.

[12] Anne-Marie Slaughter, *A New World Order* (Princeton, N.J.: Princeton University Press, 2005); Benedict Kingsbury, Nico Krisch and Richard B. Stewart, "The Emergence of Global Administrative Law", 68 *Law and Contemporary Problems* (2005), 15–61.

[13] Intergovernmental Panel on Climate Change (IPCC), *Climate Change 2007: Mitigation* (Cambridge: Cambridge University Press, 2007), at 750; Organization for Economic Cooperation and Development (OECD), *Climate Change Mitigation: What Do We Do?* (Paris: OECD, 2007), available at: http://www.oecd.org/dataoecd/30/41/41753450.pdf (last accessed on 15 June 2012),

informational instruments.[14] In practice, these instruments are applied alone or in varying combinations to different sectors, such as electricity generation, transport, buildings, and industry.[15] By diverting resources and capital away from the production of conventional goods and services, and often into costly abatement measures, these instruments can have a detrimental effect on economic growth in the short term. Over the medium and longer term, the various co-benefits of mitigation action, such as energy savings, reduced health impacts, or improved energy security, suggest that a carefully designed strategy to lower greenhouse gas emissions will generate greater benefits than costs,[16] but current political and economic decision making cycles are notorious for being myopic and providing little incentive for anticipatory governance or foresight.[17] Additionally, while the social cost of action is expected to be lower than the impacts of unabated climate change, it will nonetheless rise over time as readily available abatement options are exhausted and more costly solutions need to be explored.[18] In the context of climate change, therefore, both the rationale of policy instruments and the manner in which they are designed have

at 18–22. Pricing models date back to Arthur Cecil Pigou, *The Economics of Welfare* (London: Macmillan and Company, 1920), and notably include emissions charges and taxes set to cover the marginal damage caused by polluting activities, thereby internalizing their costs; quantity rationing, in turn, is based on work by John H. Dales, *Pollution, Property and Prices* (Toronto: University of Toronto Press, 1968), at 92–100, and W. David Montgomery, "Markets in Licenses and Efficient Pollution Control Programs", 5 *Journal of Economic Theory* (1972), 395–418, both building on Ronald H. Coase, "The Problem of Social Cost", 3 Journal of Law and Economics (1960), 1–44, and generally requires the creation of a market for tradable emission allowances, where each allowance confers the right to discharge a specified quantity of pollutants for a limited duration of time; for further details, see Thomas H. Tietenberg, *Emissions Trading: Principles and Practice* (2nd ed., Washington, DC: Resources for the Future, 2006). For a discussion of relative merits, see Martin L. Weitzman, "Prices vs. Quantities." 41 *Review of Economic Studies* (1974), 477–491.

[14] This is a very broad categorization of policy instruments, and further differentiation is possible; in 1995, for instance, the Congressional Office of Technology Assessment divided environmental policy instrument in tools without fixed targets (technical assistance, subsidies, information reporting, liability, and pollution charges), multisource tools with fixed targets (challenge regulations, tradeable emissions permits, integrated permitting), and single-source tools with fixed targets (harm-based standards, design standards, technology specifications, and product bans), see Office of Technology Assessment, *Environmental Policy Tools: A User's Guide (OTA-ENV-634)* (Washington, DC: U.S. Government Printing Office, 1995), at 81–89.

[15] In a majority of sectors, greenhouse gas mitigation will be achieved by improving the efficiency with which energy is used or by reducing its carbon intensity, see OECD, *Climate Change Mitigation*, supra, note 13, at 11, but in agriculture, forestry, and certain chemical and industrial processes where emissions are not related to energy use, different approaches – such as stabilization or expansion of carbon sinks – are applied.

[16] Especially when taking into consideration the expected costs of climate change impacts, such as extreme weather events, flooding, crop losses, vector-borne diseases, and biodiversity loss, see e.g. Congressional Budget Office (CBO), *Policy Options for Reducing CO2 Emissions* (Washington, DC: Congress of the United States, 2008), at 11.

[17] Leon Fuerth, "Forward Engagement: A New Wrinkle, in Time?", 8 *International Affairs Review* (2004), 1–5.

[18] Nicholas Stern, *The Economics of Climate Change: The Stern Review* (Cambridge: Cambridge University Press, 2006), at 63, 191.

been sensitive to economic concerns from a number of important stakeholders, prompting widespread adoption of flexible or suasive incentives alongside more coercive regulatory prescriptions.[19]

With this broad range of available instruments comes a need for reliable criteria to guide and justify selection processes between contending approaches to climate governance. While it is widely agreed that no single model can serve as a panacea for all regulatory purposes,[20] a number of criteria have gradually evolved in various academic disciplines to evaluate individual instruments and their combination in a coordinated portfolio. At a sufficient level of abstraction, the following criteria are typically proposed:

- *Environmental effectiveness*: how well does a policy instrument meet its intended environmental objective? How certain is its level of environmental impact?
- *Cost effectiveness*: can the policy achieve its objectives at a lower cost than other policies? Does it create revenue streams that can be reinvested?
- *Distributional considerations*: how does the policy impact consumers and producers? Can it be considered fair and equitable?
- *Institutional feasibility*: is the policy instrument likely to be viewed as legitimate, gain political acceptance, be adopted and ultimately implemented?[21]

While these criteria are widely advocated, albeit with slight variations,[22] it bears noting that processes of instrument choice are often complicated by the fact that individual criteria tend to compete with each other, rendering tradeoffs inevitable

[19] Limiting the economic burden requires equalization of marginal abatement costs across the economy and for each source, something price – and quantity-based instruments are said to achieve better than rigid technology standards, see William J. Baumol and Wallace H. Oates, *The Theory of Environmental Policy* (2nd ed., Cambridge: Cambridge University Press, 1988), at 177; and Nathaniel O. Keohane et al., "The Choice of Regulatory Instruments in Environmental Policy" 22 *Harvard Environmental Law Review* (1998), 313–367, at 313; as a result, conventional regulation, criticized for belonging to an "extraordinarily crude, costly, litigious and counterproductive system of technology-based environmental controls" (see Bruce A. Ackerman and Richard B. Stewart, "Reforming Environmental Law", 37 *Stanford Law Review* (1985), 1333–1365, at 1333), has been increasingly joined or supplanted by market incentives, all with an aim to "improve the command system through better balancing of regulatory costs and benefits, improved risk analysis and management and greater flexibility" (Richard B. Stewart, "A New Generation of Environmental Regulation?" 30 *Capital University Law Review* (2001): 21–182, at 21).

[20] Lawrence H. Goulder and Ian W.H. Parry, *Instrument Choice in Environmental Policy* (Washington, DC: Resources for the Future, 2008), at 2.

[21] IPCC, Mitigation, supra, note 13, at 751.

[22] Similar criteria are e.g. reported in the broader academic literature, see, for instance, Thomas Sterner, *Policy Instruments for Environmental and Natural Resource Management* (Washington, DC: Resources for the Future, 2003), at 133–134, who lists efficiency (in various forms, such as static and dynamic allocative efficiency, efficiency in the use of public funds, and transaction costs), effectiveness, fairness, effects on income distribution and other aspects related to the distribution of welfare, incentive compatibility, and political feasibility; Winston Harrington et al., "Overview: Comparing Instrument Choices", in Winston Harrington et al. (eds), *Choosing Environmental Policy* (Washington, DC: Resources for the Future, 2004), 1–22, at 5, who list effectiveness, efficiency, equity and fairness, non-intrusiveness, and public participation; or OTA,

and any selection largely dependent on specific circumstances.[23] Additionally, climate governance tends to address several market failures and seek a variety of outcomes, thus necessitating the use of more than one instrument.[24] Yet with the simultaneous operation of various instruments comes a risk of adverse interactions or even redundancies.[25] Some instruments will pursue more than one objective,[26] and the extreme uncertainties underlying causes and impacts of climate change as well as policy outcomes further complicate the evaluation of relevant instruments.[27] As the next section illustrates, similar complexities are also faced when seeking to apply evaluation criteria to international regimes; many of the considerations guiding the debate on domestic instrument choice are, however, transferable to some extent.[28]

2.2.2 *Instrument Interactions at the Domestic Level*

Growing in consecutive stages, the domestic body of rules devoted to climate policy in most jurisdictions has evolved into a comprehensive and highly diverse regulatory strategy. But as with most entities that develop over time, it has not always grown in a systematic fashion, rather adding layer upon layer to accommodate new challenges and international commitments. In recent years, for instance, the German climate strategy has been subject to growing criticism for consisting of "several barely coordinated measures and actions" whose "interaction, mutual enhancement,

---

Policy Tools, supra, note 14, at 143–147, requiring that policies be cost-effective and fair, place the least demands on government, provide assurance to the public that environmental goals will be met, use pollution prevention when possible, consider environmental equity and justice issues, be adaptable to change, and encourage technology innovation and diffusion. See also Baumol and Oates, *Theory*, supra, note 19, at 57–78; Goulder et al., *Instrument Choice*, supra, note 20, at 3–23. Of course, actual practice has often "diverged strikingly from the recommendations of normative economic theory", see Keohane et al., "Choice", supra, note 19, at 313, and will be strongly influenced by local traditions, cultures, institutions, and infrastructures, with institutional capacity especially constraining viable choices in developing countries, see Bell (2003): 22.

[23] Goulder et al., *Instrument Choice*, supra, note 20, at 2. For instance, assuring a reasonable degree of fairness in the distribution of impacts, or ensuring political feasibility, often will require a sacrifice of cost-effectiveness.

[24] Jan Tinbergen, *On the Theory of Economic Policy* (Amsterdam: North Holland, 1952).

[25] Organization for Economic Cooperation and Development (OECD), *Instrument Mixes for Environmental Policy* (Paris: OECD, 2007), at 27.

[26] William A. Knudson, "The Environment, Energy, and the Tinbergen Rule", *Bulletin of Science, Technology & Society* (2008), at 308.

[27] Martin L. Weitzman, "The Extreme Uncertainty of Extreme Climate Change: An Overview and Some Implications", Unpublished Manuscript, available at: http://www.economics.harvard.edu/faculty/weitzman/files/ExtremeUncertaintyCliCh.pdf (last accessed on 15 June 2011), at 8–10.

[28] Richard B. Stewart, "Instrument Choice", in Daniel Bodansky et al. (eds), *Oxford Handbook of International Environmental Law* (Oxford: Oxford University Press, 2007), 147–181, at 159.

and mutual cancellation" are not fully known.[29] Its instruments have been censored for "being introduced, modified or expanded in a random manner", resulting in regulatory overlap and excessive government intervention, all of which, in turn, is "stifling the market."[30]

While such verdicts mostly originate with representatives from industry and commerce, the sectors most affected by environmental and energy policies, they are not entirely unfounded: even an Advisory Council of the German federal government observed that interactions between different policies had been "insufficiently considered",[31] suggesting that the German basket of instruments for greenhouse gas mitigation deserved further attention. Generally speaking, thus, such an instrument mix can be the outcome of a carefully guided process, or merely the accidental convergence of various measures adopted by decision makers in a political system to achieve a set objective.[32]

Leaning more towards the latter category, it appears, global warming legislation has been adopted over time and in response to situational demands, sacrificing systemic coherence for a profusion of divergent terminologies and altogether various degrees of overlap, ambivalence and inconsistency. Important issues are frequently governed by executive ordinances and decrees rather than statutory law, constituting a violation of the constitutional doctrine of essentiality, which requires that substantial issues be governed by formal parliamentary acts.[33]

With energy and environmental regulation in the Member States largely initiated by Community law, many of the foregoing shortcomings can be traced back to the supranational level, where the adoption of legislation is a process strongly guided by regulatory competition between the Member States[34] and often finds its basis in a precarious compromise in the Council.

---

[29] Carsten Kreklau, Commercial Manager of the Federation of German Industries (BDI), in the *Süddeutsche Zeitung* of 17 July 2001, available at http:// www.sueddeutsche.de/deutschland/ artikel/162/9153 (last accessed on 15 June 2012): "Das gegenwärtige Instrumentarium zur Klimavorsorge besteht bereits aus vielen, kaum aufeinander abgestimmten Maßnahmen und Aktionen. Die Wechselwirkungen, die gegenseitige Verstärkung sowie die Auslöschung zwischen den bereits jetzt bestehenden Instrumenten sind noch nicht in vollem Umfang bekannt. Es geht vor allem (…) um die ungeklärten Wechselwirkungen und daraus resultierenden Begrenzungen wirtschaftlicher Tätigkeit."

[30] Wirtschaftsrat der CDU e.V., *Macht der Emissionshandel den bestehenden Instrumentenmix überflüssig?* (Berlin: Wirtschaftsrat, 2004).

[31] Wissenschaftlicher Beirat beim Bundesministerium für Wirtschaft und Arbeit, "Zur Förderung erneuerbarer Energien", 15 *Zeitschrift für Umweltrecht* (2004), pp. 400 *et sqq.*, at p. 401:

[32] Georg Hermes, "Instrumentenmix im Energieumweltrecht" in Martin Führ, Rainer Wahl and Peter von Wilmowsky (eds.), *Umweltrecht und Umweltwissenschaft: Festschrift für Eckard Rehbinder* (Berlin: Erich Schmidt, 2007), 569, at 572.

[33] This doctrine is derived from the principle of democracy contained in Article 20 (1) of the German Basic Law (*Grundgesetz für die Bundesrepublik Deutschland*) of 23 May 1949, BGBl. Part I (1949), at 1.

[34] See generally Adrienne Héritier, Christoph Knill and Susanne Mingers, *Ringing the Changes in Europe: Regulatory Competition and the Transformation of the State* (Berlin: de Gruyter, 1996), *passim.*

Looking back in time, these challenges might also find their origin in the very history of environmental legislation, which evolved from earlier rules on trade supervision and traditionally relied on a rigid system of administrative permits and control.[35] Its ambit was commonly limited to the regulation of impending threats to public safety, such as acute pollution and other perilous activities, rather than distant, elusive environmental risks.[36] Given their innate affinity to pollution prevention and control, however, measures taken to mitigate global warming were initially often assigned to the same area of law governing noise and air pollution. In Germany, for instance, a central act of legislation in this field, the Federal Ambient Pollution Control Act, mentions protection of the atmosphere amid its objectives, which is commonly understood to include the global climate.[37]

And yet, the very notion of climatic change has, by definition, originated from a precautionary outlook, seeing how it involves diffuse, cumulative manifestations of risk rather than localised and immediate danger. Unlike conventional pollutants, therefore, greenhouse gases were generally not subject to any form of management in the past, with the ability to emit greenhouse gases limited by the capacity of an installation only. Elaborating climate policies within the regulatory ambit of pollution control is, however, proving less and less viable, as legislators are compelled by economic constraints and supranational commitments to engage in a paradigmatic shift of regulatory traditions and vest flexible mechanisms and market incentives in the guise of formal law.

Unsurprisingly, significant challenges have followed from this transition for administrators and the legislature, and the latter has only succeeded in embracing a more general, preventive stance to environmental protection within the past decade. Attempts to speed up the pace of reform, for instance by supplanting traditional regulation with flexible market instruments, have often been guided by purely theoretical assumptions on the merits of a particular approach, resulting in an overly narrow focus on select mechanisms at the expense of the remaining elements in the policy architecture and the operation of the policy as a whole.[38] As with any process requiring swift adaptation to rapidly changing circumstances, the result has ultimately been characterised by no small amount of tension and outright conflicts.

[35] See Gerhard Feldhaus, "Zur Geschichte des Umweltrechts in Deutschland", in: Klaus-Peter Dolde (ed.), *Umweltrecht im Wandel* (Berlin: Erich Schmidt, 2001), 15, at 17–9; Klaus-Georg Wey, *Umweltpolitik in Deutschland: Kurze Geschichte des Umweltschutzes in Deutschland seit 1900* (Opladen: Westdeutscher Verlag, 1982), 27, at 105–27, pointing to the origins of modern pollution legislation in the area of "Gewerberecht" and its close relationship with measures to avert danger, or "Gefahrenabwehr", still found in current police legislation.

[36] Martin Winkler, "Die neue Betreiberpflicht: Klimaschutz und Emissionshandel", 14 *Zeitschrift für Umweltrecht* (2003), 395, at 395–396.

[37] Hans D. Jarass, *Bundes-Immissionsschutzgesetz* (6th ed., Munich: C.H. Beck, 2005), Section 1, annot. 4.

[38] Erik Gawel, *Umweltpolitik durch gemischten Instrumenteneinsatz* (Berlin: Duncker & Humblot, 1991), at 2.

2.2.2.1 Internal and External Conflicts – An Analytical Framework

Generally speaking, one can discern *four* categories of conflicts arising from the introduction of modern climate policies into the existing legal and constitutional order. First, there are *conflicts of objectives*, notably between environmental protection and energy market regulation. By way of illustration, the access to electricity grids and minimum feed-in rates guaranteed in many countries through rules on the promotion of renewable energy are conditional on utilisation of specified technologies, with the scope of legislation limited to generation methods defined in the law itself.

On a theoretical level, this contradicts the general commitment to free competition set out in energy market legislation, for instance European Community liberalisation rules. Likewise, the polluter pays principle adopted as a central tenet of environmental policy is inherently at odds with the requirement in many emissions trading systems to allocate a significant majority of emission allowances for free to operators under emissions trading rules.[39] Accordingly, the divergent objectives of climate policies and legislation in other issue areas are not always easy to reconcile.

Conflicts can also follow from *divergent regulatory approaches*, notably when conventional rules based on state intervention and "command and control" meet flexible policies based on the price signals of functioning markets and other financial incentives. An example for such colliding traditions can be seen in the relationship of emissions trading and many conventional ambient pollution control regimes, as the former relies on market forces to guide the standard of technology in participating installations, while the latter, in turn, tend to force rigid performance standards and emission ceilings on each individual operator. By requiring all installations – regardless of cost – to ensure a certain standard of technology, conventional regulation goes against the central premise of emissions trading, given that installations are no longer free to decide whether to acquire further allowances or invest in more efficient facilities.[40] In order to resolve this conflict, implementation of emissions trading in the European Community necessitated a legislative amendment of pollution control legislation to exempt market participants from the general performance standard.[41]

But similar tensions can also occur between two mechanisms based on the same regulatory premise, exemplified by the way emissions trading interferes with the environmental performance of certain types of renewable energy promotion. At worst, the two incentives virtually cancel each other out as a means of reducing greenhouse gas emissions, given that the generation of electricity with renewable

[39] See generally, Jonathan R. Nash, "Too Much Market? Conflict Between Tradable Pollution Allowances and the 'Polluter Pays' Principle", 24 *Harvard Environmental Law Review* (2000), 465, at 505.

[40] Hans-Joachim Koch and Annette Wieneke, "Das europäische und deutsche Anlagengenehmigungsrecht als Ordnungsrahmen eines Emissionshandels", in Hans-Werner Rengeling (ed.), *Klimaschutz durch Emissionshandel* (Cologne: Heymanns, 2001), 99, at 115.

[41] This amendment affected Council Directive 96/61/EC of 24 September 1996 concerning Integrated Pollution Prevention and Control (IPPC Directive), Official Journal L 257 of 10 October 1996, 26.

energy sources automatically increases the supply of unused allowances in the trading market and thereby disrupts the price signal required to influence corporate decisions. Moreover, the reductions achieved through renewable energy promotion could be achieved at lower cost if they were left entirely to operators participating in the market rather than a rigid promotion scheme. When this occurs, the renewable energy promotion rules ultimately subsidise CO_2 emissions originating outside of the power generation sector, rendering them an environmentally useless, but economically costly instrument.

A further illustration of conflict between two flexible instruments can be discerned in the overlap of emissions trading and voluntary declarations on climate protection by private enterprise. Under a voluntary declaration adopted by major sectors of German industry in 2000, these had pledged emissions reductions in exchange for a suspension of further regulatory measures; with the introduction of emissions trading throughout Europe, however, the federal government was bound to impose an aggregate limit on emissions for most parties to the agreement. Evidently, this did not conform with the reasoning of the earlier arrangement, although the government had no choice in the face of binding supranational commitments.

A third category of frictions can arise when implementing climate legislation in the context of *constitutional doctrines* and *fundamental rights*. On the broader level of constitutional law, the federal organisation of legislative and executive powers in many countries may impede effective elaboration and enforcement of climate policies, where a number of relevant issue fall within the purview of the federal legislator, but enforcement and administrative operationalisation, in particular, have traditionally been the prerogative of the federate provinces or states. Also, responding dynamically to changing environmental circumstances may often necessitate the delegation of legislative powers to executive bodies, whereas many national constitutions require that important issues attain the democratic legitimacy of statutory law.

Given the universal nature of global warming and the ample scope of mitigating policies, moreover, subjects may be affected in their individual rights and freedoms in manifold ways. For instance, emissions trading has been seen to be discriminatory towards sectors covered by the trading scheme, as opposed to other sectors which faced no aggregate emission limits. And altogether, with greenhouse gases traditionally subject to no form of management, the new trading system has been held to violate the established balance between individual rights and public concerns, a balance which had found its reflection in the general freedom to engage in pollutant operations subject only to a bound decision of preventive control. Emissions trading, so the argument of critics, would curtail the legal position of operators and render their ability to exercise fundamental rights dependent on a discretionary permit.[42]

And finally, tensions may arise between different *regulatory planes*, that is, divergent climate policies in domestic, supranational, and international law. What is

[42] For an overview of the arguments and their proponents, see my discussion in "European Emissions Trading and Environmental Regulation in the Member States: Irreconcilable Conflict?" in Teresa Fajardo del Castillo, Christoph Holtwisch, and Tereza Tichá (eds.), *Strengthening European Environmental Law in an Enlarged Union* (Aachen: Shaker, 2004), pp. 162 *et sqq.*

legal on the domestic plane, for instance, may conflict with precepts of supra – or international law. A salient illustration are all forms of incentives for the promotion of renewable energy sources and energy efficiency measures, as well as the free allocation of allowances to participants in the emissions trading scheme. Depending on the circumstances *in casu*, such benefits may be classified as state aid under the competition rules of the European Community[43] or as a subsidy under the Agreement on Subsidies and Countervailing Measures (SCM) administered by the World Trade Organisation.[44]

While there have been numerous efforts to reconcile separate normative environments by way of conflict or exception clauses, the tedious example of environmentally motivated trade restrictions has shown that institutions tend to prioritise their own agenda at the expense of any competing rules and objectives.[45] A second example is the admissibility of taxes or other charges on bunker fuels for aviation, which – although permissible under domestic law[46] – are precluded by anachronistic exemptions under the Chicago Convention on International Civil Aviation[47] as well as a number of bilateral agreements, formally known as "Bilateral Air Service Agreements" (BASAs).[48]

At the European level, moreover, Directive 2003/96/EC calls on Member States to "exempt … from taxation under conditions which they shall lay down for the purpose of ensuring the correct and straightforward application of such exemptions and of preventing any evasion, avoidance or abuse … energy products supplied for use as fuel for the purpose of air navigation."[49] All this has prevented legislators in

[43] See Articles 87 and 88 of the Treaty Establishing the Economic Community (EC Treaty), as amended by the Treaty of Nice Amending the Treaty on European Union, the Treaties Establishing the European Communities and Certain Related Acts, Nice, 26 January 2001, in force on 1 February 2003, OJ C 80/56 of 10 March 2001.

[44] Agreement on Subsidies and Countervailing Measures, opened for signature 15 April 1994, 1869 *United Nations Treaty Series* (1994) 14.

[45] For an overview, see Sabrina Shaw and Risa Schwartz, "Trade and Environment in the WTO: State of Play", 36 *Journal of World Trade* (2002), 129.

[46] Eckhard Pache and Joachim Bielitz, "Rechtliche Rahmenbedingungen einer Kerosinbesteuerung auf innerstaatlichen Flügen", 16 *Zeitschrift für Umweltrecht* (2004), 297–301.

[47] See article 24 of the Convention on International Civil Aviation (Chicago Convention), Montreal, 7 December 1944, in force on 4 April 1947, 15 *United Nations Treaty Series* (1944), pp. 295 *et sqq.*, elaborated by International Civil Aviation Organisation, Council Resolution on Environmental Charges and Taxes, adopted by the Council on 9 December 1996 at the 16th meeting of its 149th session, lit. 2 and 4.

[48] Members of the International Civil Aviation Organisation are required to deposit all such bilateral agreements with the Secretariat, which has compiled the roughly 3,000 BASAs in existence in a two-volume collection, ICAO, Document 9511, *Digest of Bilateral Air Transport Agreements* and *Supplement 1*.

[49] See Article 14 (1) of Council Directive 2003/96/EC of 27 October 2003 restructuring the Community Framework for the Taxation of Energy Products and Electricity, OJ 2003 L 283/51; Article 14 (2) of the Directive, however, allows Member States to limit the scope of this exemption "to international and intra-Community transport." Purely domestic flights, in other words, may be included in a kerosene taxation scheme.

several jurisdictions from implementing effective measures to contain emissions from the most rapidly growing source of greenhouse gases,[50] delaying any progress and forcing decision makers to resort to emissions trading as the only permissible measure.[51]

2.2.3 Coherence by Design: Envisioning a Domestic Climate Management Regime

Legislation implementing domestic climate policy is frequently encumbered by a number of tensions and outright conflicts. Increased harmonisation and simplification within an integrated policy framework also suggest themselves as a possible channel of improved energy and climate regulation, including better delivery of central objectives and principles to often wary addressees. Of course, a solution at the international or regional level would be preferable for various reasons, notably to lessen the concern about impacts on competitiveness and environmental efficacy. On the international plane, however, the consensus required for a sufficiently ambitious climate regime is currently absent, with the international community already facing challenges in the adoption of fairly moderate targets. At the regional level, in turn, legislative bodies tend to lack the necessary powers for comprehensive regulation of greenhouse gases, as is illustrated by the European Union, where political opinion might be more favourable than in an international setting, but the establishing treaty confers no comprehensive power to legislate climate and energy policy. With that in mind, the following sections will outline some considerations relating to the establishment of a domestic scheme to manage greenhouse gas emissions, starting with the possible sources of a legal mandate, the most important objectives, and tentative design elements.

2.2.3.1 The Legal Context – Identifying a Mandate

Before addressing the material objectives and design options of a comprehensive management regime for greenhouse gases, the current legal framework should first be assessed with a view to potential bases for such sweeping reform. In an area as sensitive as energy and climate change, after all, far-reaching policies are likely to find many linkages with fundamental tenets of constitutional law and economic

[50] Intergovernmental Panel on Climate Change (IPCC), *Special Report on Aviation and the Global Atmosphere* (Cambridge: IPCC/WMO/UNEP, 1999), especially chapter 6.

[51] See, notably, Proposal for a Directive of the European Parliament and of the Council amending Directive 2003/87/EC so as to Include Aviation Activities in the Scheme for Greenhouse Gas Emission Allowance Trading within the Community of 20 December 2006, COM(2006) 818 final.

regulation, all of which could impede the adoption of a uniform regime. At the same time, however, the legal order has gradually evolved to accommodate new and increasingly urgent environmental concerns, providing various gateways for a genuine mandate to support the adoption of a stringent climate policy architecture.

First and foremost, mitigation objectives entered by the government provide a strong foundation for comprehensive measures to meet these binding commitments, something a harmonised and consistent strategy is likely to facilitate. At the level of legal doctrine, one can point to the state objectives of environmental protection and intergenerational sustainability enshrined in constitutional documents such as the German Basic Law,[52] as well as the principle of coherence affirmed by many constitutional courts, effectively ruling out legislation that stipulates irreconcilable obligations for one and the same addressee.[53] Further support for a harmonised and consistent management scheme may be derived from the principle of integration, which has been vested in the status of positive law by the Member States of the European Community,[54] and the principle of proportionality, which could potentially impose a limit on cumulative burdens flowing from the overlap of different measures and policies.[55]

In many jurisdictions, energy and climate legislation has, to date, been based on the existing power to regulate economic activity as well as, more specifically, ambient air pollution.[56] Accordingly, there has been ample discussion whether the comprehensive management of greenhouse gases automatically incurs a violation of the fundamental right to engage in economic activity, manifested in an alleged right to use air as a resource and a medium for the absorption of emitted greenhouse gases. Indeed, in a decision on the responsibility of the state to compensate damage arising from air pollution, the German Federal Constitutional Court observed in a that "as a medium, 'air' is not subject to a management system under public law pursuant to which the holders of basic rights would generally be barred from access,

[52] See Article 20a of the German Basic Law, as amended on 27 October 1994, BGBl. Part I (1994), 3146, which reads: "[t]he state, aware of its responsibility for present and future generations, shall protect the natural sources of life within the framework of the constitutional order through the legislature and, in accordance with the law and the principles of justice, the executive and the judiciary."

[53] In its judgment of 7 May 1998 in Case 2 BvR 1991/95, Records of the Federal Constitutional Court (*BVerfGE*) (1998), 106, at 118, the Federal Constitutional Court addressed the permissibility of municipal waste and packaging charges, and found that "[t]he rule of law binds all legislative organs of the Federation and the *Länder* to coordinate their legislation in such a way as to prevent norm addressees from being confronted with countervailing rules which render the legal order contradictory" (translation by the author).

[54] See Article 6 of the EC Treaty, *supra* note 43, which reads: "Environmental protection requirements must be integrated into the definition and implementation of the Community policies and activities referred to in Article 3, in particular with a view to promoting sustainable development."

[55] On this argument, see Michael Kloepfer, *Umweltrecht* (3rd ed., Munich: C.H. Beck, 2004), Chap. 5 Annot. 284.

[56] See Article 74 (1) Nos. 11 and 24 of the German Basic Law.

and according to which use would depend on allocation by state bodies subject to their discretion."[57]

Applied to the context of climate change, such an understanding would preclude the comprehensive management of greenhouse gas emissions within an overarching framework, and would, instead, favour legislation in response to situational threats and narrowly defined issue areas. Unsurprisingly, that very approach has also been responsible for the current policy architecture, where individual policies and measures have accumulated without overall coordination, resulting in the conflicts identified in the preceding section.

Called upon to decide a challenge against the emissions trading legislation in Germany, for instance, the Federal Administrative Court has clarified that "air" could never fall within the ambit of private property, and that, instead, the rules on emissions trading merely regulate the use of property "insofar as is necessary for the general interest." In other words, the Court concluded that the emissions trading scheme was an appropriate, necessary and proportional means of protecting the global climate, and that it had merely led to the partial reorganisation of that specific area of law without infringing on the vested rights, both nationally and under Community law, of market participants.[58] Given the growing currency and media attention afforded to climate change in recent months, this perception is likely to have become more popular, providing the dogmatic basis for stringent and comprehensive management of greenhouse gas emissions in Germany.

2.2.3.2 Integrated Greenhouse Gas Management – Clinching the Objective

Any attempt to create an overarching framework for the management of greenhouse gas emissions will subsequently require the definition of uniform policy objectives. Not only is specification of a common purpose a prerequisite for the determination of substantive principles and regulatory instruments, but its very existence may also have a unifying effect on the subsequent implementation process. Clear objectives have therefore proven essential for effective governance of environmental challenges in the past.[59] Materially, however, these objectives will vary with the substantive scope afforded to the management scheme.

When deciding on the scale of the policy architecture, legislators will be called upon to make a strategic decision on its perimeters. Generally speaking, they can choose to either focus on greenhouse gas emissions and their limitation, or also include broader aspects of energy market regulation and its concurrent aims of

---

[57] See Federal Constitutional Court, Decision of 26 May 1998, Case 1 BvR 180/88, 51 *Neue Juristische Wochenschrift* (1998), 3264, at 3266 (translation by the author).

[58] See, notably, the judgment by the Federal Administrative Court *(Bundesverwaltungsgericht)*, 30 June 2005 (BVerwG 7 C 26.04), affirming that the introduction of emissions trading violated neither European fundamental rights nor the provisions of the Basic Law.

[59] Rudolf Steinberg, *Der ökologische Verfassunsstaat* (Frankfurt am Main: Suhrkamp, 1998), at 171.

energy security and an affordable, competitive energy supply. Although inherently different from mitigation policies, in turn, measures to adapt to global warming could also be included within the ambit of a management regime.

In all cases, however, substantive guidance will follow from any quantitative reduction commitments entered under international or supranational law, helping define the level of ambition that needs to be pursued with the overall management scheme. By necessity, moreover, a management scheme will have to address central aspects of the energy sector, given that achievement of the foregoing reduction targets will be conditional on a gradual transition to sustainability through improved efficiency in the exploitation of energy resources as well as in the generation, conversion, distribution, and end use of energy, but also a shift in the structure of energy sources towards increased use of renewable energy.[60]

Still, if the elaboration of a comprehensive management scheme is also meant to reduce tensions and conflicts between this scheme and other policies as well as within the scheme itself, it should aspire towards some general objectives of a systemic nature. Altogether, the management scheme should strive for the largest possible degree of consolidation and integration, ensuring the compatibility, consistency and complementarity of its various constituent policies and measures. With normative unity a central condition for the success of greenhouse gas mitigation, individual elements of this strategy must be deployed in conformity with the existing regulatory framework.[61]

By way of illustration, emissions reduction policies should be aligned with energy market rule to avoid tensions between the pursuit of a more sustainable energy supply and further market liberalisation. Ultimately, a comprehensive management scheme should avoid sending the contradictory signals relayed by current policies in place, and instead foster a high degree of harmony in its terminology, substantive goals and principles, and regulatory instruments. Another priority should be placed on curbing the excess regulation of earlier decades, reducing normative complexity and redundant bureaucratic obligations.[62]

Clear, simple and transparent norms may help reduce administrative costs and also promote identification by their addressees, thereby improving the prospects for adequate implementation. Accordingly, a comprehensive management scheme could seek to streamline mandatory procedures and consolidate permitting requirements. Given the dynamic nature of climate change and evolving responses at the regional and international plane, finally, the management scheme should be sufficiently flexible to accommodate external change. In order to safeguard the

[60] For an overview, see Martin Jänicke and Tobias Wiesenthal, "Eckpunkte und Entwicklungslinien einer nachhaltigen Energiewirtschaft", 15 *Zeitschrift für Umweltrecht* (2004), 385, at 385.

[61] See, for instance, Annex VIII No. 9 of Directive 2005/32/EC of the European Parliament and of the Council of 6 July 2005 establishing a Framework for the Setting of Ecodesign Requirements for Energy-using Products and amending Council Directive 92/42/EEC and Directives 96/57/EC and 2000/55/EC of the European Parliament and of the Council, OJ 2005 L 191/29.

[62] Michael Rodi, "Instrumentenvielfalt und Instrumentenverbund im Umweltrecht", 15 *Zeitschrift für Gesetzgebung* (2000), 231, at 234.

coherence of the overall scheme, however, future amendments should be subjected to an appropriate assessment procedure designed to identify potential impacts, as should any legislation adopted by administrative entities based on powers conferred to them.[63]

Such a Greenhouse Gas Management Act would ideally consist of a general part outlining the shared objectives, definitions, and principles, and a specific part focusing on individual sectors or issue areas, and the measures adopted within its ambit. In the general part, accordingly, the legislator could draw attention to mitigation commitments entered under international law and specify a global greenhouse gas reduction target, breaking this aggregate objective down to different sectors and activities. General principles could include a duty to take protective and preventive action against climate change, or the duty to use energy efficiently.

As for the selection of suitable instruments, the overall aim should be to arrive at a combination of different instruments capable of influencing individual and collective allocation decisions in line with the objectives defined earlier, and addressing all sources of greenhouse gas emission within the substantive and geographic scope of the Greenhouse Gas Management Act. All instruments currently in use or otherwise discussed for global warming mitigation are theoretically available, including:

- regulations and standards specifying mandatory abatement technologies or minimum requirements for pollution output;
- taxes and charges imposed on undesirable activity by a source;
- tradable permit schemes establishing a limit on aggregate emissions by specified sources and allowing trade among them;
- voluntary agreements between a government authority and one or more private parties with the aim of achieving emissions reductions beyond compliance with regulated obligations;
- subsidies and incentives awarded to an entity for performing a specified action;
- information instruments requiring public disclosure of environmentally related information, including labelling programmes and rating and certification systems; as well as
- research and development measures involving direct government funding and investment for innovative approaches to mitigation or the infrastructure of emissions reductions.[64]

---

[63] As a suitable model for such an assessment, one might refer to the legislative impact assessment required by Section 44 of the Common Rules of Procedure of the Federal Ministries (*Gemeinsame Geschäftsordnung der Bundesministerien* – GGO), 26 June 2000, Legislative and Ministerial Gazette (*GMBl*) (2000), pp. 525 *et sqq.*, as well as the creation of a special institution with the National Norm Review Committee Act (*Gesetz zur Einsetzung eines Nationalen Normenkontrollrates* – NKRG), 14. August 2006, BGBl. I (2006), pp. 1866 *et sqq.*

[64] This list is based on the draft Working Group III contribution to the Intergovernmental Panel on Climate Change Fourth Assessment Report, *Climate Change 2007: Mitigation of Climate Change* (Cambridge: Cambridge University Press, 2007), Chapter 13.1.1, available on the Internet at: <http://www.mnp.nl/ipcc/pages_media/FAR4docs/chapters/CH13_Policies.pdf> (last accessed on 1 June 2007), and is by no means comprehensive.

Further instruments might include planning and impact assessment procedures as well as liability rules and criminal sanctions, to name but a few. In order to achieve the strategic objectives of greater consolidation and integration, however, it is imperative that these instruments be carefully screened on the basis of appropriate criteria prior to their inclusion in a Greenhouse Gas Management Act, in order to avoid inconsistencies, conflicts and regulatory overlap.[65] And this is the most challenging stage in the elaboration of a suitable instrument mix. Commonly cited criteria of policy choice, such as those outlined by the Intergovernmental Panel on Climate Change in consecutive Assessment Reports, are generally too formulaic and abstract to allow for the contextuality of selection processes and the manner in which policy instruments are both formulated and implemented within a sophisticated matrix of interests, procedures and institutional mandates as well as material legal constraints.

Accordingly, criteria such as environmental effectiveness, cost effectiveness, distributional considerations and institutional feasibility may provide initial guidance, but are unable to determine the outcome of any given selection process.[66] Additional criteria, such as market conformity, administrative and transaction costs, political acceptance and legitimacy, openness to innovation, and the degree of flexibility and reflexiveness, may also prove helpful, but are equally unable to place the choice of instruments on a purely rational, objective and universally acceptable basis. In that sense, scholars and decision makers will arguably face their most important task when it comes to identifying suitable selection criteria based on the actual necessities at hand, engaging in an interdisciplinary and practically relevant discourse.[67]

2.3 Instrument Choice at the International Level

Past decades have seen an astounding proliferation of international arrangements in the area of the environment. A widespread perception that these have proven only marginally successful sparked growing interest, both institutional and academic, in the conditions and requirements of improved environmental governance. Over time, this shift in attention from the design of new international environmental arrangements to their evaluation and improvement has elicited a number of individual and collaborative research efforts across academic disciplines, producing a

[65] See, for instance, European Commission, *Green Paper on Market-Based Instruments for Environment and Related Policy Purposes*, 28 March 2007, COM(2007) 140, at pp. 8 *et seq.*

[66] Gawel, supra note 38, at 9, affirms that such theoretical criteria suffer from insufficient information on complex chains of causality, physical damage functions, persuasive valuation criteria based on contingent perception of utility, and macroeconomic costs of reallocating production factors to environmental protection, all rendering such welfare-based approaches to the description of instruments "at best a general reference system depicting ideal conditions in society" (translation by author).

[67] See, for instance, Rodi, supra note 62, at 241.

wealth of output and generating intense debate. In effect, research on the role and consequences of environmental regimes, treaties, and institutions became such a dominant part of the study of international relations at one point that it compelled a scholar to speak of a "veritable growth industry" and a "driving force" in his field.[68] Much of the resulting literature has focused on specific dimensions of regime performance, with the greatest weight being afforded to questions of effectiveness, followed by research on economic impacts, fairness, and equity.[69]

But even within these narrow categories, terms and definitions have varied greatly due to "elusive" concepts involving "daunting evaluative and analytical problems" that have given rise to much "disagreement, both in method and approach and in substantive views". Significant variations in the focus of relevant studies, as well as the distinct intellectual backgrounds and orientation of their authors, have resulted in very different approaches to the measurement of performance in terms of outputs, outcomes, and impacts. Research on the effectiveness of international environmental governance, for instance, was initially prompted by a shared concern about the ability of cooperative arrangements to influence state behavior, and hence focused on issues of regime design and improved compliance management. But definitions of what exactly constitutes "effective" governance differed widely in earlier research, with some authors merely seeking behavioral change or observable political effects, while others set the threshold higher by looking for an improvement in – or even resolution of – the situation that necessitated cooperation in the first place. Although later research has become more critical in terms of applied methods and concepts, even a recent shift to more empirical and quantitative approaches has failed to altogether eliminate some of the more persistent epistemic challenges in the study of regime effectiveness, including identification of the purpose of cooperation and of causal connections between governance systems and subsequent behavioral or physical change.

While the conceptual limitations of this line of research are thus readily apparent, the work to date reflects a sophisticated intellectual effort to determine whether international environmental cooperation plays a role in shaping collective action and social practices. Progress has been made, in particular, when it comes to distinguishing normative and utilitarian motives for state behavior and extending the perception of environmental compliance beyond binary treaty observance to a more managerial process focused on clarity, capacity, and priority, in which soft incentives and facilitation play as much a role as traditional legal coercion. More recently, scholars have responded to the rapid growth in environmental regimes by focusing on regime fragmentation and overlap, discussing options to manage conflicts and leverage synergies between multiple levels of governance and concurrent governance systems.

---

[68] Michael Zürn, "The Rise of Internavtional Environmental Politics: a Review of Current Research", 50 *World Politics* (1998), 617–649, at 649.

[69] Ronald B. Mitchell, "Evaluating the Performance of Environmental Institutions: What to Evaluate and How to Evaluate it?" in Oran R. Young, Leslie A. King, and Heike Schroeder (eds), *Institutions and Environmental Change* (Cambridge: MIT Press, 2008), 79–114.

Overall, there can be little doubt that our comprehension of international environmental cooperation has been greatly advanced, from the earliest stages of diplomatic negotiations to the final application and enforcement of individual arrangements. Nonetheless, studies of regime performance have so far failed to yield a set of clear and robust generalizations about the conditions for successful environmental governance. In particular, aspects other than compliance and effectiveness, such as economic impacts, fairness, and legitimacy, have received less systematic consideration in the absence of large, integrated research networks. Future work is likely to address such remaining gaps while further improving the clarity and transparency of analysis. Standardized definitions of key concepts, more rigorous comparison of findings across projects and disciplines, and use of advanced methods such as statistical analysis, simulations, and integrated case studies will help aggregate cumulative knowledge about the dynamics that affect regime formation and implementation. In the meantime, however, the research agenda remains heterogeneous, underscoring the earlier assertion that no single approach can capture the diverse ways of looking at international environmental cooperation, calling instead for a case by case determination of suitable evaluation criteria.

Existing surveys of alternative approaches to international climate governance have already devoted significant intellectual effort to defining generally applicable criteria for the evaluation of cooperative frameworks. What is more, they have been, to a greater or lesser extent, able to build on the cumulative insights offered by previous research on the assessment of domestic environmental policy and international environmental governance. Still, the criteria proposed in relevant literature to date are fairly heterogeneous. Only one criterion – environmental effectiveness – is common to all proposals, and even that is characterized by variations in the conceptual definition and scope. Other criteria, such as economic implications and considerations of equity, feature in a majority of studies, but again, their material content varies substantially. Comparisons across surveys become virtually impossible.

Chapter 3
Exploring the Landscape of Climate Law and Scholarship: Two Emerging Trends

Kati Kulovesi

Abstract Exploring the landscape of climate law, this chapter identifies two emerging trends increasingly visible in climate law scholarship. The first relates to the multi-layered nature of climate law. Here, the chapter argues that our understanding of the complex web of legal norms that address climate change necessitates research that also takes into consideration interactions between various sources of legal authority in regulating climate change, including their hierarchies, synergies and tensions. In addition to benefitting those implementing climate law on the ground, such an approach makes it possible to analyse the global implications of climate law, including its effectiveness and the mutual supportiveness of its various layers. The second relates to deformalization of climate law; the expanding role of non-state actors, soft law instruments and informal collaboration in global efforts to address climate change. While climate law scholarship is increasingly paying attention to this phenomenon, this chapter argues that accounting for the role of non-state actors and voluntary regulatory initiatives involves some important doctrinal challenges, including how to avoid becoming overtly descriptive and retain a normative focus.

3.1 Introduction

The emergence of the notion 'climate change law' reflects the growing volume and complexity of regulatory activity around climate change. Lawyers have begun to specialize in climate change issues, and they often do so by familiarizing themselves

---

Kati Kulovesi is Adjunct Professor in Climate Law. She holds a Ph.D. from the London School of Economics and Political Science. The author would like to thank Harro van Asselt, Tuomas Kuokkanen, Elisa Morgera, Michael Mehling and Tapio Määttä for their helpful comments on the manuscript. All mistakes and omissions remain my own responsibility.

K. Kulovesi (✉)
Department of Law, University of Eastern Finland, P.O. Box 111, 8010 Joensuu, Finland
e-mail: kati.kulovesi@uef.fi

E.J. Hollo et al. (eds.), *Climate Change and the Law*, Ius Gentium: Comparative Perspectives on Law and Justice 21, DOI 10.1007/978-94-007-5440-9_3,
© Springer Science+Business Media Dordrecht 2013

with legal norms related to climate change across jurisdictions and legal regimes. While the United Nations Framework Convention on Climate Change (UNFCCC)[1] and its Kyoto Protocol[2] have played a key role in driving and guiding the development of climate law,[3] climate change is governed and regulated at multiple levels, from the international, regional and national to the local and transnational ones, and with the involvement of diverse actors. Such regulatory diversity is understandable given the complexity of the underlying problem. As Held and Hervey indicate, the climate change challenge is "multifaceted and multi-layered" and it demands "effective policy at the level of both the nation-state and global governance."[4] The expansion of climate law can also be seen as a sign of the increasing mainstreaming of climate change; its integration into other policy domains; and its uptake by a range of organizations.[5] Thus, already in its current form, the territory of climate change law extends far beyond the UNFCCC and international law.[6]

From the scholarly perspective, climate law is still in its infancy. In line with the present book's attempt to charter the doctrinal territory of this emergent legal discipline, this chapter identifies two trends increasingly reflected in climate law scholarship and discusses their research implications. The first trend is the growing recognition that climate change, a global problem requiring local action, is governed and regulated at multiple levels. Given the transboundary nature of the underlying problem, climate law appears to have a tendency to cross legal and geographical boundaries. As a result, questions concerning the interplay between various sources of legal authority, including their hierarchies, synergies and tensions, are particularly relevant for climate law research and would arguably benefit from increased doctrinal attention. The second trend relates to deformalization; the involvement of a multitude of non-state actors in global efforts to address climate change and the increasing reliance on soft law instruments and informal collaboration.[7] Global climate change cooperation encompasses a range of local and regional initiatives,

[1] United Nations Framework Convention on Climate Change, 9 May 1992, New York, in force 21 March 1994, 31 *International Legal Materials* (1992), 849.

[2] Kyoto Protocol to the United Nations Framework Convention on Climate Change, Kyoto, 10 December 1997, in force 16 February 2005, 37 *International Legal Materials* (1998), 22.

[3] For a comprehensive overview, see Farhana Yamin and Joanna Depledge, *The International Climate Change Regime: A Guide to Rules, Institutions and Procedures* (Cambridge, UK: Cambridge University Press, 2004).

[4] David Held and Angus Hervey, "Democracy, Climate Change and Global Governance: Democratic Agency and the Policy Menu Ahead", in David Held, Angus Hervey and Marika Theors (eds), *The Governance of Climate Change: Science, Economics, Politics & Ethics* (Cornwall: Polity Press, 2011) 89, at 89.

[5] Harriet Bulkeley and Peter Newell, *Governing Climate Change* (London and New York: Routledge, 2010), at 106.

[6] For an overview of the multifaceted nature of climate change governance, see Liliana B. Andonova et al., "Transnational Climate Governance", 9 *Global Environmental Politics* (2009), 52.

[7] Ibid., at 54–56. See also Michele M. Betsill and Harriet Bulkeley, "Cities and the Multi-level Governance of Global Climate Change", 12 *Global Governance* (2006), 141, at 144; Chukwumerije Okereke, Harriet Bulkeley and Heike Schröder, "Conceptualizing Climate Governance Beyond the International Regime", 9 *Global Environmental Politics* (2009), 58.

cooperation between the public and private sectors, voluntary private sector initiatives and activities by civil society.[8] Accounting for the plural mix of regulatory initiatives around climate change involves, however, some important theoretical and ideological challenges, including how to avoid becoming overtly descriptive and retain a normative focus.

Overall, the landscape of climate law is characterised, *inter alia,* by multiple layers, overlapping sources of legal authority, deformalization and recurrent interactions between legal systems, regimes and actors involved. This regulatory complexity presents challenges for the emergent climate law scholarship.[9] Such questions are not, however, limited to the sphere of climate law. Globalization has impacted most fields of law, prompting scholars to develop new approaches focusing on themes, such as global legal pluralism,[10] global administrative law,[11] fragmentation[12] and constitutionalization[13] of international law, and so on. Given its close links with

---

[8] Bulkeley and Newell, *Governing Climate Change,* supra, note 5; Karin Bäckstrand, "Accountability of Networked Climate Governance: The Rise of Transnational Climate Partnerships", 8 *Global Environmental Politics* (2008), 74; Liliana B. Andonova, "Public-Private Partnerships for the Earth: Politics and Patterns of Hybrid Authority in the Multilateral System", 10 *Global Environmental Politics* (2010), 25; Elisa Morgera and Kati Kulovesi, "Public-Private Partnerships for Wider and Equitable Access to Climate Technologies", in Abbe Brown (ed.), *Environmental Technologies, Intellectual Property and Climate Change: Accessing, Obtaining and Protecting* (forthcoming, Edward Elgar, 2012); Kristine Kern and Harriet Bulkeley, "Cities, Europeanization and Multi-Level Governance: Governing Climate Change through Transnational Municipal Networks", 47 *Journal of Common Market Studies* (2009), 309; Betsil and Bulkeley, "Cities and the Multi-level Governance of Global Climate Change", supra, note 7.

[9] On research challenges related to transnational environmental law in general, see Elizabeth Fisher, "The Rise of Transnational Environmental Law and the Expertise of Environmental Lawyers", 1 *Transnational Environmental Law* (2011), 1 Transnational Environmental Law (2012), 43 at 45–47.

[10] For an overview, see Simon Roberts "After Government? On Representing Law without a State", 68 *Modern Law Review* (2005), 1. See also Günther Teubner, "Global Bukowina: Legal Pluralism in the World Society", in Günther Teubner (ed.), *Global Law without a State* (Aldershot: Dartsmouth, 1997), 3; Oren Perez, *Ecological Sensitivity and Global Legal Pluralism. Rethinking the Trade and Environment Conflict* (Oxford and Portland, Oregon: Hart Publishing, 2004); and Paul Schiff Berman, "Global Legal Pluralism", 80 *Southern California Law Review* (2007), 1155.

[11] Benedict Kingsbury, Nico Krisch and Richard B. Stewart, "The Emergence of Global Administrative Law", 68 *Law and Contemporary Politics* (2005), 15; Nico Krisch and Benedict Kingsbury, "Introduction: Global Governance and Global Administrative Law in the International Legal Order", 17 *European Journal of International Law* (2006), 1.

[12] *Fragmentation of International Law: Difficulties Arising from the Diversification and Expansion of International Law.* Report of the Study Group of the International Law Commission finalized by Martti Koskenniemi, UN. Doc. A/CN.4/L.682, 13 April 2006; Frank Biermann et al., "The Fragmentation of Global Governance Architectures: A Framework for Analysis", 9 *Global Environmental Politics* (2009), 14. See also Harro van Asselt, "Managing the Fragmentation of International Climate Law" in Chapter 13 of the present volume.

[13] Jan Klabbers, Anne Peters and Geir Ulfstein, *The Constitutionalization of International Law* (Oxford et al.: Oxford University Press, 2009); Jeffrey L. Dunoff, "The Politics of International Constitutions: The Curious Case of the World Trade Organization", in Jeffrey L. Dunoff and Joel P. Trachtman (eds), *Ruling the World? Constitutionalism, International Law, and Global Governance* (Cambridge et al.: Cambridge University Press, 2009), 178, at 179; Mattias Kumm,

many of the underlying questions, climate law appears to present ample opportunities to further explore and engage with these themes.

This chapter begins by illustrating in Sect. 3.2 the multi-layered nature climate change law and the diversity of actors involved. By describing the multitude of legal sources that commonly apply in parallel to a single carbon transaction under the Clean Development Mechanism (CDM), established under the Kyoto Protocol, it highlights the complexity of the regulatory landscape that lawyers and various other actors working on the CDM must navigate. Section 3.3 considers interactions between various sources of legal authority in regulating climate change, arguing that such questions are increasingly relevant for climate law research. This is partly due to the growing tendency of some actors, including the European Union (EU), to try to influence legal developments outside their territory, especially concerning climate change mitigation. However, legal norms related to climate change interact also in other ways, as the chapter shows. For example, national legislation on Green Investment Schemes in some Central and Eastern European countries complements international rules on emissions trading under the Kyoto Protocol. Finally, Sect. 3.4 addresses the trend of deformalization and the role of the private sector and other non-state actors in the field of climate change law. Non-state actors play a critical role in the battle against dangerous climate change both because they are effectively responsible for global greenhouse gas emissions and also because it has been estimated that they will be responsible for the vast majority of future financial flows to address climate change. They are also increasingly engaging in various public-private partnerships and voluntary regulatory activities around climate change. Ignoring these initiatives and the various associated soft law instruments would mean painting an incomplete picture of the landscape of climate change law. However, the argument here is that accounting for the role of the private sector and voluntary regulatory initiatives also involves some important theoretical and ideological challenges.

3.2 Mapping the Landscape of Climate Change Law

3.2.1 Role of the UNFCCC

International law has played an important role in driving the development of climate change law. In its first resolution on climate change in 1988, the United Nations (UN) General Assembly recognized climate change as "a common concern

"The Cosmopolitan Turn in Constitutionalism: On the Relationship between Constitutionalism in and beyond State", in Jeffrey L. Dunoff and Joel P. Trachtman (eds), *Ruling the World? Constitutionalism, International Law, and Global Governance* (Cambridge et al.: Cambridge University Press, 2009), 258, at 260.

of mankind."[14] It agreed that "necessary and timely action should be taken to deal with climate change *within a global framework*."[15] The UNFCCC, adopted in 1992, has attracted 195 Parties, thus becoming universal in scope. It forms the basis for a dynamic and rapidly evolving international legal regime. One of the key arguments supporting universal climate change cooperation is that it addresses the problem of free riding and reduces the costs of both mitigation and adaptation.[16] Furthermore: "No two countries will face exactly the same situation in terms of impacts or the costs and benefits of action, and no country can take effective action to control the risks that they face alone."[17] The international legal response to climate change also involves fundamental questions concerning justice, equity and fairness. Countries that have contributed least to the problem are projected to suffer the most serious consequences of climate change, especially the small island developing States, African countries and least developed countries. Given that it brings together both those responsible for the problem and those suffering its most severe consequences, my argument is that the UNFCCC enjoys a high degree of legitimacy as a negotiating forum and legal framework for addressing climate change.

Despite its significant evolution over the past 20 years, the UNFCCC regime is yet to deliver a robust legal architecture that ensures the ultimate objective enshrined in Article 2 of the Convention of avoiding dangerous anthropogenic climate change and is in line with the global long-term goal of limiting temperature increase to 2°, formally adopted by UNFCCC Parties in 2010.[18] In fact, the effectiveness of the UN climate regime and its ability to engage key countries in meaningful mitigation action have been questioned a number of times over the years.[19] Especially in the aftermath of the 2009 UN Climate Change Conference in Copenhagen, several mostly complementary venues of international climate change cooperation have emerged[20] although the argument has also been made that instead of trying to reach consensus among 195 Parties, efforts to enhance climate change mitigation should

[14] UN General Assembly Resolution, Protection of global climate for present and future generations of mankind, UN Doc. A/RES/43/53, 6 December 1988, para. 1.

[15] Ibid., para. 2. Emphasis added.

[16] Nicholas Stern, *The Economics of Climate Change. The Stern Review* (Cambridge et al.: Cambridge University Press, 2006), at 510.

[17] Ibid.

[18] UNFCCC, supra note 1, Art. 2; Decision 1/CP.16, The outcome of the work of the Ad Hoc Working Group on Long-term Cooperative Action under the Convention, UN Doc. FCCC/CP/20010/7/Add.1, 15 March 2011.

[19] See, for example, Gwyn Prins and Steve Rayner, "Time to Ditch Kyoto?", 449 *Nature* (2007), 973; and Rafael Leal-Arcas, "Top-Down Versus Bottom-Up Approaches for Climate Change Negotiations: An Analysis", 24 January 2012, available at: http://ssrn.com/abstract=1950210 (last accessed on 1 March 2012).

[20] For an overview of such claims, see Camilla Bausch and Michael Mehling, "Addressing the Challenge of Global Climate Mitigation. An Assessment of Exiting Venues and Institutions", August 2011, available at: http://library.fes.de/pdf-files/iez/08466.pdf (last accessed on 29 February 2012).

focus on major emitters.[21] Meanwhile, long-term negotiations continue under the UNFCCC, with the current deadline of concluding a new global climate treaty by 2015, to be implemented from 2020.[22] The argument here is that even accomplishing this important but challenging task, the landscape of climate change law will remain multi-layered and colourful. The following example concerning the CDM illustrates that taking into account a plural mix of legal sources is necessary even where the legal mechanism is firmly founded in an international climate treaty. Furthermore, a global climate agreement may well prove elusive concerning some sectors or actors. All this highlights the need for climate law research to take into consideration legal initiatives both within and outside the UNFCCC framework, exploring their linkages, synergies and tensions.

3.2.2 Regulation of the CDM: Multiple Layers, Diverse Actors and Deformalization

The CDM is a good way to illustrate my argument that climate law often derives from a plural mix of normative sources. Legal norms applicable to a single carbon transaction under the CDM under the CDM often originate from a variety of overlapping sources, including the Kyoto Protocol, Marrakesh Accords, CDM Executive Board, the project host country and possibly also the purchasing country. In addition, the Emission Reductions Purchase Agreement (ERPA) typically lays down a number of contractual obligations for the seller and buyer of Certified Emission Reductions (CERs), effectively creating a second legal layer, dominated by private international law and running parallel with the CDM project cycle regulated under the Kyoto Protocol.[23] CDM project participants sometimes also choose to seek compliance with voluntary standards, such as the CDM Gold Standard, that have been produced through civic regulatory initiatives. Finally, CERs themselves are typically subject to a multitude of norms, including accounting rules and practices, taxation rules, as well as rules on ownership and liability.

The legal foundation of the CDM is in Article 12 of the Kyoto Protocol. This general treaty provision did not include the necessary details for operationalizing the CDM. Instead, the text of the Kyoto Protocol tasked the Conference of the Parties serving as the meeting of the Parties to the Kyoto Protocol (COP/MOP) with elaborating "modalities and procedures with the objective of ensuring transparency, efficiency and accountability through independent auditing and verification of

[21] Leal-Arcas, "Top-Down Versus Bottom-Up Approaches for Climate Change Mitigation", supra, note 19, at 2.

[22] Decision 1/CP.17, Establishment of an Ad Hoc Working Group on the Durban Platform for Enhanced Action, UN. Doc. FCCC/CP/2011/9/Add.1, 15 March 2012, paras. 2 and 4.

[23] For critical discussion, see Anne-Marie Klijn, Joyeeta Gupta and Anita Nijboer, "Privatizing Environmental Resources: The Need for Supervision of Clean Development Mechanism Contracts?", 18 *Review of European Community and International Environmental Law* (2009), 172.

project activities."[24] Detailed rules for the CDM were subsequently adopted as part of the 2001 Marrakesh Accords and they regulate key aspects of the project cycle, including baseline development, validation and registration of the CDM project as well as verification and certification of CERs.[25] The general CDM rules adopted in Marrakesh have been subsequently complemented by decisions laying down rules for sink[26] and small-scale projects[27] under the CDM, and further COP/MOP guidance on various other issues related to the CDM.

Article 12 of the Kyoto Protocol also established the CDM Executive Board, tasked with supervising the CDM.[28] The CDM Executive Board plays an important role in regulating the CDM. It creates detailed rules through its decisions, addressed mainly at the thousands of private and public actors implementing CDM projects on the ground. This institutional structure has resulted in the evolution of a complex web of rules and requirements applicable to the CDM. It has also been argued that "the Board's decision-making practice is often not predictable, and many of its decisions have come as a surprise to project participants and technical experts."[29] After complaints from a number of stakeholders that the regulation of the CDM was quickly becoming too difficult to grasp, the COP/MOP requested the CDM Executive Board to develop a catalogue of its decisions.[30] As of 2012, this online tool remains under development by the UN Climate Change Secretariat. Thus far, a more influential initiative to categorize CDM rules has been the *CDM Rulebook*, known as the "definitive online database on CDM rules" developed and maintained by law firm Baker & McKenzie with initial funding from eight donor countries and organizations.[31] The database is updated after each meeting of the CDM Executive Board and the COP/MOP.

The CDM is also a prime example of a public-private partnership that seeks to promote climate change mitigation and sustainable development. Private entities are

---

[24] Kyoto Protocol, supra, note 2, Art. 12.7.

[25] Decision 3/CMP.1, Modalities and procedures for a clean development mechanisms defined in Article 12 of the Kyoto Protocol, UN Doc. FCCC/KP/CMP/2005/8/Add.1, 30 March 2006. The Marrakesh Accords were originally adopted by COP 7 in 2001, but their formal adoption under the Kyoto Protocol took place at COP/MOP 1 in 2005.

[26] Decision 5/CMP.1, Modalities and procedures for afforestation and reforestation project activities under the clean development mechanism in the first commitment period of the Kyoto Protocol, UN Doc. FCCC/KP/CMP/2005/8/Add.1, 30 March 2006.

[27] Decision 4/CMP.1, Guidance relating to the clean development mechanism, Annex II, Simplified modalities and procedures for small-scale CDM project activities and Decision 6/CMP.1, Simplified modalities and procedures for afforestation and reforestation project activities under the clean development mechanism in the first commitment period of the Kyoto Protocol and measures to facilitate their implementation, UN Doc. FCCC/KP/CMP/2005/8/Add.1, 30 March 2006.

[28] Kyoto Protocol, supra, note 2, Art. 12.4.

[29] Charlotte Streck and Jolene Lin, "Making Markets Work: A Review of the CDM Performance and the Need for Reform", 19 *European Journal of International Law* (2008), 409, at 410.

[30] Overall, several reform proposals for the CDM have been put forward by both states and private actors, many of which are currently being explored through the CDM Policy Dialogue, launched in 2011. More information is available at: http://cdmpolicydialogue.org/ (last accessed on 30 April 2012).

[31] The CDM Rulebook, available at: http://www.cdmrulebook.org/ (last accessed on 26 March 2012).

largely responsible for financing and implementing CDM projects on the ground. They also perform key functions in terms of ensuring compliance with the international CDM rules by validating the projects and verifying the ensuing emission reductions.[32] All this goes to show that what was originally a provision in an international treaty has become a dynamic regulatory process that involves not only sovereign states that are Parties to the Kyoto Protocol, but also the CDM Executive Board, the UNFCCC Secretariat, national CDM authorities, private sector and civil society actors, local stakeholders as well as donor countries and organizations.

The close engagement of the private sector in the governance and implementation of the CDM has posed some interesting challenges to the traditional, state-centred focus of public international law. The CDM Executive Board is an institution established under an international treaty. However, it has come to exercise authority over private actors that very much resembles administrative powers typically used by public authorities in national jurisdictions. For example, when approving and rejecting project proposals, the CDM Executive Board makes decisions with significant legal and economic implications for private actors participating in the CDM.[33] Given that it is not possible to appeal the CDM Executive Board's decisions, this aspect of the CDM is at odds with some of the key rights that are traditionally protected by domestic constitutions, including the right to a fair hearing and to effective judicial review.[34] COP/MOP 5 has consequently requested that the CDM Executive Board create an appeals procedure under the CDM and negotiations on the new appeals body are currently ongoing.[35] Given the functions that the CDM Executive Board exercises, it can be argued that "[t]he type of governance undertaken by the EB can be understood and analysed as administrative action: rule-making, administrative adjudication between competing interests, and other forms of regulatory decision-making and management."[36] This differs from the traditional, state-centred focus of international law. Likening governance of the CDM to administrative action brings to the fore links to the global administrative law project, which proceeds from the argument that:

> …we are witnessing the emergence of a 'global administrative space'; a space in which the strict dichotomy between domestic and international has largely broken down, in which administrative functions are preformed in often complex interplays between officials and institutions on different levels, and in which regulation may be highly effective despite its

[32] See Sect. 3.4. below, and also: Kati Kulovesi, "The Private Sector and the Implementation of the Kyoto Protocol: Experiences, Challenges and Prospects", 16 *Review of the European Community and International Environmental Law* (2007), 146.

[33] Streck and Lin, "Making Markets Work", supra, note 29, at 410–411; see also Ludger Gieberts and Alexander Sarac, "An Appeals Process for the Kyoto Protocol's Clean Development Mechanism", 4 *Carbon and Climate Law Review* (2010), 260, at 261.

[34] Jeffrey L. Dunoff and Joel P. Trachtman, "A Functional Approach to International Constitutionalization", in Jeffrey L. Dunoff and Joel P. Trachtman (eds), *Ruling the World? Constitutionalism, International Law, and Global Governance* (Cambridge et al.: Cambridge University Press, 2009), 3, at 17. Dunoff and Trachtman discuss this in the context of the UN Security Council actions imposing sanctions and firms suspected of involvement in terrorist activities.

[35] Decision 2/CMP.5, Further guidance related to the clean development mechanism, UN Doc. FCCC/KP/CMP/2009/21/Add.1, 30 March 2010, para. 42.

[36] Streck and Lin, "Making Markets Work", supra, note 29, at 411.

> predominantly non-binding forms… Global administrative law proposes drawing together these dispersed practices and understand them as part of a common, growing trend towards administrative-law type mechanisms for holding global regulatory governance accountable and to inquire into the challenges this set of issues poses to both domestic administrative law and international law.[37]

However, even if the implementation of the CDM involves the exercise of public authority over private actors, governance of the CDM lacks many of the checks and balances that typically play a key role in domestic administrative processes. For climate law scholarship, then, this calls for engaging in inquiries that expand the scope of legal analysis beyond the traditional doctrines of public international law to assess the legitimacy of the CDM and other novel forms of climate governance.

In addition to such doctrinal perspectives, the plural mix of legal sources applicable to the CDM may pose challenges to those implementing CDM projects on the ground. What I mean by this is that when designing and implementing a CDM project, it is necessary to take into consideration the combined and simultaneous effect of parallel legal norms originating from different sources of legal authority. The starting point is the international level, which forms the foundation for the CDM and, as we saw above, in itself includes several layers emanating from the Kyoto Protocol, COP/MOP decisions and norm-creation by the CDM Executive Board. From the practical perspective, national CDM regulations by the project host country are equally important. The international CDM rules require each non-Annex I Party to confirm both that its participation in a CDM project is voluntary and that the project contributes to its sustainable development.[38] Environmental impact assessments of CDM projects also take place in accordance with the host countries' national requirements.[39] In practice, a number of CDM host countries have developed national criteria and procedures for approving CDM projects. Complying with the host country's national CDM laws and regulations is crucial for project participants to obtain the host country's Letter of Approval, a prerequisite for registering the CDM project under the Kyoto Protocol.

In practice, most buyers of CDM credits come from the EU where the Emissions Trading Scheme (ETS) has introduced a price for greenhouse gas emissions of more than 10,000 installations.[40] Through the Linking Directive, they can use CERs

---

[37] Krisch and Kingsbury, "Introduction: Global Governance and Global Administrative Law in the International Legal Order", supra, note 11, at 1.

[38] Decision 3/CMP.1, supra, note 25, para. 40(a). The general requirement that a CDM project contributes to the host country's sustainable development is stipulated in Kyoto Protocol, supra, note 2, Art. 12.

[39] Ibid., para. 37(c).

[40] Directive 2003/87/EC of the European Parliament and of the Council of 13 October 2003 establishing a scheme for greenhouse gas emission allowance trading within the Community, OJ 2003 L 275/32; Directive 2008/101/EC of the European Parliament and of the Council of 19 November 2008 amending Directive 2003/87/EC so as to include aviation activities in the scheme of greenhouse gas emissions allowance trading within the Community, OJ 2009 L 8/3; and Directive 2009/29/EC of the European Parliament and of the Council of 23 April 2009 amending Directive 2003/87/EC so as to improve and extend the greenhouse gas emission allowance trading system of the Community, OJ 2009 L 140/63.

to comply with their emissions quota.[41] The EU has, however, introduced some stricter sustainability criteria for CDM credits under the ETS than those applicable under the Kyoto Protocol. It has, for instance, banned credits from afforestation and reforestation CDM projects and it will also prohibit CDM credits from certain industrial gas projects from May 2013 onwards.[42] Familiarity with EU climate change law is therefore highly relevant for the participants of such CDM projects that seek to generate CERs for installations included in the ETS. Furthermore, some Annex I countries within and outside the EU have enacted special national legislation laying down criteria for the approval of CDM projects.

In addition to the multiple layers of specialized CDM rules that have their origin in international, EU and national legal systems, a CDM transaction normally also raises legal questions concerning, taxes, accounting and contract law. The ERPA in particular plays a key role, especially as the vast majority of CDM projects are implemented by private actors, with limited or no involvement by an Annex I party to the Kyoto Protocol.[43] In practice, the ERPA process takes place in parallel with the official CDM project cycle and the regulation of ERPAs tends to falls under private international law rather than public international law.[44] It has been argued that the negotiation of carbon contracts and the structuring and financing of carbon transactions:

> ...requires the ability to overcome the 'disconnect' that often exists between international and national law and between private and public legal regimes, and to incorporate principles and structures, provided for in the Kyoto Protocol... into effective contracts which will bind the parties, comply with domestic law requirements, and also allow for enough flexibility to manage the constantly developing international legal framework.[45]

Some of the key elements in an ERPA include defining, *inter alia,* how various risks will be shared, the price of carbon credits, timetables for delivery and payments, as well as questions concerning liability, sanctions, applicable law and dispute resolution. An ERPA may also contain obligations related to environmental and social issues that more specific than those included in the international CDM rules and the host country's national CDM criteria. Some scholars have raised concerns over this 'dual legal nature' of the CDM cycle. Most notably, they have lamented that the private CDM contracting cycle "does not directly involve governments and is non-transparent."[46]

---

[41] Directive 2004/101/EC amending Directive 2003/87/EC establishing a scheme for greenhouse gas emission allowance trading within the Community, in respect of the Kyoto Protocol's project mechanisms, OJ 2004 L 338/18

[42] Press Release: Commission welcomes vote to ban certain industrial gas credits, IP/11/56, 21 January 2011.

[43] For interesting discussion see Soren E. Lütken and Axel Michaelowa, *Corporate Strategies and the Clean Development Mechanism: Developing Country Financing for Developed Country Commitments* (Cheltenham, UK and Northampto, MA, USA: Edward Elgar, 2008), at 110–146.

[44] Klijn, Gupta and Nijboer, "Privatizing Environmental Resources", supra, note 23, at 176.

[45] Martin Wilder, Monique Willis and Mina Guli, "Carbon Contracts, Structuring Transactions: Practical Experiences", in David Freestone and Charlotte Streck (eds), *Legal Aspects of Implementing the Kyoto Protocol Mechanisms: Making Kyoto Work* (Oxford et al.: Oxford University Press, 2005), 295, at 295–296.

[46] Klijn, Gupta and Nijboer, "Privatizing Environmental Resources", supra, note 23, at 177.

They have also highlighted "that most project developers in the developing world are inexperienced in international contracts" which "taxes their ability to participate effectively in contract negotiations and in understanding the broader ramifications of such contracts."[47] Hence, according to Klijn, Gupta and Nijboer, the "split personality of the CDM is due to it being both a public international law instrument, as well as a commercial law instrument, and is a critical legal challenge calling for solutions that reconcile these different personalities."[48]

Finally, some CDM project participants also seek to comply with voluntary regulatory initiatives designed to strengthen CDM projects' contribution to sustainable development. The best-known example is the CDM Gold Standard, established by the WWF in 2003 and currently endorsed by more than 80 non-governmental organizations worldwide.[49] The Gold Standard has been designed to "certify renewable energy and energy efficiency carbon offset projects to ensure that they all demonstrate real and permanent greenhouse gas (GHG) reductions and sustainable development benefits in local communities that are measured, reported and verified."[50] To do so, the Gold Standard project cycle involves steps that are additional to the official CDM project cycle.[51] These are sometimes turned into legally-binding obligations through the ERPA.

As this overview shows, the CDM is regulated through complex and innovative arrangements and its implementation involves a diverse mix of actors. As such, it poses challenges to both scholars researching climate change law as well as to lawyers and others implementing CDM projects on the ground. It also illustrates the multi-layered and colorful landscape of climate change law, and points towards the need to use innovative doctrinal tools and approaches in researching climate change law.

3.3 Climate Law: Interactions Between Sources of Legal Authority

3.3.1 Background: Globalization and Law

Over the past several decades, globalization has affected most areas of law.[52] For one, national legal fields have become more 'internationalized' as domestic legal

[47] Ibid.

[48] Ibid., at 181. Their suggested remedy is a supervisory body for climate change contract making.

[49] The Gold Standard website, available at: http://www.cdmgoldstandard.org/about-us/who-we-are (last accessed on 26 March 2012).

[50] Ibid.

[51] Ibid.

[52] Francis Snyder, "Economic Globalisation and the Law in the 21st Century", in Austin Sarat (ed.), *The Blackwell Companion to Law and Society* (Malden MA et al.: Blackwell Publishing, 2004), 3. Similarly, David Kennedy, "The Mystery of Global Governance", in Jeffrey L. Dunoff and Joel P. Trachtman (eds), *Ruling the World? Constitutionalism, International Law, and Global Governance* (Cambridge et al.: Cambridge University Press, 2009), 37, at 39; and Krisch and Kinsgsbury, "Introduction: Global Governance and Global Administrative Law in the International Legal Order", supra note 11, at 1.

and political developments are increasingly influenced by external factors.[53] As Snyder describes:

> Formally speaking, the sources of 'international' and 'national' norms are different, and this difference has its legal doctrinal importance in each of the two institutional and normative settings. However, the traditional distinction between 'domestic' and 'foreign,' or between 'national' and 'international,' does not often adequately capture the political origins, legal content, cultural understandings, economic assumptions, and social practices, for example the need for certain types of specialized legal professionals, of contemporary law.[54]

David Kennedy notes that most in the legal profession "thought they knew how it all worked" and legal thinking tended to be organized in relatively unproblematic categories, such as private law and public law, national law and international law.[55] Recently, however, boundaries of such categories are increasingly challenged. There are conflicting and multiplying jurisdictions, asserting the validity or persuasiveness of their rules, with no decider of last resort.[56] Kennedy also argues that specialists in every field of law "have all come to see their subject in international or comparative terms" and it is "hard to think of a legal problem that does not cross disciplinary and national boundaries."[57] Koskenniemi, in turn, draws attention to fragmentation and deformalization of international law, indicating that "traditional international law is pushed aside by a mosaic of particular laws and institutions, regimes and types of more or less formal regulation, each following its own preferences."[58]

Such developments are leaving their mark on legal scholarship and several new approaches have evolved in response to the globalization of the legal landscape.[59] They focus on themes such as constitutionalization and fragmentation of international law, the global administrative law project, and global legal pluralism.[60] According to Kennedy, public international law is in fact currently going through "a period of heightened doctrinal and methodological ferment" characterized by "disciplinary critique, confusion and rethinking".[61] Snyder, in turn, notes that many scholars are focusing on questions concerning hierarchy, coordination and multi-level governance, proceeding from the insight that different levels of governance interact, sometimes with regard to the same subject matter, sometimes with

---

[53] Snyder, "Economic Globalisation and the Law in the 21st Century", supra, note 52, at 3.

[54] Ibid.

[55] Kennedy, "The Mystery of Global Governance", supra, note 52, at 39.

[56] Ibid., at 55.

[57] Ibid., at 39.

[58] Martti Koskenniemi, "The Fate of Public International Law: Constitutional Utopia or Fragmentation", Chorely Lecture, 7 June 2006, London School of Economics and Political Science, at 13.

[59] For an overview, see Kennedy, "The Mystery of Global Governance", supra, note 52, at 43–53.

[60] See references, supra, notes 10–13.

[61] Kennedy, "The Mystery of Global Governance" supra, note 52, at 38.

regard to social life.[62] In the environmental field, scholars are increasingly interested in transnational environmental law, described by Shaffer and Bodansky in the following terms:

> The concept of transnational environmental law... is much broader than that of international environmental law. Transnational environmental law encompasses all environmental law norms that apply to transboundary activities or that have effects in more than one jurisdiction... The concept of transnational environmental law thus includes national environmental regulation that has horizontal effects across jurisdictions – for example, by providing regulatory models to other countries or by applying to or affecting the behavior of producers and consumers within them. It also includes the development of standards by private actors that have effects across borders, such as through product certification and labeling regimes. In practice, the transnational environmental law process sometimes includes international law as part of a single diachronic law-making process, but oftentimes does not.[63]

With climate law being multi-layered and characterized, *inter alia*, by overlapping sources of legal authority, deformalization, involvement of non-state actors and a high degree of specialization within the UNFCCC regime, it seems to have several links with these broader theoretical discussions and presents ample opportunities to further explore and engage with the themes reflected in these approaches.

3.3.2 Climate Law and Interaction Between Different Sources of Legal Authority

Given the global nature of the climate change problem, climate law has the tendency to cross legal and geographical boundaries. One of the arguments here is that studying interactions between various sources of legal authority in regulating climate change, including their hierarchies, synergies and tensions. including their hierarchies, synergies and tensions is necessary for analysing and understanding the combined effect of the multiple layers of climate change regulation. The relevant interactions commonly place vertically between international law and national legal systems. In many jurisdictions vertical interaction also occurs between the national and

---

[62] Snyder, "Economic Globalisation and the Law in the 21st Century", supra, note 52, at 5. As Betsill and Bulkely, "Cities and the Multi-level Governance of Global Climate Change", supra, note 7, at 149, explain: "The focus on multi-level governance emerged originally from studies of European integration, where the argument was made that the role of national governments within the EU was diminishing and that a new, multilevel system of governance was taking shape." The basic idea is that "decision-making competencies are increasingly shared between actors operating at different levels of governance" and the aim is to draw "attention to the importance of considering how political authority and processes of policymaking cross traditional divides between state and non-state actors, domestic and international spheres." Ibid.

[63] Gregory Shaffer and Daniel Bodansky, "Transnational Unilateralism and International Law", 1 *Transnational Environmental Law* (2012), 1 *Transnational Environmental Law* (2012), 31 at 32. For discussion on the concept of transnational environmental law, see also Fisher, "The Rise of Transnational Environmental Law", supra, note 9

sub-national levels. National climate regulation is also having horizontal effects across jurisdictions, for example, through unilateral climate action as well as through the diffusion of regulatory innovations.[64] Horizontal interaction also frequently takes place between specialised areas of international law. In this respect, several scholars have studied climate law against the backdrop of fragmentation of international law, raising questions concerning the mutual supportiveness of different international legal regimes from the climate change perspective, including the UNFCCC and World Trade Organization, and the UNFCCC and the Convention of Biological Diversity.[65]

The relevance of the vertical relationship has been reflected in the debate concerning the post-2012 legal architecture under the UNFCCC where one of the key questions is whether countries' mitigation commitments should be defined 'top down' through an international treaty or 'bottom up' through national legislation.[66] At the European level, questions have emerged concerning the relationship between EU climate law and its Member States' national legal systems, and also concerning the relationship between EU climate law and local regulatory initiatives.[67] Could, for example, some EU Member States implement stricter climate protection measures than those required by EU law and introduce carbon dioxide performance standards to companies included in the EU ETS?[68] Could the Mayor of London prohibit the use in London of passenger cars, which exceed the average EU emissions benchmark of 130 g of carbon dioxide per kilometre?[69]

---

[64] Shaffer and Bodansky, "Transnational Unilateralism and International Law", supra, note 63. I am also grateful for Harro van Asselt for inspiring my thinking in this regard.

[65] Harro van Asselt, Francesco Sindico and Michael Mehling, "Global Climate Change and Fragmentation of International Law", 30 *Law and Policy* (2008), 423; Margaret A. Young, "Climate Change Law and Regime Interaction", 5 *Carbon and Climate Law Review* (2011), 147; Kati Kulovesi, *The WTO Dispute Settlement System: Challenges of the Environment, Legitimacy and Fragmentation* (The Netherlands: Kluwer Law International, 2011), at 217–257; Annalisa Saravesi, "Reducing Emissions from Deforestation in Developing Countries under the UNFCCC: Caveats and Opportunities for Biodiversity", *Yearbook of International Environmental Law* (2011, forthcoming); Elisa Morgera, "Far Away, So Close: A Legal Analysis of the Increasing Interactions between the Convention on Biological Diversity and Climate Change Law", 2 *Climate Law* (2011), 85. See also the respective contributions by Harro van Asselt, Annalisa Saravesi and Elisa Morgera and myself in Part IV of the present volume.

[66] For discussion on top down and bottom up approaches, see Daniel Bodansky, "A Tale of Two Architectures: The Once and Future U.N. Climate Change Regime", March 2001, available at: http://ssrn.com/abstract=1773865 (last accessed on 26 March 2012); and Jacob Werksman and Kirk Henderson, "The Aftermath of Copenhagen: Does International Law Have a Role to Play in a Global Response to Climate Change", 25 *Maryland Journal of International Law* (2010), 142.

[67] Joanne Scott, "The Multi-Level Governance of Climate Change", in Paul Craig and Grainne de Bruca (eds), *The Evolution of EU Law,* 2nd ed. (Oxford et al.: Oxford University Press, 2011), 805, also available at: http://www.ucl.ac.uk/laws/environment/docs/hong-kong/The%20Multi-Level%20Governance%20of%20Climate%20Change%20(Joanne%20Scott).pdf (last accessed on 26 March 2012).

[68] Joanne Scott, "The Multi-Level Governance of Climate Change", 5 *Carbon and Climate Law Review* (2011), 25, at 26–27.

[69] Scott, "The Multi-Level Governance of Climate Change", supra, note 67, at 43.

In federal states, like the US, questions have surfaced concerning the compatibility of regional climate change agreements with US federalism, and also concerning the relationship between federalism and state-based climate change policies.[70] While the US federal government has lagged behind in the development of climate change law, individual states like California have taken progressive legislative steps to regulate greenhouse gas emissions. Carlarne describes how such initiatives have led to interplay between various levels of government, for instance, concerning the regulation of greenhouse gas emissions from automobile tailpipes.[71]

Horizontal effects across national jurisdictions change occur through transnational cooperation between experts and policymakers, and the diffusion of regulatory models and innovations. Dissemination of examples, concepts and models can also transpire through bilateral cooperation, including development assistance and technical cooperation.[72] There has also been discussion about linking national emissions trading schemes, and concrete plans are underway to fully link the EU ETS and the Australian emissions trading scheme by 2018. Furthermore, migration of climate law across national boundaries caused by the desire of some actors, most notably the EU, to promote climate change mitigation through regulatory schemes that seek to influence actors located outside their territory. There is in fact a rapidly growing body of research on the external dimensions of EU climate law.[73] For one, legislation included in the EU's 2009 Climate and Energy Package contains provisions that are linked to the development of international law through the UN climate negotiations.[74] Most notably, however, EU climate law includes several elements

---

[70] Cinnamon Piñon Carlarne, *Climate Change Law and Policy: EU and US Approaches* (Oxford et. al.: Oxford University Press, 2010), at 67 et seq.

[71] Ibid., at 77–85.

[72] I am grateful for Elisa Morgera for drawing my attention to this point. For detailed discussion see, Gracia Marin-Duran and Elisa Morgera, *Environmental Integration in the EU's External Relations: Beyond Multilateral Dimensions* (Oxford and Portland, Oregon: Hart Publishing, 2012).

[73] Scott, "The Multi-Level Governance of Climate Change", supra, notes 67 and 68; Kati Kulovesi, Elisa Morgera and Miquel Muñoz, "Environmental Integration and Multi-faceted International Dimensions of EU Law: Unpacking the EU's 2009 Climate and Energy Package", 48 *Common Market Law Review* (2011), 829; Sebastian Oberthür and Claire Roche Kelly, "EU Leadership in International Climate Policy: Achievements and Challenges", 43 *The International Spectator* (2008), 35; Kati Kulovesi, "Climate Change in the EU External Relations: Please Follow My Example (or I Might Force You to)", in Elisa Morgera (ed), *The External Environmental Policy of the European Union: EU and International Law Dimensions* (Cambridge University Press, forthcoming, October 2012); Kati Kulovesi, "Make Your Own Special Song even if Nobody Else Sings Along: International Aviation Emissions and the EU Emissions Trading Scheme", 2 *Climate Law* (2011), 535; Joanne Scott and Lavanya Rajamani, "EU Climate Change Unilateralism: International Aviation in the European Emissions Trading Scheme", 23 *European Journal of International Law* (2012), 469 and Biswajit Dhar and Kasturi Das, "The European Union's Proposed Carbon Equalization System: Can it be WTO Compatible?", 25 November 2009, available at: http://papers.ssrn.com/sol3/papers.cfm?abstract_id=1513231 (last accessed on 26 March 2012).

[74] For detailed analysis, see Kulovesi, Morgera and Muñoz, "Environmental Integration and Multi-faceted International Dimension of EU Law", supra, note 73.

that seek to both inspire and influence behaviour in national jurisdictions outside the EU.[75] This aspect of the EU climate law has surfaced questions concerning, for example, jurisdictional limits and the definition and permissibility of extraterritorial regulation and unilateral trade measures.[76] Concrete experiences from its implementation have also shown that such regulatory approaches, reflected most notably in the inclusion into the EU ETS of emissions from foreign airlines taking off from and landing at the EU airports, could well lead to competition between legal systems for power and influence.[77] Horizontal interaction between the UNFCCC and other specialised international legal regimes is discussed in chapters included in Part IV of this book.

3.3.2.1 Vertical Interaction: International and National Law

The vertical relationship between international and national law has played an important role throughout the history of the UNFCCC regime. Notably, the question concerning 'top-down' and 'bottom-up' approaches to climate change mitigation remains central in this regard. Different views on these two approaches originally emerged already during negotiations for the UNFCCC in the early 1990s, most prominently between the EU and US, and both approaches are reflected in the UNFCCC itself and subsequent evolution of the UN climate regime.[78] Notably, the Kyoto Protocol is based on a 'top down' legal architecture, traditionally favoured by the EU and developing countries. Accordingly, countries' emission reduction commitments are defined on the basis of an international treaty, implemented through domestic policies and measures.[79] Increasingly powerful is, however, an alternative 'bottom up' vision, advocated most notably by the US. The 'bottom up' approach relies on voluntary international mitigation pledges, made binding through national legislation and reported internationally (hence, it is also known as the pledge-and-review – approach). The rationale of this approach is that "[w]hat really matters is that the pledges reflect measurable, reportable and verifiable actions and that they are embedded in domestic law. From this perspective, the international legal character of a future climate agreement seems less important."[80] Nevertheless, the main motivation behind the bottom up approach is arguably political. It has been indicated that a bottom up approach takes into consideration national political sensitivities and complexities around climate change: "International pledges grow out of,

[75] Ibid.

[76] Kulovesi, "Make Your Own Special Song", supra, note 73, at 547–550.

[77] Ibid., 558. See also Scott and Rajamani, "EU Climate Change Unilateralism", supra, note 73, at 481 et seq.

[78] Daniel Bodansky, "A Tale of Two Architectures: The Once and Future U.N. Climate Change Regime", Arizona State University, March 2001, available at: http://ssrn.com/abstract=1773865 (last accessed on 14 March 2012), at 6.

[79] The numbers in Annex B of the Kyoto Protocol were, however, negotiated "bottom up" based on political bargaining rather than "top down" based on climate science.

[80] Werksman and Henderson, "The Aftermath of Copenhagen", supra, note 66, at 3.

and reflect, domestic policies rather than being superimposed on them. The role of the international regime is not to define what each state must do, but rather to help generate political will by raising the profile of the climate change issue and providing greater transparency."[81] The argument in favour of a 'bottom up' approach are, however, not universally accepted. Instead, in the ongoing long-term negotiations under the UNFCCC, the question concerning 'top down' and 'bottom up' approaches has therefore been politically highly sensitive. Proponents of a 'top down' approach include most developing countries and the EU, while US and some other members of the Umbrella Group have advocated a 'bottom up' approach.

In practice, the UNFCCC regime has recently taken steps from a Kyoto-type 'top up' legal structure towards a 'bottom up' legal architecture. The first step came as the (unadopted) 2009 Copenhagen Accord called on developed countries to commit to implementing quantified, economy-wide targets for 2020 and submit them for inclusion in Appendix I.[82] It also called on developing countries to implement mitigation actions and submit these for inclusion in Appendix II.[83] The 2010 Cancun Agreements subsequently followed the same approach, 'anchoring' developed and developing countries' mitigation pledges into two information documents compiled by the UNFCCC Secretariat.[84] In other words, the Cancun Agreements brought the bottom up approach into the official UNFCCC process.[85] In light of these recent developments it has been predicted that the future international climate regime "is likely to be a non-prescriptive regime based on self-selected nationally determined targets and actions, applicable in a broadly symmetrical fashion across countries, and backed not by a treaty-based compliance system, but by a robust reporting and (possibly) a review system."[86]

In context of the current configuration of a 'bottom up' approach under the UNFCCC it is important to note, however, that countries do not necessarily have in place national legislation to implement the mitigation pledges that they have communicated to the UNFCCC Secretariat. In this sense, the practical application of the pledge-and-review approach is lacking the crucial component of binding national legislation. The focus therefore shifts towards soft law.[87] Werksman and Herbertson

---

[81] Bodansy, "A Tale of Twor Architectures," supra, note 78.

[82] Decision 2/CP.15, The Copenhagen Accord, FCCC/CP/2009/7/Add.1, 30 March 2010, para. 4.

[83] Ibid., para. 5

[84] Compilation of economy-wide emission reduction targets to be implemented by Parties included in Annex I to the Convention, Revised Note by the Secretariat, UN Doc. FCCC/SB/2011/INF, 7 June 2011; Compilation of information on nationally appropriate mitigation actions to be implemented by Parties not included in Annex I to the Convention, Note by the Secretariat, UN Doc. FCCC/AWGLCA/2011/INF.1, 18 March 2010;

[85] Bodansky, "A Tale of Two Architectures", supra, note 78, at 3.

[86] Lavanya Rajamani, Jutta Brunnée and Meinhard Doelle, "Introduction: The Role of Compliance in an Evolving Climate Regime", in Jutta Brunnée, Meinhard Doelle and Lavanya Rajamani (eds), *Promoting Compliance in an Evolving Climate Regime* (Cambridge et al.: Cambridge University Press, 2012), 1, at 9.

[87] See Antto Vihma, "Analyzing Soft Law and Hard Law in Climate Change" in Chapter 7 of the present volume.

have suggested that, countries could use COP decisions and "reinvest in strengthening those aspects of the legal character of the climate change regime that are already within the UNFCCC's mandate as a legally binding treaty."[88] Reliance on soft law in climate change mitigation surfaces questions concerning, *inter alia,* effectiveness and compliance assessment. From the national law perspective, legitimacy also becomes a key consideration. For example, how much substance can be included in a COP decision before it either triggers national implementation procedures or risks violating the spirit of domestic constitutional guarantees related to democratic oversight and approval of international undertakings?

In addition to the long-standing debate on 'top down' and 'bottom up' approaches, there are interesting examples of how international and national law can interact and complement each other in the regulation of climate change law. Green Investment Schemes, for instance, illustrate how some governments may be willing to undertake stricter commitments through national legislation than under international climate treaties, and national legislation can therefore be used to enhance the environmental integrity of international climate change law.

Under Article 17 of the Kyoto Protocol, Annex I countries with legally-binding emission reduction commitments may participate in international emissions trading, provided that they comply with the eligibility criteria defined in the Marrakesh Accords.[89] One of the challenges of the Kyoto Protocol emissions trading scheme relates 'hot air,' in other words, the large amount of credits available due to the fact that emissions in several Eastern and Central European countries declined significantly from their 1990 levels as a result of economic restructuring. While the 'excess allocation' to the former communist countries was a conscious decision taken by COP 3 in 1997, it has been feared that the sale of hot air credits under Article 17 will jeopardize the environmental integrity of the Kyoto Protocol.

A legal response to the problem of 'hot air' has subsequently evolved through national legislation and other measures taken by countries concerned. Several Central and Eastern European countries, including the Czech Republic, Latvia, Hungary, Poland and Estonia, have created Green Investment Schemes.[90] The Estonian scheme, for example, is described as "a financing mechanism where finances that come from the trading of the country's CO2 quotas under the Kyoto Protocol are channelled to environmental projects and programmes that help to lower the CO2 emission."[91] In other words, countries with a Green Investment Scheme have used national legislation to ensure that international emissions trading under the Kyoto Protocol

[88] Werksman and Henderson, "The Aftermath of Copenhagen", supra, note 66, at 39.

[89] Decision 11/CMP. 1, Modalities, rules and guidelines for emissions trading, UN Doc. FCCC/KP/CMP/2005/8/Add.2, 30 March 2006.

[90] For overview, see Andreas Tuerk et al., "Working Paper: Green Investment Schemes: First Experiences and Lessons Learned", April 2010, available at: http://www.joanneum.at/climate/Publications/Solutions/JoanneumResearch_GISWorkingPaper_April2010.pdf (last accessed on 20 March 2012).

[91] Environmental Investment Center, "Green Investment Scheme", available at: http://www.kik.ee/en/kik-eng/sources-of-financing/green-investment-scheme.html (last accessed 14 February 2012).

reduces greenhouse gas emissions even if their targets in Annex B of the Kyoto Protocol would allow them to sell carbon credits without any further action to mitigate climate change. The detailed conditions for spending revenue from international emission trading are typically set forth in an Assigned Amount Unit (AAU) Purchase Agreement. Buyers can be governments or private actors authorised by their governments to participate in emissions trading in accordance with Article 17 of the Kyoto Protocol. Estonia, for example, has sold ten million AAUs to Mitsubishi Corporation.[92] According to the terms of the transaction, the proceeds will be invested to create a country-wide charging infrastructure for electric vehicles.[93] In addition, approximately 500 electric cars will be provided for the use of social workers and a grant scheme will be launched to support the purchase of electric cars by private individuals.[94] Furthermore, all owners of these electric vehicles will have to start consuming only electricity generated from renewable energy sources through a Guarantees of Origin scheme.[95] While Green Investment Schemes have developed outside the UNFCCC legal regime, their potential contribution to the problem of hot air was recognized in the decision by COP/MOP 2 to include Belarus with an emission reduction target in Annex B of the Kyoto Protocol. Accordingly, the COP/MOP welcomed that Belarus "will use any revenues generated from transfers under Article 17 of the Kyoto Protocol for further greenhouse gas emission abatement measures."[96]

There are obviously many other ways in which international and national climate change law could complement each other to increase the effectiveness and mutual supportiveness of the overall body of climate change law. One relevant area is climate finance, a key issue in the ongoing negotiations under the UNFCCC.[97] Here, developed countries' general obligation to provide climate finance under the UNFCCC could be complemented through specific legal mechanisms developed at the national level to generate climate finance. An existing example is the EU ETS and its non-binding provisions on allocating revenue from the auctioning of emission allowances.[98] Through reforms included in the EU's 2009 Climate and

[92] Estonian Government Communication Unit Press Release, "Estonia Will Promote the Use of Electric Cars under a Green Investment Scheme", 3 March 2011, available at: http://www.kik.ee/en/kik-eng/sources-of-financing/green-investment-scheme.html (last accessed 14 February 2012).

[93] Ibid.

[94] Ibid.

[95] Ibid.

[96] Decision 10/CMP.2, Proposal from Belarus to amend Annex B to the Kyoto Protocol, UN Doc. FCCC/KP/CMP/2006/10/Add.1, 2 March 2007, para. 3. According to the decision, this was still "subject to approval by the relevant authorities of the Republic of Belarus."

[97] For comprehensive overview, see Yulia Yamineva and Kati Kulovesi, "The New Legal and Institutional Framework for Climate Finance under the United Nations Framework Convention on Climate Change: A Breakthrough or an Empty Promise?" in Chapter 9 of this volume.

[98] Directive 2009/29/EC supra, note 40, Arts. 10, 10a and 10c. For discussion, see Kulovesi, Morgera and Muñoz, "Environmental Integration and the Multifaceted International Dimensions of EU Law", supra, note 73, at 855–858.

Energy Package, auctioning will gradually become the sole method of distributing allowances under the ETS.[99] The revised ETS Directive includes voluntary provisions to earmark at least 50% of the auctioning revenue for climate change mitigation and adaptation, including in developing countries.[100] In this respect, EU climate law has links to the international level, *inter alia*, through references to the Kyoto Protocol's Adaptation Fund and addressing deforestation in developing countries. While non-binding, the provisions on auctioning revenue in the context of the ETS serve to illustrate how international law and national legislation could interact vertically and complement each other in key areas, such as climate finance.

3.3.2.2 Vertical Interaction: Sub-national Initiatives

With globalization, attention is shifting towards forms of governance that take place beyond the state. Climate change cooperation is no exception. One aspect of this trend is a focus on sub-national initiatives. It has been argued that countries will be unable to meet their international climate change commitments without more explicit engagement with sub-national action.[101] Furthermore, in some countries sub-national initiatives have been key drivers for the development of climate change law. The US is the most important example in this regard.

State-led initiatives and regional cooperation have played a far more important role in the US in regulating climate change than the federal government. In her study, Carlarne describes how states, including California and New York, are "choosing to follow the footsteps of the EU to try to create robust climate change laws and policies" even if the federal government is lagging behind.[102] While noting that "states have frequently led the way for the federal government in experimenting with and promoting new environmental laws and regulations," she indicates that it is rare for them to embark on "such a widespread and coordinated campaign to develop effective environmental laws in the absence of federal leadership as in the current case of climate change governance."[103] Almost two dozen US states have some type of renewable energy obligations and over a dozen states have enacted or are in the process of enacting legislation to control greenhouse gas emissions.[104] In addition, US states are creating regional climate change partnerships,[105] also in cooperation with Canadian counterparts. These include the Western Climate

[99] Ibid.

[100] Ibid.

[101] Betsill and Bulkeley, "Cities and the Multi-level Governance of Global Climate Change", supra, note 8, at 141–142.

[102] Carlarne, *Climate Change Law and Policy,* supra, note 70, at 63.

[103] Ibid., at 61.

[104] Ibid., at 88.

[105] Ibid., at 64.

Initiative and the Regional Greenhouse Gas Initiative.[106] Carlarne concludes that state and regional programs are "now reaching a critical mass," increasing political pressure "at the national level for comparable, if not superior action."[107]

Also local governments are increasingly involved in efforts to address climate change. A prominent example is the International Council for Local Environmental Initiatives (ICLEI), a transnational network of more than 1,220 local governments from 70 different countries, representing nearly than 570 million people.[108] In 1993, ICLEI created the Cities for Climate Protection programme with five milestones to reduce greenhouse gas emissions. Many CCP member governments are taking action to mitigate climate change independently of their national governments.[109] They also interact directly across national boundaries and speak at the UN climate negotiations through the ICLEI, which has an observer status in the process.[110]

From the legal perspective, sub-national initiatives bring to the fore questions concerning competence, hierarchy and multi-level governance. As mentioned above, questions have already been raised concerning the relationship between the various state-led and regional initiatives in the US with the federal government, and concerning the relationship between EU law and local climate change initiatives.

3.3.2.3 Interaction Between National Jurisdictions

As explained above, diffusion of regulatory models and innovation across national jurisdictions is influenced by various forms of transnational cooperation between policymakers and experts. Also development cooperation and technical assistance contribute to their dissemination. However, as Shaffer and Bodansky, have recently argued, migration of legal norms also happens when "powerful states apply their national environmental standards extraterritorially, effectively imposing their standards on others."[111] Given the failure of the UNFCCC regime to steer the world on a course that avoids dangerous anthropogenic climate change, it is increasingly tempting for countries to attempt to regulate the behaviour of foreign entities and influence developments beyond their territory. This constitutes another reason for the tendency of climate law to migrate over national and legal borders.

---

[106] Ibid., at 64–65.

[107] Ibid., at 89.

[108] For more information see the ICLEI website, available at: http://www.iclei.org/index.php?id=about (last accessed on 21 March 2012).

[109] Betsill and Bulkeley, "Cities and the Multi-level Governance of Global Climate Change", supra, note 7, at 145.

[110] Ibid., at 146–147.

[111] Shaffer and Bodansky, "Transnational Unilaterlaism and International Law", supra, note 63, at 4. They highlight in particular the role of the EU and the US in creating transnational environmental law, mentioning, in particular, the EU's REACH regulation for chemicals and its scheme for genetically modified organisms and the US-prescribed methods for catching tuna and shrimp.

As also noted above, EU climate law is a prime example in this regard; it includes several elements designed not only to reduce greenhouse gas emissions in Europe but also to influence developments outside the EU. This applies equally to issues regulated under the UNFCCC regime, such as the CDM, and issues on which the international community has been unable to reach meaningful agreement, such as emissions from international aviation.[112] The external dimensions of EU climate law are linked to the EU's goal of playing a leadership role in the battle against climate change. This has been the EU's political objective since the early 1990s and it has recently been given a legal formulation in the Treaty on the Functioning of the European Union.[113] Given the size of its markets, the EU is, in theory, well-placed to use its climate law to influence developments beyond its territory:

> The EU's vast internal market… provides it with a powerful bargaining chip and gives it an excellent potential to create and alter incentives. The ability to act as a gatekeeper for those who want access to the EU market and the ability to enforce EU standards on trading partners is an extremely valuable powerful resource. The sheer scale of the internal market also means that the EU can offer and take actions that will have a dramatic environmental impact.[114]

Existing examples of external reach of EU climate change law include sustainability criteria for biofuels. To implement its 10% target for renewable energy in the transport sector by 2020, the EU adopted sustainability criteria for imported and domestically produced biofuels to ensure, *inter alia*, minimum greenhouse gas emission reductions and prevent loss of biodiversity.[115] From the conventional perspective of international law, the biofuels sustainability criteria are interesting in that they seek to influence land-use in the territory of third countries.[116] On the other hand, the implementation of the scheme relies largely on economic operators, and voluntary schemes and standards can be used and 'benchmarked' against the EU's sustainability criteria.[117]

---

[112] For comprehensive discussion, see Kulovesi, Morgera and Muñoz, "Environmental Integration and Multi-faceted International Dimension of EU Law", supra, note 73; and Kulovesi, "Climate Change in the EU External Relations", supra, note 73.

[113] According to Article 191(1) of the Treaty on the Functioning of the European Union, one of the objectives of the EU's environmental policy is to contribute to: preserving, protecting and improving the quality of the environment; protecting human health, prudent and rational utilization of natural resources; and promoting measures at international level to deal with regional or worldwide environmental problems, *and in particular combating climate change*. Emphasis added.

[114] Charles F. Parker and Christer Karlsson, "Climate Change and the European Union's Leadership Moment: An Inconvenient Truth?", 48 *Journal of Common Market Studies* (2010), 923, at 928.

[115] Directive 2009/28/EC of the European Parliament and the Council of 23 April 2009 on the promotion of the use of energy from renewable sources and subsequently repealing Directives 2001/77/EC and 2003/30/EC, OJ 2009 L 140/16, Art. 17.

[116] For discussion, see Kulovesi, Morgera and Muñoz, "Environmental Integration and Multi-faceted International Dimensions of EU Law", supra, note 73, at 877–887; Jolene Lin, "The Environmental Regulation of Biofuels: Limits of the Meta-Standard Approach", *Carbon and Climate Law Review* (2011), 34; Scott, "The Multi-Level Governance of Climate Change", supra, note 68, at 29–30.

[117] Lin, "The Environmental Regulation of Biofuels", supra, note 116, at 38–40.

Another – highly controversial – example of the transboundary influence of EU climate law is the inclusion of emissions from all flights taking off and landing in EU airports in the ETS from 2012 onwards. Emissions from international aviation are growing rapidly, threatening to cancel out the impact of climate change mitigation in other sectors. Frustrated by the lack of global progress through the International Civil Aviation Organization, the EU decided to address aviation emissions unilaterally in an attempt to inspire and influence international developments.[118] The inclusion of emissions by foreign airlines has, however, generated some strong opposition. It provoked the US to consider the "European Union Emissions Trading Prohibition Act of 2011," which passed the House of Representatives in October 2011 and would have prohibited US-based airlines from participating in the ETS if a counterpart bill passed the Senate.[119] A bill with somewhat less stringent language is expected to be adopted by the full Congress in the spring of 2012.[120] Also China has prohibited its airlines from participating in the ETS and increasing fares or imposing other charges related to the scheme, and India has instructed its airlines not to participate in the scheme.[121] The EU change law, in turn, includes some built-in legal mechanisms to consider impacts of measures taken by other jurisdictions. If, for instance, a third country adopts measures to mitigate emissions from international aviation, EU bodies will decide whether aircraft operators from the country in question should be exempt from the obligation to participate in the ETS.[122]

The external ambitions of EU climate law are increasingly attracting scholarly attention, including criticism. Alluding to "the increasing propensity of the EU to engage in climate change unilateralism," Scott and Rajamani, for example, have argued that international law and the principle of common but differentiated responsibilities in particular should constrain the global dimensions of EU climate law.[123] Despite criticism and political controversies surrounding its initiatives, it is conceivable that the EU expands the external reach of its climate law in the future – and that other countries will implement similar measures. Ideas already discussed include the inclusion of imports of energy intensive products under the ETS to avoid carbon leakage.[124] While the Commission has traditionally taken a cautious approach to the idea, the concept continues to float around, supported by France in particular. The idea of imposing measures on imports played a crucial role also in the US in

118 For comprehensive discussion, see Kulovesi, "Make Your Own Special Song", supra, note 73.

119 H.R. 2594 European Union Emissions Trading Prohibition Act of 2011.

120 "US Congress to Oppose EU Law on Aircraft Emissions", *Carbon Market Europe*, 3 February 2012.

121 BBC News, "China 'bans' airlines from joining EU carbon scheme", 6 February 2012, available at: http://www.bbc.co.uk/news/business-16901106 (last accessed on 2 March 2012); "India Confirms Boycott of EU Aviation Emissions Rule," *Bridges Weekly,* 29 March 2012, available at: http://ictsd.org/i/trade-and-sustainable-development-agenda/129985/ (last accessed on 30 April 2012).

122 Directive 2009/29/EC, supra, note 40.

123 Scott and Rajamani, "EU Climate Change Unilateralism", supra, note 73, at 481 et seq.

124 For discussion, see Kulovesi, Morgera and Muñoz, "Environmental Integration and Multi-faceted International Dimensions of EU Law", supra, note 73, at 858–862.

the currently frozen plans for a federal cap-and-trade scheme for greenhouse gas emissions. Furthermore, there has already been some analysis concerning China's growing influence in Africa and how this might affect climate law and policy. Accordingly, "China's potential to influence and assist African countries in the development of successful climate change policy and law is only just beginning" and "it would be fair to assume that China's efforts to help African countries with climate change action will substantially follow its own climate change policy and regulatory experience, and the model it has used for investment aid in Africa."[125]

For the present focus, these examples are interesting as they illustrate the growing number of linkages between different legal regimes and jurisdictions in regulating climate change, and they also show that the relationship between different legal authorities can be a dynamic one where one legal system reacts to developments in other jurisdictions. An issue of concern in this respect is, for instance, that unilateral measures implemented in one jurisdiction could lead to retaliation by other jurisdictions. This could potentially lead to competition between legal regimes for power and influence, or to 'forum shopping' and regulatory arbitrage, making the position of the private actors operating in several jurisdictions and getting caught in the legal battle uncomfortable and confusing. While there is no legal mechanism to ensure coordination and coherence between various national legal systems in regulating climate change, it is hoped that international law will play a role in taming climate change unilateralism. As Shaffer and Bodansky indicate:

> Unilateral action is not a one-step dance. It is better viewed as part of a dynamic process of action and reaction, reassessment and response, in which international law plays an uneasy role as both a check and a potential consolidator. International law needs to discipline (or, better stated, provide guidelines for) unilateral action, as part of this dynamic process. But, as with all matters, the trick is to get the balance right: there should be neither too little constraint, which would permit discriminatory and opportunistic policies, nor too much constraint, which would impede needed action.[126]

Overall, the landscape of climate law looks particularly dynamic from the perspective of interaction between legal systems and regimes. The relevance of this dimension of climate law can also be expected to increase as climate law continues to expand. Arguably, this poses challenges to both climate law scholars and practitioners who are required to take into consideration a plural mix of legal sources and understand their linkages and relationships.

3.4 Climate Law: Non-state Actors and Deformalization

The second trend in climate change law relates to the involvement of non-state actors and growing role of soft law in international climate governance. While conventional international actors, international organizations and sovereign states,

[125] Christopher Tung, "The Influence of Chinese Climate Law & Policy on Africa", 5 *Climate and Carbon Law Review* (2011), 334, at 344.

[126] Shaffer and Bodansky, "Transnational Unilateralism and International Law", supra, note 63, at 11.

continue to hold a prominent role, climate change initiatives are increasingly taking place beyond the UN climate regime and the nation state.[127] What is their relevance for climate law research and scholarship? This section first discusses public-private partnerships and self-regulation with respect to climate change. It then identifies some of the key issues for climate law research.

3.4.1 Public-Private Partnerships and Other Hybrid Initiatives

Over the years, various public-private partnerships and other hybrid forms of cooperation have emerged around climate change: "Along with inter-governmental treaty-making, the climate policy arena is characterized by civil-society led standard setting, self-regulation by transnational corporations and hybrid governance arrangements, such as multi-stakeholder partnerships".[128]

For climate change law, key public-private partnerships have been created under the UNFCCC and the Kyoto Protocol. Most notably, governance of the Kyoto Protocol's flexibility mechanisms is based on close cooperation between the public and private sectors.[129] The CDM, for example, can be characterised as a public-private partnership in which private actors participate both by implementing climate-friendly projects on the ground and ensuring the projects' compliance with the international rules adopted under the Kyoto Protocol.[130] In accordance with Article 12.5 of the Kyoto Protocol and international CDM rules, validation of CDM projects and certification of emission reductions is primarily done by Designated Operational Entities (DOEs). The current list of approximately 50 DOEs includes mostly commercial certification companies accredited by the CDM Executive Board.[131] In practice, DOEs play a critical role in ensuring the environmental integrity of the CDM and the Kyoto Protocol. As the CDM Validation and Verification Manual indicates: "The CDM is a rules-based mechanism. Therefore, it shall be the DOE's responsibility to ensure that… these rules are complied with for any project activities requesting registration as a proposed CDM project activity".[132] During the

[127] Okereke, Bulkeley and Schröder, "Conceptualizing Climate Governance Beyond the International Regime", supra, note 7, at 58.

[128] Bäckstrand, "Accountability of Networked Climate Governance", supra, note 8, at 76.

[129] For an overview, see Kulovesi, "The Private Sector and the Implementation of the Kyoto Protocol", supra, note 32, 146 et seq.

[130] Morgera and Kulovesi, "Public-Private Partnerships for Wider and Equitable Access to Climate Technologies", supra, note 8.

[131] UNFCCC, "List of DOEs", 2012, available at: http://cdm.unfccc.int/DOE/list/index.html (last accessed 1 March 2012).

[132] CDM Executive Board, *Clean Development Mechanism Validation and Verification Manual*, available at: http://cdm.unfccc.int/Reference/Manuals/accr_man01.pdf (last accessed 1 March 2012), para. 29.

validation process, DOEs are responsible for checking, *inter alia,* the critical requirement that the CDM project results in emission reductions that are "additional" to what would have been achieved in absence of the project. In verifying emission reductions, the essential task of DOEs is to ensure that the CDM project has achieved the planned emission reductions. This includes a visit to the project site to "assess that all physical features of the… CDM project activity proposed in the registered PDD [Project Design Document] are in place and that the project participants has operated the proposed CDM project activity as per the registered PDD".[133] Bearing in mind the basic idea that Annex I countries can meet a part of their emission reduction commitments under the Kyoto Protocol through offsets created under the CDM, it is clear that DOEs, in other words: private actors, are critical for the implementation of the Kyoto Protocol and ensuring its environmental integrity. One of the challenges, then, is that "[w]hile auditors in general rely on their reputation for independence and integrity to stay in business, there is less incentive to guard against reputational risks in the quasi-monopolistic environment that DOEs currently operate in."[134] In a similar vein, private actors will play an important role in verifying compliance with the EU's sustainability criteria for biofuels.[135]

While engagement of the private sector is widely seen as the CDM's greatest achievement, it is useful to bear in mind that the private sector's involvement in the CDM was neither clear nor uncontroversial from the outset. In fact, the market-based nature of the CDM continues to generate important challenges, especially for the mechanism's objective of contributing to the sustainable development of developing countries hosting the projects. During the evolution of the CDM, a market-based approach relying on private actors was initially pitted against an "interventionist" approach that would have relied on traditional development assistance from the public sector to implement CDM projects.[136] Those supporting the market-based approach argued, however, that governments should set the rules for the CDM while the private sector "which holds the capital and technology necessary to the CDM's success" should be entrusted to design the CDM projects.[137] Interventionists, in turn, were skeptical of the private sector's ability to assist non-Annex I countries to achieve sustainable development.[138] Indeed, as explained in Eni-ibukun's chapter on climate justice and the CDM in the present volume, the market-driven nature of the CDM has lead to the somewhat ironic situation "where those that are most in need

---

[133] Ibid., para. 196.

[134] Lin, "The Environmental Regulation of Biofuels", supra, note 116, at 42.

[135] See ibid. for critical assessment.

[136] Jacob Werksman, "The Clean Development Mechanism: Unwrapping the Kyoto Surprise", 7 *Review of European Community and International Environmental Law* (1998), 147, at 153.

[137] Ibid.

[138] Ibid.

of CDM projects, because of their low development levels, are actually the ones benefitting the least from the CDM."[139] All this goes to show that while private sector engagement is crucial for climate change mitigation, novel regulatory approaches, such as public-private partnerships and market-based mechanisms, also entail considerable challenges.

Under the UN climate regime, the private sector also participates in the Nairobi Work Programme on Impacts, Vulnerability and Adaptation to Climate Change along with international organizations and other public entities. To boost private sector engagement in climate change adaptation, the UNFCCC Secretariat launched in 2012 a database called Adaptation Private Sector Initiative to showcase successful strategies that businesses and communities are using to adapt to climate change, while simultaneously creating profit or avoiding losses.[140] The approximately 100 initial examples include actions by well-known global companies, such as Coca Cola, Nestlé, Levi's, Microsoft and Starbucks.[141] In launching the initiative, UNFCCC Executive Secretary Christiana Figueres underscored that "[c]limate risks which affect communities around the world are always also business risks".[142]

Outside the UNFCCC regime, one example of a public-private initiative is the Chicago Climate Exchange (CCX).[143] The CCX was initially a voluntary greenhouse gas reductions programme, which traded allowances between 2003 and 2010 and involved major corporations, utilities and financial institutions with activities in all 50 United States, 8 Canadian provinces and 16 countries.[144] Its size was estimated at around one third of the EU ETS.[145] In 2011, a new CCX Offsets Registry Programme was launched to register verified emission reductions based on a comprehensive set of established protocols.[146] The success of the CCX remains questionable, however, serving to highlight some of the challenges related to the efficacy of non-state initiatives discussed below in sect. 3.4.3.

---

139 Tomilola Eni-ibukun, "Climate Justice: The Clean Development Mechanism as a Case Study" in Chapter 10 of the present volume.

140 UNFCCC Press Release, "UNFCCC secretariat aims to help communities and businesses become climate-resilient with help of new online tool", 26 January 2012, available at: https://unfccc.int/files/press/press_releases_advisories/application/pdf/pr20122601_apsidbase.pdf (last accessed on 1 March 2012).

141 Ibid.

142 Ibid.

143 See, for example, Andonova et al., "Transnational Climate Governance", supra, note 6, at 62; Bulkeley and Newell, *Governing Climate Change,* supra, note 5, at 95.

144 CCX Fact Sheet, December 2011, available at: https://www.theice.com/publicdocs/ccx/CCX_Fact_Sheet.pdf (last accessed on1 March 2012).

145 Ibid.

146 Ibid.

3.4.2 Private Sector Engagement and Voluntary Regulatory Initiatives

The landscape of climate change law becomes even more colourful when taking into consideration regulatory initiatives around climate change launched by civic players exclusively. As such, private sector activities are crucial for climate change mitigation and one of the key objectives of climate change law is to regulate them, driving investment towards climate-friendly technologies and activities.

The private sector has been closely following climate change policy since the beginning, attempting to influence developments both internationally and within national boundaries. Initially, most private actors mobilized to stall action against climate change.[147] Subsequently, however, most have taken a more responsible stance. One of the watersheds came in December 2007 as more than 150 well-known global companies published the Bali Communiqué on Climate Change in the Financial Times, calling for a comprehensive and legally binding climate change agreement under the United Nations.[148] Networks like the World Business Council for Sustainable Development, an initiative bringing together more than 190 chief executive officers of international companies, have also been active in promoting climate change policies.

Many businesses have also begun to see climate change action as an opportunity rather than a threat.[149] A number of companies are undertaking self-regulation activities and participating in voluntary schemes and agreements to reduce greenhouse gas emissions and improve their energy efficiency. One such initiative is the Carbon Disclosure Project, which requests information annually from thousands of companies concerning their greenhouse gas emissions, energy use and other relevant issues.[150] Companies like Shell and BP have also experimented with internal emissions trading schemes. There are also various other voluntary regulatory initiatives related to carbon trading, such as the Gold Standard for the CDM, the Voluntary Carbon Standard (VCS) and the Climate, Community and Biodiversity Standard. Overall, there is a large number of partnerships and soft law initiatives seeking to address climate change.

3.4.3 Non-state Actors and Climate Law Research

Scholars of both international relations and law are increasingly interested in non-state actors, soft law and 'governance' – a notion that (in contrast to 'government')

---

[147] Bulkeley and Newell, *Governing Climate Change,* supra, note 5, at 88.

[148] An advert published in the global edition of *The Financial Times*, 30 November 2007.

[149] Bulkeley and Newell, *Governing Climate Change*, supra, note 5, at 87.

[150] For more information, see the Carbon Disclosure Project website, available at: https://www.cdproject.net/en-US/Programmes/Pages/climate-change-programs.aspx (last accessed on 21 March 2012).

includes the idea that it can take place without the state.[151] There has already been research on what is described as transnational climate governance, "a distinct form of global governance operating in a political sphere where public and private actors interact across national borders and political jurisdictions."[152] Some legal scholars have drawn attention to 'global law,' characterized as "a new body of law that emerges from various globalization processes in multiple sectors of civil society independently of laws of the nation states."[153] According to Teubner, this "fully fledged law" is distinguished from the traditional law of nation states by its peculiar characteristics: while lacking in institutional and political support, global law is "closely coupled with globalized socio-economic processes."[154]

In legal theory, accounting for the role of non-state actors points towards legal pluralism[155] and approaches challenging the traditional role of sovereign states as exclusive norm-setting institutions and emphasizing private norm-production by trade associations, professional/technical organizations, commercial arbitrators, multinational companies and other civic players.[156] Proponents of these approaches have made the argument that traditional legal theories are inadequate to grasp the increasingly multifaceted normative reality. According to Rosen-Zvi, "the world is increasingly governed by an intricate web of norm-producers, which includes international organizations, transnational bodies, states in federative systems, regions, countries, cities, national and transnational associations of subnational entities, as well as a host of private and quasi-private entities that are emerging as new types of actors on the global regulatory stage."[157] Also Perez highlights that the global economic system "is governed by multiple systems of law" and it "is not based on a coherent set of normative or institutional hierarchies. It represents, rather, a highly pluralistic mixture of legal regimes, with variable organisational and thematic structures."[158] Perez thus emphasizes the role of private legal systems, arguing that such systems are not made of the familiar sources of public international law, "but rather, are the result of (private) norm-production by trade associations, professional/technical organizations, commercial arbitrators,

[151] Andonova et al., "Transnational Climate Governance", supra, note 6, at 55.

[152] Ibid., at 68–69.

[153] Teubner, "Global Bukowina", supra, note 10, at 2.

[154] Ibid.

[155] As Koskenniemi has explained, vocabulary on legal pluralism has emerged from three different sources: the study of local laws and de facto practices in modern society; native law's coexistence with imported metropolitan laws in the context of colonialism; and globalisation. Here, the focus is on legal pluralism associated with globalisation. See Martti Koskenniemi, "Global Legal Pluralism: Multiple Regimes and Multiple Modes of Thought", Harvard, 5 March 2005, at 14.

[156] See Teubner, "Global Bukowina", supra, note 10, for comprehensive discussion.

[157] Issachar Rosen-Zvi, "Climate Change Governance: Mapping the Terrain", 5 *Carbon and Climate Law Review* (2011), 234, at 236.

[158] Perez, *Ecological Sensitivity and Global Legal Pluralism*, supra, note 11, at 7.

Multinational Enterprises and other civic players."[159] In response to such developments, it has been suggested that the scope of legal analysis should be expanded. Berman, for example, has argued that applying a pluralist framework to the global arena, "is essential if we are to more comprehensively conceptualize a world of hybrid legal spaces."[160] In the context of climate change law, Rosen-Zvi indicates that the study of climate change regulation "should go beyond traditional or even transnational regulation to encompass hybrid regulatory forms which blur the distinction between the public and the private and destabilize boundaries between the global, the national and the sub-national."[161]

Accounting for the role of non-state actors in the legal sphere comes, however, with its own challenges. For one, a project aiming to paint a comprehensive image of the complex and colourful climate governance structures involving non-state actors entails the risk that descriptiveness becomes the main objective. According to Koskenniemi:

> The problem with legal pluralism is the way it ceases to pose demands on the world. Theorists of globalisation are so enchanted by the complex interplay of the technical regimes and a positivist search for a vocabulary that would encompass all of them that they lose thus the critical point of their exercise. This is visible, for instance, in the habit of collapsing the distinction between law and regulation, a favourite technique of international relations study, and to describe law as another regime in thoroughly instrumental terms: 'legalization' as a policy-choice sometimes dictated by strategic interests.[162]

Differences between the notions of 'government' and 'governance' and 'legislation' and 'regulation' surface important questions concerning legitimacy and effectiveness; also pointing towards ideological debates surrounding neoliberalism.[163] Indeed, it was during the dawn of neoliberalism that "public regulation became anathema to powerful social forces" and the push began "for private, voluntary systems of environmental governance, as well as for public-private partnerships that might accomplish the kind of things that advocates of legal regulation had once demanded."[164] In this respect, questions can be raised concerning the effectiveness of the various climate change partnerships and their implications for legitimacy and

---

[159] Ibid., at 8.

[160] Berman, "Global Legal Pluralism", supra, note 10, at 1159.

[161] Rosen-Zvi, "Climate Change Governance: Mapping the Terrain", supra, note 157, at 234.

[162] Koskenniemi, "Global Legal Pluralism", supra, note 155, at 16.

[163] For a critical overview of neoliberalism, see David Harvey, *A Brief History of Neoliberalism* (Oxford et al.: Oxford University Press, 2005).

[164] Craig N. Murphy, "Privatizing Environmental Governance", 9 *Global Environmental Politics* (2009), 134, at 134, drawing the link to the age of Thatcher, Reagan and global neoliberalism. See also Simon Roberts, "After Government? On Representing Law without a State", 68 *Modern Law Review* (2005), 1, at 24, arguing that "it is very difficult to specify in a convincing way a secure grounding for 'law' if we try to shake it free from particular forms historically associated with the state."

democracy. Arguments supporting such informal initiatives include their flexibility, collaborative nature, speed and diverse expertise.[165] Some of the key concerns, then, include that such partnerships 'hollow out' the state, reinforce neoliberalism and accelerate privatization of environmental governance.[166] Furthermore, it is feared that they increase business influence, reinforce elite multilateralism and lead to fragmentation of global governance as well as to the retreat of state responsibility in the production of public goods.[167]

Koskenniemi has also questioned whether "informal networking by private industries, non-governmental stakeholder groups and national administrators have produced a stable basis for a formal pluralist statement."[168] Fisher, in turn, has called for "head-on engagement with extended legal pluralism," explaining that:

> By extended legal pluralism I mean the range of different legal and quasi-legal norms that can operate in transnational environmental law, whether international agreements, dispute settlement, policy-making, or negotiation. Legal pluralism is not just another word for the political science term 'governance', or for 'soft law'. Rather, it is a term that signifies that there is much that is legal in transnational environmental law but that its legal nature requires careful and nuanced analysis. To put it another way, transnational environmental law is not just politics and it does have legal content.[169]

Against this background, climate law scholarship faces the challenge of accounting for the various private sector initiatives and public-private partnerships, while retaining a normative focus.

3.5 Conclusions

Acknowledging that climate law is already a highly specialised field of legal practice, this chapter has explored the landscape of climate law and scholarship, and identified two broad trends. The first relates to the realisation that climate change is increasingly regulated at multiple levels and the various levels tend to interact and influence each other. Their hierarchies, synergies and tensions are therefore relevant for understanding the overall impact of legal norms related to climate change, including their tensions and synergies. Second, climate law is also characterized by deformalization; looking broadly, it encompasses various soft law sources and non-state actors. Accounting for the various private sector and soft law initiatives, and discerning their legal relevance while avoiding becoming overtly descriptive and losing the normative focus appear as further challenges for climate law research.

---

[165] Bäckstrand, "Accountability of Networked Climate Governance", supra, note 8, at 77.

[166] Ibid.

[167] Ibid.

[168] Koskenniemi, "Global Legal Pluralism", supra, note 155, at 14–15.

[169] Fisher, "The Rise of Transnational Environmental Law", supra, note 9, at 49.

Given the rapid evolution of the field in recent years, climate law research has tended to focus on substantive issues. This chapter has shown, however, that there are ample opportunities for climate law research to engage more closely with central themes in ongoing theoretical discussions on, inter alia, globalisation, legal pluralism, fragmentation, global administrative law, multi-level governance and transnational environmental law.

Chapter 4
Climate Change and Justice: Perspectives of Legal Theory

Felix Ekardt

Abstract A volume on climate law needs normative visions and principles to provide orientation and to line up normative requirements. This may enable to provide a comprehensive view on energy and climate topics. This contribution, while dealing with justice, gives a perspective from ethics respectively from a (re-)interpretation of national constitutions, the EU Charter of fundamental rights and the European convention on human rights in the light of sustainability. It takes us to human rights as the basic norm of any liberal democratic constitution (on national and transnational level), but criticizes the academic international law debate (unlike the practice of international law) which seems to be focused on the idea of even absolute, i.e. not subject to any balancing, environmental fundamental rights. Overall, it turns out that an interpretation of fundamental rights which is more multipolar and considers the conditions for freedom more heavily – as well as the freedom of future generations and of people in other parts of the world – develops a greater commitment to climate protection. Regarding the theory of balancing, for the purpose of a clear balance of powers the usual principle of proportionality also proves specifiable.

4.1 Theoretical Background: Ethical and Legal Considerations

Under what circumstances can we call social life "just", or the law "right"? This is the ultimate question of all thought about politics, morals, and the law. This question is also relevant when it comes to the question of how we deal with scarce energy

Felix Ekardt is a lawyer, sociologist and philosopher. He is a Professor of environmental law and philosophy of law at the University of Rostock and Director of the Research Unit Sustainability and Climate Policy in Leipzig.

F. Ekardt (✉)
Forschungsstelle Nachhaltigkeit und Klimapolitik, Könneritzstrasse 41,
04229 Leipzig, Germany
e-mail: felix.ekardt@uni-rostock.de

E.J. Hollo et al. (eds.), *Climate Change and the Law*, Ius Gentium: Comparative Perspectives on Law and Justice 21, DOI 10.1007/978-94-007-5440-9_4,
© Springer Science+Business Media Dordrecht 2013

resources and climate change, and how we balance colliding interests (for instance between contemporary and future generations). Conceptually, the term "justice" is concerned with the normative validity of a society's basic order. Thus, a normative *theory of justice* (or ethics) answers the question: how shall humans behave, and what shall the founding order look like? This question must be strictly distinguished from the question of how humans actually behave, and what the factual reasons for this action are (and what humans factually "deem right") – this is a question, respectively, of the descriptive action theory or *anthropological* theory of society.[1] A link between the theory of justice and the action theory is the equally empirical governance theory or *control theory*, i.e., the doctrine of the choice of means to effectively and factually enforce previously defined normative aims (e.g., the right to freedom from impairments to life and health), possibly after a normative balancing with other conflicting objectives (e.g., economic freedom). Such means or instruments could be for instance taxes, emissions trading systems, voluntary commitments, or regulatory prescriptions.

A volume on climate law needs normative visions and principles to provide orientation and to line up normative requirements. Only thus can it enable a comprehensive view on energy and climate topics and their relevance in societies today as well as for future generations. In the perspective of both ethics and constitutions (in international, European, and national law), the resource topic is characterized by colliding human rights: On the one hand, the freedom rights of consumers and companies; and on the other hand, rights to the elementary preconditions of freedom such as food, water, climate stability, security, energy access, a basic supply of essential resources, an absence of wars and civil wars, and so on. Generally speaking, any normative conflict can be regarded as a conflict of competing interests and thus as a balancing problem. It refers to the fundamental phenomenon of law: to find a just balance of conflicting interests.

In this chapter, climate change will be at the center of said balancing process. Since the political process has opted to promote an industrial society, allow industrial facilities, and approve traffic permits, to name but some examples, politics also knowingly accepts statistical projections of future deaths, i.e. an impairment of the right to the elementary conditions of freedom as a result of emissions of air pollutants and other detrimental impacts of permitted activities. This is done by balancing those interests with our present economic freedom to engage in production and consumption activities. The framework for legislative balancing is usually referred to as the proportionality test. Decisions by administrative authorities are mainly determined by legislative acts, and their discretion to apply a balancing test is initially (mostly) limited to the interpretation of the factual requirement of standards enacted by the legislature as an expression of its balancing assessment (if those standards leave room for interpretation).

[1] This distinction is not clear, e.g., in Jürgen Habermas, *The Theory of Communicative Action* (London: Beacon Press, 1985). Many readers, and probably the author himself, seem to attach a normative meaning to this book; the actual topic, however, is anthropology, that is: a descriptive theory of societies.

This chapter, while dealing with justice, gives a perspective from ethics and a (re-)interpretation of national constitutions, the European Charter of fundamental rights and the European convention on human rights in the light of sustainability.[2] Sustainability has been increasingly referred to as a key policy objective for the past 20 years, whether by the United Nations (UN), the European Union (EU), or national governments. It is, however, not always stringently applied. The intention of sustainability is to extend justice (and, respectively, law, morals and politics) across an intergenerational and global dimension.[3] By contrast, a common understanding is that sustainability is simply a balanced pursuit of the three pillars of environmental, economic and social issues, if necessary even without a time – or space-spanning aspect.[4] Elsewhere, it has been affirmed that this latter interpretation is at least misleading, that it adheres to expectations of unlimited economic growth which – in a physically finite world – cannot be met, and that this "pillar – perspective" is also incompatible with international law's fundamental tenets of sustainability.[5]

Hence, the subject of this chapter takes us to national, European, and international human rights as the basic norm of any liberal democratic constitution (on a national and transnational level). Human rights also form the typical core of any modern ethics. Environmental protection and intergenerational and global justice, however, are rarely addressed as guaranteed by fundamental rights in the existing legal and ethical discourse, but are rather assigned to the category of "national objectives," for instance in Article 20a of the German Constitution (Grundgesetz, GG) or, in the establishing rules of the EU, on Article 191 TFEU; or they are framed as abstract principles such as the precautionary principle or the principle of common but differentiated responsibility, thereby lacking concreteness and justitiability.

Nevertheless, it seems essential to consider fundamental rights. Unlike general objectives or abstract principles, fundamental rights not only define legal powers, but also frame legally enforceable obligations of public authority. Moreover, fundamental rights are the strongest manifestation of a liberal-democratic constitution. On a constitutional level, overcoming the economically oriented understanding of freedom

---

[2] To show that the theses of this chapter are normatively right as an ethical approach would mean to demonstrate that the principles of liberal democracy are universally right. This has been demonstrated elsewhere by previously establishing that freedom or the underlying principles of human dignity and impartiality are the universal – and sole – basis of a just basic order. For reasons of space, this is omitted here. On details, cf. Felix Ekardt, *Theorie der Nachhaltigkeit: Rechtliche, ethische und politische Zugänge – am Beispiel von Klimawandel, Ressourcenknappheit und Welthandel* (2nd edition, Baden-Baden: Nomos, 2011), §§ 3–5; similar in his basic orientation Habermas, supra, note 1; partially differing: John Rawls, *A Theory of Justice* (Cambridge, Mass.: Harvard University Press, 1971).

[3] For this understanding of the principle of sustainability (and with references to opposing views), see Ekardt, *Theorie der Nachhaltigkeit*, supra, note 2, § 1C; with a similar result (but somewhat differing arguments) cf. Konrad Ott and Ralf Döring, *Theorie und Praxis starker Nachhaltigkeit* (Marburg: Metropolis, 2004).

[4] See, e.g. Rudolf Steinberg, *Der ökologische Verfassungsstaat* (Frankfurt a.M.: Suhrkamp, 1998), at 114.

[5] Ekardt, *Theorie der Nachhaltigkeit*, supra, note 2, § 1C.

could also be the essential desideratum of a more future – and globally-oriented (thus: sustainable) legal interpretation. Furthermore, restrictions on behalf of environmental or (for instance) resource conservation in order to safeguard the conditions of individual freedom (as embodied in fundamental rights) might also be much more plausible motivationally than the usual, fairly misleading antagonism of individual self-realization versus environmental protection, as latently affirmed by national objective provisions. Incidentally, discussing human rights could even lead to a better normative justification of principles such as common but differentiated responsibility in climate policy – the discussion of historical emissions below will affirm that very clearly.

Accordingly, earlier – and even today in international law – there was often, or is respectively, a discussion about environmental fundamental rights[6] (not only with regard to future generations, of course), as environmental fundamental rights would mean a break with the traditional views diagnosed above. In the academic debate on international law (unlike the practice of international law), the idea of strong or even absolute – i.e. not subject to any balancing – environmental fundamental rights seems to be gaining support. In national debates, however, environmental fundamental rights are considered unspecific and subject to balancing; therefore they are ultimately not very helpful. Of course, the vague content of an "environmental fundamental right" would only result if one generally introduced a fundamental right "to environmental protection"; however, this author is only concerned with the question of whether a correct interpretation of fundamental and human rights (nationally or transnationally) results in greater levels of sustainability – and for instance resource and climate protection – than is often assumed.

Such an interpretation would define fundamental rights in the way they already exist in all western countries as well as in the international declarations on human rights signed by almost every state of the world, with the consequence that current policy might be in conflict with fundamental or human rights, two largely synonymous concepts. Of course, even if this issue falls within the scope of a fundamental right, the problem of necessary balancing cannot be entirely avoided. But then, this problem also applies in precisely the same way to other fundamental rights (requiring what is commonly called the "proportionality test"). Therefore, the subject of the following analysis will not be true fundamental rights "to environmental protection." At the same time, we will not limit ourselves to accepting the common assumption that basically all aspects of fundamental rights which concern environmental issues are covered by the right to life and health, which (a) includes

[6] For an outline of this common discussion, see Steinberg, *Verfassungsstaat*, supra, note 4, at 421 (explicitly criticizing "environmental fundamental rights"); Norbert Gibson, "The Right to a Clean Environment", 1 *Saskatchewan Law Review* (1990), 5; James Nickel, "The Right to a Safe Environment", 3 *Yale Law Journal* (1993), 281, at 282; on the notion of "third generation human rights", see Jack Donnelly, "Third Generation Rights", in Catherine Brölmann, René Lefeber and Marjolaine Zieck (eds.), *Peoples and Minorities in International Law* (Dordrecht: Nijhoff, 1993), 119, at 119; Pascale Kromarek (ed.), *Environnement et droits de l'homme* (Paris: UNESCO, 1987).

no provision for preventive aspects, (b) de facto favors the defensive aspect of the fundamental right over the active protection right it imparts, and (c) moreover fails to concretize environmental protection, which would be required to render it practically relevant. It is precisely this approach toward "duties of protection" (including their administrative consequences) that will be subject to criticism in the course of the following analysis.

4.2 Human Rights: Only Subordinate and Vague "Duties of Protection" with Regard to Sustainability? The Traditional Legal Point of View in Europe and Germany

It is well known that, for instance, the German constitutional and administrative courts are very reluctant to recognize environmental positions based on fundamental rights and have previously rejected corresponding claims for violations of fundamental rights on environmental protection issues.[7] They already avoid the term "protection *rights*", which would clarify that subjective, individual rights are concerned (even if they are subject to balancing with conflicting legal positions). Especially (but not only) in constitutional law cases, there is often no clear distinction between the tests of admissibility and the substantive foundation of the claim. Camouflaging the question whether a subjective, individual right exists, it thus remains unclear what the respective issue is: whether the claimant has an individual right that allows him to bring an action, or whether the underlying action is within the scope of the respective fundamental right or is an issue of restrictions of the respective fundamental right. In spite of the different outcomes, this same situation applies to abortion cases. The basis for all this is the aforementioned idea that protection rights only describe an objective, but fail to define an exact scope of protection, requiring courts to merely examine whether the protective measures taken are manifestly inadequate. However, the latter question will always be denied, since some legislative effort can be found for every objective, virtually ruling out an assessment that state action has been "manifestly inadequate." It will be elaborated later that both this result and its reasoning might deserve criticism.

From the outset, the case law of the European Court of Justice (ECJ) is hardly devoted to the issue of protection rights as such – European fundamental rights are included in the Charter of Fundamental Rights (ECFR), which has binding force since the Lisbon Treaty, and in Article 6, paragraph 1–3 of the EU Treaty.[8] So far, the ECJ has not even specifically addressed fundamental protection rights against

[7] On all the case law, see in detail Ekardt, *Theorie der Nachhaltigkeit*, supra, note 2, § 4.

[8] On the new legislation with an explicit EU Charter of Fundamental Rights, see Ekardt, *Theorie der Nachhaltigkeit*, supra, note 2, § 4 B.

the Union. Only within the Member States has it recognized the possibility of such rights. Of course, to put it provocatively, the ECJ structurally fails to adopt almost any judgment that might bind the EU in any way. It rather seems to be driven by the unspoken intention to give the EU Commission and Council ample discretion in the determination of their policies. Thus, the existing case law lacks any real reference points for the issues discussed in this article. Although the ECJ regularly requires Member States to comply with certain environmental requirements, this has nothing to do with the recognition of protection duties. It only means that the Member States are obliged to effectively implement certain environmental protection requirements adopted by the EU Commission, the Council and the Parliament. At its core, such case law is hence no more than an issue of enforcement of simple (not constitutional) European law; and it is also completely unrelated to the precise content of that law. Protection duties, however, would oblige the EU legislative bodies to act on behalf of the environmental interests of right holders, even where such action is against the legislators' will. Currently, there is no apparent example for such a right. And because of the foregoing tendency in the case law of the ECJ, it seems likely that this will not change significantly anytime soon.[9] Although Article 37 ECFR, which formally entered into force at the end of 2009, contains a commitment to environmental protection – as did the previous EU and EC Treaties – it is not designed as a fundamental right.

A similar situation applies to the European Court of Human Rights (ECtHR), which is responsible for the interpretation of the European Convention of Human Rights (ECHR), a treaty that is applicable to all European countries and is extremely similar to other international human rights treaties. Like the German Federal Constitutional Court, the ECtHR has in fact recognized obligations of states to undertake protective action based on fundamental rights, although not often, and never in an environmental protection case. Likewise, the ECtHR has granted information rights concerning environmental damages, although counterintuitively not based on the right to life and health, but on the right to privacy under Article 8 ECHR. However, all environmental cases of the ECHR are ultimately limited to ensuring that, in the course of administrative decisions, the concerns of individuals are adequately considered and, for example, the facts are weighed appropriately. This was expressed most recently in a case on mobile telecommunication. It appears that the obligation to adopt other, more effective laws on the basis of protection duties, which would trigger a larger reorientation of the social order and not merely ensure "privacy from pollutants and noise," has not been a subject of affirmative ECHR judgments so far.

In any case, the mere factual existence of case law does not per se mean that it is right. And it does not apply generally because judgments only decide a specific case,

[9] Of course, there are cases, though they are not numerous, in which the ECJ has declared EU legal acts void for formal reasons, e.g. due to a lack of legislative competence. But there does not appear to be any case in which the ECJ has ever required the EU to enact legal provisions against its legislature's will.

but do not define an abstract and general norm.[10] Thus, in the following sections, this chapter will test and analyze a somewhat altered interpretation of existing law, based on an interpretation of existing fundamental rights rather than reliance on policy considerations or suggestions of a legislative change to the catalog of fundamental rights. But what could an extended interpretation of freedom and fundamental rights that includes an intergenerational and global dimension look like in order to be more precise than the fairly vague discussion of environmental fundamental rights? Departing from what is probably a prevailing view at the domestic level, for instance in Germany, closer examination reveals that the wording and the systematic position of the fundamental concept of freedom, which is intrinsic to fundamental rights, in the ECFR, national constitutions such as the German Basic Law as well as, ultimately, the ECHR, suggest a more complex interpretation than previously assumed, which has important implications in the intergenerational context.[11] Therefore, the resulting findings can ultimately be applied to any national or transnational human rights protection effort, for instance with regard to climate change.

4.3 Intergenerational and Global Scope of Human Rights, Protecting the Conditions of Freedom, and Multipolarity of Freedom[12]

The starting point for this chapter's approach is the idea of freedom rights as classical-liberal guarantees of self-fulfillment. As far as this basic understanding goes, there is no need to criticize the prevailing view. In addition, however, freedom also has an intergenerational[13] (and global) dimension.[14]

---

[10] Laws, regulations, constitutions, etc. remain the only abstract and general norms, at least in statute law. Nevertheless it is acceptable that the practice often turns to existing judgments, because (and only) in the event that no substantial grounds be argued in favor of a change of legal opinion the burden of argumentation falls to the party challenging the existing legal opinion from previous case law (inter alia for reasons of legal certainty), cf. Robert Alexy, *Theorie der juristischen Argumentation* (2nd edition, Frankfurt a.M.: Suhrkamp, 1991); on the rationality of the application of the law and the methods of legal interpretation, see also Ekardt, *Theorie der Nachhaltigkeit*, supra, note 2, § 1 D.; Davor Susnjar, *Proportionality, Fundamental Rights, and Balance of Powers* (Leiden: Brill, 2011).

[11] The issue here is thus an interpretation of all fundamental rights. The rights of equality, which do not seem to fit, are ultimately special protections of the same freedom and thus do not contradict the following considerations.

[12] For more details and references on this subject see Ekardt, *Theorie der Nachhaltigkeit*, supra, note 2, §§ 4, 5.

[13] With a partly similar reasoning, see also Herwig Unnerstall, *Rechte zukünftiger Generationen* (Würzburg: Königshausen & Neumann, 1999), at 422; with more details, cf. Ekardt, *Theorie der Nachhaltigkeit*, supra, note 2, §§ 4, 5.

[14] To be precise, fundamental rights of future people are not current rights, but their nature is that of "pre-effects" of future rights. This, however does not or not significantly alter their relevance; see in details Unnerstall, *Rechte zukünftiger Generationen*, supra, note 13, at 52 et seq.

Why? In a nutshell[15]: for instance, young people and future generations are of course humans and hence are, or will be, protected by human rights. And this right to equal freedom must be leveraged everywhere where it is threatened – in a technological, globalized world, freedom is increasingly threatened across generations and across national borders. Therefore it is clear that fundamental rights also apply intergenerationally and globally, i.e. in favor of the likely main victims of environmental damage.

But the classical-liberal understanding of freedom, which is mainly focused on the economic freedom of those living here and now, must also be supplemented in other regards. For instance, liberties must be interpreted unambiguously in a way so as to include the elementary physical conditions of freedom – thus not only as a right to social welfare, as it was for instance recently acknowledged by the German Federal Constitutional Court, but also to the existence of a relatively stable resource base and a corresponding global climate. For without such a subsistence level – including energy access and a stable climate – and, by extension, without life and health, there is no freedom.[16] This fundamental right to the elementary conditions of freedom is explicitly provided where life and health are concerned, see, for instance, Articles 2 (2) of the German Basic Law, Articles 2 and 3 ECFR, and Articles 2 and 8 ECHR. In all other cases, it must be based on the interpretation of the general right to freedom. Contrary to the prevailing view, a literal interpretation of the ECFR reveals that Article 2 (1) of the German Basic Law has a counterpart in Article 6 ECFR in that it affords a general EU right to freedom. The same is true for Article 5 ECHR and other similarly structured bills of rights. At least elements of a general right to freedom are also indisputably included in the right to privacy under Article 8 ECHR. Based on what has been said so far, this right to life, health and subsistence also applies intergenerationally and globally, and is the subject of human rights protection e.g. against environmental damages.

"Protection of freedom where it is endangered" also means that freedom includes a right to protection (by the state) against fellow citizens (and not only in exceptional circumstances) – not only, but also for future generations. Such an understanding of the right to freedom inter alia affords protection against environmental harm which is threatening individual freedom and its conditions, for instance through climate change, *by the state and where necessary against fellow individuals.* Without that, there would be no human rights protection against intergenerational damages such as climate change, since states are not the primary emitters of greenhouse gases. The problem rather lies in the fact that states tolerate or approve e.g. greenhouse gas

---

[15] In more details on the three main arguments, cf. Ekardt, *Theorie der Nachhaltigkeit*, supra, note 2, § 4; partly cf. also Unnerstall, *Rechte zukünftiger Generationen*, supra, note 13, at 422.

[16] The international trend toward "social" rights to the various facets of minimum subsistence thus has a theoretical justification. Such a "constitution of international law" can be derived from the legal source of the "general principles of law" (cf. Article 38 of the Statute of the International Court of Justice) without recourse to, e.g., the International Covenant on Economic Social and Cultural Rights; cf. Ekardt, *Theorie der Nachhaltigkeit*, supra, note 2, § 7.

emissions by private actors. This particular idea needs to be explained in detail since it is not commonly articulated, as has been indicated above. But if fundamental rights include both a protection of freedom against the state as well as a duty of the state to protect these rights against fellow citizens, conflicts of interest of any kind must regularly be understood as multipolar (not bipolar) conflicts of freedoms (*multipolarity*); and then, it follows that such an understanding would rebut the traditional, more objective, status of fundamental rights protection (protection duties instead of protection rights, thus non-actionable duties) and the traditional imbalance between the defensive and protective side of fundamental rights, i.e. the regular elimination of protection obligations, unless there is a case of "manifest inadequacy" (understood as something which realistically never occurs, namely the complete absence of regulation in an area of law). Multipolarity would equally refute the assumption that the protective side of fundamental rights is almost entirely taken up with administrative norms, which are supposedly subject to wide legislative discretion, and is not of significant importance with regard to standing in administrative cases nor regarding the application of substantive law.

What are the arguments for multipolarity and how can these respond to certain typical counterarguments? In the following, this chapter assesses whether genuine protection rights already arise from the original scope of fundamental rights – protection rights which, in turn, would afford standing in administrative and constitutional law cases. Details regarding the subsequent balancing test (which will e.g. determine how much weight is afforded to fundamental rights when interpreting substantive administrative law, e.g. discretionary decisions, in light of those rights) will be analyzed later on. This clear distinction between the scope of fundamental rights and balancing process differs significantly from case law, which rarely clarifies whether its skepticism about (fundamental) protection rights refers to issues of standing, scope or restrictions of fundamental rights. This remains unclear even in the – ephemeral – recourse to protection rights in cases of administrative law.

First, the multipolarity of fundamental rights follows from the very idea of freedom, which lies at the center of liberal-democratic constitutions – and, as indicated in a footnote, is a philosophical necessity. Fundamental rights are elementary rights that are intended to afford protection against typical threats to freedom. Thereby, they realize the necessary *autonomy of the individual* which is embodied in the principle of dignity. This autonomy is not only threatened directly by the state, but also by private actors, whose actions are "only" approved or tolerated by the state. To dispute this statement, one would have to argue, e.g., that the construction of an industrial plant is relevant to the freedom of the operator but not to the neighboring residents' freedom. The classical-liberal thinking, in fact, tends to favor such an assumption. This view has also been adopted by the current case law. But the very purpose of a liberal state is to secure a balance of conflicts as *impartially* as possible, i.e. independent of special perspectives, and not to give precedence to a specific set of activities and ideologies (e.g. economic and commercial enterprise). All this suggests that protection rights do exist, that defense and protection are equally important, and that we should speak of protection rights, not obligations, since otherwise the equality of both categories would not be recognized.[17]

Second, the multipolarity of fundamental rights appears in limitation or balancing provisions such as Article 2 (1) of the German Basic Law or Article 52 ECFR, which are also presumed on several instances in the ECHR: as paradigmatic defining principles of liberal-democratic bills of rights, these norms also, more practically, prescribe that the freedom of action is limited by "the rights and freedoms of others." The European "constitution", here manifesting itself in the form of the ECFR and the ECHR as well as national constitutions such as the German Basic Law, thus assumes that if the state resolves specific conflicts, it deals not only with clashing interests, but also explicitly with clashing fundamental *rights*.[18]

The preceding reasoning has sought to establish (I) that, and why, there must be protection rights as part of fundamental rights and (II) that these are subjective, individual rights. Beyond that, the arguments – especially that defense and protection are mentioned side by side – also point out that (III) defense must be on an equal footing with protection.[19]

One objection that might be raised is this: such a fundamental re-interpretation of human rights in the light of sustainability could result in the will of democratic parliaments being overthrown, with "protection rights" affording far greater leeway than "defensive rights". So, does this re-interpretation of human rights undermine

[17] Incidentally, "protection" as defined in this argument can also consist in granting a benefit to an individual, such as a monetary payment to secure a minimum level of subsistence; see also Susnjar, *Balance of Powers*, supra, note 10.

[18] The third argument is the wording of provisions such as Article 1 (1) (2) of the German Basic Law, or Article 1 ECFR, which have been briefly referred to above. Public authorities shall "respect" and "protect" human dignity and also the liberties, which under Article 1 (2) of the German Basic Law ("therefore") exist for the sake of dignity, and thus must be interpreted according to its structure. This relation ("therefore") can also be found in the materials of the ECFR. In addition, the double dimension ("respect/protection") of human dignity and therefore also of the fundamental rights – given the function of dignity as a reason for all human rights which was just described – shows that freedom can be impaired by threats from various sides and that, therefore, it implies defense and protection. But most of all, the word "protect" would lose its linguistic sense if it only meant that the state shall not exercise direct coercion against the citizens (otherwise the state could simply retreat to not acting at all, instead of "protecting"). Hence norms such as Article 1 (1) of the German Basic Law and Article 1 ECFR also imply a protection against fellow citizens. And defense and protection are linguistically on equal footing there. All this implies again that there are fundamental rights of defense and protection, and that protection and defensive rights must be equally strong – and that we should speak of protection rights, not of mere protection obligations. This holds true even though (in the interests of an institutional system based on democracy and a separation of powers, which is indeed the most effective protection of freedom) this "protection" cannot be understood as a direct effect of fundamental rights among citizens, but as a claim against the state for protection (see, specifically Article 1 (3) of the German Basic Law and Article 51 ECFR). For instance, Article 1 paragraph 2 of the German Basic Law as well as the title of this section – and also the materials on the ECFR – talk about "human rights." Thus not only "some" rights are based on dignity, as one might respond, but all of them. Therefore, the structure of human rights, i.e., "equal respect and protection", applies to all and not just some human rights.

[19] In favor of an equal footing see already (but without comprehensive reasoning) Christian Calliess, *Rechtsstaat und Umweltstaat* (Tübingen: Mohr Siebeck, 2001).

democracy? In essence, that question raises the old question of the relationship between freedom and democracy. Not only lawyers, but also some philosophers think (partly by implication) that democracy has latent priority over freedom. It is correct that freedom and democracy contribute to each other. A democracy which is based on certain principles, e.g., a separation of powers, however, promises greater freedom, rationality and impartiality than a "radical" democracy. That is precisely why constitutions such as the German Basic Law are based on a separation of powers and are not structured as radical democracies. In particular, justice between generations and global justice, i.e., the freedom of young people and those living after us, are arguments against radical democracy. After all, for future generations and young people as well as those living in geographically distant locations, democracy is not an act of self-determination, but one of heteronomy. For today they are not participants in this democracy. This then leads to a democracy which is not a principle opposing freedom, but a principle resolving conflict *between* freedoms. This function makes it reasonable to have further conflict resolving institutions, e.g., courts. All this is particularly true if it can be shown that freedom may only be restricted to enhance freedom or freedom conditions – of which the elementary conditions above that were proven relevant in the context of this chapter may be subjectivized, whereas other conditions which only support freedom (such as freedom of artistic expression) however may not.

The legislature may make different choices, and the task of constitutional courts is (only) to control the framework of those decisions based on a set of balancing rules which can be derived from the liberties. The issue is always that some institution of control such as a constitutional court reviews the adherence to rules of balancing. Afterwards, the legislature may react by (partly) altering the constitution. Or the issue is that another institution of control such as a non-constitutional court assesses administrative compliance with the legislative purpose or with rules of balancing when such balancing has been deferred to the administration. Ultimately, the objective must be a deliberative process in which multipolarity supports freedom (on the one hand preventing abuses of power, on the other hand regarding democracy as a shield for freedom) and also is adequate in terms of impartiality, engaging in a "multiple-level discourse" which in turn supports rationality since it mobilizes a maximum of good reasons among the state powers.[20]

[20] First, a constitutional court may never order a judgment against a parliament stating "the legislature is required to do precisely this." On the contrary, it must always limit its decisions to statements such as: "at a minimum, you must discontinue doing this." For instance, the German Constitutional Court may not demand from the German Bundestag: "Phase out the use of coal power within four and a half years." But it may very well say: "The previous phasing out process is too slow; take a new decision on the issue until a specified date, taking into account the following factual situations, normative concerns, as well as procedural and balancing rules." Conversely, a constitutional court could rule on an action brought by an energy company: "Of course, the legislature may phase out nuclear power generation – but it must observe certain limits which it has crossed by demanding the phase out within an unreasonably short timespan." This is all the more true as the deliberative process also includes the administration and the lower courts, as just outlined by the brief introductory note on the "deferral" of the balancing test by the legislature. It allows authorities to respond to a court decision with new decisions, which then in turn are subject to judicial control. The same is true with respect to the legislator and the constitutional jurisdiction. And the legislature may also react on decisions of lower courts with legislative changes, etc.

4.4 The Case of Climate Change[21]

Now, it is possible to draw some conclusions with regard to climate change. Based on the foregoing arguments, it can also be pointed out how balancing rules derived from human rights can work in practice:

- As we have seen, freedom also has an intergenerational and global dimension, given that young people and future generations are humans and therefore are protected by human rights. Fundamental rights also apply intergenerationally and globally, i.e. in favor of the likely main victims of resource overuse, climate change, and so forth.
- Freedom rights must be interpreted unambiguously so as to include the foregoing elementary *preconditions of freedom* – and thus not only a right to social welfare in general, but also to the provision and maintenance of a relatively stable resource base, food supply, security, water supply, life-supporting functions and ecosystem services.[22] With regard to climate change, this implies: a guaranteed proper supply of food and water as well as sufficient energy access on a worldwide and intergenerational scale; a life-cycle perspective on natural resources; responsibility for maintaining life-supporting functions and services of ecosystems; and a general priority in favour of resource efficiency.
- "Protection of freedom where it is endangered" also implies that freedom includes a right to protection (by public authority) against fellow citizens (and not only in exceptional circumstances). This implies protection provided by the authorities, for example, against environmentally or socially harmful behavior that threatens freedom and its conditions, such as overuse of resources, *against fellow citizens, be these natural or legal persons.*
- In the environmental context, protection rights apply in spite of the fact that e.g. many resource problems – for instance with regard to climate change – only represent *future threats* to fundamental rights. For the scope of protection rights is already affected by such threats, not only by concrete and present encroachments. Undoubtedly, future trends are not always predictable and therefore "uncertain". However, an objection based on uncertainty would fail because impairments of fundamental rights which are "only possible" are *not* irrelevant with respect to fundamental rights, especially under the threat of irreversibility of such potential infringement. Otherwise, fundamental rights would no longer serve the very purpose of legal fundamental rights: to guarantee the protection of autonomy exactly where autonomy is threatened with impairment.

---

[21] For more details and references on this subject see Ekardt, *Theorie der Nachhaltigkeit*, supra, note 2, § 6.

[22] In liberal democracies, there are also "further" (in contrast to "elementary") preconditions of freedom such as macroeconomic stabilization, biodiversity, etc., which are extremely helpful, but not absolutely necessary to constitute freedom. Therefore, such "further" preconditions of freedom are usually seen not as human rights but as mere obligations of the public powers (without corresponding rights of individuals). This does not mean at all that these "further" conditions are not important, however – merely that individuals do not have the same degree of legal standing to require their enforcement.

The necessary balancing between all the above-mentioned aspects of sustainability-oriented human rights and the classical liberal guarantees of freedom for consumers and enterprises offers some leeway. Nevertheless, especially with regard to overuse of resources, some definite conclusions can be derived:

- A very often overlooked aspect of freedom is the polluter pays principle, which in turn follows from the principle of freedom itself. Freedom must include responsibility for the foreseeable (including environmental or social) consequences of individual behavior – even across political and temporal boundaries, and also for potentially undesirable consequences such responsibility may incur on the acting individual's life plan. The negative consequences of actions which otherwise benefit an individual (for instance, use of inexpensive resources today) must always fall back on that individual, if only by way of cost recovery for the damage created by such action. This justifies limitations of fossil fuel use and instruments that try to avoid the harmful consequences of overuse.
- Another balancing rule is that the assumptions of underlying facts must be correct. Every decision must, for instance, be based on the latest climate research in order to understand what dangers threaten the freedom of future generations. In situations of factual uncertainty such as climate change, there is also a duty to make preliminary decisions and to review them over time. Current energy and climate policy already disregards the balancing rule that its decisions shall be based on a correct factual basis: in particular, existing actions are probably erroneously deemed suitable to avoid the looming drastic problems in the future.
- Furthermore, politics has not yet taken into account in its decision making that the fundamental right of freedom also has an intergenerational and global cross-border dimension and that, therefore, legal positions of future generations and right-holders in other regions (the "proverbial Bangladeshi") need to be considered in parliamentary and legal decisions.
- The task of politics is to solve the continuous conflicts between different freedoms and, in addition, to guarantee the availability of external conditions of freedom. But generally, this does not mean that the political and democratic process has to provide an equal distribution in the sense that certain privileges – such as greenhouse gas emission rights – necessarily have to be equally distributed. Instead, the details of social distribution are subject to political discretion. However, with respect to elementary conditions of freedom, equal treatment is necessary to ensure that everyone obtains the absolute minimum required to enjoy their freedoms. For without these basic requirements such as food, water, clothing, and basic energy access, there can be no freedom to begin with. As regards food, this has direct implications for the climate problem. The "equal distribution principle" in this context is supported by two arguments:

    1. Without an equal right to the absolute minimum conditions of freedom, the latter would be of no value for the poor – and liberal constitutions as well as human rights *guarantee* equal liberties. In particular, this "equal subsistence" means two things: everyone must have a minimum level of access to resources, energy, and so on, and all must be protected from disastrous threats such as

climate change to the extent possible. Resource overuse and harmful effects such as greenhouse gas emissions caused by modern lifestyles must be reduced in absolute terms, while everyone (worldwide and also in the future) necessarily will cause at least a certain minimum level of greenhouse gas emissions (at least for food production through land use), and many around the world have not yet reached their "equal" per capita share. This makes it rather obvious to be cautious about inequalities with regard to the subject of this contribution.

2. If a collective good such as the global climate is at risk, it seems plausible to afford usage rights or the "proceeds" of an unequal distribution (such as atmospheric use) in equal parts to all persons as far as possible – for no individual can claim particular responsibility in producing that good. This second argument can also be seen as an argument *e contrario* to the polluter pays principle (which also follows from the principle of freedom). Hence, "equal wealth" (nationally or worldwide) may not be a reasonable expectation, but very probably a basic resource supply and equal greenhouse gas emission rights for all – worldwide and across generations. Incidentally, this leads to a theoretical justification of the principle of common heritage of mankind applied to geological and anthropogenic stocks.

- On a preliminary basis, a higher GHG emission rate for developing countries could be justifiable with a view to their fight against poverty (see below).
- Another important consequence of the foregoing principles is: colliding human rights call for distinct rules imposed by public authorities. Purely voluntary solutions will probably not be enough.
- On a procedural basis, the colliding human rights imply a broad participation of all stakeholders in all legislative and administrative decisions with relevance to climate change.

The implications of all this for today might be: absolute reduction of greenhouse gas emissions in industrialized countries; relative decoupling for developing countries including newly industrialising countries; minimising problem shifting between environmental media, types of resources, economic sectors, regions and generations; and driving resource productivity at a rate higher than Gross Domestic Product (GDP) growth.

4.5 The Problem of Historical Emissions

The concept of "one human, one emission right", as argued earlier on a general basis, could be amended to some degree in order to take into account historical emissions of (especially) states that form part of the Organisation of Economic Co-operation and Development (OECD). By these means, the price for emission rights could also incorporate the cost of an (inevitable) adaptation to climate change, insofar as a certain degree of climate change can no longer be prevented. As a concept, "historical emissions" take into account that OECD Member States, in particular, have been emitting vast amounts of greenhouse gases in the past 200 years which now contribute to climate change in the atmosphere. However, it would (1) not further sustainable protection of freedom by climate protection to simply allow China, India and other emerging economies another 150 years of unlimited greenhouse gas emissions, as this would compromise the living conditions of future individuals across the globe. Furthermore, (2) the OECD Member States have not necessarily acquired an "advantage" equivalent to the emitted quantity. Countries like China or India profit on their part from these "advantages", because they can reach an acceptable level of prosperity comparatively rapidly through imports of economic models and technologies that have been developed in the industrialized world. In addition, (3) taking into account "historical emissions" leads to a complex discussion as to how global historical developments in past centuries may have advantaged and disadvantaged different countries. It is therefore impossible to assign a more or less exact number of emission rights under the prospective "historical debt". Most importantly, (4) invoking historical emissions takes into account the advantages and disadvantages of deceased individuals, and considers nations as collective entities. Assuming that the foregoing approach – "only freedom and conditions of freedom" – is correct, such a collectivist perspective cannot be justified. Moreover, it raises the question whether we are really responsible for the acts of our forebears. Incidentally, the experiences with national allocation plans for emissions trading in the EU have already shown that a precise calculation of historically grown emissions is problematic for individual facilities.

All this obviously does not rule out moderate consideration of factors such as "historical emissions" and "adaptation costs" (which are, to date, only taken into account via global financial funds) when calculating the details for an international emissions trading system. Insofar as the freedom principle leads to the justification of certain equality standards and provision of certain basic needs (= fundamental conditions of freedom) and also to implementation of the polluter pays principle, for instance, these aspects can be considered e.g. when calculating the price range, and that with a minimal administrative effort.

4.6 On the Path to a Justice-Based Framework for Global Climate Governance

As shown above, the notion of "one human – one emission right" is not solely meant to be a project at the domestic level, but also an extension of the current and not very ambitious (let alone enforceable) Kyoto Protocol on a global scale after 2012. Based on the general justification provided above, the main elements of a global approach could be summarized in the following ten points:

1. In order to prevent disastrous climate change, the global per capita emissions allowance would have to be fixed and limited – and then would have to be distributed on an equal per capita basis.

2. According to the Intergovernmental Panel on Climate Change (IPCC), the per capita amount would need to be around 1 tonne of CO_2 per person annually. This would be above current emission levels in most developing countries, but far below the OECD countries' emissions.
3. If OECD countries wanted to emit more greenhouse gases, western states would have to buy emission rights from southern countries. In contrast to Kyoto, this would lead to an emission trading system between all states across the globe.
4. By these means, a reduction of greenhouse gas emissions would get started *and* funds would be mobilized for the reduction of poverty in the southern hemisphere.
5. The scheme would not have to impose the 1 tonne per capita from the outset, but could reach this goal in several stages beginning at 5 tonne per capita (which is the global emission average by now); in line with the projections of the IPCC, however, the 1 tonne level would have to be achieved by 2050.
6. Full integration of developing countries into the overall reduction obligation system should potentially be delayed by some years. Prior to that point in time, such countries could obtain additional emission rights or some form of additional payment in order to manage their reductions and adaptation.
7. Also, the sectors aviation, shipping, land use, agriculture, and deforestation would have to be fully integrated into the global emissions trading system.
8. A global institution should have the right to control emission reductions and enforce them with severe sanctions.
9. The annually decreasing aggregate number of emission certificates held by each state or group of states after international emission trading could then form the basis for a national or continental emission trading system among primary energy users (as described earlier), including an annually degressive number of certificates, annually auctioned. The basic principles of such national (or continental) distribution systems might have to be prescribed on a global level to ensure that funds really reach the socially disadvantaged (after all, many states worldwide are not democracies). Compared to existing trading systems such as the EU ETS, such a framework would possess a broader basis (primary energy), stricter goals, a lack of loopholes such as offsets, and a strictly global focus.
10. Primary energy producers or importers would have to auction certificates and pass the costs on through products, electricity and heating prices, and so on to consumers. States or regional integration organizations (such as the EU) would then distribute the auctioning revenues to all citizens on a per capita basis.

By these means, energy efficiency, renewable energy, and long-term energy security would be forced (without a highly complex "instrument mix" ordinary citizens are unable to fully understand). Western countries would partly buy certificates, but partly rely on more energy efficiency, sufficiency, and renewable energy sources, and therefore reduce their overall greenhouse emissions. Step by step, developing countries would do the same. This would stop the global "race to the bottom" with regard to climate policy. Even from a broader economic point of view, the entire concept would lead to very important advantages: it would avoid the disastrous costs of climate change; new technologies would be forced; and independence from energy imports (and rising fossil fuel prices) would increase. Emission trading would help identify the cheapest available climate protection measures, and a broad range of greenhouse gas emissions could be covered and integrated (including, for instance, emission from meet consumption or bioenergy).[23]

In developing countries, the number of transferable rights would be high initially and emission trading costs low; the opposite would apply in OECD countries. This would only be fair, as the higher per capita contribution to climate change originating from the OECD countries would be compensated, while at the same time the social justice of climate policy could be largely sustained in the same countries. Moreover, even the socially underprivileged in western countries would benefit from the financial transfers to the south, as these would stimulate the development of welfare states in the south, thereby reducing social dumping and stabilizing the western welfare state in the medium term. Furthermore, a determined attempt to combat climate change along these lines might avert the social consequences of global warming impacts in both North and South, whose severest manifestations are already emerging: migration and war for resources, such as food and water.

[23] And integration e.g. of bioenergy-caused rainforest degradation would work much more precise than by vague and incomplete "bioenergy sustainability criteria". European and national bioenergy policy is criticised in more detail by Felix Ekardt and Hartwig von Bredow, *Managing the Ecological and Social Ambivalences of Bioenergy – Sustainability Criteria Versus Extended Carbon Markets*, in: Walter Leal (ed.), *The Economic, Social, and Political Aspects of Climate Change* (Berlin: Springer, 2011), 455.

Part II
International Climate Law – Architecture and Institutions

Chapter 5
Foundations of International Climate Law: Objectives, Principles and Methods

Rowena Maguire

Abstract This chapter explores the objectives, principle and methods of climate law. The United Nations Framework Convention on Climate Change (UNFCCC) lays the foundations of the international regime by setting out its ultimate objectives in Article 2, the key principles in Article 3, and the methods of the regime in Article 4. The ultimate objective of the regime – to avoid dangerous anthropogenic interference – is examined and assessments of the Intergovernmental Panel on Climate Change (IPCC) are considered when seeking to understand the definition of this concept. The international environmental principles of: state sovereignty and responsibility, preventative action, cooperation, sustainable development, precaution, polluter pays and common but differentiated responsibility are then examined and their incorporation within the international climate regime instruments evaluated. This is followed by an examination of the methods used by the mitigation and adaptation regimes in seeking to achieve the objective of the UNFCCC. Methods of the mitigation regime include: domestic implementation of policies, setting of standards and targets and allocation of rights, use of flexibility mechanisms, and reporting. While it is noted that methods of the adaptation regime are still evolving, the latter includes measures such as impact assessments, national adaptation plans and the provision of funding.

5.1 Introduction

The purpose of this chapter is to examine the objectives, principles and methods of the international climate change regime. An understanding of the key objectives, principles and methods of the regime is essential, as all measures and policies operating within the climate change regime reinforce and build upon these primary conceptual

R. Maguire (✉)
Faculty of Law, Institute for Sustainable Resources, Queensland University of Technology, Gardens Point Campus, 2 George Street, QLD 4000, Brisbane, Australia
e-mail: rmaguire@qut.edu.au

E.J. Hollo et al. (eds.), *Climate Change and the Law*, Ius Gentium: Comparative Perspectives on Law and Justice 21, DOI 10.1007/978-94-007-5440-9_5,

© Springer Science+Business Media Dordrecht 2013

boundaries. This chapter explores the climate change regime through an examination of the United Nations Framework Convention on Climate Change (UNFCCC),[1] the Kyoto Protocol,[2] the Cancun Adaptation Framework,[3] and a number of Conference of the Parties (COP) decisions. The term climate change regime is used in this chapter to refer to the governance arrangements that exist to support the implementation of the UNFCCC. This includes a combination of laws, institutions and processes operating to assist in fulfilling the ultimate objective of the UNFCCC.

This chapter explores both mitigation and adaptation measures operating within the climate change regime. Mitigation refers to human interventions to reduce emissions of greenhouse gases from sources, or to enhance their removal by sinks.[4] By contrast, adaptation refers to adjustments in practices, process or structures which can moderate or offset the potential for damage or take advantage of opportunities created by a given change in climate.[5] Grasso examines the difference between the regimes, stating that while "[a]daptation consists in adjustment of human systems to actual or expected physical effects of climate change, variability and extreme conditions. In a broad perspective, mitigation seeks to protect natural systems against human systems whereas adaptation aims to protect the latter against nature".[6] While there has been somewhat of a disconnect between the two regimes to date, both mitigation and adaptation measures are necessary components of the climate change regime and are mutually reinforcing, and as such both worthy of equal consideration when examining international climate governance.

5.2 Objective of the Climate Change Regime

The ultimate objectives of the climate change regime is found in Article 2 of the Convention, which requires the stabilization of greenhouse gas concentrations in the atmosphere at a level that would prevent dangerous anthropogenic interference

[1] United Nations Framework Convention on Climate Change (UNFCCC), New York, 9 May 1992, in force 21 March 1994, 31 *International Legal Materials* (1992), 849.

[2] Kyoto Protocol to the United Nations Framework Convention on Climate Change, Kyoto, 10 December 1997, in force 16 February 2005, 37 *International Legal Materials* (1998), 22.

[3] The Cancun Adaptation Framework is contained within Articles 11–35 of the COP Report from the Cancun negotiations in 2010. The section dealing with the adaptation framework is titled "II Enhanced Action on Adaptation". See: Decision 1/CP.16, Cancun Agreements: Outcome of the Work of the Ad Hoc Working Group on Long-Term Cooperative Action under the Convention, FCCC/CP/2010/7/Add.1, 15 March 2011 (Cancun Adaptation Framework).

[4] Farhana Yamin and Joanna Depledge, *The International Climate Change Regime: A Guide to Rules, Institutions and Procedures* (Cambridge: Cambridge University Press, 2004), at 76.

[5] Robert T. Watson and the Core Writing Team (eds), Climate Change 2001: Synthesis Report. A Contribution of Working Groups I, II, and III to the Third Assessment Report of the Intergovernmental Panel on Climate Change (Cambridge and New York: Cambridge University Press, 2001), at 398, and Yamin and Depledge, The International Climate Change Regime: A Guide to Rules, Institutions and Procedures, supra, note 4, at 214.

[6] Marco Grasso, Justice in Funding Adaptation under the International Climate Change Regime (Netherlands: Springer, 2010), at 11.

with the climate system. Such a level should be achieved within a time frame sufficient to allow ecosystems to adapt naturally to climate change, to ensure that food production is not threatened and to enable economic development to proceed in a sustainable manner.

Article 2 provides that this objective applies not only to the Convention, but also to any related legal instrument that the COP adopts. This creates one focused objective for the regime, which is addressed through different legal polices, instruments and measures. The objective is framed as environmental quality standard.[7] It does not prohibit the emission of greenhouse gases; rather it seeks to restrict these activities when they exceed a certain threshold (that of dangerous anthropogenic interference). The objective also sets a timeline for when the environmental standard must be met, requiring that such changes take place, so as to not affect: ecosystem adaptation, food security and economic development occurring in a sustainable manner.

During the drafting process of the UNFCCC some parties (European Countries, Canada, Australia and New Zealand) sought for the adoption of an objective, which included specific targets and timetables, and initially starting with the goal of stabilizing carbon dioxide at current levels.[8] Such approaches had been used to address acid rain and ozone depletion problems and on this basis were recommended for addressing greenhouse gas emission concerns. The United States, Japan, and the former Soviet Union, however argued that setting specific targets and timetables was too rigid, given the lack of scientific certainty and that on this basis emphasis should be placed on furthering scientific research and on the development of national as oppose to international targets. Developing countries positions were also divided with the Alliance of Small Island States pushing for targets and timetables, oil rich countries questioning the science of climate change and countries in the process of industrialisation (Brazil, China and India) arguing that measures must not impinge upon their sovereign rights to development.[9] Such varying perspectives led to drafting of the current objective which stopped short of setting rigid targets and timelines but which attempted to impose an environmental quality standard as a target and apply a timeline by requiring that the environmental standards be achieved in reference to ecological factors.

Article 1 of the UNFCCC contains a number of definitions relevant to understanding the objective of the regime. No definition is provided for the key concept of "dangerous anthropogenic interference" and guidance on the definition of this concept must be sort from Intergovernmental Panel on Climate Change (IPCC) assessment reports. The closest definition provided within Article 1 related to the concept of "dangerous anthropogenic interference" is a definition of "the adverse effects of

[7] Yamin and Depledge, The International Climate Change Regime: A Guide to Rules, Institutions and Procedures, supra, note 4, at 61.

[8] Daniel Bodansky, "The History of the Global Climate Change Regime", in Urs Luterbacher and Deflet Sprinz (eds), *International Relations and Global Climate Change* (Massachusetts: MIT Press, 2001), 23, at 29.

[9] Ibid., at 31.

climate change" which are defined to mean "changes in the physical environment or biota resulting from climate change which have significant deleterious effects on the composition, resilience or productivity of natural and managed ecosystems or on the operation of socio-economic systems or on human health and welfare".

The Third Assessment Report of the IPCC (TAR) examined the concept of "dangerous anthropogenic interference". The TAR identified five broad categories of reasons for concern related to Article 2 of the UNFCCC:

1. Risks to unique and threatened systems
2. Risks from extreme climatic events
3. Regional distribution of impacts
4. Aggregate impacts
5. Risks from large scale discontinuities.[10]

The TAR did not provide a definition of dangerous anthropogenic interference, rather providing criteria and scientific assessment upon each of these criteria which could be used by policy makers in creating a definition. The Fourth Assessment Report of the IPCC (AR4) notes that the definition of "dangerous anthropogenic interference" is a complex task that can only be partially informed by science as it also involves considerations of economic, ethical and legal judgements.[11] The AR4 finds that determining the choice of a stabilization level implies a process that balances the risk of climate change against the risk that response measures will have on economic sustainability.[12] The AR4 describes the criteria of enabling economic development to proceed in a sustainable manner as a double-edged sword, hinting at the difficulty of defining and implementing the objective of the UNFCCC.[13] Given the difficulties, it is not surprising that the IPCC decided to avoid creating a specific definition.

The AR4 indirectly assists in defining the concept of the "dangerous anthropogenic interference". It does this by providing a specific temperature increase measurement. In providing this measurement the AR4 cites earlier work of the World Meteorological Organisation and the United Nations Environment Program to state that "a 2° temperature increase to be the upper limit beyond which the risk of grave damage to ecosystems and non-linear responses are expected to increase rapidly".[14]

[10] Bert Metz et al. (eds), Climate Change 2001: Mitigation. Contributions of Working Group III to the Third Assessment Report of the Intergovernmental Panel on Climate Change (Cambridge: Cambridge University Press, 2001), at 700.

[11] The AR4 focused on key vulnerabilities related to Article 2 objective. These key vulnerabilities can be broadly categories into: biological systems, social systems, geophysical systems, extreme events and regional systems. See Hans-Holger Rogner et al., "Introduction", in Bert Metz et al. (eds), *Climate Change 2007: Mitigation of Climate Change. Contribution of Working Group III to the Fourth Assessment Report of the Intergovernmental Panel on Climate Change* (Cambridge and New York: Cambridge University Press, 2007), 97, at 100.

[12] The AR4 finds that deep emissions reductions are unavoidable in order to achieve stabilisation. It also finds that climate policy can substantially reduce the risk of crossing thresholds deemed dangerous, which validates the work undertaken by the climate change regime and other leaders in climate change polices such as the European Union.

[13] Rogner et al., "Introduction", supra, note 11, at 100.

[14] Rogner et al., "Introduction", supra, note 11, at 99.

The provision of this measurement provides the climate change regime with a specific goal to work towards in meeting the ultimate objective of regime of avoiding "dangerous anthropogenic interference" with the climate system.

The work of the IPCC in providing such a definition is recognised by a number of COP decisions. The Ad Hoc Working Group on Long-Term Cooperative Action under the Convention acknowledges the AR4 finding by stating that "deep cuts in global greenhouse gas emissions are required so as to hold the increase in global average temperature below 2°C above preindustrial levels, and that parties should take urgent action to meet this long-term goal consistent with science and on the basis of equity".[15] This recognition could be used to infer that dangerous anthropocentric interference with the climate system means any change in global temperature beyond 2°C.

5.2.1 Mitigation Objectives

The mitigation objectives of the regime can be further explored by examination of the Kyoto Protocol and two tracks of working groups seeking to further Kyoto commitments and long-term cooperation under the Convention. In Article 3(1) the Kyoto Protocol defines the first commitment period of the regime as operating from 2008 to 2012. In order to understand the objectives, principles and methods of a post 2012 regime, attention is directed towards two decisions reached at the Durban (COP) negotiations in 2011. Decision COP/MOP 7 on the Outcome of the Ad Hoc Working Group on Further Commitments for Annex I Parties creates a second commitment period of the Kyoto Protocol, binding only those original parties to the Kyoto Protocol. Decision CP.17 on the establishment of an Ad Hoc Working Group on the Durban Platform for Enhanced Action[16] creates a process and timeline for creating a new legal instrument to operate from 2020 that will be applicable to all parties of the UNFCCC.

The Kyoto Protocol is the vehicle in which the mitigation obligations created within the UNFCCC are operationalized. The objective of the Kyoto Protocol can be found in Article 3 which requires at its core "greenhouse gas stabilisation and reduction commitments for industrialised (Annex I) countries meant to add up to a 5% reduction in aggregate greenhouse gas emission compared to 1990 levels".[17] The Kyoto Protocol sets individual legally binding emission reduction targets for 37 industrialised nations and the European Community in Annex B to the agreement.

[15] Decision 2/CP.17, Outcome of the Work of the Ad Hoc Working Group on Long-term Cooperative Action under the Convention, UN Doc. FCCC/CP/2011/9/Add.1, 15 March 2012, Section II, Preamble.

[16] Decision 2/CP.17, supra, note 15.

[17] Roda Verheyen, Climate Change Damage and International Law: Prevention, Duty and State Responsibility (The Netherlands: Martinus Nijhoff Publishers, 2005), at 110.

The range of targets varies with the European community adopting the most stringent reduction of 8%, while other countries such as Australia and Iceland were able to increase their emissions from 1990 levels by 8 and 10% respectively.

The delay between the drafting of the agreement in 1997 and entry into force arose from a change in domestic politics within the United States of America. The Clinton administration had signed the Kyoto Protocol, but when the George W. Bush administration took power, it expressed its intention to withdraw from the agreement.[18] Article 25 (1) of the Kyoto Protocol states the protocol shall enter into force after the date on which not less than 55 Parties to the Convention, incorporating at least 55% of total carbon dioxide emissions as at 1990 levels, have deposited their instruments of ratification, acceptance, approval or accession before the agreement entered into force. The positional change of the United States of America left the agreement in a precarious position as it now required every other Annex I party to ratify the instrument. The Kyoto Protocol entered into force on 16 February 2005 following ratification by the Russian Federation. Article 3 of the Kyoto Protocol provides that the first commitment period of the agreement operates from 2008 to 2012. During the first commitment periods, parties are required to demonstrate compliance with their individual mitigation pledges contained with Annex B of the Protocol.

The second commitment period of the Kyoto Protocol is determined by a COP decision from the Durban COP negotiations.[19] This decision determines that the second commitment period of the Kyoto Protocol commences on I January 2013 and expires either on 31 December 2017 or 31 December 2020 (the expiration date to be determined at the 2012 COP negotiations in Qatar). The objective of the second commitment period is to aim to "ensure that aggregate emissions of greenhouse gases by parties included in Annex I are reduced by at least 25–40% below 1990 levels by 2020". Annex I to the agreement sets out the new individual pledges of the Parties to the agreement. The European Union demonstrating leadership has pledged to jointly fulfil a target of a 20–30% quantified emission limitation or reduction objectives (QELROs) during the second commitment period. Pledges made by other parties such as Australia and New Zealand come with a number of caveats and conditions attached and as yet do not specify QELRO.[20] Meanwhile other parties such as Canada, Japan and the Russia Federation, have not accepted QELROs for the second commitment period, undermining the authority of the regime to deliver globally coordinated mitigation measures.

---

[18] David Freestone, "The International Climate Change Legal and International Framework: An Overview", in David Freestone and Charlotte Streck (eds), *Legal Aspects of Carbon Trading: Kyoto, Copenhagen and Beyond* (New York: Oxford University Press, 2009), 1, at 18.

[19] Decision 1/CMP. 7, Outcome of the Work of the Ad Hoc Working Group on Further Commitments for Annex I Parties under the Kyoto Protocol at its Sixteenth Session, UN Doc. FCCC/KP/CMP/2011/10/Add.1, 15 March 2012.

[20] Australia and New Zealand are prepared to consider submitting information on QELRO pursuant to domestic processes and taking into account of number of CMP and COP decisions.

The Durban COP negotiations led to the establishment of an "Ad Hoc Working Group on the Durban Platform for Enhanced Action" (Durban Platform). This process was established as it was noted with grave concern "the significant gap between the aggregate effect of Parties mitigation pledges in terms of global annual emissions of greenhouse gases by 2020 and aggregate emission pathways consistent with having a likely chance of holding the increase in global average temperature below 2°C or 1.5°C above pre-industrial levels".[21] The Durban Platform for Enhanced Action seeks to ensure that a new legal instrument that binds as many parties of the UNFCCC as possible is created. It plans to achieve this by requiring parties to start negotiations on the text of the agreement in 2012; complete the drafting of a new legal instrument by 2015; and for the new legal instrument to come into force in 2020.[22] The Durban Platform for Enhanced Action also recognises that, in order for the regime to fulfil the ultimate objective of the UNFCCC, strengthening of multilateral and rules-based regimes must be developed and implemented.

5.2.2 *Adaptation Objectives*

The adaptation regime can be understood by reference to the UNFCCC and the Cancun Adaptation Framework. The second part of the objective from Article 2 of the UNFCCC is particularly relevant to the adaptation regime. The time driven component of the objective seeks for "ecosystems to adapt naturally to climate change, to ensure that food production is not threatened and to enable economic development to proceed in a sustainable manner". Mitigation measures to reduce emissions operate with the purpose of lowering emissions so as to ensure that ecosystems can adapt naturally to climate change. Adaptation measures take a more proactive approach to ensuring that ecosystems remain in functional order by implementing measures that involve human intervention to protect or enhance vulnerable ecosystems.

The UNFCCC obliges parties in Article 4 (1) (b) to "formulate, implement, publish and regularly update national … measures to facilitate adequate adaptation to climate change". This provision creates obligations for parties to develop national measures to address domestic country specific adaptation concerns. Article 4 (1) (e) seeks to create a responsibility to assist developing countries to implement adaptation measures by directing parties to "cooperate in preparing for adaptation to the impacts of climate change, develop and elaborate appropriate and integrated plans for coastal zone management, water resources and agriculture, and for the protection and rehabilitation of areas, particularly in Africa, affected by drought and desertification, as well as by floods". The obligations in Article 4 can be read in conjunction with Article 3 (3), which deals with the precautionary principle. This provision requires parties to take precautionary measures to anticipate, prevent or

[21] Decision 2/CP. 17, supra, note 15, para. 2.

[22] Decision 2/CP. 17, supra, note 15, Art. 4.

minimize the causes of climate change and design policies and measures that take into account different socio-economic contexts. Adaptation is specifically identified within Article 3 (3) as an area of the precautionary principle has application.

While adaptation has been part of the climate regime since its inception, the development and implementation of adaptation policies and measures has been hindered by three factors: lack of agreement about the meaning, scope and timing of adaptation; limited capacity in developing countries to undertake vulnerability assessments; and bottlenecks in the availability of funding.[23] Grasso suggests that the dominant natural-science approach to climate change, based upon assessments of physical processes, is responsible for the separation of the concepts of mitigation and adaptation, consequently resulting in the climate regime focusing almost exclusively on issues of energy policy and emission control.[24] The COP Report from Cancun acknowledges the bias of the regime in the development of mitigation measures and states in paragraph 2 (b) that "*Adaptation must be addressed with the same priority as mitigation and requires institutional arrangements to support development*".[25]

The Cancun Adaptation Framework emerged from the Bali Action Plan and the work of the Ad-Hoc Working Group on Long Term Cooperative Action under the Convention. It states that "adaptation is a challenge faced by all Parties, and that enhanced action and international cooperation on adaptation is urgently required".[26] The Cancun Adaptation Framework does not specify a measurable or time bound objective. Rather the objective is stated to be the enhancement of action on adaptation through international cooperation and consideration of matters relating to adaptation under the UNFCCC.[27] The framework does however specify that "a country-driven, gender-sensitive, participatory and fully transparent approach, taking into consideration vulnerable groups, communities and ecosystems" that is based on best available science and appropriate, traditional and indigenous knowledge is to be used to integrate adaptation into relevant social, economic and environmental policies and actions, where appropriate.[28]

One of the priorities of the Cancun Adaptation Framework is to provide developing country parties particularly vulnerable to climate change with "long-term, scaled-up, predictable, new and additional finance, technology and capacity-building, consistent with relevant provisions, to implement urgent, short-, medium – and long-term adaptation actions, plans, programmes and projects at the local, national, subregional and regional levels".[29] This echoes a commitment from the UNFCCC

---

[23] Yamin and Depledge, The International Climate Change Regime: A Guide to Rules, Institutions and Procedures, supra, note 4, at 213.

[24] Marco Grasso, Justice in Funding Adaptation under the International Climate Change Regime (Netherlands: Springer, 2010), at 12.

[25] Cancun Adaptation Framework, supra, note 3.

[26] Ibid., para. 11.

[27] Ibid., paras. 12 and 13.

[28] Ibid., para. 12.

[29] Ibid., para. 18.

in Article 4 (4) that requires developed country parties to assist developing country parties that are particularly vulnerable to adverse effects of climate change by meeting the costs of adaptation arising from the adverse effects of climate change.[30] These provisions recognise the equity issues associated with adverse effects of climate change, and seek to remedy such issues though the transfer of money and other assistance.

5.3 Principles of the Climate Change Regime

Article 3 of the UNFCCC sets out a number of international environmental principles applicable to the regime. The preamble to the UNFCCC also contains references to international environmental principles.[31] The principles are to be used to guide the implementation of the instrument and assist in meeting the ultimate objective in Article 2. The principles contained within the UNFCCC are applicable to the Kyoto Protocol[32] and to all other instruments of the regime seeking to implement the ultimate objective of the convention.[33]

The international environmental principles referred to within the UNFCCC are sourced from earlier international instruments, binding acts of international institutions and customary international law. Most of the international environmental principles referred to within the UNFCCC have not reached the status of customary international law. As such, it is necessary to analyse the text of the UNFCCC in order to understand the manner, scope and application of the principles within the climate regime. There is no exhaustive list of international environmental principles, however the work of Sands can be referred to authoritatively to identify the general rules and principles of international environmental law. The following seven principles identified by Sands will be examined and their influence on the climate change regime discussed:

1. The obligation reflected in Principle 21 of the Stockholm Declaration and Principle 2 of the Rio Declaration, namely that states have sovereignty over their natural resources and the responsibility not to cause transboundary environmental damage;
2. The principle of preventive action;
3. The principle of cooperation;
4. The concept of sustainable development (encompassing the concepts of sustainable use, inter-generational equity, intra-generational equity and integration);

---

[30] Also see UNFCCC, supra, note 1, Arts. 4 (8) and 4 (9).

[31] Many of the statements in the preamble were part of earlier draft texts of the UNFCCC, which were relegated to the preamble as they were considered to be too controversial for inclusion within the Articles of the instrument. Yamin and Depledge *The International Climate Change Regime: A Guide to Rules, Institutions and Procedures,* supra, note 4, at 67.

[32] Kyoto Protocol, Preamble, supra, note 2, para. 4.

[33] UNFCCC, supra, note 1, Art. 3.

5. The precautionary principle;
6. The polluter-pays principle; and
7. The principle of common but differentiated responsibility.[34]

The principles listed above are different from substantive legal rules that are in and of themselves enforceable. Principles are used within regulatory frameworks to guide the interpretation and implementation of the obligations within the source of law under consideration. The difference between a legal rule and principle was examined in the *Gentini* case, where it was stated that:

> A 'rule'… is essentially practical and, moreover, binding… There are rules of art, as there are rules of government, while a principle 'expresses a general truth, which guides our action, serves as a theoretical basis for the various acts of our life, and the application of which to reality produces a given consequence.[35]

The inclusion of a set of guiding principles within the text of the Convention was controversial during the drafting the agreement. The United States of America, along with other developed countries, did not want to include open-ended principles within the agreement due to concerns that their inclusion would lead to the creation of additional commitments beyond those clearly spelled out within the Convention. Developing countries felt that it was essential to include a statement on principles within the articles of the text to guide the implementation of the text.[36] The final text of the agreement adopts the developing country perspective, thus creating within Article 3 a normative framework to support the implementation of the UNFCCC.[37] The principles contained within Article 3 are therefore not directly enforceable, but can be used to inform policy development and implementation modalities within the broader climate change regime. The seven international environmental general rules and principles identified by Sands will now be analysed in the context of the entire text of the UNFCCC.

5.3.1 State Sovereignty and Responsibility

The principle of sovereignty, while an essential component of the international legal order, presents difficulties in the implementation of the concept arising from the

---

[34] Philippe Sands, *Principles of International Environmental Law*, 2nd ed. (Cambridge: Cambridge University Press, 2003), at 231.

[35] *Gentini case (Italy/Venezuela)* M.C.C. (1903), J.H. Ralston and W.T.S. Doyle, Venezuelan arbitration OF 1903 ETC. (1904), 720, 725, cited in Sands, *Principles of International Environmental Law*, supra, note 34, at 233.

[36] Daniel Bodansky, "The United Nations Framework Convention on Climate Change: A Commentary", 18 *Yale Journal of International Law* (1993), 451, at 501.

[37] Yamin and Depledge, The International Climate Change Regime: A Guide to Rules, Institutions and Procedures, supra, note 4, at 66.

dual-natured character of the principle.[38] The principle of sovereignty brings with it both rights and limitations. The preamble to the UNFCCC recognises the two elements of sovereignty by recalling the wording of Principle 21 of the Stockholm Declaration and stating in paragraph 8 of the UNFCCC Preamble that States have:

> the sovereign right to exploit their own resources pursuant to their own environmental policies, and the responsibility to ensure that activities within their jurisdiction or control do not cause damage to the environment of other States or of areas beyond the limits of national jurisdiction.

Inclusion of this principle within the preamble and not within the text of the UNFCCC suggests that sovereign rights to unlimited greenhouse gas (GHG) emissions are not recognised by the agreement. The UNFCCC therefore seeks to limit state sovereign power by imposing restrictions on the level of allowable GHG emissions and as such influencing the types of activities and industries carried out within a territory. Such limitations therefore heavily impinge upon states' sovereign rights to regulate greenhouse gas emission output within their territories. It should, however be noted that the UNFCCC and Kyoto Protocol do not prohibit or remove states' rights to emit greenhouse gases, rather these instruments seek to curb the increase of such emissions within the global commons.

Some literature has focused on the responsibility component of this definition suggesting that harm suffered by countries as a result of climate change should be remedied by international law.[39] Okowa's work identifies a number of difficulties associated with implementing the principle of responsibility including: issues of retroactivity (making states liable to emission harm caused prior to the introduction of the UNFCCC); apportioning responsibility among states; apportioning responsibility for future damage; and managing the scientific uncertainty associated with such claims.[40] It is also noted that the traditional avenue for imposing responsibility to correct harm has occurred through litigation, which in the context of climate change has proved problematic.[41] The development of a loss and damage mechanism to redress harm arising from climate change provides an alternative to litigation within the climate regime. Such a mechanism does not however, impose responsibility on a particular state; rather it seeks to resolve disputes by remedying the harm suffered as a result of climate change without apportioning liability to a particular state or region.[42]

---

[38] On the different discipline understandings (international law, international relations, philosophy and economics see Melea Lewis, Charles Sampford and Ramesh Thakur, "Introduction", in Trudy Jacobsen, Charles Sampford and Ramesh Thakur (eds), *Re-envisioning Sovereignty: The End of Westphalia* (London: Ashgate Publishers, 2008), 1, at 8.

[39] See for example Richard Tol and Roda Verheyen, "State Responsibility and Compensation for Climate Change Damage – a legal and economic assessment", 32 *Energy Policy* (2004), 1109 .

[40] Phoebe Okowa, "Responsibility for Environmental Damage", in Malgosia Fitmaurice, David Ong and Panos Merkouris (eds), *Research Handbook on International Environmental Law* (United Kingdom: Edward Elgar Publishers, 2010), 303, at 304.

[41] Ibid. and on consideration of climate change see generally Brian Preston, "Climate Change Litigation (Part 1)", 5 *Carbon and Climate Law Review* (2011), 3, and Jacqueline Peel, "Issues in Climate Change Litigation", 5 *Carbon and Climate Law Review* (2011), 15.

[42] See Decision 7/CP.17, Work Programme on Loss and Damage, UN Doc. FCCC/CP/2011/9 Add.2, 15 March 2012.

5.3.2 *Principle of Preventative Action*

The principle of preventative action requires states to prevent damage to the environment and to reduce, limit, or control activities which might cause or risk such damage.[43] While the principle of preventative action is not included as principle within Article 3 of the UNFCCC, the principle of preventative action is encapsulated within the objective clause within Article 2, which requires parties to prevent dangerous anthropocentric interference with the climate system. While similar to the principle of sovereignty and responsibility, the principle of preventative action can be distinguished from the principle of sovereignty and responsibility in two ways. Firstly, the principle of preventative action requires a certain objective to be fulfilled: that of reducing environmental damage. Secondly, the preventative principle can operate to prevent a state from damaging the environment within its jurisdiction.[44] The Kyoto Protocol in Article 3(1) applies the principle of preventative action by requiring parties to reduce their overall emissions of greenhouse gasses by at least 5% below 1990 levels during the first commitment period. Such a provision seeks to reduce environmental damage and prevent parties from damaging the environment further within their jurisdictions.

5.3.3 *Principle of Cooperation*

The principle of cooperation sometimes referred to as good neighbourliness is defined in principle 27 of the Rio Declaration as requiring that "[s]tates and people shall co-operate in good faith and in a spirit of partnership in the fulfilment of the principle embodied in this Declaration and in the further development of international law in the field of sustainable development". This general principle of cooperation has evolved to include more concrete duties such as information sharing and participation in decision-making processes.[45] The UNEP Draft Principles recognise this evolution of the principle by stating in principle 7 that "exchange of information, notification, consultation and other forms of cooperation regarding shared natural resources are carried out on the basis of the principle of good faith and in the spirit of good neighbourliness".[46]

---

[43] Sands, Principles of International Environmental Law, supra, note 34, at 246.

[44] Ibid., at 246.

[45] Ibid., at 250.

[46] Draft Principles of Conduct for the Guidance of States in the Conservation and Harmonious Exploitation of Natural Resources Shared by Two or More States, United Nations Environment Programme Governing Council, XII Plenary Meeting, UN Doc. UNEP/GC/101 and Corr.1, 9 to 25 May 1978.

The principle of cooperation is found in many instances within the UNFCCC. Such repeated inclusion of this principle demonstrates an understanding of the necessity of global cooperative action in addressing climate change. The preamble in paragraph 6 calls for "the widest possible cooperation by all countries and their participation in an effective and appropriate international response". This call for cooperation could be interpreted in two ways. Firstly, it could be interpreted as requiring all parties to adopt commitments and take action in implementing mitigation and adaptation policies and activities. Or secondly it could be interpreted as merely requiring all parties to the UNFCCC to take part in the negotiation process. The first interpretation would clearly place a much heavier onus on parties, and would likely be well received by many who are frustrated by the lack of good neighbourliness conduct at recent COP negotiations.[47] The second interpretation, only requiring participation at negotiating sessions, is however reflective of state practice at COP negotiations. The ambiguity of this statement and the potentially onerous obligations that it could impart likely explain the inclusion of this statement within the preamble.

Article 4 of the UNFCCC creates binding and more specific cooperation duties.[48] The parties are requested to cooperate on a number of tasks including to:

- Promote and cooperate in the development, application and diffusion, including transfer, of technologies, practices and processes that control, reduce or prevent anthropocentric emissions[49];
- Promote and cooperate in the full, open and prompt exchange of relevant scientific, technological, technical, socio-economic and legal information related to the climate system[50];
- Promote and cooperate in education, training and public awareness related to climate change and encourage the widest participation in this process, including that of non-governmental organizations[51];
- Communicate to the Conference of the Parties information related to implementation in accordance with Article 12.[52]

---

[47] Criticism has been levelled against the UNFCCC COP process with many feeling that the process is moving too slow and that the process does not bind many of the world's highest emitters. The Copenhagen negotiations in particular attracted criticism concerning the lack of political will of the parties to reach a legally binding outcome. For an analysis of what led to failure in Copenhagen see Cameron Hepburn and Nicholas Stern, "A New Global Deal on Climate Change", 27 *Oxford Review of Economic Policy* (2011), 259, at 259–279. Also see Daniel Bodansky and Elliot Diringer, "The Evolution of Multilateral Regimes: Implications for Climate Change", 2010, available at: http://www.pewclimate.org/docUploads/evolution-multilateral-regimes-implications-climate-change.pdf (last accessed on 5 January 2012).

[48] Also note that the Kyoto Protocol creates additional cooperation obligations for parties to this agreement in Art. 1(b) and Arts. 10 (c), (d), (e), supra, note 2.

[49] UNFCCC, supra, note 1, Art. 4 (c).

[50] Ibid., Art. 4 (h).

[51] Ibid., Art. 4(i)

[52] Ibid., Art. 4 (j). The requirements of Article 12 will be discussed in greater detail further on in this chapter.

5.3.4 The Concept of Sustainable Development

The concept of sustainable development originates from the 1987 Brundtland Report where the concept was described as "development that meets the needs of the present without compromising the ability of future generations to meet their own needs".[53] The concept of sustainable development has become the one of the key goals of international environmental and developmental regimes, though implementation of the concept remains elusive.[54] Sands identifies four principles of the concept of sustainable development:

1. The need to preserve natural resources for the benefit of future generations (the principle of inter-generational equity);
2. The aim of exploiting natural resources in a manner which is sustainable or prudent or rational or wise or appropriate (the principles of sustainable use);
3. The equitable use of natural resources, which implies that use by one state must take account of the needs of other states (the principle of equitable use or intra-generational equity);
4. The need to ensure that environmental considerations are integrated into economic and other development plans, programmes and projects, and that development needs are taken into account in applying environmental objectives (the principle of integration).[55]

The principle of inter-generational equity[56] is recognised in the final line of the preamble and in Article 3(1) of the UNFCCC. The final line in the preamble states that the parties to this convention are "determined to protect the climate system for present and future generations". Similarly, Article 3(1) is broad in coverage stating that "[t] he Parties should protect the climate system for the benefit of present and future generations". The most contentious issue of the principle is defining the

[53] World Commission on Environment and Development, *Our Common Future* (Oxford: Oxford University Press, 1987), at 43.

[54] For instance, a major summit – the United Nations Conference on Sustainable Development (Rio+20) – will take place in June 2012 in Rio de Janeiro, Brazil. This event marks the 20th anniversary of the Rio Declaration 1992, signed at the Earth Summit in Rio. The objective of the Conference is to secure renewed political commitment for sustainable development, assess the progress to date and the remaining gaps in the implementation of the outcomes of the major summits on sustainable development, and address new and emerging challenges. The Conference will focus on two themes: (a) a green economy in the context of sustainable development and poverty eradication; and (b) the institutional framework for sustainable development. For the complete agenda and further background information, see on the Internet http://www.uncsd2012.org (last accessed on 25 March 2012).

[55] Sands, Principles of International Environmental Law, supra, note 34, at 253.

[56] For further background on the concept of inter-generational equity see Edith Brown Weiss, "Our Rights and Obligations to Future Generations for the Environment", 84 *American Journal of International Law* (1990), 198.

nature and extent of the rights of future generations.[57] The current incorporation of the principle of inter-generational equity within the UNFCCC does not create any specific rights or duties; it merely recognises that future generations have an interest in the natural environment.

Applying the principle of sustainable use requires the adoption of a standard that sets out the rate of use or exploration of specific natural resources, as opposed to relying on their preservation for future generations as an outcome.[58] The standard of sustainable development within the UNFCCC can be found in the objectives clause of Article 2, where it is stated that stabilisation of greenhouse gases must occur within a "timeframe sufficient to allow ecosystems to adapt naturally to climate change, to ensure that food production is not threatened and to enable economic development to proceed in a sustainable manner". As discussed above, the AR4 report of the IPCC suggests that sustainable use in the context of climate change involves limiting temperature increases to a maximum of 2° in order to meet the ecosystem, food production and economic standards of the UNFCCC.[59]

The principle of equitable use/intra-generational equity are based upon notions of fairness with regards to the access and use of the environment and enjoyment of the environment. The concept of equity also allows for a consideration of how to share the benefits and burdens of environmental protection and or environmental harm (for example pollution, water scarcity). There are parallels between the principles of equitable use/intra-generational equity and theories of distributive and environmental justice.[60] The principle of equitable use/intra-generational equity gives recognition to the fact that the poorest of the poor in the world (including poor people in prosperous societies) are going to be the groups worst hit by climate change.[61] This principle can be considered from two perspectives:

> An international perspective: examining the inequities that arise between the distribution of environmental benefits and burdens between different countries; and
>
> A country level perspective: examining the inequities which arise between different community groups and stakeholders within a specific region in distributing environmental benefits and burdens.

[57] Peter Doherty, "What Do We Owe to Future Generations?", in Helen Sykes (ed.), *Future Justice* (Albert Park, Vic.: Future Leaders, 2010), 21 and more generally see generally Laura Westra, *Environmental Justice and the Rights of Unborn and Future Generations: Law, Environmental Harm and the Right to Health* (United Kingdom: Earthscan, 2006).

[58] Sands, Principles of International Environmental Law, supra, note 34, at 257.

[59] The most recent IPCC report is the 4th Assessment Report, available at http://www.ipcc.ch/publications_and_data/publications_and_data_reports.shtml (last accessed on 25 March 2012).

[60] See generally on the topic of environmental justice David Schlosberg, *Defining Environmental Justice: Theories, Movements, and Nature* (Oxford: Oxford University Press, 2007); Klaus Bosselman and Benjamin J. Richardson, *Environmental Justice and Market Mechanisms: Key Challenges for Environmental Law and Policy* (London: Kluwer Law, 1999).

[61] Maxine Burkett, "Just Solutions to Climate Change: A Climate Justice Proposal for a Domestic Clean Development Mechanism", 56 *Buffalo Law Review* (2008), 169, at 177.

The climate change regime has focused on the first perspective, that of the inequities that arise between developed and developing countries in the climate change context. Climate change is perceived as raising ethical and justice issues. It has been stated that "[c]limate change is … chiefly an issue of (in)justice, since it has been caused by rich nations and poses risks upon the poor, who are the least responsible and the most vulnerable to the damages and risk associated with it".[62] The UNFCCC recognizes indirectly that developed countries are largely responsible for global emissions by directing "developed country parties to take the lead in combating climate and adverse effects thereof" in Article 3(1). The Cancun Adaptation Framework, however recognizes directly "that the largest share of historical global emissions of greenhouse gases originate in developed countries, and that owing to this historical responsibility developed country Parties must take the lead."[63]

The UNFCCC also guides parties in Article 3(2) to recognize:

> The specific needs and circumstances of developing country Parties, especially those that are particularly vulnerable to the adverse effects of climate change, and of those Parties, especially developing country Parties, that would have to bear a disproportionate or abnormal burden under the Convention, should be given full consideration.

Article 4 (8) of the UNFCCC identifies a number of groups likely to suffer from the adverse impacts of climate change and implementation of response measures. These groups include: small island countries, countries with low-lying coastal areas, countries prone to natural disaster, drought, and desertification, countries with fragile ecosystems, countries with economies highly dependent on income associated with consumption or trade in fossil fuels, and land-locked and transit countries. Within its preamble, the Cancun conference report notes "Resolution 10/4 of the United Nations Human Rights Council on human rights and climate change, which recognizes that the adverse effects of climate change have a range of direct and indirect implications for the effective enjoyment of human rights and that the effects of climate change will be felt most acutely by those segments of the population that are already vulnerable owing to geography, gender, age, indigenous or minority status, or disability". As such the Cancun Adaptation Framework calls for parties to reduce vulnerability and assist in building resilience in countries with urgent and immediate needs.[64]

The principle of integration requires for environmental considerations to be taken into account in economic and development activities. Article 3 (4) of the UNFCCC directs parties to ensure that "policies and measures to protect the climate system against human-induced change … be integrated with national development programmes, taking into account that economic development is essential for adopting

---

[62] Chukwumerije Okereke and Heike Schroeder, "How Can the Objectives of Justice, Development and Climate Change Mitigation be Reconciled in the Treatment of Developing Countries in a Post-Kyoto Settlement?", Background Paper for the DSA-DFID Policy Forum on Climate Change and International Development, University of Greenwich, 2 June 2008, at 1.

[63] Cancun Adaptation Framework, supra, note 3, Art. 8.

[64] Ibid., Art. 11.

measures to address climate change". Environmental impact assessments and strategic environmental assessments are the tools used at the state level to integrate environmental considerations into development and planning decisions. Concerns about the preservation of economic development have led to the development of market-based mechanisms (such as emission trading schemes and carbon taxes) to regulate domestic carbon emissions.

5.3.5 The Precautionary Principle

The precautionary principle seeks to provide assistance in the development and implementation of international environmental law when there is scientific uncertainty. The principle includes considerations of risk prevention, cost effectiveness, ethical responsibilities towards the earth and the shortcomings of human understanding.[65] Principle 15 of the Rio Declaration provides a definition of the principle: "where there are threats of serious or irreversible damage, lack of full scientific certainty shall not be used as a reason for postponing cost-effective measures to prevent environmental degradation". Differing views exist as to the meaning of the principle with some believing that it provides a justification for early intervention, while others view the principle as hampering human activity and creating a system of over-regulation.[66] From a legal perspective the crucial component of the principle is a requirement to take positive action to protect the environment, prior to the existence of scientific evidence detailing specific harm. The principle is also proactive in nature and operates to prevent unsustainable or degrading environmental practices as opposed to the majority of reactionary processes used in environmental regulation.

The precautionary principle is included within Article 3(3) of the UNFCCC and requires parties to:

> take precautionary measures to anticipate, prevent or minimise the causes of climate change and mitigate its adverse effects. Where there are threats of serious or irreversible damage, lack of full scientific certainty should not be used as a reason for postponing such measures, taking into account that policies and measures to deal with climate change should be cost-effective so as to ensure global benefits at the lowest possible cost.

Findings from the Stern Review suggest that immediate adoption and implementation of the precautionary principle is required.[67] The complicating factor in implementing the precautionary principles arises in the determination of what is considered to be a cost-effective solution. The preamble to the UNFCCC provides some guidance

---

[65] Minna Pyhälä, Anne Brusendorff and Hanna Paulomäki, "The Precautionary Principle", in Malgosia Fitmaurice, David Ong and Panos Merkouris (eds), *Research Handbook on International Environmental Law* (United Kingdom: Edward Elgar Publishers, 2010), 203, at 203.

[66] Sands, Principles of International Environmental Law, supra, note 34, at 267–268.

[67] Nicholas Stern, *The Economics of Climate Change: The Stern Review* (Cambridge: Cambridge University Press, 2007).

on cost effectiveness in paragraph 17 by stating that "various actions to address climate change can be justified economically in their own right and can also help in solving other environmental problems".

Assessments about cost effectiveness are value judgments based on "the amount of damage that is acceptable; and the costs that society is willing to pay to reduce or lower the risk of such damage".[68] The global consensus on the level of acceptable climate change is a temperature increase of 2°.[69] Determining the cost that the global community is willing to pay in order to reduce global warming can be understood by reference to the mitigation pledges provided during and after the Copenhagen (2009) and Cancun (2010) climate negotiations.[70] Assessments carried out by the United Nations Environment Program suggest that the emission pledges reached at Copenhagen will not be sufficient to prevent more than a 2° global temperature rise.[71] A challenge therefore in implementing the precautionary principle within the climate regime will be to increase mitigation pledges and the associated implementation of such pledges in order to align with the agreed level of acceptable climate damage.

5.3.6 The Polluter Pays Principle

The polluter pays principle requires that individuals, states or corporations engaging in polluting or hazardous activities that cause damage to the environment should be held responsible for the consequences of their action.[72] The polluter pays principle is defined in principle 16 of the Rio Declaration as:

> National authorities should endeavour to promote the internalisation of environmental costs and the use of economic instruments, taking into account the approach that the polluter should, in principle, bear the costs of pollution, with due regard to the public interest and, without distorting international trade and investment.

The UNFCCC does not incorporate the polluter pays principle. A number of climate change suits are being bought which in effect seek to implement the polluter

[68] Pyhälä, Brusendorff and Paulomäki, "The Precautionary Principle", supra, note 65, at 215.

[69] Global acceptance refers to the agreed upper temperature increases agreed within the climate change regime. See Decision 2/CP.17, supra, note 15, para. 2.

[70] Compilation of Economy-wide Emission Reduction Targets to Be Implemented by Parties included in Annex I to the Convention, FCCC/SB/2011/INF.1/REV.1, 7 June 2011.

[71] Kelly Levin and Murray Ward, "The Emissions Gap Report", 2010, available at: http://www.unep.org/publications/ebooks/emissionsgapreport (last accessed on 24 February 2012).

[72] Priscilla Schwartz, "The Polluter-pays Principle", in Malgosia Fitmaurice, David Ong and Panos Merkouris (eds), *Research Handbook on International Environmental Law* (United Kingdom: Edward Elgar Publishers, 2010), 243.

pays principle.[73] National courts have taken a cautious approach with such cases and it seems that there is a preference from both the judiciary and the literature surveying these decisions for legislative responses to be created dealing with the imposition of liability for harm and the specification of the appropriate remedy for such harm (i.e. who should receive payment, on what basis and for what purpose). The evolution of cases in this area is a response to the lack of legislative responses to climate change at both the international and national levels giving effect to a polluter pays type obligation.[74]

5.3.7 *The Principle of Common But Differentiated Responsibility*

The concept of common but differentiated responsibility adopts a substantive approach to justice by recognising that different groups before the law require different rights and responsibilities. As such, the principle recognises:

- The common responsibility of countries to protect the environment;
- The differing contributions of countries to climate change; and
- The differing inabilities of countries to prevent, reduce and control the threat of climate change.[75]

The principle therefore, recognises the historical differences in the contribution of developed and developing countries to climate change and the difference in their respective economic and technical capacity to respond to these problems.[76] This concept was defined and brought to life through the 1992 Rio

[73] *Connecticut v American Electrical Power Company Inc,* Judgement, 20 June 2011, 406F.Supp. 2d, at 265; *Korsinsky v U.S. EPA*, Judgement, 29 September 2005, No 05–859 (NRB), 205 U.S. Dist LEXIS 21778; *California ex rel Brown v General Motors Corporation*, Judgement, 17 September 2007, U.S. Dist LEXIS 68547. For a discussion of all of these cases see Theodore J. Boutrous and Dominic Lanza, "Global Warming Tort Litigation: The Real Public Nuisance", 80 *Ecology Law Currents* (2008), 80.

[74] The Kyoto Protocol aims at an international solution to this problem. However, any climate policy measures would still have to be implemented at the national level. Germany and the European Union are acting as forerunners in international climate change policy. Michael Grubb, "Seeking Fair Weather: Ethics and the International Debate on Climate Change", 71 *International Affairs* (1995), 463.

[75] Angela Williams, "Promoting Justice within the International Legal System: Prospects for Climate Refugees", in Benjamin J. Richardson, Yves Le Bouthillier, Heather McLeod-Kilmurray, Stepan Wood (eds.), *Climate Law and Developing Countries: Legal and Policy Challenges for the World Economy* (Cheltenham: Edward Elgar, 2009), 84, at 90.

[76] Kati Kulovesi and Maria Gutierrez, "Climate Change Negotiations Update: Process and Prospects for a Copenhagen Agreed Outcome in December 2009", 18 *Review of European Community and International Environmental Law* (2009), 229, at 236.

Declaration.[77] The principle was also defined and explained in Article 3(1) of the UNFCCC[78] which states that:

> "The parties should protect the climate system for the benefit of present and future generations of humankind, on the basis of equity and in accordance with their common but differentiated responsibilities and respective capabilities. Accordingly, the developed country Parties should take the lead in combating climate change and the adverse effects thereof.

The Kyoto Protocol is explicitly based on the Common but Differentiated Responsibility Principle.[79] As stated by Honkonen[80] "the burden-sharing agreement under the Kyoto Protocol was a remarkable achievement, paving the way to country specific commitments in international environmental cooperation". Specifically, the Protocol demonstrated the applicability of the Common but Differentiated Responsibility principle through its operational provisions, for example, by excluding non-Annex 1 countries (which mainly consist of developing countries) from binding emissions reduction obligations. Future climate instruments will change the way in which the responsibility burden is shared, with the Durban Platform for Enhanced Action seeking to create some form of mitigation obligation for all parties to the UNFCCC by 2020.

The principle of common but differentiated responsibility also features prominently within adaptation policies of the regime. Paragraph 14 of the Cancun Adaptation Framework, the key provision of the framework defining the parameters of domestic adaptation policies and measures recognizes the application of the principle of common but differentiated responsibility in the implementation of such measures. The principle of common but differentiated responsibility is a central pillar in the climate change regime. The ability of the principle to recognize historical acts

[77] See Rio Declaration on Environment and Development, Rio de Janeiro, 14 June 1992, A/CONF.151/5/Rev.1, Vol.1, Annex 1. Most notably, see Principle 6, which states that the "special situation and needs of developing countries, particularly the least developed and those most environmentally vulnerable shall be given special priority. International actions in the field of environment and development should also address the interests and needs of all countries." See also Principle 7, which states that "States should cooperate in a spirit of global partnership to conserve, protect and restore the health and integrity of the Earth's ecosystem. In view of the different contributions to global environmental degradation, States have common but differentiated responsibilities. The developed countries acknowledge the responsibility that they bear in the international pursuit of sustainable development in view of the pressures their societies place on the global environment and of the technologies and financial resources they command".

[78] See Art. 4(i) of the UNFCCC, which outlines the national and regional development priorities, objectives and circumstances supra, note 1. See also Art. 8, which explains that parties shall give full consideration as to what actions are necessary under the convention specifically in relation to the needs and concerns of developing country parties, supra, note 1.

[79] See UNFCCC, supra, note 1, Art. 10.

[80] Tuula Honkonen, "The Principle of Common But Differentiated Responsibility in Post 2012 Climate Negotiations", 18 *Review of European Community & International Environmental Law* (2009), 257, at 259. See also M. Bothe, "The United Nations Framework Convention on Climate Change – An Unprecedented Multilevel Regulatory Challenge", 63 *Zeitschrift für ausländisches öffentliches Recht und Völkerrecht* (2005), 239, at 252.

and the current capacity of parties to respond to climate change will see this principle continuing to have great application in all future climate change policies and measures.

5.4 Methods of the Climate Change Regime

The UNFCCC establishes the procedural manner in which the climate change regime is to evolve and operate. Article 7 establishes the Conference of the Parties (COP), which serves as the supreme body of the Convention. The COP is charged with: reviewing the implementation of the Convention; making any decision necessary to promote the effective implementation of the Convention; and to review any related instruments adopted by the COP.[81] The COP is comprised of representatives of all governments that are parties to the Convention, who meet annually to progress implementation of the Convention.[82] Each party has one vote and regional economic integration organisations (for example the European Union) are able to vote collectively.[83] While the regime seeks to make decisions on all matters of substance by consensus, when efforts to reach consensus have been exhausted and no agreement has been reached the decision shall at last resource be taken by a two third majority vote.[84] The COP negotiations of the climate change regime have attracted criticism from the media and some academics for failing to deliver on measures to address climate change. The urgency in which climate change measures need to be undertaken, versus the cost implications of implementing such measures divides parties opinions on the types of measures to be adopted at COP negotiations. The Kyoto Protocol demonstrates that parties to UNFCCC are willing to act in the absence of consensus in order to implement measures to meet the objective of the regime.

5.4.1 Mitigation Regime

The mitigation regime is based on the principles of preventative action and common but differentiated responsibility. Mitigation by its very nature seeks to prevent damage from arising. And yet, the principle of common but differentiated responsibility is applied in the Kyoto Protocol by establishing binding obligations for industrialized countries only. The UNFCCC and Kyoto Protocol both advocate for developed

[81] UNFCCC, supra, note 1, Art. 7(2).

[82] Ibid., Art. 7(4).

[83] Organizational Matters: Adoption of Rules and Procedures, FCCC/CP/1996/2, 22 May 1996, rule 41.

[84] Ibid., rule 42.

nations to show leadership in implementing mitigation measures within their own territories first; along with providing assistance to developing countries in order for them to implement mitigation measures.[85] The mitigation regime uses a combination of methods to achieve this purpose and the ultimate objective of the Convention. These methods include:

- The obligation to implement policies and measures at the domestic level to limit anthropocentric emissions of greenhouse gases;
- The setting the targets and allocation of assigned amount units;
- The application of the Kyoto Protocol flexibility mechanisms: Joint Implementation; the Clean Development Mechanism; and Emissions Trading, which can be used to assist in meeting the targets imposed by the regime; and
- The submission of reports outlining emission reduction activities that comply with the guidelines and methodologies developed by the Intergovernmental Panel on Climate Change.

The UNFCCC creates specific obligation for parties in Article 4(2). Article 4(2)(a) provides that parties "shall adopt national policies and take corresponding measures on the mitigation of climate change, by limiting its anthropogenic emissions of greenhouse gases and protecting and enhancing its greenhouse gas sinks and reservoirs". Article 2(a) of the Kyoto Protocol outlines a number of areas in which parties are requested to implement or elaborate policies and measures. This list includes a wide range of measures such as: energy efficiency; protection of sinks; sustainable forms of agriculture; renewable forms of energy; progressive reduction or phasing out of market imperfections, fiscal incentives, tax or duty exemptions that run counter to objective of Convention; encouragement of reforms aimed at reducing emissions of greenhouse gases; and measures to limit emissions in the transport sector. Proactive parties could use this list of measures as a checklist to ensue that measures to reduce emissions greenhouse gases occurs in all relevant sectors. Article 10 of the Kyoto Protocol affirms the existing commitments under Article 4 of the UNFCCC and requests parties to formulate cost effective programs to improve the quality of local emission factors taking into account their common but differentiated responsibilities and national development priorities.[86] The wording of Article 10 is reflective of the objective of the regime by its reference to cost effective measures, which could be perceived as watering down the commitments by specific inclusion of this criterion.

Article 3(1) of the Kyoto Protocol sets individual emission targets for Annex I parties, the details of which are set out in Annex B of the Protocol. Each party is allocated a quota of assigned amount units, which are calculated pursuant to their quantified emission limitation and reduction commitment (QELRO). Assigned amount units are the currency used within the regime and represent the carbon dioxide equivalent of all gases covered within the regime (carbon dioxide, methane, nitrous

[85] See for example UNFCCC, supra, note 1, Art. 4(a) and (b).

[86] Ibid., Art. 10 (a).

oxide, hydrofluorocarbons, perfluorocarbons and sulphur hexafluoride).[87] Parties are obligated under Article 3(1) to ensure that they do not exceed their allocation of assigned amount units during the first commitment period (2008–2012).

The flexibility mechanisms were designed to lower the overall costs associated with meeting the emission targets contained in Annex B of the Protocol. The flexibility mechanisms of the Kyoto Protocol are considered to be innovative in nature by creating a means in which the objective of the UNFCCC can be realised through the participation of a broader range of parties than Annex I members alone.[88] The mechanisms provide a degree of flexibility by allowing for emission reduction activities that occur outside the territory of an Annex I party to count towards their emission reduction commitments as required by Article 3 of the Protocol. However, parties are not able to invest solely in emission reduction activities that take place outside of their jurisdiction; there is a requirement under all three mechanisms that participation within these mechanisms can only be used to supplement emission reduction measures occurring within the territory of the state whose emissions are in questions.

Article 6 of the Kyoto Protocol gives life to wording contained in Article 4(2) (b) of the UNFCCC, which foreshadows Annex I parties meeting their emission reduction objectives individually or jointly. Joint Implementation (JI) allows for parties to transfer or acquire from another Annex I Emission Reduction Units (ERUs). Joint implementation is seen as providing a cost effective means of meeting QELRO, as it allows parties to source ERUs from regions where the costs associated with emission reduction are lower. There are some requirements that must be met in order for ERUs to be issued. The ERUs must come from a project activity and must meet additionality requirements[89]; parties are prevented from acquiring ERUs if they are in breach of their reporting obligations under Article 5 and 7 of the Protocol[90]; and the acquisition of ERUs must be supplemental to domestic actions.[91] Liability for ensuring the validity of the ERUs passes to the party acquiring the interest, who must ensure that the JI project has been verified before completing a transaction or risk being liable for the costs of remediation activities.[92]

The Clean Development Mechanism (CDM) evolved from a proposal made by Brazil that sought to establish a non-compliance fund, the proceeds of which would be used to fund mitigation and adaptation activities in non-Annex I countries.[93]

---

[87] The gases covered by the regime are set in Annex A of the Kyoto Protocol, supra, note 2.

[88] Freestone, "The International Climate Change Legal and International Framework: An Overview", supra, note 18, at 12.

[89] Kyoto Protocol, supra, note 2, Art. 6 (1) (a) and (b).

[90] Ibid., Art. 6 (1) (c).

[91] Ibid., Art. 6 (1) (d). The Kyoto Protocol does not define supplementarity, but the European Union has decided that this means at least 50% of domestic policies and measures.

[92] Ibid., Art. 6 (4).

[93] See for further discussion on this point Roda Verheyen, *Climate Change Damage and International Law: Prevention, Duty and State Responsibility* (Netherlands: Martinus Nijhoff Publisers, 2005), at 113.

The Clean Development Mechanism as described in Article 12 instead operates as a project fund that allows Annex I parties to invest in emission reduction activities in developing countries. The term used to describe the currency of CDM transactions is Certified Emission Reduction units (CERs). The CDM creates an avenue for sustainable development to take place in developing countries, which provides benefits to the country in which the project takes place and which also assists in meeting the ultimate objective of the Convention. The CDM has a number of procedural requirements such as: an Executive Board that oversees all transactions[94]; designated operational entities (DOEs) who certify that participation in CDM activities is voluntary, that real and measurable long term mitigation benefits are achieved as a result of CDM activities, and that reductions are additional to any that would have occurred in absence of the project.[95] A share of proceeds from CDM activities is to be used to cover the administrative costs associated with verification as well as assisting developing country parties particularly vulnerable to adverse effects of climate change to meet the costs of adaptation.[96]

Article 17 deals with the concept of emission trading and allows parties to participate in emission trading for the purpose of fulfilling their Article 3 commitments. This Article provides the basis for the existence of a global emission trading scheme, with participation contingent upon the trade of emission units being supplemented by domestic activities. Emissions trading systems have been established at the domestic level by the European Union, several states and provinces in North America, New Zealand, and other jurisdictions as a means of meeting Article 3 commitments.

A stringent reporting framework is established by the UNFCCC and the Kyoto Protocol to measure emission reduction activities and enhancement of sinks. Article 4(2) (b) of the UNFCCC requested parties to communicate detailed information on their mitigation policies and measures implemented at the domestic level within the first 6 months of Convention being in force. Article 12 of the UNFCCC provides guidance on the reporting requirements requesting: a national inventory of anthropogenic emissions by sources and removal by sinks of greenhouse gases; a general description of the steps taken by the parties; a detailed description of the policies and measures that it has adopted to reduce sources of emissions and to enhance sinks or reservoirs; and a specific estimate of the effects that the policies and measures adopted will have on anthropogenic emissions.

The Kyoto Protocol requires parties to establish national systems that estimate sinks and sources of greenhouse gas emissions, and prescribes that such estimates should be based upon guidelines and methodologies developed by the Intergovernmental Panel on Climate Change and the Subsidiary Body for Scientific and Technological Advice.[97] Article 7 of the Kyoto Protocol requires parties to submit their annual

[94] Kyoto Protocol, supra, note 2, Art. 12 (4)

[95] Ibid., Art. 12 (5).

[96] Ibid., Art. 12 (8).

[97] Kyoto Protocol, supra, note 2, Art. 5. The IPCC revised 1996 guidelines for national greenhouse gas inventories are the current guidelines prescribed by the regime.

inventories along with supplementary information to demonstrate compliance with Article 4 (2) (a) of the Convention. Article 8 provides that expert teams will provide a through and comprehensive technical assessment of all aspects of the implementation of the reports submitted by parties.

5.4.2 *Adaptation Methods*

The adaptation regime is evolving with consideration of equity as a central theme. The principle of intra-generational equity has particular relevance in relation to adaptation measures and policies. The methods of the adaptation regime are not as established as those of the mitigation regime. The Cancun Adaptation Framework was established in 2010 and does not create legally binding commitments for the parties.[98] Rather, it sets out a program to enhance adaptation action by the parties. Paragraph 14 of the Cancun Adaptation Frameworks sets forth the activities and parameters of the regime. The measures in paragraph 14 can be categorised as follows:

- Conduct impact vulnerability and adaptation assessments[99];
- develop national and subnational adaptation plans and strategies and implement prioritised adaptation projects and programmes under the plan[100];
- strengthen institutional capacity to implement adaptation activities in the areas of water resources; health; agriculture and food security; infrastructure; socio-economic activities; aquatic ecosystems and coastal zones[101];
- enhance climate change related disaster risk reduction strategies that pay regard to the Hyogo Framework for Action. Examples of activities include: early warning systems; risk assessment and management; develop risk transfer and risk share mechanisms such as insurance; and increase public awareness and education concerning climate adaptation[102];
- coordinate measures dealing with climate change induced displacement, migration and planned relocation at the national, regional and international levels[103];
- improve climate-related data collection, modelling and knowledge systems, and improve research and technologies associated with adaptation activities in developing countries.[104]

The Durban COP negotiations led to the advancement of implementation of the Cancun Adaptation Framework through: the establishment of modalities for the Adaptation Committee; definition of activities under the work program on loss and

[98] Cancun Adaptation Framework, supra, note 3, para. 13.

[99] Ibid., para. 14 (b).

[100] Ibid., para. 14 (a) .

[101] Ibid., para. 14 (c) and (d).

[102] Ibid., para. 14 (e) and (h).

[103] Ibid., para. 14 (f).

[104] Ibid., para. 14 (g), (h) and (i).

damage[105]; and the development of modalities and guidelines for national adaptation plans. The developments concerning the adaptation committee and the guidelines on national adaptation plans will be briefly canvassed before moving on to discuss the issue of funding, which is central to the operation of the adaptation regime.

The Adaptation Committee will be comprised of 16 members and will include: 2 members of each of the 5 United Nations regional groups; 1 member from a small island developing state; 1 member from a least developed country party; 2 members from parties included in Annex I to the Convention; 2 members not included in Annex I to the Convention.[106] This is a fairly balanced committee, though it could be improved by giving the two seats to non-Annex I members to other least developed nations in order to increase their representation and say in how measures affecting their country are progressed. The Adaptation Committee will coordinate and manage linkages with all relevant bodies within the climate change and other relevant international institutions working on climate related adaptation and report annually to the COP on its progress and recommendations for action.[107] This is a particularly important task in the realm of adaptation, as climate related disaster risk reduction activities and adaptation initiatives currently take place in a fragmented manner across a number of international institutions, without any one regime playing an authoritative or lead role. The Adaptation Committee has been requested to develop a 3-year plan for its wok that includes milestones, activities, deliverables, and resource requirements, which is to be approved at the COP 18 negotiations in 2012.[108] This 3-year plan will provide a strategic framework of action that is currently lacking within the adaptation regime.

The initial guidelines for the formation of national adaptation plans by least developed country parties are located in the Annex of the COP Decision on National Adaptation Plans.[109] The guidelines envisage a four-stage cycle of planning. Parties have been requested to trial on a voluntary basis the implementation of these guidelines and provide feedback on the usefulness and ways in which these guidelines can be improved.[110] The elements of national adaptation plans are summarized below:

Laying the groundwork and addressing gaps:

- Identifying gaps and weakness in enabling environments;
- conducting assessment on climate change impacts; and

[105] The decision of the Work Program on loss and damage is a decision that outlines a number of meetings and that commissions reports as they relate to this topic. See Decision 7/CP.17, Work Program on Loss and Damage, UN Doc. FCCC/CP/2011/9/Add.2, 15 March 2012.

[106] Decision 7/CP.17, supra, note 15 para. 101.

[107] Ibid., paras. 99 and 100.

[108] Ibid., para. 97.

[109] Decision 5/CP.17, National Adaptation Plans, UN Doc. FCCC/CP/2011/9/Add.1, 15 March 2012.

[110] Ibid., paras. 29 and 39.

- conducting comprehensive assessment of development needs and climate vulnerability.

Preparatory work:

- Conducting assessment of medium and long term adaptation needs;
- integrating climate change adaptation into national and sectoral planning;
- participatory consultation and communication and awareness raising.

Implementation strategies:

- Prioritize programs and strengthening institutional capacity to implement
- Reporting, monitoring and review

One of the primary challenges of the adaptation regime is the provision of funding to carry out the activities envisaged by the regime. Article 11 of the UNFCCC establishes a financial mechanism to provide financial resources and technology transfer to assist in the implementation of the convention. The Global Environmental Facility (GEF) has been entrusted with the operation of the financial mechanism of the Convention.[111] There are three different funding entities within the regime: the GEF funds (see below), the Adaptation Fund operating under the Kyoto Protocol, and the Green Climate Fund. The finance for each of these three entities comes from different components of the regime, though it seems less then ideal to continue adaptation funding in such a fragmented approach. This may well be something that the Adaptation Committee considers early in its program, due to potential savings on administrative costs of an integrated fund and increased efficiency outcomes that could be achieved by avoiding duplication and overlapping adaptation initiatives.

The GEF has three funds for adaptation activities: the GEF Trust Fund providing support for vulnerability and adaptation assessments; The GEF Least Developed Country Fund; and the GEF Special Climate Change Fund. The Adaptation Fund under the Kyoto Protocol is financed by 2% of CDM proceeds and focuses on adaptation measures for those particularly vulnerable to the adverse effects of climate change. The Cancun COP negotiations in 2010 led to the development of the Green Climate Fund. The purpose of the Green Climate Fund is to provide new and additional resources approaching USD 30 billion for the period from 2010 to 2012, with a balanced allocation between adaptation and mitigation and priority for the most vulnerable developing countries such as least developed countries, small island developing states, and Africa.[112]

[111] Art. 11 (1) of the UNFCCC allows for the operation of the financial mechanism to be entrusted to one or more existing international entities, supra, note 1.

[112] Report of the Conference of the Parties on its sixteenth session, FCCC/CP/2010/7/Add.1, 15 March 2011, para. 95.

5.5 Conclusion

This chapter has explored the foundations of the international climate change regime by analysing the objectives, principles and methods of mitigation and adaptation policies and measures of the regime. Achieving the ultimate objective of the regime will remain challenging, as considerations of economic growth continue to displace environmental objectives and concerns about community vulnerability and functionality. The application of all seven principles discussed in this chapter would assist in meeting this objective and in adjusting the way in which economic considerations are weighed against environmental and social considerations. The precautionary principle, in particular, has great potential to change the way in which economic interests are prioritised over other relevant considerations.

The methods of the mitigation regime are considered advanced by international environmental law standards. The stringent reporting requirements along with the existence of innovative flexibility mechanisms show that serious consideration is being given to the issue of global greenhouse gas mitigation. The same level of innovation needs to be directed towards creating an adaptation regime that is supportive of mitigation measures, while also obtaining prominence of its own accord. Further integration between mitigation and adaptation measures and polices is needed given the reinforcing and supporting character of these measures.

In conclusion, attention should be directed to paragraph 10 of Cancun COP report, which provides a vision for the way in which the ultimate objective can be obtained. It states "that addressing climate change requires a paradigm shift towards building a low-carbon society that offers substantial opportunities and ensures continued high growth and sustainable development, based on innovative technologies and more sustainable production and consumption and lifestyles, while ensuring a just transition of the workforce that creates decent work and quality jobs."[113]

[113] Ibid., para. 10.

Chapter 6
Alternative Venues of Climate Cooperation: An Institutional Perspective

Camilla Bausch and Michael Mehling

Abstract Climate change is now widely recognized as a political priority, yet different views exist on how to shape an international response. Following serious setbacks in the climate negotiations, this question has also grown to encompass the adequacy of different venues and institutions to address the challenge of climate change mitigation. Applying a diverse set of metrics, this chapter assesses the structure and achievements of a number of existing and proposed fora for international climate cooperation. It starts with the United Nations Framework Convention on Climate Change and the Kyoto Protocol, and proceeds to survey other venues focused on climate change. Rather than identifying one single panacea for climate change mitigation, the analysis shows that different approaches to climate cooperation evidence distinct strengths and shortcomings, typically accompanied by correlating trade-offs; and that a balanced combination of approaches may be needed to address the climate mitigation challenge. Unsurprisingly, the analysis also affirms that the global objective of effective climate change mitigation cannot be reached in any of the venues unless participants raise their political ambition.

Camilla Bausch works in the Senior Management and as a Senior Fellow at the Ecologic Institute, Berlin; Michael Mehling is President of the Ecologic Institute, Washington DC, and an Adjunct Professor at Georgetown University.

C. Bausch (✉)
Ecologic Institute, Pfalzburger Str 43/44, 10717, Berlin, Germany
e-mail: camilla.bausch@ecologic.eu

M. Mehling
Ecologic Institute, 1630 Connecticut Ave. NW, Suite 300, 20009, Washington, DC, USA
e-mail: michael.mehling@ecologic-institute.us

E.J. Hollo et al. (eds.), *Climate Change and the Law*, Ius Gentium: Comparative Perspectives on Law and Justice 21, DOI 10.1007/978-94-007-5440-9_6,
© Springer Science+Business Media Dordrecht 2013

6.1 Introduction

Climate change has been widely recognized as an urgent problem by the international community. Views differ widely, however, about the shape of an adequate international response. Closely related is the question of the suitable institutions to drive international climate cooperation. Following the controversial United Nations (UN) climate summit in Copenhagen convened in December 2009, questions relating to the venue of global climate efforts have been discussed with particular vigor.

Although many policymakers and stakeholders are calling for equal weight to be given to adaptation and mitigation, the following analysis will focus on the latter; this is not a value judgment on the relative importance of mitigation and adaptation, but rather a reflection of the need for focus in an issue area of considerable complexity. The chapter will begin by providing an overview and assessment of international policy fora currently engaged in mitigation, covering venues with different degrees of formality, institutional capacity, and specificity of mandate. Building on this survey and subsequent analysis, the chapter draws a series of conclusions, with consideration given to options and challenges for more effective cooperation going forward.

For the underlying analysis, the imperative of keeping average global temperature increases below two degrees Celsius (2°C) above preindustrial levels, an objective that has been decided upon by the international community,[1] will be used as the central point of reference. Existing and proposed venues for international climate cooperation will be measured against this objective, based on past achievements and future potential.

6.2 Assessing Current Venues of Climate Cooperation

Since the onset of international climate cooperation, the negotiations conducted under the auspices of the UN have surely been the most prominent venue for addressing mitigation efforts. There is, however, a multiplicity of other fora – be they climate-specific or general, technical or broadly political, formal and centralized or informal and decentralized – which have also sought to address the mitigation challenge or aspects thereof. To shed light on the question of whether and how such venues might complement or even replace parts of the UN negotiations, the following sections will outline a selection of the most prominent venues and their goals and achievements with respect to the overarching aim of reaching the 2°C goal.[2]

---

[1] See, most recently, Decision 2/CP.17, Outcome of the work of the Ad Hoc Working Group on Long-term Cooperative Action under the Convention, UN Doc. FCCC/CP/2011/9/Add.1, 15 March 2012, recital 3 of Sec. II.A.

[2] Katherine Michonski and Michael A. Levi, *Harnessing International Institutions to Address Climate Change* (New York, NY: Council on Foreign Relations, 2010) discuss a number of bodies not dealt with here, such as the United Nations Environment Programme (UNEP), the Food and Agriculture Organization (FAO) or the International Atomic Energy Agency (IAEA).

6.2.1 Classical Multilateralism: The United Nations

In the area of climate change, the UN has operated mainly through a framework convention, the UNFCCC, and subsequent rules elaborated within the institutional setup created by that treaty.[3] Although different facets of the climate change issue have also been addressed by the UN General Assembly, the UN Security Council or the UN Environment Program (UNEP),[4] the center of gravity is clearly the work done under the auspices of the UNFCCC and its Kyoto Protocol. In addition and almost accidentally, the Montreal Protocol on Substances that Deplete the Ozone Layer has had considerable relevance for the mitigation of GHGs.

6.2.1.1 The UNFCCC and Its Kyoto Protocol

The UN climate regime is credited with high legitimacy and near universal membership. Still, the past two decades of its operation have been anything but smooth, notably as states have turned to negotiating new rules with potentially significant implications for the current and future global economy. Indeed, progress and stagnation exemplify the potential, the difficulties, and the cumbersome nature of UN efforts.

What Has the Regime Achieved with Regard to Mitigation?

Any attempt to survey the success of the UN climate regime depends on what one defines as success. If measured by the degree to which the regime has compelled parties to mitigate GHG emissions in accordance with the recommendations set out by the scientific community, the UNFCCC and the Kyoto Protocol clearly have not

---

[3] For discussion of the principles and objective of the UN climate regime, see in this volume Rowena Maguire, Foundations of International Climate Law: Objectives, Principles and Methods.

[4] See e.g. UN Security Council, Presidential Statement on Possible Security Implications of Climate Change, UN Doc. S/PRST/2011/15, 20 July 2011, available at http://www.un.org/News/Press/docs/2011/sc10332.doc.htm (last accessed on 24 June 2012); General Assembly, Resolution, "Protection of Global Climate for Present and Future Generations of Humankind", A/RES/65/159 December 2010, available at http://unfccc.int/resource/docs/2011/un/eng/ares65159.pdf (last accessed 24 June 2012); for more details on UNGA and UN Security Council see: Camilla Bausch, "The Power of Plurality – On the Different Fora Contributing to the Post-2012 Negotiations", in Michael Rodi and Michael Mehling (eds), *Bridging the Divide in Global Climate Policy: Strategies for Enhanced Participation and Integration* (Berlin: Lexxion, 2009), 53; For the climate-related work of UNEP, refer to the UNEP website: http://unep.org/climatechange (last accessed on 15 June 2012). On the role of UNEP in global environmental governance, as well as on the UNEP reform debate, see: Marianne Beisheim and Nils Simon, "Neuer Schwung für die Reform der internationalen Umweltgovernance", 37 *SWP-Aktuell* (2010).

risen to expectations. There is broad agreement that the efforts, actions, commitments and pledges under the regime remain insufficient to even come near the objective of limiting global warming below 2°C above pre-industrial levels.[5]

Also, an assessment of parties' compliance with their mitigation commitments under the Kyoto Protocol yields a mixed picture. Not taking into account parties which abstained from ratification altogether, such as the United States, only a subset of parties seems to be on track to meeting their obligations for the first commitment period (2008–2012). One party, Canada, has effectively ignored its emission reduction target, showing that even binding agreements have their limitations when political will is absent. Following the Durban climate summit of 2011, Canada formally announced its withdrawal from the Kyoto Protocol.

While the Protocol marked an important step in climate cooperation, its practical effect has thus been described as narrow, thin, or even ultimately symbolic by critics.[6] Neither the Protocol nor the UNFCCC is currently adequate to the ultimate objective of preventing "dangerous anthropogenic interference" with the climate system. Negotiations on mitigation and on the future of the regime have proven to be so cumbersome and protracted, that – combined with the diverging and partly antipodal positions of the parties involved – the prospects for adoption of sufficiently ambitious decisions in time to prevent dangerous climate change are dire.

Nevertheless, while success in terms of short-term emission reductions and adequate mid-term commitments may currently be difficult to establish, other aspects of the regime can indeed be considered important achievements for broader climate cooperation and global mitigation efforts. For one thing, with one of the largest memberships of any multilateral treaty, the UNFCCC has undoubtedly been very successful in securing near-universal endorsement of the need for climate cooperation.

Also, the UN climate regime has triggered comprehensive processes through which to address broader questions of a future climate regime. In doing so, the UN negotiations have helped raise the political profile of climate change and its mitigation to the level of heads of state and government. Furthermore, it was also due to the dynamics of the UN climate negotiations that the issue of climate change has been placed on the agenda of other fora such as the Group of Eight (G8). As a result

---

[5] This was acknowledged even at the UN climate summit in Durban, South-Africa, in 2011. Decision 1/CP.17 on the Establishment of an Ad Hoc Working Group on the Durban Platform for Enhanced Action, UN Doc. FCCC/CP/2011/9/Add.1, 15 March 2012, spells this out in its preambular language: "Noting with grave concern the significant gap between … Parties' mitigation pledges … and … pathways consistent with having a likely chance of holding the increase of … temperature below 2 degrees C or 1.5 degrees C above pre-industrial levels." Many scientific analyses support this finding, including the UNEP "Emission Gap Report", UNEP, *The Emissions Gap Report – Are the Copenhagen Accord Pledges Sufficient to Limit Global Warming to 2°Celsius or 1.5°Celsius? A Preliminary Assessment* (November 2010).

[6] David G. Victor, *The Collapse of the Kyoto Protocol and the Struggle to Slow Global Warming* (Princeton, NJ: Princeton University Press, 2001); Ruth Greenspan Bell, "The Kyoto Placebo", 24 *Issues in Science and Technology* (2006), 28; Christoph Böhringer and Carsten Vogt, "Dismantling of a Breakthrough: The Kyoto Protocol as Symbolic Policy", 20 *European Journal of Political Economy* (2004), 597.

of this dynamic, new actors at different levels, from different backgrounds and a broad range of agencies have become involved.

It has also – directly and indirectly – triggered action at the national and regional level. The European Union has regularly referenced its international obligations when designing its climate policy. In fact, its emissions trading system might never have been implemented in the absence of the emission reduction obligations and the possibility of using the flexibility mechanisms[7] under the Protocol. Furthermore, political leaders often want to use the UN climate summit to announce or launch new climate protection initiatives, such as, for instance, the South African Renewable Initiative in 2011. The opportunity to attract visibility at a global summit helps mitigation opportunities be accelerated and embraced.

More technical aspects of the regime have facilitated a better understanding of the scope and nature of the climate challenge; for instance, the reporting obligations currently imposed under the UNFCCC and the Kyoto Protocol have greatly increased transparency and knowledge of emissions trends in different jurisdictions and, by extension, at the global level. Further increasing such understanding and transparency will be crucial for comprehending and steering global mitigation efforts successfully.

At a practical level, moreover, the UNFCCC and the Kyoto Protocol have resulted in the creation of an infrastructure with its own resources and highly diverse expert staff that currently no other international institution or initiative focused on climate change can match. The UNFCCC secretariat alone, with a staff of several hundred experts, brings a pool of technical knowledge to the climate process that would be very difficult to build up in any other institution or venue.[8] Also, given its nearly two decades of evolution, the UNFCCC has been able to build up an institutional memory (including, for instance, a vast documentation database) and professional routines that, again, would take years to develop in another setting or forum.

One of the most evident outcomes of the Kyoto Protocol has been the creation of a carbon market, especially through the Clean Development Mechanism (CDM). Such a market for mitigation units would not have been possible without the binding Quantified Emission Limitation and Reduction Objectives (QELROs) and the comparatively advanced compliance regime created with the Protocol. Leaving aside justified concerns about the environmental integrity of certain mitigation projects and high transaction costs, the CDM regime has far exceeded initial expectations in terms

[7] "Flexibility mechanisms" is a collective term applied to three instruments aiming at efficient GHG emission reduction: international Emissions Trading (ET), the Clean Development Mechanism (CDM), and Joint Implementation (JI), see Kyoto Protocol, articles 6, 12 and 17, respectively. For an introduction into the mechanisms see Camilla Bausch et al., "Efficient Climate Policy through Flexible Mechanisms", in Michael Rodi (ed), *Between Theory and Practice: Putting Climate Policy to Work, Vol. 1* (Berlin: Lexxion, 2008), 9.

[8] According to the UNFCCC Secretariat, its staff of "around 500 international civil servants works towards the UNFCCC's goals, guided by the Convention's 194 and the Protocol's 190 Parties." See UNFCCC, "Fact sheet: UNFCCC Secretariat", available at http://unfccc.int/files/press/backgrounders/application/pdf/unfccc_secretariat.pdf (last accessed on 12 June 2012).

of the investment it has attracted. It helped to deploy climate-friendly technologies, bringing about greater involvement by the private sector, and identifying a number of pathways for mitigation. While the other project based mechanism – Joint implementation (JI) – has not been as prominent and successful, both mechanisms – together with domestic or regional efforts such as the EU emissions trading system – have helped establish an understanding of the potential and pitfalls of carbon pricing.

Furthermore, capacity-building efforts under the UNFCCC and the Kyoto Protocol have contributed to disseminating knowledge about the challenges of global warming and possible solutions around the world.

In conclusion, the UN regime has been able to deliver outcomes of considerable importance. At the same time, it has not yet proven its capacity to live up to the challenges at hand, with regard to both urgency of action and the requisite level of ambition. Understandably, this has not bred a sense of unwavering confidence in the UN as the single most important forum for solving the climate crisis. Two questions, therefore, invariably arise: first, can more be expected from the UN climate regime in the future, and second, are there any convincing alternatives?

What Can Be Expected in the Future?

One of the defining characteristics of decisions taken at the UN level is the need for consensus, which – at least on issues involving widely divergent interests and lack of political will – has tended to allow agreement only on sufficiently watered down compromises. Some stakeholders have therefore called for the adoption of rules of procedure that would allow majority voting, thereby preventing individual parties from blocking an overwhelming majority of countries prepared to move forward. However, past experience suggests that an agreement to introduce general majority voting (in addition to the few cases already foreseen in the Convention and its Protocol) will not be achieved in the near future.

Without even entering into the discussion over the corresponding advantages and disadvantages, it can therefore be expected that consensus will remain the default mode of decision making in the UN climate regime for the foreseeable future.[9] Furthermore, a change in decision-making procedures alone cannot solve the impasse whenever divisions are held by influential parties, such as China and the US. Indeed, the lack of a shared vision between parties is one of the most decisive stumbling blocks for progress.

Looking ahead, the decisions of the climate summits in Cancún (2010) and Durban (2011) outline a broad work program, which may serve as a basis for prog-

---

[9] Nevertheless, the Cancún summit saw parties exploring new ways of dealing with the consensus requirement. Despite Bolivia explicitly opposing to the decision taken, the Presidency declared consensus. This approach shows the shaping power of political will, albeit potentially at the risk of undermining the perceived legitimacy of the outcomes, see Antto Vihma, *A Climate of Consensus: The UNFCCC Faces Challenges of Legitimacy And Effectiveness* (Helsinki: FIIA, March 11, 2011), at 2 seqq.

ress. The Durban Platform decision[10] launched a workplan to close the ambition gap with "a view to ensuring the highest possible mitigation efforts by all parties". Also, parties adopted a new process for a future regime to be decided by 2015 and be implemented from 2020. Both the roadmap and the workplan will prove useful to negotiating a future regime.

Science tells us, however, that global emissions will need to peak by 2020 at the latest if the 2°C goal is to be achieved. As it appears unlikely that the current workplan alone will be unable to deliver on this objective, the UN regime is evidently in need of any synergies it might mobilize in the short and medium term and most likely also beyond 2020. And that is where additional initiatives, venues and institutions can play an important role.

6.2.1.2 The Montreal Protocol

To address the problem of stratospheric ozone depletion, the international community has adopted an international regime comprising the 1985 Vienna Convention for the Protection of the Ozone Layer and its 1987 Montreal Protocol on Substances that Deplete the Ozone Layer. A total of 196 Parties have ratified the Protocol, including the United States. It is widely considered one of the most successful multilateral environmental agreements, in terms of the number of parties and its verified progress towards safeguarding the ozone layer, promoting North–south cooperation, and building robust institutions.

The Montreal Protocol determines the phase-out of the production and consumption of several groups of ozone-depleting substances, most prominently chlorofluorocarbons (CFCs). Both developed and developing country parties are subject to reduction and phase-out obligations, although the timetables to complete the phase-out are more generous for developing countries. Phase-out schedules can be revised on the basis of periodic scientific and technological assessments.

The Montreal Protocol and Greenhouse Gas Mitigation

Looking only at the issues relevant to climate change, the Montreal Protocol has had both positive and negative material repercussions for GHG mitigation. On the one hand, the Protocol is acknowledged for its contributions to climate protection because many substances phased out under the Protocol are also powerful GHGs.[11] Furthermore, the Montreal Protocol may have triggered – as a side-effect – energy efficiency improvements for refrigeration and air conditioning appliances.

[10] Decision 1/CP.17 on the Establishment of an Ad Hoc Working Group on the Durban Platform for Enhanced Action, UN Doc. FCCC/CP/2011/9/Add.1, 15 March 2012.

[11] According to UNEP calculations, the Montreal Protocol is projected to have reduced GHG emissions by 11 billion tonnes of CO_2-equivalent emissions by 2010; it remains unclear to the authors whether the negative interplay has been taken into account in these calculations.

At the same time, however, there have also been negative side-effects.[12] The Montreal Protocol directly and indirectly promoted the use of hydrochlorofluorocarbons (HCFCs) and hydrofluorocarbons (HFCs) as substitutes for ozone-depleting substances, despite the fact that these are potent GHGs. The disruptive effect of the Montreal Protocol's promotion of HCFCs and HFCs on the climate change regime triggered scientific and technical cooperation between both regimes, starting in 1998.

In 2007, the parties to the Montreal Protocol finally embraced the objective of climate protection in their work, and agreed to an accelerated phase-out of HCFCs. Stable and sufficient funding possibilities for developing countries, including priority funding for climate friendly alternatives, were a crucial element of the agreement on accelerated action.[13] Owing to the special "adjustment procedure" applicable, the 2007 agreement entered into force automatically, without any need for ratifications, in May 2008.

However, the net climate effect is likely to be impeded by the fact that several of the most promising HCFC substitutes are again GHGs – most importantly HFCs which are controlled under the Kyoto Protocol. Nothing in the agreement reached under the Montreal Protocol restricts the use of these substitutes.

The Montreal Protocol as a Future Driver for GHG Mitigation?

Despite its mixed impact on climate change mitigation, some observers have portrayed the Montreal Protocol as being more successful in terms of mitigation than the Kyoto Protocol.[14] After the Copenhagen climate summit, reports about meetings under the Montreal Protocol referred to a "more proven tool" to fight climate change than the UN climate regime itself. The less politicized atmosphere under the Montreal Protocol, coupled with its more streamlined procedures (especially with regard to amendments), were seen as important advantages; its success in protecting the ozone layer is considered evidence of its capabilities.

Certainly, the Vienna Convention and Montreal Protocol offer some useful lessons for mitigation. Interestingly, their success cannot be explained simply by reliance

---

[12] For a comprehensive overview, see Sebastian Oberthür and Kelly Yasuko Matsumoto, "Managing Policy Contradictions between the Montreal and Kyoto Protocols: The Case of Fluorinated Greenhouse Gases", in: Sebastion Oberthür and Olav Schram Stokke (eds), *Institutional Interplay and Global Environmental Change: Interplay Management and Institutional Complexes: State of the Art and Perspectives* (Cambridge, MA: MIT Press, 2009).

[13] Philip Drost, *Multilateral Environmental Agreements 2008: State of Affairs and Developments, ed.* (Utrecht: Eleven International Publishing, 2008), at 211.

[14] See, for example, Jessica Leber, "Emissions: Decades-old Global Pact Morphs into Potent Climate Treaty", ClimateWire, 26 November 2008; see also Guus J.M. Velders et al., "The Importance of the Montreal Protocol in Protecting Climate", 104 *Proceedings of the National Academy of Sciences* (2007), 4814.; John M. Broder, "Experts Point to a More Proven Tool to Fight Warming", *International Herald Tribune*, 9 November 2010, at 2.

on an alternative governance paradigm relative to the Kyoto Protocol: both treaty regimes are highly formal and legally binding, with centralized institutions and a strong compliance mechanism.

Arguably, the success of the Montreal Protocol makes a case for the benefits of more narrowly focused regimes: substantively, the ozone regime is limited to phasing out a limited group of industrial chemicals. Where ozone-depleting chemicals are found only in narrowly defined contexts of economic activity and daily life, the climate challenge pervades nearly every facet of modern society.

But this potential advantage of the ozone regime also would seem to rule out its suitability for broader climate mitigation beyond the substances it currently covers: as soon as for example the disproportionately more common GHGs CO_2 and methane are included in the scope of the Montreal Protocol, politics and divisions are likely to find their way into its governance processes, ushering in the same diplomatic challenges that have slowed down progress under the climate regime. Finally, the amendment procedures which currently allow the Montreal Protocol regime to adjust to new challenges comparatively swiftly would probably no longer be acceptable to parties if the regime's substantive scope was broadened. Attempting to shift the issue of climate protection in its entirety to this forum might thus not only fail to deliver the intended success, but also put at risk the efficiency of the Montreal Protocol.

6.2.2 *New Impetus from Outside the UN?*

Growing recognition of climate change as a political priority as well as the divisions limiting progress under the UN climate negotiations have prompted the international community to explore other venues for climate cooperation. Ironically, both climate laggards and frontrunners have shown interest in such alternative fora to further their respective aims.

6.2.2.1 The MEF: An Example for an Alternative Climate-Specific Venue

Soon after the Kyoto Protocol entered into force in 2005, political forces critical of the UN climate process – such as the US – sought to divert attention from this emerging venue: Therefore, the Major Economies Meeting on Energy Security and Climate Change (MEM) was initiated by President George W. Bush in 2007.[15] It took extensive international pressure for the MEM to reposition itself as a forum contributing to the UNFCCC process. But criticism remained. For many, the MEM was initiated primarily as a political instrument to draw major emitting developing countries out of the Group of 77 (G77) voting bloc in which they have frequently

[15] More details on the MEM and the MEF can be found in Bausch, "The Power of Plurality", supra, note 4, at 47 et sqq.

been organized within the UNFCCC regime.[16] Furthermore, the MEM was criticized "for pushing an agenda of voluntary measures to combat global warming, as opposed to mandatory caps on emissions",[17] and for falling short with regard to tangible results.[18]

Building on the MEM, US President Barack H. Obama launched the Major Economies Forum on Energy and Climate Change (MEF) in 2009. In contrast to the MEM, the MEF has sought to create renewed momentum in international climate cooperation, clearly emerging in favor of the UN process, to which it aims to contribute.[19] The MEF is intended to facilitate "dialogue among major developed and developing economies" and to "advance the exploration of concrete initiatives and joint ventures that increase the supply of clean energy while cutting GHG emissions."[20] Participation extends to 17 major economies, jointly accounting for approximately 80 % of global GHG emissions.[21] The Obama administration has also tried to establish the MEF as an international forum rather than a US initiative. Accordingly, meetings were hosted not only by the US, but also in other countries such as the UK, Italy, and Mexico. Nevertheless, the US has remained the single most influential country shaping the MEF profile, as reflected by the clear majority of meetings taking place in the US.[22]

Activities Concerning Climate Protection

Participating MEF countries have focused largely on general mitigation needs and technology cooperation, although attention has increasingly shifted to include other issues, such as climate finance.[23] In 2009, in particular, MEF parties made considerable

---

[16] See also Andrew Light and Nina Hachigian, "Rise of the Green Dragon?", available at http://www.americanprogress.org/issues/2009/04/rise_green_dragon.html (last accessed on 12 June 2012).

[17] Available at http://www.greenpeace.org/international/press/reports/bush-mem (last accessed on 12 June 2012).

[18] In 2008, MEM participants produced a "Declaration of Leaders Meeting of Major Economies on Energy Security and Climate Change", available at http://www.mofa.go.jp/policy/economy/summit/2008/doc/doc080709_10_en.html (last accessed on 12 June 2012). Chinese news criticized the MEM as "fruitless": Zhang Jin, "No Progress on Carbon Emission Cuts at MEF Meeting", 27 May 2009, China Radio International, available at http://english.cri.cn/6966/2009/05/27/1461s488215.htm (last accessed on 10 June 2012).

[19] Hillary Rodham Clinton, "Remarks at the Major Economies Forum on Energy and Climate", 27 April 2009, available at http://www.state.gov/secretary/rm/2009a/04/122240.htm (last accessed on 10 June 2012).

[20] State Department, "Major Economies Forum on Energy and Climate", available at http://www.state.gov/r/pa/prs/ps/2009/04/122097.htm (last accessed on 10 June 2012).

[21] These are: Australia, Brazil, Canada, China, the European Union, France, Germany, India, Indonesia, Italy, Japan, Korea, Mexico, Russia, South Africa, the United Kingdom, and the United States.

[22] For a list of events, see http://www.majoreconomiesforum.org/meetings (last accessed on 10 June 2012).

[23] For a summary of past meetings, see http://www.majoreconomiesforum.org/past-meetings (last accessed on 10 June 2012).

efforts to use this forum to advance the climate agenda in preparation for the Copenhagen summit. In July 2009, the Heads of State and Government of the participating jurisdictions convened for a "Leaders Meeting", where they adopted a political declaration embracing the 2°C goal.[24] For Copenhagen, this was an important signal, which was further strengthened by a clear alignment of efforts with the G8 (see below).

In addition, this meeting launched a Global Partnership for low-carbon and climate-friendly technologies aimed at increasing and coordinating public sector investments in research, development, and demonstration of these technologies, with a view to "doubling such investments by 2015."[25] Using the Global Partnership as a starting point, US Secretary of Energy Steven Chu launched a Clean Energy Ministerial (CEM) with slightly expanded membership.[26] At the inaugural meeting in Washington DC in 2010, ministers from 24 countries launched 11 technology-focused initiatives to accelerate the transition to greater energy sustainability, for instance by promoting the rapid deployment of electric vehicles and supporting the market for renewable energy and carbon capture technologies. For the most part, these initiatives aim to coordinate efforts and improve the exchange of best practices; some are accompanied by funding pledges from participant countries, and some set in motion processes to elaborate technical and policy guidance. What they do not specify, however, is mitigation commitments for individual countries or emissions reduction objectives for the group as a whole.

Assessment and Outlook

An exclusive focus on climate change, as well as its composition, afford the MEF an interesting position in the mitigation debate. It has, however, been criticized for lacking transparency and, perhaps more importantly, for being a vehicle of US foreign policy objectives. In the past years, the MEF has arguably been useful as a complement to the negotiations under the UNFCCC. It has given participating states an additional venue for sharing views, identifying common interests, and addressing potential or existing conflicts, all in a less formal atmosphere and with fewer actors than under the auspices of the UNFCCC. Still, while contributing to the overall debate and providing a potentially useful format with the Global Partnership and its technology forum, the MEF has not triggered significant breakthroughs on mitigation.

---

[24] Chair's Summary, L'Aquila, July 10, 2009, available at http://www.g8italia2009.it/static/G8_Allegato/Chair_Summary,1.pdf (last accessed on 10 June 2012).

[25] See http://www.majoreconomiesforum.org/the-global-partnership (last accessed on 10 June 2012).

[26] Participants at the launch were Australia, Brazil, Canada, China, Denmark, the European Commission, Finland, France, Germany, India, Indonesia, Italy, Japan, Korea, Mexico, Norway, Russia, South Africa, Spain, Sweden, the United Arab Emirates, the United Kingdom, and the United States.

The activities of the MEF are purely political in nature.[27] It is unsuitable for facilitating anything but informal political arrangements.

Moreover, the MEF neither has the financial resources nor the staff to take over the various functions currently performed by the UNFCCC Secretariat. In effect, it is still perceived by some as an initiative driven by the United States, something that the spin-off Clean Energy Ministerial underscores; as such, it would most likely not be accepted as a legitimate forum for more comprehensive, let alone formal negotiations – neither by participants themselves, nor by the vast majority of members of the international community not included in the MEF process.[28] Last but not least, both the domestic climate policy impasse of its main proponent, the US, and the unclear future of the MEF more generally constrain its potential political weight. In hindsight, it appears that the MEF itself had its strongest political traction in 2009 in the lead-up to the Copenhagen climate summit. In the end, the MEF might contribute to advancing the future climate protection agenda – possibly with a focus on technology – but for the highlighted reasons is unlikely to become a driving force for meaningful mitigation action.

6.2.2.2 Alternative Venues with Broader Agendas

Group of Eight (G8)

The Group of Eight (G8) industrialized nations is a forum for the governments of eight developed nations in the northern hemisphere.[29] Each year, the G8 process culminates in a summit of the Heads of State and Government of the participating countries. In preparation for this summit, several meetings at ministerial level are convened. The Presidency – which rotates annually – sets the agenda, hosts the summit, and determines which ministerial meetings will take place. Nevertheless, the presidency will typically have to secure sufficient buy-in from the other G8 member states to ensure a constructive and fruitful debate. The G8 summits aim primarily to send political signals and set trends, and do not produce binding results.

Unlike the aforementioned MEM and MEF, the G8 is a forum not focusing on climate issues in particular. With climate change rising on the political agenda, the G8 presidencies have, however, addressed the issue with varying energy. From the perspective of climate protection the G8 is interesting not only in that it assembles

---

[27] Brazil and India, for instance, explicitly opposed the outcome of the Leaders Meeting being framed as a negotiated communiqué, arguing that negotiation of the elements of a climate deal should be left to the UNFCCC, see Teriete, "Major Economies Meet in Mexico – Many Good Ideas in their Text, But All in Square Brackets", 24 June 2009, available et http://blogs.panda.org/climate/2009/06/24/major-economies-meet-in-mexico-%E2%80%93-many-good-ideas-in-their-text-but-all-in-square-brackets (last accessed on 10 June 2012).

[28] It should be noted, however, that *no* such attempts are currently apparent to extend the mandate and role of the MEF.

[29] Its members are currently Canada, France, Germany, Italy, Japan, Russia, the United Kingdom and the United States.

particularly influential states and economies, but also in that these countries have particularly high absolute, per capita and historical emissions.

Activities on Climate Change

Since the 2005 summit hosted by the United Kingdom, climate change has featured as a more or less prominent issue on the annual agenda of the G8. Under the Russian G8 Presidency in 2006, climate protection was less of a priority, while in 2007, the German Presidency once again placed great emphasis on the objective of climate mitigation, championing agreement on the importance of the UN climate process.[30] Convening in Japan in 2008, the G8 expressed its determination to reach agreement on the goal of halving global GHG emissions by 2050[31]; it also addressed important issues related to climate finance.

In 2009, the year of the UNFCCC Copenhagen Summit, the G8 aligned their efforts with the Major Economies Forum (see above).[32] Convening in Italy, G8 leaders went so far as to agree on a reduction goal of 80 % or more for developed countries by 2050.[33] In the aftermath of the failed UN Copenhagen Summit, however, the 2010 G8 summit in Muskoka, Canada, again brought little progress.[34] Likewise, in 2011, the French G8 Presidency reportedly had to be pressured into agreeing on a climate text which showed little tangible content. Accordingly, the Deauville summit declaration[35] reflects limited progress with respect to mitigation needs.

More recently, at the 2012 summit in Camp David, Maryland, governments endorsed action on short-lived climate pollutants. They backed an initiative to reduce such pollutants, the Climate and Clean Air Coalition to Reduce Short-lived Climate Pollutants, launched earlier that year by a group of states and UNEP.[36]

---

[30] Chair's Summary, 8 June 2007, available at http://www.g-8.de/Content/EN/Artikel/__g8-summit/anlagen/chais-summary,templateId=raw,property=publicationFile.pdf/chairs-summary.pdf (last accessed on 10 June 2012).

[31] G8 Hokkaido Toyako Summit Leaders Declaration, Hokkaido Toyako, 8 July 2008, Points 22 to 35, available at http://www.mofa.go.jp/policy/economy/summit/2008/doc/doc080714__en.html (last accessed on 10 June 2012).

[32] Chair's Summary, L'Aquila, Italy, 10 July 2009, at 5, available at http://www.g8italia2009.it/static/G8_Allegato/Chair_Summary,1.pdf (last accessed on 10 June 2012).

[33] Chair's Summary, L'Aquila, Italy, 10 July 2009, available at http://www.g8italia2009.it/static/G8_Allegato/Chair_Summary,1.pdf (last accessed on 10 June 2012).

[34] For details see Chair's Summary, Muskoka, Canada, 25–26 June 2010, available at http://www.canadainternational.gc.ca/g8/summit-sommet/2010/muskoka-declaration-muskoka.aspx?lang=eng (last accessed on 10 June 2012).

[35] The declaration can be accessed here: http://www.g20-g8.com/g8-g20/g8/english/the-2011-summit/declarations-and-reports/declarations/renewed-commitment-for-freedom-and-democracy.1314.html (last accessed on 10 June 2012).

[36] Camp David Declaration, Camp David, Maryland, United States, 18–19 May 2012, available at http://www.whitehouse.gov/the-press-office/2012/05/19/camp-david-declaration (last accessed on 10 June 2012).

Additionally, the G8 nations reaffirmed their strong support for efforts to phase-out fossil fuel subsidies – a topic also addressed by the G20 (see below) and by many of the most recent submissions by the parties to the UNFCCC.[37]

Despite a welcome dynamic on very specific issues, the G8 – like all the above-mentioned fora – reflects the difficulty of advancing mitigation efforts in the absence of political will.

Assessment and Outlook

The G8 forum assembles some of the largest economies and emitters around the globe. And yet, in recent years, both economic power and emissions growth have been shifting at the international level, a development that can be expected to accelerate. Furthermore, as the G8 is primarily a high-level forum for the exchange of ideas and opinions, it has very limited – if any – capacity to adopt operational decisions comparable to the formal decisions taken under the UN regime. In its current shape, the G8 lacks the institutional and technical expertise needed to promote comprehensive mitigation policies, let alone a new legal regime. While each G8 member has skilled personnel at the domestic level, and such experts could support the staff responsible for the G8 summit, there are probably challenges with regard to free capacities. Furthermore, there is currently no specialized Secretariat to support work carried out by the G8.

For the time being, therefore, what the G8 is able to deliver is nothing more – and nothing less – than a forum to facilitate and foster political will and provide political guidance at the highest level. Thus, in practice, the G8 will be most effective when it triggers broader processes and, in doing so, facilitates high-level debates. To date, the G8 has had its greatest impact on the climate debate by making the issue a topic dealt with by heads of state and government. For the small group of developed states constituting its membership, moreover, the G8 has provided a useful forum for political engagement and exchange. At the same time, the political declarations emerging from the G8 summits have done little to promote actual mitigation. This underscores the limitations of a body addressing a great breadth of (sometimes changing) issues, a body without a negotiating mandate, and without recourse to a professional secretariat and its financial resources and staff.

Overall, the influence and political weight of the G8 are declining as other economies grow and new powers emerge, a trend reflected in the establishment of the G8+5 format and the recent ascendancy of the G20 (see the next sections). Also, continued questioning of its legitimacy further weakens the G8. At this point, some observers have even suggested that the G8's work only continues because of the "illusion that this community of values can achieve something significant."[38]

[37] A compilation of the submissions, dated 28 March 2012, can be accessed here: http://unfccc.int/resource/docs/2012/adp1/eng/misc01.pdf (last visited on 24 June 2012).

[38] Frankfurter Allgemeine Zeitung, 27 May 2011, "Magerer Ertrag in Deauville", available at http://www.faz.net/f30/common/Suchergebnis.aspx?term=eingespielten+Rhythmus+und+der+Illusion&x=0&y=0&allchk=1 (translation by authors; last accessed on 10 June 2012).

But even an unexpectedly ambitious G8 could not reach the 2°C goal on its own, as it excludes other major emerging emitters.[39]

Group of Eight and Major Emerging Economies (G8+5)

A somewhat institutionalized extension of the G8 is the G8+5 group, which was formed in 2005. This format adds China, India, Brazil, Mexico and South Africa to parts of the G8 talks. In 2007, their inclusion was solidified by the *Heiligendamm Process* (HDP), which launched a topic-driven, non-negotiating dialogue "on an equal footing"[40] between the G8 and the "+5" countries. It reflected the acknowledgment that the inclusion of the five largest emerging economies would be necessary to address global challenges.

This structured dialogue aimed to complement "the work in other multilateral and regional institutions".[41] One of the four topic areas of this dialogue was energy, with a focus on energy efficiency – areas that are decisive for climate protection. In 2009, parties agreed on the first ever G8/G5 joint declaration "Promoting the Global Agenda",[42] which refers to climate change without affording it a particularly prominent role. The concluding report of the HDP,[43] however, addresses the issue of energy and efficiency in some detail (with a focus on retrofitting coal-fired power plants, energy efficient and sustainable buildings, and renewable energy).

After the 2-year HDP, the Heiligendamm L'Aquila Process (HAP) was created with a 2-year mandate, leading up to the French Deauville Summit in 2011. The HAP was to broaden the range of topics addressed (with energy remaining one of the core issues)[44] and aimed to produce more concrete results. It specifically sought to "explore further possibilities for producing spillovers from the HAP to other forums of international cooperation."[45] However, the partners did not follow up on their own plans. There was neither a published interim report in 2010, nor a con-

[39] This is why some see the G8 as an inadequate forum. See, for example, Sascha Müller-Kraenner, "Weichenstellung statt Katerstimmung – Nach dem Kopenhagen Debakel braucht Europa eine effektive Klimastrategie", 66 *Internationale Politik* (2010), 2–7.

[40] As underlined in paragraph 2 of the concluding document of the HDP, available at http://www.oecd.org/dataoecd/4/53/43288908.pdf (last accessed on 10 June 2012).

[41] As underlined in paragraph 2 of the concluding document of the HDP, available at http://www.oecd.org/dataoecd/4/53/43288908.pdf (last accessed on 10 June 2012).

[42] Available at http://www.oecd.org/dataoecd/61/59/43299158.pdf (last accessed on 10 June 2012).

[43] Available at http://www.oecd.org/dataoecd/4/53/43288908.pdf (last accessed on 10 June 2012).

[44] The HAP agenda is available at http://www.ioc.u-tokyo.ac.jp/~worldjpn/documents/texts/summit/20090709.O2E.html (last accessed on 10 June 2012).

[45] See http://www.oecd.org/site/0,3407,en_21571361_40549151_1_1_1_1_1,00.html (last accessed on 10 June 2012); see also, for example, Ulrich Benterbusch, OECD Director of the HDP, outlining the way forward: http://wn.com/the_heiligendamm_process_extending_the_g8-g5_dialogue (last accessed on 10 June 2012).

cluding report in 2011. No joint statement or advanced agenda gave any positive indication of the continuation and intensification of the G8+5 cooperation.[46]

In retrospective, it appears that the disappointing Copenhagen summit with its controversial debate between emerging economies and industrialized countries may have undermined any impulse to further the climate issue in the G8+5 setting. Admittedly, this body brings together some of the largest current and future emitters of GHGs, including some of the most powerful countries on the globe, and – last but not least – many of the countries whose divisions have made it so difficult to reach agreement at the UNFCCC level. But given the setting – less formal even than the G8 and despite the reiteration of the "equal footing" with a "guest role" only for the "+5" states – it is unlikely that any strong outcome, let alone anything of a more formal or even binding nature, can be achieved. Countries such as China are no longer willing to participate at a side-table. In addition, given the shortcomings of the G8 as elaborated above, there seem to be few to no opportunities for achieving a breakthrough on mitigation issues under the G8+5. In some ways, the G8+5 constellation may thus have been a transition stage reflecting the evolving geopolitical landscape, culminating in the ascendancy of yet another venue with greater developing country participation: the G20.

Group of Twenty (G20)

Since 1999, the Group of Twenty (G20) has brought together high-level public representatives from 20 large economies.[47] Originally created in the wake of the 1997 Asian Financial Crisis to convene major advanced and emerging economies to help stabilize the global financial market, the G20 has also proceeded to address broader economic and related environmental issues. However, its mandate remains focused on international economic development.[48] The G20 primarily aims to send political signals and set trends, not to produce binding results. Unlike the UNFCCC, there is no intention to establish strong institutions, such as a permanent secretariat.

---

[46] As a side observation, it bears noting that, in Deauville, the G8 also issued a joint declaration with Africa on shared values and responsibilities, which did not however mention climate at all and barely touched on the issue of renewable energy, while underlining the importance of access to energy, see http://www.g20-g8.com/g8-g20/g8/english/the-2011-summit/declarations-and-reports/declarations/shared-values-shared-responsibilities-g8-africa.1320.html (last accessed on 10 June 2012).

[47] The G20 comprises: Argentina, Australia, Brazil, Canada, China, France, Germany, India, Indonesia, Italy, Japan, Mexico, Russia, Saudi Arabia, South Africa, Republic of Korea, Turkey, United Kingdom, United States of America and the European Union, which is represented by the rotating Council presidency and the European Central Bank. Initially, the G20 convened the finance ministers and Central Bank governors of these states, but more recently, G20 summits have also attracted heads of state and government.

[48] Available at http://www.g20.org/about_what_is_g20.aspx (last accessed on 10 June 2012).

The G20 claims a "high degree of representativeness and legitimacy on account of its geographical composition (members are drawn from all continents) and its large share of global population (two-thirds) and world Gross National Product (GNP) (around 90 %)."[49] But poor countries still see their interests inadequately represented, giving rise to questions about the legitimacy of the G20.[50]

In 2008, the G20 gained in overall political importance as a result of the challenges raised by the global financial and economic crisis. This rise in power can be expected to continue,[51] partly at the expense of the G8. As a group, the G20 is not as homogenous in its interests, backgrounds, structures and value systems as the G8. This makes it more difficult to find common positions – including in the field of climate change mitigation.

Activities on Climate Change

Following the Pittsburgh Summit of 2009, the G20 finance ministers were tasked with taking forward work in nine areas, including a "framework for strong, *sustainable*, and balanced "growth" and "energy" security and climate change." On the latter, the G20 has focused primarily on questions of how to finance global mitigation efforts. At the 2009 Pittsburgh summit, states announced their intention to "rationalize and phase out over the medium term inefficient fossil fuel subsidies that encourage wasteful consumption" as a means of helping to protect the climate, improve energy efficiency and transition to a green economy.[52] What was missing, however, was a clear time schedule for this phase-out, or the adoption of a binding agreement on this issue. Instead, the G20 requested the respective ministers to prepare implementation strategies and timeframes, and called on the IEA, the OECD and other institutions to report on such subsidies and suggest remedies. Dialogue, submissions,

---

[49] Official website of the French 2011 G20 presidency, FAQs: http://www.g20.org/about_faq.aspx (last accessed on 10 June 2012).

[50] See also Joy A. Kim, who points out that this perspective on climate governance "is neither desirable nor useful", "Polycentric Governance of Climate Change in the Post-Copenhagen Era: The Role of the G20", Conference Paper submitted to the 2nd Global Conference on Environmental Governance and Democracy – Strengthening Institutions to Address Climate Change and Advance a Green Economy, Yale, 2010, at 1, 10.

[51] This impression is supported, for example, by Frankfurter Allgemeine Zeitung, "Magerer Ertrag in Deauville", 27 May 2011, available at http://www.faz.net/f30/common/Suchergebnis.aspx?term=eingespielten+Rhythmus+und+der+Illusion&x=0&y=0&allchk=1 (last accessed on 10 June 2012); Political indications for this trend are manyfold, for example: Deutsch-Chinesisches Gemeinsames Kommuniqué zur umfassenden Förderung der Strategischen Partnerschaft, July 2010, Recital 9, available at http://www.auswaertiges-amt.de/cae/servlet/contentblob/334836/publicationFile/50199/100718-DeutschChinesisches_Kommunique.pdf (last accessed on 10 June 2012).

[52] G20, Leaders' Statement: The Pittsburgh Summit, 24–25 September 2009, available at http://www.pittsburghsummit.gov/mediacenter/129639.htm (last accessed on 20 June 2012).

research and national implementation strategies followed,[53] and publication of research and strategy papers enhanced the transparency of the issue.[54]

Following the disappointment of the 2009 Copenhagen summit, some perceived a growing role for the G20, as all of the G20 members had agreed to or associated themselves with the Copenhagen Accord. But this new-found interest did not necessarily match the wishes of the parties at the table; different G20 members from both developing and industrialized countries showed a keen interest in keeping the issue off the table, preferring to see it negotiated under the auspices of the UN or not at all. In Seoul in 2010, G20 nations thus only reiterated their "commitment to take strong and action-oriented measures and remain fully dedicated to UN climate change negotiations".[55] The summit declaration limits itself to a general commitment to "achieving a successful, balanced result that includes the core issues of mitigation, transparency, finance, technology, adaptation, and forest preservation," and focusing on a number of actions related to green growth. During the 2011 French presidency, some parties showed an interest in pushing the climate finance issue more prominently onto the G20 agenda. Other parties, such as India and China, however, have resisted such approaches, fearing that this would dissolve the distinction between parties and party groups as established under the UN regime. At the latest summit in Los Cabos, Mexico, in 2012, the G20 affirmed its support for the phase out of fossil fuel subsidies and generally endorsed the concept of a "green economy", but otherwise did not propose specific actions in the area of climate change mitigation. A G20 Study Group on Climate Finance was, however, created.[56]

Assessment and Outlook

If key nations such as China, India, and the US fail to support the notion of addressing climate change mitigation through the G20, it will be difficult or even impossible to meaningfully advance the issue through this forum. Having said that, the G20 is an interesting venue with regard to both its members' emission profiles and their political and economic power. Its members announced in 2009 that the group would replace the G8 as the main economic council of wealthy nations.[57] While such a

[53] For details, see David Runnalls, "Fossil Fuel Subsidies and the G20", in John Kirton and Madeline Koch (eds), *G8 & G20: The 2010 Canadian Summits* (London: Newsdesk Media, 2010), 164, at 164.

[54] IGO-4, Analysis *of the Scope of Energy Subsidies and Suggestions for the G-20 Initiative* (Paris et al.: IEA/OPEC/OECD/World Bank Joint Report, 2010), at 5.

[55] G20 Seoul Summit, Leaders' Declaration, 11–12 November 2010, available at http://www.g20.org/Documents2010/11/seoulsummit_declaration.pdf (last accessed on 20 June 2012).

[56] See G20 Los Cabos Summit, Leaders' Declaration, 18–19 June 2012, available at http://g20mexico.org/images/stories/docs/g20/conclu/G20_Leaders_Declaration_2012.pdf (last accessed on 20 June 2012), paras. 70–72.

[57] Available at http://www.whitehouse.gov/the_press_office/Fact-Sheet-Creating-a-21st-Century-International-Economic-Architecture (last accessed on 20 June 2012). It should be mentioned, however, that some researchers see a risk that the G20 will exhaust themselves; the trend then would rather be to integrate new countries in the G8 format.

transition clearly cannot occur overnight, the G20 could be very influential for the climate mitigation debate if it applies its growing political weight to the topic.

Conceptually, the G20 is of interest because – while it is a high-level forum – its original mandate has afforded the group a more applied focus compared to the G8; consequently, the G20 would also seem more suited to addressing some of the more technical issues raised by climate cooperation. Furthermore, the G20 assembles all major emitters and much of global economic power and could thus successfully address the mitigation challenge at the global level. But the primary focus of G20 nations remains financial stability and economic growth. The absence of a strong impulse from the Copenhagen Summit certainly has not helped to trigger political support to widen the agenda. Furthermore, the G20 does not engage in formal negotiations geared towards a binding outcome.

Nevertheless, some observers assume that "[i]n the post-Copenhagen era, climate governance is likely to take a polycentric approach and the G20 could play a critical role in setting the direction of the green economy and addressing climate change."[58] While the G20 may not yet be a driver of mitigation efforts, it could be instrumentalized and grow into such a role. The ability of the G20 to act swiftly "through its highly informal institutional set-up and flexible coordination tools without heavy obligations"[59] could prove useful in advancing the mitigation agenda. A first step could be to address aspects of the mitigation challenge which have a financial dimension, and installing corresponding working groups. Work on fossil fuel subsidies, for instance, has been a step into the right direction, and the Study Group on Climate Finance could likewise prove potentially helpful.

Considering the current G20 structures, any leadership on climate issues will, of course, depend greatly on the respective G20 presidency. In addition, the presidency will need strong backing from a broader group of G20 members. In essence, the question thus really becomes: which country or group of countries could be willing and successful in taking the lead in transforming the G20 into a major force for climate mitigation? While there have been some more or less timid attempts to push the agenda – for example by Germany – , no country has yet shown strong leadership. But even if one or more G20 members were to champion the cause of mitigation, it still remains unlikely that such an attempt would be welcomed and supported by members such as China or Brazil.

Again: Considering the current G20 mandate centered on global financial and economic priorities, and the states it comprises, it seems unlikely that the G20 will become a driving force for mitigation efforts in the near future.[60] Like the G8, it furthermore lacks a robust institutional framework and the designated staff and

[58] Kim, "Polycentric Governance", supra, note 50, with further references to arguments of some that the "expansion of its agenda beyond global economic governance … is a means for the group to further develop and solidify its status in the future" – which could be an additional incentive for heads to put climate change prominent on the G20 agenda.

[59] Kim, "Polycentric Governance", supra, note 50.

[60] Trevor Houser, *A Role for the G-20 in Addressing Climate Change?* (Washington, DC: Peterson Institute for International Economics, 2010).

resources needed to engage on the technicalities of climate protection at the same level as the UN climate negotiations. In the longer run, however, and subject to a corresponding surge in political will, the G20 could reinvent itself and create a process through which mitigation efforts might be taken to a next level. It appears very probable that, if the countries assembled under the G20 do agree on the way forward on climate protection, there could also be progress under the roof of the UN. Some have warned, however, that any attempt to reshape the architecture of the G20 would also entail the risk of losing its specific strength – which is to react swiftly and flexibly on an informal basis in topic-specific coalitions to new and pressing global problems.[61]

Organisation for Economic Cooperation and Development (OECD)

The OECD was established in 1961 and currently counts 34 member countries, including mainly industrialized economies, but also some economies in transition. Other emerging economies, such as Brazil, South Africa, India and China, are included in OECD activities through an "Enhanced Engagement" program.[62] Broadly speaking, the mission of the OECD is to promote policies that will "improve the economic and social well-being of people around the world."[63] It has an Environment Directorate to provide governments with an "analytical basis to develop policies that are effective and economically efficient."[64]

While the OECD can contribute to the development of legislation, typically through so-called "OECD Acts" prepared by its numerous Committees and adopted by its Council, such acts are mainly limited to non-binding recommendations, declarations, and understandings. Under its constitutive treaty, however, the OECD Council also has the ability to adopt binding decisions and enter into international agreements with states and other international organizations.[65] In practice, the OECD has additionally acquired great importance through its publications and databases. For its work, the OECD can draw on ample resources, including an annual budget

[61] Kim, "Polycentric Governance", supra, note 50.

[62] See OECD, "Members and Partners", available at www.oecd.org/document/25/0,3746,en_36734052_36761800_36999961_1_1_1_1,00.html (last accessed on 20 June 2012).

[63] See OECD, "Our Mission", available at www.oecd.org/pages/0,3417,en_36734052_36734103_1_1_1_1_1,00.html (last accessed on 20 June 2012).

[64] See OECD, "Environment Directorate", available at www.oecd.org/department/0,3355,en_2649_33713_1_1_1_1_1,00.htm (last accessed on 20 June 2012).

[65] See Article 4 of the Convention on the Organisation for Economic Co-operation and Development, Paris, 14 December 1960, available at www.oecd.org/document/7/0,3746,en_2649_201185_1915847_1_1_1_1,00.html (last accessed on 20 June 2012): "Article 5: In order to achieve its aims, the Organisation may: (a) take decisions which, except as otherwise provided, shall be binding on all the Members; (b) make recommendations to Members; and (c) enter into agreements with Members, non-member States and international organisations."

in excess of EUR 300 million and a secretariat staff of approximately 2,500. For the issue of climate protection, the expertise and resources concentrated in the OECD provide an opportunity to complement and facilitate international cooperation with the systematic compilation and assessment of information. OECD information has already been of great value to the UNFCCC secretariat when drafting its assessment of financial flows.[66]

Nevertheless, it appears very unlikely that the OECD, with its comparatively technical focus and limited membership, would become a venue for formal negotiations on a new climate regime. While the OECD theoretically has sufficient resources to facilitate negotiations, the highly political nature of climate cooperation would undermine the neutrality and objectivity with which the OECD is currently credited; and of course, given its membership, any formal arrangements would exclude emerging economies like the BASIC countries[67] and the vast group of developing countries. It comes as no surprise, therefore, that nobody currently foresees a leading role for the OECD in climate protection.

6.3 Many Venues, Few Achievements, Dire Prospects?

6.3.1 Necessary Conditions for a Breakthrough

After the momentous Copenhagen summit, many stakeholders and media reports suggested changing the political forum to address climate change and the mitigation challenge. For them, the UN process had become too cumbersome, with too many parties in the room stalling an already arduous decision-making process. A change of venue, they argued, would circumvent these problems and offer better prospects to address climate change.[68]

Indeed, while the international community has agreed in several contexts that it wants to limit global warming to below 2°C above preindustrial levels, it is clearly far from reaching this objective. While the Cancún summit, in particular, but also the Durban summit were each able to create a positive dynamic and produce some results, they did not deliver on mitigation targets adequate to prevent dangerous climate change.

Current emission trends suggest that only a limited number of countries – the major present and future emitters from the developed and the developing world – are

[66] UNFCCC, "Investment and Financial Flows to Address Climate Change", 2007, available at http://unfccc.int/files/cooperation_and_support/financial_mechanism/application/pdf/background_paper.pdf (last accessed on 20 June 2012); and the respective update: "Investment and Financial Flows to Address Climate Change: An Update", UN Doc. FCCC/TP/2008/7, 2008.

[67] BASIC = Brazil, South Africa, India and China.

[68] See also Daniel Bodansky, *A Tale of Two Architectures: The Once and Future U.N. Climate Change Regime* (Phoenix, Az.: Arizona State University, 2011), at 18; Michonski and Levi, *Harnessing International Institutions,* supra, note 2, at 1–3.

needed to successfully protect the climate. While acknowledging that other countries might be able and willing to contribute to collective mitigation efforts, in the final analysis, success will hinge on these major emitters. Which emitters eventually have to be on board can vary, depending on who is willing to contribute what level of emission reductions. But despite a limited margin or "gray area", the 15 to 20 parties that must be on board can be easily identified.[69]

From a practical point of view, one might thus argue that a forum can successfully tackle climate change mitigation if it, at the minimum, fulfills the following criteria:

- all major current and future emitters are participants;
- it has sufficient resources, time and expertise to deal with the complex issues at hand;
- it is able to ensure transparency, both procedurally, but also with regard to efforts and emissions;
- it can facilitate agreement on mitigation (including commitments or pledges) and take relevant decisions (such as mechanisms to incentivize compliance);
- it has a firm political will to act swiftly to achieve the 2°C goal; and
- it reflects a common vision on *how* this should be achieved.

The criteria applied in this chapter are explicitly based not on specific schools of thought in a specific discipline, such as international relations theory, but on the observed characteristics of the different venues and institutions described in Sect. 6.2, and their ability to impact the practical achievement of the mitigation objectives serving as a benchmark in this chapter.[70]

None of the venues analyzed fulfills all the abovementioned criteria. Alternatively, one might rely on an even more limited group of countries – which then would have to be highly influential – to come to a common understanding and then create sufficient political dynamics for spillover effects. If, for example, cooperation between China and the United States were to solidify, as occasionally proposed, creating a more formal Group of Two (G2), any agreement by these two powerful nations to meet the mitigation challenge might be able to trigger a landslide within the broader international community. The emergence of such a group pushing mitigation ambition is, however, currently not in sight. But pushing the agenda by forming strong alliances across boundaries might still be a strategy worth considering in the future.

---

[69] See supra, Section 2.2.

[70] Other, more theoretical criteria for the assessment and classification of international climate policy frameworks have been proposed by Robert N. Stavins, Joseph E. Aldy, Scott Barrett, "Thirteen Plus One: A Comparison of Global Climate Policy Architectures", 3 *Climate Policy* (2003), 373; Valentina Bosetti et al., *Modeling Economic Impacts of Alternative International Climate Policy Architectures: A Quantitative and Comparative Assessment of Architectures for Agreement*, Discussion Paper 08–20 (Cambridge, MA: Harvard Project on International Climate Agreements, 2008); Daniel Bodansky, *International Climate Efforts Beyond 2012: A Survey of Approaches* (Arlington, VA: Pew Center on Global Climate Change, 2004).

6.3.2 Why Different Venues Are Needed, and Associated Risks

Evidently, the UN climate process fulfills almost all the criteria mentioned above. With regard to the difficulties of consensus decision-making in a forum representing almost every country, Cancún and – perhaps to a lesser extent – Durban have shown that the UN climate regime might be able to deal with this challenge more easily than some observers thought. And yet, the UN regime has also shown the limits of its capabilities, most of all its questionable capacity to act swiftly. What seems to be lacking is political will in some quarters, and also a common vision on the way forward.

But the UN regime is evidently not the only game in town. Recent years have seen the emergence of a multitude of venues addressing – to a greater or lesser extent – the challenge of mitigation. This is as much a product of the complexities and far-reaching implications of climate change, as it is a reflection of different powers seeking to curb or accelerate progress within the UN climate regime. In a sense, this fragmentation of governance structures follows a similar trend in the broader context of international cooperation, where competing policy architectures operate side by side on a broad range of issue areas. Drawing on experience in the global marketplace for goods and services, competition and specialization might be seen as beneficial, helping to promote an issue and deliver faster, more efficient and ultimately more effective solutions.

However, while some degree of differentiation may appear inevitable and even useful, the existence of different fora to address what essentially remains one connected challenge does not automatically translate into improved cooperation and stronger mitigation. Overlap of mandates and activities can lead to redundancies, tensions, or even inconsistencies, along with an inefficient use of already scarce resources. Initiatives with similar objectives can even undermine each other in their work – especially when they are instrumentalized for that specific purpose.

Fortunately, it no longer appears that any of the major international fora addressing climate mitigation are directly counteracting each other. Earlier, that diagnosis may have been less tenable, when groups such as the MEM were initially seen as attempts by the previous US administration to create a counterbalance to the UN climate negotiations. However, any such attempts largely ceased as a result of international pressure and changing political leadership, and possibly also due to the fact that the mainstream climate negotiations have changed in nature and approach, as discussed earlier. While this may not preclude renewed attempts to undermine an ongoing process, for the time being, no such efforts are apparent.

But even where such conscious efforts to frustrate the operation of rival regimes are not apparent, the existence of alternative fora may give rise to "forum shopping", with parties favoring whichever venue is most likely to further their priorities and interests. Furthermore, too many venues might undermine each other due to the constraints imposed on public budgets and the limited time of decision-makers. This has already been in evidence, with high-level meetings scheduled too close to each other time-wise, but too far from each other geographically, preventing some ministers from attending. Such risks arising from regime fragmentation suggest an additional and potentially important benefit of a common framework for climate cooperation, such as the one currently maintained by the UN climate regime.

Aside from the UN climate treaties and the Montreal Protocol, none of the initiatives mentioned above has a mandate for formal negotiations on emission reduction objectives, let alone a new climate treaty; nor are they likely to be given one anytime soon. Indeed, depending on how they are framed, their practical value may lie more in:

- providing an arena for less formal interaction between large emitters that have often diametrically opposed positions in the formal negotiations – for instance, such states can take advantage of the informal setting of such alternative fora to explore contentious issues without the pressure of needing to advocate rigid diplomatic positions;
- moving the climate issues to a higher political level in order to obtain guidance and support for the respective common approaches;
- assembling a group of parties who share certain values and characteristics to develop common ideas, visions or projects;
- advancing certain – primarily technical – issues more quickly to provide practical examples and lighthouse projects.

The numerous initiatives shaping international climate policy can be characterized by virtue of their scope and mandate. Such initiatives can be assigned to two categories: initiatives that are explicitly focused on climate change, and initiatives addressing climate change as only one among many issues in their substantive portfolio. In both cases, a forum may address the entire breadth of climate change mitigation, or only individual aspects. Many initiatives allow for a group of interested parties which share a certain set of common values to come together, while excluding or limiting participation by particularly difficult actors.

6.3.3 Climate-Specific Initiatives and Their Contributions

Venues focused on the climate challenge can again be subdivided into those which specifically aim to inform the UN climate process, and those which seek to address the issue without relating themselves to the UNFCCC or the KP. Among the larger initiatives, those contributing to the UN negotiations outweigh those which do not specifically link their work to that of the UN. Attempts to divert attention away from the UN climate regime, such as the earlier MEM, or steer it in another direction, have not succeeded in gaining sufficient political weight to shape the climate agenda. This shows that a majority of countries agree on the general direction and governance approach of the UN climate negotiations, and that they are willing to invest political capital and resources to defend the UN climate regime against attempts to sideline it. While this willingness might have lessened somewhat after the Copenhagen summit, it generally still seems to prevail.

A second line of distinction relates to the scope of efforts. Climate initiatives contributing to the UN climate negotiations can be differentiated into broader approaches, often at the highest political level, which cover a comprehensive range of issues and seek to further the debate and understanding of the UN process more generally (for instance, the MEF Summit); and more technical approaches – partly emerging from the former – which tackle, often at an expert level, specific technical issues and aim to become a platform to test strategies and instruments in an isolated issue area (such as REDD+ and MRV). The Cartagena Dialogue seems to integrate both such aspects while being composed of likeminded parties only.

Often, achieving broad acceptance on mitigation-related issues is a critical and difficult challenge within the UN climate regime. This applies all the more in the wake of the Copenhagen summit, where many parties felt that they had not been heard adequately. Success in helping to identify robust policy approaches while avoiding a situation in which individual parties block results or impulses from such initiatives will remain a sensitive issue. Some of the more technical initiatives can be seen as a bottom-up approach to contribute to the UN climate process, which is still largely top-down in nature. They provide a space for experts to convene without being encumbered by diplomatic considerations, and afford an opportunity to share and build knowledge and elaborate common understandings or even standards where formal negotiations on the same issues are momentarily stalled for political reasons. Some of these initiatives are also bolstered by having been given access to considerable financial resources. With availability of financing and often significant political support, these initiatives can develop a dynamic of their own. One – not necessarily desirable – effect may be the creation of path dependencies and constituencies with intrinsic interests. While such initiatives would still contribute to the UN climate process, their work might become influenced by new and independent institutional considerations, and may also reduce the openness of actors to engage in alternative routes. At the broader level of climate negotiations, moreover, these initiatives will inevitably be aligned with the interests of some countries more than others; their support may thus become perceived as a political bargaining chip. In the end, therefore, technical initiatives may not remain entirely free from politics, and hence carry the risk of becoming encumbered by the same impasses that characterize the formal negotiations. At worst, they may even divert attention and resources away from the latter.

For now, however, this risk has not materialized, and it remains to be seen whether the international community can successfully confront it. On balance, cooperation on technical issues alongside the UNFCCC and its Kyoto Protocol negotiations is more likely to help further the cause of mitigation and achieving the 2°C objective than to detract from it. Indeed, given the urgency of mitigation and the often long lead times of different abatement options, any initiative that facilitates the exploration of options and pilot projects, even if only involving a smaller group of participants and without a formalized, central governance structure, has great potential utility as an instrument to accelerate subsequent action on a larger scale. Informal cooperation may also prove instrumental in establishing bridges between developed and developing countries, building trust and a deeper understanding of the issues at hand. For certain technical issues, these initiatives may thus offer a way to circumvent the cumbersome decision-making process under the UN climate regime, while not abandoning the process as such. By aligning themselves with the formal negotiations and their respective topics, they retain a degree of legitimacy that other fora might not be able to muster. Accordingly, if such venues prove successful, the future will likely bring an increase in their number and political significance.

6.3.4 Broader Initiatives and Their Future Role

A number of important venues have brought climate mitigation into their broader agendas, and can be expected to continue doing so in the future, including the G8, the G8+5, and the G20. In the past, these high-level fora – and especially the G8 – have proven useful to reaffirm positions and allow discussion of relevant climate issues at the highest political level in groups of influential countries. In the event that a corresponding political will emerges, these venues – and especially the G20, with its broader membership – could even drive the global agenda by agreeing on some of the more controversial issues with respect to climate protection.

Lacking a specific climate focus, however, and the resources and technical expertise to address complex mitigation issues in great depth, such fora are not suited to governing technical details or providing specific guidance on issues such as the carbon market. Moreover, due to numerous factors relating to how agendas are defined and decisions are made, the role of these venues – and especially the G20 – in climate policy has not yet become sufficiently established to afford them the status of a reliable and stable forum for climate protection. Likewise, while specialized agencies such as the OECD and the IEA possess significant technical expertise, their lack of an explicit mandate for climate cooperation also precludes an advanced role as a forum for concerted mitigation.

In the case of the G8, for instance, the 2008 presidency assigned climate mitigation a distinctly lower priority than previous presidencies. Based on the outcomes of the latest summits, the G8 and its extended formation, the G8+5, have at best played a flanking role in mobilizing political will for actual mitigation commitments. Limited to political declarations that do not provide much of an operational roadmap, the main benefit of these fora lies in their ability to foster discussion and awareness of the mitigation challenge. However, they cannot establish the institutional framework or provide the in-depth technical outcomes needed to operationalize and implement political visions. Considering its institutional setup, with a focus on the industrialized world and a broad mandate which does not prioritize climate change, it is unlikely that the G8 – even in the G8+5 formation – will be able to achieve a major breakthrough for global mitigation efforts. This applies all the more given the G8's declining importance relative to the rapidly emerging economies and the fact that it does not comprise all major emitters.

Given its composition and increased political weight, however, there is the potential for a growing role for the G20. Unlike the G8, the G20 has established a track record of more specific objectives and activities, most recently in the area of fossil fuel subsidy reform. Again, this body is limited to political outcomes without direct legal effect, and its activities so far have focused strongly on global financial challenges. Although the G20 with its current mandate can thus make an important contribution to specific aspects of the mitigation challenge, it is unlikely to drive broader mitigation efforts in the near future. If, however, the G20 acquires greater political weight and if key players invest political capital into moving mitigation further into the focus of its activities – something that is not apparent at this time – the G20 could become an important player in the medium term. To what extent and under what conditions a country such as China, which has so far shown a preference for negotiating within the G77 block and under the auspices of the UN rather than in smaller fora (let alone a G2 setting), would be willing to embrace such an approach remains to be seen.

At any rate, a stronger role for the G20 would also require that some internal governance questions be addressed, such as the issue of agenda setting for summits, the need for more formal outcomes, and extended internal support structures. At this point, there are clear indications that some developed and some developing countries are reluctant to endow the G20 with further institutional resources and a broader mandate. Developments in this regard might depend, *inter alia*, on how the UN climate negotiations progress in the coming years, and how much of a political vacuum might be felt in the area of mitigation. Considering the resources and capacity concentrated in the UN climate regime – including the expert support provided by roughly 500 staff at the UNFCCC secretariat – it seems unlikely that the G20 will be able to address all the issues dealt with in the UN process any time soon; also, while any progress on mitigation is likely to attract support in the broader international community, it still may invite questions of legitimacy if an informal forum with 20 participants takes on a central role on an issue – climate change – affecting the entire international community, and particularly impacting many of the countries not participating in the G20. It is more likely that the G20 would choose to focus on very specific aspects of mitigation – especially questions with a financial dimension, such as climate finance or tax issues – and promote the international agenda in that way.

6.3.5 The Future of the UN Climate Regime

After suffering a serious legitimacy crisis in recent years,[71] the UN climate regime appears to have recovered some confidence and support in the international community, despite the fact that reform of its cumbersome voting rules remains unlikely,

[71] See Vihma, *A Climate of Consensus*, supra, note 8.

if not impossible. Its increased politicization in recent years is reflected in the appointment by both Mexico and South Africa of their respective foreign minister as president of the COP. It remains to be seen whether such an approach will help to avoid clashes between parties, as witnessed in Copenhagen. In any case, it underlines a new set of interests and dynamics driving the climate process.

The UN climate regime has long been described as "having no alternative", in large part due to the perception of legitimacy instilled by the nearly universal membership of the UNFCCC. Following the problematic conduct and outcome of the Copenhagen climate summit in 2009, critics of the UNFCCC process had a unique opportunity to call for a fundamental departure from the paradigm it represents.[72] However, by the end of the climate summit in Cancún one year later, that very process was described as revitalized. In terms of the ability to legitimize policy decisions, foster better understanding of global emissions trends, define necessary action, and ensure its implementation, there is currently indeed no alternative to what the UN climate process can deliver. Despite much room for further improvement, its highly developed regime for the measuring, reporting, and verification of emissions from industrialized countries, now complemented by the emergence of more stringent processes for developing-country emissions, are ample evidence of this institutional capacity.

It is also probably the only forum able to ensure some degree of transparency with regard to global emission trends and reduction ambitions. And no other forum could currently bring similar infrastructure, expertise, and broad support to bear on the climate challenge. Given the urgency of swift progress on mitigation, moreover, it stands to reason that political capital and financial resources should not be invested in establishing entirely new institutions or building up the capacity of alternative institutions at record speed, but rather to draw and build on the demonstrated capacities of the existing UN regime.[73] It follows that the UN climate regime will remain the centerpiece of climate cooperation for the foreseeable future, and will probably play a critical role in taking forward the global mitigation effort.

Nevertheless, the UN climate regime also has a number of important limitations. With regard to mitigation, it has proven too slow and cumbersome to live up to the urgency of the issue. At the Cancún summit, some of the most challenging issues – including long-term emission reduction targets, the inflection point for GHG emissions, the specific distribution of mitigation efforts, and the legal nature of any related commitments – have once more been deferred to future negotiations. And the Durban Summit has barely shown progress with respect to mitigation, as have the mitigation sessions in Bonn and Bangkok in 2012.

Furthermore, there seems to be a trend to turn away from the top-down approach with binding targets and a strong compliance regime. Instead, a more informal, bottom-up "pledge-and-review" approach has emerged and has been strengthened

---

[72] See, for instance, the discussion by Robert N. Stavins, *Options for the Institutional Venue for International Climate Negotiations,* Issue Brief 2010–3 (Cambridge, MA: Harvard Project on International Climate Agreements, 2010).

[73] Michonski and Levi, *Harnessing International Institutions,* supra, note 2, at 3.

by the Copenhagen Accord, as well as by the Cancún Agreements. Development of a new compliance regime under the UNFCCC, with elements of facilitation and enforcement, is not yet in sight. While the last word on this matter has yet to be spoken, and some important actors – including the EU – are still advocating a future regime with internationally binding targets, current trends point in another direction. Coupled with the timing challenge, political realities in a number of important states indicate that the UN climate regime is not suited, or at least not sufficient, to deal with the mitigation challenge on its own.

Going forward, thus, the UN regime may again fall prey to unrealistic expectations, which ultimately could undermine support.[74] Unlike any of the other fora mentioned above, the UN climate regime has to live up to the exceedingly high expectations of stakeholders – including many parties and the broader public. While it is important to acknowledge that the UN climate regime is currently the only forum with potential to deliver a comprehensive and robust policy framework with an adequate compliance regime, it would also be unrealistic to hope for a sweeping breakthrough on these challenges anytime soon.

Instead, the near – and mid-term focus will probably have to be limited to a step-by-step process, with openness to "soft" bottom-up elements such as mitigation pledges, slowly creating fertile ground for the longer term vision expounded by climate scientists.[75] Only the future can show how ambitious these steps will be. Positive developments, such as the announcement by several Latin American parties that they intend to strengthen their current mitigation pledges, are offset by negative signals from major industrialized emitters resisting a second commitment period under the Kyoto Protocol. The latest formal negotiations in Bonn and Bangkok in 2012 have given no reason to hope for unexpected breakthroughs at future climate negotiations heading toward 2015.

If the expectations revived after Cancún are again disappointed, the debate about the appropriate institutional venue will gain new momentum; but, as yet, no natural successor could simply replace the political credibility and institutional resources provided by the UN climate regime, underscoring the importance of careful expectation management.[76] Given the scientific imperative of prompt and steadily rising mitigation efforts, parties will have to balance realism and ambition to identify creative solutions.

Frontrunners may be needed to exemplify progress on emissions mitigation without sacrificing other interests, such as economic stability or prosperity, and successful regional and national initiatives could help to inspire global action and

[74] See Houser, A Role for the G-20 in Addressing Climate, supra, note 68.

[75] Daniel Bodansky, *A Tale of Two Architectures: The Once and Future U.N. Climate Change Regime* (Phoenix, Az.: Arizona State University, 2011), at 18; Daniel Bodansky and Elliot Diringer, *The Evolution of Multilateral Regimes: Implications for Climate Change* (Arlington, VA: Pew Center on Global Climate Change, 2010); Michonski and Levi, *Harnessing International Institutions,* supra, note 2; Robert O. Keohane and David G. Victor, *The Regime Complex for Climate Change*, Discussion Paper 2010–33 (Cambridge, MA: Harvard Project on International Climate Agreements, 2010), at 3–4.

[76] Houser, A Role for the G-20 in Addressing Climate, supra, note 68.

create a positive dynamic for political will and ambition. For many actors, however, the ultimate objective remains a legally binding instrument that involves all major economies and allows for the pursuit of the 2°C target – and under the current circumstances, this objective can be achieved only within the UN climate regime.

6.4 Conclusion: What Does All This Add Up To?

Clearly, the proliferation of negotiation and cooperation venues alone will not help solve the climate challenge. It may be tempting, therefore, to conclude that more is not automatically better. But, as so often, the reality is more complex. Trade-offs do exist between the characteristics of different fora. Greater inclusiveness, for instance, may increase legitimacy, but it will also typically reduce the pace and flexibility of negotiation processes; more formal engagement and legally binding outcomes may solidify expectations and instill trust, but they may moderate the level of ambition participants are willing to commit to; and the list goes on.

As a result, different initiatives can play important roles in their own right and will ideally complement each other, but no single forum will prove a panacea for the mitigation challenge. Ultimately, the outcomes of climate cooperation will only be as good as the willingness of parties to act. Regardless of which venue emerges as the main arena of mitigation efforts, and of whether the future climate architecture is driven more by bottom-up or top-down approaches, if the level of ambition is insufficient, the international community will fail to achieve the 2°C objective. Given the realities outlined earlier, the UN climate regime might not be able to fully deliver on the mitigation challenge.[77] However, for many reasons also described throughout this study, none of the existing alternatives are currently in a position to meet the challenge by themselves. Nevertheless, if harnessed correctly, they may contribute to defining suitable pathways, finding solutions, and increasing the ambition to achieve the 2°C objective.

Assuming that all the venues discussed in this chapter remain active in the area of climate change, they can undeniably contribute in the short term to mitigation at different levels: some with respect to the political will of leaders of selected countries (for example, the G8), others with regard to the evaluation and elaboration of technical solutions for specific issues (for example, MRV Partnership). Furthermore, while certain fora assemble actors with very controversial opinions (for example, G20), others are based on a symmetry of political objectives and expectations (for example, the Cartagena Dialogue). Accordingly, while a venue such as the G20 can help to bridge differences, the Cartagena Dialogue can be an engine for more ambitious

[77] Aware of this very real possibility, the UN climate regime has already decided to launch a comprehensive review, starting in 2013 and set to conclude by 2015, the year when global emissions should peak.

efforts. It seems likely that particularly ambitious countries will increasingly form alliances to establish themselves as frontrunners, spearheading mitigation efforts and showcasing opportunities.[78]

When it comes to providing a comprehensive framework for climate change mitigation, however, the UN climate regime is currently the only realistic option. If the UN climate negotiations were to collapse altogether, the burden of mitigation may be shifted to other high-profile venues, such as the G20. But such a transition would need to be accompanied by substantial governance changes under strong political pressure and extreme time constraints. What is more, the loss of institutional resources may indeed be one of the most consequential setbacks entailed by a failure of the UN climate regime. If failure in the UN climate regime occurs more gradually, the burden may be distributed in a more systematic manner between different complementary venues, such as the Montreal Protocol, the G20, and various technical initiatives.[79] Regardless of which forum might ultimately be favored by the international community in this event, success on such a complex challenge will not come overnight. However, it would be premature to anticipate a failure of the UN climate regime.

Indeed, achieving the required mitigation efforts would currently appear to be less a matter of the venue or institution; rather, the diversity of interests among major emitters – irrespective of the forum they are engaged in – is what is currently stalling significant progress. While different institutions can provide for more or less cumbersome rules on decision-making, advocate different levels of ambition, or address relevant issues at the level of experts or of heads of state and government, none will be able to eradicate the current divisions among major players. In the end, it does not matter which forum is chosen to address the mitigation challenge if parties do not bring with them sufficient will to act; and even the best regime design will not achieve the necessary mitigation levels if it is not followed up with robust implementation. Both aspects are strongly contingent on the domestic politics of parties. But that also means that national leaders with a strong vision and a will to act have a unique opportunity to advance our collective efforts on one of the most complex challenges ever to face humankind.

[78] Leading by example will also be an issue relevant to the political discussion within the EU, which is currently reflected in the debate on the – 30 % emission reduction target for 2020.

[79] With this implication Robert N. Stavins, *Options for the Institutional Venue for International Climate Negotiations,* Issue Brief 2010–3 (Cambridge, MA: Harvard Project on International Climate Agreements, 2010).

Chapter 7
Analyzing Soft Law and Hard Law in Climate Change

Antto Vihma

Abstract There is a great deal of variety in the international environmental agreements that have mushroomed in past decades. These legal arrangements can be placed on a continuum from hard law – *precise* and legally binding treaties that *oblige* a behavioural change with *delegated* enforcement bodies – to the softest of soft law, with its vague, aspirational goals and no delegation or institutional follow-up. The legalization continuum is a more insightful starting point for analyzing international agreements than 'bottom-up' vs. 'top-down' or 'pledge-and-review' vs. 'targets-and-timetables' that are often suggested by reports and policy papers. When applying the legalization lenses to the UN climate regime, two big trends emerge. There is a notable turn toward soft law in developed country commitments in climate mitigation. In the meantime, the UN regime is becoming harder by providing greater transparency of climate actions of all major economies.

7.1 Introduction

The recent climate meetings have witnessed no shortage of political drama and many of the central quarrels have included a strong legal perspective. The UN Climate Change Conference in Copenhagen, famously, concluded in a messy final plenary, "taking note" of the Copenhagen Accord, and was followed by months of blame game. In Cancún, a package of decisions was adopted by stretching the

Antto Vihma has completed his Ph.D. in the University of Helsinki, Faculty of Social Sciences, analyzing effectiveness and legitimacy of international agreements. He currently works as a Research Fellow at the Finnish Institute of International Affairs.

A. Vihma (✉)
The Finnish Institute of International Affairs, Kruunuvuorenkatu, 4, 00160 Helsinki, Finland
e-mail: antto.vihma@fiia.fi

E.J. Hollo et al. (eds.), *Climate Change and the Law*, Ius Gentium: Comparative Perspectives on Law and Justice 21, DOI 10.1007/978-94-007-5440-9_7,
© Springer Science+Business Media Dordrecht 2013

definition of "consensus" further than ever before in the climate regime. At the latest session of the Conference of the Parties (COP) in Durban the limelight of the final political struggle was the issue of legal form, this time for a mandate to negotiate an agreement beyond the year 2020. Since COP 13 in Bali in 2007 the UN climate regime has experienced legal-political turmoil; this reality calls for policy relevant analysis of the characteristics of different legal options, their limitations and possibilities.

The increasing reliance on global regulation through diverse types of legal arrangements, a phenomenon also referred to as the "legalization" of international affairs,[1] has been interpreted as a necessary corollary to globalization.[2] There is a great deal of variety in the international agreements that have mushroomed in past decades. Driven by the realist challenge to prove the ability of international law to exert influence on nation states, much of the research has focused on the international agreements in their legally binding treaty form with enforcement ('hard law') such as the World Trade Organization, as well as on economically powerful organizations such as the International Monetary Fund and the World Bank.[3] Since the early 1990s, however, increased attention has also been paid to the more amorphous, non-legally binding ('soft law') instruments. There is a growing body of research that studies private authority, networks, transnational standard-setting with non-state actors, and other profoundly soft modes of global governance.[4]

How should we approach soft law? As outlined above, when analyzing the current global response to climate change, we are confronted with several critical puzzles regarding international law in general. Even a cursory glance at the empirical world of global governance shows that there is considerable diversity in the legal characteristics of international agreements. How to analyze this diversity in a way that is both academically solid and relevant to the political debates in climate change? The aim of this chapter is to contribute to our understanding on this vital question.

There are broadly three main alternatives to study the legal characteristics of international agreements at the nexus of IR theory,[5] including rationalist, social

---

[1] Kenneth Abbott et al., "The Concept of Legalization", 54 *International Organization* (2000), 401.

[2] Abram Chayes and Antonia H. Chayes, *The New Sovereignty: Compliance with International Regulatory Agreements* (London: Harvard University Press, 1995); John Braithwaite and Peter Drahos, *Global Business Regulation* (Cambridge: Cambridge University Press, 2000).

[3] Xinuan Dai, *International Institutions and National Policies* (Cambridge: Cambridge University Press, 2007), at 7.

[4] See for example Braithwaite and Drahos, *Global Business Regulation*, supra, note 2; Dinah Shelton (ed), *Commitment and Compliance: The Role of Non-binding Norms in the International Legal System* (Oxford: Oxford University Press, 2000), Steve Bernstein and Benjamin Cashore, "Can Non-State Global Governance be Legitimate? An Analytical Framework", 1 *Regulation & Governance* (2007), 347; Julia Black, "Constructing and Contesting Legitimacy and Accountability in Polycentric Regulatory Regimes", 2 *Regulation & Governance* (2008), 137.

[5] This grouping is by no means exhaustive list of perspectives that legal scholars use in studying international cooperation as a whole. Many other schools of thought and theoretical debates exist and are influenced by other disciplines such as anthropology, sociology, political philosophy and history.

constructivist, and critical formalist.[6] The orientation in this essay is influenced to a great extent by rationalist scholars and especially the special issue of *International Organisation* that provided a definition of "legalization" and kick-started the ensuing academic debate on that approach.[7] The legalization school argues that international agreements can be placed on a continuum from hard law – *precise* and legally binding treaties that *oblige* a behavioural change with *delegated* enforcement bodies – to the softest of soft law, with its vague, aspirational goals and no delegation or institutional follow-up.[8] From this point of view it is possible to further analyse the politics of institutional choice in the fragmented international legal order.

In the field of environmental politics, major multilateral agreements are commonly expressed in legally binding treaty form as "conventions" and "protocols" to those conventions.[9] For example the ozone regime,[10] biological diversity regime[11] and climate regime[12] include provisions for signature, ratification, accession, approval, and withdrawal recognized by international treaty law and customary law as a means of formalizing the consent of a state to be bound. These treaties have been complemented with soft law that exists *outside* their umbrella and soft law that exists *within* these regimes. In global climate governance, there has been a broader trend of States negotiating minilateral, non-legally binding agreements outside of the United Nations Framework Convention on Climate Change (UNFCCC). This trend gained momentum around 2005 and includes agreements that focus on the implementation of activities[13] as well as agreements that focus on political declarations

[6] Several analysts such as Shaffer and Pollack would call these critics of soft law "legal positivists". However, some notable critical scholars such as Koskenniemi do not sit well with legal positivism, as he constantly emphasizes that his goal is not to promote positivist formalism, which could mask or neutralize political choices and conflicts. For this reason I adopt the term "critical formalism" to describe these viewpoints. See Gregory Shaffer and Mark Pollack, "Hard Law vs. Soft Law: Alternatives, Complements and Antagonists in International Governance", 94 *Minnesota Law Review* (2010), 706.

[7] See seminal articles by Abbott et al., "Concept of Legalization", supra, note 1; Kenneth Abbott and Duncan Snidal, "Hard and Soft Law in International Governance", 54 *International Organization* (2000), 421.

[8] Abbott et al., "The Concept of Legalization", supra, note 1.

[9] Jacob Werksman and Kirk Herbertson, "The Aftermath of Copenhagen: Does International Law have a Role to Play in a Global Response to Climate Change?", 25 *Maryland Journal of International Law* (2010), 109; Farhana Yamin and Joanna Depledge, *The International Climate Change Regime: A Guide to Rules, Institutions and Procedures* (Cambridge: Cambridge University Press, 2004).

[10] The Vienna Convention for the Protection of the Ozone Layer, Vienna, 22 March 1985, in force 22 September 1988, 26 *International Legal Materials* (1986), 1529.

[11] The Convention on Biological Diversity, Rio de Janeiro, 5 June 1992, in force 29 December 1993, 31 *International Legal Materials* (1992), 818.

[12] The Framework Convention on Climate Change, Rio de Janeiro, 9 May 1992, in force 21 March 1993, 31 *International Legal Materials* (1992), 849.

[13] For example the Asia-Pacific Partnership on Clean Development and Climate (APP), see Harro Van Asselt, "From UN-ity to diversity? The UNFCCC, the Asia-Pacific Partnership, and the Future of International Law on Climate Change", 1 *Carbon and Climate Law Review* (2007), 17; Sylvia Karlsson-Vinkhuyzen and Harro Van Asselt, "Introduction: Exploring and Explaining the Asia-Pacific Partnership on Clean Development and Climate", 9 *International Environmental Agreements* (2009), 195.

and guidance.[14] Second, soft law arrangements have emerged from within the UN climate regime, as the prospects for a legally binding protocol to include other parties than European countries have become more and more daunting for the 2012–2020 period in the climate regime. Also the mandate for the post-2020 agreement, preliminarily decided in the 2011 Durban meeting, may well yield an outcome that is considerably softer than the Kyoto Protocol architecture (see below). Furthermore, already the operationalization of the legally binding Kyoto Protocol relied considerably on the decisions of the Conference of Parties serving as the Meeting of the Parties to the Kyoto Protocol (COP/MOP), which can be seen as a type of soft law.[15]

I will first provide an overview to the concepts of hard and soft law, and from there move on to present the main theoretical insights to these concepts. The focus is on the legalization approach and the critique it has faced during the last decade. Applying the legalization approach, I present some insights into major trends within the UNFCCC negotiations. Lastly, some conclusions are drawn.

7.2 Analyzing Soft and Hard Law

7.2.1 The Legalization Continuum

For the term hard law, which naturally was not referred to as 'hard' before the emergence of soft law, there are relatively widely accepted legal definitions, including on its sources (treaty and custom), and the implication of the general obligation imposed on states (*pacta sunt servanda*) to follow them. The well-established category of international customary law emerges from state practice and is in many cases not explicitly designed or formally codified. Therefore, it is excluded from further discussion in this essay.

The early discussions on soft law in the international sphere date back to the late 1970s. At that time the term was usually placed in quotation marks.[16] By the late

[14] For example the Major Economies Forum/Meeting on Energy Security and Climate (MEF), several G8 and G20 meetings, and numerous regional forums, see Antto Vihma, "Friendly Neighbor or Trojan Horse? Assessing the Interaction of Soft Law Initiatives and the UN Climate Regime", 9 *International Environmental Agreements* (2009), 239.

[15] The Marrakesh Accords operationalized some of the key aspects of the Kyoto Protocol after long – lasting negotiations in 2001, related to, for example, reporting, verification and compliance. See Decisions 2-14/CP.7, The Marrakesh Accords, UN Doc. FCCC/CP/2001/13/Add.1, 21 January 2002.

For a recent discussion on the properties of COP decisions, see for example Antto Vihma, "A Climate of Consensus: The UNFCCC Faces Challenges of Effectiveness and Legitimacy", 75 *Finnish Institute of International Affairs Briefing Papers* (2011).

[16] For example Rene-Jean Dupuy, "Declaratory Law and Programmatory Law: From Revolutionary Custom to 'Soft Law'" in Robert Akkerman et al. *Declarations on Principles: A Quest for Universal Peace* (Leiden: Sijthoff, 1977), 247.

1980s and early 1990s, the concept had gained momentum,[17] and the analysis not only recognized and described an empirical phenomenon, but also reflected on its implications against the binary terms of legal formalism (legal/illegal; binding/non-binding), and to the great disappointment of formalist-oriented scholars, "in doing so ended up rejecting the binary code of law altogether".[18] Also in the contemporary literature the term soft law is on many occasions defined in binary terms, and usually in terms of what it is *not*. Soft laws are not legally binding by themselves, they are not in treaty form, and they do not belong to the category of customary law.[19]

The essential issue on which there is considerable disagreement is whether it is possible and/or useful to make this kind of binary divide between hard and soft law. In the formal view an international agreement is either international law or it is not; if a "soft norm meets the requirements of the doctrine of sources of international law, it is hard law".[20] Several analysts in a leading volume on soft law edited by Shelton subscribe to this viewpoint, in which soft law means "normative agreements that are not legally binding".[21] Another perspective is presented in the same volume by Chinkin, who frames soft and hard law in a hierarchy in a descending "hardness" of laws, including legal soft law (imprecise hard law); secondary or delegated soft law (which includes the "statements and practice that develop around a treaty to supplement or correct the text"); and non-legal soft law (resolutions, declarations, the output of intergovernmental conferences, etc.).[22]

The continuum approach to international legalization, of which Chinkin's categories are a variant, is supported by many rationalist scholars. For these analysts, the whole sanctity of "bindingness" in international law is a somewhat misleading hyperbole.[23] International legalization offers decision-makers many shades of grey

[17] Christine Chinkin, "The Challenge of Soft Law: Development and Change in International Law", 38 *International and Comparative Law Quarterly* (1989), 850.

[18] Jan Klabbers, "Reflections on Soft International Law in a Privatized World", XVI *Finnish Yearbook of International Law* (2007), 313, at paragraph II.

[19] Dinah Shelton, "Introduction: Law, Non-law and the Problem of 'Soft Law'", in Shelton (ed.) *Commitment and Compliance*, supra, note 4.

[20] Jonathan L. Charney, "Commentary: Compliance with International Soft Law" in Shelton (ed.), *Commitment and Compliance*, supra, note 4, at 115.

[21] Shelton, "Introduction", supra, note 18; Wolfgang Reinicke and Jan Martin Witte, "Interdependence, Globalization, and Sovereignty: The Role of Non-binding International Legal Accords", in Shelton (ed.), *Commitment and Compliance,* supra, note 4.

[22] Chinkin also includes an unnamed category in her study, which encompasses the norms that are developed without the involvement of states. Some scholars would not include these in the term soft law, while others consider such private regulation as a central part of international soft law. The realm of "private" soft law – which in itself can range from very precise, elaborate and enforced rules to vague principles or codes of conduct – is not addressed here. See Christine Chinkin, "Normative Development in the International Legal System" in Shelton (ed), *Commitment and Compliance*, supra, note 4, at 27.

[23] See Charles Lipson, "Why are Some International Agreements Informal?", 45 *International Organization* (1991); Abbott et al., "The Concept of Legalization", supra, note 1; Abbott and Snidal, "Hard and Soft Law", supra, note 6; Kal Raustiala, "Form and Substance in International Agreements", 99 *American Journal of International Law* (2005); Shaffer and Pollack, "Hard Law vs. Soft Law", supra, note 5.

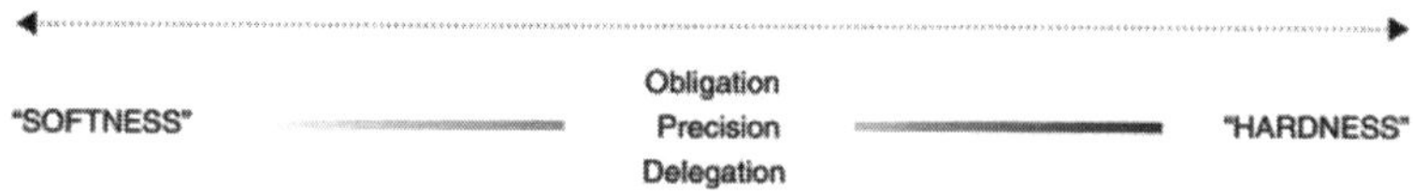

Fig. 7.1 The figure illustrates the continuum of international legalization, based on the criteria of precision, obligation, and delegation (Figure 7.1 is from Sylvia I. Karlsson-Vinkhuyzen and Antto Vihma, "Comparing the Effectiveness and Legitimacy of Global Hard and Soft Law: An Analytical Framework", 3 *Regulation & Governance* (2009), 400, at 402)

instead of a clear black and white distinction between binding and non-binding, and this is not to be seen normatively as a bad thing. A case in point is that traditional and formal hard law treaties may be so generally worded as to be devoid of legal content – the category referred to by Chinkin as "legal soft law".[24] Also many constructivists, as well as "realist" legal scholars, emphasize the "law-in-action" as opposed to "law-in-books", noting that domestic laws also vary considerably in their real-life bindingness, that binary distinctions are not useful,[25] and even accuse formalist scholars of being guilty of "elite ignorance" and "non-knowledge of the social".[26]

In their seminal article Abbott, Keohane, Moravcsik, Slaughter and Snidal elaborate on their factors of "hardness" for international law.[27] They pronounced three criteria: providing binding obligation, precise wording, and a degree of delegation. If international agreements are weakened along these dimensions, they enter the realm of soft law. *Obligation* means that the behaviour of actors under the treaty is subject to change and scrutiny under the general rules, procedures, and discourse of international law. *Precision* indicates that "rules unambiguously define the conduct they require, authorize, or proscribe", a particularly important feature of law at the global level, as laws and rules are usually created consensually by states and interpreted afterwards by those same states. *Delegation* gives a third party some level of authority to implement and interpret the rules and to resolve disputes.[28]

The critical formalist viewpoint regards international law as a clear, binary choice between something binding, which is law, and something non-binding, which is not law. In this view, the concept of soft law and its characteristics are not interesting,[29]

[24] Chinkin, "Normative Development", supra, note 22.

[25] David Trubek, Patrick Cotrell and Mark Nance, "soft Law, Hard Law, and EU Integration" in Joanne Scott and Gráinne de Búrca (eds), *New Governance and Constitutionalism in Europe and the US* (Oxford: Hart Publishing, 2006).

[26] Peter Goodrich, "Law-Induced Anxiety: Legalists, Anti-Lawyers and the Boredom of Legality", 9 *Social & Legal Studies* (2000), at 150.

[27] Abbott et al., "The Concept of Legalization", supra, note 1; Abbott and Snidal, "Hard and Soft Law", supra, note 7.

[28] Abbott et al., "The Concept of Legalization", supra, note 1, at 401.

[29] Jan Klabbers, "The Redundancy of Soft Law", 65 *Nordic Journal of International Law* (1996), 167.

and/or not desirable.[30] However, from an *ex ante* point of view of the actors, differences in legal characteristics offer the decision-makers room for manoeuvring, as different legal characteristics involve different costs and benefits. International agreements usually have both hard and soft elements[31] and do not constitute "sharp dichotomous choices" for decision-makers but rather "choices of various strategies, or combinations of strategies".[32] Hard and soft law may also act as mutually supportive or as competing strategies; law is not only a facilitator of social order, but also a tool advanced by actors looking to fulfil their aims.[33] This is especially the case in the sphere of global governance, with its fragmented nature, lack of clear hierarchy and lack of a final institutional arbiter (a supreme court). Also, soft law regimes may be "hardened", for example with links to other regimes, while hard law regimes may be "softened" with ambiguous paragraphs or decisions.

The exploration of the hard and soft law landscape results in a long continuum, "from hard law through varied forms of soft law, each with its individual mix of characteristics".[34] Hard law and soft law are useful concepts as end points of the continuum, but a systematic and policy-relevant evaluation of law needs to pay attention to the diversity along the continuum (Fig. 7.1). However, while negotiators draft legal arrangements of descending or ascending hardness, there is one strong element which is *not* amenable to flexibility: the decision on whether to conclude a ratifiable treaty or not. This is a profoundly binary decision.[35] If the intergovernmental negotiations result in a ratifiable treaty, it will go through procedures which are determined in the national constitutions.

In the climate regime, one of the main arguments for governance by COP decisions is that they do not, *an sich*, require ratification. However, in several countries, the content of an international instrument – rather than its name or formal status – determines the legal procedures through which it must be transposed into national legislation. This means that for many countries than the more substance is put into COP decisions, the more likely they are to require ratification in accordance with national constitutional requirements. In some key countries, however, the form and name of the agreement might make a remarkable difference. Globally the most crucial

---

[30] See Jan Klabbers, "The Undesirability of Soft Law", 67 *Nordic Journal of International Law* (1998), 381. The normative argument is centred on the notion that increasing reliance on soft law represents a shift of power from legal institutions to "administrative power" in the EU context, namely to the European Commission.

[31] Richard Bilder, "Beyond Compliance: Helping Nations Cooperate", in Shelton (ed.), *Commitment and Compliance*, supra, note 4.

[32] John Kirton and Michael Trebilcock, "Introduction: Hard Choices and Soft Law in Sustainable Global Governance" in John Kirton and Michael Trebilcock (eds), *Hard Choices, Soft Law: Voluntary Standards in Global Trade, Environment and Social Governance* (Cornwall: Ashgate, 2004), 3.

[33] Shaffer and Pollack, "Hard Law vs. Soft Law", supra, note 5.

[34] Karlsson-Vinkhuyzen and Vihma, "Comparing the Legitimacy and Effectiveness", supra, note 34, at 401.

[35] I would like to thank Professor Timo Koivurova for emphasizing this point in our correspondence.

implication of the ratification requirement probably is the advice and consent procedure and the two thirds majority by which the US Senate has to consent to international agreements signed by the US executive branch. This has effectively prevented most environmental treaties from being implemented into US domestic legislation. A recent report sheds light to the status of ten pending environmental treaties – half signed by Democratic presidents and half signed by Republican presidents – which have been pending an *average* of 13 years, awaiting ratification.[36]

Focusing the analysis on the legal form only does not seem to provide answers to critical puzzles. Many practitioners and academics alike assume that a legally binding form would have a positive effect on compliance; however, available evidence does not univocally support this proposition.[37] From a rationalist point of view, states and other international actors utilize hard law to order their relations, because it helps to reduce transaction costs, strengthen the credibility of their commitments, expand their available political strategies, and resolve problems of incomplete contracting. The hard law path, however, comes at a significant cost, as hard law restricts actors' behaviour and sovereignty.[38] The rationalist paradigm sees that the advantage of soft law is that it is less costly in terms of the sovereignty of states – a key theme in this literature is that soft law involves less negotiation costs, as states exercise more caution in drafting hard law due to greater consequences of a subsequent violation.[39] The other side of the coin is that soft law arguably represents a less credible commitment to the issue at hand than hard law. In the face of serious global risks such as climate change, many would argue that the virtue of credible commitments is worth significant costs, given the nature and limitations of the non-hierarchical and fragmented international legal order. One virtue could the formality of reciprocal expectations, that could, perhaps, build trust and thus enable greater ambition. Secondly, formalizing substantive and procedural rights and duties could in turn elevate the position of smaller actors, which in the case of climate change could also have implications on the ambition level. Many civil society groups such as environmental NGOs fear that soft law arrangements can be used cynically, to "take the heat off" political leaders, allowing symbolic but empty promises to substitute for real action.[40]

The constructivist paradigm has focused on "appropriate behaviour", which is intimately connected to the construction of the identities of states. Changes in state behaviour can thus occur through processes of socialization and the expansion of norms, ideas and principles. Constructivist-oriented legal scholars quite frequently

[36] Mary Jane Angelo et al., *Reclaiming Global Environmental Leadership: Why the United States Should Ratify Ten Pending Environmental Treaties,* Center for Progressive Reform White Paper #1201 (2012), available at: http://www.progressivereform.org/articles/International_Environmental_Treaties_1201.pdf (last accessed on 23 February 2012).

[37] Helmut Breitmeier, Oran Young and Michael Zürn, *Analyzing International Environmental Regimes: from Case Study to Database* (Cambridge: MIT Press, 2006).

[38] Abbott and Snidal, "Hard and Soft Law", supra, note 6.

[39] Lipson, "Why are Some International Agreements Informal?", supra, note 23; Abbott and Snidal, "Hard and Soft Law", supra, note 7; Kirton and Trebilcock, "Introduction: Hard Choices and Soft Law", supra, note 32; Shaffer and Pollack, "Hard Law vs. Soft Law", supra, note 6.

[40] See for example Vihma, "Friendly Neighbor or Trojan Horse?", supra, note 144, at 250.

take this kind of approach to international law, whether discussing "the active role of the regime in modifying preferences",[41] "internationalization processes" that work over time,[42] or "the compliance pull" of international law that is deemed legitimate.[43] From this perspective, the soft law approach might have advantages in promoting norm diffusion and learning, allow a wide spectrum for deliberation in governing,[44] and generate shared norms and a sense of common purpose and identity.[45] As constructivists analyze international law in terms of values and the formation of state identities, the world no longer needs to be thought of in terms of power and interest. This idealist perspective alone, I would argue, makes the analysis susceptible to classic criticisms which realist scholars of international relations originally levelled at international law.[46]

This chapter suggests a middle path between constructivist and rationalist paradigms.[47] I follow the argument that there is evidence that decision-making related to creating and complying with international law is influenced by drivers from both paradigms,[48] as the interest-based and normative strategies are deeply intertwined.[49] In the case of environmental regimes, utilitarian motives and normative motives are most often both at work, and simultaneously so.[50] Furthermore, both approaches can be improved on the ground "by carefully incorporating the arguments made by the other" in the analysis.[51] It is also worth re-emphasizing that the continuum approach resonates well with the view of the practitioners, namely the negotiators who craft multiple wordings ascending in various ways from "binding" to "non-binding" language. This *ex-ante* viewpoint of the multilateral negotiations is strikingly different from the binary distinction picture painted by some critical formalist scholars of academic literature.[52]

---

[41] Chayes and Chayes, *The New Sovereignty,* supra, note 2.

[42] Harold Koh, "Why do Nations Obey International Law?", 106, *Yale Law Journal* (1997), 2599.

[43] Thomas Franck, *The Power of Legitimacy Among Nations* (Oxford: Oxford University Press, 1990), 312.

[44] Trubek et al., "Soft Law, Hard Law, and EU Integration", supra, note 25, at 3.

[45] Shaffer and Pollack, "Hard Law vs. Soft Law", supra, note 6, at 3.

[46] See Martti Koskenniemi, "Turn to Ethics in International Law", available at: http://www.helsinki.fi/eci/Publications/Koskenniemi/Ethics.pdf (last accessed on 22 February 2012); also, see Martti Koskenniemi, "The Lady Doth Protest too Much: Kosovo, and the Turn to Ethics in International Law", 65 *The Modern Law Review* (2002), 159.

[47] Karlsson-Vinkhuyzen and Vihma, "Comparing the Effectiveness and Legitimacy", supra, note 34, at 405.

[48] See, for example, Oran Young, *The Institutional Dimensions of Environmental Change* (Cambridge: The MIT Press, 2002).

[49] Abbott and Snidal, "Hard and Soft Law", supra, note 7.

[50] Oran Young, Leslie King and Heike Schroeder, *Institutions and Environmental Change: Principle Findings, Applications, and Research Frontiers* (Cambridge: The MIT Press, 2008).

[51] Abbott and Snidal, "Hard and Soft Law", supra, note 6, at 422.

[52] The *ex-post* view is more at home in a situation where a judge faces the decision in a court on whether a given instrument is binding or not. However, this view should not be simplified to the extreme either, see Shaffer and Pollack, "Hard Law vs. Soft Law", supra, note 6, at 12.

7.2.2 *Criticism of Legalization*

The legalization continuum approach has sparked notable critical comments from two opposing camps – the formalist-oriented legal scholars who guard the sanctity of the binary character of law,[53] and several social constructivist legal scholars.[54]

The constructivist critique is aimed at the "narrow conception of law", rooted in "positivism," "formalism," and "Western tradition".[55] For many constructivists, law is a very broad concept, and in the end, law is "whatever people recognize and treat as law through their social practices".[56] According to this critique, focusing on legalization variables leads to diminished attention paid to important topics such as legitimacy, from which international law gets its "force" in a non-hierarchical system, and customary law, as well as the process of law. The constructivist scholars call for more focus on identities as generators of interest, and research which illuminates how identities are shaped through social interaction.[57]

Certainly, taking a profoundly sociological view on law, the legalization approach can also seem formal and alien to the developments on the ground, in the real world where law operates which is what most analysis is ultimately interested in. I also share the constructivist critics' view that legitimacy is a central concept, and furthermore, it is deeply intertwined with questions of effectiveness and compliance. In spite of this, I argue, the legalization approach highlights important aspects about the making and implementation of international law, and serves as a useful starting point for analysis.

The second branch of criticism stems from the directly opposing group to constructivist perspectives, namely scholars emphasizing critical formalism. There certainly is no love lost in Koskenniemi's assessment of constructivist research, which he sees as "returning to analyses of international politics in terms of its rights and wrongs, good and evil" and celebrating "moral enlightenment of a new world, a universal liberal *Gemeinschaft*".[58]

---

[53] Martti Koskenniemi, "International Law: Between Fragmentation and Constitutionalism", available at: http://www.helsinki.fi/eci/Publications/Koskenniemi/MCanberra-06c.pdf (last accessed on 22 February 2012). Most themes Koskenniemi touches upon in this key presentation feature in his collection of essays, Martti Koskenniemi, *The Politics of International Law* (Oxford: Hart Publishing, 2011). See also Klabbers, "The Undesirability of Soft Law", supra, note 30; Klabbers, "Reflections on Soft International Law", supra, note 18.

[54] Jutta Brunnée and Stephen Toope, "International Law and Constructivism: Elements of an Interactional Theory of International Law", 39 *Columbia Journal of Transnational Law* (2000), 19; Jutta Brunnée and Stephen Toope, "Interactional International Law", 3 *International Law Forum* (2001), 186; Marthe Finnemore and Stephen Toope, "Alternatives to "Legalization": Richer Views of Law and Politics", 55 *International Organization* (2001), 743.

[55] Finnemore and Toope, "Alternatives to 'Legalization'", supra, note 55.

[56] Brian Tamahana, *A General Jurisprudence of Law and Society* (Oxford: Oxford University Press, 2001).

[57] Brunnée and Toope, "Interactional International Law", supra, note 55.

[58] Koskenniemi, "Turn to Ethics", supra, note 47, at 22.

The formalist critique is usually aimed at 'soft law' in general, not only the legalization continuum approach. Although this critical viewpoint admits that soft law "may seem useful at first sight", as soon as it is to be applied it collapses into either hard law or no law at all. Soft law is like balancing a coin on its edge; it looks good for a moment, "but as soon as you start to spend it, it will fall heads or tails", so no continuum really exists.[59] The accusation by Klabbers is that soft law typically gets applied like hard law – especially in the EU context – with the difference being that it does not have to be accepted by "domestic democratic bodies" like formal treaties.[60] Again, in the EU context, this means a power shift towards the bureaucratic initiatives of the European Commission.[61] The backdrop is about power: once you give up formalism, a chaotic state prevails, who shouts loudest wins, and legal concepts and regimes cannot be systematically analyzed.[62]

It seems that some of the critical formalist views are laced with an overwhelming nostalgia for a more stable and simple point in time, when rules were clear, knowledge was uniform, and the road ahead was well laid out. The critique is concerned that international law is "no longer taken seriously" but is a policy option among others[63] (was it really ever anything else?), that soft law enables today's power-holders to escape "democratic scrutiny" (more than international affairs did before?), and compares "fragmented order" to the times when there was no international regulation whatsoever, for instance, for environmental problems.[64]

From a broader but equally critical perspective, soft law is claimed to represent fragmentation and managerialism, which leads to erosion of international law.[65] From this viewpoint Koskenniemi presents a masterful and critical reflection on the fragmentation of international law, leading to "imperial and solipsistic" sub-systems, which threaten the universalism international law ought to highlight.[66] A fundamentally "managerial" approach has emerged as international law comes to us in separate boxes, and serves an instrumental purpose for particular values, interests and preferences, such as the "European project", "trade project" or "environmental project".[67] Koskenniemi acknowledges that the fragmentation goes further than the differentiated soft-hard characteristics of law, and emphasizes that each

[59] Klabbers, "The Undesirability of Soft Law", supra, note 30, at 382.

[60] Klabbers, "Reflections on Soft International Law", supra, note 18, paragraph IV.

[61] Klabbers, "The Undesirability of Soft Law", supra, note 30.

[62] Klabbers, "Reflections on Soft International Law", supra, note 18. See for example paragraph II, "Any definition, or even any broader concept of soft law, has so far proved highly elusive", and "if everything is law, nothing is".

[63] Klabbers, "Reflections on Soft International Law", supra, note 18, at paragraph V.

[64] Klabbers, "Reflections on Soft International Law", supra, note 18, at paragraph II; Koskenniemi, "International law", supra, note 51.

[65] Koskenniemi, "International Law", supra, note 51; see also Martti Koskenniemi, "The Politics of International Law – 20 years later" in Koskenniemi, *The Politics of International Law*, supra, note 54.

[66] Koskenniemi, "International Law", supra, note 54.

[67] Koskenniemi, "International Law", supra, note 54, paragraph 8.

subsystem of international law has a different objective, different ethos and a different "structural bias", no matter what its legal characteristics are.[68] However, from the point of view of his critique, it is evident that soft law is a way to spread these specialized projects and their differentiation further, with a quintessentially managerial approach to law.[69] International lawyers are taught to speak of "regimes" instead of institutions and of "regulation" instead of rule, to change the language of government to "governance", responsibility to "compliance" and lawfulness to "legitimacy". Ultimately, international law becomes drained of law.[70]

From a practice-oriented perspective this critique is problematic. First, as Koskenniemi naturally acknowledges, the empirical reality is that specialized regimes are commonplace in contemporary international law, but he argues that this is not "natural and inevitable" as many others would suggest.[71] It is certainly true that the sub-areas of international law do not automatically arise from the "nature of things", and that most real-world events and cases relate to, for example, environmental law, trade law and human rights law simultaneously. However, it is still hard to escape the increasing inevitability of such specialization and division of labour, even though its origins are a social construct and have to do with "powerful interests".[72] While agreeing with many of the problems raised by Koskenniemi, his criticism is on a higher level of abstraction than this essay, suggesting that international law should return to the "culture of formalism" and "constitutional mindset",[73] universality in a Kantian sense, law as a language for the critique of power. His view is openly normative, the world as it ought to be, and, in contrast to this edited volume, he also highlights that he is ultimately not interested in architectural questions.[74] This critique thus offers few applicable tools for analyzing the architectural issues in the contemporary international legal landscape in empirical terms.

Finally, both the constructivist critique and the perspectives that emphasize formalism are connected to the broad theme of legitimacy. Constructivists caution that in the legalization approach effectiveness overrides legitimacy and, in a way, so does the broader version of formalist critique. Kantian deontological reasoning cited by Koskenniemi requires the decision-maker to focus on the morality of actions themselves, without "making principles subordinate to the end", without deriving justification from the consequences, as values and purposes

---

[68] Koskenniemi, "International Law" supra, note 54; see also Martti Koskenniemi, *From Apology to Utopia: The Structure of International Legal Argument* (Cambridge: Cambridge University Press, 2005).

[69] Koskenniemi is fiercely critical of the "deformalisation" of international law. See for example Koskenniemi, "International Law", supra, note 54, paragraphs 17 and 21.

[70] Koskenniemi, "International Law", supra, note 54, paragraphs 20 and 21.

[71] Koskenniemi, "International Law", supra, note 54, paragraph 9.

[72] Koskenniemi, "International Law", supra, note 54, paragraph 9.

[73] Martti Koskenniemi, *The Gentle Civilizer of Nations: The Rise and Fall of International Law 1870–1960* (Cambridge: Cambridge University Press, 2001).

[74] Koskenniemi, "International Law", supra, note 54, paragraphs 20 and 25.

represent *hubris* and *Schwärmerei*.[75] The opposing rationalist camp follows the "consequentialist" or "welfarist" paradigm that acknowledges the priority of good over process.[76] In this rationalist view, justice is seen firstly as a matter of outcome; a political and legal decision can produce injustice, however fair the procedure is. It is in this instrumental way that specialized regimes and projects – with their faults and biases – currently are justified as legitimate.[77] In the traditional view, legitimacy is crucial in achieving state compliance and thus effectiveness. But equally importantly, effectiveness is a component of legitimacy, as the lack of acceptable performance undermines the legitimacy of the norm in the long term. This argument has been widely noted in sociology, but has not been internalized by many analysts of international regimes.[78]

7.2.3 Hard Law-Soft Law Dynamics

The dynamic of legal characteristics operates over time. For example, an initially soft agreement may earn high enough legitimacy to be turned into hard law.[79] At the level of practical politics within the field of global environmental governance, the soft-hard law dynamics are at play in the process of operationalizing softer framework conventions into harder legal instruments and decisions through multilateral negotiations. This approach has been adopted, for example, with the ozone regime, the biodiversity regime, and the climate regime.[80]

Framework conventions in international environmental law are formal, ratifiable and legally binding treaties. However, framework conventions typically do not contain clear, detailed, or specific rules that could be implemented in domestic legislation in a straightforward manner. In contrast with the generality of framework conventions, the protocols or other legal instruments developed within their regime, as well as decisions adopted by the decision-making bodies established by the regime, typically provide rules and mechanisms that are very specific.[81]

---

[75] Koskenniemi, "International Law", supra, note 54; Immanuel Kant, *The Critique of Pure Reason* (1781), available at: http://ebooks.adelaide.edu.au/k/kant/immanuel/k16p/part1.2.html (last accessed on 22 February 2012).

[76] See discussion in Jekwu Ikeme, "Equity, Environmental Justice and Sustainability: Incomplete Approaches in Climate Change Politics", 13 *Global Environmental Change* (2003), 195.

[77] See Frtitz W. Scharpf, *Governing in Europe: Effective and Democratic?* (New York: Oxford University Press, 1999); Robyn Eckersley, "Ambushed: The Kyoto Protocol, the Bush Administration's Climate Policy and the Erosion of Legitimacy", 44 *International Politics* (2007), 308.

[78] See also Eckersley, "Ambushed: The Kyoto Protocol", supra, note 78.

[79] Shelton, "Introduction", supra, note 19.

[80] See supra, notes 10, 11 and 12.

[81] See for example Yamin and Depledge, *The International Climate Change Regime*, supra, note 9.

As is well illustrated by the legalization approach, international agreements are very varied rather than a dichotomy of two categories with different forms. Consequently, the term soft law can also be used to refer to soft provisions in "hard law" instruments. Abbott and Snidal elaborate on how states can limit their legal obligation through "hortatory language, exceptions, reservations and the like".[82] In practice, the soft provisions mean vague and flexible formulations in treaty texts, such as mandating a party to take "such actions as it deems necessary"[83] or to act in a certain manner "as appropriate".[84] These types of provisions are also referred to as "escape clauses",[85] or more to the point, "non-decisions",[86] or "elements of non-commitment in commitment".[87] Soft provisions in treaties exist parallel to hard ones, but their vagueness leaves it up to states to decide how to implement the provision. In spite of their vagueness, the principle of *pacta sunt servanda* applies and a treaty remains "binding" on paper, even if the chances of actual effective implementation of the provision in question are reduced by the generality of the obligation.

Soft law can bring parties to the negotiation table – under a framework convention – and involve parties in a process that leads to harder international obligations in the future.[88] From this dynamic viewpoint, the soft provisions within a regime are left open for future negotiations. Regime critics have also raised this issue in order to point out the respective lack of formalism in contemporary treaties. To agree on a framework convention is, in practice, also to agree to continuous negotiations, contextual deal striking, and bargaining of experts; and as laws do not spell out the conditions of their application in their entity, the management of a regime will have to take place by open-ended standards.[89]

Constructivist-oriented scholars tend to view the framework convention approach positively, claiming that it may catalyze the dialogic process of norm-building.[90] Rationalists do not take a stand on whether this is the case, but conclude that the framework convention approach is fruitful at least in cases of technical uncertainty, where states can facilitate information generation and common understanding via

---

[82] Kenneth Abbott and Duncan Snidal, "Pathways to Cooperation", in Eyal Benvenisti and Moshe Hirsch (eds), *The Impact of International Law on International Cooperation: Theoretical Perspectives* (Cambridge: Cambridge University Press, 2004), at 50.

[83] Article 5, North Atlantic Treaty Organization, Washington, 4 April 1949, in force 24 August 1949, 34 *United Nations Treaty Series,* 243.

[84] Article 4.5, UNFCCC, supra, note 12.

[85] Lavanya Rajamani, "From Berlin to Bali and Beyond: Killing Kyoto Softly?", 57 *International and Comparative Law Quarterly* (2008), 909.

[86] Joyeeta Gupta, *The Climate Change Convention and Developing Countries: From Conflict to Consensus?* (Dordrecht: Kluwer Academic Publishers, 1997), 249.

[87] Michael Glennon, *Constitutional Diplomacy* (Princeton: Princeton University Press, 1990).

[88] Jutta Brunnée and Stephen Toope, "Environmental Security and Freshwater Resources: Ecosystem Regime Building", 91 *The American Journal of International Law* (1997).

[89] Koskenniemi, "International Law", supra, note 54, paragraph 15 and paragraph 25.

[90] Trubek et al., "soft Law, Hard Law, and EU Integration", supra, note 25; Braithwaite & Drahos, *Global Business Regulation*; Brunnée and Toope, *Legitimacy and Legality in International Law: An Interactional Account* (Cambridge: Cambridge University Press, 2010).

clarified costs and benefits.[91] The ozone regime – the Vienna Convention and its Montreal Protocol – has served as a positive example for scholars from both paradigms, both for norm internalization and monitoring, as well as cooperative research, transparency and information exchange.

The main legal-political challenge is agreeing on whether normative principles and rules are overlapping or not – does a Conference of the Parties decision or a protocol "change" or "operationalize" the framework convention? This is the delicate balancing act that the negotiations within a framework convention call for. As noted already in the 1980s by Krasner, changes in rules and decision-making procedures are "changes *within* the regime, provided the principles and norms are unaltered, whereas changes in principles and norms are changes *of* the regime".[92] There is a dynamic approach built into the regimes as the substantive obligations can change along the way, based on the progress of the negotiations as well as input from external processes, such as increased scientific insights into the problem that needs to be addressed, or, at least ideally, changes in the respective responsibilities and capabilities of states to address the problem. Sometimes the uncertainties related to the operationalization are also used cynically to slow down the negotiations. There are cases in which parties assume positions that are contradictory to the basic understanding of a system of negotiating a protocol and decisions on the basis of a framework convention.[93]

International law is also influenced by horizontal interaction between hard law and soft law.[94] The commonplace viewpoint of the literature is that hard and soft law act as complements, that hard law can generate secondary or delegated soft law,[95] or that hard law linkages can indirectly harden soft law.[96] The complementary assumption has also been claimed to be biased, as Shaffer and Pollack conclude that "the scholarship has failed to address how, when and why hard law and soft law operate as antagonists".[97] Their viewpoint is not completely original, however, as some earlier literature already suggests that emerging principles of soft law can soften existing

[91] Abbott and Snidal, "Hard and Soft Law", supra, note 7; Shaffer and Pollack, "Hard Law vs. Soft Law", supra, note 6.

[92] Stephen Krasner, *International Regimes* (Ithaca, NY: Cornell University Press, 1983), at 3.

[93] See Antto Vihma, "India and the Global Climate Governance: Between Principles and Pragmatism", 20 *Journal of Environment & Development* (2011), 69.

[94] The vertical interaction between levels of governance is also a case in point. The vertical dynamics include international soft law, which can "harden" at lower levels of governance; for example, when a principle from a soft international declaration is elaborated into a more binding instrument nationally or regionally. See, for example, Jeremy Wates, "The Aarhus Convention: A Driving Force for Environmental Democracy", 2 *The Journal for European & Environmental Planning Law* (2005).

[95] Chinkin, "Normative Development in the International Legal System", supra, note 17.

[96] Sylvia Karlsson, Multilayered Governance: Pesticides in the South – Environmental Concerns in a *Globalised World* (Linköping: Linköping University, 2000).

[97] Shaffer and Pollack, "Hard Law vs. Soft Law", supra, note 6, at 2.

hard law by undermining its legitimacy.[98] Although not framed in soft law-hard law terminology, the environmental regime theory is also well informed of the possible antagonist relationship between different international legal arrangements.[99]

In the world of climate governance it seems reasonable to assume that the multitude of climate governance arrangements is not simply a somewhat uncoordinated group of peacefully co-existing institutional processes – instead, these elements might be used to create overlaps and to interact with intentional synergetic or disruptive consequences.[100] The latter is the case if the institutional overlaps result from deliberative efforts of interested parties to pursue their own objectives by creating competitive arenas, and/or opening up opportunities for strategic behaviour for those who have less interest in the problem.[101] Based on these premises, the multitude of processes in global climate governance calls for analysis of the positions of relevant actors and mechanisms through which the influence could occur, perhaps in the footsteps of the insightful analysis by Shaffer and Pollack. Such work on the interaction between soft and hard law is, however, beyond the scope of this essay.[102]

7.2.4 Legalization Insights to the Climate Regime

The legalization approach emphasizes the benefits and costs of different legal characteristics and thus a rationalist perspective. But clearly law also engages normative considerations. It requires commitment to a background set of legal norms – the "engagement in established legal processes and discourse"[103] or "the practice of legality"[104] – and provides opportunities for parties to epitomize normative values. Normative processes and interests enable laws to be effective, and also constrain the success of law. The key message of Abbott and Snidal is that the form and content of international laws are parts of the same package, the muscle of international law, into which the legalization continuum – with its variables of obligation, precision and delegation – offers an insightful analytical approach. Several other approaches are suggested to categorize between different types of law in the policy literature,

[98] Christine Chinkin, "The Challenge of Soft Law: Development and Change in International Law", 38 *International and Comparative Law Quarterly* (1989), 850.

[99] See for example Frank Biermann et al., "The Fragmentation of Global Governance Architectures: A Framework for Analysis", 9 *Global Environmental Politics* (2009), 14; Sebastian Oberthür and Thomas Gehring (eds), *Institutional Interaction in Global Environmental Governance* (Cambridge: The MIT Press, 2006); Young, *The Institutional Dimensions*, supra, note 49.

[100] Biermann et al., "The Fragmentation of Global Governance Architectures", supra, note 100.

[101] Young, *The Institutional Dimensions*, supra, note 49, at 112–113.

[102] See, for instance, the chapter by Camilla Bausch and Michael Mehling in this volume.

[103] Abbott and Snidal, "Hard and Soft Law", supra, note 7, at 425.

[104] Brunnée and Toope, "Legitimacy and Legality", supra, note 91.

such as "top-down vs. bottom-up" and "pledge-and-review vs. targets-and-timetables".[105] These approaches, however, run a notable risk of being misleading. First, in the policy discourse of the UN climate regime, "pledge-and-review" is usually used to imply a very soft architecture.[106] However, reviewing policies, pledges, targets or obligations is actually an ambitious task for international law. The UN climate regime itself illustrates that many countries are extremely sensitive about allowing such measures to be taken.[107] Second, "top-down" architecture, on the other hand, risks sounding idealistic and lacking credibility in the community of practitioners. It suggests that the international community would be able to agree on the needed aggregate amount of emissions reductions and then divide the pie to different parties via negotiations. This picture is far from the reality of policy making, where countries' emissions targets are adopted "bottom-up", agreed upon by the domestic constituencies, and then communicated to the international arenas. The role of international negotiations considering ambition is not irrelevant but more subtle than "top-down" – it is to provide a framework of reference for the domestic politics of emissions reductions. Examples of this include the collective sense of the level of effort in Kyoto negotiations, and the 2° target and 450/550 pm targets discussed and debated in various international fora in recent years.

So what type of insights can the legalization approach give to the contemporary developments in the climate regime? The decision on legal form was one of the main political struggles in Durban COP-17. The conference resulted in Parties launching a process titled "The Durban Platform on Enhanced Action" to negotiate "a Protocol, another legal instrument or agreed outcome with legal force under the Convention applicable to all".[108] The negotiations are scheduled to adopt a decision in 2015 and implement it from 2020 onwards. The compromise language "agreed outcome with legal force" cobbled together by US and Brazilian negotiators to solve a political stand-off between the EU and India does not reflexively signal a ratifiable instrument.[109] However, it makes a ratifiable treaty the most likely and widely expected form of the outcome for the post-2020 period.[110]

As suggested by the legalization continuum, the ratifiable versus non-ratifiable form should not be the only criterion when evaluating the legal dimension of the UN

---

[105] For a very recent example see Daniel Bodansky, "A Tale of Two Architectures: The Once and Future UN Climate Regime", available at http://ssrn.com/abstract=1773865 (last accessed on 22 February 2012).

[106] See for example "Greenpeace Guide to Kyoto, Bali, APEC, the G8 and Major Emitters Meeting", Greenpeace Briefing, available at: http://www.greenpeace.org/usa/Global/usa/report/2007/11/greenpeace-guide-to-kyoto-bal.pdf (last accessed on 22 February 2012).

[107] These difficulties are featured, for example, in Vihma, "India and the Global Climate Governance", supra, note 94.

[108] Decision 1/17.CP, Establishment of an Ad Hoc Working Group on the Durban Platform for Enhanced Action, UNFCCC, UN Doc. FCCC/CP/2011/9/Add.1, 15 March 2012, available at: http://unfccc.int/resource/docs/2011/cop17/eng/09a01.pdf (last accessed on 22 February 2012).

[109] Rajamani, "Deconstructing Durban", *Indian Express*, 15 December 2011.

[110] Rajamani, "Deconstructing Durban", *Indian Express*, 15 December 2011.

climate regime. Building on the analysis of Rajamani, there are at least two broader tendencies that are traceable at least from Bali COP-13 onwards, namely i) softer *obligation* and less *delegation* on developed country commitments and ii) harder *obligation, delegation* and *precision* on major developing country reporting and transparency. These tendencies are determining the parameters of the 2012–2020 climate regime and may well be a strong influence from 2020 onwards as well. Moreover, even if the legal form becomes "hard" as in a ratifiable treaty for the post-2020, the character of the commitments for developed countries is likely to be softer than under KP and their form more self-selected.

First, the move towards a soft law approach in the post-2012 era for developed countries not parties to the second commitment period of the Kyoto Protocol is quite evident if we use the Kyoto Protocol itself as a yardstick. There is no facilitative or punitive compliance mechanism on the Convention track of the Bali Action Plan. The mitigation by developed countries will be subject to International Assessment and Review (IAR) procedures,[111] but the scope of the "assessment" is unclear, namely whether the assessment concern the adequacy of data, adequacy of targets, or adequacy of performance.[112] In comparison with the Kyoto Protocol's architecture,[113] the Convention track is softer than the KP in all aspects of the legalization continuum.

The evolution towards soft law has also taken place *within* the Convention track since the Bali meeting in 2007, as pointed out by Rajamani.[114] The Cancún and Durban decisions use the language of "targets" instead of "commitments" like the Bali Action Plan,[115] and similarly they "promote comparability" instead of "ensuring comparability".[116] Furthermore, the Cancún and Durban outcomes essentially re-emphasize the pledges countries submitted under the Copenhagen Accord, but do this in a non-legal manner, by "taking note" of these pledges, collected in an information document.[117] The main point is not that the pledges are nationally determined and then submitted to the international sphere – many would say that also

[111] Decision 1/CP.17, supra, note 109.

[112] Rajamani, "Deconstructing Durban", *Indian Express*, 15 December 2011.

[113] Articles 5, 7 and 8, Kyoto Protocol to the United Nations Framework Convention on Climate Change, Kyoto, 10 December 1997, in force 16 February 2005, 37 *International Legal Materials* (1998), 22.

[114] Rajamani, "From Berlin to Bali and Beyond", supra, note 86; Rajamani, "The Cancun Climate Agreements: Reading the Text, Subtext and Tea Leaves", 60 *International and Comparative Law Quarterly* (2011), 499.

[115] Decision 1/CP.16, The Cancun Agreements: Outcome of the work of the Ad Hoc Working Group on Long-term Cooperative Action under the Convention, UNFCCC, UN Doc. FCCC/CP/2010/7/Add.1, 15 March 2011, paragraphs 36–38; Decision 1/CP.13, The Bali Action Plan, UNFCCC, UN Doc. FCCC/CP/2007/6/Add.1, 14 March 2007, paragraph 1 (b) (i).

[116] Decision 1/CP.16, supra, note 116, paragraph 44; decision 1/CP.13, supra, note 116, paragraph 1 (b) (i).

[117] Information documents have no legal status in the process, but are commonly used for example as a way to distribute the list of participants.

the Kyoto commitments were in essence nationally determined and in that way "bottom-up" as noted above – but that these pledges have different conditions, base years and caveats.[118] They are thus lacking common accounting and comparability metrics.

Second, the mitigation actions of developing countries are internationalized through increased reporting requirements and a process of International Consultation and Analysis (ICA). Although ICA is to be conducted "in a manner that is non-intrusive, non-punitive and respectful of national sovereignty",[119] it is a significant development on the Convention track, and has been subject to notable political controversy, both in the UNFCCC negotiations as well as in the national sphere of major developing countries, for example in the Indian Parliament.[120] Since the launch of the UN climate regime the developing countries have faced virtually no transparency requirements with any degree of international delegation: the National Communications have not been regular, they have not been designed in accordance with international guidelines, and they have been allowed to use ancient data. Comparing this long time *status quo* to the biennial reporting with 4 years old data and an ICA procedure, all envisioned in the Cancún and Durban decisions, shows a significant step forward in the hardness of the 2012–2020 climate regime.

While agreeing to the main conclusions of Rajamani, it seems that in some parts of her analysis, the formal and political meanings of "developing countries" overlap slightly. Formally, as is well known, there is very little differentiation among the developing countries (non-Annex I countries) in the climate regime. From this perspective it is plausible to conclude growing parallelism among developed and developing countries. Politically, however, the pressure for parallelism in reporting and legal form has not been on "developing countries" but on certain major economies, China ahead of others. Secondly, Rajamani takes a firm stand that in Bali the (major) developing countries agreed only to measure, report and verify internationally supported actions. However, most developed countries had an interpretation of the Bali Action Plan that the transparency requirements covered also unsupported domestic actions.[121] At least the paragraph in question has been open to different interpretations and quite a lot of political controversy, as witnessed already in the Bali final plenary, where India inserted a carefully placed comma to the text and South Africa clarified their interpretation of the text to overcome objections from the US.[122]

---

[118] See, for example, submissions from the US, available at: http://unfccc.int/files/meetings/cop_15/copenhagen_accord/application/pdf/unitedstatescphaccord_app.1.pdf (last accessed on 22 February 2012).

[119] Decision 1/CP.16, supra, note 116, paragraph 63.

[120] For example Lokh Sabha of the Indian Parliament, 21 December 2009 (transcript on file with author).

[121] Decision 1/CP.13, supra, note 116, Article 1 b (ii).

[122] Bali COP-13, final plenary, 15 December 2007.

7.3 Discussion and Conclusions

Crafting legal arrangements is a central tool in global governance, whether we look into the issue areas of trade, security, human rights, or the environment. The types of norms that have been generated during the past decades have very different legal characteristics. The aim of this essay is not to have the final word on the strengths and weaknesses of hard and soft law, but to suggest a way for further analysis that would be academically rigorous as well as politically relevant.

The law that is designed as an instrument of global governance can be placed on a continuum from ideal hard law – precise and legally binding treaties with delegated enforcement bodies – to the softest of soft law, with its vague, aspirational goals and little or no institutionalized follow-up. The legalization continuum, I argue, is a more insightful starting point for analyzing international agreements than "bottom-up" vs. "top-down" or "pledge-and-review" vs. "targets-and-timetables" that are often suggested by reports and policy papers.

To date, little work has been done on combining constructivist and rationalist paradigms in the analysis of hard and soft law,[123] although there seems to be considerable value in incorporating arguments from both paradigms into the research framework. I argue that we should remain agnostic as to which theoretical camp most accurately captures the true nature of hard and soft law and their relevant qualities, and approach the question on a contextual basis. In sum, different legal characteristics have advantages and drawbacks in different contexts, whether framed in rationalist or constructivist terms. The qualities of global hard and soft law are largely based on specific, political and functional questions, such as the North–south politics, the domestic/foreign policy interface, and the institutional interaction. It seems that the issues of legal character, effectiveness and legitimacy cannot and should not be solved in an abstract or general way. This echoes the views of Young, King and Schroeder, who summarized the literature on environmental regimes and recommended a "diagnostic approach" to designing specific institutions rather than "a search for design principles or generalizations" applicable to the full range of international environmental agreements.[124] The debate should be firmly grounded in the context of a particular policy domain, its incentives, discourses and operational capacities. An almost inescapable context for the effectiveness and legitimacy of global environmental governance is, however, the North–south politics, which have received relatively scant attention in some more theoretical analyses of global law and its implications.

The UN climate negotiations can be framed as efforts to operationalize the soft law of the framework convention into decisions or legal instruments, with a greater degree of obligation, precision, and delegation. From this legalization perspective there indeed seems to be a notable drive towards soft law within the climate

[123] Trubek et al., "Soft Law, Hard Law, and EU Integration", supra, note 25.

[124] Young et al., "Institutions and Environmental Change", supra, note 51, at 3.

change regime. This is not only due to the form of the agreement that is relying on COP-decisions that "take note" of parties' actions and not a ratifiable protocol, but to the broader tendency towards less obligation, precision and delegation for developed country parties in mitigation. However, there is another broad tendency, which is scaling up the transparency requirements of (major) developing countries. For the first time a delegated and precise reporting system is being agreed upon, and although the first report is framed in voluntary terms, this is not a minor development. The legal form of the post-2020 agreement is likely to continue to draw the big headlines, but in the meantime, the UN regime is becoming "harder" by providing greater transparency of climate actions of all major economies. The caveat is that while mitigation commitments and transparency are central issues in the climate negotiations, they do not paint a complete picture of the regime. Alongside them there are many other interesting developments, including Parties' commitments in long-term financing and the evolution of technology, adaptation and REDD+ mechanisms. There are signs of a process of stronger institutionalization and hence delegation in these areas, with new decision making bodies with a mandate from the COP and limited membership. Simultaneously, the common design standards are lacking in market mechanisms and, above all, the commitments that generate demand for credits. There remains a need to look closer into these specific issue areas, as well as the regime functions as a whole.

In the context of growing parallelism, I would be tempted to argue that a trade-off between hard law characteristics and effectiveness of the regime may well be present. The political context of parallelism and the drive towards "hard law" outcomes will make states hyper-cautious about what they commit to, potentially leading to decreased ambition, and possibly, an absence of a major player such as the US or China or Russia. This is especially the case where the legal form of the obligation is concerned, but may well surface also with more innocent attempts to delegate authority away from parties to the international sphere.

One suggestion to unravel the complex dynamic of effectiveness and legitimacy is to focus on enhanced decision-making in the UNFCCC.[125] The legal vacuum of unadopted Rules of Procedure and pushing the limits of "consensus" do not seem like sustainable strategies.[126] The idea of voting has recently been floated by several scholars.[127] This is often justified by highlighting the problems of a consensus-based decision-making structure: "*Moving the climate change agenda forward multilaterally among 195 parties to the UNFCCC is proving to be a serious challenge […] The turn today toward a multipolar world indicates that approaches based on consensus*

[125] Antto Vihma and Kati Kulovesi, "Strengthening the Global Climate Change Negotiations", *Nordic Council of Ministers Working Paper* (forthcoming, 2012).

[126] Vihma, "Climate of Consensus", supra, note 15.

[127] "One of the core findings of our research program is that the current consensus principle as it is being implemented in the climate negotiation, but also in many other international environmental negotiations, is obsolete." Professor Frank Biermann, interview with Deutsche Welle 27 March 2012, available at: http://www.dw.de/dw/article/0,,15840057,00.html (last accessed on 25 May 2012).

are unlikely to produce results".[128] While procedural reforms do not offer "low hanging fruits" in the short term, strengthening the basis of decision-making for the future of global climate governance would be a productive exercise that could, in time, contribute to a positive cycle of increased legitimacy and effectiveness.

In general, the constructivist point of view does not see hard law, or credible compliance and enforcement systems, as key motivators for states in international regimes, as measuring the utilitarian value of compliance and non-compliance is not the central issue. For rationalists, regimes as "information providers" have been a centrepiece for research,[129] as compliance mechanisms in a broader sense begin with observability. More empirical research is needed on which functions can be effectively covered with soft law and which would require a hard law approach. These insights would, in turn, feed back into the more theoretical debates between rationalists, constructivists and the critical formalist scholars.

In Koskenniemi's view, the practice-oriented approach and emphasis on the contextual – as argued for in this essay – can turn international law into an apologist deference to power. In his work, "apology" has at least two distinct meanings, namely referring to international law as being descriptive of what states do, and international law as reflecting the wishes or values of its subjects (which might not be "good").[130] From the viewpoint of this essay, which is more open to rationalist argumentation than Koskenniemi's deconstruction, only the first is a concern. It is, in essence, the classic realist challenge. Future research would duly benefit from answering the call by examining international agreements with a legalization approach, from a broad and practice-oriented perspective.

[128] Rafael Leal-Arcas, "Top-Down versus Bottom-Up Approaches for Climate Change Negotiations: An Analysis", 6 *IUP Journal of Governance and Public Policy* 6 (2011).

[129] Many scholars have discussed these issues, see for example Xinuan Dai, *International Institutions and National Policies* (Cambridge: Cambridge University Press, 2007).

[130] Notably, the classic critique presented by Koskenniemi is not only about international law being apologetic, but about being caught between the destructive dynamics of apology and utopia. See Koskenniemi, *From Apology to Utopia*, supra, note 69.

Chapter 8
Compliance and Enforcement in the Climate Change Regime

Meinhard Doelle

Abstract This chapter tracks the work of the compliance committee under Kyoto Protocol since the operationalization of the Kyoto compliance system in 2006. The basic elements of the compliance system, including its facilitative and enforcement branches, are described. Key issues brought before the committee between 2006 and 2012 are reviewed. In particular, the effectiveness of the more active enforcement branch is assessed through the first seven issues of implementation brought before the branch. The case against Greece, the first matter considered by the branch, is considered in detailed, followed by an assessment of issues raised in the six subsequent cases. Finally, some opportunities to strengthen the Kyoto compliance system are identified.

8.1 Introduction

The Kyoto compliance system has long been recognized as a testing ground for compliance theory.[1] While compliance theorists actively debated the relative merits of self-interest and norm-building in motivating countries to meet their international

---

Meinhard Doelle is a Professor at the Schulich School of Law, Dalhousie University, Halifax, Canada (Mdoelle@dal.ca), where he currently serves as Associate Director of the Marine & Environmental Law Institute. Parts of this chapter are based on Meinhard Doelle, "Early Experience with the Kyoto Compliance System: Possible Lessons for MEA Compliance System Design", 1 *Climate Law* (2010), 237.

[1] See, for example, Jutta Brunnée, "Promoting Compliance With Multilateral Environmental Agreements", in Jutta Brunnée, Meinhard Doelle and Lavanya Rajamani, *Promoting Compliance in an Evolving Climate Change Regime* (Cambridge: Cambridge University Press, 2012).

M. Doelle (✉)
Schulich School of Law, Dalhousie University,
6061 University Avenue, Halifax, NS, Canada
e-mail: moelle@ac.ca

E.J. Hollo et al. (eds.), *Climate Change and the Law*, Ius Gentium: Comparative Perspectives on Law and Justice 21, DOI 10.1007/978-94-007-5440-9_8,
© Springer Science+Business Media Dordrecht 2013

commitments, negotiators of the Kyoto compliance system strove to develop a compliance system that would be capable of building norms and facilitating compliance while at the same time deterring parties that might be tempted make a calculated choice not to comply. The result of these negotiations was the Kyoto compliance system, including its facilitative and enforcement branches.[2]

The Kyoto compliance system is enabled in Article 18 of the Kyoto Protocol.[3] It was negotiated over a 4-year period following the signing of the Protocol. The resulting Compliance Procedures were then formally adopted by way of a decision of the Conference of the Parties serving as the Meeting of the Parties to the Kyoto Protocol (COP/MOP) at its first meeting in Montreal in 2005.[4] This was followed, after some initial experience, by Rules of Procedure developed by the compliance committee and adopted by the COP/MOP.[5]

The compliance committee established under the Compliance Procedures has functioned since 2006 in the form of a plenary, a bureau, and two branches. One branch, the facilitative branch, serves to facilitate countries efforts to comply with obligations under the Kyoto Protocol. The other branch, the enforcement branch, serves to impose consequences in case of non-compliance with specific obligations.

The plenary of compliance committee consists of the members of the facilitative and enforcement branches. The chairs and vice-chairs of the two branches constitute the bureau. Each branch is composed of one member from each of the five regional groups of the United Nations, one member representing small island States, and two members each from Annex I countries and Non-Annex I countries. An alternate is appointed for each member of the committee in case a member is unavailable. Decisions are to be made by consensus whenever possible. In case consensus is not possible, a majority of three-quarters is required for any decision of the committee or one of its branches. In addition, decisions by the EB require the support of a majority of both Annex I and non-Annex I members.

See also Jutta Brunnée, "A Fine Balance: Facilitation and Enforcement in the Design of a Compliance for the *Kyoto Protocol*", 13 *Tulane Environmental Law Journal* (2000), 223; Meinhard Doelle, *From Hot Air to Action? Climate Change, Compliance and the Future of International Environmental Law* (Toronto: Carswell, 2005); Peggy Rodgers Kalas and Alexia Herwig, "Dispute Resolution under the *Kyoto Protocol*", 27 *Ecology Law Quarterly* (2000), 53; and David G. Victor, "Enforcing International Law: Implications for an Effective Global Warming Regime", 10 *Duke Environmental Law and Policy* (1999), 147.

[2] Sebastian Oberthür and René Lefeber, "Holding Countries to Account: The Kyoto Protocol's Compliance System Revisited After Four Years Of Experience", 1 *Climate Law* (2010), 133. See also René Lefeber and Sebastian Oberthür, "Key Features of the Kyoto Protocol's Compliance System" in Jutta Brunnée, Meinhard Doelle & Lavanya Rajamani, *Promoting Compliance in an Evolving Climate Change Regime* (Cambridge: Cambridge University Press, 2012).

[3] Kyoto Protocol to the United Nations Framework Convention on Climate Change, 10 December 1997, UN Doc. FCCC/CP/1997/L.7/add. 1, 37 *International Legal Materials* (1998), 22, Art. 18.

[4] Annex to Decision 27/CMP.1 on Procedures and Mechanisms Relating to Compliance under the Kyoto Protocol, 92, UN Doc. FCCC/KP/CMP/2005/8/Add.3, 30 March 2006.

[5] See Annexes to Decisions 4/CMP.2 and 4/CMP.4 on the Compliance Committee, UN Doc. FCCC/KP/CMP/2006/Add.1, 2 March 2007, 17, and UN Doc. FCCC/KP/CMP/2008/11/Add.1, 19 March 2009, 14.

The plenary is responsible for reporting to the COP, and for the overall administration of the compliance process. The bureau receives and reviews questions of implementation brought to the compliance committee and determines which branch of the compliance committee is responsible for responding to the issues raised. The facilitative branch is generally responsible for assisting Parties in their efforts to meet their commitments under the Kyoto Protocol. This includes providing advice, and otherwise facilitating compliance with respect to Articles 5 and 7 of the Protocol.

With respect to Articles 5 and 7, the mandate of the facilitative branch overlaps with that of the enforcement branch, which has a mandate to determine compliance and impose consequences of non-compliance with these provisions. In addition to providing advice on Articles 5 and 7, the facilitative branch has the exclusive mandate to address questions of implementation with respect to supplementarity under Articles 6, 12, and 17, Article 3.14 dealing with effects of mitigation measures on developing countries, and reporting on demonstrable progress under Article 3.2.

The jurisdiction of the enforcement branch is limited to provisions that have a clear link to the emissions reduction target under Article 3.1. In addition to the emission reduction obligation itself, this includes accounting and reporting obligations necessary to determine a Party's emissions and mitigation efforts, and the eligibility to participate in emissions trading, joint implementation and the clean development mechanism. All other commitments under the Kyoto Protocol are subject to facilitation, but not subject to enforcement.

Decisions of the enforcement branch regarding compliance with Article 3.1 will generally follow the review of the final reports submitted by a Party under Article 8 at the end of the commitment period, which are expected to be concluded in 2014.[6] Before a determination of noncompliance is made at this point, Parties will be provided with an opportunity to come into compliance by purchasing the necessary credits from another Party. Under Part XIII of the compliance annex, a Party may purchase credits for compliance purposes up to 100 days after the expert review process for the commitment period under Article 8 is declared by the conference of the Parties to be concluded.[7]

The importance of the distinction between the two branches becomes apparent in light of the consequences applied by each of the branches. The facilitative branch, under part XIV of the Annex on compliance, can apply the following consequences:

- Provision of advice and facilitation of assistance;
- Facilitation of financial and technical assistance, including technology transfer and capacity building;
- Formulation of recommendations to a Party on what could be done to address concerns about a Party's ability to comply with its obligations.[8]

---

[6] Until the end of the first commitment period, the focus of the enforcement branch will be on compliance with accounting and reporting rules under Articles 5 and 7.

[7] Compliance Procedures, supra, note 4, Part XIII, Additional Period for Fulfilling Commitments at 74.

[8] Ibid., in particular Part XIV, Consequences Applied by the Facilitative Branch at 75.

The enforcement branch has the power to apply the following consequences:

- Declaration of noncompliance;
  - Requiring a Party to submit a compliance action plan, which would include an analysis of the causes of non-compliance, measures to be taken to return to compliance, and a timetable for implementing the measures;
    - Suspending a Party's eligibility to use the mechanisms, if a Party is found not to meet one of the eligibility requirements;
- In case of failure to meet its emissions reduction target under Article 3.1, deducting from the Party's assigned amount for the second commitment period 1.3 times the amount of excess emissions from the first commitment period.[9]

The most substantive consequence of not meeting the first commitment period target, therefore, is the reduction of the assigned amount in the second commitment period.[10]

The compliance process is generally initiated by referring questions of implementation to the bureau for a determination of which branch has jurisdiction. There are three ways issues can come before the compliance committee: as a result of a review of a country's submissions by an expert review team under Articles 5 and 7, at the initiative of a Party that realizes it requires assistance in meeting one of its obligations, or at the request of another Party that questions compliance of a Party with one of its obligations.[11]

The process described is generally open to the public and reasonably transparent. However, there are provisions in the compliance agreement[12] that can reduce or eliminate the transparency of the process to a point where it risks losing its credibility. There are broad powers, for example, to prevent information from being made public until after the conclusion of the process. Similarly, there is provision for the hearings of the enforcement branch to take place in private. These powers have generally not been exercised to date except for deliberations of the committee, which have generally been held in private.

The agreement provides for an appeal process, but grounds for appeal are limited to due process issues. The Conference of the Parties (COP) serves as the appeal body, and decisions being appealed stand pending the appeal. This is designed to ensure that the appeal process, which can take some time given that the COP generally only meets once a year, is not used as a way to delay application of consequences of non-compliance.[13]

---

[9] Ibid., Part XV, Consequences Applied by the Enforcement Branch at 75.

[10] Also referred to as borrowing or restoration.

[11] Compliance Procedures, supra, note 4, para. 3, at 70.

[12] Ibid., paras. 4–6 at 70; Ibid., Part IX, Procedures for the Enforcement Branch, para. 2 at 71; Ibid., Part X, Expedited Procedures for the Enforcement Branch, para. 1 at 72.

[13] The appeal process has been utilized once to date, in a case involving Croatia discussed below.

In summary, the Kyoto compliance system operates through its facilitative and enforcement branches, a plenary, and a bureau. Compliance issues can be referred either by a party or by an Expert Review Team (ERT). Matters referred are to be allocated by the bureau to the appropriate branch. The compliance procedures include detailed rules on the composition and functions of the two branches, the bureau, and the plenary. The compliance procedures furthermore outline the general process to be followed, and the powers of each branch. The rules of procedure, supplemented by working arrangements adopted by the plenary, detail the process implemented to give effect to the compliance procedures.[14]

The entry into force of the Kyoto Protocol was delayed until 2005.[15] As a result, the work of the compliance committee did not get underway until 2006, only 2 years before the start of the first commitment period.[16] This delay had particular implications for the work of the facilitative branch (FB), given that one of its main tasks was to assist parties in preparing for a range of obligations. Many of the obligations subject to facilitation in some way related to parties' commitments to report and to reduce their greenhouse gas emissions for the first commitment period. The opportunity to actively work with parties to assist in this process was, as a result of the delay, reduced by 6 years.

It is therefore not surprising that the FB has been relatively inactive. The enforcement branch (EB), on the other hand, has been relatively busy dealing with the estimation, reporting, and verification of emissions and credits of Annex I parties. The focus of the EB initially was on compliance with rules under Articles 5 and 7 for initial eligibility to trade under the Kyoto mechanisms. It has since transitioned into the second phase of its work, the ongoing compliance with rules under Articles 5 and 7. The third phase of its work, compliance with parties' emission-reduction obligations for the first commitment period, is not expected to start until well over a year after the end of the commitment period.

When it was first negotiated in 2001, the general expectation of parties was that the Kyoto compliance system would serve the climate change regime for a long time. While this is still possible, the future of the Kyoto compliance system is very much uncertain as a result of the ongoing negotiations of the post-2012 climate change regime. At the time of writing, it is unclear to what extent the regime will continue to be built around binding emission-reduction commitments and whether or under what circumstances those commitments will be subject to international

---

[14] It is worth noting that the application of the current compliance system under Kyoto is focussed on developed countries, but there are elements that could be utilized for developing-country parties in the future. An interesting question in reviewing the current system, therefore, would be what adjustments would have to be made to expand the application of this kind of compliance system to address monitoring, reporting, and verification involving developing countries.

[15] See Meinhard Doelle, "The Kyoto Protocol; Reflections on its Significance on the Occasion of its Entry into Force", 27 *Dalhousie Law Journal* (2005), 556.

[16] The first commitment period under the Kyoto Protocol started on 1 January 2008 and runs until 31 December 2012.

enforcement. It remains to be seen, therefore, how much the experience to date will be considered and built upon in the design of the compliance system of the emerging climate change regime.

The uncertainty over the future of the climate change regime also has immediate implications for the current compliance system. First, the requirement that parties make up missed emission reductions in the subsequent commitment period is an effective enforcement tool only if there are subsequent commitment periods. The closer we come to the end of the first commitment period before the post-2012 regime is finalized, the greater will be the temptation of parties at risk of missing their emission-reduction target to either reject a second commitment period altogether or to incorporate the expected consequence into their second commitment period targets. With every passing year of uncertainty, the risk of parties not taking the work of the compliance committee seriously increases. Much of the work of the EB is yet to come, and the uncertainty over the future of the climate regime is at risk of increasingly affecting its work.[17]

Regardless of the future of the Kyoto compliance system, much of its work is on issues that will continue to be important both for the climate change regime and for other multilateral environmental agreements. While it is impossible to make accurate predictions about the future of the climate change regime after 2012, it is nevertheless valuable to reflect on the experience with the Kyoto compliance system, whether for improvements to the Kyoto compliance system itself or for MEA compliance more generally.

8.2 The Facilitative Branch

Until the Kyoto compliance system was designed, facilitation had been the dominant approach to compliance in Multilateral Environmental Agreements (MEAs).[18] MEAs offer a rich experience with facilitation, though not in the context of the rigorous reporting and review requirements in Articles 5, 7, and 8 of the Kyoto Protocol.[19] One would expect the experience with facilitation under the Kyoto compliance system to offer new insights into reporting and review, as well as on the more general experiment with the combination of facilitation and enforcement.

---

[17] Already, Canada has withdrawn from the Kyoto Protocol and Japan has indicated that it is not going to take on a second commitment period target. Both parties are among those considered most likely to struggle to meet their first commitment period targets. The federal government in Canada, in fact, had previously declared in 2007 that it would not meet its target.

[18] See Jane Bulmer, "Compliance Regimes in Multilateral Environmental Agreements", in Jutta Brunnée, Meinhard Doelle and Lavanya Rajamani (eds.), *Promoting Compliance in an Evolving Climate Change Regime* (Cambridge: Cambridge University Press, 2012).

[19] See, for example, Montreal Protocol on Substances that Deplete the Ozone Layer, 16 September 1987, amended at London on 29 June 1990, amended at Copenhagen on 25 November 1992, amended at Vienna in 1995, amended at Montreal on 17 September 1997, and amended at Beijing on 3 December 1999, in force 1 January 1989, 1522 *United Nations Treaty Series* (1989), 3.

The FB was expected to play an important role in the Kyoto compliance system, both as an early-warning system for compliance matters which ultimately might be subject to enforcement, and to deal with the range of commitments not subject to the jurisdiction of the EB. Whether it lived up to expectations is considered in this section.

The only substantive matter referred to the FB to date has been a submission filed by South Africa in its capacity as chair of the G-77/China bloc. The submission was filed with respect to Austria, Bulgaria, Canada, France, Germany, Ireland, Italy, Latvia, Liechtenstein, Luxembourg, Poland, Portugal, Russia, Slovenia, and the Ukraine. The focus of the submission was to bring to the attention of the FB a number of instances of late filing of reports on demonstrable progress by Annex I countries toward meeting their emission-reduction targets. The letter submitted by South Africa reads in part as follows:

> South Africa, as Chairman of the Group of 77 and China, on behalf of the Group of 77 and China, is submitting a question of implementation to the Compliance Committee, for consideration by the Facilitative Branch. ... This question of implementation is raised against those Parties who have not provided their reports demonstrating progress, even after a period of nearly six months from the January 1 deadline.[20]

The submission requested the branch to investigate the alleged violations and to consider whether they were indicative of potential non-compliance with more substantive requirements, such as Article 3.1 of the Kyoto Protocol. The FB decided not to proceed against Latvia and Slovenia as both countries had submitted the required documentation by the time the branch met to consider the submission. This decision not to proceed was approved with two abstentions and one vote against.[21]

With respect to the other parties, the members of the branch could not agree on whether the submission in the form of a letter from South Africa on behalf of the Group of 77 and China properly brought the matter before the compliance committee. The dispute was over the requirement that questions of implementation be brought by a party or by an Expert Review Team. The branch was split on whether the submission by South Africa was properly filed by a party. As a result, the FB was not able to make a preliminary decision to proceed or not to proceed.

The branch failed to comply with the requirement to make a preliminary decision within 3 weeks of the referral of a question of implementation, and reported this failure to the compliance committee.[22] The Rules of Procedure approved by the

---

[20] Letter submitted by South Africa: CC 2006-1-1/FB, available at http://unfccc.int/kyoto_protocol/compliance/facilitative_branch/items/3786.php (last accessed on 8 April 2012).

[21] See Decision not to proceed against Slovenia CC-2006-14-2/Slovenia/FB and Decision not to proceed against Latvia CC-2006-8-3/Latvia/FB , available at http://unfccc.int/kyoto_protocol/compliance/facilitative_branch/items/3786.php (last accessed on 8 April 2012).

[22] See Report to the Compliance Committee on the Deliberations in the Facilitative Branch relating to the Submission entitled "Compliance with Article 3.1 of the Kyoto Protocol" (Party concerned: Canada), CC-2006-3-3/FB, available at: http://unfccc.int/kyoto_protocol/compliance/facilitative_branch/items/3786.php (last accessed on 8 April 2012).

COP/MOP in Nairobi in 2006 now clarify the process for making submissions of this kind. To date, no further referrals have been made to the FB, either by a party of by an ERT.[23]

It is noteworthy that the FB has not had any opportunity to facilitate compliance with emission-reduction targets of Annex I parties. Appling the FB process to Canada, for example, would have been an interesting test of facilitation for Annex I parties with respect to their emission-reduction targets. A possible trigger for the work of the FB with respect to Canada would have been the so-called demonstrable progress report or its national communications.[24] There might have been value in providing the FB with the opportunity to schedule a meeting or some form of consultation with a party that may be at risk of missing its target based on the demonstrable progress report filed.

In 2010, the FB branch took a modest step toward proactively facilitating compliance. The FB initiated contact with Monaco with respect to Monaco's delay in submitting its fifth national communication. In the letter, the FB offers facilitation and advice, and seeks a response from Monaco.[25] There has been some further communication between the FB and Monaco resulting from this initial letter, suggesting that Monaco has accepted the FB's role in this regard.[26]

8.3 The Enforcement Branch

The work of the enforcement branch of the Kyoto compliance system is of particular interest because it is the first time that an MEA has taken enforcement seriously. The EB has to date been confronted with seven questions of implementation related to a party's compliance with its Kyoto commitments. The cases involve Greece,

---

[23] The immediate concern raised by the South Africa submission was the split between Annex I and Non-Annex I parties on this issue. The broader concern is the difficulty of bringing matters before the FB. The fact that no party was willing to follow up the South Africa submission on its own is telling in this regard. It suggests a fear of reprisal by individual parties.

[24] Clare Breidenich and Daniel Bodansky, "Measurement, Reporting and Verification in a Post-2012 Climate Agreement" (2009 Pew Center on Global Climate Change), available at: http://www.pewclimate.org/docUploads/mrv-report.pdf (last accessed on 8 April 2012), at 15, where the authors discuss the difference in rigour of the reporting obligations for inventories and reporting on mitigation measures. The requirements for inventories are much more specific, making it much more likely that an ERT would trigger the compliance process for inventories than for mitigation measures including progress toward commitment-period targets. Clear standards for reporting on mitigation measures would be an essential foundation for more effective facilitation and enforcement of compliance with mitigation commitments.

[25] See report on decision to send letter to Monaco, available at: http://unfccc.int/kyoto_protocol/compliance/facilitative_branch/items/3786.php (last accessed on 8 April 2012).

[26] See 10th Meeting of the FB, 11–12 October 2011, "Provisions Related to Facilitation: Advice and Facilitation", available at: http://unfccc.int/kyoto_protocol/compliance/facilitative_branch/items/3786.php (last accessed on 8 April 2012).

Canada, Croatia, Bulgaria, Romania, Ukraine, and Lithuania. All cases to date have had to follow the expedited procedures in section X of the Compliance Procedures, set up to ensure the time-sensitive issue of eligibility to utilize the Kyoto mechanisms is dealt with in an expedited manner. Section X provides shorter timelines than the general procedures and establishes specific rules for the reinstatement of eligibility to participate in the mechanisms.[27]

The case against Greece is reviewed in detail, as it offered the first opportunity to observe the functioning of the EB. As such, it provides a good opportunity to illustrate the general process followed by the EB. The other six cases are drawn upon to highlight new issues they raise about the functioning of the EB. Significant changes to the process in these subsequent cases that signal an evolution of the process are also identified.[28]

8.3.1 *Proceedings Against Greece*

This case represents the first question of implementation brought before the EB. As noted above, a question of implementation can be brought before the compliance committee either by an ERT or by a party; the bureau determines whether it comes within the jurisdiction of the EB, the FB, or both. Once the EB receives a question of implementation from the bureau, it conducts a preliminary review of the issue raised and makes a determination on whether to proceed. If the EB decides to proceed, the party under investigation is informed of this decision. It then has the right to request a hearing and make written submissions. The party can also request under section VIII(6) of the compliance procedures that information be kept private until the conclusion of the proceedings. The EB will usually hear from the party, the ERT, any other party, as well as from any independent experts it feels are needed to resolve the issue raised. The EB can also request specific information from the party under investigation, and can consider submissions from non-parties. There are set timelines for the major steps in the process.

After the hearing, the EB makes a preliminary finding as to whether the party is in compliance. The party has an opportunity to comment on the preliminary finding. If it does not, the preliminary finding stands as the final decision of the EB. If the party submits comments on the preliminary finding, the EB issues a final decision in light of the comments filed. The EB has to give reasons for its decisions. A finding of non-compliance will result in a range of consequences depending on

[27] The expedited procedures can take a maximum of 17 weeks, whereas the general procedures can take up to 36 weeks.

[28] For a more detailed assessment of the first four cases before the EB, see Meinard Doelle, "Experience with the Kyoto Compliance System", in Jutta Brunnée, Meinhard Doelle and Lavanya Rajamani, *Promoting Compliance in an Evolving Climate Change Regime* (Cambridge: Cambridge University Press, 2012).

the nature of the violation. A key part of the process is the preparation of a compliance plan within 3 months of the determination of non-compliance, with regular updates thereafter on the implementation of the compliance plan. Key substantive requirements for the compliance plan are set out in Section XV (2) of the Compliance Procedures.

The case against Greece was the first opportunity to test this process. It resulted from the ERT's review of the initial report filed by Greece and from the ERT's in-country review of Greece's national system for the estimation of emissions and the preparation of information required under Article 7 of the Kyoto Protocol.[29] The ERT summed up the situation as follows:

> The ERT concludes from the information contained in the initial report and the additional information received during and after the in-country review that the national system of Greece does not fully comply with the guidelines for national systems under Article 5, paragraph 1 of the Kyoto Protocol (decision 19/CMP.1) and the guidelines for the preparation of the information required under Article 7 of the Kyoto Protocol (decision 15/CMP.1). In particular, the ERT concludes that the maintenance of the institutional and procedural arrangements; the arrangements for the technical competence of the staff; and the capacity for timely performance of Greece's national system is an unresolved problem, and therefore lists it as a question of implementation.[30]

The ERT report was received by the compliance committee on 31 December 2007. It was allocated by the bureau to the enforcement branch on 7 January 2008. On 22 January 2008, the EB decided unanimously, by way of an electronic system for taking decisions outside of a conventional meeting, to proceed with the case against Greece.[31] A number of steps followed in short order. Greece was informed of the decision to proceed. It requested a hearing and filed a written submission in February 2008.[32] The EB requested expert advice from members of the ERT and from independent experts. The request for expert advice included a list of specific questions to be addressed by the experts.

A hearing of the EB was held in March of that year, followed by a preliminary finding of non-compliance.[33] Greece filed further written submissions in response to the preliminary finding. At a further meeting of the EB in April, the preliminary finding was confirmed. No submissions were filed by non-parties. Once the EB

---

[29] Kyoto Protocol, supra, note 3, Arts. 5, 7.

[30] See Report of the Review of the Initial Report of Greece: CC-2007-1-1/Greece/EB, 8 January 2008, par. 244, available at: http://unfccc.int/kyoto_protocol/compliance/enforcement_branch/items/5455.php (last accessed on 8 April 2012). See also par. 5–10 of the ERT Report, including table 1.

[31] Decision on Preliminary Examination: CC-2007-1-2/Greece/EB, available at: http://unfccc.int/kyoto_protocol/compliance/enforcement_branch/items/5455.php (last accessed on 8 April 2012).

[32] Written Submission of Greece: CC-2007-1-5/Greece/EB, 26 February 2008, available at: http://unfccc.int/kyoto_protocol/compliance/enforcement_branch/items/5455.php (last accessed on 8 April 2012).

[33] See Preliminary Finding: CC-2007-1-6/Greece/EB, 6 March 2008, available at: http://unfccc.int/kyoto_protocol/compliance/enforcement_branch/items/5455.php (last accessed on 8 April 2012).

made its finding of non-compliance, the process shifted to the consequences of non-compliance and Greece's efforts to remedy the problems identified. Greece filed two successive compliance plans and made a formal request to the EB for eligibility to use the Kyoto mechanisms. This process took until November 2008, when the EB decided that Greece had come into compliance.

The key steps in this process are now considered in more detail.

8.3.1.1 First Hearing Regarding Greece

The third meeting of the EB served as the first hearing in the case against Greece.[34] It was held on 4 and 5 March 2008, in accordance with Rule 9 of the Rules of Procedure. Most of the meeting was held in public, but the deliberations on the preliminary finding were held in private. The public portions of the hearing are accessible by webcast. Greece did not seek to prevent disclosure of information to the public, nor did the EB.

Substantively, the focus of the question of implementation raised by the ERT was on the transition of the role of "technical consultant" from the National Observatory of Athens (NOA) to the National Technical University of Athens (NTUA). Greece appears to have relied heavily on the NOA in establishing its national system. While the ERT had no concerns with the work done by the NOA, the heavy reliance on an outside consultant raised concerns about the capacity of the government officials responsible for the national system. It also raised concerns about the decision to switch consultants from the NOA to the NTUA. Throughout the EB proceedings, there was disagreement over the actual extent of the responsibility of the technical consultant. At least some of the ERT members were of the view that the consultant had overall responsibility for Greece's national system and that government officials lacked the capacity to oversee the work of the consultant.[35] Greece took the position that the responsibility throughout rested with the responsible Ministry, not with either the old or the new technical consultant.

Knowledge transfer was a central concern for the ERT, both with respect to the transfer from the NOA to the NTUA and for possible future transfers of responsibility. A key problem appears to have been that the description of the organizational structure, and the role of the consultant in maintaining Greece's national system ignored the fact that the consultant's responsibility was to be transferred from NOA to NTUA. Greece's response appears to have been that it would ensure the transition would take place properly, but without providing the detail necessary to satisfy the ERT with respect to knowledge transfer.[36]

---

[34] The meeting was held on 4–5 March 2008 in Bonn. The webcast is available at: http://unfccc.int/kyoto_protocol/compliance/enforcement_branch/items/5455.php (last accessed on 8 April 2012).

[35] See webcast of 4–5 March 2008 meetings in Bonn. The webcast is available at: http://unfccc.int/kyoto_protocol/compliance/enforcement_branch/items/5455.php (last accessed on 8 April 2012).

[36] The capacity of the new responsible entity was an issue at least in principle, in that the ERT was not able to verify its capacity during the in-country review.

The new system was first explained by Greece in its written submission to the EB. The experts invited to the March 2008 meeting of the EB, some of whom were members of the original ERT, seemed pleased with the new system as described by Greece, but felt that the capacity of the new Greek team (consisting of new Ministerial staff and the NTUA) could not be assessed based on the submission. The invited experts, including the ERT members, felt that a further in-country review was required to confirm the capacity of the new team.

The timing of the transition of responsibility had not enabled the ERT members to meet with the NTUA who had taken over responsibility for the maintenance of the national GHG inventory system after the ERT's in country visit. Therefore, the ERT members felt that they could not conclude that the maintenance of Greece's national system was in good hands with the NTUA. The concern appeared in part to be a result of discussions with the original technical consultant involved, the NOA, during the in-country review. NOA staff had indicated that they had not been engaged in any knowledge transfer to NTUA.

The contract between the Ministry for the Environment, Physical Planning and Public Works of the Government of Greece and NOA ended in April 2007. An agreement between the Ministry and the NTUA to take over as technical consultant was not reached until December 2007. In the interim, the Ministry had sole responsibility for the maintenance of the national system. During the course of the EB hearings, Greece indicated that it had increased the capacity of the Ministry by hiring six new staff, that the new technical consultant would play a less prominent role than the previous consultant, and that a workshop would be held to ensure knowledge transfer from NOA to NTUA.

The key issue in the end seemed to be whether another full in-country review or some other process (such as a modified in-country review, a centralized review, or a desk review) was needed to ensure that there was now capacity to manage the inventory going forward. In this regard, Greece pointed out that if the transition had happened after a successful initial in-country review, the transition would have triggered a desk review, not another in-country review. This raised the question for the EB whether in light of the ERT's findings (including the finding that the NOA process had been adequate and that the problem really had to do with the transition), there was still a need for an in-country review of the national GHG inventory system.

In-session documents, including working drafts of reports and decisions were not available from the UNFCCC website, and requests for these documents were denied, making it difficult at times to follow the work of the EB in detail through the webcasts.[37] Electronic communications among members of the EB were also not available, even though the EB did conduct some of its formal business electronically to reduce travel time and cost.[38] No observers registered to attend the March 2008 meeting of the EB.[39]

---

[37] E-mail communication requesting these documents is on file with the author.

[38] Such as the preliminary decision to proceed made on 22 January 2008.

[39] The author was the first registered observer at the April 2008 meeting of the EB.

The process used by the EB is not an adversarial process with both sides represented and the EB playing the role of judge. Neither the UNFCCC secretariat not the ERT is playing the role of prosecutor. This suggests that members of the EB need to take a pro-active role in bringing out and exploring critical issues. It is clear from the webcast that some members of the EB were more comfortable with this role than others.

8.3.1.2 Preliminary Finding

After the public hearing, the EB went into a private session for its deliberations. The result was a preliminary finding of non-compliance. Reasons for the decision are somewhat limited. The following are the key provisions of the preliminary finding:

16. The information submitted and presented has not been sufficient for the enforcement branch to conclude that the question of implementation has now been fully resolved. Additional information is required that specifically addresses whether and how the national system is maintained through transitions. The enforcement branch agrees with the expert advice provided that a further in-country review of Greece's new national system, in conjunction with a review of an annual inventory report generated by this national system, is required for the enforcement branch to assess present compliance with the guidelines.
17. The enforcement branch determines that Greece is not in compliance with the guidelines for national systems under Article 5, paragraph 1, of the Kyoto Protocol (decision 19/CMP.1) and the guidelines for the preparation of the information required under Article 7 of the Kyoto Protocol (decision 15/CMP.1). Hence, Greece does not yet meet the eligibility requirement under Articles 6, 12 and 17 of the Kyoto Protocol to have in place a national system in accordance with Article 5, paragraph 1, of the Kyoto Protocol and the requirements in the guidelines decided thereunder.
18. In accordance with section XV, the enforcement branch applies the following consequences:

    (a) Greece is declared to be in non-compliance.
    (b) Greece shall develop a plan referred to in paragraph 1 of section XV and submit it within 3 months to the enforcement branch in accordance with paragraph 2 of section XV. The plan should demonstrate measures to ensure the maintenance of the national system through transitions and include appropriate administrative arrangements to support an in-country review by the expert review team of the new national system of Greece, coordinated by the secretariat in conjunction with a review of an annual inventory report generated by this national system.

(c) Greece is not eligible to participate in the mechanisms under Articles 6, 12 and 17 of the Protocol pending the resolution of the question of implementation.

19. These findings and consequences take effect upon confirmation by a final decision of the enforcement branch.[40]

8.3.1.3 Written Submissions

On 8 April 2008, Greece filed a written submission in response to the preliminary finding of the EB.[41] The main point made in the submission is that regardless of the difficulties at the time of the ERT review, the transition in Greece was complete as of the date of the 8 April submission. According to the submission, the Ministry had improved its capacity, the new technical consultant had been hired, and the workshop between the NOA and NTUA had been held. Greece stated, moreover, that it had submitted its new inventory. Greece took the position that the quality of the new inventory should answer any question about its national system and that in these circumstances it would be inappropriate to hold up its access to the Kyoto mechanisms for the purpose of conducting an in-country review.

The submission filed by Greece also raised a question about the consistency in ERTs' approaches to referrals to the EB. The submission made the point that many of the issues raised regarding Greece had been raised by other ERTs in other initial reviews conducted for other parties without raising questions of implementation. Greece argued that as a matter of consistency, therefore, these issues should not delay Greece's eligibility to use the mechanisms.

8.3.1.4 Further Hearing

The main purpose of the second hearing on 16 and 17 April was to review the preliminary finding in light of the comments from the party.[42] The EB considered whether Greece's submissions warranted any change to the preliminary decision or whether it should be adopted as final. The Chair clarified at the outset that Greece was not yet required to comply with the terms of the preliminary decision. Specifically, Greece was not yet required to submit a compliance action plan on how Greece would bring its national system into compliance. Greece was required to act only if the preliminary finding of non-compliance were affirmed through a final decision.

At the April 16 hearing, the EB went through Greece's April 2008 submission in detail to consider whether the submission warranted a change to the preliminary finding.

---

[40] See Preliminary Finding: CC-2007-1-6/Greece/EB, available at: http://unfccc.int/kyoto_protocol/compliance/enforcement_branch/items/5455.php (last accessed on 8 April 2012).

[41] Further Written Submission of Greece: CC-2007-1-7/Greece/EB, available at: http://unfccc.int/kyoto_protocol/compliance/enforcement_branch/items/5455.php (last accessed on 8 April 2012).

[42] The meeting was held on 16–17 April 2008 in Bonn. The webcast is available at: http://unfccc.int/kyoto_protocol/compliance/enforcement_branch/items/3785.php (last accessed on 8 April 2012).

Concerns raised by EB members focused on the fact that there was no information on how Greece would better prepare for the eventuality of another transition in the future, and that experts at the April hearing continued to take the position that some form of in-country review would be needed to confirm that the new team (consisting of the new Ministry staff and NTUA) had the capacity and had effected the transfer of the relevant knowledge to properly maintain the national system.

8.3.1.5 Final Decision

The final decision of the EB, released on 17 April 2008, confirms the preliminary finding of non-compliance as well as the consequences identified in the preliminary finding.[43] As of the date of the final decision, Greece was declared to be in non-compliance, was required to submit a compliance plan within 3 months, and was declared ineligible to participate in the mechanisms.

The decision was not unanimous. Unfortunately, there are no reasons given for the one dissenting vote. At the October 2008 meeting of the compliance committee, the plenary clarified that, in the future, members of either branch who cast a dissenting vote will be able to provide an explanation in the report of the meeting, but that that explanation will not be part of the decision. It remains to be seen whether members will avail themselves of this opportunity in a meaningful way.

8.3.1.6 Greece's Compliance Plan

In accordance with the 17 April decision of the EB, Greece filed its compliance plan on 16 July 2008.[44] The plan contemplated an in-country review to take place in September 2008, and otherwise indicated that Greece's current system was adequate to address the concerns expressed by the EB in the 17 April decision.

At the meeting of the EB on 6 and 7 October 2008, Greece's compliance plan was reviewed and found to be inadequate in addressing the issues raised in the April decision and the requirements in Section IV(2) of the Compliance Procedures. In particular, the branch noted that the document contained no plan on how to improve future transitions of responsibility for components of its national system. The report was also found to be inadequate in its form, in that it did not specifically respond to each of the issues raised in the April decision. Furthermore, the EB clearly did not accept Greece's position that everything was in order and that the in-country review was the only event that stood in the way of having its eligibility reinstated.

---

[43] See Final Decision: CC-2007-1-8/Greece/EB, 17 April 2008, available at: http://unfccc.int/kyoto_protocol/compliance/enforcement_branch/items/5455.php (last accessed on 8 April 2012).

[44] See Plan Pursuant to Final Decision CC-2007-1-9/Greece/EB, 16 July 2008, available at: http://unfccc.int/kyoto_protocol/compliance/enforcement_branch/items/5455.php (last accessed on 8 April 2012).

The EB confirmed that it could not make a final decision about Greece's state of compliance without access to the written report from the ERT on its follow-up in-country review of Greece's national system in September. There was some discussion at the EB's October 2008 meeting about the time-delay in reviewing the compliance plan submitted by Greece, and there was general agreement that in the future the EB should endeavour to respond within 4 weeks. The EB noted that the Rules of Procedure with respect to the review of compliance plans were inadequate, and proposed amendments.[45]

8.3.1.7 Final Resolution

Greece filed a revised compliance plan on 27 October 2008. The matter was resolved on 13 November 2008, when the EB, on a request by Greece, decided to grant it eligibility to participate in the mechanisms. The decision is based on the written report of the ERT following its in-country review in September 2008 and the revised compliance plan. The ERT report concluded that Greece had made considerable improvements in the implementation of its national system, and that it had addressed the EB's and ERT's concerns about future transitions in responsibility for maintaining its national system. The revised compliance plan was found to be in compliance with the formal requirements set out in the EB's 17 April decision. On this basis, Greece was found to be in compliance and it was declared eligible to use the Kyoto mechanisms.

8.3.2 *Subsequent Proceedings Before the EB*

There have been six further questions of implementation brought before the EB, one against Canada and five against eastern European countries. All seven cases have followed the same basic process, though subsequent cases have benefitted from the rules of procedures developed in 2007 and more generally from the experience gained by the enforcement branch over time. Key issues that arose out of the subsequent six cases are briefly outlined in this section of the chapter.

8.3.2.1 Canada

At the heart of the question of implementation before the EB with respect to Canada was a delay in establishing Canada's national registry. A national registry is a computerized system used to track holdings of greenhouse gas credits, and is a

[45] See Compliance Committee 2008 Annual Report, UN Doc. FCCC/KP/CMP/2008/5, Annex I, available at: http://unfccc.int/kyoto_protocol/compliance/plenary/items/3788.php (last accessed on 8 April 2012)

requirement for all Annex I parties. The question of implementation did not extend to any actual accounting of emissions. Canada's declared intention not to meet its emission reduction target by the end of 2012 was not before the EB.[46]

Canada's approach in its written and oral submissions was not to dispute the question of implementation raised, but to point out that the problem had been addressed and that the registry was now in place. Canada took the position that it was in compliance at the time of the hearing, and that there was therefore no point in the EB proceeding further with the question of implementation raised. The EB agreed not to proceed, but to Canada's displeasure made a point of noting Canada's past non-compliance.

The key new issue raised by the proceedings against Canada was whether it is appropriate for the EB to make reference to past non-compliance of a party or whether, in the instant case, it should have simply found Canada to be in compliance because the registry was established by the time the hearings were held. It seems clear that if the EB is to serve its role of motivating parties to comply by bringing instances of non-compliance to the attention of the public, being able to bring attention to past non-compliance may be a valuable tool.

8.3.2.2 Croatia

In this case the issues raised by the ERT centered on an attempt by Croatia to add 3.5 megatonnes of CO_2 eq. to its assigned amount. The ERT concluded that the 3.5 Mt enlargement was not in accordance with modalities established under decision 13/CMP.1 and raised this as an issue of implementation. Croatia claimed this amount based on a recognition of Croatia's special circumstances by the COP prior to Croatia joining the Kyoto Protocol.

The EB concluded that the flexibility provided for in Articles 4.6 of the UNFCCC and 3.5 of the Protocol does not extend to additions to the assigned amount. Furthermore, the recognition of Croatia's special circumstances in 7/CP.12 was made by the UNFCCC's COP, and not by the Protocol's COP/MOP. Thus the EB concluded that there was no basis on which Croatia could claim special treatment for the determination of its assigned amount under the rules of the Protocol.[47]

The EB essentially decided that any recognition of special circumstances under the UNFCCC had to be confirmed by the COP/MOP to be applicable to Croatia's Kyoto obligations. The EB also concluded that the flexibility for EITs under the Kyoto Protocol is limited to the choice of base year. Decision 7/CP.12 is based on the Convention, which allows for more flexibility with respect to EITs. Essentially, the EB acknowledged Croatia's special circumstances, but concluded that it was up to the COP/MOP to consider these circumstances and take appropriate action in light of the more limited flexibility under the Kyoto Protocol.

---

[46] See Oberthür, "Holding Countries to Account", supra, note 2, at 154.

[47] See Final Decision: CC-2009-1-8/Croatia/EB, 26 November 2009, available at: http://unfccc.int/kyoto_protocol/compliance/enforcement_branch/items/5456.php (last accessed on 8 April 2012).

The decision of the EB was appealed by Croatia, representing the first time the compliance mechanism's appeal provisions have been used. Croatia's Notice of Appeal includes the following grounds: violation of Article 31, paragraphs 1, 2, and 3(b), as well as Article 32 of the Vienna Convention on the Law of Treaties; improper application of Article 3, paragraph 5, of the Kyoto Protocol; violation of COP and COP/MOP decisions and provisions of the Kyoto Protocol; violation of the equal-treatment principle; and violation of the procedures and mechanisms relating to compliance, in particular: indication of information relevant to the decision; the right to respond; and the independence, impartiality, and conflict-of-interest principles.[48] Croatia withdrew its appeal in 2011, before a final decision from the COP/MOP.

8.3.2.3 Bulgaria

The case against Bulgaria was triggered as a result of a question of implementation raised in the 2009 ERT report on Bulgaria.[49] In this case, the ERT concluded that Bulgaria's national system did not operate in accordance with the Guidelines for National Systems for the Estimation of Emissions by Sources and Removals by sinks under Article 5, paragraph 1, of the Kyoto Protocol.[50] The ERT was also not satisfied with institutional arrangements and arrangements for technical competence of staff within the national system involved in the inventory-development process.[51]

The problems identified were not new; nevertheless, in-country and desk reviews carried out in previous years had not resulted in issues of implementation with respect to these ongoing problems. The ERTs had, of course, fulfilled their role in facilitating compliance by making recommendations for improvements to Bulgaria's national system, but had chosen not to engage the compliance committee until 2010.

The matter was referred to the EB, which followed the same general process established in previous cases. Bulgaria filed a detailed submission prior to the EB's preliminary finding of non-compliance. After a hearing on 10 May 2010, the EB issued its preliminary finding, essentially confirming the findings of the ERT with respect to Bulgaria's inventories of emissions and sinks, particularly with respect to institutional arrangements and staff. The conclusions of the EB were confirmed in its final decision rendered at the conclusion of the meeting of the EB on 28 June 2010. The focus of the decision was the requirement of a compliance plan, regular updates, and a further in-country review. Compliance with the decision of the EB took Bulgaria until February, 2011, when its eligibility to use the Kyoto mechanisms was re-instated.

---

[48] See Notice of Appeal filed by Croatia, available online at http://unfccc.int/resource/docs/2010/cmp6/eng/02.pdf (last accessed on 8 April 2012).

[49] See Report of the Review of the Initial Report of Bulgaria, UN Doc. FCCC/ARR/2009/BGR, available at: http://unfccc.int/kyoto_protocol/compliance/questions_of_implementation/items/5538.php (last accessed on 8 April 2012).

[50] Decision 19/CMP.1, Guidelines for national systems under Article 5, paragraph 1, of the Kyoto Protocol, UN Doc. FCCC/KP/CMP/2005/8/Add.3, 30 March 2006.

[51] Bulgaria ERT Report, supra, note 49, at para. 194.

The extent of Bulgaria's difficulties and the length of time it has already taken to try to resolve them are perhaps the key aspects of this case. The fact that this matter had not previously come before either the FB or the EB must be considered a shortcoming of the process, even if ERTs have been diligent in working with Bulgaria to resolve these issues. This is a case where both branches could have been engaged. The EB in its findings limited itself more or less to a finding of non-compliance and to identifying a process of determining when Bulgaria has come into compliance. While this is entirely appropriate for the EB, it would seem that Bulgaria is in need of more detailed advice on the steps it needs to take. The EB could have referred the matter to the FB to fill this gap.

This case also serves to illustrate the problem of triggering primarily through ERT referral. It seems clear that ERTs reviewing Bulgaria's inventory system have had concerns and have identified problems for some time, but decided, until 2010, not to raise them as questions of implementation. The first in-country review likely should have resulted in referral to the FB or perhaps even the EB. If the FB had the responsibility to review each ERT report and the power to initiate proceedings on its own, it would seem likely that this would have led to a pro-active approach to this matter years earlier. The end result is that this ongoing problem first came before the EB with only 2 years left before the end of the commitment period.

8.3.2.4 Romania, Ukraine and Lithuania

The three most recent cases involving Romania, Ukraine and Lithuania arise out of the ERT reviews of the parties' 2010 annual submission. As a result of these reviews, the ERT raised questions of implementation regarding the national system of each of these parties. The process followed in each case was similar to the one used in previous cases. Romania, Ukraine and Lithuania were each found to be in non-compliance and were asked to submit a compliance plan.

At the time of writing, only Ukraine had submitted its compliance plan and had completed the implementation of the compliance plan to the satisfaction of the EB. As a result, Ukraine had its eligibility re-instated in March, 2012. With respect to Romania, the EB had accepted the compliance plan submitted by Romania, and was awaiting a second update on its implementation. Lithuania had not yet filed its compliance plan.

8.4 Observations on the Experience to Date

Overall, the Kyoto compliance system has performed remarkably well given the circumstances, particularly with respect to the requirements for initial eligibility and the establishment of national systems and inventories by Annex I parties. Considerable facilitation appears to have occurred at the ERT level with respect to the requirements of Articles 5, 7, and 8. It seems that the threat of formal proceedings before the compliance committee has been an effective motivator for parties to cooperate with ERTs.

A few key shortcomings of the Kyoto compliance system can nevertheless be identified based on the experience to date. They are discussed below. The most obvious is that the triggers for proceedings before the two branches have proven to be inadequate. The adequacy of the "consequences" has also been brought in doubt, largely by Canada's declared intention not to work toward its emission-reduction target, but also, as explained below, by the inactivity of the FB. Transparency is a third key area. The role of key actors in the compliance system is also briefly discussed.[52]

8.4.1 Triggering

Triggering is perhaps the most obvious shortcoming of the current system. The compliance system allows self-triggering by parties, party to party triggering, and triggering by ERTs. The self-trigger has not been used. The party-to-party trigger was attempted once and failed. The limitation of self-triggering and party-to-party triggering was, of course, recognized in the design of the system. Triggering by ERTs was offered as the solution. This solution will likely prove adequate with respect to the emission-reduction obligations at the end of the first commitment period. It is less clear that it has been adequate in allowing the compliance committee to act early to encourage compliance in a pro-active, preventative manner.

The ERT process has not been an adequate triggering process to date. In particular, the triggering of proceedings before the FB has been practically non-existent, in spite of clear evidence of numerous concerns and violations under the jurisdiction of the FB. The most notable example is the inability of either branch of the compliance system to take any action in response to Canada's declared intention as early as 2007 not to meet its emission-reduction target. The stakes for the compliance system were particularly high with respect to Canada, as its position struck at the core of the Kyoto Protocol. Yet, the compliance committee could not act, and had to rely on the threat of the ultimate consequences to be applied in 2015 as the only tool within the system to encourage Canada to change its position.[53]

[52] For a more detailed discussion on the lessons learned from the Kyoto compliance system, see Meinhard Doelle, Jutta Brunnee and Lavanya Rajamani, "Conclusion: Promoting Compliance in An Evolving Climate Regime", in Jutta Brunnee, Meinhard Doelle and Lavanya Rajamani, *Promoting Compliance in an Evolving Climate Regime* (Cambridge: Cambridge University Press, 2012).

[53] In light of Canada's recent decision to withdraw from the Kyoto Protocol altogether, one might be inclined to take the view that even with an appropriate trigger, there was nothing the compliance committee could have done to convince an unwilling party to change its position. In the end, however, this is an unanswered question. Would proceedings before the compliance committee have an impact on the position of the Canadian government? Would it affect its relationship to other parties? Would it affect the credibility of the government domestically? Would it affect the domestic debate on this issue?

8.4.2 Consequences

On the enforcement side, the main question regarding consequences is whether the experience to date warrants a reconsideration of the adequacy of the ultimate consequences of non-compliance. Leaving aside the immediate problem of the uncertain future of the Kyoto Protocol and the fact that targets for a second commitment period are uncertain as a result, what could be done to increase the likelihood that parties acting out of short-term self-interest are nevertheless motivated to comply with their obligations?

One step forward would be a safeguard against using the compliance process to continuously borrow from future commitment periods. Parties could be prevented from borrowing in two sequential commitment periods, and instead be required to pay a financial penalty. The 1.3 rate could be increased in case of repeated failure to meet emission-reduction targets. The compliance action plan could be made subject to more rigorous international review and approval for repeat offenders. An international compliance fund could be reconsidered as a means of preventing repeated borrowing, particularly in light of the need to finance mitigation and adaptation in developing countries.[54] Such a compliance fund could, for example, require payment for each ton of carbon missed at a rate equal to or higher than the cost of achieving the reductions during the commitment period, and make the funds available to non-Annex I parties for mitigation or adaptation purposes.

On the facilitative side, the main issue regarding consequences is whether the FB should have access to concrete tools and resources to assist parties in their effort to meet commitments, particularly with respect to tracking of emissions, sinks, credits, and reporting. The FB should be able to offer help in the form of funding and expertise, certainly in the context of EITs. In situations where facilitation extends to developing countries, this becomes even more important. Providing the FB with such tools may encourage less developed parties that experience compliance difficulties to self-report to the branch.

8.4.3 Transparency of the Process

When the compliance system was negotiated, there were legitimate concerns that transparency had been weakened in the late stages of the negotiations with the inclusion of section VIII(6), which allows information to be kept from the public until the conclusion of the proceedings on request by the party being investigated at the discretion of the EB.[55] It is encouraging that this mechanism has not been used, and

[54] For a discussion of the consideration of a compliance fund in the negotiations of the Kyoto compliance system, see Doelle, *From Hot Air to Action*, supra, note 1, at 60.

[55] See Doelle, *From Hot Air to Action*, supra, note 1, at 136.

that the committee and its two branches have made considerable efforts toward transparency. Examples include webcasting proceedings other than deliberations on decisions, a straightforward mechanism for observers to attend public meetings, and full access to all key documents on the UNFCCC website.[56] Nevertheless, a few transparency issues have arisen from the experience to date.

One limitation of the current process is that public proceedings frequently make reference to working documents that are not publicly accessible, making it difficult to follow the discussions taking place. In order for the webcasts to truly create transparency, working documents that are the subject of discussion should be provided, unless there is an overriding reason why they cannot be made available to the public.

A second issue relates to the increasing use of electronic means of communication, in place of meetings. While this practice should be encouraged, exchanges by electronic means that otherwise would be public should be made publicly available. In essence, e-mail exchanges should be treated like in-person meetings—they should be public unless there is a reason to keep them confidential. Currently, the form instead of the substance of communication dictates whether information is accessible.

A third issue relates to the level of detail offered in annual reports and decisions of the committee and its branches. The EB has gradually provided more detail in its decisions, and this trend should be encouraged and continued. More detailed reasons can help fill in some of the gaps left by the inaccessibility of working documents and e-mails.

8.4.4 Roles of Key Actors

The ERT process generally appears to be working well. It is, however, not consistently bringing issues of implementation before the compliance committee. This has been a concern from the time of the first case against Greece. The case against Bulgaria would seem to reinforce the point. Consistency is clearly an issue for the ERT process. Whether the review by ERTs, in particular through in-country reviews, is sufficiently detailed and frequent for the credibility and integrity of the reporting system is unclear based on the experience to date. It may be worth considering complementary ways to review and verify emissions and credits, such as through direct engagement of civil society in reporting methodological issues.[57]

[56] Surprisingly, to date no submission has been made by civil society, and there have only been very few registered observers.

[57] For example, there could be a formal process through which civil society could be encouraged to register to review and publically comment on ERT reports. These comments could then be considered by the appropriate branch, and could potentially even feed into a branch-based triggering process.

As a group, the members of the compliance committee appear to have served the process reasonably well. There are few indications of voting along party lines.[58] The expertise of members appears to vary, resulting in some members being very engaged while others seeming to limit their involvement to a narrow range of issues. It is noteworthy that some members appear to have technical expertise, whereas others seem to have legal expertise. To deal with technical issues, the EB has made extensive use of outside experts, being careful to draw on ERT members and independent experts. Legal issues, however, have not been resolved through the use of outside experts. This may need to be rethought if legal disagreements continue to arise within the EB.[59] One solution would be to provide the compliance committee with access to independent legal advice.

The COP/MOP has to date been relatively unengaged with the work of the compliance committee. This may be partly due to its focus on the post-2012 negotiations. As a general rule, this may be a good thing, as it will limit political interference in the work of the committee. In defining an appropriate role for the COP/MOP, timing, the number of parties, and the political nature of the COP/MOP all need to be taken into consideration. Its role as the ultimate overseer of the process without much direct involvement generally seems appropriate.

The role of secretariat has been the subject of some discussion within the EB. The secretariat has been resistant to requests from members of the EB to provide preliminary analysis of cases that come before it. The impartiality of the secretariat and the independence of the compliance committee appear to be the main reasons. On balance, it would seem that the secretariat's approach has generally been appropriate. Limits in the capacity, resources, and expertise of members of the branches should be addressed directly, rather than blurring the line between the secretariat and the members of the compliance committee. However, a review of ERT reports for consistency by the secretariat would seem appropriate.

8.5 Conclusion

Much of the focus of the work of the compliance committee to date has been on developing and testing its basic rules of procedure. The seven cases before the EB, and the case brought by South Africa on behalf of the G-77/China before the FB, stand out as the main sources of experience with the Kyoto compliance system to date. These are early days for the Kyoto compliance system, and one would be well

[58] The South Africa submission to the FB on behalf of the G-77, and the one abstention on the final decision in the Croatia case are perhaps worth noting here. One issue to watch in this regard are the voting rules, which can serve to encourage block voting along Annex I/non-Annex I lines.

[59] One prominent example was a discussion of the Plenary in 2007 on an issue related to the timing of early eligibility. The webcast of the October 2007 annual meeting is available at http://unfccc.int/kyoto_protocol/compliance/plenary/items/3788.php (last accessed on 8 April 2012).

advised not draw firm conclusions about the effectiveness of the compliance system based on this limited experience. Nevertheless, it is clear that the EB is off to a promising start. At the same time, the experience does suggest that the compliance system is underutilized. A number of issues, ranging from delays in reporting to methodological issues and Canada's decision to abandon its emission-reduction obligation, have either not come before the branches or have not done so in a timely manner.

Overall, the Kyoto experiment to combine facilitation and enforcement shows considerable promise. The main task ahead is to encourage more and better facilitation, and to adjust the consequences as needed. The good news is that the experience to date suggests that enforcement can and does encourage constructive facilitation, even if the facilitation to date has been carried out by ERTs rather than the FB. On the enforcement side, the process seems to be reasonably effective, efficient, and fair. There are still details to be worked out, but the current system offers a strong basis to work from.

Part III
International Climate Law – Cross-Cutting Issues

Chapter 9
The New Framework for Climate Finance Under the United Nations Framework Convention on Climate Change: A Breakthrough or an Empty Promise?

Yulia Yamineva and Kati Kulovesi

Abstract In this chapter, we first show that the framework for climate finance under the United Nations Framework Convention on Climate Change (UNFCCC) has been controversial, fragmented and insufficient to meet the growing financing needs of developing countries for adaptation and mitigation. We then describe and analyze the reformed framework for climate finance under the UNFCCC. We argue that the establishment of the Green Climate Fund constitutes an important milestone and progress has also been made in other respects. However, long-standing divides and mistrust between developed and developing countries have shaped the negotiations and continue to be reflected in their outcomes (and non-outcomes). This, together with the lack of clarity over long-term sources of finance, casts shadows over the future effectiveness of the new framework.

9.1 Introduction

Climate finance has rapidly evolved into a critical area of international climate policy and law. One of the reasons is that it cuts across other key elements of international climate change cooperation, namely adaptation, mitigation and technology.

Yulia Yamineva holds a Ph.D. in International Studies from the University of Cambridge. She is currently a postdoctoral researcher at the Department of Law, University of Eastern Finland. She followed the UNFCCC negotiations in 2007–2009 for the Earth Negotiations Bulletin and was a consultant on climate finance based at the UNFCCC Secretariat in 2010–2011. Kati Kulovesi holds a Ph.D. from the London School of Economics and Political Science. She is Adjunct Professor in Climate Law at the Department of Law, University of Eastern Finland.

Y. Yamineva (✉) • K. Kulovesi
Department of Law, University of Eastern Finland, P.O. Box 111, 80101 Joensuu, Finland
e-mail: yulia.yamineva@uef.fi; kati.kulovesi@uef.

E.J. Hollo et al. (eds.), *Climate Change and the Law*, Ius Gentium: Comparative Perspectives on Law and Justice 21, DOI 10.1007/978-94-007-5440-9_9,
© Springer Science+Business Media Dordrecht 2013

To avoid dangerous anthropogenic climate change, considerable new investments are needed in the coming decades to turn the course of the global economy and reduce developing country emissions below business-as-usual growth projections. Developing countries will also need funding to adapt to the unavoidable impacts of climate change. This is particularly true for the most vulnerable countries, such as small island developing States (SIDS), least developed countries (LDCs) and Africa.

Under the United Nations Framework Convention on Climate Change (UNFCCC),[1] finance for developing countries is provided through the Convention's financial mechanism. This constitutes, however, just one of several sources of climate finance, which include a range of multilateral and bilateral sources, as well as the private sector. Estimates show that climate finance through the UNFCCC has been modest in comparison to bilateral funding and private sector investments; it is estimated at less than US$0.3 billion annually for climate change mitigation.[2] The Clean Development Mechanism (CDM) under the Kyoto Protocol[3] plays a special role, providing an additional US$3–10 billion annually in the form of payments for carbon credits.[4] Bilateral assistance for climate-related purposes is estimated at US$5.8 billion a year, while climate funds outside the Convention provide around US$2 billion, and other multilateral assistance is estimated at about US$3 billion a year.[5] Key multilateral sources for climate funding outside the UNFCCC include the Climate Investment Funds, established in 2008, and UN agencies and other international bodies. Private sector finance in the form of investments is estimated at US$35 billion a year, most of which goes for renewable energy.[6] As for adaptation, estimates of current funding range from US$1 to 4 billions per year, with funding through the specialized adaptation funds constituting less than 10%.[7]

It is clear that funding currently available under the UNFCCC and the Kyoto Protocol is insufficient compared to the investment and financial flows needed to address climate change.[8] Additional external funding for climate change mitigation and adaptation will be particularly important for developing countries in sectors that depend on government investment and financial flows.[9] Reform of the Convention's financial mechanism, mobilization of new resources and strengthening

[1] United Nations Framework Convention on Climate Change, New York, 9 May 1992, in force 21 March 1994, 31 *International Legal Materials* (1992), 849. (UNFCCC).

[2] Susanna Olbrisch et al., "Estimates of Incremental Investment for and Cost of Mitigation Measures in Developing Countries", 11 *Climate Policy* (2011), 970, at 981, Table 6.

[3] Kyoto Protocol to the United Nations Framework Convention on Climate Change, Kyoto, 10 December 1997, in force 16 February 2005, 37 *International Legal Materials* (1998), 22.

[4] Olbrisch et al., "Estimates of Incremental Investment for and Cost of Mitigation Measures in Developing Countries", supra, note 2, at 981.

[5] Ibid.

[6] Ibid.

[7] Joel Smith et al., "Development and Climate Change Adaptation Funding: Coordination and Integration", 11 *Climate Policy* (2011), 987, at 990–992.

[8] UNFCCC, Investment and Financial Flows to Address Climate Change (October 2007), para. 5.

[9] Ibid., para. 10.

the governance structures for climate finance have therefore been key themes in the ongoing negotiations on strengthening the UNFCCC regime, launched by the thirteenth session of the Conference of the Parties (COP) in Bali in 2007.

While there is general agreement among UNFCCC Parties on the need to increase financial flows to address climate change, considerable differences remain concerning many of the critical details. Some of the most divisive issues have included the respective roles of the private sector and public funding from developed countries. Climate finance negotiations have tended to surface long-standing divides between developed and developing countries, including a mutual lack of trust[10] and dissatisfaction with Annex I and non-Annex I countries' respective levels of commitment under the UNFCCC.[11] The basic dynamic has been that developing countries highlight developed countries' historical responsibility for climate change and their better capacity to act. Developed countries in turn, have resisted detailed funding commitments, especially those requiring public funding from their governments.

The negotiations have been further complicated by the fact that in the two decades following the adoption of the UNFCCC, several developing countries have experienced rapid economic growth leading to a significant increase in their greenhouse gas emissions. The world is also changing in terms of developing countries' economic and financial capacities. Meanwhile, many developed countries are struggling to cope with diminishing industrial production, aging populations and unsustainable amounts of sovereign debt. While it has been argued that the current financial crisis should not be allowed to interfere with much-needed reforms to the financial mechanism of the UNFCCC,[12] it will undoubtedly affect discussions concerning the scale of public finance by developed countries. At the same time, the world's poorest countries remain desperately vulnerable with their populations, economies and infrastructure increasingly exposed to the adverse impacts of climate change – a problem they have not caused and are unable to prevent. The current system for climate finance is inadequate to meet even the most urgent needs of these countries. These broader issues should be kept in mind when focusing on the legal and governance framework for climate finance under the UNFCCC.

In this chapter, we will start by describing the basic legal and institutional framework for climate finance under the UNFCCC prior to the recent reforms. We argue that the framework used to be fragmented, controversial and insufficient to meet the growing funding needs of developing countries to address climate change. We then proceed to describe progress made during the long-term negotiations, highlighting agreements at COP 15 in Copenhagen, COP 16 in Cancun and COP 17 in Durban. Notably, the decision by COP 16 in 2010 to establish the Green Climate Fund

---

[10] Richard B. Stewart, Benedict Kingsbury and Bryce Rudyk, "Climate Finance for Limiting Emissions and Promoting Green Development", in Richard B. Stewart, Benedict Kingsbury and Bryce Rudyk (eds), *Climate Finance: Regulatory and Funding Strategies for Climate Change and Global Development* (New York and London: New York University Press, 2009), 3, at 15.

[11] See similarly Luis Gomez-Echeverri and Benito Müller, "The Financial Mechanism of the UNFCCC: A Brief History", ECBI Policy Brief, April 2009, at 1.

[12] Ibid., at 4.

and the commitment by developed countries at COP 15 in 2009 to collectively mobilize US$100 billion of long-term climate finance annually by 2020 are important milestones.[13] We argue that the recent advances lay foundations for an improved governance structure for climate finance under the UNFCCC. However, we will also highlight the main sticking points in the negotiations where considerable challenges remain, especially concerning the critical issue of funding sources and measuring, reporting and verifying (MRV) financial support. We conclude by arguing that the emergent new climate finance framework under the UNFCCC holds significant promise. However, there are no guarantees that it will live up to the high expectations. Crucially, the question of funding sources remains unanswered and is likely to be affected by the financial crisis and sovereign debt problems of developed countries at least in the short-term.

9.2 Pre-2012 Legal and Governance Framework for Climate Finance Under the UNFCCC

9.2.1 *Principles and Basic Commitments*

The basic obligation for developed countries to provide financial assistance to developing countries to address climate change is contained in Article 4.3 of the UNFCCC and affirmed in Article 11 of the Kyoto Protocol. According to Article 4.3 of the UNFCCC, countries listed in Annex II of the Convention "shall provide new and additional financial resources" to developing countries "to meet the agreed full costs" of implementing their general commitments and reporting obligations under the Convention. The Article further states that Annex II countries shall provide finance to developing countries to cover "the agreed full incremental costs" of implementing mitigation measures. It identifies the need for adequacy and predictability of finance, and the importance of appropriate burden-sharing among developed countries. Further obligations are contained in Article 4.4 of the Convention,[14] which requires Annex II Parties to assist particularly vulnerable developing countries with adaptation costs, and in Article 4.5 concerning technology transfer.[15]

[13] Decision 1/CP.16, The Cancun agreements: outcome of the work of the Ad Hoc Working Group on Long-term Cooperative Action under the Convention, UN Doc. FCCC/CP/2010/7/Add.1, 15 March 2011.

[14] UNFCCC, supra, note 1, Art. 4.4 requires Annex II Parties to "assist developing countries that are particularly vulnerable to the adverse effects of climate change in meeting costs of adaptation to those adverse effects." Article 4.8 of the UNFCCC requires Parties to give "full consideration" to actions that are needed under the UNFCCC to meet the needs and concerns of developing countries arising from the adverse effects of climate change. Such actions can relate to funding, insurance and the transfer of technology.

[15] Ibid., Art. 4.5 requires Annex II Parties to take "all practicable steps to promote, facilitate and finance technology transfer to other Parties, in particular developing country Parties, to enable them to implement the Convention's provisions."

The legal structure of financial commitments in the UNFCCC reflects the principle of common but differentiated responsibilities and respective capabilities.[16] Annex II countries with financial obligations towards developing countries are those that were members of the Organization for Economic Cooperation and Development (OECD) in 1992 when the Convention was adopted. The term 'developing country' has not been defined in the Convention, leading to the question whether all non-Annex I parties fall under the definition.[17] The group of non-Annex I countries includes a wide range of countries and not all of them consider themselves to be developing countries or associate themselves politically with the Group of 77 and China.[18] The most notable examples include Chile, Mexico, the Republic of Korea, Israel and the Group of Countries of Central Asia, Caucasus, Albania and Moldova. In practice, UNFCCC Parties have not managed to agree on a definition of a 'developing country' and all non-Annex I countries are, in theory, eligible for funding through the Convention's financial mechanism.[19] Not all of them have, however, availed themselves of this opportunity. Furthermore, Annex I countries undergoing the process of transition to a market economy (EITs) also receive financial support through the UNFCCC. Although the Convention initially envisaged that funds would assist developing countries, in practice, assistance to EITs has subsequently moved higher up on the climate policy agenda – despite resistance by developing countries.[20]

9.2.1.1 Controversies Over the Scale of Funding

The scale of funding for developing countries under the UNFCCC has been a highly controversial issue over the past two decades. The controversies originate from the fact that the Convention does not specify the level of resources to be provided for developing countries by Annex II Parties.[21] Article 11.3(d) of the UNFCCC requires the COP to work with the operating entity of the financial mechanism to determine "in a predictable and identifiable manner of the amount of funding necessary and the conditions under which that amount shall be periodically reviewed." Developing countries initially proposed that the COP – rather than Annex II parties or the Global Environment Facility (GEF) – should assess the scale of resources needed by developing countries to implement their commitments under the Convention.[22] Developed

---

[16] Ibid., Art. 3.1.

[17] Farhana Yamin and Joanne Depledge, *The International Climate Change Regime: A Guide to Rules, Institutions and Procedures* (Cambridge: Cambridge University Press, 2004), at 272.

[18] Ibid., at 274.

[19] Ibid., at 273.

[20] Ibid., at 265.

[21] Ibid., at 267.

[22] Ibid.

countries, being traditionally reluctant to undertake stringent and specific international financing commitments, opposed the idea. The compromise was a provision in the Memorandum of Understanding between the GEF and the COP, according to which the GEF Council and the COP will jointly determine the aggregate GEF funding requirements for the purposes of the Convention.[23] COP 3 subsequently adopted further guidance on this, indicating that prior to each GEF replenishment, the COP is to make an assessment of the amount of funds necessary to assist developing countries.[24] The assessment should take into account: the agreed full costs of preparing national communications; agreed full incremental costs of implementing other developing country commitments; information from the GEF on approved projects and programmes as well as those turned down due to lack of resources; and other funding sources available for implementation of the Convention.[25] These assessments have not, however, led to concrete numbers being agreed by the COP.[26] In practice, therefore, "it has been the GEF Secretariat and the Trustee who make assessments and funding scenarios on the basis of estimates of what the donors are willing to contribute."[27]

9.2.1.2 Controversies Over Definitions of "New and Additional" Funding and "Incremental Cost"

As explained above, Article 4.3 of the Convention requires that funding be "new and additional." The Convention does not define, however, the concept of "new and additional" finance leaving room for diverse interpretations. The emphasis on additionality reflects developing countries' fear that official development assistance (ODA) will be repackaged as climate-related finance. However, the additionality of developed country financing contributions is a difficult question both politically and technically.[28] On the political side, to measure the additionality of financial support, countries need to negotiate a common baseline and agree on what counts as climate finance.[29] Even if this politically sensitive problem is resolved,

[23] Ibid.

[24] Ibid., at 268. See also Decision 12/CP.3, Annex to the Memorandum of Understanding on the determination of funding necessary and available for the implementation of the Convention, UN Doc. FCCC/CP/1997/7/Add.1, 25 March 1998.

[25] Ibid.

[26] See, for example, Decision 9/CP.10, Assessment of funding to assist developing countries in fulfilling their commitments under the Convention, UN Doc. FCCC/CP/2004/10/Add.1, 19 April 2005, para. 1, which merely provides that the assessment report prepared by the Secretariat, in collaboration with the GEF, constitutes an input from the COP to the negotiations on the fourth GEF replenishment.

[27] Gomez-Echeverri and Müller, "The Financial Mechanism of the UNFCCC", supra, note 11, at 6.

[28] Yamin and Depledge, *The International Climate Change Regime*, supra, note 17, at 276.

[29] Ibid.

a further, more technical issue persists as to how effectively collect financial data from various government departments and international organizations.[30]

Some COP decisions contain hints on as to what can be considered as a baseline for new and additional finance. For instance, in the context of the establishment of the Special Climate Change Fund (SCCF) and the Least Developed Country Fund (LDC Fund), COP 7 indicated that "there is a need for funding, including funding that is new and additional to contributions which are allocated to the climate change focal area of the Global Environment Facility and to multilateral and bilateral funding, for the implementation of the Convention."[31] Further, another decision by COP 7 on funding under the Kyoto Protocol included a similar definition.[32] The language of these two decisions suggests that "new and additional" climate finance refers to the increase of funds in relation to those provided in the context of the implementation of the Convention. However, this definition does not seem to have been adopted by all Parties. Annex I Parties have been required to report in their national communications on "new and additional" financial resources provided to developing countries as well as on how they determined such resources as "new and additional."[33] As the latest compilation and synthesis of the fifth national communications shows, developed country Parties have used different approaches to defining "new and additional" finance.[34] While several of them referred to the increase in climate-related funding over a reporting period, others defined "new and additional finance" in relation to the pledges made in the Bonn Agreements on the implementation of the Buenos Aires Plan of Action.[35] Some countries, including the United Kingdom, suggested that the "new and additional" nature of their contributions should be measured against the target of providing 0.7% of their gross national income in ODA by 2013.[36]

Also the notion of 'incremental cost' in Article 4.3 of the UNFCCC has generated controversy over the years. As Yamin and Depledge indicate, the concept "raises politically sensitive issues about the sustainable development pathways developing

[30] Ibid.

[31] Decision 7/CP.7, Funding under the Convention, UN Doc. FCCC/CP/2001/13/Add.1, 21 January 2002, para. 1(a).

[32] Decision 10/CP.7, Funding under the Kyoto Protocol, UN Doc. FCCC/CP/2001/13/Add.1, 21 January 2002, para. 1(a).

[33] Guidelines for the Preparation of National Communications by Parties Included in Annex I to the Convention, UN Doc. FCCC/CP/1999/7, 16 February 2000, paras. 50–56. (Guidelines for Annex I National Communications).

[34] Compilation and synthesis of fifth national communications. Note by the secretariat. Addendum. Financial resources, technology transfer, vulnerability, adaptation and other issues relating to the implementation of the Convention by Parties included in Annex I to the Convention, UN Doc. FCCC/SBI/2011/INF.1/Add.2, 20 May 2011, paras. 45–47.

[35] Decision 5/CP.6, The Bonn Agreements on the implementation of the Buenos Aires Plan of Action, UN Doc. FCCC/CP/2001/5, 25 September 2001.

[36] Compilation and synthesis of fifth national communications, supra, note 34.

countries can and should follow."[37] This is because the notion of 'incremental costs' aims to exclude baseline costs and limit financing under the Convention to the benefits of the global environment.[38] This means that the activity eligible for financing must be compared to a baseline scenario and the activity that it will replace or make redundant.[39] Or, in the words of Yamin and Depledge, "the incremental costs of a project are the difference in costs between doing a project that achieves national goals but does not give global environmental benefits and doing one that does not."[40] While for many mitigation projects, it is a fairly straightforward task to calculate the incremental cost of a cleaner technology, this is not the same for adaptation projects which often have only local benefits.[41]

Since the adoption of the Convention, the COP and the GEF have attempted to clarify the notion of incremental costs and simplify its application.[42] In practice, the GEF has played an important role in operationalizing the concept through its funding decisions, including through the highly controversial Resource Allocation Framework, a system for allocating funds between countries based on country performance and potential to generate global environmental benefits. This system did not take into account vulnerability to impacts of climate change and adaptation needs and was described by some as a "*de facto* effort by the GEF to fine-tune its interpretation of the concept of incremental costs, based on the World Bank indicators designed without consulting the UNFCCC COP."[43] The GEF recently replaced the Resource Allocation Framework with a new policy that takes into consideration countries' GDP in the allocation of funds.

9.2.2 Financial Mechanism and Special Funds

Financial assistance to developing countries under the UNFCCC is provided through the financial mechanism. In addition, the Convention recognizes that financial assistance to developing countries can also be provided through bilateral, regional and other multilateral channels. The basic provisions concerning the financial mechanism can be found in Article 11 of the Convention. Accordingly, the financial mechanism provides financial resources "on a grant or concessional basis." The mechanism is to function "under the guidance of" the COP and "be accountable" to it. Article

[37] Yamin and Depledge, *The International Climate Change Regime,* supra, note 17, at 278.

[38] Charlotte Streck, "Ensuring New Finance and Real Emission Reduction: A Critical Review of the Additionality Concept", 5 *Carbon and Climate Law Review* (2011), 158, at 162.

[39] Ibid.

[40] Yamin and Depledge, *The International Climate Change Regime*, supra, note 17, at 280.

[41] Erik Haites, Development Perspectives for a Post-Copenhagen Climate Financing Architecture (OECD, 2010), at 12–13.

[42] Yamin and Depledge, *The International Climate Change Regime,* supra, note 17, at 278.

[43] Gomez-Echeverri and Müller, "The Financial Mechanism of the UNFCCC", supra, note 11, at 3.

11 also states that the operation of the financial mechanism can be trusted to one or more international entities. Article 21 of the Convention designated the GEF as the interim operating entity of the Convention's financial mechanism and COP 4 later confirmed the GEF's role, subject to review every 4 years.[44]

UNFCCC Parties have subsequently established four funds. The SCCF and the LDC Fund are both managed by the GEF. The Adaptation Fund, however, is managed by the Adaptation Fund Board (AFB) under the authority and guidance of Conference of the Parties serving as meeting of the Parties to the Kyoto Protocol (COP/MOP). In 2010, COP 16 established the Green Climate Fund, managed by the Green Climate Fund Board, and designated it as an operating entity of the Convention's financial mechanism. The following paragraphs provide an overview of climate funding provided through the GEF Trust Fund, LDC Fund, SCCF and the Adaptation Fund. The new Green Climate Fund will be discussed in Sect. 9.3.

9.2.2.1 GEF Trust Fund

As noted above, the GEF has been an operating entity of the Convention's financial mechanism since 1994. It is therefore responsible for providing grant and concessional funding for developing countries to assist them with incremental costs of climate change mitigation projects, and full costs of preparation of national communications to the UNFCCC. The GEF has also funded several pilot and demonstration projects on adaptation to impacts of climate change through its *Strategic Priority on Adaptation*, which is now closed. The GEF also operates the financial mechanism for other multilateral environmental agreements (MEAs), such as the Convention on Biological Diversity,[45] Stockholm Convention on Persistent Organic Pollutants[46] and the United Nations Convention to Combat Desertification.[47]

The GEF is an international financial organization, which is in many ways different from other international institutions. The legal and institutional structures underpinning the GEF are rather loose and complex.[48] The GEF Assembly, consisting

---

[44] Decision 3/CP.4, Review of the financial mechanism, UN Doc. FCCC/CP/1998/16/Add.1, 25 January 2005, para. 1.

[45] Convention on Biological Diversity, Rio de Janeiro, 5 June 1992, in force 29 December 1993, 34 *International Legal Materials* (1992), 822.

[46] Stockholm Convention on Persistent Organic Pollutants, Stockholm, 22 May 2001, in force 17 May 2004, 40 *International Legal Materials* (2001), 532.

[47] United Nations Convention to Combat Desertification in Those Countries Experiencing Serious Drought and/or Desertification, Particularly in Africa, Paris, 17 June 1994, in force 26 December 1996, 33 *International Legal Materials* (1994), 1328.

[48] For detailed overview, see Jacob Werksman, "Consolidating Global Environmental Governance: New Lessons from the GEF?", 2003, available at: www.environmentalgovernance.org/cms/wp-content/uploads/docs/dialogue/oct03/papers/Werksman%20GEF.pdf (last accessed on 30 March 2012).

of all countries that have agreed to participate in the GEF, meets every 3 years to review the GEF's performance and negotiate its replenishment.[49] The GEF Council is the main governing body and makes decisions within the GEF.[50] It has 32 members representing participants' constituency groups, with 16 members from developing countries, 14 members from developed countries and two members from EITs.[51] The GEF Council meets twice a year for 3 days to develop, adopt and evaluate GEF programs.[52] The Council operates by consensus.[53] However, there are also formal voting rules, which place the emphasis on donor countries. Accordingly, the GEF Council could, as a last resort, take decisions through double weighted majority voting system, requiring an affirmative vote from 60% of the total number of participating countries and 60% of total contributions.[54]

The GEF's role has been controversial and divided developed and developing countries since preparations for the 1992 UN Conference on Environment and Development in Rio de Janeiro.[55] Developed countries hoped to avoid a proliferation of funds with overlapping and conflicting mandates, thus preferring to use the GEF for all MEAs.[56] Developing countries, in turn, would have preferred to establish MEA-specific financial mechanisms under the direct authority of each relevant COP.[57] They objected to the GEF because of the lack of transparency over its workings and a governance structure dominated by donor governments[58] which was in "sharp contrast" to consensus-based decisions-making procedures under MEAs, including the UNFCCC.[59] Some also saw the GEF's consolidation of financing functions as an attempt to limit the amount of overall funding that might otherwise be available under MEAs.[60]

The compromise reached under the UNFCCC and the Convention on Biological Diversity was a distinction between the financial mechanism, governed by the COP, and the international entity (entities) designated to operate the financial mechanism.[61] From this follows that the GEF is not the same thing as the Convention's financial

[49] Ibid., at 5.

[50] GEF Council website, available at: http://www.thegef.org/gef/council (last accessed on 30 March 2012).

[51] Ibid.

[52] Ibid.

[53] Ibid.

[54] Instrument for the Establishment of the Restructured Global Environment Facility (October 2011), para. 25(c).

[55] Yamin and Depledge, *The International Climate Change Regime,* supra, note 17, at 265.

[56] Ibid.

[57] Ibid.

[58] Ibid.

[59] Werksman, "Consolidating Global Environmental Governance", supra, note 48, at 6.

[60] Ibid.

[61] Yamin and Depledge, *The International Climate Change Regime,* supra, note 17, at 265–266.

mechanism but "an" entity trusted with its operation. The financial mechanism, in turn, can be defined as "the totality of legal, institutional and procedural arrangements that regulate and make possible the flow of financial resources mandated by the Convention."[62]

The GEF is funded through voluntary contributions from donor countries who pledge money every 4 years to the GEF Trust Fund through a process known as the GEF Replenishment. While the GEF receives most of its funds from industrialized countries required to provide financial assistance to developing countries under the UNFCCC, countries that have not been listed in Annex II of the UNFCCC have also made pledges to the GEF.[63] Negotiations on the Fifth GEF Replenishment were completed in May 2010, with a total of US$4.34 billion pledged by 35 donors for 2010–2014.[64] Approximately US$1.4 billion will be programmed under the agreed climate change mitigation strategy.[65]

The GEF is serviced by an independent secretariat, based at the World Bank. It has three implementing agencies, the UN Environment Programme (UNEP), UN Development Programme (UNDP) and World Bank, which help developing countries to design and implement eligible projects. Between 1991 and June 2011, the GEF has supported efforts on climate change mitigation, adaptation, and enabling activities, financing 914 projects with US$3.84 billion in 156 developing countries and economies in transition.[66] These projects have attracted US$21.8 billion in co-financing.[67] The emphasis in financial assistance has clearly been on mitigation as the GEF has funded 755 mitigation projects with US$3.39 billion.[68]

The GEF currently allocates funds according to a policy its Council adopted in 2010, known as the System for Transparent Allocation of Resources (STAR). The STAR policy essentially allows determining a country allocation for biodiversity, climate change, and land degradation based on a system of indices agreed by the GEF Council. Such indices cover country performance, benefits for the environment and the country's GDP.[69] The previous allocation policy known as the Resource Allocation Framework was heavily criticised by developing countries for ignoring the most vulnerable countries. Taking into account a country's national GDP in allocating funds is a new feature of the GEF allocation policy meant to address those concerns and include support for the poorest countries.

The relationship between the COP and the GEF is governed by Article 11.1 of the UNFCCC and the Memorandum of Understanding (MoU) between the GEF and

---

[62] Ibid., at 283.

[63] Gomez-Echeverri and Müller, "The Financial Mechanism of the UNFCCC", supra, note 11, at 5.

[64] Report of the Global Environment Facility to the Conference of the Parties, UN Doc. FCCC/CP/2011/7, 19 September 2011, at 2.

[65] Ibid.

[66] Ibid., at 1.

[67] Ibid.

[68] Ibid.

[69] System for Transparent Allocation of Resources, GEF Policy Paper, GEF/P.3, 24 June 2010.

the COP.[70] Accordingly, the COP decides on the relevant climate change policies, programme priorities and eligibility criteria for funding.[71] The task of approving concrete projects has been left to the GEF as an operating entity.[72] Under the UNFCCC, developing countries have been highly critical of the GEF's provision of financial support due to its donor-dominated decision-making, arguing that it fails to take into account the needs and interests of country-recipients and difficulties in accessing funding from the GEF. As elaborated by Mace:

> The UNFCCC requires that the GEF follow guidance provided by the COP. Nevertheless, the GEF Council takes decisions that have a substantial impact on the way funding is allocated by the GEF within the climate change focal area. At the UNFCCC COP, each State party theoretically has an equal vote. The same is not true in the GEF.[73]

Others have noted that the relationship between the COP and the GEF has been "disappointing due as much to the failure of the COP to provide explicit guidance… as it has been due to the vested interested represented on the GEF Council and the COP."[74] As it will be explained below, developing countries' dissatisfaction with the GEF motivated the creation of independent boards to govern the Adaptation Fund and the new Green Climate Fund.

9.2.2.2 Special Climate Change Fund

COP 7 created the SCCF in 2001 to address special needs of developing countries under the UNFCCC regime in the areas of: adaptation; technology transfer; energy, transport, industry, agriculture, forestry and waste management; as well as economic diversification.[75] Funding by the SCCF is complementary to funding from the GEF Trust Fund. Adaptation is its highest priority and unlike other multilateral sources, adaptation funding from the SCCF is open to all developing country parties under the UNFCCC.[76]

The operationalization of the SCCF has suffered from tensions among developing countries concerning prioritization of activities to be funded, and among donors and developing countries on issues, such as full-cost funding and scope of activities to be funded.[77] Unlike the GEF Trust Fund, the SCCF is not based on periodic

[70] Decision 2/CP.12, Memorandum of Understanding between the Conference of the Parties and the Council of the Global Environment Facility, UN Doc FCCC/CP/1996/15/Add.1, 29 October 1996.

[71] Ibid. See also UNFCCC, supra, note 1, Art. 11.1.

[72] Gomez-Echeverri and Müller, "The Financial Mechanism of the UNFCCC", supra, note 11, at 5.

[73] M.J. Mace, "Funding for Adaptation to Climate Change: UNFCCC and GEF Developments since COP 7", 14 *Review of European Community and International Environmental Law* (2005), 225, at 229.

[74] Gomez-Echeverri and Müller, "The Financial Mechanism of the UNFCCC", supra, note 11, at 4.

[75] Decision 7/CP.7, supra, note 31, para. 2.

[76] Report of the Global Environment Facility to the Conference of the Parties, supra, note 64, at 27.

[77] Mace, "Funding for Adaptation to Climate Change", supra, note 73, at 236.

replenishment by donors but it relies on voluntary contributions. These have been inadequate to the date[78] making the role of the SCCF in climate-related assistance to developing countries limited. By the summer of 2011, the SCCF had mobilized US$130.1 million for adaptation projects and programs in non-Annex I countries.[79] It had approved 32 projects funding, with 2 projects completed and 17 projects having started implementation on the ground.[80]

9.2.2.3 Least Developed Country Fund

COP 7 also established the LDC Fund in 2001 to address the special needs of LDCs under the UNFCCC. The Fund relies on voluntary contributions and by June 2011, US$415.5 million had been pledged to the LDC Fund.[81] The LDC Fund's priorities focus on the preparation and implementation of National Adaptation Programmes of Action (NAPAs). Preparation of NAPAs is a process designed for LDCs to identify their urgent and immediate needs with regards to adaptation and to formulate priority actions and projects to address those needs.

Since its inception, the LDC Fund has provided funding for the preparation of 48 NAPAs, and 45 of these have been completed.[82] The LDC Fund has therefore subsequently shifted its focus from NAPA support to NAPA implementation. By 2011, 47 projects and programs in 39 countries had been approved for funding, totaling US$178.6 million and leveraging US$826.43 million in co-financing.[83] However, funding for NAPA implementation and other aspects of the LDC Work Programme constitutes an important challenge. The LDC Expert Group has estimated that at least US$1.6 billion would be needed to implement all the priority projects identified in NAPAs.

9.2.2.4 Adaptation Fund

Created by COP 7 in 2001,[84] the Adaptation Fund operates under the Kyoto Protocol with the objective of funding concrete adaptation projects and programmes in developing countries. The legal basis for its establishment is Article 12.8 of the Kyoto Protocol, requiring COP/MOP to ensure that a share of proceeds from CDM project activities is used to assist developing countries that are particularly vulnerable to

---

[78] GEF Evaluation Office, Assessment of the SCCF, October 2011, at 8.

[79] Ibid.

[80] Ibid.

[81] Report of the Global Environment Facility to the Conference of the Parties, supra, note 64, at 25.

[82] Ibid.

[83] Ibid.

[84] Decision 10/CP.7, supra, note 32.

the adverse impacts of climate change to meet the costs of adaptation. Against this background, the Adaptation Fund is exceptional in that its resources are largely independent of donors' contributions. The Fund is also an innovative institution because it pioneered a direct access by developing countries to financial resources and due to the large role which developing countries play its governance.

The operating entity supervising and managing the Adaptation Fund is the AFB.[85] It functions under the *authority* and *guidance* of the COP/MOP, which makes decisions on the Fund's overall policies. The AFB is "*fully accountable*" to COP/MOP.[86] The Board consists of 16 members, with two representatives from each of the five UN regional groups,[87] one representative from small island developing States and LDCs respectively, two other Annex I representatives and two other non-Annex I representatives.[88] Members of the AFB are nominated by their respective constituencies and elected by COP/MOP.[89] The AFB has legal capacity, which enables it to enter contractual relationships and fund projects directly rather than through an intermediary.[90]

COP 13 agreed on an interim arrangement whereby the GEF provides secretariat services for the Adaptation Fund and the World Bank serves as its trustee. A review of these interim institutional arrangements began in 2011 and is expected to conclude in 2012. A review report recommended considering a new approach with respect to the Adaptation Fund secretariat in order to bring about organizational independence, management control, transparency and accountability.[91] No pressing needs were identified requiring altering the existing arrangements concerning the trustee.[92] Negotiations on these issues are ongoing.

As explained above, the Adaptation Fund is mainly financed through the carbon market. Its primary source of funding is a 2% levy on Certified Emission Reductions (CERs) from CDM projects. In practice, the World Bank conducts CER sales through exchange trades on a daily basis as well as through over-the-counter transactions and auctions in order to provide a predictable flow of resources for the Adaptation Fund, consistent with guidance from the COP/MOP[93] and the CER

[85] Decision 1/CMP.3, Adaptation Fund, UN Doc. FCCC/KP/CMP/2007/9/Add.1, 14 March 2008, para. 3.

[86] Ibid., para. 4.

[87] The five UN regional groups are: Africa, Asia, Latin America and the Caribbean, Central and Eastern Europe, as well as Western Europe and Others.

[88] Decision 1/CMP.3, supra, note 85, para. 6.

[89] Ibid., para. 8.

[90] COP/MOP 4 decided that the AFB should be conferred such legal capacity as necessary for the execution of its functions. See Decision 1/CMP.4, Adaptation Fund, UN Doc. FCCC/KP/CMP/2008/11/Add.2, 19 March 2009, para. 11. The German Parliament conferred legal capacity on the Adaptation Fund Board on 8 February 2011.

[91] Report of the Adaptation Fund Board, Addendum: Review of the Interim Arrangements of the Adaptation Fund, UN Doc. FCCC/KP/CMP/2011/6/Add.1, 22 November 2011, para. 22.

[92] Ibid., para. 26.

[93] Decision 1/CMP.3, supra, note 85, para. 28 gives guidance on monetization.

Monetization Guidelines adopted by the AFB.[94] Between May 2009 and August 2011, the World Bank, as trustee, had sold 9.7 million CERs, generating revenues of US$166.[95] The estimates of potential resources available for the Fund in 2012 are in the range between US$187 million and US$223 million.[96] This means that funding for adaptation activities through the Adaptation Fund is currently more significant than through other channels under the UNFCCC. In addition, the Fund can also receive contributions from governments, private sector and individuals. In August 2011, donations to the Adaptation Fund, predominantly from governments, amounted to US$86 million.[97]

One of the exceptional features of the Adaptation Fund is that developing countries may choose whether to access financial resources directly or indirectly through implementing and executing entities chosen by developing country governments.[98] The option for direct access was pushed through by developing countries during intense negotiations at COP/MOP 4.[99] The direct access modality addresses recipient country concerns over difficulties experienced when accessing funds through implementing agencies and associated high administrative costs. In order to submit a project proposal and access the funds, Parties must meet the criteria adopted by the AFB[100] in accordance with the principle of sound financial management.[101] These criteria also apply to regional and multilateral implementing agencies. To ensure that entities receiving resources from the Adaptation Fund meet the detailed fiduciary standards concerning the use, disbursement and reporting of funds, the AFB has created an Accreditation Panel with three independent experts and two Board members.[102] The panel makes recommendations to the AFB concerning the accreditation of national, regional and multilateral implementing agencies.[103]

The Adaptation Fund focuses on concrete adaptation projects and programmes in developing countries. Its funding is based on *Strategic Priorities, Policies and Guidelines of the Adaptation Fund*, adopted by COP/MOP 4.[104] In 2010, the

---

[94] Ibid.

[95] Report of the Adaptation Fund Board, UN Doc. FCCC/KP/CMP/2011/6, 22 November 2011, para. 33.

[96] Financial Status of the AF Trust Fund, AFB/EFC.8/7, 14 February 2012, Annex 2.

[97] Ibid.

[98] Decision 1/CMP.3, supra, note 85, para. 29.

[99] "Summary of the Fourteenth Conference of Parties to the UN Framework Convention on Climate Change and Fourth Meeting of Parties to the Kyoto Protocol", *The Earth Negotiations Bulletin* 12(395), 15 December 2008.

[100] Decision 1/CMP.3, supra, note 85, para. 30.

[101] Decision 5/CMP.2, Adaptation Fund, UN Doc. FCCC/KP/CMP/2006/10/Add.1, 2 March 2007.

[102] Website of the Adaptation Fund, available at: http://www.adaptation-fund.org/about/accreditation-panel (last accessed 2 April 2012).

[103] Ibid.

[104] Decision 1/CMP.4, supra, note 90, Annex IV.

Adaptation Fund Board approved its first two projects.[105] The first is a direct-access project in Senegal, aiming to combat coastal erosion exacerbated by climate change and sea level rise.[106] The second project, seeking reduce the vulnerability of about 13,000 poor households in the Tegucigalpa region by improving water management, was submitted by Honduras through the UNDP.[107] During the latest reporting period in 2010–2011, the AFB approved nine more funding proposals for a total of US$55.4 million.[108] It also approved ten project concepts for a total of US$59.4 million.[109] These projects relate to agriculture, coastal management, disaster risk reduction, food security, rural development, infrastructure, and water resources management.[110]

9.3 Reforming the Framework for Climate Finance Under the UNFCCC

9.3.1 Climate Finance in the Post-2012 Negotiations

The overview in the previous section indicates that the climate finance framework under the UNFCCC has been controversial, fragmented and insufficient to meet the growing funding needs by developing countries for adaptation and mitigation. Climate finance has therefore been a critical issue in the ongoing long-term negotiations under the UNFCCC. These negotiations formally began at COP 13 in 2007 with the adoption of the Bali Action Plan and establishment of the Ad Hoc Working Group on Long-term Cooperative Action under the Convention (AWG-LCA).[111] Finance became one of the key pillars of the future climate regime as the AWG-LCA was assigned the task of addressing "enhanced action on the provision of financial resources and investment to support action on mitigation and adaptation and technology cooperation."[112]

The initial deadline for completing negotiations under the Bali Action Plan was COP 15 in 2009. During the 2 years of intense pre-Copenhagen negotiations, discussions on financial arrangements were both fruitful and divisive. At the heart of the debate were principles, the scale and sources of funding, as well as mechanisms

[105] Report of the Adaptation Fund Board, UN Doc. FCCC/KP/CMP/2010/7, 4 November 2010, para. 23.

[106] Ibid., para. 24.

[107] Ibid.

[108] Report of the Adaptation Fund Board, UN Doc. FCCC/KP/CMP/2011/6, 22 November 2011, para. 23.

[109] Ibid., para. 25.

[110] Ibid., para. 26.

[111] Decision 1/CP.13, The Bali Action Plan, UN Doc. FCCC/CP/2007/6/Add.1, 14 March 2008.

[112] Ibid., para. 1(e).

for the delivery of funds.[113] Early in the process, Mexico proposed to establish a world climate change fund, or green fund, on mitigation, adaptation and technology transfer,[114] and this proposal was subsequently modified and developed further. Overall, the idea of streamlining the existing, multiple funding sources and establishing a new umbrella fund under the Convention enjoyed wide support among the Parties. A number of innovative proposals also floated around on where and how to obtain the necessary finance to assist developing countries. These included imposing a global levy on all fossil fuel emissions and auctioning emission allowances to finance adaptation.[115] Some of these proposals proved highly controversial, for instance, a proposal for a levy on airline and shipping emissions.

Despite unprecedented publicity and high-level participation, COP 15 in Copenhagen did not manage to complete negotiations on key issues under the AWG-LCA, let alone produce a global and comprehensive climate agreement. Instead, the Conference became immersed into various procedural scandals and its outcome cast doubt over the effectiveness of multilateral efforts to address climate change.[116] The key outcome, the Copenhagen Accord was not formally adopted; instead the COP agreed to "take note" of the Accord,[117] reflecting a lack of consensus in the room during the final night of the negotiations.[118] Nevertheless, the Accord was supported by an overwhelming majority of Parties[119] and, among other things, included several important milestones for a future climate finance framework under the UNFCCC. These included an agreement to establish the Copenhagen Green Climate Fund as an operating entity of the Convention's financial mechanism.[120] The Accord also

---

[113] For a review of the situation immediately before Copenhagen, see Kati Kulovesi and María Gutíerrez, "Climate Change Negotiations Update: Prospects for a Copenhagen Agreed Outcome in December 2009", 18 *Review of European Community and International Environmental Law* (2009), 229, at 241.

[114] Ideas and Proposals on the Elements Contained in Paragraph 1 of the Bali Action Plan. Submissions from Parties, UN Doc. FCCC/AWGLCA/2008/MISC.2, 14 August 2008, paras. 40–45.

[115] "Summary of the Fifth Session of the Ad Hoc Working Group on Long-Term Cooperative Action and the Seventh Session of the Ad Hoc Working Group on Further Commitments for Annex I Parties under the Kyoto Protocol: 29 March – 8 April 2009", *The Earth Negotiations Bulletin* 12(407), 8 April 2009.

[116] For analysis, see Lavanya Rajamani, "The Making and Unmaking of the Copenhagen Accord", 59 *International and Comparative Law Quarterly* (2010), 825; Per Meilstrup, "The Runaway Summit: The Background Story of the Danish Presidency of COP 15, the UN Climate Change Conference", *Danish Foreign Policy Yearbook* (2010), 113; Daniel Bodansky, "The Copenhagen Climate Change Conference: A Post-Mortem", 104 *American Journal of International Law* (2010), 230.

[117] Decision 2/CP.15, Copenhagen Accord, UN Doc. FCCC/CP/2009/11/Add.1, 30 March 2010.

[118] For an overview of the final plenary, see "Summary and Analysis of the Copenhagen Climate Change Conference: 7–19 December 2009", *The Earth Negotiations Bulletin* 12 (459), 22 December 2009.

[119] Decision 2/CP.15, supra, note 117. The chapeau of the Copenhagen Accord annexed to the decision lists countries supporting the document.

[120] Ibid., para. 10.

contained a collective commitment by developed countries to provide US$30 billion of fast-track climate finance for developing countries in 2010–2012 with a balanced allocation between mitigation and adaptation, and agreement to prioritize adaptation finance for the most vulnerable developing countries.[121] Developed counties also undertook to mobilize jointly US$100 billion of climate finance a year by 2020 from both public and private source.[122] The Copenhagen Accord would have also established a high-level panel to study funding sources.[123]

While reflecting progress on finance, the Copenhagen Accord remained outside the formal UNFCCC regime. Furthermore, it did not contain sufficient details for operationalizing the new finance framework. Negotiations on finance thus continued in 2010 under the AWG-LCA and also informally.[124] UN Secretary-General Ban Ki-moon also convened the High-level Advisory Group on Climate Finance, tasked with analysing funding sources. Building on progress in negotiations throughout 2010, COP 16 in Cancun then resulted in the adoption of the Cancun Agreements, a package of decisions bringing key elements of the Copenhagen Accord, including those on finance, formally under the UNFCCC regime.[125] It is fair to argue that the Cancun Agreements significantly modified the institutional framework for climate finance under the UNFCCC. First, they formally established the Green Climate Fund as an operating entity of the financial mechanism of the Convention "to support projects, programmes, policies and other activities in developing countries, using thematic funding windows."[126] The Cancun Agreements also indicated that the new Fund would be managed by a Board with equal representation by developed and developing countries.[127] A Transitional Committee was created to complete the Fund's design by COP 17.[128] Second, the Cancun Agreements created a new Standing Committee to assist the COP in governing the financial mechanism of the Convention.[129] They also recognized developed countries' commitments in the Copenhagen Accord concerning fast-track and long-term climate finance.[130] While no specific conditionalities were attached to long-term finance, the Cancun Agreements mention that the developed country commitment is made "in the context of meaningful mitigation actions and transparency on implementation."[131] This

[121] Ibid., para. 8.

[122] Ibid.

[123] Ibid., para. 9.

[124] Asheline Appleton and Kati Kulovesi, "A Summary Report of the Geneva Dialogue on Climate Finance", 5 September 2010, available at: http://www.iisd.ca/download/pdf/sd/ymbvol179num1e.pdf (last accessed on 31 March 2012).

[125] Decision 1/CP.16, supra, note 13.

[126] Ibid., para. 102.

[127] Ibid., para. 103.

[128] Ibid., para. 109.

[129] Ibid., para. 112.

[130] Ibid., paras. 95 and 98.

[131] Ibid., para. 98.

implies that developing countries are expected to take meaningful mitigation action and provide information on those actions internationally.

With regard to principles of climate finance the Cancun Agreements mostly reiterated the Convention's provisions. Decision 1/CP.16 promises developing countries "scaled up, new and additional, predictable and adequate funding, taking into account the urgent and immediate needs of developing countries that are particularly vulnerable to the adverse impacts of climate change."[132] The Cancun Agreements stopped short of providing clarity on the sources of this new funding, and the commitment thus remains open to various interpretations. They merely state that "funds provided to developing country Parties may come from a wide variety of sources, public and private, bilateral and multilateral, including alternative sources."[133] In the negotiations preceding the Cancun Agreements, there had been proposals to define the notion of new and additional finance more precisely. In particular, developing countries favoured defining the term as an increase over ODA or as a percentage of GDP.[134] Developed countries, on the other hand, preferred the more vague approach already taken in the Convention.

As it will be explained below, the new climate finance framework under the UNFCCC was developed further a year later at COP 17 in Durban with agreement on the governing instrument of the Green Climate Fund[135] and a work programme on long-term finance in 2012.[136] Given the Durban outcome, the Green Climate Fund and the Standing Committee will begin to operate in 2012. Our argument here is that the Cancun Agreements and Durban outcome, with the preceding Copenhagen Accord, opened a new chapter under the UNFCCC on climate finance and governance. However, many important challenges remain despite the recent positive developments. In particular, the question concerning sources of long-term finance remains both undefined and controversial. Most developing countries insist that the funding should mainly come from public sources in developed countries and be channelled through the UNFCCC. This would essentially mean budgetary commitments from the developed world. The argument of developing countries is that such an approach would be fair in the context of developed countries' historical responsibility for climate change. Other funding sources, such as multilateral and bilateral funding outside the UNFCCC, private sector and innovative financing could be used, but only to complement public funding. Developed countries, on the other hand, are keen to avoid strong and prescriptive language on public funding, advocating instead a significant role for the private sector and innovative sources. They argue that is unrealistic to expect the public sector to provide the necessary scale of resources.

---

[132] Ibid., para. 97.

[133] Ibid., para. 99.

[134] Negotiating text. Note by the secretariat, UN Doc. FCCC/AWGLCA/2010/14, 13 August 2010.

[135] Decision 3/CP.17, Launching the Green Climate Fund, UN Doc. FCCC/CP/2011/9/Add.1, 15 March 2012.

[136] Decision 2/CP.17, Outcome of the Work of the Ad Hoc Working Group on Long-term Cooperative Action under the Convention, UN Doc. FCCC/CP/2011/9/Add.1, 15 March 2012, para. 127.

There has been little alignment on this controversial matter over years. The Durban outcome means that through the new work programme on long-term finance, discussions on this controversial issue will intensify.

In light of these general developments in the ongoing negotiations on climate finance since 2007, we will provide a more detailed overview of the major outcomes, starting with fast-start finance, long-term financing and MRV of financial support. We will then discuss the Green Climate Fund, Standing Committee as well as the new registry for NAMAs, which seeks to facilitate matching financial, technological and capacity-building support for NAMAs.

9.3.2 Fast-Start Finance

The pledge by developed countries in Copenhagen to mobilize US$30 billion of fast-start finance in 2010–2012 can be seen as an attempt to build trust and generate good will in the interim period before a new international architecture for long-term climate finance is operational. The weakness of the Copenhagen Accord was, however, that it did not define what type of financial support counts as fast-start finance, merely stating that the funding should be new and additional,[137] which, as we have seen above, is an ambiguous and controversial expression. Developing countries and non-governmental organizations argued that much of the fast-start finance provided in 2010 was neither new nor additional. However, their calls for formalised reporting on fast-start finance under the Convention were met with the resistance from developed countries since the Copenhagen Accord was legally outside of the UNFCCC. As a compromise, in Cancun, COP 16 took note of the funding provided for climate projects and invited developed countries to submit information annually to the UNFCCC Secretariat on financial resources provided, including on how developing countries accessed those funds.[138] No details were provided, however, as to what information donor countries should include in their submissions.

The first round of submissions on fast-start finance took place in May 2011[139] and additional updates were provided in late 2011 prior to COP 17.[140] Due to the lack of a common reporting format, the information provided varies considerably concerning the amount of detail, reporting periods and currency as well as interpretation of what constitutes new and additional finance. Types of support reported by developed

[137] Decision 2/CP.15, supra, note 117, para. 8.

[138] Decision 1/CP.16, supra, note 13, para. 96.

[139] Submissions on information from developed country parties on the resources provided to fulfil the commitment referred to in decision 1/CP.16, paragraph 95. Note by the Secretariat, UN Doc. FCCC/CP/2011/INF.1, 15 August 2011.

[140] For a web site with updated information on fast-start finance, see UNFCCC, "Fast-start Finance," available at: http://unfccc.int/cooperation_support/financial_mechanism/fast_start_finance/items/5646.php (last accessed on 2 April 2012).

countries include, for example, Australian contributions to the LDC Fund, Germany's grant to build solar power plants in Brazil and Japan's loan support to improve energy access in Kenya and Tanzania.[141] Such differences make any aggregation of data difficult and the findings on fast-start finance highly approximate. While it does not appear possible to conclude how much money has actually been delivered in 2010, analysis by the World Resources Institute in November 2011 suggests that of the total pledges for 2010–2012, "US$16.23 billion has been requested and/or budgeted by the executive bodies" of donor countries.[142]

The data for 2011 shows that most of the fast-start funding came from public sources; however, several countries also reported private sector finance in the form of investments in clean energy, renewable energy and other sectors, clean technology transfer and public-private partnerships.[143] Donor countries use both bilateral and multilateral channels for the delivery of finance, and the funds are provided in the form of grants, concessional lending, technical cooperation and other instruments. Although it is difficult to determine whether the current fast-start finance provides for balanced allocation of funds between mitigation and adaptation as initially agreed, many countries reported an increase in financing for adaptation activities in developing countries.

In sum, the commitment on fast-start finance could boost climate finance and investment flows at a critical moment in the battle against climate change. It holds potential to enhance trust between developed and developing countries in the UN negotiations concerning a topic that has been divisive over the past two decades. However, as shown above, the lack of common approaches and a reporting format for fast-track funding delivered makes aggregation of information across donor countries highly difficult. In other words, the first experiences of developed country reporting on their fast-track finance highlighted the already known problems with the MRV of financial support under the UNFCCC. Given this, the implementation of the fast-start financing pledge risks deepening divides between developed and developing countries over such issues as new and additional resources, public versus private financing, delivery channels and forms of financing.

9.3.3 Long-Term Finance

There are two critical questions with regard to long-term climate finance:

- How much future financing will be needed for climate change mitigation and adaptation in developing countries? and

[141] Submissions on information from developed country parties on the resources provided to fulfil the commitment referred to in decision 1/CP.16, para. 95, supra, note 139.

[142] Kristen Stasio et. al., "Summary of Developed Country 'Fast-Start' Climate Finance Pledges", World Resources Institute, November 2011, available at:

http://www.wri.org/publication/summary-of-developed-country-fast-start-climate-finance-pledges (last accessed on 2 April 2012).

[143] Submissions on information from developed country parties on the resources provided to fulfil the commitment referred to in decision 1/CP.16, para. 95, supra, note 139.

- Where will this money come from and can the current levels of support be scaled up to meet those needs?

What is clear from the outset is that the current financial support for developing countries within and outside of the UNFCCC framework falls significantly short of what will be necessary in the coming years to avoid dangerous climate change and for developing countries to adapt to its unavoidable consequences.

A number of estimates are available concerning the scale of financial needs by developing countries to help them address climate change.[144] However, none of them allow accurately determining how much support will actually be required.[145] Recent analyses differentiate between incremental costs and incremental investments needed for mitigation actions in developing countries.[146] Incremental investments are simpler to calculate but differ from actual costs of mitigation because investments for many mitigation measures, such as energy efficiency and some renewables, have a high and quick return. The available estimates only inform, but do not determine, the support required for developing countries, which is likely to be in between the estimates of incremental costs and incremental investments.[147] In its ground-breaking technical paper from 2007, the UNFCCC estimated that additional investment and financial flows of around US$177 billion will be necessary in 2030 for mitigation in non-Annex I countries.[148] McKinsey estimated that incremental investments of US$659 billion in 2030 will be necessary for mitigation activities in developing countries, while the estimate by the International Energy Agency amounts to US$377 billion in 2030.[149] Few estimates of costs of mitigation actions exist: for instance, McKinsey estimates those at US$175 billion in 2030.[150] On adaptation, the 2007 UNFCCC technical paper estimated that additional investment of US$28–67 billion per year will be necessary by 2030 for adaptation activities in developing countries,[151] while the World Bank calculated that adaptation costs will be approximately US$80–90 billion annually by 2030.[152] To sum up, the financing

---

144 For synthesis of available estimates for mitigation, see Olbrisch et al., "Estimates of Incremental Investment for and Cost of Mitigation Measures in Developing Countries", supra, note 2, at 970–986. On adaptation, see Smith et al., "Development and Climate Change Adaptation Funding: Coordination and Integration", supra, note 7, at 987–1000.

145 Erik Haites, "Climate Change Finance – editorial", 11 *Climate Policy* (2011), 963, at 964.

146 Olbrisch et al., "Estimates of Incremental Investment for and Cost of Mitigation Measures in Developing Countries", supra, note 2, at 971.

147 Ibidem.

148 UNFCCC, Investment and financial flows to address climate change, supra, note 8, at 175, table IX-64.

149 Olsbrisch et al., "Estimates of Incremental Investment for and Cost of Mitigation Measures in Developing Countries", supra, note 2, at 974.

150 Ibid., at 975, table 2.

151 UNFCCC, *Investment and Financial Flows to Address Climate Change*, supra, note 9, para. 26 of the executive summary.

152 Smith et. al., "Development and Climate Change Adaptation Funding: Coordination and Integration", supra, note 7, at 989.

needs of developing countries for climate change mitigation and adaptation amount to hundreds of billions of US$ annually and therefore are broadly consistent with the developed country pledge for long-term financing under the UNFCCC.

Recent assessments of their financing needs for mitigation and adaptation by several developing countries under the UNFCCC are also worth noting in this context. They were part of the 2010 National Economic, Environment and Development Study (NEEDS) for Climate Change Project, requested by the Subsidiary Body for Implementation (SBI) with the aim of providing information on financing needs of non-Annex I Parties to implement mitigation and adaptation measures.[153] The ten participating countries, including Costa Rica, Egypt, Indonesia and Mali, assessed both their current financial flows for climate-related policies and resources needed in the short term and long term. The project was based on a bottom up and country-driven approach meaning that the participating countries relied on their own priorities and methodologies. The estimated short- and long-term costs of mitigation by these countries range between US$45 million and US$33.01 billion, while costs of adaptation to climate change range between US$161.5 million and US$20.69 billion.[154] Given that the countries participating in the assessment did not apply a common methodology, any aggregation across countries is highly approximate. Nevertheless, the results provide some perspective on the future needs of these countries to address climate change. The approach also reflects developing countries' preference for country ownership in setting priorities and assessing financial needs. The needs assessments are also relevant in the context of the new Standing Committee under the UNFCCC. Among other functions, the Committee should prepare a biennial assessment of funding needs, drawing on a number of sources which include "information provided by Parties on assessments of their needs."[155]

Concerning funding sources, the Copenhagen Accord included agreement to establish a high-level panel under the guidance of, and accountable to, the COP to study "the contribution of the potential sources of revenue, including alternative sources of finance, towards meeting this goal."[156] Given that the Copenhagen Accord was not formally adopted, the UN Secretary-General proceeded in 2010 to establish a High-Level Advisory Group on Climate Change Financing to study potential sources of finance. The Advisory Group, consisting of prominent figures from governments, international financing institutions, academia and the UN, was not an exact replica of the panel envisaged in the Copenhagen Accord and, most notably, it was not overseen by the COP. Yet, the objective of its work was essentially the same. In its report at the end of 2010, the Advisory Group came to the

---

[153] Report of the Subsidiary Body for Implementation on its 28th session, held in Bonn from 4 to 13 June 2008, UN Doc. FCCC/SBI/2008/8, 11 July 2008, para. 30.

[154] Synthesis report on the National Economic, Environment and Development Study (NEEDS) for Climate Change Project, Note by the secretariat, UN Doc. FCCC/SBI/2010/INF.7, 24 November 2010.

[155] Decision 2/CP.17, supra, note 136, para. 121(f).

[156] Decision 2/CP.15, supra, note 117, para. 9.

conclusion that "it is challenging but feasible to meet" the pledge for US$100 billion annually for long-term financing made in the Copenhagen Accord.[157] On sources of financing, the Group was rather opaque concluding that funding should come from "a wide variety of sources, public and private, bilateral and multilateral, including alternative sources of finance, the scaling up of existing sources and increased private flows."[158] Given the unclear mandate of the Advisory Group, its findings did not have a direct influence on the course of the UNFCCC negotiations but they still provide a useful analytical base for the discussion.

In Durban, Parties were again unable to agree on the controversial issue of long-term funding sources. Instead, they established a special work programme on long-term financing to analyse options for the mobilisation of resources from various sources as well as relevant financing needs of developing countries.[159] Such analytical work should be based on findings of the Advisory Group and the report on mobilising climate finance prepared at the request of G20 finance ministers and should take into account lessons learned from fast-start financing.[160] The work programme will include workshops on which the programme's co-chairs are requested to prepare a report for consideration at COP 18.[161] It is difficult to predict the effectiveness of the work programme at this point. However, in light of the continuing financial crisis in the EU and elsewhere, it could easily become an avenue for keeping negotiations on long-term funding sources alive rather than bringing about a real change.

9.3.4 MRV of Support

Measuring, reporting and verifying financial support for climate-related actions is key to ensuring the effectiveness of the international framework for climate finance. Without clear rules on MRV, it will be impossible to determine how much finance flows from developed to developing countries and for what activities. *Measuring* of climate finance relates to the question of what kind of support counts as climate finance. Here, the definition of "new and additional" finance is pivotal. *Reporting* relates to what kind of information countries communicate to the UNFCCC on finance, including the technical but important questions concerning reporting guidelines and a common reporting format. The element of *verification* is more complex as it relates to verifying whether climate finance reported by countries complies with agreed principles for the provision of financial support. In the context of diametrically opposite interpretations of such principles, verification of climate finance

[157] Report of the Secretary-General's High-Level Advisory Group on Climate Change Financing, 5 November 2010, at 5.

[158] Ibid.

[159] Decision 2/CP.17, supra, note 136, paras. 127–131.

[160] Ibid., para. 130.

[161] Ibid., paras. 127 and 129.

is also a highly political question. In addition, there is currently no clarity on which body would undertake such verification function and how this would be linked, if at all, to compliance.

Prior to the reforms agreed by COP 16 in 2010, Annex II countries were only required to report in their periodic national communications on the "new and additional" financial resources they had provided to developing countries and how they had determined such resources to be "new and additional."[162] In particular, Parties were required to report on financial support provided through bilateral and multilateral channels, including GEF funds, provide to the most vulnerable countries to address the adverse effects of climate change as well as on their definition of new and additional finance. However, Annex I countries tended to use different reporting approaches resulting divergences over definitions of new and additional finance, reporting periods and currency, which made any aggregation of data difficult.[163] Annex I national communications were subject to in-depth review according to agreed guidelines, which mostly addressed issues of completeness of information.[164] Findings from each round of Annex I national communications were compiled and synthesised by the UN Climate Change Secretariat in a report considered by the SBI.

The Cancun Agreements included agreement on a system of enhanced reporting on the provision of financial, technology and capacity building support to developing countries. Alongside other information on mitigation actions, this support is now part of biennial reports of developed countries.[165] Parties also decided to improve the reporting guidelines for national communications by Annex I Parties, "including the development of common reporting formats and methodology for finance, in order to ensure that information provided is complete, comparable, transparent and accurate."[166] Furthermore, Parties decided to enhance guidelines for the review of information on support in national communications.[167] A work programme for the development of relevant guidelines and modalities was also established. As result, COP 17 was able to adopt the UNFCCC biennial reporting guidelines for developed country Parties.[168] The first reports are due by 1 January 2014.[169] Concerning a common reporting format, Parties requested the Subsidiary Body for Scientific and Technological Advice (SBSTA) to develop methodologies for reporting information on financial support, taking into account existing international methodologies, and lessons learnt in preparing the first biennial reports, with a view of recommending a decision to COP 20.[170]

---

162 Guidelines for Annex I National Communications, supra, note 33.

163 Compilation and synthesis of fifth national communications, supra, note 34.

164 Guidelines for Annex I National Communications, supra, note 33.

165 Decision 1/CP.16, supra, note 13, para. 40 (a).

166 Ibid., para. 41.

167 Ibid., para. 42.

168 Decision 2/CP.17, supra, note 136, Annex I, paras. 16–20.

169 Ibid., para. 13.

170 Ibid., para. 19.

Information on financial support alongside other information provided in biennial reports will be subject to international assessment and review.[171] However, information on the provision of support is only subject to technical review but not to multilateral assessment.[172] It is presumed therefore that verification will be conducted through a biennial assessment of climate finance flows by the Standing Committee.

Overall, the new guidelines for MRV of climate finance address many of the shortcomings of the previous reporting system, such as inconsistency in reporting approaches and incompleteness. However as noted above, the development of a common reporting format is still pending. More importantly, the fundamental question of what finance is "new and additional" remains unresolved.

9.3.5 Green Climate Fund

One of the most notable achievements of the Cancun Agreements was that they formally established the Green Climate Fund, an idea that evolved on the basis of the Mexican proposal and was initially mentioned in the unadopted Copenhagen Accord. In Cancun, parties were also able to specify the main funding principles and basic governance arrangements for the Fund, and to outline a process for completing its design.[173] During 2011, a Transitional Committee worked to finalize the Green Climate Fund in accordance with terms of reference agreed in Cancun.[174] The Committee consisted of 40 members with 15 from developed countries and 25 from developing countries. While the Transnational Committee was ultimately unable to reach consensus on its recommendations, COP 17 was nevertheless able to agree on a decision operationalizing the Fund and approving its governing instrument.[175] As a result, attention has shifted towards making the new Fund fully operational. The Green Climate Fund can be seen as a key milestone in the evolution of the legal and institutional framework for climate finance. While it remains undecided how much of the US$100 billion of annual long-term climate finance will eventually flow through the Fund, the Green Climate Fund promises to become a remarkable international institution for climate finance.

According to the Governing Instrument of the Green Climate Fund, the Fund operates in accordance with the principles and provision of the Convention.[176] The purpose of the Fund is "to make a significant and ambitious contribution to the global efforts towards attaining the goals set by the international community to

[171] Decision 2/CP.17, supra, note 136, Annex II, para. 4.

[172] Ibid., para. 5.

[173] Decision 1/CP.16, supra, note 13, paras. 102–111.

[174] Ibid., Appendix III.

[175] Decision 3/CP.17, supra, note 135, Annex containing the Governing Instrument of the Green Climate Fund.

[176] Ibid., para. 3.

combat climate change."[177] It will do so through providing new, additional, adequate and predictable financial resources and catalysing both public and private climate finance to support developing countries in limiting or reducing their greenhouse gas emissions, and adapting to the impacts of climate change.[178] The Fund will be sourced through contributions by developed countries but it may also receive funds through other channels, public and private, and through alternative sources.[179]

The Fund is designated as an operating entity of the financial mechanism of the Convention.[180] Further arrangements will have to be concluded between the COP and the Green Climate Fund to ensure that the Fund is *accountable to* and functions *under the guidance* of the COP.[181] In this sense, the format resembles the relationship between the COP and the GEF whereby the COP provides annual guidance to the GEF, which takes appropriate action and reports annually back to the COP. During the negotiations, the relationship between the Green Climate Fund and the COP constituted a source of contention between developed and developing countries in the negotiations. Developing countries argued that the Fund should operate under the *authority* of the COP rather than just under its guidance, implying a greater role for the COP – similar to the arrangements on the Adaptation Fund under the Kyoto Protocol. The final compromise in Cancun was that although the Fund operates under the guidance of the COP, developing countries had a larger representation on the Transitional Committee that designed the Fund's details. Furthermore, they also received equal representation on the Green Climate Fund Board governing the Fund.[182] Representation in the Green Climate Fund Board is based on the UN regional groupings and representatives from SIDS and LDCs, who are selected within their constituencies.[183]

Finally, designating the Green Climate Fund as an operating entity to the financial mechanism of the Convention raises questions concerning its relationship to the current operating entity, which is the GEF. The COP has yet to clarify the respective roles of, and the relationship between the GEF and the Green Climate Fund in operating the Convention's financial mechanism.

Any developing country is eligible to receive support from the Green Climate Fund, which will cover "agreed full and agreed incremental costs for activities to enable and support enhanced action on adaptation, mitigation (including REDD-plus), technology development and transfer (including carbon capture and storage), capacity-building and the preparation of national reports by developing countries."[184]

---

[177] Ibid., para. 1.

[178] Ibid., para. 3.

[179] Ibid., paras. 29–30.

[180] Ibid., para. 4.

[181] Ibid., para. 6.

[182] Decision 1/CP.16, supra, note 13, para. 103.

[183] Ibid.

[184] Decision 3/CP.17, supra, note 135, Annex containing the Governing Instrument of the Green Climate Fund, para. 35.

At this point it remains unclear when the Fund will use full-cost funding and when it will only provide funding for incremental costs. What has been agreed is that funding will be provided in the form of grants and concessional lending and other instruments through specific funding windows, of which there will be two during the initial stage, namely mitigation and adaptation.[185]

Questions concerning access to funding were pivotal in the negotiations that led to the establishment of the Green Climate Fund. The outcome was that developing countries will be able to access the Fund directly through accredited sub-national, national and regional entities as well as through accredited international entities.[186] This is an innovative element in the Fund's design and aims to address developing countries' concerns and their difficulties in obtaining funds from existing sources, including the GEF. As such, the direct access modality has already been tested through the Adaptation Fund. It has also been agreed that the Green Climate Fund Board will ensure balanced allocation of funding between mitigation and adaptation.[187] For adaptation, the Board should take into account "the urgent and immediate needs of developing countries that are particularly vulnerable to the adverse effects of climate change, including LDCs, SIDS and Africa" through the use of minimum allocation floors.[188]

A strive towards a country-driven approach is another characteristic of the Fund which will "promote and strengthen engagement at the country level through effective involvement of relevant institutions and stakeholders."[189] In particular, countries may designate a national authority, which can both recommend proposals for funding in line with national climate strategies and which will be consulted on other proposals their consistency with national strategies.[190]

Compromises reached in Cancun also included agreement that the World Bank will serve as the interim trustee of the Green Climate Fund, administering the Fund's assets "only for the purpose of, and in accordance with the relevant decisions of the Green Climate Fund Board."[191] The operation of the Fund will be supported by an independent secretariat.[192] At the time of writing this chapter, the UNFCCC was in process of receiving nominations for members of the Green Climate Fund Board as well as expressions of interest from countries willing to host the Fund. A Memorandum of Understanding between the COP and the Fund is expected to be concluded at COP 18.

[185] Ibid., paras. 37 and 54.

[186] Ibid., paras. 45–49.

[187] Ibid., para. 50.

[188] Ibid., para. 52.

[189] Ibid., para. 31.

[190] Ibid., para. 46.

[191] Decision 1/CP.16, supra, note 13, paras. 105–107.

[192] Ibid., para. 108.

9.3.6 Standing Committee

The Cancun Agreements also established a Standing Committee under the COP to work on "improving coherence and coordination in the delivery of climate change financing, rationalization of the financial mechanism, mobilization of financial resources and measurement, reporting and verification of support provided to developing country Parties."[193]

The idea of a specialised, coordinating body on climate finance under the Convention was based on a strong rationale. There was general agreement among Parties that the current system for climate finance, with its multiple funding institutions and channels, was fragmented and that there was a need to promote a more coherent and coordinated approach. Institutionally, the UNFCCC regime appeared as the logical framework to achieve this through a new platform for communication and exchange of information. Furthermore, with the Green Climate Fund, the Convention's financial mechanism itself came to include two operating entities, strengthening the need for ensuring coherence. The question also had to be addressed as to which body under the Convention would perform the function of MRVing financial support to developing countries. At the political level, developing countries also feared that the COP would not have sufficient power over the operation of the Green Climate Fund and recipient countries could become side-lined on the Fund's Board. They therefore promoted establishing a strong specialised body on climate finance under the COP to ensure that the Fund as an operating entity of the financial mechanism would remain under its guidance. The Standing Committee was created in response to such concerns.

COP 17 reached agreement on the composition and role of the Standing Committee.[194] The Committee will consist of 20 members with equal representation from Annex I and non-Annex I Parties.[195] The members must have the necessary experience in the areas of climate change, development and finance.[196] The Committee will hold its meetings twice a year and develop further modalities for observer participation.

To address fragmentation of climate finance flows, it was agreed that the Standing Committee would organise a forum for communication and exchange of information among bodies and institutions relating to climate finance, including those outside of the Convention.[197] The Committee will also provide guidance to the COP concerning operating entities of the Convention's financial mechanism, make recommendations on how to improve their work in terms of coherence, efficiency and effectiveness, and provide expert contribution into periodic reviews of the financial

[193] Ibid., para. 112.

[194] Decision 2/CP.17, supra, note 136, Annex VI.

[195] Ibid., para. 1.

[196] Ibid., para. 2.

[197] Decision 2/CP.17, supra, note 136, para. 121(a).

mechanism.[198] The Standing Committee is also tasked with ensuring the necessary coordination with the SBI and thematic bodies of the Convention.[199] In relation to MRV of support, the Committee will prepare a biennial assessment of climate finance flows on the basis of: national communications and biennial reports by both developed and developing countries; information in the NAMA registry; information by developing countries on the assessment of their needs; reports by operating entities of the financial mechanism; and information available from other entities providing climate finance.[200] It remains unclear to what extent this assessment will contribute to the verification of climate finance given the lack of agreement on key principles and concepts, such as "new and additional," among Parties. The danger thus exists that such assessments will become another venue for political power play, highlighting divisions between donor and recipient countries.

9.3.7 Support for NAMAs

Support in the form of finance, capacity building and technology transfer can be vital for developing countries to take mitigation actions. COP 16 decided that developed countries should provide increased support for the preparation and implementation of NAMAs in developing countries and for enhanced reporting on those actions.[201] The Cancun Agreements also set up a registry to record NAMAs seeking international support and facilitate matching those actions with available finance, capacity building and technology support.[202] At COP 17 Parties specified that the registry will be developed as a web platform and participation will be voluntary.[203] On the support side, developed countries, operating entities of the Convention's financial mechanism and other donors were invited to submit information on support available and/or provided for NAMAs.[204] On the needs side, developing countries were invited to submit information on individual NAMAs seeking international support.[205] It is envisaged that the registry will facilitate matching action and support by providing enhanced information to donors and recipients. In addition, the registry also increases transparency of developing country mitigation actions by recording individual NAMAs which do not seek international support.

Although it is too early to determine the effectiveness of the NAMA registry, doubts can be expressed regarding the added value of such arrangement. With the newly established Green Climate Fund and its funding window for mitigation, and

[198] Ibid., para. 121(c) – (e).

[199] Ibid., para. 121(b).

[200] Ibid., para. 121(f).

[201] Decision 1/CP.16, supra, note 13, para. 52.

[202] Ibid., paras. 53–59.

[203] Decision 2/CP.17, supra, note 136, para. 45.

[204] Ibid., para. 48.

[205] Ibid., para. 46.

other entities providing climate finance for mitigation actions in developing countries, the registry with its informative role appears to be rudimentary. Further work is also needed to clarify how its role in recording support provided by developed countries will be linked to MRV of support and feed into a biennial assessment of climate finance flows by the Standing Committee.

9.4 Conclusions: A Breakthrough or an Empty Promise?

Long-term negotiations under the UNFCCC have continuously disappointed those hoping for a comprehensive, legally-binding and meaningful agreement. Yet, in the past 5 years, progress has been made on many important issues. Notably, the negotiations launched in Bali in 2007 have resulted in a number of important reforms to the climate finance architecture under the UNFCCC.

In this chapter, we have argued that the scale of climate finance has been a key controversy in the history of UNFCCC regime. Without agreement on a concrete figure, it has been difficult to evaluate Annex II countries' support for non-Annex I Parties. At COP 15, developed countries finally placed a tangible figure on the table, committing to mobilising jointly US$100 billion per year by 2020 to assist developing countries in climate change mitigation and adaptation. While not everyone agreed with it, the figure had been influenced by several assessments of developing countries' future funding needs. It was therefore an important step forward.

Negotiations under the Bali Action Plan have also succeeded in strengthening the institutional framework for climate finance under the UNFCCC. Most notably, agreement has been reached to establish the Green Climate Fund and the Standing Committee. The figure above illustrates how the new institutional framework will function as a result of these reforms (Fig. 9.1).

Our conclusion is that the Green Climate Fund holds important potential to improve the framework for climate finance under the UNFCCC. While it remains unclear how much of the US$100 billion of annual climate finance expected by 2020 will flow through the Fund, it promises to become a remarkable financing institution. The governance of the new Green Climate Fund also incorporates innovative features.

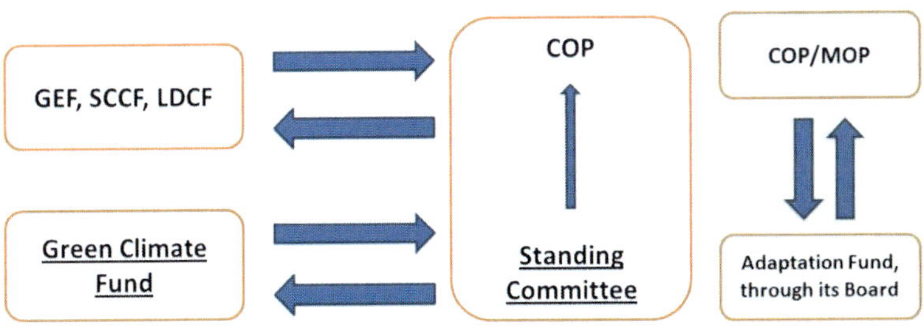

Fig. 9.1 Institutional framework for climate finance under the UNFCCC and Kyoto Protocol. The underlined Green Climate Fund and Standing Committee are new elements in the institutional structure

In contrast to the highly controversial GEF, developed and developing countries have equal representation on the Green Climate Fund Board. In the future, developing countries will thus be better placed to influence funding decisions in relation to the Convention's financial mechanism. This will hopefully help bridge some of the deep and long-standing divides between developed and developing countries. Direct access and a country-driven approach are other distinctive characteristics of the Green Climate Fund, making it stand out form among other funding institutions.

Attempts have also been made to address fragmentation of the international framework for climate finance both within and outside the UNFCCC. The new Standing Committee has been tasked with improving coherence and coordination in the delivery of international climate finance. Concrete steps will involve a forum for different bodies and institutions as well as biennial assessments of climate finance flows. The Standing Committee's function to prepare biennial assessments will seek to provide a complete picture on financial support to address climate change. Also the MRV framework for climate finance under the UNFCCC has been strengthened. At the heart of these reforms lie the new reporting guidelines for developed countries, requiring extended information on financial support provided. This will make it possible to collect information on financial flows in a more efficient and effective manner. Reporting arrangements for fast-start finance and the NAMA registry also serve the goal of improved MRV of financial support.

Despite these important reforms, several core issues have yet to be resolved. Although the Green Climate Fund is currently being operationalized, it does not have money. Without adequate resources, the Fund will become nothing more than an empty promise. Experiences with the SCCF and the LDC Fund during the past decade demonstrate that funding arrangements that are not backed by adequate financial resources risk become inefficient. Moreover, even if pledged, the money does not necessarily materialize, as evidenced by the failure of most developed country governments to comply with the 0.7% target for ODA from GDP.[206] In the case of the Green Climate Fund, the stakes are high, not least because developing countries' future needs for international assistance amount to hundreds of billions of dollars. If the US$100 billion annual commitment fails to materialize, serious consequences will follow both for the credibility of UNFCCC finance framework and most importantly, for the battle against dangerous climate change. However, against the backdrop of a serious financial crisis, deep budget cuts and gruelling austerity measures, the political will of developed country governments to channel public resources to developing countries is far from strong.

Thus far, Parties have not made any meaningful progress towards resolving the highly controversial issue of long-term finance and what roles the public and private sectors as well as innovative sources are expected to play in the provision of climate finance. The work programme on long-term finance established in Durban can achieve some progress on these issues. Ultimately, however, it is the political

[206] See, for instance, Net official development assistance from Development Assistance Committee and other OECD members in 2011 – preliminary data for 2011, available at http://www.oecd.org/dataoecd/44/13/50060310.pdf (last accessed 30 April 2012)

will of countries that will determine whether and how the issue moves forward. Given how deeply divisive the question of public and private funding is, it might be more constructive to focus discussions on mechanisms of generating funds through innovative means rather than trying to determine the relative contribution of public and private sectors to multilateral climate-related assistance to developing countries.

As this chapter shows, a tectonic divide between developed and developing countries has shaped the negotiations, and the legal and institutional framework for climate finance under the UNFCCC. Developing countries have been deeply dissatisfied with the GEF as an operating entity of the financial mechanism and with the ineffectiveness of SCCF and the LDC Fund. Developing countries' ambition for greater involvement and "having a say" in the decision-making over financial assistance have critically influenced the recent reforms to the Convention's financial architecutre. The reformed climate finance architecture reflects a compromise between these two camps while non-outcomes echo the most politically sensitive issues. Clarity on unresolved issues should be achieved speedily as its absence will hinder any progress made in the delivery of funds to developing countries in the future.

Chapter 10
Climate Justice: The Clean Development Mechanism as a Case Study

Tomilola Eni-ibukun

Abstract Justice considerations are now almost inextricably linked to the climate change discourse because of the recognition that global injustice and inequity are evident in the climate change problem, from its causes to its impacts. Consequently, the climate change regime contains a range of provisions, tools and measures to promote justice in the regime. One such tool is the Clean Development Mechanism (CDM), which gives developing countries the opportunity to contribute to climate change mitigation and also provides them with sustainable development benefits. However, the CDM itself is beset with its own justice issues, specifically distributive justice issues. This chapter focuses on the distributive justice issues of the CDM. It defines what distributive justice in the CDM means, examines what it should look like, and identifies the main causes for the lack of distributive justice in the CDM.

Tomilola Eni-ibukun holds a Ph.D. in law from the University of Dundee. She currently works as an independent legal consultant in the field of environmental law. The author is grateful to Elizabeth Kirk of the University of Dundee for reading the draft of this chapter and giving useful advice and comments.

T. Eni-ibukun (✉)
Sustainability Global Ltd., 8C Loganlee Terrace,
DD4 9AB Dundee, UK
e-mail: tomilola@gmail.com

E.J. Hollo et al. (eds.), *Climate Change and the Law*, Ius Gentium: Comparative Perspectives on Law and Justice 21, DOI 10.1007/978-94-007-5440-9_10,
© Springer Science+Business Media Dordrecht 2013

10.1 Introduction

The problem of climate change raises issues of equity and justice,[1] particularly with regard to its causes and impacts. This is because those that have contributed the least to climate change face most of its impacts. Climate change is historically attributable to the developed world and developed countries have benefitted the most from the activities that caused the problem.[2] However, developing countries, which have historically contributed the least to climate change,[3] are expected to be the most affected by it.

The impacts of climate change are expected to be quite severe, and the Intergovernmental Panel on Climate Change (IPCC) notes that these impacts would be greater in developing countries than in developed countries, concluding that climate change will likely exacerbate income inequalities between and within countries.[4] Developing countries also have lower capacity to adapt. This goes to another issue of justice – the ability or capacity to address the problem of climate change. Developed countries, with their greater resources and technological advancement, are generally recognised as having greater capacity to address climate change than developing countries.

The issue of historical responsibility for climate change also gives rise to another justice consideration. Developing countries argue that developed countries have had many years to develop, and that in their development process, have caused the current climate change problem; and that developing countries in turn need to increase their energy use in order to achieve development and alleviate poverty in

---

[1] Equity and justice are used interchangeably in this chapter, as appropriate. The ordinary dictionary meaning of equity includes definitions like "justice according to natural law or right, freedom from bias or favouritism, or something that is equitable", The Merriam-Webster Dictionary, available at: http://www.merriam-webster.com/ (last accessed on 1 March 2012); "the quality of being fair and impartial" *The Concise Oxford English Dictionary,* 11th ed. (Oxford: Oxford University Press, 2008); "fairness," and "justice," Samantha Hepburn, *Principles of Equity and Trusts*, 2nd ed. (Sydney/London: Cavendish Publishing Pty Limited, 2001), at 3; and "that which is just or right," Leslie Curzon, *Equity & Trusts*, 2nd ed. (London: Cavendish Publishing, 1996), at 1.

[2] See generally on the science and effects of climate change, Barrie Pittock, *Climate Change: Turning up the Heat* (London: Earthscan, 2005); John Houghton, *Global Warming: The Complete Briefing*, 3rd ed. (Cambridge: Cambridge University Press, 2004); and Mohan Munasinghe and Rob Swart, *Primer on Climate Change and Sustainable Development* (Cambridge: Cambridge University Press, 2005).

[3] Although this is still true, in terms of current emissions, some developing countries have overtaken or are overtaking developed countries and there is therefore a call for such developing countries to undertake appropriate mitigation actions.

[4] Samuel Fankhauser et al., "Vulnerability to climate change and reasons for concern: a synthesis", in James McCarthy et al. (eds), *Climate Change 2001: Impacts, Adaptation, and Vulnerability: Contribution of Working Group II to the Third Assessment Report of the Intergovernmental Panel on Climate Change* (Cambridge: Cambridge University Press, 2001), at 916.

their countries.[5] This is one of the reasons why developing countries have resisted attempts to cap their emissions. Responsible development, however, should not be taken to mean unrestricted freedom to continue to produce greenhouse gas (GHG) emissions. Any consumption that leads to GHG emissions should be done in light of the need for 'sustainable' development – defined as "development that meets the needs of the present without compromising the ability of future generations to meet their own needs."[6]

Finally, there is the issue of intergenerational equity. The impacts of climate change will exceed the impacts that are being seen today, and will continue to be felt far into the future – by those persons that did nothing to contribute to the problem and perhaps will not even enjoy the same benefits of industrialisation being enjoyed by the developed world today.

Due to these issues, justice considerations were key considerations in the design of the climate change regime, which contains a range of provisions and mechanisms aimed at ensuring justice.[7] One of the mechanisms adopted to help deliver justice under the regime is the Clean Development Mechanism (CDM). The CDM is, however, a flawed mechanism. One of the major problems with the CDM is that the distribution of projects under the CDM is generally regarded as inequitable.

This chapter focuses on this particular flaw. It examines what "distributive justice" in the context of the CDM means and ascertains the main reasons why the CDM has been unable to achieve this.

10.2 The CDM and Justice

First, a brief explanation of the role the CDM should play in delivering justice. The CDM is a market-based mechanism established by Article 12 of the Kyoto Protocol.[8] Under the CDM, projects or programmes of activities can generate Certified Emission Reductions (CERs) through activities implemented in developing countries that result in lower GHG emissions than would otherwise have been produced.

[5] See United Nations Framework Convention on Climate Change, New York, 9 May 1992, in force 21 March 1994, 31 *International Legal Materials* (1992), 851, para. 22 of the Preamble, which recognises that developing countries need access to resources and that their energy consumption will grow, in order to achieve sustainable social and economic development, albeit taking account of the possibilities for achieving greater energy efficiency and for controlling GHG emissions.

[6] See "Our Common Future, Chapter 2: Towards Sustainable Development", in *Our Common Future: Report of the World Commission on Environment and Development*, UN Doc A/42/427, 4 August 1987, Annex, para. 1.

[7] On climate change and justice generally, see Friedrich Soltau, *Fairness in International Climate Change Law and Policy* (Cambridge: Cambridge University Press, 2009); and Eric A. Posner and David Weisbach, *Climate Change Justice* (Princeton: Princeton University Press, 2010).

[8] Kyoto Protocol to the United Nations Framework Convention on Climate Change, Kyoto, 11 December 1997, in force 16 February 2005, 37 *International Legal Materials* (1998), 32.

The two main objectives of the CDM are to contribute to sustainable development in developing countries and to contribute to climate change mitigation through the GHG emission reductions achieved by the projects. Generally, in relation to developing countries, the CDM aims to assist developing countries to achieve sustainable development and also to contribute to the ultimate objective of the Convention[9] through the reduction in their GHG emissions achieved by the CDM projects.[10] In relation to developed countries, the CDM provides them with flexibility and cost-effective opportunities to comply with their emission reduction commitments under the Kyoto Protocol.[11]

As already noted above, the CDM is a key mechanism for achieving justice within the climate change regime. In recognition of their limited responsibility for the climate problem and their limited capability to address it, developing countries do not have emission reduction commitments under the Kyoto Protocol.[12] However, through the CDM, they are given the opportunity to contribute to climate change mitigation. In addition, in recognition of their need for sustainable development, CDM projects are required to contribute to sustainable development in developing countries. In this way, the CDM attempts to ensure justice in the treatment of developing countries within the climate change regime.[13]

Although the CDM is one of the key justice mechanisms within the climate change regime, the CDM itself, in its operation and implementation, also has justice problems, specifically distributive justice problems. The first CDM project was registered in 2004, and there are now more than 5,600 projects in the CDM pipeline, including over 3,800 registered projects.[14] Although there are currently 128 developing countries that are eligible to participate in the CDM,[15] only 73 countries do.

---

[9] The ultimate objective of the Convention is to stabilise GHG concentrations in the atmosphere at a level that would prevent dangerous human interference with the climate system. See UNFCCC, supra, note 5, Art 2.

[10] Kyoto Protocol, supra, note 8, Art. 12.2.

[11] Ibid., Art. 3.1 and Annexes A and B. Accordingly, developed countries are required to ensure that their total emissions of certain greenhouse gases (carbon dioxide, methane, nitrous oxide, hydrofluorocarbons, perfluorocarbons and sulphur hexafluoride) do not exceed their allowed emission levels. The aim is to reduce their overall emissions of these gases by at least 5% below 1990 levels in the first commitment period of the Kyoto Protocol, which runs from 2008 to 2012.

[12] Ibid., Arts. 3 and 10.

[13] The CDM is a good example of the implementation of the principle of common but differentiated responsibilities, which is one of the justice principles of the climate change regime. On the common but differentiated responsibilities principle in the climate change regime, see UNFCCC, supra, note 5, Arts. 3.1, 3.2 and 4.1.

[14] Statistics correct as of 30 January 2012. See CDM, "CDM in Numbers", available at: http://cdm.unfccc.int/Statistics/index.html (last accessed on 1 March 2012).

[15] This refers to those countries that have fulfilled the CDM participation requirements, which are: Kyoto Protocol ratification; designation of a national authority; and confirmation of voluntary participation. See Decision 3/CMP.1, Modalities and procedures for a clean development mechanism as defined in Article 12 of the Kyoto Protocol, FCCC/KP/CMP/2005/8/Add.1, 30 March 2006, Annex, paras. 28–30.

Of this number, just two countries – China and India – account for 67% of all projects, and China, India, Brazil and Mexico together account for 76% of all CDM projects. This skewed distribution continues at the regional level, with Asia and the Pacific region hosting 82% of projects, Latin America and the Caribbean hosting 15%, and Africa hosting just over 2% of all CDM projects.[16]

In 2001, countries highlighted the need to promote equitable distribution of CDM projects.[17] At the first Conference of the Parties serving as the meeting of the Parties to the Kyoto Protocol (COP/MOP) in 2005, countries again identified addressing the issue of equitable distribution of CDM projects as one of their roles.[18] At most of the subsequent COP/MOPs, Parties have addressed the need to ensure an equitable distribution of CDM projects, and have taken various actions, all aimed at achieving this goal.[19] However, the goal remains elusive, and the distribution of CDM projects, both among countries and among regions, still appears to be inequitable. Although the number of registered CDM projects has multiplied, the distribution of projects among countries has not changed much and the same four countries, namely India, China, Brazil and Mexico, have been consistently dominating the CDM market.[20] This is therefore obviously a justice problem in the CDM.

---

[16] All statistics are correct as of 30 January 2012. See CDM, "Registered project activities by host party", available at: http://cdm.unfccc.int/Statistics/Registration/NumOfRegisteredProjByHost PartiesPieChart.html (last accessed on 1 March 2012).

[17] See Decision 17/CP.7, Modalities and procedures for a clean development mechanism as defined in Article 12 of the Kyoto Protocol, FCCC/CP/2001/13/Add.2, 21 January 2002, Preamble, para. 6.

[18] Decision 3/CMP.1, supra, note 15, Annex, para. 4(c).

[19] See for example Decision 7/CMP.1, Further guidance relating to the clean development mechanism, FCCC/KP/CMP/2005/8/Add.1, 30 March 2006, para. 32; and Decision 2/CMP.5, Further guidance relating to the clean development mechanism, FCCC/KP/CMP/2009/21/Add.1, 30 March 2010, paras. 47–50. For a detailed discussion of the actions that have been taken with the CDM regime to address the problem of inequitable distribution of projects, see Tomilola Akanle, "Distributive Justice in International Law: Can the CDM Achieve an Equitable Geographic Distribution of Projects?", Ph.D. thesis on file at the University of Dundee, (2011), at 189–238.

[20] As of March 2007, the distribution of projects among the top 4 CDM hosts was: India (33%), China (8%), Brazil – (16%) and Mexico – (13%). In January 2008, the distribution was as follows: India: 33%; China: 16%; Brazil: 12% and Mexico: 11%. In July 2010, it was China (40%), India (22%), Brazil (7%) and Mexico (5%). In April 2011, the distribution was: China (44%), India (21%), Brazil (6%) and Mexico (4%) (all statistics obtained by the author from the CDM website at the relevant times). In January 2012, the distribution was: China (47%), India (20%), Brazil (5%) and Mexico (4%). The significance of these statistics is not so much that it is the same four countries that are the top CDM hosts. Much more significant is that although there has been some fluctuation in their percentage shares, they still host by far the majority of all CDM projects – the distribution has not levelled out. These four countries were hosting 70% of the 516 registered CDM projects as of March 2007, 72% of the 850 projects as of January 2008, 75% of the 2,312 registered projects as of August 2010, 76% of the 2,970 registered projects as of April 2011, and 76% of the 3,815 registered projects as of January 2012. The growth in the number of CDM projects has not led to a percentage increase in the number of projects hosted by other countries or a significant increase in the number of countries participating in the CDM. Instead, the status quo has mostly been maintained.

10.3 Meaning of Justice in the CDM

Countries have been highlighting the need for distributive justice within the CDM even before the first CDM project was registered.[21] However, what distributive justice in the CDM means has never been defined. Although a lot of effort has gone into achieving what countries refer to as "equitable distribution" of CDM projects,[22] and countries have, for many years, been addressing the apparent problem of the inequitable geographic distribution of projects, the exact nature of the problem has never been defined. There is no description of what an equitable distribution should be, so efforts to achieve this goal essentially amount to efforts to achieve an uncertain goal.

While there is a broad range of literature on distributive justice in general, far less has been written on the issue of distributive justice within the CDM or equitable distribution of CDM projects. It is generally regarded as given that the distribution of projects is inequitable and the focus is usually on determining the reasons for the inequitable distribution of projects, rather than on defining "equitable distribution." So the question is 'what is the meaning of equitable distribution of CDM projects/ distributive justice under the CDM?'

There are many different approaches to distributive justice broadly speaking. Firstly, there are various theories of distributive justice, such as egalitarianism, utilitarianism and Rawls's difference principle.[23] Generally, these theories can be regarded as 'outcome-based' approaches, as they would often result in set outcomes, regardless of the specific situation under consideration. For instance, egalitarianism requires equal distribution, whatever the circumstances surrounding the distribution, such as the specific circumstances of the recipients of the benefits.[24] Utilitarianism holds that a just outcome is the one that results in the greatest overall utility and maximises the happiness of society as a whole.[25] It gives no weight to individual happiness and only considers collective happiness, and would, for

[21] In 2001, when establishing the rules to govern the CDM, countries recognised the need to promote equitable distribution of projects. See Decision 17/CP.7, supra, note 16, Preamble, para. 6.

[22] "Equitable distribution" is the specific term used within the CDM regime to refer to distributive justice.

[23] See generally on egalitarianism, Ronald Dworkin, *Sovereign Virtue: Equality in Theory and Practice* (Cambridge: Harvard University Press, 2000); and Andrew Mason (ed.), *Ideals of Equality* (Oxford: Blackwell Publishers, 1998). On utilitarianism, see Jeremy Bentham, *An Introduction to the Principles of Morals and Legislation* (Kitchener: Batoche Books, 2000) (originally published 1781); and John Stuart Mill, *Utilitarianism* (London: Electric Book Company, 2001). On the difference principle, see John Rawls, *A Theory of Justice* (Cambridge: Harvard University Press, 1971); and John Rawls, *Political Liberalism,* expanded edition (New York: Columbia University Press, 2005).

[24] See Felix E. Oppenheim, "Egalitarianism as a descriptive concept", in Louis P. Pojman and Robert Westmoreland (eds), *Equality: Selected Readings* (Oxford: Oxford University Press, 1997), at 56; and Mason, *Ideals of Equality*, supra, note 23, at 3.

[25] See James W. Harris, *Legal Philosophies*, 2nd ed. (London: Butterworths, 1997), at 41.

instance, require an individual to sacrifice their own happiness, regardless of their circumstances, if this would increase the overall collective happiness.[26]

Thus, in these theories, specific circumstances are often disregarded[27] and distributive justice would require that the same formula be applied to all cases, irrespective of relevant circumstances. This approach can be contrasted with that used in international law. The approach to distributive justice in international law appears to be what can be called a "process-based" approach. Distributive justice is usually seen as the outcome of a process that takes certain relevant issues into consideration.[28] For instance, under the law of international watercourses,[29] a just outcome is achieved when factors such as the needs and uses of States, as well as the geographic and hydrographic factors of the watercourses, are taken into consideration.[30] In the case of the delimitation of maritime borders, account must be taken of circumstances such as the existence of islands, coastal configurations and proportionality, in order to reach an equitable outcome.[31]

These factors must be specific to the issue under consideration and should not be generalised. When these factors are fully taken into consideration, the outcome of this process would be considered just or equitable and distributive justice would be achieved. There is therefore no 'one-size-fits-all' equitable outcome. It is this approach, the one used in international law generally, that this chapter also adopts for the CDM. Consequently, equitable distribution of CDM projects can be regarded as the result of a process that takes certain relevant factors into consideration, rather than as a set or pre-determined outcome.[32] Following from this conclusion, the question is what the relevant factors in relation to the CDM are.

As already noted above, the relevant factors to be considered vary depending on the specific regime in question. Under the CDM regime, it is not necessary to go far

---

[26] See Howard Davies and David Holdcroft, *Jurisprudence: Texts and Commentary*, *Commentary* (London: Butterworths, 1991), at 219.

[27] For instance, both utilitarianism and egalitarianism do not require consideration of relevant circumstances.

[28] See generally, Akanle, "Distributive Justice in International Law", supra, note 19, at 131–136.

[29] One of the basic rules governing the use of, or access to, shared watercourses is the requirement for equitable and reasonable sharing of the watercourses. See *Gab ikovo-Nagymaros Project (Hungary/Slovakia)*, Judgment, 25 September 1997, *ICJ Reports* (1997), at 54. See also Convention on the Law of the Non-Navigational Uses of International Watercourses, New York, opened for signature 21 May 1997, not yet in force, 36 *International Legal Materials* (1997), 703 (Watercourses Convention).

[30] The Watercourses Convention does not expressly define "equitable and reasonable" use. Instead, it outlines some of the factors for determining whether a use is equitable and reasonable. According to Article 6, to achieve equitable and reasonable use, account should be taken of all relevant factors and circumstances, some of which are identified in the Article (6).

[31] See generally *Continental Shelf (Libyan Arab Jamahiriya/Malta)*, Judgment, 3 June 1985, *ICJ Reports* (1985), at 39–40; and David Freestone et al. (eds), *The Law of the Sea: Progress and Prospects* (Oxford: Oxford University Press, 2006), at 150–159.

[32] A set or pre-determined outcome under the CDM would be something to the effect that all countries should host the same number of projects, that countries should each host x number of projects, and such like.

to determine what should be the relevant factors to be considered. The CDM was established to achieve two objectives: to reduce greenhouse gas emissions and to contribute to developing countries' sustainable development.[33] The relevant factors that should be considered can therefore be distilled from these two objectives: countries' potential to achieve GHG emission reductions and their need for sustainable development.

Countries' potential to achieve GHG emission reductions can be referred to as their GHG emission reduction potential. This is a relevant factor because the CDM aims *inter alia* to help developing countries contribute to the ultimate objective of the UNFCCC to stabilise GHG concentrations in the atmosphere and to assist developed countries to comply with their emission reduction commitments.[34] Countries' emission reduction potential and the realisation of this potential determine how much countries can contribute to these objectives of the CDM. Uruguay, which emits about 45 million tonnes of carbon dioxide (CO_2) equivalent annually, cannot be expected to host the same number of projects as Indonesia, which emits in excess of 2 billion tonnes of CO_2 equivalent annually.[35] Consequently, a country which produces very little GHG emissions may not have much in the way of potential CDM projects and should not be expected to host more CDM projects than it has the potential for. The argument here is that to ensure that the CDM objective of GHG emission reduction is achieved, countries' emission reduction potential, determined by their GHG emission levels, must be taken into consideration.

The reference to potential is often a reference to emission reduction potential. However, another kind of potential that should also be considered is the sustainable development potential of countries. Since sustainable development is one of the objectives of the CDM, it is not sufficient to only consider the emission reduction potential and opportunities for cost-effective emission reductions in countries, as these only measure one of the objectives of the CDM – its objective to promote cost-effective emission reductions. The objective of contributing to sustainable development is equally important. Consequently, countries' need for sustainable development, or their sustainable development potential, should also be considered. Sustainable development potential can be taken to refer to how far along the development path a country is, considering its current development level. Countries that are less developed have greater sustainable development potential and greater need, and presumably, need the sustainable development benefits of the CDM more than those countries that are more developed.

In conclusion, an equitable geographic distribution of CDM projects is a distribution among countries based on their GHG emission reduction potential and their sustainable development potential. A distribution that is the result of the consideration of these two elements can then be regarded as just or equitable.

---

[33] See the discussion in Sect. 10.2 above.

[34] Kyoto Protocol, supra, note 8, Art. 12.

[35] For countries' emissions data, World Resources Institute, "World Resources Institute's Climate Analysis Indicators Tool (CAIT) Version 7.0.", 2010, available at: http://cait.wri.org/ (last accessed on 27 January 2012).

10.4 How Just Is the Current Geographic Distribution of CDM Projects?

This section examines whether the current distribution of CDM projects among countries is equitable, using the meaning of equitable distribution described above. It uses the relevant factors, emission reduction potential and sustainable development potential, to calculate countries' CDM potential and provide an outline of what the distribution of CDM projects should be. It then compares the current distribution of projects to this ideal, with the aim of determining whether or not the current distribution fits this ideal.

All developing countries produce GHG emissions, and therefore, all have the potential to reduce their emissions. It is however unlikely that every country will be able to host as many projects as it has the potential to, largely due to practical issues, specifically the size of the CDM market. As of January 2012, the CDM generated over 560 million CERs annually, which is equivalent to annual reductions of 560 million tonnes of CO_2 equivalent.[36] Annual developing country GHG emissions for 2005 are estimated to be about 25 billion tonnes of CO_2 equivalent, which means that annually, only about 2.2% of developing countries' emissions are being reduced through the CDM.[37]

Countries' emissions data is available from the World Resources Institute's Climate Analysis Indicators Tool (CAIT).[38] The emissions data for 2005, which is the year with the most comprehensive record of all GHG emissions for all countries, will be used.[39]

---

[36] See CDM, "CDM in Numbers", available at: http://cdm.unfccc.int/Statistics/index.html (last accessed on 31 January 2012).

[37] The estimated demand for CERs has been steadily falling. See Alexandre Kossoy and Philippe Ambrosi, *State and Trends of the Carbon Market 2010* (Washington, DC: World Bank, 2010), at 55–59. The bulk of this demand is from the European Union, which accounts for about 70% of demand. See page 55. However, supply too is expected to fall, due, among other things, to the revised EU Emissions Trading Scheme (EU ETS) Directive, which provides that CERs from new projects registered after 2012 will only be accepted into the EU ETS if the projects are in LDCs. See Council Directive 2009/29/EC amending Directive 2003/87/EC so as to improve and extend the greenhouse gas emission allowance trading scheme of the Community, OJ 2009 L 140/63, Article 11a(4), and infra, note 92. It is difficult to estimate with any kind of precision, the demand and supply of CERs in the post-2012 period, as these depend on several factors, such as the emission reduction commitments of developed countries, and rules for the use of CERs to meet these commitments. See generally, World Bank, *State and Trends of the Carbon Market 2011* (Washington, DC: World Bank, 2011), at 47–68.

[38] World Resources Institute, "World Resources Institute's Climate Analysis Indicators Tool (CAIT) Version 7.0.", 2010, available at: http://cait.wri.org/ (last accessed on 27 January 2012).

[39] CAIT contains the GHG emissions of most countries and can help with calculating a country's potential for GHG emission reductions. However, the available data has some shortcomings. The total CO_2 emissions data for all countries is available up to 2006. For non-CO_2 emissions (such as methane and nitrous oxide), this data is only available up to 2005 and is not available for all countries. In addition, for some countries, their emissions data from land use, land-use change and forestry activities is also not available. However, the CAIT database contains the most up to date and comprehensive information found.

To measure countries' sustainable development potential, this section uses the UN Development Programme's (UNDP) Human Development Index (HDI). Countries with low HDI are considered to have greater sustainable development potential due to their low human development levels. The HDI measures the average achievements in a country in three basic dimensions of human development: a long and healthy life (health), access to knowledge (education) and a decent standard of living (income).[40] The basic use of the HDI is to rank countries by level of "human development." The HDI has not been generally accepted as a measure of human development and has been criticised for, *inter alia,* not including environmental indicators in its assessment.[41] Nonetheless, it is widely used and is regarded as a more complete assessment of a country's development than, for example, gross domestic product (GDP) or gross national product (GNP), because it assesses not only economic, but also social development. The latest HDI data available is for 2011, and it is this data that is used in this section.[42] HDI data is available for all eligible developing countries with the exception of the Democratic People's Republic of Korea (North Korea).

10.4.1 CDM Potential and the Current Geographic Distribution of Projects

Using the meaning of equitable distribution described above, this section calculates countries' CDM potential and compares this potential to the current distribution of projects. A three-step process is employed to calculate countries' CDM potential.

Firstly, countries are classified according to their GHG emission reduction (ER) potential, using their 2005 emissions data. For ease of analysis, countries are divided into five categories, representing the ER potential of each category: 1 billion tonnes and over (very high); 500 million–1 billion tonnes (high); 100 million–500 million tonnes (medium); 1–100 million tonnes (low); under 1 million tonnes (very low).

---

[40] See UNDP, "Frequently Asked Questions (FAQs) about the Human Development Index (HDI)", available at: http://hdr.undp.org/en/statistics/hdi/ (last accessed on 27 January 2012).

[41] See Mark McGillivray, "Measuring development? The UNDP's Human Development Index", 5 *Journal of International Development* (1993), 183–192; and Ambuj D. Sagara and Adil Najam, "The Human Development Index: A Critical Review", 25 *Ecological Economics* (1998), 249–264.

[42] See UNDP, "Human Development Index and Its Components", 2011, available at: http://hdr.undp.org/en/media/HDR_2011_EN_Table1.pdf (last accessed on 21 January 2012). Although it is possible to use the 2005 HDI data in order to be consistent with countries' GHG emissions data, the 2011 data is a more accurate measurement of countries' current development levels than the 2005 data. As the purpose of this section is not to compare countries' sustainable development potential to their GHG emission reduction potential, but to carry out a comparison among countries, this author determines that it is better in this situation to be accurate.

Each country grouping is then assigned a value, as follows: Very High ER Potential (5); High ER Potential (4); Medium ER Potential (3); Low ER Potential (2); and Very Low ER Potential (1).

Secondly, countries are classified according to their sustainable development (SD) potential, using their 2011 HDI values. This requires further explanation. In UNDP's pre-2010 classifications, countries were classified into four groups, based on values, as follows: Low HDI (0 to 0.499); Medium HDI (0.500 to 0.799); High HDI (0.800 to 0.899); and Very High HDI (0.900 to 1.000).[43] However, using this classification, almost double the number of countries fell into the medium HDI group, compared to the number in the other groups. Because so many countries fell in the medium HDI group, compared to the other groups, and to make it easier to analyse the data more precisely, this section further splits the medium HDI group into two. To achieve this, UNDP's low HDI group is renamed "very low HDI" and UNDP's medium HDI group is split into two equal groups.

Therefore, the groups and values used to classify countries are as follows: very low HDI = very high SD potential (0–0.499); low HDI = high SD potential (0.500–0.649); medium HDI = medium SD potential (0.650–0.799); high HDI = low SD potential (0.800–0.899); very high HDI = very low SD potential (0.900–1.000). As a result, this section classifies countries into five groups according to their HDI, which also enables cross comparison with the data on developing country GHG emissions, where countries are also categorised into five categories. Currently, however, no developing country falls in the "very high HDI" category.

Each country grouping is assigned a value (the same used for ER potential), as follows: Very High SD Potential (5); High SD Potential (4); Medium SD Potential (3); Low SD Potential (2); and Very Low SD Potential (1).

The third step is to calculate countries' complete CDM potentials. To do this, a simple arithmetic calculation is done using the values assigned to each country in steps 1 and 2, and adding these numbers to show overall out of ten, what each country's potential is.

---

[43] Since 2010, countries are now divided into four roughly equal quartiles, as follows: low, medium, high and very high HDI. In this classification system, the cut-off point for each category does not depend on countries' HDI values. Rather, countries are simply grouped into roughly equal quartiles, and the cut-off point depends on the number of countries to be included in each quartile, regardless of the HDI values of the countries. The result of this is that two countries with the same HDI value could fall into different categories. For instance, although Tunisia, Jordan and Algeria all have the same HDI value of 0.698, Tunisia is categorised into the high HDI category and Jordan and Algeria into the medium category, essentially because with Tunisia, the number of countries to be included in the high HDI category was completed and so the next countries (starting from Jordan) were classified in the next (medium HDI) category. It is the opinion of this author that the previous classification system (of using absolute values) is a better system, as it will ensure that all countries with the same or similar values fall in the same categories. Consequently, it is this system that this section uses in classifying countries according to their sustainable development potential. See, for instance, the 2007 HDI, in UNDP, "Human Development Report 2009: Summary", 2009, available at: http://hdr.undp.org/en/media/HDR_2009_EN_Summary.pdf (last accessed on 21 January 2012), at 12.

Countries are then classified again into five groups to show what the distribution of CDM projects among countries should look like, based on their CDM potential. The categories and values used are as follows: Very High CDM Potential (9–10); High CDM Potential (7–8); Medium CDM Potential (5–6); Low CDM Potential (3–4); and Very Low CDM Potential (1–2). This is compared to the current geographic distribution of projects (as of 30 January 2012), to show whether or not this distribution is just or equitable.

All this data is presented in Table 10.1 below. It is important to note that this table is intended only as a rough representation of countries' CDM potential. It cannot, and is not intended to, be used to determine exactly how many projects countries should host compared to other countries. Instead, the purpose of this table is to provide a guide as to which countries should be performing well under the CDM, due to their ER and SD potentials taken together. The ultimate objective is to use this information to reach a conclusion about whether or not those countries that should be performing well in the CDM are the ones actually performing well and if not, to ascertain the possible reasons for this. However, the exact number of projects that a particular country can or should host will depend on the country's own ER and SD potentials.

This is particularly so because, due to countries' varying ER potential, the number of projects they can host will also vary. Therefore, countries which, according to Table 10.1, have the same CDM potential, are not necessarily expected to host the same number of projects. For example, although Guinea-Bissau and Iran have the same CDM potential value of 7, this does not mean both countries should host the same number of projects. While Guinea-Bissau has a low ER potential, Iran has a high potential and this necessarily affects the numbers of projects these countries can host. This however does not change the fact that Guinea-Bissau should be performing well under the CDM because it has a high CDM potential, considering both its ER and SD potentials. Because Guinea-Bissau has a high SD potential, it should receive priority or preferential treatment to facilitate its participation in the CDM. However, the precise meaning of "well," in terms of exact number of projects, will depend on the country's ER potential and how many projects it can actually host. And because Guinea-Bissau currently hosts no project, it is obvious that it should be doing much better than it is currently doing.

10.4.2 Analysis of the Distribution of CDM Projects

Table 10.1 above shows clearly the countries with the highest CDM potential. Forty-nine countries fall into the category of those with very high and high CDM potential, comprising countries from all regions. Within this category are those already hosting the largest number of projects, such as India, China and Brazil. On the other hand, only about 60% (29 countries) currently host projects and this hosting is extremely skewed. It ranges from China hosting 1,800 projects, to Iran hosting 7 and Mali hosting 1. The other 20 countries, such as Angola, Myanmar and Mozambique, host no projects at all.

Table 10.1 Countries' CDM potential and the current geographic distribution of projects

	Country	Emission reduction potential	Sustainable development potential	CDM potential	No of registered projects
1.	India	Very high (5)	High (4)	9 (Very high)	776
2.	Indonesia	Very high (5)	High (4)	9 (Very high)	75
3.	China	Very high (5)	Medium (3)	8 (High)	1,800
4.	Dem. Republic of the Congo	Medium (3)	Very high (5)	8 (High)	2
5.	Nigeria	Medium (3)	Very High (5)	8 (High)	5
6.	Zambia	Medium (3)	Very high (5)	8 (High)	1
7.	Angola	Medium (3)	Very High (5)	8 (High)	0
8.	Brazil	Very high (5)	Medium (3)	8 (High)	201
9.	Cameroon	Medium (3)	Very High (5)	8 (High)	2
10.	Myanmar	Medium (3)	Very high (5)	8 (High)	0
11.	Sudan	Medium (3)	Very high (5)	8 (High)	0
12.	Tanzania	Medium (3)	Very high (5)	8 (High)	1
13.	Bangladesh	Medium (3)	High (4)	7 (High)	3
14.	Benin	Low (2)	Very high (5)	7 (High)	0
15.	Burkina Faso	Low (2)	Very high (5)	7 (High)	0
16.	Burundi	Low (2)	Very high (5)	7 (High)	0
17.	Cambodia	Medium (3)	High (4)	7 (High)	5
18.	Chad	Low (2)	Very high (5)	7 (High)	0
19.	Côte d'Ivoire	Low (2)	Very high (5)	7 (High)	3
20.	Gambia	Low (2)	Very high (5)	7 (High)	0
21.	Guinea	Low (2)	Very high (5)	7 (High)	0
22.	Guinea-Bissau	Low (2)	Very high (5)	7 (High)	0
23.	Iran	High (4)	Medium (3)	7 (High)	7
24.	Liberia	Low (2)	Very high (5)	7 (High)	1
25.	Malawi	Low (2)	Very high (5)	7 (High)	0
26.	Mali	Low (2)	Very high (5)	7 (High)	1
27.	Mozambique	Low (2)	Very high (5)	7 (High)	0
28.	Niger	Low (2)	Very high (5)	7 (High)	0
29.	Pakistan	Medium (3)	High (4)	7 (High)	13
30.	Rwanda	Low (2)	Very high (5)	7 (High)	3
31.	Senegal	Low (2)	Very high (5)	7 (High)	2
32.	Sierra Leone	Low (2)	Very high (5)	7 (High)	0
33.	Togo	Low (2)	Very high (5)	7 (High)	0
34.	Djibouti	Low (2)	Very high (5)	7 (High)	0
35.	Haiti	Low (2)	Very high (5)	7 (High)	0
36.	Egypt	Medium (3)	High (4)	7 (High)	10
37.	Lesotho	Low (2)	Very high (5)	7 (High)	0
38.	Madagascar	Low (2)	Very high (5)	7 (High)	1
39.	Mauritania	Low (2)	Very high (5)	7 (High)	1
40.	Mexico	High (4)	Medium (3)	7 (High)	136
41.	Nepal	Low (2)	Very High (5)	7 (High)	4
42.	Papua New Guinea	Low (2)	Very High (5)	7 (High)	5
43.	Philippines	Medium (3)	High (4)	7 (High)	57
44.	South Africa	Medium (3)	High (4)	7 (High)	20
45.	Uganda	Low (2)	Very High (5)	7 (High)	9

(continued)

Table 10.1 (continued)

	Country	Emission reduction potential	Sustainable development potential	CDM potential	No of registered projects
46.	Uzbekistan	Medium (3)	High (4)	7 (High)	13
47.	Viet Nam	Medium (3)	High (4)	7 (High)	94
48.	Zimbabwe	Low (2)	5 (Medium)	7 (High)	0
49.	Yemen	Low (2)	Very High (5)	7 (High)	0
50.	Algeria	Medium (3)	Medium (3)	6 (Medium)	0
51.	Bhutan	Low (2)	High (4)	6 (Medium)	2
52.	Bolivia	Medium (3)	Medium (3)	6 (Medium)	4
53.	Ghana	Low (2)	High (4)	6 (Medium)	0
54.	Kenya	Low (2)	High (4)	6 (Medium)	6
55.	Lao	Low (2)	High (4)	6 (Medium)	1
56.	Swaziland	Low (2)	High (4)	6 (Medium)	0
57.	Thailand	Medium (3)	Medium (3)	6 (Medium)	64
58.	Argentina	Medium (3)	Medium (3)	6 (Medium)	25
59.	Botswana	Low (2)	High (4)	6 (Medium)	0
60.	Colombia	Medium (3)	Medium (3)	6 (Medium)	38
61.	Ecuador	Medium (3)	Medium (3)	6 (Medium)	17
62.	Equatorial Guinea	Low (2)	High (4)	6 (Medium)	0
63.	Guatemala	Low (2)	High (4)	6 (Medium)	11
64.	Guyana	Low (2)	High (4)	6 (Medium)	1
65.	Honduras	Low (2)	High (4)	6 (Medium)	21
66.	Kyrgyzstan	Low (2)	High (4)	6 (Medium)	0
67.	Moldova	Low (2)	High (4)	6 (Medium)	4
68.	Morocco	Low (2)	High (4)	6 (Medium)	8
69.	Namibia	Low (2)	High (4)	6 (Medium)	0
70.	Nicaragua	Low (2)	High (4)	6 (Medium)	6
71.	Tajikistan	Low (2)	High (4)	6 (Medium)	0
72.	Peru	Medium (3)	Medium (3)	6 (Medium)	26
73.	Republic of Korea	High (4)	Low (2)	6 (Medium)	67
74.	Saudi Arabia	Medium (3)	Medium (3)	6 (Medium)	0
75.	Syria	Low (2)	High (4)	6 (Medium)	3
76.	Solomon Islands	Low (2)	High (4)	6 (Medium)	0
77.	Comoros	Very Low (1)	Very High (5)	6 (Medium)	0
78.	Armenia	Low (2)	Medium (3)	5 (Medium)	5
79.	Azerbaijan	Low (2)	Medium (3)	5 (Medium)	1
80.	Belize	Low (2)	Medium (3)	5 (Medium)	0
81.	Bosnia & Herzegovina	Low (2)	Medium (3)	5 (Medium)	0
82.	Oman	Low (2)	Medium (3)	5 (Medium)	0
83.	Dominican Republic	Low (2)	Medium (3)	5 (Medium)	2
84.	El Salvador	Low (2)	Medium (3)	5 (Medium)	6
85.	Eritrea	Low (2)	Very high (5)	5 (Medium)	0
86.	Ethiopia	Low (2)	Very high (5)	5 (Medium)	1
87.	Fiji	Low (2)	Medium (3)	5 (Medium)	2
88.	Gabon	Low (2)	Medium (3)	5 (Medium)	0
89.	Georgia	Low (2)	Medium (3)	5 (Medium)	2
90.	Jamaica	Low (2)	Medium (3)	5 (Medium)	1

(continued)

Table 10.1 (continued)

	Country	Emission reduction potential	Sustainable development potential	CDM potential	No of registered projects
91.	Jordan	Low (2)	Medium (3)	5 (Medium)	3
92.	Mongolia	Low (2)	Medium (3)	5 (Medium)	3
93.	Paraguay	Low (2)	Medium (3)	5 (Medium)	2
94.	Sri Lanka	Low (2)	Medium (3)	5 (Medium)	7
95.	Suriname	Low (2)	Medium (3)	5 (Medium)	0
96.	Tunisia	Low (2)	Medium (3)	5 (Medium)	2
97.	Turkmenistan	Low (2)	Medium (3)	5 (Medium)	0
98.	Albania	Low (2)	Medium (3)	5 (Medium)	1
99.	Bahamas	Low (2)	Medium (3)	5 (Medium)	0
100.	Costa Rica	Low (2)	Medium (3)	5 (Medium)	8
101.	Cuba	Low (2)	Medium (3)	5 (Medium)	2
102.	Cape Verde	Very low (1)	High (4)	5 (Medium)	0
103.	Lebanon	Low (2)	Medium (3)	5 (Medium)	0
104.	Macedonia	Low (2)	Medium (3)	5 (Medium)	1
105.	Malaysia	Low (2)	Medium (3)	5 (Medium)	105
106.	Mauritius	Low (2)	Medium (3)	5 (Medium)	1
107.	Montenegro	Low (2)	Medium (3)	5 (Medium)	0
108.	Panama	Low (2)	Medium (3)	5 (Medium)	8
109.	Serbia	Low (2)	Medium (3)	5 (Medium)	0
110.	Libya	Low (2)	Medium (3)	5 (Medium)	0
111.	Trinidad and Tobago	Low (2)	Medium (3)	5 (Medium)	0
112.	United Arab Emirates	Medium (3)	Low (2)	5 (Medium)	5
113.	Uruguay	Low (2)	Medium (3)	5 (Medium)	7
114.	Barbados	Low (2)	Medium (3)	5 (Medium)	0
115.	Kuwait	Low (2)	Medium (3)	5 (Medium)	0
116.	Bahrain	Low (2)	Low (2)	4 (Low)	0
117.	Chile	Low (2)	Low (2)	4 (Low)	52
118.	Maldives	Very low (1)	Medium (3)	4 (Low)	0
119.	Samoa	Very low (1)	Medium (3)	4 (Low)	0
120.	Antigua and Barbuda	Very low (1)	Medium (3)	4 (Low)	0
121.	Cyprus[a]	Low (2)	Low (2)	4 (Low)	8
122.	Grenada	Very low (1)	Medium (3)	4 (Low)	0
123.	Israel	Low (2)	Low (2)	4 (Low)	22
124.	Malta[b]	Low (2)	Low (2)	4 (Low)	0
125.	Qatar	Low (2)	Low (2)	4 (Low)	1
126.	Saint Lucia	Very low (1)	Medium (3)	4 (Low)	0
127.	Singapore	Low (2)	Low (2)	4 (Low)	2
128.	Democratic People's Republic of Korea	Medium (3)	NA	NA	0

Source: Author

Source of project data: UNFCCC: CDM in Numbers (January 2012)

[a]Annex I to the UNFCCC has been amended to include Cyprus. This amendment will take effect from 1 January 2013 or on a later date. This means from the entry into force of this amendment, Cyprus will no longer be eligible to host new CDM projects. See Decision 10/CP.17, Amendment to Annex I to the Convention, FCCC/CP/2011/9/Add.2, 15 March 2012

[b]Annex I to the UNFCCC has been amended to include Malta, which means that the country is now no longer eligible to host new CDM projects. See Decision 3/CP.15, Amendment to Annex I to the Convention, FCCC/CP/2009/11/Add.1, 30 March 2010

This skewed distribution cannot be explained solely by countries' GHG emission levels. Although the countries that are currently performing well are among those with the highest ER potential,[44] many of the countries that also have relatively high potential are underperforming[45] particularly when compared to other countries in the same category[46] or those in a lower category.[47]

It also cannot be explained by countries' SD potential, as the current distribution of CDM projects does not match with that required by this factor. The groups of countries with the with high and very high SD potential are actually hosting the least number of projects, with most of them not hosting any project. Therefore considering both the ER and SD potentials of countries, neither of these elements explains the current distribution of CDM projects.

Table 10.1 shows also that many of the countries, such as Mexico, the Philippines, Thailand and Viet Nam, currently performing very well under the CDM are not among those with the highest CDM potential. In fact, countries like Israel, Malaysia and Chile have among the lowest CDM potential, but relatively high number of projects. This again cannot be explained by either their ER potential (low) or their SD potential (low or medium). In relation to those countries with higher CDM potential, the conclusion must be that this distribution is not equitable.

It is therefore reasonable to conclude that the current geographic distribution of CDM projects is inequitable and the reason for this inequity cannot be found solely in countries' ER or SD potential. Therefore, in order to address the problem of the inequitable geographic distribution of projects, it is necessary to ascertain the cause(s) of the problem, so that efforts can be targeted at these causes. This is what the next section sets out to do: to identify the main reasons for the inequitable geographic distribution of CDM projects.

10.5 Barriers to Distributive Justice in the CDM

Some of the barriers to participation in the CDM and equitable distribution of projects are internal to the countries involved, and include barriers that would affect any kind of investment. Examples of such internal barriers are corruption, lack of security, poor governance structures, conflict and political instability, all of which lead to high investment risks.[48] These internal barriers to investment are beyond the ability of the CDM regime to address, and so will not be discussed in this section, as modifications to the

[44] For example, China, Brazil, Indonesia, India, Mexico and the Republic of Korea are the countries with the highest GHG emissions and they are among the countries with the largest number of CDM projects.

[45] Such as Iran (seven projects), Nigeria (five projects) and Cambodia (five projects).

[46] Such as the Philippines (57 projects) or Malaysia (105 projects).

[47] Such as Chile (52 projects).

[48] See generally Matthias Busse and Carsten Hefeker, "Political Risk, Institutions and Foreign Direct Investment", 23 *European Journal of Political Economy* (2007), 397; and Chantal Dupasquier and Patrick N. Osakwe, "Foreign direct investment in Africa: Performance, challenges, and responsibilities", 17 *J. Asian Economics* (2006), 241.

CDM regime at the international level cannot address these barriers. However, many other barriers stem from the institutional makeup of the CDM itself and are issues that the international CDM regime can address, such as lack of capacity and lack of financing opportunities. These CDM barriers are the focus of this section.

Nevertheless, before moving on, it is useful to show that these internal barriers are not the key or sole reasons for the inequitable distribution of CDM projects. Statistics show that although many countries do have internal barriers to investment, this has not stopped some of them from performing well under the CDM. In addition, some of the countries that are actually doing well in terms of their internal governance structures are under-performing under the CDM. For instance, Botswana, Cape Verde, Mauritius, the United Arab Emirates, Uruguay and Qatar are performing relatively well in terms of the World Bank's governance indicators,[49] which are: voice and accountability,[50] political stability, government effectiveness, regulatory quality, rule of law and control of corruption.[51] Nevertheless, these countries are not doing well under the CDM: Botswana and Cape Verde do not host any project; Mauritius and Qatar host just one; and the United Arab Emirates hosts five. The Republic of Korea and Israel, whose good governance rankings are similar to these countries', host 67 and 22 projects, respectively. Mexico and the Philippines have much worse rankings, but they host 136 and 57 projects, respectively. China, which hosts almost half of all registered projects, ranks low compared with many other countries, such as Brazil (201), South Africa (20), Bhutan (2) and Lesotho (0), but this has not stopped it from being the single largest CDM host country and doing far better than these other countries. Even though some of these differences can be explained by the varying levels of ER potential and/or SD potential in these countries, not all of them can. For example, South Africa has greater ER and SD potentials than the Philippines,[52] and South Africa's governance ranking by the World Bank is higher than that of the Philippines, but while South Africa hosts 20 projects, the Philippines hosts nearly three times this number – 57 projects.[53]

---

[49] For the World Bank good governance indicators, see World Bank, "The Worldwide Governance Indicators (WGI) project", 2011, available at: http://info.worldbank.org/governance/wgi/index.asp (last accessed on 30 January 2012).

[50] These refer to the perception of how much a country's citizens are able to participate in selecting their government, as well as freedom of expression and association, and a free media. See Daniel Kaufmann, Aart Kraay and Massimo Mastruzzi, "The worldwide governance indicators: methodology and analytical issues", September 2010, available at: http://siteresources.worldbank.org/INTMACRO/Resources/WPS5430.pdf (last accessed on 6 February 2012), at 4.

[51] These countries are performing well for most, though not necessarily all of the statistics. But in comparison to other developing countries, they *are* performing very well.

[52] In absolute values. In the classification in Table 10.1, both have the same ER and SD potential rankings (as these rankings cover a range of absolute values).

[53] All governance statistics are for 2010 (the latest available). See World Bank, "Access governance indicators", 2011, available at: http://info.worldbank.org/governance/wgi/sc_country.asp (last accessed on 6 February 2012). Even computing beyond 2010, the conclusion remains that governance is not the key barrier to equitable distribution. For example, comparing South Africa's and the Philippines' governance indicators for 2007, 2008 and 2009, South Africa has consistently ranked higher, but the Philippines is still performing better under the CDM.

These statistics suggest that while internal structures and barriers may play a part in determining the distribution of CDM projects, there are other, probably more important, considerations that investors look out for, and these internal barriers are not the overriding barrier to CDM participation. Therefore, this section briefly outlines some of the barriers to distributive justice in the CDM, including the key barriers.

10.5.1 Lack of Capacity and Local Expertise

There are two elements to hosting CDM projects which may impact on the distribution of projects: the general investment/project element; and the CDM-specific element. CDM-specific issues arise out of the need to comply with the CDM modalities and procedures when developing and implementing CDM projects.[54] General investment issues are those that would affect normal investments, not just CDM projects, and relate to the underlying project. They include the legal and regulatory framework for investment within the host developing country, and the available infrastructure, such as transportation and telecommunications facilities. Lack of capacity in these two areas, that is, lack of CDM-specific and general investment capacity, has been identified as a barrier to CDM hosting and equitable distribution of projects.[55]

The primary reason why lack of capacity, particularly project development capacity, constitutes a barrier to CDM hosting and the equitable distribution of projects is the unilateral nature of many CDM projects. In the unilateral CDM structure, developing country entities themselves develop, finance and implement projects, rather than with developed country support.[56] As a result of this, those that lack the capacity to develop and implement projects are under-performing in the CDM market.

This capacity barrier to distributive justice in the CDM primarily undermines the 'SD potential' factor for achieving equitable distribution because the countries with the lowest human development and greatest SD potential are often those with the least capacity. It also undermines the 'ER potential' factor because many of the countries that lack the capacity to effectively participate in the CDM and are

---

[54] The CDM modalities and procedures are provided by the Conference of the Parties serving as the Meeting of the Parties to the Kyoto Protocol (COP/MOP) and the CDM Executive Board. These include the modalities and procedures for undertaking activities such as those relating to selecting the project methodologies, preparing the necessary project documentation such as the project design documents, and registering the project activities. See "Rules and References", available at: http://cdm.unfccc.int/Reference/index.html (last accessed on 27 February 2012).

[55] See Ann E. Prouty, "The Clean Development Mechanism and its Implications for Climate Justice", 34 *Columbia Journal of Environmental Law* (2009), 513, at 523; Sanja Lutzeyer, "Climate trading: the clean development mechanism and Africa", 12 *Stellenbosch Economic Working Papers* (2008), 1, at 27; and Emily Boyd et al., "The clean development mechanism: an assessment of current practice and future approaches for policy", 2007, available at: http://www.tyndall.ac.uk/sites/default/files/wp114.pdf (last accessed on 15 January 2012), at 23.

[56] See the discussion of the unilateral CDM structure as a barrier to equitable distribution, in Sect. 10.5.5 below.

therefore affected by this capacity barrier, such as least developed countries (LDCs) and sub-Saharan African countries among others, do have sufficient ER potential to participate in the CDM.[57]

10.5.2 *Finance and Cost-Related Barriers*

Lack of funding has been identified as a major barrier to the equitable distribution of CDM projects, or as a barrier to the hosting of projects by certain groups of countries, such as LDCs and African countries.[58] As with most projects, the funding required for CDM projects can be divided into: funding for the project transaction costs; and funding for the underlying project.

10.5.2.1 Transaction Costs

CDM project transaction costs include the cost of identifying potential CDM projects, identifying potential partners and negotiating the CDM contract, as well as the costs involved in the approval process, such as those associated with establishing baselines, proving additionality, validation, registration and verification of the project. They also include the share of proceeds and registration fees required by the Kyoto Protocol.[59]

Transaction costs are a barrier to local developers who cannot access the funds required to pay the transaction costs associated with the development of CDM projects. As these costs are incurred upfront, project developers would require some financing to cover the costs, which could be quite substantial. UNEP estimates the costs incurred during the CDM planning phase as ranging from US$18,500 to US$610,000, depending on various things such as the complexity and scale of the project.[60] The need for host country project developers to bear the bulk of these transaction costs would generally only arise in the case of unilateral projects, where

---

[57] Most of these countries fall in the medium ER potential category and others fall in the low potential category. For example, Angola (no project), Zambia (one project), Tanzania (one project), and Nigeria (five projects), all have medium ER potential and very high SD potential. See the classification of countries according to their emission reduction in Table 10.1 above.

[58] See UNEP and Ecosecurities, *Guidebook to Financing CDM Projects* (Roskilde: UNEP, 2007), at 3 and 7.

[59] Kyoto Protocol, supra, note 8, Art. 12.8 provides that a share of the proceeds of CDM projects should be used to cover administrative expenses, as well as to assist in meeting the cost of adaptation in developing countries. The share of proceeds to support adaptation in developing countries is 2% of CERs issued. See Decision 17/CP.7, supra, note 17, para. 15(a). The share of proceeds to cover administrative expenses, including the registration fee, is US$0.10 per CER issued for the first 15,000 tonnes of CO_2 equivalent and US$0.20 per CER issued for any amount in excess of 15,000 tonnes. See Decision 7/CMP.1, supra, note 19, para. 37.

[60] See, UNEP and Ecosecurities, *Guidebook to Financing CDM Projects*, supra, note 58, at 56.

the host country entity itself undertakes and finances all the preliminary elements of the CDM project.[61] However, even in the case of bilateral projects, the host country project developer may still have to bear some of the transactions costs, such as negotiation costs.

This barrier to equitable distribution particularly affects those countries with the lowest human development and greatest SD potential and therefore undermines the 'SD potential' factor for achieving equitable distribution.[62] It also, obviously, undermines the 'ER potential' factor, because some of these countries with the greatest SD potential that are unable to effectively participate in the CDM also have ER potential.[63] Their inability to participate effectively therefore means that their potential is not being adequately exploited under the CDM.

10.5.2.2 Implementation Costs

These refer to the actual or direct cost of producing the goods, as opposed to the transaction costs, which are the costs associated with organising production. Under the CDM, the implementation costs would include the project construction costs, such as purchasing the plant and equipment, and the project operating costs, such as the cost of maintenance and other running costs.[64]

Lack of underlying finance for the project has been identified as a major barrier to CDM participation, particularly for those smaller developing countries that do not have strong financial institutions. For example, Sieghart, commenting on the Yemeni experience, states that "some buyers offer to assist with the designing of the project. However, transaction costs are not perceived as the major financial barrier by project developers. Developers face difficulties in procuring underlying finance due to a deficiency of domestic capital and both to country-specific and CDM-specific risks."[65]

The original expectation of the CDM was that it would attract foreign investment, and that this foreign investment would provide financing for the actual CDM project, beyond the purchase of CERs generated from the projects. If this original expectation was generally the case, local project developers would only have to secure foreign developed country counterparts to invest in the projects and this investment would cover the implementation costs of the project, in exchange for the

[61] See the discussion on unilateral CDM projects below.

[62] See, for example, Jane Ellis and Sami Kamel, "Overcoming Barriers to Clean Development Mechanism Projects", May 2007, available at; http://www.oecd.org/dataoecd/51/14/38684304.pdf (last accessed on 12 February 2012), at 32–33, where the authors state that transactions costs are a barrier faced by many project developers, especially for small-scale projects, and in poor developing countries.

[63] See note 57 above for examples of such countries.

[64] See generally for the financing requirements of CDM projects, UNEP and Ecosecurities, *Guidebook to Financing CDM Projects,* supra, note 58.

[65] Lia C. Sieghart, "Unilateral clean development mechanism – an approach for a least developed country? The case of Yemen", 12 *Environmental Science and Policy* (2009), 198, at 201.

CERs generated by the project.[66] However because of the prevalence of unilateral CDM projects and pure CER purchase-type transactions, the norm has become that local developers source local financing for the underlying projects and then secure foreign developed country counterparts to purchase the CERs generated by the projects. This is a problem for many countries that do not have well-developed financial institutions, and for those that even where these institutions exist, local financiers are reluctant to invest in CDM projects because of a lack of understanding of its operation and because of its greater risk compared to other kinds of projects. In these situations, local developers have difficulty sourcing the required financing for the underlying projects locally and depend on foreign investment, which is often not forthcoming because of the preference for simply purchasing CERs.[67]

Just like the transaction costs barrier, this implementation costs barrier undermines the 'SD potential' and 'ER potential' factors for achieving equitable distribution of CDM projects.

10.5.3 Preference for Large-Scale Projects

The size of CDM projects has been identified as a barrier to the distribution of projects. Specifically, this has been highlighted as investors' preference to invest in projects that will generate a minimum quantity of CERs. This is partly in order to ensure that considering the transaction costs of the project, the quantity of CERs generated is enough to make the project worthwhile.[68] Linked to this barrier is the relatively low level of industrial development in some countries, resulting in limited opportunities for large-scale projects. Because the CDM seeks to assist developed countries to meet their Kyoto targets in a cost-effective way, investors will consider cost-effectiveness in determining the attractiveness of any CDM project. Some investors have a minimum project size they will invest in. For example, the World Bank requires the volume of emission reductions to be generated from a project to be large enough to make a project viable, and states that for example, a small-scale project should generate at least 50,000 tonnes of CO_2 equivalent annually.[69]

---

[66] Here, the developing country would for example benefit from the use of renewable energy, capacity building, technology transfer and other sustainable development benefits arising from the project.

[67] See Gregor Pfeifer and Geoff Stiles, "Carbon finance in Africa – a policy paper for the Africa Partnership Forum", 2008, available at: http://www.africapartnershipforum.org/dataoecd/40/15/41646964.pdf (last accessed on 16 February 2012), at 17; Axel Michaelowa, "Unilateral CDM – can developing countries finance generation of greenhouse gas emission credits on their own?", 7 *International Environmental Agreements: Politics, Law and Economics* (2007), 17, at 17; and Sieghart, "Unilateral clean development mechanism", supra, note 67, at 202.

[68] See the discussion of transaction costs in Sect. 10.5.2.1 above.

[69] See World Bank, "Minimum Project Requirements", available at: http://web.worldbank.org/WBSITE/EXTERNAL/TOPICS/ENVIRONMENT/EXTCARBONFINANCE/0,,contentMDK:21844766~menuPK:5220728~pagePK:64168445~piPK:64168309~theSitePK:4125853,00.html (last accessed on 16 February 2012).

However, although this has been identified as a barrier by some authors,[70] the number of small-scale projects that have been registered and that are in the pipeline belies this claim. As of 30 January 2012, of the 3,815 registered CDM projects, 1,627 (43%) were small-scale and 2,188 (57%) were large-scale projects.[71] This means that even if investors do prefer large-scale projects in order to minimise cost and maximise cost-effectiveness, small-scale projects are still being developed and registered at almost the same rate as large-scale projects. A likely explanation for this is that most small-scale projects are unilateral, the host countries themselves are almost solely responsible for the projects, which involve no foreign investment, and therefore what foreign investors want does not directly affect the rate of developing and implementing such projects.

The barrier presented by many investors' preference for large-scale projects mainly undermines the ER potential factor for achieving equitable distribution. This is because when countries with limited opportunities for large-scale projects are ignored or overlooked, their ER potential (even though this potential can only be tapped primarily through small-scale projects) is basically lost and is not exploited under the CDM. In addition, it also undermines the SD potential factor because many of the countries with the lowest human development and greatest SD potential have limited opportunities for large-scale projects and are thereby affected by this barrier.

10.5.4 The Market-Based Nature of the CDM

As highlighted above, the CDM is a market-based mechanism. Developed or developing country entities can invest in these projects[72] and the resulting CERs can either be traded or used directly by the developed country participant (to comply with its emission reduction commitment).

Although the CDM was created as a mechanism that would both generate cost-effective emission reductions and contribute to sustainable development,[73] the

---

[70] See Alan Silayan, "Equitable distribution of CDM projects among developing countries", 255 *Hamburg Institute of International Economics Report* (2005), 1, at 23–24; Prouty, "The Clean Development Mechanism and its Implications for Climate Justice", supra, note 57, at 523; and Ben Pearson, "Market failure: Why the Clean Development Mechanism Won't Promote Clean Development", 15 *Journal of Cleaner Production* (2007), 247, at 250.

[71] See CDM, "Registered project activities by scale", available at: http://cdm.unfccc.int/Statistics/Registration/RegisteredProjByScalePieChart.html (last accessed on 30 January 2012).

[72] If investment comes from developing country entities, the projects are referred to as unilateral, and if from developed country entities, they are either bilateral or multilateral, depending on the number of developed country entities involved in the project.

[73] Kyoto Protocol, supra, note 8, Art. 12. See also Prouty, "The Clean Development Mechanism and its Implications for Climate Justice", supra, note 55, at 522.

very nature of the CDM as a market-based instrument is preventing it from achieving these objectives equitably among developing countries. The nature of the CDM means that apart from the necessary environmental constraints,[74] normal market considerations, such as risk and cost, largely dictate the location of projects. Investors are generally more interested in lower cost and risk projects, with the cost of a CDM project and the profit to be derived from it being the major considerations.[75] Added to the problem is the fact that the sustainable development element of the CDM, unlike its GHG emission reduction element, has no monetary value, and is therefore not factored into the cost or profit of the CDM.[76] There is no market incentive to promote sustainable development and no particular benefit to investors from investing in projects with high sustainable development benefits.[77] Because of this, for investors, who are considering cost and profit, the GHG reduction element is usually the paramount consideration. It is partly because the market-based nature of CDM projects that the size of projects, which partly determine the profit to be achieved from projects, and cost-related issues also constitute barriers to equitable distribution.

The consequence is that those developing countries that are rapidly industrialising, with the attending industries, high emission levels, institutions and possibly project experience or existing foreign direct investment, are better placed to host CDM projects, and investors will therefore gravitate towards these countries. This is compounded by the CDM no longer being used purely as a compliance tool by developed country entities, but also as a profit-generating mechanism. This means that although many public and private entities invest in the CDM in order to use the CERs generated to meet their emission reduction commitments or to comply with environmental regulations in their jurisdictions, many invest in the CDM in order to

---

[74] Such as rules to ensure that projects result in real, measurable, and long-term benefits related to the mitigation of climate change, and that reductions in emissions are additional to any that would occur in the absence of the certified project activity. See Kyoto Protocol, supra, note 8, Art. 12.

[75] See Sieghart, "Unilateral clean development mechanism", supra, note 65, at 199; and Harrie Oppenoorth et al., "The Bali guide on CDM: towards a sustainable CDM", November 2007, available at: http://www.snm.nl/pdf/klimaattopbali_brochure_bali_guide_def_webversie_copy.pdf (last accessed on 12 January 2012), at 20.

[76] According to the CDM rules, the host developing countries are responsible for determining that projects will contribute to their sustainable development. The host country is required to confirm that the CDM project activity assists it in achieving sustainable development. See Decision 3/CMP.1, supra, note 15, para 40(a) of the Annex. Also the host entity usually provides in the project design document, an explanation of the sustainable development contributions of the project. Beyond this, there is no regulation or rule concerning what this means or should constitute. The regulatory tools that have been developed (such as tools for assessing the additionality of the project) are mainly focused on calculating the emission reductions achieved by the project, and not measuring the sustainable development benefits it provides.

[77] See Christoph Sutter and Juan Carlos Parreño, "Does the current clean development mechanism (CDM) deliver its sustainable development claim? An analysis of officially registered CDM projects", 84 *Climatic Change* (2007), 75, at 89.

trade the CERs generated and make profit from such trade.[78] Because of this, these entities would not only go for projects that cost the least, they will in particular go for projects that can generate the greatest profit, and most likely follow the normal foreign direct investment trends.

Although these issues are doubtless relevant and should be considered, the important point is that market-based indicators are only suitable for one element of the CDM – the GHG emission reduction element. The sustainable development element of the CDM must also be considered if the CDM is to actually achieve its dual objectives but these indicators do not compute this element. It is not suggested that the CDM should no longer operate as a market, or have market characteristics. However, it is essential that to ensure achievement of both objectives of the CDM, while investors consider market factors in selecting host countries and projects, they also consider sustainable development factors, such as countries' needs and sustainable development potential. A combination of the two, rather than just the cost-effectiveness factor, should guide investors' choices.

The barrier to equitable distribution presented by the market-based nature of the CDM mainly undermines the "SD potential" factor for achieving equitable distribution. This is because by not considering countries' SD potential, investors are not adequately considering the specific circumstances of those countries with low development levels. If countries' SD potential was actually considered by investors, then, it follows that those countries with the highest SD potentials (because of their low development levels) would be preferred over those countries with less SD development potential, or at least that they would have the opportunity to participate more effectively in the CDM. In addition, because these countries with the greatest SD potential also have ER potential, this barrier also undermines the "ER potential" factor for achieving equitable distribution.

10.5.5 The Unilateral CDM Structure

The above discussions show that many of the barriers to equitable distribution, such as lack of capacity, lack of financing and other cost-related barriers, constitute barriers to equitable distribution of projects primarily because of the unilateral nature of many CDM projects. Hence, one of the major barriers to equitable distribution of

[78] For example, as of November 2010, Ecosecurities was the largest CDM investor/CER purchaser, with a share of about 12% of all registered CDM projects. See the CDM Pipeline, 1 November 2010. Ecosecurities is however not a compliance buyer, but a CER trader, and is in the business of "sourcing, developing and trading emission reduction credits." See Ecosecurities, "Who we are", 2010, available at: http://www.ecosecurities.com/Home/EcoSecurities__the_carbon_market/Who_we_are/default.aspx (last accessed on 1 March 2012). See the CDM Pipeline (available at: www.cdmpipeline.org (last accessed on 1 March 2012)) for an analysis of all CDM projects and the official CDM website (available at: http://www.cdm.unfccc.int (last accessed on 1 March 2012)) for CDM statistics.

CDM projects is the predominance of unilateral CDM projects in the CDM market. In the unilateral CDM structure, the CDM project is developed and implemented by local project developers with financing obtained usually from local investors/financial institutions, and the resulting CERs are then sold to developed countries, developed country private entities or market traders. The key element here is that the purchaser of CERs does not invest in the underlying project – the only finance provided is for the purchase of the CERs.

These commodity-style purchase transactions are possibly the most common form of CDM projects.[79] The CDM is rapidly moving away from the envisaged foreign investment and involvement-based mechanism to one which mainly involves local developers and financiers. This dominance of unilateral CDM projects constitutes a barrier to equitable distribution because unilateral projects require the hosts to have sufficient financial and technical capacity to undertake such projects. Developing countries that lack such capacity are unable to implement unilateral CDM projects and are consequently sidelined in the CDM market. This ability to unilaterally host projects is not in itself inequitable. On the contrary, it is very beneficial particularly to those developing countries that have the capacity to unilaterally develop and implement projects, and that can also raise the necessary financing. For instance, the host countries would be able to focus on projects that align with their sustainable development objectives, rather than those projects that are more financially-attractive to a developed country sponsor.

The disadvantage arises specifically because both unilateral and bilateral CDM projects compete in the same market and for the same developed country entities. There is a finite demand for CDM projects/CERs. If there is preference for unilateral projects over bilateral projects, then the demand for bilateral projects ultimately will be reduced. And because unilateral projects currently dominate the market, the share of bilateral projects is inevitably reduced.

Many countries, especially LDCs and other poor developing countries, rely on foreign investment and capacity building to be able to develop and host projects. These countries lack the financial and technical capability to exploit their CDM potential and will thus be unable to enjoy the sustainable development benefits (such as direct investment, capacity building and technology transfer) the CDM is meant to contribute to. And yet, they are likely to be those most in need of these benefits because of their low human development.[80] Consequently, this barrier created by the unilateral CDM structure mainly undermines the "SD potential" factor for achieving equitable distribution. However, as already highlighted several times, because those countries with the greatest SD potential also have ER potential, this barrier also undermines the "ER potential" factor.

---

[79] Because project proponents are not required to disclose their source and style of funding, it is not possible to determine precisely how the market is divided among the various structures available. It is possible that although in some PDDs, it is not stated that the foreign entity is investing directly in the project, or that a contract has been signed for the purchase of CERs, that this is actually the case.

[80] See the classification of countries according to their SD potential/human development levels in Table 10.1 above.

10.6 Analysis of Barriers

Regarding whether the CDM regime can support an equitable distribution of projects, the answer is that two of the main elements of the CDM regime constitute the main barriers to equitable distribution of projects. These two elements in fact lead to those countries with the greatest SD potential being unable to effectively participate in the CDM. These elements are the market-based nature of the CDM and the prevalence of unilateral CDM projects in the CDM market. Because of the prevalence of unilateral CDM projects and the availability of CERs for purchase, developed country entities have less of an incentive to directly invest in CDM projects, with the attendant risks and financial commitments required. However, even where developed country entities invest directly in projects, they prefer to transact with larger, rapidly-industrialising developing countries, mainly because of their greater potential and financial and technical capacity, to the detriment of the smaller, less-industrialised developing countries, who are often those with the lowest human development and greatest SD potential.[81] This means that even bilateral projects will often by-pass smaller developing countries because of the market-based nature of the CDM.

Various actions have been undertaken within the CDM regime to address the inequitable distribution of CDM projects, such as: the Nairobi Framework, which aims to increase the African region's participation in the CDM, primarily through capacity building[82]; initiatives to reduce the transaction costs of projects, for LDCs or countries generally, such as fee exemptions for LDCs,[83] and the provision of loans to countries hosting fewer than 10 CDM projects[84]; and a registration fee exemption and simplified modalities for small-scale projects.[85] However, there are currently no initiatives which address the two main barriers of the market-based nature of the CDM and the prevalence of unilateral CDM projects. This is unfortunate, because as noted above, these two barriers are the main reasons why the current distribution of CDM projects is inequitable. It is probably because of this that these initiatives have not been particularly successful in ensuring a more equitable distribution of projects.[86]

---

[81] As noted above, internal barriers such as lack of good governance cannot completely explain the distribution of CDM projects.

[82] See CDM, "Regional Distribution – Nairobi Framework", available at: http://cdm.unfccc.int/Nairobi_Framework/index.html (last accessed on 28 March 2012).

[83] See Decision 17/CP.7, supra, note 17, para. 15(b) and Decision 2/CMP.3, Further Guidance Relating to the Clean Development Mechanism, FCCC/KP/CMP/2007/9/Add.1, 14 March 2008, para. 31.

[84] See Decision 2/CMP.5, supra, note 19, paras. 49–50; and Decision 3/CMP.6, Further Guidance Relating to the Clean Development Mechanism, FCCC/KP/CMP/2010/12/Add.2, 15 March 2011, para. 64 and Annex III.

[85] See Report of the 37th Meeting of the CDM Executive Board, Annex 20, para. 4; and Decision 4/CMP.1, Guidance Relating to the Clean Development Mechanism, FCCC/KP/CMP/2005/8/Add.1, 30 March 2006, Annex II. For a more detailed discussion of these and other initiatives, see Akanle, "Distributive Justice in International Law", supra, note 19, at pages 189–238.

[86] See note 20 above for a history of the distribution of projects, which highlights that the same four countries have consistently been hosting the majority of projects.

As it currently operates therefore, the CDM regime, with its market-based nature, primacy of market forces and the prevalence of unilateral CDM projects, does not and will probably be unable to support an equitable distribution of CDM projects.

10.7 Recommendations

There are various ways of addressing these barriers to equitable distribution. For instance, to address the capacity barrier, the most obvious solution is capacity building. To have the greatest impact, this should, ideally, be targeted specifically at those countries with the lowest human development and greatest SD potential. The capacity building should also respond to specific capacity needs, such as those identified through a comprehensive study of countries' capacity.

To help overcome the barrier created by the market-based nature of the CDM, investors can be required to take countries' SD potential into consideration when selecting countries to invest in. This should go beyond considering the SD potential of projects, as this could just lead to more sustainable projects in the same countries already dominating the market.[87] Instead, in keeping with the factors to be considered for achieving distributive justice under the CDM, countries' SD potential should be considered and preference given to those countries with the greatest potential. When investing in countries, investors should consider why that particular country is the most appropriate, given its human development level and SD potential.

Considering that there is no real benefit to investors of taking sustainable development potential into consideration, especially when this may necessitate investing in countries that can only produce less profitable projects, there is a risk that a requirement of this sort may drive investors away from the CDM market because it will diminish their opportunities to generate profits relative to costs. However, having a market that has a different focus may result in a different outcome. In other words, the CDM market needs the right focus. It needs to focus, not on maximising profit and minimising risk and cost, but on ensuring achievement of its environmental objectives of reducing GHG emissions and promoting sustainable development, which should be done equitably among developing countries. Although making profit and reducing risk and cost could be part of the focus of the market, it should not be, as it is now, the primary focus.

One way of addressing this issue is by promoting the practice of socially-responsible investing within the CDM – investing in a way that incorporates social, environmental or ethical criteria with financial objectives.[88] For socially-responsible investors,

[87] There is nothing wrong with this. The issue is that those countries that are underperforming should also have the chance to fulfil their CDM potential.

[88] See Peter Waring and Tony Edwards, "Socially responsible investment: explaining its uneven development and human resource management consequences", 16 *Corporate Governance: An International Review* (2008), 135, at 135.

making a return on their investments, though an important aim, is not the overriding concern.[89] The situation with CDM investors should be similar – the overriding concern of CDM investors should not be making profit, but achieving the objectives of the CDM, which are to achieve GHG emission reductions and promote sustainable development, rather than to generate profit for investors.

An excellent example of this is the initiative by the European Commission in relation to LDCs. The Commission has decided that in Phase III of the European Union (EU) Emissions Trading Scheme (ETS), which will run from 2013 to 2020, the only new CDM projects which will be automatically eligible for inclusion in the Scheme are CDM projects implemented in LDCs.[90] Specifically, CERs generated from new CDM projects registered in LDCs from 2013 onwards will be automatically accepted into the EU ETS, whereas CERs from new projects in non-LDCs will only be accepted if there is an agreement for this purpose between the Commission and the relevant country.[91]

This is a good example of the kind of preferential treatment that should be given to countries with the lowest human development and highest SD potential, as it would help ensure that they fulfil their CDM potential. It is also important to ensure that the CDM market does not falter because of this kind of provision. While there is an obvious need to increase the participation of LDCs in the CDM market, it is also important to ensure that other countries, which are not LDCs, also have the opportunity to host projects. The key thing is that the ER and SD potentials of all countries should be considered and they should be given the opportunity to host projects, according to these potentials. In addition, it is not sufficient to limit automatic eligibility to LDCs – what is needed is for developed countries to actively source and finance projects in the developing countries that are currently underperforming in the CDM, which have both the ER and SD potentials to perform better.

To address the problem of the prevalence of unilateral projects/financial assistance, the most obvious solution is to require that a specific percentage of all registered projects must be bilateral in the real sense and, where the projects are multilateral,[92] they should be funded by the multilateral investor, rather than by the host country entity itself. This can be done, for instance, by having a requirement that x% of registered projects should be bilaterally-funded, or that x% of CERs used by developed countries to fulfil their emission reduction objectives should be obtained from bilaterally-funded projects. These options however may not directly improve the participation of those countries with the greatest need, as developed countries, in complying with the options, may simply increase their investments in the countries already performing well under the CDM. That is why these options should be used

---

[89] Ibid.

[90] Such CERs will be accepted into the EU ETS until 2020 or until these countries have entered into an agreement with the EU for this purpose, whichever is earlier. See Council Directive 2009/29/EC amending Directive 2003/87/EC so as to improve and extend the greenhouse gas emission allowance trading scheme of the Community, OJ 2009 L 140/63, Art. 11a(4).

[91] Ibid., Art. 11a(5).

[92] That is, where they involve several developed country entities whether acting directly or through a fund, such as the various World Bank carbon funds.

in conjunction with that proposed as a solution to the market-based nature of the CDM – requiring investors to consider countries' SD potential, to encourage them to increase their investments in those countries with the lowest human development.

10.8 Conclusion

It is ironic that those countries that are most in need of CDM projects, because of their low development levels, are actually the ones benefitting the least from the CDM. It is also ironic that they are benefitting the least precisely for the reason they are most in need – because of their low development levels. They lack the technical and financial capacity to implement CDM projects, and as most CDM projects are undertaken by developing countries themselves, this has formed an effective barrier to prevent the poorest countries from participating in the CDM.

Although there appears to be a possible solution to the problem of the prevalence of unilateral projects in the CDM market, the barrier created by the market-based nature of the CDM does not appear to be as easy to overcome. If the concept of socially-responsible investing is introduced into the CDM market, with an emphasis on effectively considering the sustainable development objective of the CDM and ensuring that more countries are able to participate in the CDM, this could reduce the focus of the market on financial incentives and refocus the market more effectively on the CDM's environmental objectives of promoting sustainable development (and GHG emission reductions) equitably among developing countries. However, there is possibly no legal solution to effectively ensure consideration of this, although investors or groups can voluntarily adopt the socially-responsible investing approach to ensure that those countries that are underrepresented in the CDM, particularly those with the greatest need, are helped to increase their level of CDM participation.

So far, this chapter has focused on the current structure and operation of the CDM, and the barriers preventing an equitable distribution of projects. As highlighted above, the Kyoto Protocol first commitment period comes to an end in 2012,[93] and countries were meant to finalize considerations for the second commitment period in 2009.[94] They were however unable to meet this deadline, spurring concern about

---

[93] Supra, note 11.

[94] The GHG emission reduction commitments contained in the Kyoto Protocol (in Annex B) must be achieved by the end of the first commitment period which runs from 2008 to 2012 (Protocol, Art. 3.1). The Protocol does not contain the commitments for subsequent periods, but provides in Art. 3.9 that consideration of these commitments shall be initiated by 2005. During the 11th Conference of the Parties (COP 11) in December 2005, the *Ad Hoc* Working Group on Further Commitments for Annex I Parties under the Kyoto Protocol (AWG-KP) was established. Its aim is to determine what commitments developed countries will take on post-2012, and how they will meet those commitments. See Decision 1/CMP.1, Consideration of Commitments for Subsequent Periods for Parties Included in Annex I to the Convention under Article 3, paragraph 9, of the Kyoto Protocol, FCCC/KP/CMP/2005/8/Add.1, 30 March 2006, paras. 2–3. Countries decided to conclude this work and forward their conclusions to COP/MOP 5 in December 2009. See Report of the Ad Hoc Working Group on Further Commitments for Annex I Parties under the Kyoto Protocol on its Resumed Fourth Session, FCCC/KP/AWG/2007/5, 5 February 2008, para. 22(c).

the future of the Kyoto Protocol and its instruments, including the CDM. However, at COP/MOP 7 in December 2011, countries established the second commitment period, which will start on 1 January 2013 and end in 2017 or 2020.[95] The CDM (and the other Kyoto Protocol instruments) will continue to operate during this period.[96]

In addition, some organisations have made efforts to ensure that the CDM market would continue to operate, even if countries had been unable to reach agreement regarding the Protocol's second commitment period before the end of the first. For instance, the European Commission will continue to accept CERs from CDM projects implemented in the LDCs into the EU ETS.[97] Also, the World Bank's Umbrella Carbon Facility has put up new funding of €68 million (US$89 million) for CERs generated after 2012.[98] It is therefore probably accurate to say that the CDM has a future, and the issues raised in this chapter are equally relevant to the operation of the CDM during the Kyoto Protocol second commitment period.[99]

Countries have also "defined" a new market mechanism, operating under the guidance and authority of the Conference of the Parties, which developed countries can use to meet part of their mitigation targets or commitments. They have also undertaken to "maintain and build upon" the existing Kyoto Protocol flexibility mechanisms, which includes the CDM.[100] This means that the new market mechanism

---

[95] See Decision 1/CMP.7, Outcome of the Work of the Ad Hoc Working Group on Further Commitments for Annex I Parties under the Kyoto Protocol at its Sixteenth Session, FCCC/KP/CMP/2011/10/Add.1, 15 March 2012, paras. 1–2.

[96] See Decision 3/CMP.7, Emissions Trading and the Project-Based Mechanisms, FCCC/KP/CMP/2011/10/Add.1, 15 March 2012, para. 1. See also CDM, "Frequently Asked Questions", available at: http://cdm.unfccc.int/faq/index.html (last accessed on 26 March 2012).

[97] It will also accept CERs from CDM projects implemented in non-LDC countries with which it enters into an agreement for this purpose. See supra, note 90.

[98] See World Bank, "World Bank ups funding for post-2012 credits", 13 January 2011, available at: http://wbcarbonfinance.org/docs/World_Bank_ups_funding_for_post-2012_credits.pdf (last accessed on 28 March 2012); and World Bank, "Umbrella Carbon Facility T2", available at: http://wbcarbonfinance.org/Router.cfm?Page=UCFT2&ItemID=53224&FID=53224 (last accessed on 28 March 2012).

[99] Although the CDM market will continue to operate, it is still uncertain just how much demand there will be for CERs. See the discussion at supra, note 37.

[100] See Decision 2/CP.17, Outcome of the work of the Ad Hoc Working Group on Long-term Cooperative Action under the Convention, FCCC/CP/2011/9/Add.1, 15 March 2012, para. 83 and Preamble to Part E, para. 4. This new mechanism may be created under the Convention. It is however more likely to be created under the new international agreement which countries are currently negotiating, which is intended to come into effect and be implemented from 2020. See Decision 1/CP.17, Establishment of an Ad Hoc Working Group on the Durban Platform for Enhanced Action, FCCC/CP/2011/9/Add.1, 15 March 2012, para. 4.

Countries have requested the *Ad Hoc* Working Group on Long-term Cooperative Action under the Convention to conduct a work programme to elaborate modalities and procedures for the new mechanism, with a view to recommending a decision to COP 18. Parties and admitted observer organisations have been invited to submit their views on possible modalities and procedures, including their positive and negative experiences with existing approaches and mechanisms, as well as lessons learned. See Decision 2/CP.17, para. 85. This would provide a good opportunity for countries to ensure that the lessons from the operation of the CDM are taken into account when designing the new mechanism.

will likely be modelled on the CDM. It is very important that the new mechanism not repeat the mistakes of the CDM in order not to face the same criticisms faced by the CDM.[101] Instead, when designing this new mechanism, countries should learn from the CDM and ensure that the new mechanism is a better, improved mechanism. For instance, in order to give the new mechanism a better chance of achieving distributive justice, some of the recommendations in this chapter could be implemented, such as limiting the percentage of unilateral projects that can be registered.[102]

Nevertheless, there does not appear to be very much that can be done to address the problems created by the current design of the CDM as a market-based instrument. And the new mechanism "defined" by parties is also intended to be a market-based mechanism. There is no real incentive that can be given to investors to make it really worth their while to take sustainable development into consideration and there is the risk that requiring them to do so may drive investors away from the market. Addressing the problem of unilateral CDM projects should go some way in correcting the skewed distribution of projects. There is however the very real possibility that if investors cannot purchase enough CERs and need to invest directly in projects, they will simply do this in the countries where it makes the best market sense. So the problem may not be solved at all. The question that this chapter cannot shy away from therefore is whether the CDM can achieve distributive justice and whether there is any point in continuing efforts to achieve this. Should the CDM continue to attempt to achieve sustainable development and GHG emission reductions equitably among countries? Or should it be streamlined to be simply a market mechanism to achieve cost-effective emission reductions, with no significance attached to where the reductions are achieved?

In reality, CERs are issued for emission reductions achieved in countries, and not for sustainable development contributions. This is how it has to be in order to maintain the environmental integrity of the CDM, considering that these CERs are then used to offset the emission reduction objectives of developed countries. The final conclusion is that the CDM regime, given its market-based nature, may not be able to achieve a truly equitable distribution of projects, and that there is no legal solution to this problem. The only option would be to accept that the CDM cannot continue to operate as a simple market mechanism and to introduce regulations that are not

---

[101] For some of these criticisms, see Charlotte Streck, "Expectations and Reality of the Clean Development Mechanism: A Climate Finance Instrument between Accusation and Aspirations", in Richard Stewart, Benedict Kingsbury and Bryce Rudy (eds), *Climate Finance: Regulatory and Funding Strategies for Climate Change and Global Development* (New York: New York University Press, 2009), 67, at 67–75; and Pearson, "Market Failure: Why the Clean Development Mechanism Won't Promote Clean Development", supra, note 70, at 249.

[102] As noted above, there are several criticisms of the CDM, and there is a lot of literature on how the CDM should be reformed in the post-2012 period. The suggestions contained in such literature could also be useful in the design of a new market mechanism. See for instance, Emily Boyd et al., "Reforming the CDM for Sustainable Development: Lessons Learned and Policy Futures", 12 *Environmental Science and Policy* (2009), 820.

really suitable for a typical market mechanism,[103] but which would help the CDM to achieve its dual objectives. On the other hand, countries could simply accept that there will be no truly equitable distribution of projects and turn their attention elsewhere in efforts to contribute to sustainable development in developing countries.

[103] Such as requiring investors to invest directly in certain countries, or requiring them to take countries' sustainable development potential into consideration. See the recommendations in Sect. 10.7.

Chapter 11 Legal Aspects of Climate Change Adaptation

Jonathan Verschuuren

Abstract This chapter provides an overview of the current state of affairs with regard to the legal aspects of adaptation to climate change. After a brief introduction into the relationship between adaptation and law, I will discuss the international legal obligations with regard to adaptation that exist under the umbrella of the UNFCCC. The remainder of this chapter will show the impact of adaptation on the fields of law involved, particularly law related to marine and coastal adaptation, water management, biodiversity conservation, planning and land use and buildings and infrastructure, energy and telecommunications, and migration. In fact, there are hardly any policy fields and associated laws that are not impacted by climate change. The big question that arises is whether all of the pieces of legislation associated with these policy fields are suited to facilitate adaptation measures. Is, perhaps, adaptation of the law required, and, if so: what needs to be changed?

11.1 Introduction

Climate change is here to stay, at least for the time being. Even halting all greenhouse gas emissions today (which, of course, is highly unlikely) would only lead to first observable changes for the better after 30–40 years. So we have to adapt to the changing climate. The Working Group II report of the Intergovernmental Panel on Climate Change (IPCC) concludes that adaptation will be necessary to address impacts resulting from the warming which is already unavoidable due to past

Jonathan Verschuuren is a professor of International and European Environmental Law at Tilburg University.

J. Verschuuren (✉)
Tilburg Sustainability Center, Tilburg Law School,
5000 Tilburg, Netherlands
e-mail: j.m.verschuuren@tilburguniversity.edu

E.J. Hollo et al. (eds.), *Climate Change and the Law*, Ius Gentium: Comparative Perspectives on Law and Justice 21, DOI 10.1007/978-94-007-5440-9_11,
© Springer Science+Business Media Dordrecht 2013

emissions. For some impacts, namely those that already show or will show in the very near future, adaptation is *the only available and appropriate response*, according to the IPCC.[1] Such impacts for instance are:[2]

- increased water availability in some regions, and
- decreased water availability and increasing drought in others
- increased ecosystem changes (species shifting their natural range) and risk of extinction of species
- negative impacts on a small scale for poor farmers and fishers
- increased damage from floods and storms
- increased burden from malnutrition and infectious diseases and a changed distribution of some disease vectors such as the mosquito's vectoring malaria dengue.

The drafters of the 1992 UN Framework Convention on Climate Change (UNFCCC) treated mitigation and adaptation as equally important. The UNFCCC mentions adaptation as one of the policies and measures to mitigate the adverse effects of climate change in Article 3(3). Under the commitments listed in Article 4 of the UNFCCC, several deal with adaptation. Adaptation, however, received much less attention than mitigation in the 20 years that followed the signing of the UNFCCC. This not only goes for policy and legal measures taken on the basis of the UNFCCC and regional and national implementation thereof, but also for academic research.

Perhaps the lack of attention is one of the reasons why adaptation measures are still in their infancy. It appears to be much more difficult to devise and implement adaptation policies than it is to devise and implement mitigation policies. Partly this is due to the fear that attention afforded to adaptation measures will make it more difficult to get mitigation measures adopted and implemented. Partly, the problem with adaptation is that you cannot always take one-size fits all measures. Some of the consequences of climate change, particularly sea level rise, are similar in all regions of the world. Mostly, however, the consequences of climate change may dramatically differ from one region to another. Some regions suffer from droughts, whereas others face increased floods. This may even be so within one country, such as is the case for instance in large countries such as the US, China and Australia. It may even be so that within the same region, periods of extreme droughts are followed by a period of intense rainfall causing floods. The biggest problem, however, is an inherent difficulty of adaptation law and policy. Whereas mitigation measures primarily can be implemented in one policy field (that of environmental law), adaptation measures have to be implemented through a wide range of policies such as water, marine and coastal, fisheries, biodiversity, energy, building and construction, agriculture, telecommunications, infrastructure, etc. It requires an immense, coordinated effort, to adapt all the policies and laws in these fields of government in such a way

[1] Neil Adger et al., "Summary for Policymakers", in Martin L. Parry et al. (eds), Climate Change 2007: Impacts, Adaptation and Vulnerability. Contribution of Working Group II to the Fourth Assessment Report of the Intergovernmental Panel on Climate Change (Cambridge and New York: Cambridge University Press, 2007), 18.

[2] Ibid., see Table SPM-1 of the WGII summary for policymakers, at 15.

that society at large will be prepared for the climatic changes and associated extreme weather events that we are going to experience in the course of the twenty-first century, with a peak expected between 2050 and 2080.

This chapter provides an overview of the current state of affairs with regard to the legal aspects of adaptation to climate change. After a brief introduction into the relationship between adaptation and law (Sect. 11.2), I will discuss the international legal obligations with regard to adaptation that exist under the umbrella of the UNFCCC (Sect. 11.3). The remainder of this chapter will show the impact of adaptation on the policy fields and the related laws involved. There are hardly any policy fields that are not impacted by climate change. Hence, adaptation measures in these fields are required. The big question that arises is whether all of the pieces of legislation associated with these policy fields are suited to facilitate adaptation measures. Is, perhaps, adaptation of the law required, and, if so: what needs to be changed?

11.2 General Introduction to Adaptation Law

11.2.1 What Is Adaptation?

Adaptation is generally, quite loosely, defined as the process of adjusting to climate change and its impacts.[3] This definition is sometimes criticized for its somewhat reassuring connotation.[4] Species, including humans, have always adapted to changes in their natural environment, and the human species has proven to be particularly good at that. "We are found in a wide diversity of physical environments and thrive under a range of climatic characteristics", even though, "in geological terms, the human presence on the Earth has been exceptionally brief".[5] Adaptation to climate change, however, is different from what we are used to because, among others, the changes that are occurring now cannot be considered to be natural, are very complex and diverse and are not just gradual changes, but may be sudden, drastic changes as well. Therefore, there are ecological, individual and cultural limits to adaptation.[6] Hence, it is sometimes concluded that we should not talk "just" about

[3] Ben Orlove, "The Past, the Present and Some Possible Futures of Adaptation", in W. Neil Adger, Irene Lorenzoni and Karen L. O'Brien (eds), *Adapting to Climate Change. Thresholds, Values, Governance* (Cambridge: Cambridge University Press, 2009), 131.

[4] Ibid, at 131–163.

[5] Donald R. Nelson, "Conclusions: Transforming the World", in W. Neil Adger, Irene Lorenzoni and Karen L. O'Brien (eds), *Adapting to Climate Change. Thresholds, Values, Governance* (Cambridge: Cambridge University Press, 2009), 491.

[6] W. Neil Adger et al., "Adaptation now", in W. Neil Adger, Irene Lorenzoni and Karen L. O'Brien (eds), *Adapting to Climate Change. Thresholds, Values, Governance* (Cambridge: Cambridge University Press, 2009), 1.

adaptation, but about "transformational changes", that require us to reform the basis on which we think about the world.[7]

Although it is debatable whether we are doing just that when discussing legal aspects of climate change adaptation, I do feel that lawyers are not underestimating the changes that are necessary to our legal system to facilitate and accommodate the necessary adaptation measures. Let us just have a brief look at the measures that are suggested in, what probably is the most advanced adaptation plan that exists today, the 2010 report by the New York City Panel on Climate Change (NYPCC). This plan gives a good overview of the measures that should be taken in a modern metropolitan area like New York City. They include, for instance[8]:

- Zoning, environmental, water and waste regulations to manage precipitation, flooding and stormwater, to ensure sustained water supply and to protect wastewater infrastructure;
- Zoning and land use regulations dealing with sea level rise and storm surges;
- Laws and regulations aimed at facilitating increased energy demand during heat waves and designating public buildings that can serve as emergency cooling centers in heat waves;
- Laws and regulations enabling the development of new power generation resources, and encouraging siting of emergency power generators to supply energy when there is a peak in demand (during heat waves) or in case of a weather induced electric outage;
- Laws and regulations requiring upgrading underground energy and telecommunications infrastructure to withstand flooding and sea level rise;
- Various environmental impact assessment regulations requiring assessment of the consequences of the proposed activity on adaptation to climate change;
- Amending energy, building, and sewer codes to adapt buildings to high wind conditions, flooding, and high temperatures;
- Requirements on green or energy-smart landscaping, leading to energy consumption reductions in buildings, storm water retention and tree shading;
- Requirements on adapting the transportation infrastructure to deal with flooding, saltwater damage, increased power demands and power outages, overheating of subway platforms, increased stress on infrastructure because of higher temperatures;
- Air quality requirements dealing with more frequent periods of elevated concentrations of ground-level ozone, as well as elevated ozone concentrations;
- Requirements on waste management sites and brownfields to prevent containment leaking in case of flooding and sea level rise.
- Improving laws and regulations concerning emergency preparedness to be able to deal with storm surges, inland flooding and heat waves during summer.

---

[7] Nelson, supra, note 5, at 497.

[8] As discussed by Edna Sussman et al., "Climate Change Adaptation: Fostering Progress Through Law and Regulation", 18 *N.Y.U. Environmental Law Journal* (2010), 55.

This list just deals with the specific situation in New York City. Other cities around the globe will face other challenges. Coastal defense systems may have to be reinforced or even re-created from the start, for which land reclamation may be in order or expropriation of current land owners; negative impacts on coastal habitats will have to be minimized or mitigated. Parts of a city may even be moved to floating platforms. Other cities may not have the capacity and funds to adopt the kind of the measures New York City is adopting and may have to divert to much more drastic measures such as abandoning parts of the city or the entire city altogether. The same goes for rural areas that are below or on sea level or rural communities that have to cope with increasing droughts in addition to other stresses such as extreme poverty, land degradation, or a large number of HIV/AIDS infections. Bangladesh and parts of southern Africa are the cases in point for situations like this. Mass displacement and mass migration both internally and across border are direct consequences, which in turn may lead to food shortage, lack of adequate housing and jobs, violence between migrants and the existing population. Current international refugee law clearly is not equipped at all to deal with these so called climate refugees and their problems.

Outside of the city, a wide range of adaptation measures are indeed necessary. In order to protect biodiversity, policies should be aimed at making protected areas climate proof by making sure that these areas are large, robust, stable and interconnected enough to adapt to the changing climate. Protected areas should be able to live through occasional flooding, wild fires, and extreme weather events, such as heavy storms. They should have enough variety in habitat types to host new species in search of a more suitable new climate zone. Current biodiversity law, both at the domestic and at international level, is hardly able to provide for these kind of measures. In addition, in most legislatures, biodiversity law is aimed at specific "flag species". As a consequence of climate change, though, species are appearing and disappearing within just a few years, rendering the idea of having a specific flag species for a certain site useless.

Measures in the field of planning law and water law may be necessary to deal with the increased risk of inland flooding. Lands may be designated to serve as controlled flooding areas (or flood control reservoirs) to protect more sensitive parts of the land against flooding. Land may also have to be designated for fresh water storage, so as to have buffer capacity in times of drought. In those regions with increased winter precipitation and increased risk of drought during summer, such as will be the case in northwestern Europe, farmers may have to store fresh water on their agricultural lands in winter, to be used in summer. It is obvious that here, agricultural policies and water policies are closely interlinked, as water in controlled flooding areas, which often are agricultural lands, can be used for this purpose. Farmers may even engage in recreational activities in summer with water thus stored on their lands.

The agricultural sector also will have to adjust crop variety depending on the changing climate and weather conditions. In those areas of the world where local agriculture is essential for local food production, the government will have to intervene in order to keep or restore food security. The kind of measures that have to be taken vary greatly throughout the world. Farmers may have to shift to crops that are better suited in a wetter, or dryer, or warmer climate, or crops that are better suited in an environment of increasing salinity. More or less the same goes for the forestry sector. The fisheries sector will have to get used to regular adjustments of fish quota due to climate change induced decreased fish stocks. The surplus of water on agricultural lands described above, may foster the introduction of aquaculture on agricultural lands.

Planning law and building law are among the fields of law that are applied by government authorities seeking to reduce potential harms from wild fires in those areas that are prone to increasing risks of bush fires.

Climate change is not only impacting government policies in the fields mentioned above, but also has a direct impact on private actors and thus on private law. Insurance companies are assessing the consequences of climate change for their line of business, the tourist industry is shifting its focus to new, yet to be developed areas, and law firms are setting up climate litigation divisions as the number of law suits is likely to increase with the increasing damage by climate change and with increasing adaptation costs. With the increase of court cases, it is clear that litigation law, insurance law, and even property law is affected by climate law and is, in a way, adapting to climate change as well.

We can, therefore, conclude that climate change adaptation necessitates a thorough revisiting of the law as it is. A transformation of the law is necessary, as much as a transformation of societies as a whole.

11.2.2 Links Between Adaptation and Mitigation

Adaptation cannot be regarded in isolation from mitigation. Not only are both important elements of any climate change policy, they also are positively and negatively interlinked. First, many adaptation measures equally serve mitigation goals. Afforestation, reforestation, preserving and restoring mangroves, with the goal to protect the land against flooding, against landslides following intense rainfall, or against negative impact of storms, are all measures that can as well be part of a mitigation strategy. The same goes for adaptation measures in the field of biodiversity conservation. Creating corridors between protected areas, and enlarging protected areas through restoration leads to an increase in vegetation, and thus to additional carbon uptake. This is even true for the construction of green buildings and green roofs: not only are they naturally cooler and thus an effective measure against heat waves, they also lead to less energy consumption and, again, to additional carbon uptake. In fact, any measure to reduce the amount of energy consumption to avoid an energy fall out during heat waves is a mitigation measure as well.

The second link between adaptation and mitigation is a negative one. Adaptation measures can be harmful to the climate in the sense that they lead to a) more greenhouse gas emissions, or b) a reduction of carbon uptake. The examples are obvious. Installing air conditioners to combat the heat lead to more energy consumption and thus, if the energy comes from a coal fueled energy installation, to higher emissions. Replacing a natural coastal habitat by a large seawall to combat sea level rise and storm surges, leads to a loss of natural carbon uptake. This example, coincidentally, shows that an adaptation measure in one policy field may be detrimental to adaptation policy in another field, as this large seawall that replaces natural coastal habitat probably will lead to a loss of biodiversity adaptation opportunities.

Mitigation measures, on the other hand, can be harmful for adaptation as well. Afforestation in arid and semi-arid regions strongly reduces water yields and thus has a negative impact on local agriculture and biodiversity. Switching to hydropower may reduce irrigation options for farmers and thus deprive them of adaptation opportunities.

It is clear that adaptation and mitigation policies have to be developed together, so that they are mutually beneficial.[9]

11.2.3 Dealing with Risks and Uncertainty

Although thousands of scientists around the globe have been researching climate change for many years now, we still are faced by many uncertainties. Uncertainties as to the exact nature and intensity of the changes that we can expect, the timeframe within which they are to be expected, and the locations that will be hardest hit. Uncertainties present themselves in two ways: (a) uncertainties caused by our lack of knowledge, and (b) uncertainties caused by randomness inherent to the phenomenon at hand. In literature on risks, these are usually described as epistemic risks (a) and aleatory risks (b). Both types of risks are relevant in climate adaptation law and policies.

Obviously, risks inherent to the randomness of floods, storms and other events influenced by climate change will never disappear. We can only predict that in certain areas storm intensity will increase, rainfall will increase etc., without being able to tell exactly when and where a storm will hit and how strong that storm will be.

Uncertainties as a consequence of a lack of knowledge do gradually become smaller because of increasing scientific knowledge as to what is happening to atmospheric and climatic processes due to higher levels of greenhouse gases in the

[9] See extensively chapter 18 of the IPCC's Working Group II report: Richard J. T. Klein et al., "Inter-relations between Adaptation and Mitigation", in Martin L. Parry et al. (eds), *Climate Change 2007: Impacts, Adaptation and Vulnerability. Contribution of Working Group II to the Fourth Assessment Report of the Intergovernmental Panel on Climate Change* (Cambridge and New York: Cambridge University Press, 2007), 745, at 745–777.

Earth's atmosphere. We must, however, not be so optimistic as to think that this type of uncertainty will eventually disappear altogether, at least not in the foreseeable future. Our lack of knowledge still is quite large, especially because as a consequence of climate change, *everything* is changing. Increasing carbon emissions lead to temperature rise, which leads to a great number of subsequent changes in weather patterns. Many of these changes probably do not occur in a linear way. Instead, it becomes increasingly clear that there are many tipping point effects and non-linear effects. Such effects are much more difficult to predict. Paradoxically, increasing scientific knowledge thus has lead to an increase in the epistemic risks. We now better know what we do not know. Our lack of knowledge also concerns the effect of these changes on human behavior and on that of the other living organisms on Earth. How are living organisms, including human beings, going to adapt to these changes once they occur, gradually or suddenly? In addition to all of this, we should keep in mind that scientists will keep disagreeing on certain aspects of their findings. This is simply what scientists do: they question findings in order to establish a theory on the basis of which the most accurate predictions can be made. Policy makers have to deal with this "tug of war among competing theories and qualifications of theories."[10]

Kleindorfer shows how both types of risks are tightly intertwined with choice. A homeowner or a business might consider options such as insurance or mitigation before the fact in order to either reduce or pay for losses resulting from an extreme weather event. In order to reach such a decision, the homeowner of business may want to gain knowledge to reduce the epistemic risk. Knowledge thus is a fundamental input to rational choice under uncertainty.[11] The same processes take place at the macro level of state authorities making climate adaptation policies and regulations. Authorities are faced with the huge challenge of regulating for an uncertain future and with dealing with "low probability, high consequence events". It requires a proactive and long term approach of policy makers and regulators; an approach in which they can rely less on front-end methodologies,[12] such as environmental impact assessments, regulatory impact assessments, and cost benefit analyses.[13] Laws and regulations have to leave room for long term decision-making under uncertainty and give the decision-makers the tools to still reach legitimate decisions.

[10] Paul Kleindorfer, "Interdependency of Science and Risk Finance in Catastrophe Insurance and Climate Change", 18 January 2010, available at: http://papers.ssrn.com/sol3/papers.cfm?abstract_id=1538161 (last accessed on 25 February 2012), at 10.

[11] Ibid., at 12–13.

[12] J.B. Ruhl, "Climate Change Adaptation and the Structural Transformation of Environmental Law", 40 *Environmental Law* (2010), 363.

[13] Climate change decision-making truly shook up discussions on cost benefit analyses because it forces us to think about how to value the benefits for future generations in today's currency. See extensively Richard L. Revesz and Matthew R. Shahabian, "Climate Change and Future Generations", 15 August 2010, available at: http://papers.ssrn.com/sol3/papers.cfm?abstract_id=1666423 (last accessed on 25 February 2012).

11.2.4 Climate Ethics

Like the whole international climate change debate, discussions on adaptation have a strong ethical dimension too. Many claim that developed states, by not taking the lead in an aggressive mitigation scheme, place the burden of climate change, almost entirely caused in developed countries, on the shoulders of people living in poor countries and people belonging to the next generation. This process is labeled "an ethical failure."[14] Developed countries and future generations are facing huge adaptation challenges which they either cannot afford, or which are technically unfeasible. In 2007, the UNFCCC estimated the total global adaptation costs in 2030 to amount to $49–171 billion per annum, of which $27–66 billion would accrue in developing countries.[15] A 2009 evaluation of these calculations showed that the actual costs may very well be a factor 2 or 3 higher.[16] As discussed below, therefore, current international negotiations mainly focus on generating funds for developing countries to meet these costs. Nevertheless, there exists also a strong body of literature arguing that climate law and policy should focus on the environmental problem of climate change instead of connecting climate change policies to developmental policies.[17] It may not come as a surprise that this approach is criticized in an equally strong body of literature.[18]

11.3 Adaptation in the International Climate Regime

11.3.1 Introduction

It is fair to say that adaptation law originates at the international level. Both the 1992 United Nations Framework Convention on Climate Change,[19] and the 1997 Kyoto Protocol[20] comprise various obligations for the parties to these international

---

[14] Stephen M. Gardiner, *A Perfect Moral Storm: The Ethical Tragedy of Climate Change* (Oxford: Oxford University Press, 2011).

[15] UNFCCC, Investment and Financial Flows to Address Climate Change (Bonn: UNFCCC, 2007).

[16] Martin Parry et al., "Assessing the Costs of Adaptation to Climate Change. A Review of the UNFCCC and other Recent Estimates", 2009, available at: http://pubs.iied.org/pdfs/11501IIED.pdf (last accessed on 25 February 2012).

[17] For example, Eric A. Posner and David Weisbach, *Climate Change Justice* (Princeton, NJ: Princeton University Press, 2010).

[18] For example: Rosemary Lyster, "Towards a Global Justice Vision for Climate Law in a Time of 'Unreason'", June 2011, available at: http://papers.ssrn.com/sol3/papers.cfm?abstract_id=1943818 (last accessed on 25 February 2012).

[19] United Nations Framework Convention on Climate Change (UNFCCC), New York, 9 May 1992, in force 21 March 1994, 31 *International Legal Materials* (1992), 849.

[20] Kyoto Protocol to the United Nations Framework on Climate Change, Kyoto, 10 December 1997, in force 16 February 2005, 37 *International Legal Materials* (1998), 22.

agreements to adopt and implement adaptation policies. The documents also impose upon developed countries the duty to, both financially and practically, assist developing countries with their adaptation actions. In the international arena, most attention is going to the latter issue, as will be obvious from the overview of the most important adaptation related actions both under the UNFCCC (Sect. 11.3.2) and under the Kyoto Protocol (Sect. 11.3.3).

11.3.2 International Adaptation Law Under the UNFCCC

11.3.2.1 Adaptation in the UNFCCC

The primary objective of the UNFCCC, as laid down in Article 2, is to achieve stabilization of greenhouse gas concentrations in the atmosphere at a level that would prevent dangerous anthropogenic interference with the climate system. Such a level should be achieved within a time frame sufficient to allow ecosystems to adapt naturally to climate change, to ensure that food production is not threatened and to enable economic development to proceed in a sustainable manner. Mitigation, according to the objective of the UNFCCC, is supposed to be successful so that ecosystems, food production and the economy are more or less automatically kept as they were before.

Fifteen years later, the IPCC sounded the alarm bell. Adaptation is necessary to address impacts resulting from the warming which is already unavoidable due to past emissions. For impacts that already show or will show in the very near future, adaptation is *the only available and appropriate response*, according to the IPCC in 2007.[21]

Fortunately, we did not have to start from scratch after these alarming words of the IPCC had been published. As a matter of fact, other than the objective of the UNFCCC cited above suggests, the 1992 convention focuses not just on combating climate change, but also on combating the adverse effects of climate change. This is obvious from the principles of the UNFCCC, laid down in Article 3. Developed country Parties, for instance, should take the lead in combating the adverse effects of climate change.[22] Developing country Parties that are particularly vulnerable to the adverse effects of climate change should be given full consideration.[23] Precautionary measures should be taken to mitigate the adverse effects of climate change.[24]

[21] Adger et al., "Summary for Policymakers", supra, note 1, at 18.

[22] UNFCCC, supra, note 20, Art. 3(1).

[23] Ibid., Art. 3(2).

[24] Ibid., Art. 3(3).

Under the commitments listed in Article 4 of the UNFCCC there are several commitments that deal with adaptation. Parties to the Convention have to:

- formulate, implement, publish and regularly update national and regional programmes containing measures to facilitate adequate adaptation to climate change[25];
- cooperate in preparing for adaptation to the impacts of climate change; develop and elaborate appropriate and integrated plans for coastal zone management, water resources and agriculture, and for the protection and rehabilitation of areas, particularly in Africa, affected by drought and desertification, as well as floods[26];
- take climate change considerations into account in the relevant social, economic and environmental policies and actions, and employ appropriate methods, for example impact assessments, with a view to minimizing adverse effects on the economy, on public health and on the quality of the environment, of projects or measures taken by them to adapt to climate change[27];
- promote and cooperate in research (scientific, technological, technical, socio-economic and other) intended to further the understanding and to reduce the remaining uncertainties regarding the economic and social consequences of various response strategies,[28] as well as exchange information on this[29]; and
- promote and cooperate in education, training and public awareness related to climate change in general, thus including adaptation issues.[30]

In addition to this impressive list of adaptation duties, even further obligations have been imposed on developed country parties. They also have to help developing countries with their adaptation policies by:

- assisting developing countries that are particularly vulnerable to the adverse effects of climate change in meeting the costs of adaptation to those adverse effects[31]; and by
- taking all practicable steps to promote, facilitate and finance the transfer of, or access to, environmentally sound technologies and know how to developing country Parties, to enable them to implement the provisions of the UNFCCC[32]; this obligation, thus, also relates to the adaptation provisions mentioned above. Hence, the transfer of technology not only applies to mitigation technologies, but also to technologies and know-how necessary to implement adaptation measures.

[25] Ibid., Art. 4(1)(b).

[26] Ibid., Art. 4(1)(e).

[27] Ibid., Art. 4(1)(f).

[28] Ibid., Art. 4(1)(g).

[29] Ibid., Art. 4(1)(h).

[30] Ibid., Art. 4(1)(i).

[31] Ibid., Art. 4(4).

[32] Ibid., Art. 4(5).

An especially interesting provision on adaptation commitments is Article 4(8), in which all Parties are compelled to focus on the specific needs and concerns of developing country Parties arising from the adverse effects of climate change and the impact of the implementation of response measures. A whole range of countries especially vulnerable to the effects of climate change is mentioned here: small island countries, countries with low-lying coastal areas, countries with arid and semi-arid areas, forested areas and areas liable to forest decay, countries with areas prone to natural disasters, countries with areas liable to drought and desertification, countries with areas of high urban atmospheric pollution, countries with areas with fragile ecosystems, including mountainous ecosystems, countries whose economies are highly dependent on income generated from the production, processing and export, and/or on consumption of fossil fuels and associated energy-intensive products, and landlocked and transit countries.

All subsequent provisions of the UNFCCC, on such issues as research and systematic observation, education, training and public awareness, the financial mechanism, etc., apply to both mitigation and adaptation.[33] The two subsidiary bodies that are established by the UNFCCC,[34] the Subsidiary Body for Scientific and Technological Advice (SBSTA), and the Subsidiary Body for Implementation (SBI), both are competent to deal with adaptation issues.

11.3.2.2 National Adaptation Programmes of Action

Many countries have developed national adaptation programmes pursuant to Article 4(1)(b) of the UNFCCC. Developing countries, however, appeared to be largely unable to draft such programmes. Since such a programme is considered to be the first step towards financing and implementation of adaptation measures, the conference of the parties, in its 2001 session in Marrakesh, adopted guidelines for the preparation of National Programmes of Action (NAPAs)[35] to assist the least developed countries in the process,[36] and established a Least Developed Country Fund (LDCF).[37] This has been quite successful. By 2011, 45 least developed countries had submitted their NAPAs, containing detailed priority policies to respond to their urgent and immediate needs to adapt to climate change.[38] Each NAPA reports on the urgent

[33] Ibid., Arts. 5 and seq.

[34] Ibid., Arts. 9 and 10.

[35] Not to be confused with NAMAs, Nationally Appropriate Mitigation Actions.

[36] Decision 28/CP.7, Guidelines for the preparation of national adaptation programmes of action, FCCC/CP/2001/13/Add.4, 21 January 2002.

[37] Decision 7/CP.7, Funding under the Convention, UN Doc. FCCC/CP/2011/13 Add.1, 15 March 2012.

[38] The secretariat of the UNFCCC makes all of these available through their website, available at: http://unfccc.int/cooperation_support/least_developed_countries_portal/submitted_napas/items/4585.php (last accessed on 25 February 2012).

and immediate needs for which further delay could increase vulnerability to climate change or lead to increased costs at a later stage.

The 2010 Nepal NAPA, for instance, formulates extensive policies to increase communities' resilience in the field of water management, agriculture, disaster management, forest and ecosystem management, public health, and others.[39] Associated with these policies is an impressive list of over a hundred specific climate adaptation options for agriculture and food security, for the water sector, for the energy sector, for forests and biodiversity, for public health, for urban settlements and infrastructure, and to address climate-induced disasters. Concrete measures range from simple practical measures such as tree planting around farm lands and water resources in mountainous regions[40] and better rain water collection,[41] to complex legal measures such as a zoning program to adapt to water induced disasters and the associated activation of an inundation committee,[42] and the enforcement of planning regulations and building codes in urban areas incorporating climate change dimensions.[43]

After submission to the secretariat, the NAPA enters the so called GEF cycle which allows for funding of the projects identified in the NAPA under the LDCF, which is administered by the Global Environment Facility (GEF). The GEF is a financial organization instituted by governments, international organizations such as the World Bank and UNEP, NGOs and the private sector, to fund environmental projects in developing countries.[44]

11.3.2.3 The Buenos Aires Programme of Work on Adaptation and Response Measures and the Nairobi Work Programme on Impacts, Vulnerability and Adaptation to Climate Change

Since the adoption of the UNFCCC in 1992, adaptation has been an issue during several Conferences of the Parties (COP), particularly those held in Buenos Aires in 2004, and in Cancun in 2010. COP10 in Buenos Aires led to the adoption of the Buenos Aires Programme of Work on Adaptation and Response Measures, which is a rather weak attempt to push adaptation, especially in developing countries, forward. Developing countries were requested to make use of the NAPA opportunities mentioned above, developed countries were asked to make available additional funding for adaptation measures under NAPAs, and the GEF was requested to step

---

[39] Government of Nepal, "National Adaptation Programme of Action (NAPA) to Climate Change", 2010, available at: http://unfccc.int/essential_background/library/items/3599.php?rec=j&priref=7329#beg (last accessed on 25 February 2012).

[40] Ibid., at 69.

[41] Ibid., at 68.

[42] Ibid., at 70.

[43] Ibid., at 77.

[44] For more information, see the GEF's website, available at: http://www.thegef.org (last accessed on 25 February 2012).

up its activities in the implementation phase of the NAPAs. In addition to this, the Buenos Aires programme of work initiated additional information gathering and capacity-building measures. Finally, the SBSTA was requested to develop a 5-year programme of work on scientific, technical and socio-economic aspects of impacts, vulnerability and adaptation to climate change.[45]

This 5-year programme was adopted 1 year later during COP11 in Montreal,[46] and has as its main objective to assist the parties, especially developing countries, to improve their understanding of impacts, vulnerability and adaptation, and to make informed decisions on practical adaptation actions. The SBSTA operationalized this again 1 year later during COP12 in Nairobi,[47] hence its current name: Nairobi Work Programme on Impacts, Vulnerability and Adaptation to Climate Change (NWP) "Understanding Vulnerability, Fostering Adaptation". This programme has been running ever since and acts as a coordination mechanism for a wide variety of capacity-building and information dissemination activities by some 200 organizations, including private companies such as insurance firms and consultancy firms, universities and other research institutes, development banks and other financial organizations, NGOs, and charity organizations.[48]

11.3.2.4 The Cancun Adaptation Framework and the Adaptation Committee

It was only in 2010 that the COP placed adaptation high on its agenda. In the Cancun Agreements, adopted in 2010, the conference dealt with adaptation first (before mitigation), stating that enhanced action on adaptation was urgently needed.[49] The conference decided to establish the Cancun Adaptation Framework under which all parties to the UNFCCC will[50]:

- Plan, prioritize and implement adaptation actions;
- Assess adaptation actions, including economic, social and environmental evaluation of adaptation options;

[45] Decision 1/CP.10, Buenos Aires Programme of Work on Adaptation and Response Measures, FCCC/CP/2004/10/Add. 1., 19 April 2005.

[46] Decision 2/CP.11, Five-year Programme of Work of the Subsidiary Body for Scientific and Technological Advice on Impacts, Vulnerability and Adaptation to Climate Change, UN Doc. FCCC/CP/2005/5/Add. 1, 30 March 2006.

[47] Report of the Subsidiary Body for Scientific and Technological Advice on its Twenty-Fifth Session, UN doc. FCCC/SBSTA/2006/11, 1 February 2007, at 5–13.

[48] Information on activities and organizations involved can be found at the UNFCCC's website, available at: http://unfccc.int/adaptation/nairobi_work_programme/items/3633.php (last accessed on 25 February 2012).

[49] Decision 1/CP.16, The Cancun Agreements: Outcome of the work of the Ad Hoc Working Group on Long-term Cooperative Action under the Convention, UN Doc. FCCC/CP/2010/7/Add. 1., 15 March 2011.

[50] Ibid.

- Strengthen institutional capacities for adaptation;
- Build resilience of socio-economic and ecological systems;
- Enhance climate change related disaster risk reduction strategies, taking into consideration early warning systems, risk assessment and management, and sharing and transfer mechanisms such as insurance;
- Take measures to enhance understanding, coordination and cooperation with regard to climate change induced displacement, migration and planned relocation;
- Research, develop and diffuse technologies, practices and processes for adaptation;
- Strengthen information, education and public awareness;
- Improve climate related research in order to help provide information to decision-makers at the national and regional levels.

The Adaptation Committee (AC) was established to promote the implementation of all this.[51]

11.3.3 *International Adaptation Law Under the Kyoto Protocol*

11.3.3.1 Adaptation in the Kyoto Protocol

Five years after the adoption of the UNFCCC, the Kyoto Protocol mainly focused on the reduction of greenhouse gas emissions and not on adaptation. Given the stage of international discussions on climate change at that time, this was to be expected: a substantial first step towards greenhouse gas emission reductions was within reach. This explains why the Kyoto Protocol's provisions on adaptation are a bit meager.

The Protocol does work out a little further the general provisions of the UNFCCC mentioned above. It states that the programmes to facilitate adequate adaptation to climate change have to concern the energy, transport and industry sectors as well as agriculture, forestry and waste management, and that adaptation technologies as well as spatial planning are important elements in such adaptation policies.[52] In addition, the Kyoto Protocol regulates that information on such national adaptation policies has to be integrated into the communication that has to be submitted to the secretariat of the UNFCCC.[53] From a practical point of view, the most important provision on adaptation in the Kyoto Protocol probably is a section in the provision on the Clean Development Mechanism, stating that money generated through CDM projects should be used to assist developing countries to meet the

[51] Ibid. During the Durban COP in 2011, composition of, and modalities and procedures for, the Adaptation Committee were adopted. It yet has to develop its work plan.

[52] Kyoto Protocol, supra, note 21, Art. 10(b)(i).

[53] Ibid., Art. 10(b)(ii).

costs of adaptation.[54] It was this provision that, 4 years later, formed the basis for the establishment of the Adaptation Fund.

11.3.3.2 Adaptation Fund and Other Funds

The Adaptation Fund was established during COP7 of the UNFCCC in Marrakesh in 2001,[55] but it operates under the Kyoto Protocol. Its aim is to finance concrete adaptation projects in developing country parties to the Kyoto Protocol, as well as some specific adaptation activities that are not limited to developing countries. The latter is a consequence of another decision taken in Marrakesh which aimed at the following activities [56]:

- Starting to implement adaptation activities promptly where sufficient information is available to warrant such activities, in the areas of water resource management, land management, agriculture, health, infrastructure development, fragile ecosystems, including mountainous ecosystems, and integrated coastal zone management;
- Improving the monitoring of diseases and vectors affected by climate change, and related forecasting, and improving disease control and prevention;
- Supporting capacity-building for preventive measures, planning, preparedness and management of disasters relating to climate change, including contingency planning, in particular, for droughts and floods in areas prone to extreme weather events;
- Strengthening and, if necessary, establishing national and regional information networks for rapid response to extreme weather events.

It was not until the Conference of the Parties serving as the meeting of the Parties to the Kyoto Protocol (COP/MOP) first convened in 2005 in Montreal that the Adaptation Fund materialized.[57] In the next few years, various decisions were taken to operationalize the Fund, first, in 2006, by adopting its guiding principles,[58] then, in 2007, more importantly, by instituting the Adaptation Fund Board (AFB) as the operating entity of the Fund.[59] Although the COP/MOP remains in control over the Adaptation Fund, the AFB does have considerable power over the actual

---

[54] Ibid., Art. 12(8).

[55] Decision 10/CP.7, Funding under the Kyoto Protocol, UN Doc. FCCC/CP/2001/13/Add.1, 21 January 2002.

[56] Decision 5/CP.7, Implementation of Article 4, paragraphs 8 and 9, of the Convention (decision 3/CP.3 and Article 2, paragraph 3, and Article 3, paragraph 14, of the Kyoto Protocol), UN Doc. FCCC/CP/2001/13/Add.1, 21 January 2002.

[57] Decision 28/CMP.1, Initial guidance to an entity entrusted with the operation of the financial mechanism of the Convention, for the operation of the Adaptation Fund, UN Doc. FCCC/KP/CMP/2005/8/Add.4, 30 March 2006.

[58] Decision 5/CMP. 2, Adaptation Fund, UN Doc. FCCC/KP/CMP/2006/10/Add.1, 2 March 2007.

[59] Decision 1/CMP. 3, Adaptation Fund, UN Doc. FCCC/KP/CMP/2007/9/Add.1, 14 March 2008. On the functioning of the Adaptation Fund, see in more detail Ralph Czarnecki and Kaveh Guilanpour, "The Adaptation Fund after Poznań", 3 *Carbon and Climate Law Review* (2009), 79.

funding process.[60] The AFB is a legal entity under German law,[61] with involvement of both the GEF and the World Bank, acting as secretariat and trustee respectively.[62] The Adaptation Fund is financed by a fixed share of 2% of the proceeds of all CDM project activities (2% of the Certified Emission Reductions, or CERs, issued for a CDM project to be precise), as well as individual donations by countries.

Finally, in 2010, almost 10 years after the decision was taken to establish an adaptation fund, the AFB took its first decision to fund an adaptation project, namely an adaptation project to coastal erosion due to sea level rise in Senegal.[63]

Besides the Adaptation Fund, there are two more funds, both of which operate under the UNFCCC. Like the Adaptation Fund, the Special Climate Change Fund (SCCF) and the LDCF were established during COP7 in Marrakesh in 2001.[64] The SCCF is a general fund, aimed at financing activities, programmes and measures relating to climate change in a number of areas, one of which is adaptation (the others being transfer of technology, energy, transport, industry, agriculture, forestry, waste management, and activities that assist developing countries to diversify their economies). The LDCF, as it name already indicates, is specifically aimed at financing projects in the least developed countries, particularly, as indicated above in Sect. 11.3.2.2, those related to their NAPAs. Both funds are operated by the GEF, and their councils have joint meetings. They both mainly rely on voluntary contributions from Annex I states.

Examples of projects funded under the joined funds are for instance a project to promote the implementation of national and transboundary integrated water resource management that is sustainable and equitable given expected climate change in Swaziland, a project integrating climate change risks into water and flood management by vulnerable mountainous communities in the Greater-Caucus region of Azerbaijan, and a project to promote a value chain approach to adaptation in agriculture in Ghana.[65]

With the proliferation of funds, there is a clear risk of overlap and of inefficiency. The GEF, involved in all of them, as well as in their own GEF Trust Fund, which initially also had funding of adaptation measures as one of its main goals,[66] is aware of that. In their 2010 Revised Programming Strategy on Adaptation to Climate Change, the GEF Council tries to show how the different adaptation related funds

[60] Its rules of procedure were adopted in 2008, Decision 1/CMP.4, Adaptation Fund, UN Doc. FCCC/KP/CMP/2008/11/Add.2., 19 March 2009.

[61] Decision 5/CMP. 6, Report of the Adaptation Fund Board, UN Doc. FCCC/KP/CMP/2010/12/ Add.1, 15 March 2011.

[62] MOU's with both institutions have been concluded to formalize this involvement. Decision 1/CMP. 4, UN Doc. FCCC/KP/CMP/2008/11/Add.2, supra, note 62, paras. 11 and 12.

[63] Information on the projects funded under the Adaptation Fund is available from the AFB's website, available at: http://www.adaptation-fund.org (last accessed on 25 February 2012).

[64] Decision 7/CP.7, Funding under the Convention, UN Doc. FCCC/CP/2001/13/Add.1, 21 January 2002.

[65] UN Doc. GEF/LDCF/SCCF.9/JointSummary, 18 November 2010.

[66] Through its Strategic Priority on Adaptation, a 50 million USD allocation inside of the GEF Trust Fund. As of 2010, the SPA was merged with the SCCF/LDCF, see the GEF website, available at: http://www.thegef.org/gef/adaptation (last accessed on 25 February 2012).

are all complimentary.[67] The arguments used are not particularly strong though. The fact that the LDCF and the SCCF have a mandate under the UNFCCC and the Adaptation Fund under the Kyoto Protocol, for instance, is but a formal argument. The fact that their revenues come from different sources is not a strong argument either, nor is the argument that the LDCF and the SCCF, other than the Adaptation Fund, are aimed at other issues than adaptation as well.

The biggest problem of the funds, however, is the lack of available funds. In the 42 NAPAs that had been submitted by 2009, urgent adaptation measures have been laid down of a total cost of around 2 billion USD.[68] In 2010, however, the LDCF received a total of pledges of 131.5 million USD, and the SCCF a total of 49.8 million USD from Annex I states. The largest contributors for both funds in 2010 were the United States and Germany, both with donations of around 50 million USD in total (for both funds together). Current funding is totally inadequate to address just the most urgent and immediate adaptation needs of the least developed countries. It is, therefore, not surprising that the least developed countries raise this issue in every COP, and not without success. At the Copenhagen and Cancun meetings in 2009 and 2010, pledges were made by developed countries to provide new and additional resources approaching USD 30 billion for the period 2010–2012, with a balanced allocation between adaptation and mitigation, and with prioritizing funding for adaptation for the most vulnerable developing countries, such as the least developed countries, small island developing states and Africa.[69] In addition to that, the developed countries even committed to the goal of mobilizing jointly 100 billion USD per year by 2020 to address the needs of developing countries.[70] These funds should be channeled through again a new fund, the Green Climate Fund (GCF), established in Cancun in 2010.[71] The set up of this new fund, however, saw a slow start with the first meeting of the Transitional Committee, entrusted with the task to draft the operating documents of the GCF, postponed to late April 2011 because of lack of agreement of the composition of this committee. Although this had been resolved by the Durban COP in 2011, the GCF still has to be further operationalized.

11.4 Adaptation in the Various Policy Fields

As already stated above (Sect. 11.2), adaptation touches upon almost everything. It should therefore not come as a surprise that many fields of law are to be addressed. In this section, laws related to the most important policy fields affected by the impact

[67] Revised Programming Strategy on Adaptation to Climate Change for the Least Developed Countries Fund (LDCF) and the Special Climate Change Fund (SCCF), UN Doc. GEF/LDCF.SCCF.9/4/Rev.1, 19 October 2010, at 7–9.

[68] See the LDC Expert Group, The Least Developed Countries. Support Needed to Fully Implement National Adaptation Programmes of Action (NAPAs) (Bonn: UNFCCC, 2009), at 17.

[69] FCCC/CP/2010/7/Add.1, supra, note 50, at 16.

[70] Ibid., at 17.

[71] Ibid.

of climate change, will be briefly reviewed. The main purpose of this section is to give an idea of the kind of legal arrangements that have to be in place in order to facilitate adaptation measures. Within the context of this chapter, it is not possible to extensively deal with all legal issues involved in all of these policy fields.[72] We focus our attention on marine and coastal adaptation (Sect. 11.4.1), water management adaptation (Sect. 11.4.2), adaptation in planning and land use law, buildings and infrastructure (Sect. 11.4.3), adaptation and energy and telecommunications (Sect. 11.4.4), adaptation and migration (Sect. 11.4.5), adaptation and biodiversity (Sect. 11.4.6). Section 11.4.7 briefly mentions other relevant fields of laws.

11.4.1 Marine and Coastal Adaptation

Sea level rise, disappearing islands, melting ice in the arctic and Antarctic, increased storm intensity, shifting ranges of fish stocks and ocean acidification have a profound impact on marine and coastal law, especially in such fields as maritime jurisdictional claims, fisheries, marine biodiversity, and coastal defences. Let us briefly look into these fields.

Sea level rise leads to receding low-water baselines. As a consequence, maritime zones will also retreat, leading to a potential loss of maritime jurisdiction of coastal states. This would then have strong economic consequences, as the right over natural resources is threatened. This is even more so in cases where islands determine maritime jurisdiction. Once islands start to disappear altogether, huge jurisdictional changes will be apparent. It has been suggested that there may be a need for a new process or understanding regarding either the fixing of normal baselines and/or maritime limits. This might develop through State practice, with coastal States choosing particular charts for maritime jurisdictional purposes or simply declaring the location of the limits of their maritime claims. Multilateral negotiations probably are inevitable.[73] The consequences of the disappearance of small island states, obviously, are huge. Responses to such a loss of statehood that have been suggested are the acquisition of "new" territory, merger or confederation with a non-inundation threatened State, and the creation of a new legal category of "deterritorialised" State.[74]

The transformation of the Arctic Ocean from an ice-covered sea to a seasonally ice-free sea, has many legal consequences, especially in the field of international law. States in this region are adapting to the new situation by extending their jurisdiction on the basis of Art. 76 of the UN Convention on the Law of the Sea (UNCLOS), thus aiming to get control of natural resources in the area (especially

[72] For such an extensive overview, see: Jonathan Verschuuren (ed.), *Research Handbook on Climate Adaptation Law* (Cheltenham: Edward Elgar Publishers, 2012).

[73] Clive Schofield, "Shifting Limits? Sea Level Rise and Options to Secure Maritime Jurisdictional Claims", 3 *Carbon and Climate Law Review* (2009), 405.

[74] Ibid., at 415.

fossil fuels) and of water ways.[75] A military build-up is emerging, and non-arctic nations are getting involved as well. It is expected that the Arctic Council, consisting of eight Arctic nations with the involvement of indigenous peoples organizations, will have to lead this process. The Arctic Council, however, has no regulatory authority. Some existing international law can be used as well, especially rules administered by the International Maritime Organization,[76] the North-East Atlantic Fisheries Commission (established under Art. 118 UNCLOS), and rules under the 1992 Convention for the Protection of the North-East Atlantic (OSPAR). In due course, however, it is expected that the adoption of a new arctic (polar) treaty, perhaps along the lines of the Antarctic Treaty, is inevitable.[77]

Ocean warming and increases in the amount of CO_2 dissolved in the ocean leads to a loss of habitat forming species (e.g. corals, giant kelp, sea grasses and mangroves), declines in ocean productivity and shifts in the geographic distributions and latitudinal ranges of marine organisms. This has a profound impact on fisheries and on people depending on fish for food or livelihood.[78] Habitat restoration and the largescale designation of marine protected areas are necessary to help marine biodiversity adapt to their changing environment. In the EU, the Marine Strategy Framework Directive and the Birds and Habitats Directives offer a legal framework for these measures. On the basis of these instruments, areas will have to be closed for fishing to help stocks to recover. Fisheries laws, such as those designed to implement the Common Fisheries Policy in the EU, will have to be used to regulate the redistribution of fisheries rights. Scientists have shown that we can expect a large-scale redistribution of global catch potential, with an average of 30–70% increase in high-latitude regions and a drop of up to 40% in the tropics.[79]

Coastal defence systems are under pressure because of sea level rise and increased storm intensity. With ineffective defence systems in place, coastal erosion and inundation of low lying areas is to be expected. Low lying coastal cities are especially vulnerable and receive a lot of attention in the academic and public debate. Within the context of various international legal instruments, there is growing attention for taking adaptation measures in coastal areas. In a 2009 Report, the OSPAR Commission reviewed the national adaptation policies of its member states and urged states to integrate adaptation measures into Integrated Coastal Zone Management (ICZM)

---

[75] Paul Arthur Berkman and Oran R. Young, "Governance and Environmental Change in the Arctic Ocean", 324 *Science* (2009), at 339.

[76] Already in 2002, the IMO established the "Guidelines for ships operating in ice-covered arctic waters".

[77] Berkman and Young, "Governance and Environmental change in the Arctic Ocean", supra, note 76, at 340.

[78] John D. Koehn et al., "Climate Change and Australian Marine and Freshwater Environments, Fishes and Fisheries: Synthesis and Options for Adaptation", 62 *Marine and Freshwater Research* (2011), 1148; Alexander Proelss and Monika Krivickaite, "Marine Biodiversity and Climate Change", 3 *Carbon and Climate Law Review* (2009), 437.

[79] William W. L. Cheung et al., "Large-scale Redistribution of Maximum Fisheries Catch Potential in the Global Ocean under Climate Change", 16 *Global Change Biology* (2010), 24.

and marine spatial planning, as facilitated by both the OSPAR Convention and the EU's Marine Strategy Framework Directive and the Water Framework Directive.[80] In addition, coastal adaptation is one of the key issues integrated in the North-East Atlantic Environmental Strategy, adopted at the 2010 OSPAR ministerial conference.[81] Also in 2010, the UN General Assembly encouraged states to develop means of marine and coastal adaptation under the UNCLOS and other relevant international frameworks.[82] The UNCLOS secretariat itself produced a general document on oceans and climate change, in which some attention is paid to adaptation by stating that the resilience of coastal ecosystems such as mangroves, salt marshes and sea grasses should be enhanced, that integrated conservation and managements measures to protect marine species should be adopted and that the vulnerability of coastal communities, particularly in developing countries, should be reduced by capacity building and transfer of technology.[83]

The EU has adopted binding legislation forcing its member states to develop coastal and estuarine adaptation law. These requirements stem from the 2000 Water Framework Directive (WFD),[84] the 2007 Floods Directive,[85] the 2008 Marine Strategy Framework Directive (MSFD),[86] as well as a series of non-legally binding policy documents in which member states are urged to incorporate adaptation in their coastal management.[87] EU member states like the UK and the Netherlands have already prepared new legislation that enables the authorities to take far-reaching adaptation measures in coastal areas.[88] A recent comparative study showed that an integrated approach to coastal adaptation law is currently needed to lay the foundations for the required long-term strategy.[89] Such an approach would establish processes by which adaptation objectives are agreed for each part of the coast; ensure land use

---

[80] OSPAR Commission, "Assessment of Climate Change Mitigation and Adaptation", 2009, available at: http://www.ospar.org (last accessed on 15 February 2012), at 22–28.

[81] The North-East Atlantic Environment Strategy, Strategy of the OSPAR Commission for the Protection of the Marine Environment of the North-East Atlantic 2010–2020 (OSPAR Agreement 2010–2013), Annex 25, available at: http://www.ospar.org (last accessed on 15 February 2012).

[82] Resolution 64/71 (114), UN Doc. A/RES/64/71, 12 March 2010.

[83] Oceans and Climate Change, "Division for Ocean Affairs and the Law of the Sea of the UN", 2010, available at: http://www.un.org/Depts/los/oceans_climate_change/oceans_climate_change_7_september_2010.pdf (last accessed on 15 February 2012).

[84] Directive 2000/60/EC establishing a Framework for Community Action in the Field of Water Policy, OJ 2000 L 327.

[85] Directive 2007/60/EC on the Assessment and Management of Flood Risks, OJ 2007 L 288.

[86] Directive 2008/56/EC establishing a Framework for Community Action in the Field of Marine Environmental Policy, OJ 2008 L 164.

[87] Such as the 2002 Recommendation 2002/413/EC concerning the implementation of Integrated Coastal Zone Management in Europe, OJ 2002 L 148, and the 2009 White Paper, Adapting to Climate Change: Towards a European Framework for Action, COM(2009)0147 final.

[88] Jonathan Verschuuren and Jan McDonald, "Towards a Legal Framework for Coastal Adaptation: Assessing the First Steps in Europe and Australia", 1 *Transnational Environmental Law* (forthcoming 2012).

[89] Ibid.

planning that can accommodate future change and does not expose new communities to risk; integrate coastal adaptation with biodiversity and coastal zone policy; allocate regulatory responsibility in a way that promotes subsidiarity and consistency; and ensures that funds are available for future measures.

11.4.2 Water Management Adaptation

The impact of climate change on river basins has already been apparent for a number of years. Already in 2007, the UNECE finds that "most basins experience an impact of climate change on water quantity (e.g. decreasing water resources availability and extreme hydrological events, including severe floods and long-lasting droughts). With a reduction in precipitation of up to 30% over the last decade, water resources availability, for example, is decreasing in river basins in the discharge area of the Mediterranean Sea. The effects of climate change on the ecological regime of rivers are also becoming visible in transboundary basins in Central Asia, where the rise in air temperatures leads to significant melting of glaciers, resulting in noteworthy changes of the river's hydrological and ecological regimes. Thus, climate change adaptation measures in water management and water-depended activities and services (e.g. agriculture, forestry, water supply, hydropower generation) are needed in the entire UNECE region."[90] The same region saw an increase in the average number of annual disastrous weather and climate-related events by about 65% between 1998 and 2007,[91] many of which impacted water and wastewater services systems.

In the area of water management, we have the relative advantage that, almost universally, an integrated river basin approach is applied.[92] Such a holistic approach to all interrelated water issues is particularly relevant to water adaptation measures.[93]

[90] UNECE, Our Waters: Joining Hands across Borders. First Assessment of Transboundary Rivers, Lakes and Groundwaters (New York and Geneva: UNECE, 2007), at 29.

[91] European Environment Agency, Impacts of Europe's Changing Climate, 2008 Indicator-based Assessment, EEA Report 04/2008, 2008, at 169.

[92] Based upon the UNECE Convention on International Watercourses and Transboundary Lakes, Helsinki, 17 March 1992, 31 *International Legal Materials* (1992), 1312 and the UN Convention on the Law of the Non-navigational Uses of International Watercourses, New York, NY, 21 May 1997, 36 *International Legal Materials* (1997), 700.

[93] There is a wealth of academic papers in which this is acknowledged, see for instance: Craig A. Arnold, "Law's Adaptive Capacity and Climate Change's Impacts on Water", 5 *Environmental and Energy Law and Policy Journal* (2010), v. Craig A. Arnold, "Adaptive Watershed Planning and Climate Change", 5 *Environmental and Energy Law and Policy Journal* (2010), 417; Richard T. Kingsford, "Conservation Management of Rivers and Wetlands under Climate Change: A Synthesis", 62 *Marine and Freshwater Research* (2011), 217. Dan Tarlock, "Four Challenges for International Water Law", 23 *Tulane Environmental Law Journal* (2009), 369; David N. Cassuto and Romulo S. R. Sampaio, "Water Law in the United States and Brazil – Climate Change & Two Approaches to Emerging Water Poverty", 35 *William & Mary Environmental Law and Policy Review* (2011), 371; Poh-LingTan, "Adaptation Measures for Water Security in a Changing

Such measures may have to be aimed at preventing floods and minimising flood damage, as well as at mitigating the impact of droughts, even in one geographic area. In 2009, the COP to the UNECE Water Convention adopted the Guidance on Water and Adaptation to Climate Change,[94] prepared by the Task Force on Water and Climate.[95] These elaborate guidelines also take an integrated river basin approach as a starting point for adaptation measures. Integrated river basin planning, therefore, should plan for extreme events. Flood prevention and mitigation policies can make use of a wide range of instruments, such as the designation of water retention areas that are used for controlled flooding (thus releasing the pressure on other, for instance more densely populated areas), or the widening of the river bed (thus enhancing the capacity of the river to discharge larger amounts of water to the ocean), and/or the creation and management of controlled water levels through dams and other technical measures. In cities, storm water management is an important element of any water policy. Increased precipitation will put a greater pressure on storm water systems, which may lead to pollution in case the sewage system was not adapted to deal with an increased amount of storm water.[96]

Water scarcity and draughts have to be addressed in integrated river basin plans as well. Water scarcity occurs where there are insufficient water resources to satisfy long-term average requirements, whereas draughts are caused by a temporary decrease of the average water availability due to rainfall deficiency. Storing water on agricultural lands for later use, crop changes to less water intensive crops, or to crops that are resistant to salination, are examples of measures that can be integrated into river basin plans.[97]

Finally, it should be noted that mitigation measures may have a negative impact on water resources. Carbon dioxide capture and storage, for instance, may lead to groundwater degradation in case of leakages, while large scale hydro-electric power may have a negative impact on river ecosystems and fisheries. Hence the need to integrate mitigation and adaptation policies in the field of water management.

Climate: Policy, Planning and Law", in Tim Bonyhady, Andrew Macintosh and Jan McDonald (eds), *Adaptation to Climate Change. Law and Policy* (Sydney: The Federation Press, 2010), 135; Barbara Cosens and Mark Williams, "Resilience and Water Governance: Adaptive Governance in the Columbia River Basin", 2011, available at: SSRN: http://ssrn.com/abstract=1942587 (last accessed on 15 February 2012).

[94] Meeting of the Parties to the Convention on the Protection and Use of Transboundary Watercourses and International Lakes Working Group on Integrated Water Resources Management, see UN Doc. UNECE/MP.WAT/29, 4–5 May 2011, at 10.

[95] Guidance on Water and Adaptation to Climate Change, ECE/MP.WAT/30, 15–16 December 2009.

[96] In 2010, the WHO and the UNECE issued Luciana Sinisi and Roger Aertgeerts, *Guidance on Water Supply and Sanitation in Extreme Weather Events* (Copenhagen: WHO/UNECE, 2010).

[97] See UNECE Convention on International Watercourses and Transboundary Lakes and UN Convention on the Law of the Non-navigational Uses of International Watercourses, supra, note 93.

11.4.3 Adaptation in Planning and Land Use Law, Buildings and Infrastructure

Land use law and spatial planning law are important instruments to adapt to a variety of climate change impacts. Most research in this field shows that adaptation and mitigation go hand in hand here.[98] As already mentioned above, land use law will have to be used to avoid unwanted development in areas prone to flooding, or even to remove existing buildings from such areas (in which case expropriation of property owners by be necessary as well). The same also goes for areas that need to be used for increased precipitation run off and stormwater management. These areas are not just at or below street level. Buildings may also be used for such purposes as reducing run off for instance through the use of green roofs (buildings partially or completely covered with vegetation and soil). Under the phrase of "green cities" or "green buildings", comprehensive plans are developed to design inner cities and its buildings in such a way that they both play a role in adaptation and mitigation. Next to applying land use law and zoning law, building requirements have to be set as well in building and/or construction codes.[99]

Planning and land use law are also important instruments for transport infrastructure adaptation. Both underground and surface transportation systems may be in areas prone to flooding and thus may have to be adapted. Elevated railwaylines are, for instance, are less vulnerable to flooding than non-elevated ones. Increased energy demand during heat waves may have a negative impact on trains and subways (see also Sect. 11.4.4 below). Material failure can be an issue during heat waves or during severe storms. Waterways can be affected both by droughts and by high water levels. All of these issues will have to be addressed in a transport infrastructure adaptation policy. Again, adaptation and mitigation are closely linked here. Smart neighbourhood planning so as to avoid urban sprawl and excessive transport needs and bicycle friendly land-use, are advocated as a means to both reduce the pressure on vulnerable transportation infrastructure and to mitigate GHG emissions by the transport sector at the same time.

As will be further discussed below, planning law is also vital to create connectivity between natural areas, thus helping biodiversity to adapt to the changing climate (see Sect. 11.4.6). Planning and building law may also play a role in policies

---

[98] See among others: Anne Leitch, Ben Harman and Marcus B. Lane, "From Blueprint to Footprint: Climate Change and the Challenge for Planning", in Tim Bonyhady, Andrew Macintosh and Jan McDonald (eds), *Adaptation to Climate Change. Law and Policy* (Sydney: The Federation Press, 2010), 63; John R. Nolon and Patricia. E. Salkin, "Integrating Sustainable Development Planning and Climate Change Management: A Challenge to Planners and Land Use Attorneys", 63 *Planning & Environmental Law* (2011), 3; Sussman et al., "Climate Change Adaptation: Fostering Progress Through Law and Regulation", supra, note 8, at 63–77.

[99] See extensively, Sussman et al., "Climate Change Adaptation: Fostering Progress Through Law and Regulation", supra, note 8, at 97–103 and Edna Sussman, "Reshaping Municipal and County Laws to Foster Green Building, Energy Efficiency, and Renewable Energy", 16 *N.Y.U. Environmental Law Journal* (2008), 1.

designed to reduce fire threats in areas prone to wild fires, for instance by regulating vegetation free zones around public buildings like schools, planning exit roads, setting building requirement aimed at making buildings better fire resistant.

11.4.4 Adaptation and Energy and Telecommunications

Energy demand is expected to sharply rise in those areas with increased heat waves during summer as the use of air conditioning will increase. Long periods of warm weather in summer may also lead to a shortage of cooling water for energy production. This may lead to energy fall out. That may also happen when floods or storms damage power infrastructure. Designing and planning for a robust power infrastructure, thus, is needed. Again, land use law, building law environmental permits for installations, but also water management law, will have to be applied to achieve that.[100] In literature, this goal usually is combined with the goal to reduce the use of energy as part of a mitigation strategy. Smart grids are often seen as the most important means to achieve both adaptation and mitigation targets in the energy sector.[101] The smart grid is an ICT-based transmission and distribution network, which provides for a better integration of renewable energy sources into the grid. Lyster shows that the Smart Grid gives utilities an enhanced ability to identify the location of a failure and quickly re-route electricity to locations where demand is most critical. "This could occur during times of climate change-induced crisis, or peak demand, and prevents outages through proactive diagnosis of the grid and its individual elements. Importantly, it enhances the ability of the grid to continue to provide power following a catastrophic event and to support vital emergency responses as well as military, economic and social activities during a crisis."[102] There are many legal aspects involved in the roll out of smart grids, not just in field of environmental, energy and competition law, but also concerning the protection of consumer privacy.

On a final note, it should be mentioned that for this sector, there are several opportunities as well, such as a higher potential for wind energy as well as for solar energy, whereas the overall demand in energy in a large parts of the world will decrease because of milder winters.

As far as telecommunications are concerned, similar adaptive measures will be necessary. Floods and severe winds may damage telecommunication infrastructure, whereas, at the same time, such events, simultaneously, may trigger a peak in telecommunications traffic. Telecommunications black outs can be caused by

---

[100] See among others, Dirk T. G. Rübbelke et al., "Impacts of Climate Change on European Critical Infrastructures: The Case of the Power Sector", 2010, available at: http://www.bc3research.org (last accessed on 15 February 2012).

[101] Zhen Zhang, "Smart Grid in America and Europe: Similar Desires, Different Approaches", 149 *Public Utilities Fortnightly* (2011).

[102] Rosemary Lyster, "Smart Grids: Opportunities for Climate Change Mitigation and Adaptation", 36 *Monash University Law Review* (2010), 173.

this, severely hampering disaster relief operations. Hence, regulating for network reliability is required.[103]

11.4.5 Adaptation and Migration

A lot of attention is focused on climate induced migration. Severe climatic events such as floods, as well as a gradual loss of fresh water availability or agricultural resources, can trigger human migration. Research shows that migration usually is caused by a multitude of factors, climate change being one of those. Such migration may be internal migration or cross border displacement. The increased pressure caused by the immigrants on the host communities can lead to food and water shortages, housing problems, unemployment and ultimately to violence and armed conflict.[104] Most legal research, so far, has aimed at identifying the most suitable international law approaches to dealing with cross border migration. Current refugee law is not applicable to environmental refugees as they are not forced to move for political reasons or human rights violations, hence suggestions to either come up with a new stand-alone international law instrument, or with the adoption of a protocol on climate refugees within the UNFCCC framework.[105] Research aimed at the role of international law in the prevention of armed conflicts as a consequence of climate change still has to emerge.

11.4.6 Adaptation and Biodiversity

Biologists have been observing many changes to biodiversity caused by climate change for years. They advocate making protected areas climate proof by making sure that these areas are large enough and stable enough to adapt to the changed climate. Protected areas should be able to live through flooding in winter, wild fires in the summer, storm damage and should have enough variety in habitat types to host new species. This for many protected areas means enormously intensified protection measures are to be taken, for instance by enlarging sites or connecting existing sites into one much larger site.[106] To prepare for the ecological impacts of climate

[103] See Sussman et al., "Climate Change Adaptation: Fostering Progress Through Law and Regulation", supra, note 8, at 114–115.

[104] Idean Salehyan, "From Climate Change to Conflict? No Consensus Yet", 45 *Journal of Peace Research* (2008), 315.

[105] Bonnie Docherty and Tyler Giannini, "Confronting a Rising Tide: A Proposal for a Convention on Climate Refugees", 33 *Harvard Environmental Law Review* (2009), 349; Brendan Gogarty, "Climate-change Displacement: Current Legal Solutions to Future Global Problems", 21 *Journal of Law, Information and Science* (2011), 1.

[106] For an overview see Saja Erens, Jonathan Verschuuren and Kees Bastmeijer, "Adaptation to Climate Change to Save Biodiversity: Lessons Learned from African and European Experiences", in Benjamin Richardson et al. (eds), *Climate Law and Developing Countries: Legal and Policy Challenges for the World Community* (Cheltenham: Edward Elgar, 2009), 206.

change, a landscape approach is needed in which existing protected areas are enlarged and secured and ecological corridors between areas are protected and restored, thus establishing a real ecological network that is resilient to future change. "Connectivity" is the key word in many biodiversity adaptation policies.[107]

In Europe, most attention focuses on the Birds – and Habitats Directives.[108] While there is some criticism as to the soft legal nature of some of its provisions,[109] the Habitats Directive does, in Article 10, propose that member states, in their land-use planning and development policies, maintain and develop features of the landscape of major importance for wild fauna and flora, thus improving the ecological coherence of the EU Natura 2000 network. Article 10 in particular mentions features which, by virtue of their linear and continuous structure (such as rivers with their banks or the traditional systems for marking field boundaries) or their function as stepping stones (such as ponds or small woods), are essential for the migration, dispersal, and genetic exchange of wild species. Most authors conclude on the basis of the combined duties arising from a variety of provisions in both EU Directives and relevant international conventions, such as the CBD, the Convention on Migratory species and the Bern Convention on the Conservation of European Wildlife and Natural Habitats, that there already exists a legal obligation within the EU to establish corridors and other measures to help biodiversity adapt to climate change.[110] Recent case law by the European Court of Justice seems to point in the same direction.[111]

11.4.7 Other Areas of Law

Besides the above obvious fields of law that play a major role in climate change adaptation, there are more areas to cover. Pollution control legislation must be used to control additional or altered environmental effects of emissions. Higher temperatures, for example, lead to elevated concentrations of ground level ozone, which is harmful to human health as well as to the vegetation. Hazardous waste facilities can be

---

[107] The Work Programme on Protected Areas adopted at the 2004 COP of the Biodiversity Convention, for instance, sets as a target for 2015 the integration of all protected areas and protected area systems into the wider land – and seascape by applying the ecosystem approach and taking into account ecological connectivity and the concept of ecological network, COP 7 Decision VII/28, Protected Areas, UN Doc. UNEP/CBD/COP/DEC/VII/28, 13 April 2004, Arts. 8(a) to (e).

[108] Directive 2009/147/EU on the Conservation of Wild Birds, OJ 2010 L 20/7, and Directive 92/43/EEC on the Conservation of Natural Habitats and of Wild Flora and Fauna, OJ 1992 L 206/7.

[109] Jonathan Verschuuren, "Climate Change: Rethinking Restoration in the European Union's Birds and Habitats Directives", 28 *Ecological Restoration* (2010), 431.

[110] Arie Trouwborst, "Conserving European Biodiversity in a Changing Climate: The Bern Convention, the European Union Birds and Habitats Directives and the Adaptation of Nature to Climate Change", 20 *Review of European Community and International Environmental Law* (2011), 62; An Cliquet et al., "Adaptation to Climate Change: Legal Challenges for Protected Areas", 5 *Utrecht Law Review* (2009), 158.

[111] Case C-404/09, European Commission v. Kingdom of Spain [2011] ECR I-0000.

damaged during extreme weather events and floods, which may lead to pollution of large areas outside the waste facility. The impact of the 2011 tsunami on the Fukushima nuclear power plant shows that nuclear contamination of large areas may occur in extreme weather events. Tightened environmental rules for such installations are required. The instrument of environmental impact assessment is a promising tool for assessing the possible consequences of climate change on an installation.[112]

Disaster prevention and mitigation, and disaster response are other important areas of law. Basically, the entire disaster management system has to be employed to prevent and mitigate disasters and to respond to disasters, for instance caused by extreme weather events, floods, wild fires etc.[113]

Public health is affected by climate change in a great number of ways, examples of which are increased hospitalizations and deaths caused by heat spells, increased respiratory problems caused by elevated ozone levels also during heat spells, increased numbers of victims of vector-borne infectious diseases, and casualties from severe weather events and floods (including mental problems caused by such events). Although most attention has been focussed on emergency preparedness-law has to be ready so that emergency services are able to deal with a sudden influx of additional patients-, attention is now shifting to handling the substantial, slowly emerging, intensification of more routine health threats that are expected to be seen as an impact of climate change, both domestically[114] and internationally.[115]

Last, but certainly not least, mention should be made of the huge adaptation efforts needed in agriculture. Changing weather patterns, increasing draughts in some areas, and too much water in others (or in the same area, but in different parts of the year) and new plant diseases and pests, require changes in agricultural practices such as the choice of crops and the use of pesticides. Agricultural law can facilitate farmers to implement such changes.[116] As with most of the topics addressed in this chapter, adaptation measures in this field are strongly connected to mitigation actions. Low carbon agriculture is increasingly gaining popularity as this variety of measures both help farmers adapt to the changing climate, while at the same time

---

[112] Caleb W. Christopher, "Success by a Thousand Cuts: The Use of Environmental Impact Assessment in Addressing Climate Change", 9 *Vermont Journal of Environmental Law* (2007–2008), 549.

[113] Among others, Tim Bonyhady, "The Law of Disasters", in Tim Bonyhady, Andrew Macintosh and Jan McDonald (eds), *Adaptation to Climate Change: Law and Policy* (Sydney: The Federation Press 2010), 265.

[114] See Lindsay F. Wiley, "Adaptation to the Health Consequences of Climate Change as a Potential Influence on Public Health Law and Policy: From Preparedness to Resilience", 15 *Widener Law Review* (2009–2010), 483.

[115] Lindsay F. Wiley, "Moving Global Health Law Upstream: A Critical Appraisal of Global Health Law as a Tool for Health Adaptation to Climate Change", 22 *Georgetown International Environmental Law Review* (2010), 439.

[116] Robert W. Adler, "Balancing Compassion and Risk in Climate Adaptation: U.S. Water, Drought and Agricultural law", 64 *Florida Law Review* (2012), 201; Richard Munang and Johnson N. Nkem, "Using Small-scale Adaptation Actions to Address the Food Crisis in the Horn of Africa: Going Beyond Food Aid and Cash Transfers", 3 *Sustainability* (2011), 1510.

they reduce GHG emissions, which, depending on mitigation laws applicable, may lead to financial benefits for them as well.

The forestry sector will be confronted with similar impacts, such as reduced growth rates, changes in wood quality and quantity, increased pests, increased competition by exotics and increased fires. Changes in forest management systems are in order, and again, law can play a role to facilitate these changes.[117] Again, adaptation and mitigation go hand-in-hand, which is particularly visible in mechanisms for educing emissions from deforestation and forest degradation (such as REDD+): vegetation promoted by REDD+ not only mitigates GHG emissions, but also reduces erosion, protects biodiversity, and may provide a range of ecosystem services to local communities. Much, however, depends on how the measures are designed, as evidence shows that REDD+ may also lead to negative impacts on adaptation, for instance for biodiversity.[118]

11.5 Conclusion

More and more attention is focused on adaptation. It is clear that huge efforts are needed on a variety of areas to help the world adapt to the changing climate. In this chapter, the international law with regard to adaptation has been described and assessed, as well as domestic legal issues involved. It is obvious that a wide variety of measures needs to be taken using legal instruments from water law, marine law, planning and land use law, building law, biodiversity law, agriculture law, etc. etc. In fact, almost all the laws that we use to facilitate basic economic and societal processes may have to play a role in the transformational changes that we have to start. In some fields, such as coastal adaptation and biodiversity, such transformations have already started, at least in some countries. In general, however, it must be concluded that adaptation law is still in its infancy. Sometimes existing laws can be used to foster adaptation, more often, laws will first have to be adapted themselves before they provide the right instruments to foster adaptation. Internationally, much attention is focused on developing countries. Financial instruments are currently being developed at the international level to assist these countries with their adaptation efforts. Legal assistance will be required as well, given the enormous scale of the transitions needed. In developed countries, knowledge on how to create the most suitable legal environment for adaptation has to grow as well. It seems that international cooperation and exchange of ideas and experiences would facilitate such a process.

[117] D.L. Spittlehouse and Richard B. Stewart, "Adaptation to Climate Change in Forest Management", 4 *BC Journal of Ecosystems and Management* (2003), 1.

[118] Andrew Long, "Taking Adaptation Value Seriously: Designing REDD to Protect Biodiversity", 3 *Carbon and Climate Law Review* (2009), 314.

Chapter 12
Climate Change and Human Rights

Timo Koivurova, Sébastien Duyck, and Leena Heinämäki

Abstract This chapter examines the inter-relationship between human rights and climate change, a linkage that has been given little attention, but whose importance is likely to grow in the coming years. Some aspects of the relationship between climate change and human rights have been selected, especially those that have emerged as having most potential in influencing climate change governance. We will identify how climate change, with its dramatic consequences, impacts the enjoyment of human rights and has already led to a human rights petition against the United States. We will, then, turn to the implications of human rights to the functioning of the climate change regime, such as how the emerging rights to participate in environmental decision-making are reflected in the negotiation process of defining the elements of the current climate change regime. More difficult question on whether human rights can or even should influence the future design of the climate change regime will be examined. The concluding remarks will focus on evaluating the pros and cons of using human rights in the struggle against climate change impacts and the influence of human rights on the design and operation of the climate change regime.

Dr. Timo Koivurova, Research Professor, Northern Institute for Environmental and Minority Law, Arctic Centre, University of Lapland; Sébastien Duyck, Doctoral Candidate and Researcher, Northern Institute for Environmental and Minority Law, Arctic Centre, University of Lapland; Dr. Leena Heinämäki, Researcher, Northern Institute for Environmental and Minority Law, Arctic Center, University of Lapland.

T. Koivurova (✉) • S. Duyck • L. Heinämäki
Northern Institute for Environmental and Minority Law, University of Lapland,
Arctic Centre, 96101, Rovaniemi, Finland
e-mail: timo.koivurova@ulapland.fi; sebastien.duyck@ulapland.fi; leena.heinamaki@ulapland.fi

E.J. Hollo et al. (eds.), *Climate Change and the Law*, Ius Gentium: Comparative Perspectives on Law and Justice 21, DOI 10.1007/978-94-007-5440-9_12,
© Springer Science+Business Media Dordrecht 2013

12.1 Introduction

There is hardly an environmental issue that has captured as much or more global attention as global warming and the associated climate change. The Fourth Assessment Report of the Intergovernmental Panel on Climate Change (IPCC), released in 2007, warned that climate change is unequivocal and accelerating.[1] Over the last century, the global average temperature has increased by 0.74°C, which constitutes the largest and fastest warming trend in world history.[2] It is predicted that it will increase by 1.1–6.4°C.[3] It is also estimated that climate change will, among other impacts, increase the severity of droughts, land degradation and desertification, the intensity of floods and tropical cyclones, the incidence of malaria and heat-related mortality and decreasing crop yield and food security.[4]

Despite states' inertia to adequately respond to this phenomenon, climate change – with good reason – has been characterized as "the defining human development issue of our generation."[5] As expressed in the UN Human Development Report, climate change differs from other problems facing humanity, because of its ability to challenge us to think differently in many ways. Above all, it forces us to consider what it means to live as a part of an ecologically interdependent human community. Climate change serves as a reminder of what humanity shares: planet Earth. All nations and all people share one atmosphere.[6]

Notwithstanding the urgency of the climate change problem, and despite clear human rights risks, governments have only very recently awoken to a discussion on the human rights dimensions of climate change and the potential role that human rights law, principles, and institutions could play in responding to climate change.[7] A notable example in this direction is the United Nations (UN) Human Rights Council's consensus resolution 2009, which was adopted by a total of 88 UN member states and encouraged the greater involvement of human rights expert bodies in the UNFCCC process.[8]

---

[1] Rajendra K. Pachauri and Andy Reisinger (eds), *Climate Change 2007: Synthesis Report. Contribution of Working Groups I, II and III to the Fourth Assessment Report of the Intergovernmental Panel on Climate Change* (Cambridge and New York: Cambridge University Press, 2007), at 30.

[2] Ibid., at 30.

[3] Ibid., at 23.

[4] Ibid., at 26.

[5] United Nations Development Programme, "Human Development Report 2007–2008, Fighting Climate Change: Human Solidarity in a Divided World", 2007, available at: http://hdr.undp.org/en/media/HDR_20072008_EN_Complete.pdf (last accessed on 25 February 2012), at 16.

[6] Ibid., at 17.

[7] Siobhán McInerney-Lankford, Mac Darrow, and Lavanya Rajamani, *Human Rights and Climate Change: A Review of the International Legal Dimensions* (Washington, DC: The World Bank Study, 2011), at 55.

[8] United Nations Human Rights Council (HRC) Resolution 10/4, Human Rights and Climate Change, UN Doc. A/HRC/10/L.11, 12 May 2009.

In regard to the implications of climate change on human rights, the resolution notes:

> [Noting that] climate change-related impacts have a range of implications, both direct and indirect, for the effective enjoyment of human rights including, inter alia, the right to life, the right to adequate food, the right to the highest attainable standard of health, the right to adequate housing, the right to self-determination and human rights obligations related to access to safe drinking water and sanitation, and recalling that in no case may a people be deprived of its own means of subsistence.[9]

The resolution also serves as a reminder of how the implications of climate change mostly affect those who already find themselves in vulnerable situations due to factors that include, among others, geography, poverty, gender, age, indigenous or minority status, and disability.[10]

There are various ways in which human rights and climate change may be regarded as being inter-related. Climate change influences the enjoyment of human rights and human rights, in turn, should affect how the climate change regime functions and could function. However, as human rights have not played a strong role in how climate change is perceived as a politico-legal problem, this is not inevitable. It was with the pioneering of the 2005 Inuit human rights petition, against the United States, that climate change was concretely framed as a human rights problem. It was primarily this first international climate change litigation that spurred interest in the role of human rights in combatting climate change.

This chapter will examine selected aspects of the relationship between climate change and human rights and will particularly concentrate on those that have emerged with the most potential in influencing climate change governance. We will begin by identifying how climate change, with its dramatic consequences, impacts the enjoyment of human rights. We will focus on two core human rights – to life and health – and will generally examine how international instruments and environmental rights jurisprudence began paying attention as to how these rights are violated by environmental pollution, including climate change. As an example of this approach, we will briefly demonstrate how the Inuit's developed a human rights petition against the United States because of the latter's climate policy.

Human rights are drawn into – or at least should be, as they have not yet played a strong role – the discussion on how the climate change regime functions today and should be reformed for the future. The second section, examines the implications of

---

[9] Ibid.

[10] Ibid. Already a year before, indeed, the Council had made a resolution "Human rights and climate change", recognising the implications of climate change on the enjoyment of human rights, see HRC Resolution 7/23, Human rights and climate change, UN Doc. A/HRC/RES/7/23, 28 March 2008. In implementation of that resolution, the Council prepared and submitted a study on the relationship between climate change and human rights. Report of the Office of the High Commissioner for Human Rights [hereinafter OHCHR] on the relationship between climate change and human rights, UN Doc. A/HRC/10/61, 15 January 2009. This concerned was also reaffirmed by the more recent resolution of the Council on Human Rights and Climate Change, HRC Resolution 18/22, UN Doc. A/HRC/18/L.26/Rev.1, 28 September 2011.

human rights on the function of the climate change regime, such as how emerging participatory rights in environmental decision-making are reflected in the negotiation process and defines elements of the current climate change regime. The case study on palm oil plantations in the Bajo Aguan illustrates the risk that climate policies may infringe on local communities' rights. We then consider how procedural rights, and human rights more broadly, are to be exercised in the implementation of the climate change regime at the domestic level.

Thereafter, we approach a more challenging question on whether human rights can or should influence the future design of the climate change regime. There are various perspectives as to how and, on which basis the climate change regime should be developed in order to meet the vast challenges ahead. In the chapter's third section, we examine a human rights approach to climate change that is skilfully argued by Professor Caney.[11]

The final section's concluding remarks will focus on evaluating the pros and cons of using human rights in the struggle against climate change impacts and in having human rights influence the design and operation of the climate change regime. As noted, although it appears as though human rights should play various roles in climate change related decision-making, this is still a very incipient development. Consequently, it is useful to consider the strengths and weaknesses of human rights in climate change governance.

12.2 Climate Change Impacts on the Enjoyment of Human Rights

Despite the fact that the inherent connection between human rights and climate change has, thus far, not been widely addressed by international forums, the general linkage between human rights and the environment has been widely discussed since the 1970s.[12] Although this debate has not led to a wide-spread recognition of an independent human right to decent or satisfactory environment,[13] it has spurred a lot

---

[11] Simon Caney, "Climate Change, Human Rights and Moral Thresholds", in Stephen Humphreys (ed.), *Human Rights and Climate Change* (Cambridge: Cambridge University Press, 2010), 69.

[12] When looking at the evolution of the concept of "a human right to the environment", it can be seen as dating back to the United Nations Stockholm Declaration on Human Environment 1972, which provides that: "Man has the fundamental right to freedom, equality and adequate conditions of life, in an environment of a quality that permits a life of dignity and well-being, and he bears a solemn responsibility to protect and improve the environment for present and future generations." Declaration of the UN Conference on the Human Environment, UN Doc. A./CONF. 48/14/Rev.1, 16 June 1972, principle 1.

[13] There have been attempts at different forums to create an instrument that would explicitly recognize a human right to the environment. In 1994, Special Rapporteur Fatma Zohra Ksentini delivered her Final Report on Human Rights and the Environment to the UN Sub-Commission on Prevention of Discrimination and Protection of Minorities (Draft Principles on Human Rights and the Environment, Annex I, UN Doc. E/CN.4/Sub.2/1994/9, 6 July 1994). The Report included a Draft Declaration of Principles on Human Rights and the Environment, inter alia, stating that "all persons have the right to a secure, healthy and ecologically sound environment".

of action where existing human rights are regarded as important in protecting the environment, also against climate change consequences. So, despite the fact that a specific universal right to the environment may not have been recognized explicitly[14], many other substantive human rights, such as the right to life, health, and property, or procedural rights, such as participatory rights or the right to effective remedies, have been applied by human rights monitoring bodies in an environmental context.

Before examining how human rights to life and health are applied by monitoring bodies in cases concerning environmental interference and in relation to human rights violations considering the likely impacts of climate change on the enjoyment of human rights, it is useful to define our meaning of "environmental rights" – the possibility of formulating claims relating to the environment in terms of human rights.[15] There is no general definition of environmental rights, but the concept includes rights that belong to both general human rights law and the instruments of international environmental law.[16]

Many international human rights bodies, including those with the authority to hear complaints or resolve disputes, have acknowledged environmental issues in one way or another. These institutions commonly appear to support the idea that environmental degradation may affect human rights in demonstrable ways. However, the precepts and analyses upon which these bodies have acted and articulated the connection between human rights and the environment vary. Despite the fact that no single standard or analytical tool exists for the evaluation of environmental issues within the human rights doctrine, there is a legal precedent for considering these issues within the global institutional framework and, more concretely, region by region.[17]

[14] On the other hand, there are a few regional instruments that explicitly recognize the right to environment; See, Additional Protocol to the American Convention on Human Rights, San Salvador, 17 November 1988, 28 *International Legal Materials* (1989), 156, at 161. American Convention on Human Rights, San José, 22 November 1969, in force 18 July 1978, 114 *United Nations Treaty Series* (1978), 123, Art. 11; The African Charter on Human and Peoples' Rights, 27 June 1981, in force 21 October 1986, 21 *International Legal Materials* (1982), 58, Art. 24.

[15] See generally, for example, Alan E. Boyle and Michael R. Anderson (eds), *Human Rights Approaches to Environmental Protection* (Oxford: Clarendon Press, 1996); and Dinah Shelton, "Environmental Rights", in Philipp Alston (ed.), *Peoples' Rights* (Oxford: Oxford University Press, 2001), 189.

[16] See, for instance, Shelton, "Environmental Rights", supra, note 15; Dinah Shelton, "Environmental Rights in Multilateral Treaties Adopted between 1991 and 2001", 32 *Environmental Policy and Law* (2002), 70.

[17] Michelle T. Leighton, From Concept to Design: Creating an International Environmental Ombudsperson, *Legal and Normative References: Environmental Human Rights* (Berkeley, Ca.: The Nautilus Institute for Security and Sustainable Development, 1998), at 12.

12.2.1 Climate Change Impact on the Right to Life

The right to life is often considered to be one of the most fundamental human rights. International human rights treaties and customary international law affirm states' obligation to not undertake acts that harm or threaten human life. The right to life is guaranteed by nearly all major human rights instruments. For example, Article 3 of the Universal Declaration of Human Rights states, "[e]veryone has the right to life, liberty and security of person."[18] In a similar vein, Article 6 of the International Covenant on Civil and Political Rights (CCPR) states that "every human being has the inherent right to life."[19]

Many treaties, including the CCPR, attempt to clarify the content of the right to life by only prohibiting the "arbitrary deprivation" or "intentional deprivation" of life. However, in relation to this right, states' obligations extend beyond the requirement of arbitrary or intentional deprivation of life. There appears to be a general understanding that the right to life itself requires a precautionary approach by governments, which means that government officials must prevent harm or threats to human life in cases where they may be foreseen.[20]

In its General Comment No. 6, the UN Human Rights Committee has also stated that the right to life "has been too often narrowly interpreted. The expression "inherent right to life" cannot properly be understood in a restrictive manner, and the protection of this right requires that states adopt positive measures."[21]

In regard to the environmental dimension of the right to life, the UN Human Rights Committee has indicated that state obligations to protect the right to life may

---

[18] The Universal Declaration of Human Rights, Paris, 10 December 1948, UN Doc. A/810, Art. 3.

[19] The International Covenant on Civil and Political Rights, New York, 16 December 1966, in force 23 March 1976, 999 *United Nations Treaty Series* (1976), 302. Additionally, this right has been recognized in the United Nations Convention on the Rights of the Child, New York, 20 November 1989, in force 2 September 1990, 1577 *United Nations Treaty Series* (1990), 3, Art. 6; The European Convention on Human Rights and Fundamental Freedoms, Rome, 4 November 1950, in force 3 September 1953, 213 *United Nations Treaty Series* (1951), 222, Art. 2; the African Charter on Human and Peoples' Rights, supra, note 13, Art. 4: the American Convention on Human Rights, supra, note 13, Art. 4; and the American Declaration of the Rights and Duties of Man, Bogotá, 2 May 1948, OEA/Ser.L.V/II.82 doc.6 rev.1 (1992), Art. 1.

[20] For instance the Inter-American Court of Human Rights has interpreted the obligation as requiring states to exercise their power in a manner that legally ensures the full enjoyment of human rights, including preventing, investigating and punishing any violation of the rights provided by the American Convention on Human Rights. See, e.g., Valesquez Rodriguez Case, Judgement, Inter-Am. Ct. H.R., App. VI, OAS/Ser.L/V/III.19, doc. 13, at 70–71. In its Report on the Situation of Human Rights in Ecuador, the Inter-American Commission on Human Rights stated that "the right to have one's life respected is not … limited to protection against arbitrary killing". Inter-American Commission on Human Rights, Report on the Human Rights Situation in Ecuador, OEA/Ser.L/V/II.96, 24 April 1997, available at: cidh.org/countryrep/ecuador-eng/index%20-%20 ecuador.htm (last accessed on 25 February 2012), at chapter 8.

[21] Human Rights Committee, General Comment No. 6, The Right to life (Art.6), UN Doc./A/37/40, 30 April 1982, available at: www.unhchr.ch/tbs/doc.nsf/(Symbol)/84ab9690ccd81fc7c12563ed00 46fae3?Opendocument (last accessed on 25 February 2012), item 5.

include positive measures designed to reduce infant mortality and protect against malnutrition and epidemics.[22] *In E.H.P. v. Canada,*[23] a case concerning the storage of radioactive waste near the claimants' home, the UN Human Rights Committee said the case raised "serious issues with regard to the obligation of States parties to protect human life."[24]

The Inter-American Commission on Human Rights has also recognized the environmental dimension of the right to life by laying down that:

> The realization of the right to life, and to physical security and integrity is necessarily related to and in some ways dependent upon one's physical environment. Accordingly, where environmental contamination and degradation pose a persistent threat to human life and health, the foregoing rights are implicated.[25]

Furthermore, when discussing the connection between the physical environment and the right to life, the Inter-American Commission concluded that environmental degradation can "give rise to an obligation on the part of a state to take reasonable measures to prevent the risk to life associated with environmental degradation."[26] The Commission noted that human rights law "is premised on the principle that rights inhere in the individual simply by virtue of being human", and that environmental degradation, "which may cause serious physical illness, impairment and suffering on the part of the local populace, [is] inconsistent with the right to be respected as a human being."[27]

The Inter-American Commission has also dealt with the right to life in a petition brought by the Yanomami community against the Brazilian government.[28] In the petition, the Commission explicitly recognized that environmental degradation can violate the right to life. In that case, the Brazilian government constructed a highway through Yanomami territory and authorized the exploitation of the territory's resources. These actions led to an influx of non-indigenous people who brought contagious diseases that spread to the Yanomami, resulting in disease and death.[29] The Commission found that, among other things, the government's failure to protect the integrity of Yanomami lands violated the Yanomami's rights to life, liberty, and personal security, which are guaranteed by Article 1 of the American Declaration

---

[22] Ibid., para. 5.

[23] In E.H.P. v. Canada, a group of Canadian citizens alleged that the storage of radioactive waste near their homes threatened the right to life of present and future generations. E.H.P. v. Canada, Communication No. 67/1980, UN Doc. CCPR/C/OP/1, 1984, at 20, para. 8. Also available at: http://www1.umn.edu/humanrts/undocs/html/67-1980.htm (last accessed on 25 February 2012).

[24] Ibid., para 8.

[25] Report on the Human Rights Situation in Ecuador, supra, note 19.

[26] Ibid.

[27] Ibid.

[28] Case of Yanomami Indians, Judgement, 1985, Case 7615 (Brazil), Inter-Am. C.H.R., OEA/Ser.L/V/II.66 doc. 10 rev. 1. Also available at: www.cidh.org/annualrep/84.85eng/Brazil7615.htm (last accessed on 25 February 2012).

[29] Ibid., under the section "Background", para. 3.

of the Rights and Duties of Man.[30] In this statement, the Commission importantly connected the interference with the lands of indigenous peoples to a violation of their right to life.

Another regional human rights body, the European Court of Human Rights, has found a direct connection between the right to life and environmental interference. Both *Öneryildiz v. Turkey*[31] and the case of *Budayeva and Others v. Russia*[32] were brought before the European Court on the basis of Article 2 (a right to life) and Article 1 of a Protocol (a right to property) to the European Human Rights Convention. In the former case, in addition to the right to life, the Court also found a violation of the right to property.[33]

The first case concerned a vast waste-collection site in Turkey, which was established in opposition to the Environmental Act and the Regulations on Solid-Waste Control. In April 1993, a methane explosion occurred at a site that was near the slum dwelling area. The explosion was followed by a mudslide, which was caused by pressure and led to the death of 39 people.[34] The Court found that the administrative and municipal authorities knew or ought to have known that there was a real and immediate risk to people. Therefore, they had a positive obligation, under Article 2 of the Convention, to take necessary and sufficient preventive measures to protect those individuals. Hence, the Court "unanimously [held] that there has been a violation of Article 2 of the Convention in its substantive aspect, on account of the lack of appropriate steps to prevent the accidental death of nine of the applicant's close relatives."[35] Additionally, the Court stated that "there has also been a violation of Article 2 of the Convention on its procedural aspect, on account of the lack of adequate protection by law safeguarding the right to life."[36]

---

[30] Ibid., under the section "The Inter-American Commission on Human Rights, resolves", para. 1.

[31] Case of Öneryildiz v. Turkey, Judgement, 41 EHRR (2004), at 20.

[32] Case of Budayeva and Others v. Russia, Judgement, EHRR 15339/02, 21166/02, 20058/02, 11673/02 and 15343/02 /2008.

[33] In an environmental context, the right to property has been applied quite extensively by the Inter-American Human Rights Court. See, for instance, The Mayagna (Sumo) Awas Tingni Community v. Nicaragua, Judgement, 31 August 2001, Inter-Am. Ct. HR., (Ser.C), No. 79; Sawhoyamaxa Indigenous Community v. Paraguay, Judgement, 29 March 2006, Case 0322/2001, Report No. 12/03, Inter-Am. C.H.R., OEA/Ser.L/V/II.118 Doc. 70 rev.; Yakye Axa Indigenous Community v. Paraguay, Judgement, 24 August 2010, Case 12.313, Report No. 2/02, Inter-Am. C.H.R., Doc. 5 rev. 1, at 387; Moiwana Community v. Suriname, Judgement, 15 June 2005, Series C No. 124; Saramaka People v. Suriname, Judgement, 28 November 2007, Inter-American Court of Human Rights, (ser. C). No. 172. For an extensive analysis of the protection of the property rights of indigenous peoples in international instruments, see Nigel Bankes, "The Protection of the Rights of Indigenous Peoples to Territory through the Property Rights Provisions of International Regional Human Rights Instruments", 3 *The Yearbook of Polar Law* (2011), 57.

[34] For an analysis, see Malgosia Fitzmaurice, *Contemporary Issues in International Environmental Law* (Cheltenham: Edward Elgar, 2009), at 201.

[35] Item 1 of the considerations of the merits, available at: http://www.elaw.org/node/5566 (last accessed on 25 February 2012).

[36] Item 2, see Ibid.

The facts of the latter case are very similar to those of the first and relate to the alleged negligence of Russian authorities in mitigating the result of the mudslide, which resulted in both deaths and the destruction of property. Similarly, the Court found the violation of substantive and procedural aspects of the right to life.[37]

It is indisputable that humanity's ultimate survival is indelibly linked to the state of the Earth's environment.[38] Human rights monitoring bodies have also increasingly recognized situations where environmental destruction may seriously affect human life. It is undeniable that anthropogenic climate change violates the right to life. To start with climate change is projected to result in increasingly severe weather occurrences, including tornadoes, hurricanes, storm surges, and floods, which may lead to a direct loss of life.[39]

In these cases, one cannot specify the individuals that will suffer in advance. However, this does not weaken the argument that the actions in question undermine human rights. Examples of storm-surge flooding in Bangladesh[40] and extreme heat waves in Chicago[41] or Western Europe are merely examples of the climatic changes leading to a considerable increase in deaths.[42]

In its 2009 report on climate change and human rights, the OHCHR, based on the 2007 IPCC assessment, states:

> A number of observed and projected effects of climate change will pose direct and indirect threats to human lives. IPCC…projects with high confidence an increase in people suffering from death, disease and injury from heat waves, floods, storms, fires and droughts. Equally, climate change will affect the right to life through an increase in hunger and malnutrition and related disorders impacting on child growth and development, cardio-respiratory morbidity and mortality related to ground-level ozone. Climate change will exacerbate weather-related disasters which already have devastating effects on people and their enjoyment of the right to life, particularly in the developing world. For example, an estimated 262 million people were affected by climate disasters annually from 2000 to 2004, of whom over 98 per cent live in developing countries.[43]

---

[37] See Items 2 and 3 of the consideration of the merits, available at: http://cmiskp.echr.coe.int/tkp197/viewhbkm.asp?sessionId=69032692&skin=hudoc-en&action=html&table=F69A27FD8FB86142BF01C1166DEA398649&key=28166&highlight=Budayeva (last accessed on 25 February 2012).

[38] See generally Prue Taylor, *An Ecological Approach to International Law, Responding to Challenges of Climate Change* (London: Routledge, 1998).

[39] Special Report: Managing the Risks of Extreme Events and Disasters to Advance Climate Change Adaptation, Summary for Policy Makers (2011).

[40] R.F. Mclean and Alla Tsyban, "Coastal Zones and Marine Ecosystems", in James J. McCarthy et al. (eds), *Climate Change 2001: Impacts, Adaptation, and Vulnerability. Contribution of Working Group II to the Third Assessment Report of the Intergovernmental Panel on Climate Change* (Cambridge and New York: Cambridge University Press, 2001), at 366–367.

[41] Jonathan A. Patz et al., "The Potential Health Impacts of Climate Variability and Change for the United States: Executive Summary of the Report of the Health Sector of the U.S. National Assessment", 108 *Environmental Health Perspectives* (2000), 367, at 370.

[42] For a general overview, See Andrew Haines et al., "Climate Change and Human Health: Impacts, Vulnerability, and Mitigation", 369 *The Lancet* (2006), 2101.

[43] OHCHR, supra, note 9, paras. 22–23.

Climate change, by redrawing the maps of water availability, food security, disease prevalence, population distribution, and coastal boundaries has the potential to exacerbate insecurity and violent conflict on a potentially large scale.[44] While threats to life are more immediate in some countries and regions than in others, a recent report by the U.S. Center for Naval Analyses argues that climate change acts as a threat multiplier in already fragile regions, exacerbating conditions that lead to failed states and breed terrorism and extremism, and concluded that "projected climate change poses a serious threat to America's national security."[45]

12.2.2 Climate Change Impacts on the Right to Health

The right to health,[46] similar to the right to life, is guaranteed by many widely accepted international human rights instruments.[47] The International Covenant on Economic, Social and Cultural Rights (CESCR) recognizes the right to the "highest

[44] See, McInerney-Lankford, Darrow and Rajamani, *Human Rights and Climate Change: A Review of the International Legal Dimensions*, supra note 6, at 13. See also, Oli Brown and Alec Crawford, "Rising Temperatures, Rising Tensions: Climate Change and the Risk of Violent Conflict in the Middle East", 2009, available at: http://www.iisd.org/pdf/2009/rising_temps_middle_east.pdf (last accessed on 26 February 2012). United Nations Environment Programme "From Conflict to Peace-Building: The Role of Natural Resources and the Environment", 2009, available at: http://postconflict.unep.ch/publications/pcdmb_policy_01.pdf (last accessed on 26 February 2012); Nick Mabey, "Delivering Climate Security: International Security Responses to a Climate Changed World", 2008, available at: http://www.tandf.co.ul/journals/spissue/rwhi-si.1.asp (last accessed on 25 February 2012); Brahma Chellaney, "Climate Change and Security in Southern Asia: Understanding the National Security Implications", 152 *Royal United Services Institute Journal* (2007), at 63.

[45] Center for Naval Analyses Corporation, "National Security and the Threat of Climate Change", 2007, available at: http://securityandclimate.cna.org/report/National%20Security%20and%20 the%20Threat% 20of%20Climate%20Change.pdf (last accessed on 25 February 2012). See also Douglas V. Johnson, "Global Climate Change: National Security Implications", 2007, available at: http://www.strategicstudiesinstitute.army.mil/pubs/display.cfm?pubID=779 (last accessed on 26 February 2012).

[46] The World Health Organization, which addresses health concerns in a variety of cultural and social contexts, defines health as "a state of complete physical, mental and social well-being, and not merely the absence of disease or infirmity". Constitution of the World Health Organization, New York, 22 July 1946, in force 7 April 1948, Official Records of the World Health Organization, Vol. 2, preamble, at 100. The definition and application of the universal right to health, then, must account for the complex interplay of physical, mental and social experiences and circumstances, and the varying cultural and social norms used to evaluate them. Michael F. Willis, "Economic Development, Environmental Protection, and the Right to Health", 9 *Georgetown International Environmental Law Review* (1996), 195–220, at 197.

[47] The Universal Declaration of Human Rights, supra note 17, Art. 25 (1); The European Social Charter, 18 October 1961, revised in May 1996, in force 1999, 529 *United Nations Treaty Series* (1997), 89, Art. 11; The American Declaration of the Rights and Duties of Man, supra, note 13, Art. XI; The Additional Protocol to the American Convention on Human Rights in the Area of Economic, Social and Cultural Rights ("Protocol of San Salvador"), supra, note 13, Art. 10.

attainable standard of physical and mental health."[48] The Committee on Economic, Social and Cultural Rights states that this right is indispensable for the enjoyment of other human rights.[49]

As interpreted by the CESCR and other authoritative or adjudicatory bodies, the substantive content of this right includes timely and appropriate health care, access to safe and potable water, adequate sanitation, an adequate supply of safe food, nutrition and housing, healthy occupational and environmental conditions, and access to health-related education and information.[50] Furthermore, the Committee goes on to state that victims of violations of the right to health should have access to remedies at both national and international levels and should be entitled to adequate reparation.[51] These are considered to be the basic determinants of health that, according to the assessment of the World Health Organization (WHO), will be placed at risk due to climate change.[52]

The only United Nations' human rights treaty whose text directly refers to environmental issues in relation to the right to health is the United Nations Convention on the Rights of the Child. While recognizing the right of the child to the enjoyment of their highest attainable standard of health, state parties, *inter alia*, shall take appropriate measures to combat disease and malnutrition, "taking into consideration the dangers and risks of environmental pollution."[53]

In the context of the state reporting procedure, the UN Committee on the Rights of the Child has issued observations calling for better compliance with Article 24(2)(c) in some of its Concluding Observations. For instance, it recommended that Jordan "take all appropriate measures, including through international cooperation, to prevent and combat the damaging effects of environmental pollution and contamination of water supplies on children and to strengthen procedures for inspection."[54] The Committee also expressed concern regarding South Africa and

---

[48] The International Covenant on Economic, Social and Cultural Rights, New York, 16 December 1966, in force 3 January 1976, 993 *United Nations Treaty Series* (1996), Art.12.

[49] The Committee on Economic, Social and Cultural Rights 3 [hereinafter CESCR], General Comment No. 14, The right to the highest attainable standard of health, Un. Doc. E/C.12/2000/4 11 August 2000, paras. 1 and 3.

[50] Ibid, para. 11. For a discussion on the sources and content of this right, see, P. Hunt, Report of the Special Rapporteur on the Right of Everyone to the Highest Attainable Standard of Physical and Mental Health, UN Doc. E/CN.4/2003/58, 2003, paras. 10–36.

[51] Ibid., para. 59.

[52] World Health Organization, "Protecting Health from Climate Change", 2008, available at: http://www.who.int/world-health-day/toolkit/report_web.pdf (last accessed on 26 February 2012), at 6.

[53] CRC, supra note 19, Art. 24.

[54] UN Committee on the Rights of the Child, Concluding Observations on Jordan, UN Doc. CRC/C/15/Add.125, at para. 50. Also available at: www1.umn.edu/humanrts/crc/jordan2000.htm (last accessed on 15 January 2012). See UN Committee on the Rights of the Child, Concluding Observations on South Africa, UN Doc. CRC/C/15/Add. 122, 26 January 2000, at para. 30. Also available at: http://www1.umn.edu/humanrts/crc/southafrica2000.html (last accessed on 26 February 2012).

"the increase in environmental degradation, especially as regards air pollution."[55] It recommended that South Africa fight environmental degradation, particularly air pollution, by facilitating "the implementation of sustainable development programmes to prevent environmental degradation, especially as regards air pollution."[56]

Many multilateral environmental agreements acknowledge and address the impact that environmental harms may have on human health.[57] The UNFCCC, in its definition of adverse effects of climate change, includes "significant deleterious impacts on human health and welfare", and requires Parties to account for, *inter alia*, health impacts in relevant social, economic, and environmental policies.[58]

The close relationship between environmental integrity and health has been recognized by various studies concerning international human rights. UN Special Rapporteur on human rights and the environment of the then UN Sub-Commission on Prevention of Discrimination and Protection of Minorities,[59] Fatma Zohra Ksentini, identified the right to health as a fundamental right and analyzed the effects of the environment on it.[60] After studying various international human rights documents and national constitutions, she concluded that, under customary international law, "everyone has a right to the highest attainable standard of health."[61] Furthermore, she came to the conclusion that "in the environmental context, the right to health essentially implies feasible protection from natural hazards and freedom from pollution."[62]

---

[55] See UN Committee on the Rights of the Child, Concluding Observations on South Africa, supra, note 54, at para. 30.

[56] Ibid.

[57] See e.g. Convention on Long-Range Transboundary Air Pollution, Geneva, 13 November 1979, in force 16 March 1983, 18 *International Legal Materials* (1979), 1442; Basel Convention on the Control of Transboundary Movements of Hazardous Wastes and their Disposal, Basel, 22 March 1979, in force 5 May 1992, 28 *International Legal Materials* (1979), 656; Rotterdam Convention on the Prior Informed Consent Procedure for Certain Hazardous Chemicals and Pesticides in International Trade, Rotterdam, 10 September 1998, in force 24 February 2004, 38 *International Legal Materials* (1999), 1; The Stockholm Convention on Persistent Organic Pollutants, Stockholm, 23 May 2001, in force 17 May 2004, 40 *International Legal Materials* (2001), 532. See also Need to Ensure a Healthy Environment for the Well-being of Individuals, G.A. Res. 45/94, UN Doc. !/ RES/45/94, 14 December 1990.

[58] United Nations Framework Convention on Climate Change, United Nations Framework Convention on Climate Change, New York, 9 May 1992, in force 21 March 1994, 31 *International Legal Materials* (1992), 849, Art. 1(1).

[59] The body has changed its name and is now the Sub-Commission on Promotion and Protection of Human Rights.

[60] Fatma Z. Ksentini, "Review of Further Developments in the Fields with Which the Sub-Commission Has Been Concerned: Human Rights and the Environment", UN Doc. E/CN.4/ Sub.2/1994/9, 6 July 1994, paras. 176–187.

[61] Ibid., para. 176.

[62] Ibid. Other rapporteurs of the UN have also found connections between environmental degradation and the right to health. The United Nations' Special Rapporteur on the right to health, Paul Hunt, noted that the right to health gives rise to an obligation on the part of a State to ensure that environmental degradation does not endanger human health. See P. Hunt, Right of Everyone to the

The Inter-American Commission on Human Rights has also recognized the close relationship between environmental degradation and the right to health, especially in the context of indigenous peoples. In the *Yanomami case*, aside from the right to life, the Commission recognized that harm to people, resulting from environmental degradation, violated their right to health in Article XI of the American Declaration.[63] Additionally, in *the Belize Maya case*, the Commission noted that indigenous people's right to health and well-being was so dependent on the integrity and condition of indigenous land that "broad violations" of indigenous property rights essentially impacted the health and well-being of the Maya.[64]

The European Court of Human Rights and its case-law as regards Article 8 of the European Human Rights Convention (a right to home and privacy) has established a close connection to the right to health and well-being. There are several cases in the Court that relate to the environmental interference causing health-related problems to applicants.[65] Based on the study concerning the Court's jurisprudence, Boyle concludes that states have a positive duty to take appropriate measures to prevent industrial pollution or other forms of environmental nuisance from seriously interfering with health or the enjoyment of private life or property.[66]

There is, by now, an extensive literature pointing to the severe health impacts of anthropogenic climate change.[67] For example, the Fourth Assessment Report of the IPCC notes that anthropogenic climate change will increase the number of people suffering from disease and injury as a result of heat waves, floods, storms, fires, and droughts; increase the range of malaria in some places, while decreases in

---

Highest Standard of Physical and Mental Health: Addendum, Mission to Peru, UN Doc. E/CN.4/2005/51/Add.3, 2005, para. 54. A reference to human rights generally in relation to the environment was also made by Special Rapporteur Rodolfo Stavenhagen of the UN Commission on Human Rights, who took particular account of indigenous peoples. He concluded that "the effects of global warming and environmental pollution are particularly pertinent to the life changes of Aboriginal people in Canada's North, a human rights issue that requires urgent attention at the national and international levels, as indicated in the recent Arctic Climate Impact Assessment". Rodolfo Stavenhagen, Human Rights and Indigenous Issues: Report of the Special Rapporteur on the Situation of Human Rights and Fundamental Freedoms of Indigenous People, Addendum, UN Doc. E/CN.4/2005/88/Add. 4, 15 December 2004, para. 94.

[63] Yanomami Indians case, under the section "The Inter-American Commission on Human Rights, resolves", supra, note 28, para. 1.

[64] Belize Maya, *Judgement, 2004,* Case 12.053, Report No. 40/04, Inter-Am. C.H.R., OEA/Ser.L/V/II.122 Doc. 5 rev.1, at 727, paras 154–156.

[65] See, for instance, Lopez Ostra v. Spain, Judgement, 20 EHRR (1994), at 277; Guerra v. Italy, Judgement, 26 EHRR (1998), at 357; Fadeyeva v. Russia, Judgement, ECHR (2005), at 376; Raynor and Powell v. United Kingdom, Judgement, 12 EHRR (1990), at 355; Taskin v. Turkey, Judgement, ECHR 46117/99; Tatar v. Romania, Judgement, ECHR, 67021/01. See also two Hatton cases: Hatton and others v. United Kingdom, Judgement, 2 October 2001 and 8 July 2003.

[66] Alan E.. Boyle, "Human Rights and the Environment: A Reassessment", 2010, available at: http://www.unep.org/environmentalgovernance/LinkClick.aspx?fileticket=GccCLN-brmg%3D&tabid=.... (last accessed on 26 February 2012), at 16. Originally published in 18 *Fordham Environmental Law Review* (2008), 471.

[67] Caney, supra, note 11, at 79.

others; increase the burden of diarrhoeal diseases; increase cardio-respiratory morbidity associated with ground-level ozone; and increase the number of people at risk of dengue.[68] Thus, human-induced climate change clearly results in a variety of threats to the human right to health.[69]

The IPCC also predicts that adverse health impacts will be greatest in low-income countries. Across all countries, "the urban poor, the elderly and children, traditional societies, subsistence farmers, and coastal populations" are at greatest risk.[70] Health equity is also at risk, as are prospects for achieving the health-related Millennium Development Goals.[71] Overall, negative health effects will be disproportionately felt in Sub-Saharan Africa, South Asia, and the Middle East.[72] In the Arctic area, as described in relation to the Inuit Petition examined in the following chapter, climate change is already having health-related impacts on indigenous peoples living on the lands.

12.2.3 The Inuit Petition as an Example of a Human Rights Approach to Climate Change

To date, global climate change has most intensively been felt in the Arctic area. Over the past few decades, the average Arctic temperature has risen twice as much as the average global temperature.[73] The United Nations Intergovernmental Panel on Climate Change (IPCC) has predicted that Arctic temperatures will increase by 5–7° by 2099, while the Earth's temperature is predicted to rise by 1.8–4°.[74]

The Arctic Climate Impact Assessment (ACIA) – a comprehensive international evaluation of Arctic climate change and its impact undertaken by hundreds of

---

[68] Pachauri and Reisinger (eds), *Climate Change 2007*, supra, note 1, at 48.

[69] Caney, supra, note 11, at 80.

[70] Ulisses Confalonieri et al., "Human Health", in Martin L. Parry et al. (eds), *Climate Change 2007: Contribution of Working Group II to the Fourth Assessment Report of the Intergovernmental Panel on Climate Change* (Cambridge and New York: Cambridge University Press, 2007), at 393. See also Nicholas Stern, "Stern Review on the Economics of Climate Change", 2006, available at: http://www.hm-treasury.gov.uk/independent_reviews/stern_review_economics_climate_change/sternreview_index.cfm (last accessed on 26 February 2012), part II, chapter 3.

[71] See, McInerney-Lankford, Darrow and Rajamani, *Human Rights and Climate Change, A Review of the International Legal Dimensions*, supra, note 7, at 16. See also World Health Assembly, "Resolution on Climate Change and Health", 24 May 2008, available at: http://www.who.int/gb/ebwha/pdf_files/A61/A61_R19-en.pdf (last accessed on 26 February 2012).

[72] OHCHR (2009), supra, note 10, para. 32.

[73] Arctic Climate Impact Assessment (ACIA) (Cambridge: Cambridge University Press, 2005).

[74] See summary for policymakers of the synthesis report of the fourth assessment report of the Intergovernmental Panel on Climate Change, Rajendra K. Pachauri and Andy Reisinger (eds), *Contribution of Working Groups I, II and III to the Fourth Assessment Report of the Intergovernmental Panel on Climate Change – Synthesis Report*, supra, note 1. Also available at: http://www.ipcc.ch/ (last accessed on 26 February 2012).

scientists – points to dramatic changes in the Arctic environment and Arctic indigenous peoples' nature-based lifestyle as a result of global climate change. According to the ACIA, over the next 100 years, climate change is expected to accelerate and will contribute to major physical, ecological, social, and economic changes, many of which have already begun.[75]

Many environmental changes, studied and predicted by the ACIA, are already having a direct impact on Arctic indigenous peoples' traditional lifestyle. Indigenous peoples throughout the Arctic area depend on the land and sea for food and income as well as traditional activities, including hunting, fishing, gathering, and reindeer herding, which are vitally important for indigenous society and culture.[76] The hunting culture of many Arctic indigenous peoples is particularly endangered. However, climate change will also affect other traditional livelihoods.

For Arctic indigenous peoples, global climate change is an important human rights issue due to their traditional, nature-based way of life, which is often considered to be the crux of the culture of indigenous peoples.[77] For this reason, in 2005, Sheila Watt-Cloutier, the former president of the Inuit Circumpolar Council (ICC), an organization representing Inuit peoples in four Arctic states,[78] filed a petition against the United States at the Inter-American Commission on Human Rights, for the damage caused to the Inuit and their rights as a result of global climate change.[79] This chapter briefly explores the petition with the aim of showing how the present impacts of climate change are already making Arctic indigenous peoples particularly vulnerable by infringing on many of their important human rights.

According to the Inuit petition, the impact of climate change caused by acts and omissions of the United States, violates the Inuit's fundamental human rights, which are protected by the American Declaration of the Rights and Duties of Man and other international instruments. At that time, the petition served as a reminder that

---

[75] See generally ACIA, supra, note 73.

[76] Ibid., at 4.

[77] Indigenous peoples often live in the most vulnerable ecosystems, such as in areas of high biological diversity or in the stark arctic regions. According to estimates made in 1990, around 200 million of the world's 300 million indigenous peoples live in vulnerable ecosystems. See Report of the Commission on Human Rights at its forty-sixth session E/1990/22-E/CN.4/1990/94, 1990, at 8.

[78] Alaska (USA), Canada, Greenland (Denmark) and the Russian Federation.

[79] The Inuit petition was submitted by Sheila Watt-Cloutier, the president of the ICC at the time, "with the support of the Inuit Circumpolar Conference", on behalf of all the Inuit of the Arctic regions of the United States and Canada; it is signed by 62 people in addition to Watt-Cloutier. See the Petition to the Inter American Commission on Human Rights Seeking Relied From Violations Resulting from Global Warming Caused by Acts and Omissions of the United States, 7 December 2005, available at: http://www.inuitcircumpolar.com/files/uploads/icc-files/FINALPetitionICC.pdf (last accessed on 1 February 2012), at 1. According to the rules of procedure of the Commission, any person, group of persons or non-governmental entity may submit a petition as long as the petition involves an alleged violation of a human right recognized under the IAHR regime. See Rules of Procedure of the Inter-American Commission on Human Rights, approved by the Commission at its 109th Session, 8 December 2000, in force 1 May 2001.

the United States is the world's largest contributor to global warming, which has a damaging effect on the Inuit. As the world's leading consumer of energy, both historically and at the time of the petition, the United States is responsible for the largest amount of cumulative emissions of any state.[80]

The Inuit petition greatly relies on the ACIA and uses the assessment as a scientific basis. The petition points out that, because average annual Arctic temperatures are increasing more than twice as fast as temperatures in the rest of the world, climate change has already seriously impacted the Arctic. This includes the deterioration of ice conditions, a decrease in the quantity and quality of snow, changes in the weather and weather patterns, as well as a transfigured landscape as permafrost melts at an alarming rate, which causes slumping, landslides, and severe erosion in some coastal areas.[81] For instance, in the Shishmaref village in Alaska, many of the houses owned by local Inuit have been badly damaged and partly fallen into the sea due to erosion and a rise in the sea-level.[82] Inuit observations and scientific studies consistently document many types of environmental changes. Importantly, the ACIA contains a chapter related to indigenous traditional knowledge and indigenous peoples' observations on climate change.[83]

According to the petition, several principles of international law guide the application of human rights issues in this case. Most directly, the US membership in the Organization of American States and its acceptance of the American Declaration of the Rights and Duties of Man oblige it to protect the rights of the Inuit.[84] The petition alleges that diverse impacts of climate change violate several human rights, such as the rights to the benefits of culture, to property, to the preservation of health, life, physical integrity, security and a means of subsistence, and to residence, movement, and inviolability of the home.[85]

The petition's legal starting point is that indigenous peoples' human rights are to be interpreted in the context of indigenous culture, which requires the protection of their land and environment.[86] The petition points out that, in applying rights contained in the American Declaration to indigenous peoples, both the Inter-American

---

[80] The Inuit petition, supra, note 79, at 103.

[81] Ibid., at 2.

[82] See BBC News, David Willis, "Sea Engulfing Alaskan village", available at: http://news.bbc.co.uk/1/hi/world/europe/3940399.stm (last accessed on 26 February 2012).

[83] ACIA, "The Changing Arctic: Indigenous Perspectives", supra, note 73, at 61–98.

[84] The Inuit petition, supra, note 79, at 5. The United States is not a party to the American Convention on Human Rights, so the Convention cannot be applied to this case. The American Declaration is nevertheless regarded as having become a legally binding instrument through so-called double-incorporation. See Thomas Buergenthal, "The Revised OAS Charter and the Protection of Human Rights", 69 *American Journal of International Law* (1975), 828. Additionally, the Inter-American Commission has regarded the Inter-American Declaration as legally binding in its case law. See Douglass Cassel, "Inter-American Human Rights Law, Soft and Hard", in Dinah Shelton (ed.), *Commitment and Compliance: The Role of Non-binding Norms in the International Legal System* (Oxford: Oxford University Press, 2000), 393, at 397.

[85] The Inuit petition, supra, note 79, at 5.

[86] Ibid., at 70.

Human Rights Court and the Commission have repeatedly emphasized the need to account for the unique context of indigenous culture.[87] The Commission has stated that, by interpreting the American Declaration as safeguarding the integrity, livelihood and culture of indigenous peoples through the effective protection of their individual and collective human rights, the Commission respects the very purpose underlying the Declaration, which, as expressed in the preamble, recognizes that "it is the duty of man to preserve, practice and foster culture by every means within his power."[88] Furthermore, the Commission has stated that "indigenous peoples maintain special ties with their traditional lands, and a close dependence upon the natural resources provided therein – respect for which is essential to their physical and cultural survival."[89]

According to the petition, the lives and culture of the Inuit demonstrate that indigenous peoples' human rights are inseparable from their environment. Therefore, the preservation of the Arctic environment is "one of the distinct protections required for the Inuit to fully enjoy their human rights on an equal basis with all peoples."[90] The petition claims that states, thus, have an international obligation to not degrade the environment to an extent where it threatens the culture, health, life, property, or ecological security of indigenous peoples.[91]

The petition serves as a reminder that the Inuit and their culture have developed over thousands of years in relation and response to the Arctic's physical environment.[92] The Inuit have, thus, developed an intimate relationship with their surroundings, using their understanding of the Arctic environment to develop tools, techniques, and knowledge that has enabled them to subsist on their scarce environmental resources.[93] All aspects of Inuit life depend on Arctic ice, snow, land, and weather conditions. The petition even goes so far as to argue that "the subsistence harvest is essential to the continued existence of the Inuit as a people."[94]

---

[87] Ibid. The petition refers to many cases that will be dealt with in this section.

[88] Case of Mary and Carrie Dann, Judgement, 27 December 2002, Report No. 75/02, Case 11.140, Inter-Am. C.H.R., 2002, para. 131, quoting the American Declaration of the Rights and Duties of Man. Also available at: http://www.cidh.oas.org/annualrep/2002eng/USA.11140.htm (last accessed on 26 February 2012).

[89] Report on the Human Rights Situation in Ecuador: Human Rights Issues of Special Relevance to the Indigenous Inhabitants of the Country, OEA/Ser.L/V/II.96, 27 April, 1997. Available at: http://cidh.org/countryrep/ecuador-eng/index%20-%20ecuador.htm (last accessed on 26 February 2012).

[90] The Inuit petition, supra, note 79, at 72.

[91] Ibid.

[92] The petition refers to Margie A. Gibson and Sally B. Schullinger, *Answers from the Ice Edge: The Consequences of Climate Change on Life in the Bering and Chukchi Seas* (Anchorage, Alaska: Arctic Network & Greenpeace, 1998), at 6.

[93] The petition refers to the ACIA Overview report *Impacts of a Warming Arctic: Arctic Climate Impact Assessment* (Cambridge: Cambridge University Press, 2004), at 16.

[94] The petition refers to the ACIA Overview report, supra, note 93, at 94; Alaska Native Science Commission, "National Subsistence Technical – Planning Meeting for the Protection of Traditional & Tribal Life-ways", 15 April 2003, available at: http://www.nativescience.org/pubs/reports.htm (last accessed on 25 February 2012); Alaska Regional Assessment Group, *The Potential Consequences of Climate Variability and Change* (Fairbanks: The Center for Global Change and Arctic System Research, 1999).

The Inter-American Commission on Human Rights has clearly recognized that environmental interference with indigenous peoples' lands may lead to the infringement of their human rights.[95] So, in principle, at the Commission, it may be assumed that the consequences of climate change could be considered to be an issue of human rights.

On November 16, 2006, the Commission rejected the Inuit petition, stating that "the information provided does not enable us to determine whether the alleged facts would tend to characterize a violation of rights protected by the American Declaration."[96] Following a request of the petitioners, the Inter-American Commission decided to hold a public hearing to gather more evidence on the link between global warming and human rights. However, the petitioners' request modestly states, "[w]e are aware that the Commission has dismissed that petition and do not seek here to reopen that decision."[97] Now that several years have passed, it appears to be obvious that the Commission has decided not to proceed with the case. The pros and cons of the Inuit petition will be discussed in Sect. 12.4.

12.3 Human Rights Influence on the Function and Design of the Climate Change Regime

The situation of Northern Honduras provides a dramatic example of the manner in which climate mitigation measures may potentially clash with the enjoyment of basic human rights. The background of the case is as follows. The region of Bajo Aguan is the location of a longstanding land claim dispute caused by the government's illegal sale of land, previously allocated to peasants, to private entrepreneurs as a result of the agrarian reform in the 1990s. Within the context of a general insecurity in the country since 2009, the situation resulted in a conflict between peasants

---

[95] In the planning state of the petition, the ICC was trying to determine whether there might be other suitable bodies for the petition. In the beginning of 2003, the Executive Council of the ICC issued a resolution pondering the issue. The resolution mentions in particular two states, the Russian Federation and the United States, which had not at that time ratified the Kyoto Protocol. See ICC, ICC Executive Council Resolution 2003–1, available at: http://www.inuit.org/index.asp?lang=eng&num=244 (last accessed on 15 February 2012). Russia, unlike the United States, has ratified the Optional Protocol to the CCPR, so in principle the Russian Inuit could have brought an individual communication to the UN Human Rights Committee. Importantly, however, the Russian Federation ratified the Kyoto Protocol before the Inuit took the legal action against the United States, so a claim against the Russian Federation was no longer so topical (The Russian Federation ratified the Kyoto Protocol on 5 November 2004).

[96] Letter from Ariel E. Dulitzki, Assistant Executive Secretary to Paul Crowley, Legal Representative for Sheila Watt-Cloutier, 16 November 2006, available at: http://graphics8.nytimes.com/packages/pdf/science/16commissionletter.pdf (last accessed on 26 February 2012).

[97] Letter from Sheila Watt-Cloutier, Martin Wagner and Daniel Magraw, "Request for a Hearing on the Relationship between Global Warming and Human Rights", 15 January 2007, available at: http://www.ciel.org/Publications/IACHR_Letter_15Jan07.pdf (last accessed on 26 February 2012).

claiming their land rights and private security firms protecting the interest of large palm oil plantations. As a result, between January 2010 and the winter of 2011, 42 people – primarily peasants, but also journalists – have been assassinated in the context of the conflict. During its field visit in May 2010, the Inter-American Commission on Human Rights already expressed "its concern over the involvement of the armed forces in matters related to citizen security; as such matters should be the exclusive purview of the civilian law enforcement."[98] Military presence was, however, increased in the region in August 2011. On 24 October 2011, the Inter-American Commission on Human Rights held a hearing on the situation in Bajo Aguan, during which the situation was described by the petitioners as "the most severe repression and aggression against peasant communities in a sub-region in Central America in the past 15 years."[99]

In July 2011, the Aguan biogas project, which produces fuel from the output of local palm oil plantations, was registered by the Executive Board (EB) of the Clean Development Mechanism (CDM). This decision was reached despite the involved firms' direct link to cases of alleged murders. Hence, the project's registration under UNFCCC flexibility mechanisms, will provide a financial incentive for the continuation of the dispute between entrepreneurs and local peasants. This issue prompted non-governmental and institutional stakeholders to call for a reform of the CDM in order to ensure that the respect of human rights becomes a basic requirement for all projects.[100]

This unfortunate example highlights that the convention's implementation does not guarantee the respect of human rights but may sometimes lead to a gross violation of the rights of local communities. Three approaches may be relied upon in order to mitigate this risk. Firstly, the effective enjoyment of stakeholders' procedural rights should be guaranteed.[101] In the design of the climate regime, the exercise of these rights provides an opportunity for civil society representatives to highlight the risks and flaws inherent in the development of the climate regime. Secondly, the exercise of these rights in the adoption of mitigation and adaptation policies ensures that the measures adopted in the implementation of the convention do not infringe on the specific rights of local communities. Thirdly, the necessity to respect substantive human rights, when fulfilling obligations resulting from the climate regime, may be explicitly addressed by the working bodies of the regime.

---

[98] Preliminary Observations of the Inter-American Commission on Human Rights on its visit to Honduras, OEA/Ser.L/V/II., Doc. 68, 18 May 2010, para. 120.

[99] See report, Claire Grandison, "Human Rights Situation in the Bajo Aguan, Honduras", 28 October 2011, available at: http://hrbrief.org/2011/10/human-rights-situation-in-the-bajo-aguan-honduras/ (last accessed on 26 February 2012).

[100] See for instance the European Parliament resolution of 16 November 2011 on the climate change conference in Durban (COP 17), P7_TA-PROV(2011)0504, para. 28.

[101] See Svitlana Kravchenko, "Procedural Rights as a Crucial Tool to Combat Climate Change", 38 *Georgia Journal of International and Comparative Law* (2010), 635.

12.3.1 Procedural Rights in the Intergovernmental Process

Based on the guidelines for participation of observers under the UNFCCC, public participation allows "vital experience, expertise, information and perspectives from civil society to be brought into the process to generate new insights and approaches" and "promotes transparency."[102] The situation in the Bajo Aguan region proves, in an extreme manner, that some of the climate regime's design mechanisms, here the CDM, may lead to consequences that were obviously unexpected. In light of this tragedy and judging by some of the parties' current disarray, one may expect that their negotiating teams, had they anticipated the consequences of the terms adopted in 2001, would not have accepted the existing modalities and procedures. In this case, one of the flaws of the current modalities and procedures is the fact that substantial time might pass between the consultation of local stakeholders and the registration of a project by the EB, thus potentially allowing a knowledge gap. Effective access to the negotiation process and the stakeholders' adequate participation may, in many instances, provide a warning when the proposed decisions may risk infringing on the rights of local communities.

The Convention provides that all parties have the obligation to "encourage the widest participation in this process, including that of non-governmental organizations."[103] The general extent of observers' rights to partake in the UNFCCC is defined by the Conference of the Parties.[104] The category of observers includes governments who are not party to the convention (or to the Kyoto Protocol, in relation to processes established under the protocol), intergovernmental organizations, and stakeholders.

The Subsidiary Bodies consider observers' consultation of and participation to the intergovernmental process on a cyclical basis.[105] Such discussion typically lasts over several sessions and includes a first session that is dedicated to a general statement by interested parties, a consultation round of both parties and stakeholders, via submissions or a workshop, and a final decision by the subsidiary body, as well as a possible endorsement of the decision by the COP itself.

Due to the lack of preparedness of the secretariat, in comparison to the unprecedented level of attendance and a historically high participatory rate of heads of

[102] UNFCCC, "Guidelines for the Participation of Representatives of Non-governmental Organizations at Meetings of the Bodies of the United Nations Framework Convention on Climate Change", 2003, available at: http://unfccc.int/files/parties_and_observers/ngo/application/pdf/coc_guide.pdf (last accessed on 25 February 2012), at 3.

[103] UNFCCC, supra, note 56, Art. 4.1(i).

[104] Ibid., Art. 7.6.

[105] The Subsidiary Body for Scientific and Technological Advice (SBSTA) was mandated to consider this issue as part of its agenda until 1997, after which this fall under the competence of the Subsidiary Body for Implementation (SBI) according to the division of labour between the two bodies. Report of the Subsidiary Body for Scientific and Technological Advice on the work of its 7th session held in Bonn from 25 February to 28 February 1997, FCCC/SBSTA/1997/4, 29 November 1997, para. 37(b).

state and government officials, COP 15 in 2009 presented a particular logistical challenge. Once they had arrived at the venue, thousands of delegates who had pre-registered in accordance with the established procedures were unable to obtain their accreditation badge and participate in the conference. Furthermore, during the conference's remaining 3 days, the secretariat decided to severely limit access to the conference (to approximately 2% of the civil society representatives during the final 48 h) by communicating this decision merely a few hours prior to its implementation. This situation led to an unparalleled denial of the rights of duly accredited members of the public to attend and participate at an intergovernmental meeting and led the Executive Secretary to express his personal regret for this situation.[106] Consequently, civil society's participation has been the subject of intergovernmental negotiations, under the COP and the Subsidiary Body for Implementation (SBI), with the objective of enhancing participation.[107] Additionally, the secretariat has conducted internal reviews and taken concrete steps in preventing such a situation at future sessions. The secretariat has also established the practice of meeting with all constituencies' representatives in order to discuss the modalities and challenges associated with the participation of observers in the process.

12.3.1.1 Access to Negotiations

The observer status is necessary for intergovernmental and non-governmental organizations that are interested in attending meetings and participating in the process. Non-governmental organizations, as well as non-UN intergovernmental bodies, interested in partaking in the negotiating process may submit an application to the secretariat in order to be admitted as an observer organization. In order to be admitted, organizations must demonstrate that they are "qualified in matters covered by the Convention."[108] The first Conference of the Parties, serving as the Meeting of the Parties to the Kyoto Protocol (COP/MOP), decided to extend the arrangements related to the participation of observer organizations to the meetings under the Kyoto Protocol of the Convention.[109] Once accredited, an organization may then nominate representatives to attend any negotiation meetings. There is no formal limit on the number of delegates that each organization may originally

[106] Message from the Executive Secretary to the Designated Focal Points of observer organizations, YdB/SD/HP/MEN, 23 February 2010.

[107] Report of the Subsidiary Body for Implementation on its 32nd session held in Bonn from 31 May to 9 June 2010, UN Doc. FCCC/SBI/2010/10, 25 August 2010, para. 167.

[108] UNFCCC, supra, note 58, Art. 7.6 and Draft Rules of Procedure of the COP and its Subsidiary Bodies, applied provisionally, UN Doc. FCCC/CP/1996/2, 22 May 1996, Rule 7(1).

[109] Decision 17/CP.9, Arrangements for the First Session of the Conference of the Parties serving as the Meeting of the Parties to the Kyoto Protocol, UN Doc. FCCC/CP/2003/6/Add.2, 30 March 2004, and Decision 36/CMP.1, Arrangements for the Conference of the Parties serving as the Meeting of the Parties to the Kyoto Protocol at its First Session, UN Doc. FCCC/KP/CMP/2005/8/Add.4, 30 March 2006, para. 2. (c).

nominate for a particular meeting. However, in order not to exceed the physical capacity of the venues, the secretariat has implemented since 2010 a quota system through which it allocates a specific number of accreditation badges to each organization proportionally to the number of delegates that they originally accredited.[110]

According to the procedures' draft rules, the COP meetings are held in public unless otherwise decided.[111] The rules of procedure provide that the subsidiary bodies' meetings are to be held in private, but an interpretative footnote comments on this rule, providing that it is to be interpreted in a manner that permits "duly accredited observers to participate in "private" meetings."[112]

However, the main bodies' plenary sessions, established under the convention (and the protocol), are primarily dedicated to the session's ceremonial opening, to crosscutting stocktaking, and to the final negotiations during the last hours of each session.[113] Most of the negotiations take place during the sessions in thematic groups established under one or several of the main bodies. Open-ended "contact groups" are the most formal manner of discussions. Observers may attend the contact group meetings unless a third of a session's present parties request the opposite.[114] The presiding officers also have the authority to close a contact group to observers at any given time. In recent years, governmental delegations have complained about their inability to attend all formal meetings taking place simultaneously. Consequently, current rules concerning the scheduling of sessions foresee that only six meetings may be scheduled in parallel, with only two of those as either plenary meetings or contact group.[115]

Hence, most meetings scheduled during the negotiation sessions are, however, organized as informal working groups. These groups allow for more flexible

---

[110] In response to concerns expressed by civil society and parties delegates on the impact of the seize of the venue for the participation of observers, the SBI also "encouraged hosts of future sessions of the COP and the CMP to consider, in their planning and organization, the size of the venue and the need to facilitate the participation of all Parties and admitted observer organizations". Report of the SBI (2010), supra, note 107, para. 166. Representatives under the age of 18 years old can be registered at the discretion of the secretariat, which allows their participation only for specific event and with additional requirements. See Guidelines for Participation, supra, note 99, section A, para. 4.

[111] Rules of Procedures, supra, note 108, rule 30. In practice logistical constraints in the implementation of this rule have been addressed through the use of webcasts and screening of the proceedings of the main sessions of the COP in parallel conference room in order to accommodate a large number of participants.

[112] For the reference to a prior discussion by the Intergovernmental Negotiating Committee on this issue, see Report of the Committee on its Eight Session, A/AC.237/41, paras. 105 and 106(c).

[113] For an example of the use by presiding officers of diverse degree of openness towards observers in their meetings, see Joanna Depledge, *The Organization of Global Negotiation: Constructing the Climate Change Regime* (London: Earthscan, 2005), at 218.

[114] Decision 18/CP.4, Attendance of Intergovernmental and Non-governmental Organizations at Contact Groups, UN Doc. FCCC/CP/1998/16/Add.1, 25 January 1999, para. 1.

[115] Report of the SBI, supra, note 107, para. 164.

procedural rules and enable more open discussions between negotiators.[116] In the absence of a formal recommendation on the access of observers to informals within the climate change regime, the meetings' facilitators adopted in most cases a default practice in refusing access to observers. The SBI considered this issue and recommended in June 2011 that the first and last informals shall be open to observers in case the agenda item under discussion is not the object of a contact group, parties retaining the right to close any such meeting.[117]

Outside the main negotiation sessions, workshops are organized in order to facilitate discussion regarding the negotiations' technical aspects or in order to foster a more open exchange of views on new approaches. These intersessional workshops typically only involve a limited number of parties and do not constitute an integral part of the official process. The presence of observers at these meetings is particularly relevant as their expertise and perspective may promote new thinking in the discussions.[118] As they are organized on an ad-hoc basis, participation rules may vary at the discretion of the chair of the subsidiary body conveying the workshop and depending on its nature and substance. In 2002, the SBI requested that the chairs of the subsidiary bodies and workshops, as well as the secretariat, "promote transparency and observer participation, while safeguarding the effectiveness of workshops" and adapt the number of observers attending based on the nature of each workshop.[119] More recently, the SBI called for observers' enhanced participation in workshops and invited the meetings' chairs to "make greater use of observer input"[120] and "invite, time permitting, observer organizations to make presentations."[121]

12.3.1.2 Access to Information

In climate change negotiations, NGOs have access to official documents in a similar manner as governmental delegations. Documents distributed in negotiating rooms are distributed to civil society delegates once all parties are provided with the text. Official documents are also made available on the webpage of the convention as

---

[116] Farhana Yamin and Joanna Depledge, *The International Climate Change Regime: A Guide to Rules, Institutions and Procedures* (Cambridge: Cambridge University Press, 2004), at 453.

[117] Report of the Subsidiary Body for Implementation on its 34th session, held in Bonn from 6 June to 17 June 2011, UN Doc. FCCC/SBI/2011/7, 19 September 2011, para. 167.

[118] Yamin and Depledge, *The International Climate Change Regime*, supra, note 116, at 462.

[119] Report of the Subsidiary Body for Implementation on its 17th Session, held in New Delhi from 23 October to 1 November 2002, UN Doc. FCCC/SBI/2002/17, 13 February 2003, paras. 50(c) and (d). In practice, the later request is managed through the involvement of the constituencies, which are often expected to nominate a maximum of one or two representatives among their rank for a given workshop.

[120] Report of the SBI, supra, note 107, para. 178 (a) ii.

[121] Ibid, para. 176.

soon as they are released. In the past, the default practice regarding access to non-official documents distributed in closed meetings, such as the latest non-papers proposed by facilitators, was to not release them to stakeholders.[122] Since 2010, this practice has been reversed so that civil society representatives also have access to informal negotiating texts except when parties or presiding officers explicitly decide otherwise. In order to increase the transparency of the process and in order to allow those who are not attending a meeting to follow the discussions, the secretariat has increasingly utilized webcasts.[123]

In her assessment of the COPs' legitimacy and the importance of transparency in the body's proceedings, Brunnée noted the role played by the online streaming of these meetings in this context, as well as the publication of semi-official reports "Earth Negotiation Bulletin" by the non-governmental organization IISD.[124]

12.3.1.3 Public Participation

Stakeholders are invited to make interventions in the plenary sessions of the convention's main working bodies. In order to channel the perspective of all stakeholder groups, while limiting the number of interventions, one intervention is traditionally invited from each constituency that is recognized by the secretariat. Yamin and Depledge described this right as the implementation of the right to participate provided in the rule of procedures.[125] This participatory right is, however, limited. It is not guaranteed in relation to all the working bodies and statements addressing the Subsidiary Bodies or the Ad-Hoc Working Groups are most of the time at the discretion of the chair and often conditioned by the availability of time. The SBI recently invited presiding officers to "seek opportunities" for such interventions when time allows.[126] In these bodies, chairs may invite general statements or requests

[122] For a classification of the various types of official and non-official documents, see Depledge, *The Organization of Global Negotiation*, supra, note 113, table 11.1.

[123] Good practice and challenges for public participation in international forums: Report prepared by the secretariat in cooperation with the Chair of the Task Force on Public Participation in International Forums, ECE/MP.PP/2011/10, 9 March 2012, para. 40. Webcasts currently covers plenary sessions of the main working bodies of the Convention, and some of the special events and workshops. Webcasts are also used to cover some sessions of the meetings of the Clean Development Mechanism Executive Board and the Joint Implementation Joint Committee. In its conclusions on the enhancement of the participation of observers, the SBI has recently noted this practice and requested the secretariat, "subject to the availability of resources and where appropriate, to increase the number of meetings that are webcast", Report of the SBI, supra, note 107, para. 178(e)ii.

[124] Jutta Brunnée, "COPing with Consent: Law-Making Under Multilateral Environmental Agreements", 15 *Leiden Journal of International Law* (2002), at 45.

[125] See Rules of Procedures, supra, note 108, rule 7(2), providing that observers may, upon invitation of the President, participate without the right to vote in the proceedings of any session in matters of direct concern to the body or agency they represent, unless at least one third of the Parties present at the session object.

[126] Report of the SBI, supra, note 107, para. 178(a)ii.

the stakeholders to more specifically address one of the discussed agenda items.[127] Such an intervention occurs on an ad-hoc basis upon the invitation of the meeting's facilitator, and allows for more meaningful participation as it enables delegates to directly respond to ongoing proceedings.

Written submissions are often invited by working bodies in between sessions in order to provide views and information that are useful for an upcoming discussion.[128] In 2004, the SBI agreed that the calls for submission would be extended to stakeholders "where appropriate and on the understanding that such submissions would not be issued as official documents, but would be made available on the secretariat web site."[129] In 2011, in responding to concerns expressed by NGO representatives on the lack of accessibility of their submission to the UNFCCC website, the SBI requested that the secretariat, when feasible, "post submissions from observer organizations on the UNFCCC website in a way that makes them accessible to Parties."[130] During the sessions, written materials may only be distributed at organizations' exhibits or, if submitted in advance, at a dedicated counter. The distribution of all other written material is officially prohibited.[131] Observer organizations are also allowed to organize a more visual demonstration within the venues of the negotiations in order to attract the attention of the negotiators and/or the media on a given issue under negotiation. In order to be authorized, these actions must be registered in advance, must respect the rules defined by the participatory guidelines, as well as other requirements indicated by the secretariat.[132]

---

[127] In the past, observers were requested to submit their interventions in advance to facilitate interpretation. Taking into account concerns expressed about the difficulties to address most recent issues on the agenda due to this rule, this practice was suspended in 2011. In more limited cases, civil society delegates are sometimes invited to contribute directly to the discussions of contact groups.

[128] The faculty to provide written submission is the only form of participation authorized for non-accredited organizations as calls for submissions might in exceptional cases be open to any relevant stakeholder when explicitly provided by a working body, see for instance, Article 6 of the Convention: Draft conclusions proposed by the Chair, UN Doc. FCCC/SBI/2011/L.6, 15 June 2011, para. 2.

[129] Report of the Subsidiary Body for Implementation on its 12th Session, held in Bonn from 16 June to 25 June 2004, UN Doc. FCCC/SBI/2004/10, 31 August 2004, para. 104.

[130] Report of the SBI, supra, note 107, para. 178(d).i.

[131] UNFCCC, "UN Security Guidelines related to Media Actions, Distribution of Publicity Materials, and Use of UN Emblem at the UNFCCC Conferences", available at:

http://unfccc.int/files/parties_and_observers/ngo/application/pdf/un_security_guidelines.pdf (last accessed on 25 February 2012), at 1.

[132] These guidelines for participation were established by the secretariat based on the general UN guidelines and in consultation with NGOs representatives. NGOs have raised concerns, for instance, against the systematic prohibition of actions naming the World Bank, as potentially constitutive of harassment. See Report of the Compliance Committee on its 35th meeting, ECE/MP.PP/2011/10, 9 March 2012, para. 111. The UN security and the secretariat retain the authority to exclude provisionally or definitely any delegates or organizations breaching the codes of conduct. Guides for Participation, supra, note 102.

In order to provide additional informal opportunities, expertise, and input, the UNFCCC secretariat established additional channels for participation that are based on the experience of other UN fora. Accredited organizations can apply in advance in order to obtain an exhibit within the conference venues. The possibility to organize side events during the session provides an additional means for stakeholders to share their views. While governments can also apply for side events and exhibits, the UNFCCC secretariat guarantees that a minimum number of both are allocated to civil society organizations in order to enhance their participation.

12.3.1.4 The Role of the Aarhus Convention and Its Task Force on Public Participation in International Forums

The UNECE Convention on Access to Information, Public Participation in Decision-Making and Access to Justice in Environmental Matters (hereinafter the "Aarhus Convention")[133] constitutes the most advanced international agreement providing procedural rights in environmental matters, to date. While most of the convention's obligations apply at the domestic level, the convention also considers the importance of the implementation of its principles in international forums.[134] This provision has led to the adoption of the Almaty Guidelines on Promoting the Application of the Principles of the Aarhus Convention in International Forums[135] and to the creation of a dedicated Task Force.[136] Considering that over 40 UNFCCC parties are also party to the Aarhus Convention, their obligations under the latter convention are relevant in the context of the definition of the role and rights of stakeholders in the climate regime.

In June 2010, the Secretariat of the Aarhus Convention organized a specific discussion with governmental representatives, stakeholders, and a liaison officer of the UNFCCC secretariat on the case study of promoting the principles of the Aarhus Convention in the lead up to, during and after the United Nations Climate Change Conference 2009, Copenhagen. The case study concluded with 13 concrete recommendations.[137] In its June 2010 synthesis report on ways to enhance the engagement

[133] Convention on Access to Information, Public Participation in Decision-Making and Access to Justice in Environmental Matters, Aarhus, 25 June 1998, in force 30 October 2001, 38 *International Legal Materials* (1999), 515.

[134] Ibid., Art. 3.7.

[135] Almaty Guidelines on Promoting the Application of the Principles of the Aarhus Convention in International Forums, UN Doc. ECE/MP.PP/2005/2/Add.5, 20 June 2005, at 4.

[136] Decision II-4, Promoting the Application of the Principles of the Aarhus Convention in International Forums, ECE/MP.PP/2005/2/Add.5, 20 June 2005, para. 5, and renewed mandate Decision III-4, Promoting the Application of the Principles of the Aarhus Convention in International Forums, UN Doc. ECE/MP.PP/2008/2/Add.6, 13 June 2008, para. 2.

[137] Excerpts from the Chair's Summary of the Workshop on Experiences of promoting the application of the principles of the Aarhus Convention in international forums, UN Doc. ECE/MP.PP/WG.1/2011/3, 25 January 2011, Annex, para. 8. Three of these recommendations were included in the report of the chair of the task force, which was taken note by the Working Group of the Parties. Decisions and Major Outcomes as adopted by the Working Group at its twelfth meeting, Aarhus Convention WGP-12/Inf.5, 2 July 2010, Item 5 (b) e-g.

of observer organizations, the UNFCCC secretariat noted that many submissions from parties and non-governmental organizations referred to the Almaty guidelines and the recommendations of the June 2010 workshop.[138] In spring 2011, the Task Force organized a second workshop, on the theme of "Making Aarhus work in international forums", with a session that was fully dedicated to the promotion of the Aarhus Convention's principles in the context of the 2010 Cancun Climate Conference.

On the basis of the consultation with intergovernmental organizations and of the information shared during its meetings, the secretariat of the Aarhus Convention prepared a report on "Good practice and challenges for public participation in international forums."[139] The report quotes the UNFCCC procedures as examples of good practices in terms of public participation and access to information several times. However, it notes challenges regarding the volume of participating NGOs, the need for members' capacity building, as well as limits to the freedom of expression and to peaceful assembly at meetings of the climate change regime.[140]

12.3.2 *Procedural Rights in the Implementation of the Convention*

12.3.2.1 Procedural Rights in Domestic Climate Policies

In relation to their climate policies, the Convention refers to the obligation of parties to:

> [p]romote and facilitate at the national and, as appropriate, subregional and regional levels, and in accordance with national laws and regulations, and within their respective capacities:
>
> (ii) public access to information on climate change and its effects;
> (iii) public participation in addressing climate change and its effects and developing adequate responses.[141]

The Kyoto Protocol reiterates the duty of its parties to "cooperate in and promote at the international level" public access to information.[142] A country-driven work programme was adopted in 2002 in order to facilitate the cooperation and implementation of the Article 6 of the convention.[143] While the work programme

[138] Draft conclusions proposed by the SBI Chair, supra, note 135.

[139] UNECE Report on Good practice and challenges for public participation in international forums, supra, note 120.

[140] Ibid., paras. 100, 110 and 111.

[141] FCCC Art. 6(a). Article 6 also addresses education, public awareness, training and international cooperation.

[142] Kyoto Protocol to the United Nation Framework Convention on Climate Change, Kyoto, 10 December 1997, in force 16 March 1998, 37 *International Legal Materials* (1998), 22, Art. 10(e).

[143] Decision 11/CP.8, New Delhi Work Programme on Article 6 of the Convention, UN Doc. FCCC/CP/2002/7/Add.1, 28 March 2003, at 23.

is primarily focused on other themes covered by Article 6, it suggests that parties could, in the implementation of their obligations under the Convention, take the following steps:

(h) Seek opportunities to disseminate widely relevant information on climate change;
(i) Seek input and public participation, including participation by youth and other groups, in the formulation and implementation of efforts to address climate change and encourage the involvement and participation of representatives of all stakeholders and major groups in the climate change negotiation process;
(j) Inform the public about causes of climate change and sources of greenhouse gas emissions, as well as actions that can be taken at all levels to address climate change.[144]

Parties are also invited to report, in their national submissions, of activities undertaken in the frame of this programme and to highlight challenges and best practices. The work programme was renewed in 2007 and the definition of its scope, in relation to public access to information and public participation, was then further defined.[145] The adoption of a new version of the work programme is expected at the COP18 in 2012.[146] However, contrary to the recognition of the fundamental nature of procedural rights, it is striking that the language used in Article 6 of the Convention and Article 12 of the Kyoto Protocol is not mandatory. These provisions have, nevertheless, served as a legal basis for domestic litigation.[147] The Office of the High Commissioner for Human Rights (OHCHR) noted that access to information and public participation are of "key importance in efforts to tackle climate change."[148] It further reaffirmed that the right to participate in decision-making is implied in the right to "take part in the conduct of public affairs" protected by the International Covenant on Civil and Political Rights,[149] and the rights of children and indigenous peoples under specific conventions.[150] The importance of public participation is acknowledged in the Convention's text, which broadly defines the modalities for the participation of observers to the intergovernmental process.[151]

[144] Ibid., para. 15.

[145] Decision 9/CP.13, Amended New Delhi Work Programme on Article 6 of the Convention, UN Doc. FCCC/CP/2007/6/Add.1, 14 March 2008, paras. 14 and 15.

[146] See Draft Conclusions of the SBI chair, supra, note 126.

[147] See for instance the Ukrainian NGO "Environment People Law" that asked its domestic courts to force the government to make publicly available information regarding its climate change policies, available at: http://epl.org.ua/en/environment/climate-change/cases/ (last accessed on 25 February 2012).

[148] Report of the OHCHR on the relationship between climate change and human rights, supra, note 10, paras. 78 and 79.

[149] Ibid., para. 79, referring to ICCPR, Art. 25.

[150] United Nations Declaration on the Rights of Indigenous Peoples, UN Doc. A/RES/61/295, 13 September 2007, Art. 19 and CRC, supra, note 19, Art. 12.

[151] FCCC, Art. 7.6.

12.3.2.2 Procedural Rights in Flexibility Mechanisms: The Example of CDM

This subsection only addresses the issue of the exercise of procedural rights in the project cycle established under the CDM.[152] The effective exercise of procedural rights of stakeholders is particularly important in the context of the CDM due to its particularly layered governance structure of the delegation of administrative authorities. Firstly, the COP, which directly represents the parties, delegates management authority to the Executive Board of the CDM, an intergovernmental body especially created. The CDM Executive Board then delegates some regulatory functions, including the tasks of validation and verification of CDM projects, to private certifiers known as Designated Operating Entities (DOEs). In international governance, the distance between elected national representatives and agents, to which regulatory authority is delegated, affects the governance models legitimacy.[153] This delegation raises additional legitimacy issues given the discrepancy between the absence of a formal status of non-state actors as subjects of international environmental law in comparison to the increasing functions performed by these actors.[154] Activities, undertaken by private regulators in the context of the CDM, are susceptible to indirectly affect the rights of third parties and local communities as exemplified by the Bajo Aguan case where their decisions may provide positive incentives for economic actors to engage in activities that potentially infringe on the rights of a local community's members. In this context, all stakeholders' adequate enjoyment of their procedural rights is particularly important to the legitimacy of the governance of the CDM.

Access to Information

In order to be formally accredited under the DOE status, a certifier is required to make information related to its internal governance, functioning, and expertise publicly available.[155] The CDM Executive Board is responsible for maintaining a publicly available and updated list of the status of all DOE, where each DOE is responsible for the maintenance of an updated list of all projects for which it has been contracted.[156] The communication of additional project-specific information is

---

[152] Discussion of aspects related to access to information and public participation in the governance of the Clean Development Mechanism has voluntarily been omitted from this chapter for the sake of brevity.

[153] Daniel Esty, "Good Governance at the Supranational Scale: Globalizing Administrative Law", 115 *Yale Law Journal* (2006), 1502.

[154] See on this issue Asher Alkoby, "Non-State Actors and the Legitimacy of International Environmental Law", 3 *Non-State Actors and International Law* (2003), at 25.

[155] Decision 3/CMP.1, Modalities and Procedures for a Clean Development Mechanism as defined in Article 12 of the Kyoto Protocol, Annex, UN Doc. FCCC/KP/CMP/2005/8/Add.1, 30 March 2006, para. 1(g).

[156] Ibid., para. 27 (f).

also required from certifiers in relation to every project for which they have been contracted. The procedures of the CDM generally state that the DOE should make all information publicly available provided by the project participants, except information marked as confidential.[157] The procedures further define the content of this general obligation with regard to both stages at which the DOE may intervene. In relation to a validation phase, the DOE is responsible to make both the project design documents and its validation report available to the public.[158]

Public Participation

Stakeholders' consultations, in relation to specific projects, are organized via two different processes and various geographic scopes. The primary participatory channel for local stakeholders to the project cycle consists of their involvement in local consultations, which the project participants are requested to organize during the design of the project. The report of this consultation should be included in the information transmitted by the project participants to the DOEs during the validation of the project.[159] The registration documents should include the project participant's description of the steps taken in inviting public comments, as a summary of the comments, and a report on how the received comments have been evaluated.

On a global level, stakeholders and UNFCCC-accredited organizations are invited to provide comments through the "global stakeholders consultations" managed by the DOE. Global stakeholders consultations are initiated by the DOE's publication of the project design document's non-confidential elements, after which stakeholders are invited to submit comments.[160] The DOE validation report must provide information on how each comment has been duly addressed throughout the process.[161] However, once the project has been registered, there is no formal and automatic opportunity for stakeholders to play a role in relation to the DOE's verification of the emissions reduction resulting from a project. Stakeholders can only then attempt to persuade parties to the project or members of the CDM Executive Board to trigger the review of the request for issuance of Certified Emissions Reductions. Such a review may, however, only address cases of fraud, malfeasance, and the incompetence of the DOE.[162]

---

[157] Ibid., para. 27 (h).

[158] Ibid., para. 40. This requirement is however more limited at the verification phase as DOEs are only required to make publicly available its verification report. Ibid., para. 62 (h).

[159] Ibid., para. 37 (b).

[160] Ibid., para. 40 (c).

[161] See Procedures for processing and reporting on validation of Clean Development Mechanism project activities (Version 03), CDM-EB-50, Annex 48, para. 12.

[162] Ibid., para. 65.

Access to a Review Process

The right to an adequate access to judicial remedies constitutes the third pillar of procedural rights. The absence of procedures enabling individuals and private entities to directly challenge decisions reached by intergovernmental institutions is a rising issue in public international law due to the multiplication of instances in which decisions, taken at the international level, directly target individuals.[163] Currently, only parties may appeal to the decision of the COP/MOP to the enforcement branch of the compliance committee,[164] while such mechanism do not currently exist for decisions made by its subsidiary body.[165]

The CDM Modalities and Procedures request DOEs to establish internal review procedures and to make "their procedures for handling complaints, appeals and disputes" publicly available.[166] In addition to these internal procedures, the CDM Executive Board also established an external process for handling complaints against the DOEs, to which any stakeholder who has participated at a global consultation may appeal.[167] In such a case, the Executive Board would organize a contradictory procedure, which could eventually lead to the suspension of the DOE.

The issue of the need for a legal standing for various actors in a review process was already identified by academics as a potential issue before the adoption of the Marrakech Accords.[168] At the COP15, the COP/MOP requested that the CDM Executive Board proposes, in consultation with stakeholders, procedures for appeals

[163] Charlotte Streck and Jolene Lin note the examples of the UN Security Council with regards to individuals directly affected by individual sanctions, and the Court of Arbitration for Sport in relation to decisions adopted by the World Anti-Doping Agency. Charlotte Streck and Jolene Lin, "Making Markets Work: A Review of CDM Performance and the Need for Reform", 19 *European Journal of International Law* (2008), 428. For further analogies with other administrative review processes established in relation to decisions adopted by international institutions, see also the elements drawn from six other international mechanisms by the FCCC secretariat, Procedures, mechanisms and institutional arrangements for appeals against the decisions of the Executive Board of the clean development mechanism, UN Doc. FCCC/TP/2011/3, 17 May 2011. The processes concerned are mentioned in para. 11.

[164] Decision 27/CMP.1, Procedures and Mechanisms Relating to Compliance under the Kyoto Protocol, UN Doc. FCCC/KP/CMP/2005/8/Add.3, 30 March 2006, annex, section VII.

[165] Christiana Figueres and Charlotte Streck, "A Post-2012 Vision for the Clean Development Mechanism", in David Freestone and Charlotte Streck (eds), *Legal Aspects of Carbon Trading: Kyoto, Copenhagen and Beyond* (Oxford: Oxford University Press, 2009), at 575.

[166] CDM Modalities and Procedures, supra, note 155, Appendix A, para. 1(g)vi. The standards for the accreditation of DOEs also contains a second reference to such processes, requesting from applicant entities to communicate their procedures to allocate responsibility in relation to the handling of complaints, Ibid., para. 1(e). The CDM accreditation standards for DOEs further develop on the content of these requirements, elaborating on each of these three types of contention processes. Clean Development Mechanism Accreditation Standard for Operational Entities (Version 02), CDM-EB-56, 17 September 2010, Annex 1, para. 133.

[167] Ibid., Procedure for accrediting Operational Entities by the Executive Board of the Clean Development Mechanism, (Version 10.1), Annex 2, Appendix 3.

[168] Peggy R. Kalas and Alexia Herwig, "Dispute Resolution under the Kyoto Protocol", 27 *Ecology Law Quarterly* (2000), at 121.

against CDM Executive Board decisions "that are brought by stakeholders directly involved, defined in a conservative manner."[169] While this request does not specifically defines the scope of the legal standing in this process, the wording used indicates that the COP/MOP expects that the review processes' procedures would strike a balance between the processes' openness and the safeguard of the mechanism's effectiveness. The CDM Executive Board developed a recommendation for an appeal procedure of its decisions, which interpreted the reference to "stakeholders directly involved, defined in a conservative manner" as only encompassing the economic entities involved in a project.[170] Due to a lack of consensus among the parties on the proposal contained in this recommendation, negotiations continued in 2011.[171] A technical paper, issued by the secretariat, noted that the decision, over the scope of the legal standing. would have implications "for issues such as environmental integrity, legitimacy and confidence as well as for caseload, the efficiency of the appeal mechanism and the possibility of vexatious or frivolous claims."[172] Contrary to the draft procedures proposed by the Executive Board, the paper also refers to the role that stakeholders could play in the process, suggesting that the appeal mechanism may have the possibility to solicit, at its own discretion, views from stakeholders previously engaged in the project cycle.[173] Since parties could not agree on the terms of this appeal at the COP17, a decision on the scope of the appeal offered to the CDM Executive Board decision remains pending and may be decided in 2012.

12.3.3 The Respect of Substantial Human Rights in Climate Policies

The discussion regarding the need for the respect of substantial human rights in the implementation of the framework convention is a more recent phenomenon. Concrete examples, such as the case of the Bajo Aguan palm oil project, have raised the regime participants' awareness of the potential importance of considering this issue. In the case of the Bajo Aguan, the project was approved by the CDM Executive Board despite NGOs having brought to its attention the resulting violation of local communities' rights. Nonetheless, the Executive Board certified the project on the

---

[169] Decision 2/CMP.5, Further Guidance Relating to the Clean Development Mechanism, UN Doc. FCCC/KP/CMP/2009/21/Add.1, 30 March 2010, para. 42

[170] 2010 Annual Report of the EB to the CMP, UN Doc. FCCC/KP/CMP/2010/10, 3 November 2010, Annex II.

[171] Decision 3/CMP.6, Further guidance relating to the clean development mechanism, UN Doc. FCCC/KP/CMP/2010/L.8, 10 December 2010, para. 18.

[172] UNFCCC, Technical Paper: Procedures, Mechanisms and Institutional Arrangements for Appeals against the Decisions of the Executive Board of the Clean Development Mechanism, UN Doc. FCCC/TP/2011/3, 17 May 2011, para. 131.

[173] Ibid, para. 132.

basis that the project respected all requirements defined by the modalities and procedures of the CDM at the time of the submission of the registration documents. The project applicants' respect for human rights does not constitute a criteria that the CDM Executive Board is currently mandated to consider in its decision-making processes. It also lacks the capacity to review or withdraw certification and to suspend the transfer of credits if such circumstances are brought to its knowledge. The Kyoto Protocol definition of the CDM as projects assisting non-Annex 1 parties to achieve sustainable development[174] could provide an avenue for the introduction of a human rights criterion in the definition of projects' requirements. However, CDM modalities and procedures provide that national authorities are competent in interpreting the notion of "contribution to sustainable development."[175]

The risk that climate change mitigation and adaptation measures may infringe on the exercise of human rights was only acknowledged by the COP in 2010. The Cancun agreements provide that the COP "emphasizes that Parties should, in all climate change-related actions, fully respect human rights."[176] This provision was echoed by the Human Rights Council, which "urged States to take human rights into consideration when developing their environmental policies."[177] A more specific safeguard was defined in the context of the Reduction of Emissions from Deforestation and forest Degradation (REDD), the Cancun Agreements calling for the [r]espect for the knowledge and rights of indigenous peoples and members of local communities, by taking into account relevant international obligations, national circumstances and laws, and noting that the United Nations General Assembly has adopted the United Nations Declaration on the Rights of Indigenous Peoples.[178]

12.3.3.1 Role of the HRC in Raising Awareness on Interlinkages at the UNFCCC

The Human Rights Council has periodically considered the inter-linkage between human rights and climate change since 2008.[179] The Council has placed a particular emphasis on working together with the UNFCCC secretariat and in informing

---

[174] Kyoto Protocol, Art. 12.2.

[175] CDM Modalities and Procedures, supra, note 155, para. 40(a).

[176] Decision 1/CP.16, The Cancun Agreements: Outcome of the work of the Ad Hoc Working Group on Long-term Cooperative Action under the Convention, UN Doc. FCCC/CP/2010/7/Add.1, 15 March 2011, para. 8. For an insider account of the negotiations of such right-based language in the climate change process, see for instance Kravchenko, "Procedural Rights as a Crucial Tool to Combat Climate Change", supra, note 101.

[177] HRC Resolution 16/11, Human Rights and the Environment, UN Doc. A/HRC/RES/16/11, 12 April 2011. The resolution also noted the human right language contained in the Cancun Agreements.

[178] Decision 1/CP.16, supra, note 176, Appendix 1. 2(c), Guidance and safeguards for policy approaches and positive incentives on issues relating to reducing emissions from deforestation and forest degradation in developing countries; and the role of conservation, sustainable management of forests and enhancement of forest carbon stocks in developing countries.

[179] See HRC Resolution 7/23, supra, note 10.

UNFCCC parties of its own proceedings. The Council's resolutions request the Office of the High Commissioner on Human Rights to consult the UNFCCC secretariat when collecting information.[180] It also repeatedly requested that the OHCHR to release the outcomes of the discussions and workshops organized by the Human Rights Council on this issue to the UNFCCC COP.[181] These efforts have partly resulted in the Ad hoc Working Group on Long-term Cooperative Action taking note of the resolution of the 10/4 Council and quoting, in the outcome document's preamble, the recognition by the Council that the adverse effects of climate change have a range of direct and indirect implications for the effective enjoyment of human rights and that the effects of climate change will be felt most acutely by those segments of the population that are already vulnerable owing to geography, gender, age, indigenous or minority status and disability.[182]

On the other hand, the Human Rights Council also recognized the role of the UNFCCC in contributing to the protection of human rights. In its report on the implication of climate change for the exercise of human rights, the Council noted that effective international cooperation to enable the "full, effective and sustained implementation of the UNFCCC in accordance with the provisions and principles of the Convention is important in order to support national efforts for the realization of human rights implicated by climate change-related impacts."[183]

12.4 Human Rights as Influencing the Design Principles of a Possible New Climate Regime

There are also scholars who argue that human rights should have a bearing on the climate change regime and the currently standing regime cannot stabilize "greenhouse gas concentrations in the atmosphere at a level that would prevent dangerous anthropogenic interference with the climate system", the objective of the climate regime.[184] Perhaps, in this respect, Professor Caney has made the most credible argument.[185] Firstly, he identifies that there is no scientific uncertainty regarding the radical impact of climate change on human rights: some of these have already occurred, some will concretize in time. Although there may be uncertainty as to how climate change will violate some human rights, Caney focuses on the most modest and widely accepted interpretations of human rights – to life, health, and subsistence – and how these are and will be violated by climate change. For example,

[180] Ibid., para. 1.

[181] HRC Resolution 10/4, supra, note 8, para. 2 and HRC Resolution 18/22, supra, note 10, para. 4(b).

[182] Decision 1/CP.16, supra, note 176, preamble.

[183] HRC Resolution 18/22, supra, note 10.

[184] FCCC Art. 2.

[185] Caney, "Climate Change, Human Rights and Moral Thresholds", supra, note 11.

he notes that there may be controversies surrounding the human right to life, but not in terms of its very core – all persons have a human right to not be arbitrarily deprived of their life – as prescribed by the International Covenant on Civil and Political Rights (ICCPR).[186] As he argues, this formulation of the right to life does not make the possibly contentious claim that each person has a positive right to have their life saved from all kinds of threats because it only insists on arbitrary loss of life.[187]

Caney points to the continuing loss of life caused by climate change, but especially those projected by science. He contrasts a human rights approach to various versions of cost-benefit analysis on how to justly allocate burdens over climate change. The general problem with cost-benefit models in re-designing the way that mitigation burdens should be allocated is that they have a hard time moving beyond nation-states because, even if allocation were to take place on a per capita basis, it is impossible to account for large differentiation within states; surely, it is unjust and unreasonable to expect poor Indians to shoulder the same mitigation burden as the wealthier ones. Although there is potential in making this differentiation – like Henry Shue's distinction between subsistence/survival emissions (GHG's that are used to fulfil basic human needs must be differentiated from those aiming to perpetuate luxurious lifestyles) – this model remains based on Gross Domestic Product (GDP) and, thus, cannot examine vulnerability beyond or within states, even if it can differentiate between nation-states.[188] Caney's "human rights as thresholds" requires differentiation between human beings within states, as climate change will not affect the enjoyment of basic human rights of all, but only some people within countries and in many different ways. The human rights approach, thus, has the potential to differentiate the burdens of mitigation and adaptation both between nation-states and within them in a more nuanced manner.

It is important to note that, in Caney's approach, those suffering from climate change driven human rights violations, have a right to compensation in the case that their human rights are violated. This is a missing aspect of the current climate change regime, which only emphasizes adaptation to climate change consequences and avoids the discussion of responsibility from damage caused by climate change.[189]

---

[186] FCCC, Art. 6 (1).

[187] Caney, "Climate Change, Human Rights and Moral Thresholds", supra, note 11, at 76.

[188] Henry Shue, "Subsistence Emissions and Luxury Emissions", 15 *Law & Policy* (1993), 39; another well-known proposal is so-called Contraction and Convergence (C&C), proposed originally by the Global Commons Institute. The idea is first that future total of greenhouse gas emissions from human sources is decreased over time to near zero-emissions within a specified time-frame (contraction). To achieve this, global per capita average of emissions arising under the contraction rate is chosen (convergence), which thus varies in accordance with states per capita emissions. See GCI, "Contraction and Convergence: Climate Justice without Vengeance", available at: http://www.gci.org.uk/contconv/cc.html (last accessed on 25 February 2012).

[189] However, the new Work Programme on Loss and Damage established by the Cancun agreement does consider some of these issues. It does not address state responsibility as such but only "approaches to address loss and damage associated with climate change impacts in developing countries that are particularly vulnerable", Decision 1/CP.16, supra, note 176, para. 26.

This has led e.g. small island states to make declarations to the effect that their participation in the UNFCCC and the Kyoto Protocol does not mean that they renounce their rights under general international law to invoke state responsibility over environmental damage caused by climate change.[190]

12.5 Concluding Remarks – Evaluation

As reviewed in this chapter, there are many ways that human rights and climate change are interconnected. Yet, as Stephen Humphrey's rightly notes, the climate change regime appears to avoid the use of human rights language and climate change does not figure into the human rights discourse.[191] There are many reasons for this – for instance, the well-documented phenomenon of fragmentation whereby various sub-disciplines of international law increasingly function independently of each other – but the more interesting question is whether it is desirable or not to have human rights and the climate change (regime) to more actively interact. It is useful to focus on evaluating the pros and cons of this inter-relationship in the order that we have studied each question: whether it makes sense to use human rights in the struggle against climate change impacts, as manifested by the Inuit Petition; what, if any, should be the consequences of human rights to the functioning of the climate change regime; and, whether human rights should, in effect, guide the development of a new type of international policy to combat climate change.

12.5.1 Evaluation

The Inuit petition showed the strengths and weaknesses of using a human rights petition to combat climate change. Even if the Inuit petition was clearly well prepared, it appears that despite the Inter-American Commission's fairly innovative manner of construing human rights requirements, finding that the US is infringing on the Inuit's various human rights via its irresponsible climate policy was a perspective that was just too extraordinary for the Commission. Even in the unlikely case

[190] For instance, Nauru made the following declaration upon ratification of the Kyoto Protocol: "Nauru declares its understanding that the ratification of the Kyoto Protocol shall in no way constitute a renunciation of any rights under international law concerning State responsibility for the adverse effects of climate change", see at UNFCCC, "Declarations and Reservations by Parties – Kyoto Protocol: Nauru", available at: http://unfccc.int/kyoto_protocol/status_of_ratification/items/5424.php (last accessed on 25 February 2012).

[191] Stephen Humphreys, "Conceiving Justice: Articulating Common Causes in Distinct Regimes", in Stephen Humphreys (ed.), *Human Rights and Climate Change* (Cambridge: Cambridge University Press, 2009), 299.

that the petition had been found meritorious, it would have hardly achieved its aim: the effective protection of the rights of the Inuit. It is difficult to see how the Commission could so dramatically affect the climate change policy of the United States that the major climatic changes, that are already beginning to take place in the Arctic, would not threaten the traditional livelihoods of the Inuit. The case also demonstrates – at least currently – that traditional human rights mechanisms cannot be effectively used in protecting the rights of indigenous, as well as other peoples and individuals, from global environmental interference, such as climate change. We must also be critical of the potential consequences of the Inuit petition; given that it did not even proceed to the merits stage in the Inter-American Commission, it cannot be ruled out that it has also had a disillusioning effect on the use of human rights in the fight against climate change.

On the other hand, even if it presently seems as though it is difficult to address climate change concerns via human rights petitions, cases such as the Inuit petition importantly challenge human rights bodies to open up new ways of thinking and interpreting the articles of human rights instruments that were not originally created to handle the complex impacts of global climate change. In a similar vein, research on human rights and climate change increased greatly after the Inuit petition, which particularly showed that this is a possible course of action.

For the major victims of climate change – such as the Inuit – the most significant and immediate consequence of their human rights legal strategy was not winning their case. By making their legal claims against the worst polluters public, victims are able to improve their position in an effort to combat climate change. The climate regime involves a great number of actors and decision-making structures. The publication of their legal claims allowed Inuit to reinforce their activities in the climate regime and to obtain a louder voice in the global regime. For example, the ICC consciously brought its human rights petition to the public eye during its drafting phase and organized press meetings during the climate change regime's Conference of the Parties. They also openly advocated for human rights and other legal actions to be taken all over the world in order to combat climate change.

The ICC's actions demonstrate how to effectively challenge both the basic rules prescribed by the climate regime and the structure upheld by international law as a society of states. By raising the human rights petition against the United States, the Inuit expanded society's notion of who is entitled to participate in the fight against climate change. Through their consolidated agency, the Inuit also brought their plight – the death of their culture – into the public eye, which is not easily achieved. This message arguably challenged the climate regime's view that climate change is a problem that we can control and manage. The ICC petition may have also opened a new era for climate litigation with the office of the High Commissioner for Human Rights noting the importance of "the recognition of the extraterritorial obligations of States [which] allows victims of … dangerous climate change, to have access to remedies."[192]

[192] Analytical study on the relationship between human rights and the environment, Report of the OHCHR, UN Doc. A/HRC/19/34, 16 December 2011, para. 72.

Turning to the question as to what, if any, should be the consequences of human rights to the functioning of the climate change regime? As demonstrated in Sect. 12.3, it is clear that particularly the Rio principle on access to information, public participation in environmental decision-making, and access to justice is already gradually encroaching on the climate change regime and domestic climate policies. This is a welcome development, as not only climate change affects vulnerable communities' enjoyment of human rights. Still, response policies may also potentially constitute a threat for the protection of these rights. Even if this is the case, they are rarely openly seen as hard-core human rights that must be accounted for, but rather as accommodating diverse interests in managing climate change. It appears to be evident that it is desirable to have the procedural principles of human rights play a more forceful role in the climate change regime, also because these principles have more clearly matured as universally applicable human rights principles.[193]

Finally, we will ask whether human rights should guide the development of a new type of international policy to combat climate change. It seems clear that Simon Caney's human rights approach would have a lot to give if the climate change regime would engage in real soul-searching, which is yet to happen. The brilliance of Caney's argument is – in effect – that the design principles for a new climate regime may already be found in existing universally valid human rights law and that these design principles also make sense: they enable us to nuance the way climate mitigation and adaptation burdens are allocated between nation-states and within them; they also introduce the missing element of compensation for damage resulting from climate change impacts to the design of possible new international policy of climate policy and law. The other side of the coin is, of course, that Caney's ideas are very far from the current reality of the climate change regime.

Caney not only goes against the cost-benefit analysis models studied above, but also argues that the human rights approach has a lot more to offer in combating climate change than a security-oriented approach, an approach that may also be regarded as a viable alternative.[194] However, Caney understands the security-oriented approach too narrowly as he argues that this approach "gives us reason to be concerned about climate change only if, because, and to the extent that, it results in violent conflict."[195] This is a constricted reading of the security-oriented approach, as it does not consider the long-standing discourse on the securitization of environmental problems, particularly in the case of climate change.[196]

[193] A good overview is in McInerney-Lankford, Darrow, and Rajamani, *Human Rights and Climate Change*, supra, note 7, at 32–36.

[194] See, e.g. Timo Koivurova, "International Legal Avenues to Address the Plight of Victims of Climate Change: Problems and Prospects", 22 *Journal of Environmental Law & Litigation* (2007), 267.

[195] Caney, "Climate Change, Human Rights and Moral Thresholds", supra, note 11, at 85–86.

[196] See Bill McSweeney, *Security, Identity and Interests: a Sociology of International Relations* (Cambridge: Cambridge University Press, 1999), 45; see also Lorraine Elliott, "Expanding the Mandate of the United Nations Security Council", in W. Bradnee Chambers and Jessica F. Green (eds), *Reforming International Environmental Governance: from Institutional Limits to Innovative Reforms* (Tokyo: United Nations University Press, 2005), 204.

Climate change could also be framed anew as a collective security problem, as opposed to an environmental problem, with a corresponding soft welfare approach to its solution. It may well be that only in re-framing climate change – likely the biggest collective security problem faced by humanity – and understanding it as a matter of collective security, will stronger response measures follow. We must acknowledge that the climate change regime has failed to deliver and we are faced with gloomy future scenarios. We may, of course, defend the present climate regime as the only viable alternative. However, if it continues to act as a façade for inaction, providing states the excuse to argue that they are combatting climate change while they are not, then it is important to seriously examine other perspectives and possibilities of framing and solving climate change as a politico-legal problem.

Unfortunately, at the moment, there are only weak signs that such a "climate change securitization" is taking place. There are also no strong signs that human rights will determine our response to climate change in the manner that Caney insightfully outlines. After the 2011 Climate Conference in Durban, governments have agreed to a new timeframe for negotiations, thus, postponing the roadmap that was originally agreed upon in Bali and failing to address the urgency of the climate response urged by scientists. At least, at the moment, the managerial approach of the current climate change regime seems to prevail. Yet, when the consequences of climate change become more manifest, it is likely that some actors will increasingly choose other approaches to combat climate change. Let us hope that we still have time for this.

Part IV
International Climate Law – Sectoral Issues

Chapter 13
Managing the Fragmentation of International Climate Law

Harro van Asselt

Abstract This chapter focuses on the fragmentation of international law related to climate change and the interactions between the relevant legal regimes. It examines various management strategies with a view to enhancing synergies and mitigating conflicts between climate-related international legal regimes. The chapter starts with an overview of the ongoing debate on the fragmentation of international law. It then identifies the features of international climate lawmaking and implementation that constrain the usefulness of well-known legal techniques for avoiding and resolving conflicts. The chapter moves on to show how institutional cooperation between political bodies and bureaucracies may lead to enhanced coherence between the climate change regime and other legal regimes, while arguing that such a strategy will also encounter specific concerns related to their legitimacy. The chapter concludes by highlighting the need to apply various strategies for managing the fragmentation of international climate law, and identifies areas for further inquiry in this regard.

13.1 Introduction

This chapter examines the international legal response to climate change by placing the United Nations climate regime in the context of the broader international regulatory environment. It aims to highlight the fragmented international legal order that is

---

Harro van Asselt is a postdoctoral research fellow at the Stockholm Environment Institute. He is also a visiting research associate with the Environmental Change Institute at the University of Oxford and a visiting researcher with the Institute for Environmental Studies at VU University Amsterdam. The author would like to acknowledge funding from the European Commission (Intra-European Fellowship, CLIMATEGOV; contract 253090).

H. van Asselt (✉)
Stockholm Environment Institute, Kräftriket 2b, SE 106 91 Stockholm, Sweden
e-mail: harro.vanasselt@sei-international.org

E.J. Hollo et al. (eds.), *Climate Change and the Law*, Ius Gentium: Comparative Perspectives on Law and Justice 21, DOI 10.1007/978-94-007-5440-9_13,
© Springer Science+Business Media Dordrecht 2013

relevant for addressing climate change, and to suggest ways of managing interactions between the relevant international legal regimes. Building on the emerging body of literature on the fragmentation of international law, it discusses the opportunities for, as well as the limitations of, addressing the relationship between different areas of international law related to climate change through specific legal and institutional strategies.

The chapter proceeds from the assumption that it is not possible to find a single, comprehensive legal response to the problem of climate change. The rationale lies in the very nature of the problem – climate change can be conceived as a 'wicked problem' *par excellence*.[1] This means, among others, that there is no exclusive definition of what the problem is. Is the climate change problem, for instance, essentially about reducing greenhouse gas emissions, or phasing out fossil fuels, or is the problem more profound than that: Is it about the insistence on economic growth?[2] Similarly, there is no simple 'solution' to the climate change problem, as 'solving' the climate change challenge will depend on how one defines the problem in the first place. Any proposed solution will thus be influenced by one's value judgement; actors promoting diverging values and ideologies are likely to advocate different responses to the climate change problem. Moreover, solutions to wicked problems like climate change may well have ripple effects, potentially causing new problems in their wake. Climate change is also characterized by specific traits that make it a 'super' wicked problem.[3] First, the causes and impacts of, and responses to climate change cut across all sectors of the society. Various human activities and societal sectors contribute to the growing concentrations of greenhouse gases in the atmosphere. At the same time, the same activities and sectors may also be affected by the impacts of climate change. Second, climate change requires an urgent response if the goal is to avoid large-scale, irreversible impacts.[4] Third, responding to climate change is complicated by the fact that international and national decision-makers cannot fully control the choices of actors that are relevant for addressing climate change.[5] Fourth, climate change is a transboundary problem, and may indeed be "the greatest collective action problem the international community has yet faced."[6] This enhances tensions between countries, especially because those who are in the

[1] Horst W.J. Rittel and Melvin M. Webber, "Dilemmas in a General Theory of Planning", 4 *Policy Sciences* (1973), 155, at 160–169.

[2] For an excellent discussion of different framings of the climate change problem, see Mike Hulme, *Why We Disagree about Climate Change: Understanding Controversy, Inaction and Opportunity* (Cambridge, UK: Cambridge University Press, 2009).

[3] Kelly Levin et al., "Playing It Forward: Path Dependency, Progressive Incrementalism, and the 'Super Wicked' Problem of Global Climate Change", paper presented at the International Studies Association Convention, Geneva, 28 February–3 March 2007, at 4–9; Richard J. Lazarus, "Super Wicked Problems and Climate Change: Restraining the Present to Liberate the Future", 94 *Cornell Law Review* (2009), 1153, at 1159–1183.

[4] Levin et al., "Playing It Forward", supra, note 3, at 8–9.

[5] Ibid., at 9.

[6] Daniel H. Cole, "Climate Change and Collective Action", 61 *Current Legal Problems* (2008), 229, at 232.

best position to take action have little incentive to do so.[7] Fifth, climate change has an undeniable intertemporal dimension: to mitigate impacts in the future, action now is needed.[8] Finally, the problem is characterized by various levels of scientific uncertainty, including uncertainty regarding the future development of greenhouse gas emissions as well as the impacts (and associated costs) of climate change in the long term.

By its nature, the climate change problem thus covers a broad range of narrowly defined issue areas, and its resolution inevitably requires a variety of responses. The implication at the international level is that issues relevant for the climate change problem are governed by a multitude of legal regimes with overlapping jurisdictions.[9] For instance, the simple facts that some greenhouse gases are also ozone depleting substances, and that the substitutes for some of these substances are in turn greenhouse gases, inevitably means that the international legal regime for ozone layer depletion, notably the Vienna Convention for the Protection of the Ozone Layer[10] and its Montreal Protocol,[11] is relevant for tackling climate change.[12] Similarly, because of the intricate connections between climate change mitigation and adaptation on the one hand, and biodiversity loss on the other, international biodiversity law, particularly the Convention on Biological Diversity (CBD)[13] may affect the response to climate change, and may itself be affected by climate policies.[14]

---

[7] Levin et al., "Playing It Forward", supra, note 3, at 9.

[8] Lazarus, "Super Wicked Problems and Climate Change", supra, note 3, at 1174–1176.

[9] I adopt the definition proposed by Margaret Young (which is in turn adapted from the consensus regime definition proposed by Stephen Krasner): "regimes are sets of norms, decision-making procedures and organisations coalescing around functional issue-areas and dominated by particular modes of behaviour, assumption and biases." Margaret A. Young, "Introduction: The Productive Friction Between Regimes", in Margaret A. Young (ed.), *Regime Interaction in International Law: Facing Fragmentation* (Cambridge: Cambridge University Press, 2012), 1, at 11. See also Stephen D. Krasner, "Structural Causes and Regime Consequences: Regimes as Intervening Variables", in Stephen D. Krasner (ed.), *International Regimes* (Ithaca: Cornell University Press, 1983), 1, at 2.

[10] Convention on the Protection of the Ozone Layer, Vienna, 22 March 1985, in force 22 September 1988, 26 *International Legal Materials* (1987), 1529.

[11] Protocol on Substances that Deplete the Ozone Layer, Montreal, 16 September 1987, in force 1 January 1989, 26 *International Legal Materials* (1987), 1550.

[12] Sebastian Oberthür, Claire Dupont and Yasuko Matsumoto, "Managing Policy Contradictions Between the Montreal and Kyoto Protocols: The Case of Fluorinated Greenhouse Gases", in Sebastian Oberthür and Olav Schram Stokke (eds), *Managing Institutional Complexity: Regime Interplay and Global Environmental Change* (Cambridge: The MIT Press, 2011), 115.

[13] Convention on Biological Diversity, Rio de Janeiro, 5 June 1992, in force 29 December 1993, 34 *International Legal Materials* (1992), 822.

[14] See, for instance, Rüdiger Wolfrum and Nele Matz, *Conflicts in International Environmental Law* (Berlin: Springer, 2003); Imke Sagemüller, "Forest Sinks under the United Nations Framework Convention on Climate Change and the Kyoto Protocol: Opportunity or Risk for Biodiversity?", 31 *Columbia Journal of Environmental Law* (2006), 189; Harro van Asselt, "Integrating Biodiversity in the Climate Regime's Forest Rules: Options and Tradeoffs in Greening REDD Design", 20 *Review of European Community and International Environmental Law* (2011), 139; Harro van Asselt, "Managing the Fragmentation of International Environmental Law: Forests at the Intersection of the Climate and Biodiversity Regimes", 44 *New York University Journal of International Law and Politics* (2012, forthcoming).

Climate change and climate policy are also closely connected with economic activities, such as international trade and investment. Therefore, international economic law, including the law of the World Trade Organization (WTO), is of importance in the response to climate change.[15] Furthermore, international climate law may interact with other areas of international law, including the law of the sea,[16] human rights law,[17] and the law on transboundary air pollution.[18]

In other words, international law on climate change is characterized by a certain degree of fragmentation. The phenomenon of regulatory fragmentation is clearly not limited to the issue of climate change. Indeed, over the past decade, fragmentation of international law has moved from the periphery to the centre of international legal debate. The increasing specialization of international law had already been noted by early observers in the 1980s,[19] but discussions on the subject intensified at the turn of the millennium, when fragmentation was included in the work program of the International Law Commission (ILC).[20] The ILC released its report on fragmentation in 2006, providing an impressive overview of the various questions

[15] See, for instance, Ludivine Tamiotti et al., Trade and Climate Change: A Report by the United Nations Environment Programme and the World Trade Organization (Geneva: WTO Secretariat, 2009); Tracey Epps and Andrew Green, *Reconciling Trade and Climate: How the WTO Can Help Address Climate Change* (Cheltenham: Edward Elgar, 2010); Fariborz Zelli and Harro van Asselt, "The Overlap Between the UN Climate Regime and the World Trade Organization: Lessons for post-2012 Climate Governance", in Frank Biermann, Philipp Pattberg and Fariborz Zelli (eds), *Global Climate Governance Beyond 2012: Architecture, Agency and Adaptation* (Cambridge: Cambridge University Press, 2010), 79.

[16] See, for instance, Karen N. Scott, "The Day After Tomorrow: Ocean CO2 Sequestration and the Future of Climate Change", 18 *Georgetown International Environmental Law Review* (2005), 57; Meinhard Doelle, "Climate Change and the Use of the Dispute Settlement Regime of the Law of the Sea Convention", 37 *Ocean Development and International Law* (2006), 319.

[17] Stephen Humphreys (ed.), *Human Rights and Climate Change* (UK: Cambridge University Press, 2009); Edward Cameron, "Human Rights and Climate Change: Moving from an Intrinsic to an Instrumental Approach", 38 *Georgia Journal of International and Comparative Law* (2010), 673; Ole W. Pedersen, "The Janus-Head of Human Rights and Climate Change: Adaptation and Mitigation", 80 *Nordic Journal of International Law* (2011), 403.

[18] See, for instance, Erika Rosenthal and Robert Watson 2011, "Multilateral Efforts to Reduce Black Carbon Emissions: A Lifeline for the Warming Arctic?", 20 *Review of European Community and International Environmental Law* (2011), 3.

[19] See notably Bruno Simma, "Self-Contained Regimes", 16 *Netherlands Yearbook of International Law* (1985), 845.

[20] Fragmentation of International Law: Difficulties Arising from the Diversification and Expansion of International Law, Report of the Study Group of the International Law Commission finalized by Martti Koskenniemi, UN. Doc. A/CN.4/L.682, 13 April 2006, para. 729; On the fragmentation of international law see, for instance, Martti Koskenniemi and Päivi Leino, "Fragmentation of International Law? Postmodern Anxieties", 15 *Leiden Journal of International Law* (2002), 553; Matthew Craven, "Unity, Diversity and the Fragmentation of International Law", 14 *The Finnish Yearbook of International Law* (2003), 3; Gerhard Hafner, "Pros and Cons Ensuring From Fragmentation of International Law", 25 *Michigan Journal of International Law* (2004), 849; Joost Pauwelyn, "Bridging Fragmentation and Unity: International Law as a Universe of Inter-Connected Islands", 25 *Michigan Journal of International Law* (2004), 903; Eyal Benvenisti and George W. Downs, "The Empire's New Clothes: Political Economy and the Fragmentation of

raised by the increasing specialization and diversification of international law.[21] The report shows how conflicts may arise between special and general international law, as well as between different branches of international law, and reviews various legal techniques for avoiding and resolving conflicts of norms and regimes.[22]

The ILC report points to possible tensions between different branches of international law, and recommends that "increasing attention will have to be given to the collision of norms and regimes and the rules, methods and techniques for dealing with such collisions."[23] However, only a handful of the studies responding to this call have focused on international environmental law.[24] This is surprising, as the proliferation of international legal instruments is one of the key features of the development of international environmental law over the past decades. In 1993, Edith Brown Weiss already discussed the possible consequences of "treaty congestion" in international environmental law, identifying "operational inefficiency" as a key concern.[25] While the multiplication of international environmental agreements has certainly not been ignored in the period since, and has received particular attention in the context of discussions on reforming international environmental governance,[26]

International Law", 60 *Stanford Law Review* (2007), 595; Alexandra Khrebtukova, "A Call to Freedom: Towards a Philosophy of International Law in an Era of Fragmentation", 4 *Journal of International Law and International Relations* (2008), 51; Margaret A. Young (ed.), *Regime Interaction in International Law: Facing Fragmentation* (Cambridge: Cambridge University Press, 2012).

[21] ILC, Fragmentation of International Law, supra, note 20.

[22] Ibid.

[23] Ibid., para. 493.

[24] Exceptions include Harro van Asselt, Francesco Sindico and Michael A. Mehling, "Global Climate Change and the Fragmentation of International Law", 30 *Law & Policy* (2008), 423; Harro Van Asselt, "Legal and Political Approaches in Interplay Management: Dealing with the Fragmentation of Global Climate Governance", in Sebastian Oberthür and Olav Schram Stokke (eds), *Managing Institutional Complexity: Regime Interplay and Global Environmental Change* (Cambridge: The MIT Press, 2011), 59; Margaret A. Young, "Climate Change Law and Regime Interaction", 4 *Carbon and Climate Law Review* (2011), 147; Margaret A. Young, *Trading Fish, Saving Fish: The Interaction between Regimes in International Law* (Cambridge: Cambridge University Press, 2011); Cinnamon Piñon Carlane, "Good Climate Governance: Only a Fragmented System of International Law Away?", 30 *Law &Policy* (2008), 450; Karen N. Scott, "International Environmental Governance: Managing Fragmentation Through Institutional Connection", 12 *Melbourne Journal of International Law* (2011), 177.

[25] Edith Brown-Weiss, "International Environmental Law: Contemporary Issues and the Emergence of a New Order", 81 *Georgetown Law Journal* (1993), 675, at 697–702. See also Bethany Lukitsch Hicks, "Treaty Congestion In International Environmental Law: The Need For Greater International Coordination, Comment", 32 *University of Richmond Law Review* (1999), 1643; Donald K. Anton, "Treaty Congestion' in Contemporary International Environmental Law", in Shawkat Alam et al. (eds), *Routledge Handbook of International Environmental Law* (London: Routledge, 2012, forthcoming).

[26] See, for instance, Steinar Andresen, "Global Environmental Governance: UN Fragmentation and Co-ordination", in Olav Schram Stokke and Øystein B. Thommessen (eds), *Yearbook of International Co-operation on Environment and Development 2001/2002* (London: Earthscan, 2001), 19; Steven Bernstein and Maria Ivanova, "Institutional Fragmentation and Normative Compromise in Global Environmental Governance: What Prospects for Re-embedding?", in Steven Bernstein and Louis W. Pauly (eds), *Global Liberalism and Political Order: Towards a New Grand Compromise?* (Albany: State University of New York Press, 2007), 161.

only limited attention has been paid to (the effectiveness of) strategies for managing the fragmentation of international environmental law.[27]

Against this backdrop, this chapter aims to provide insights into strategies for managing fragmentation international climate change law by examining the potential of legal and institutional strategies with a view to enhancing synergy and mitigating conflict between various legal regimes. The chapter is structured as follows: Section 13.2 provides an introduction to the debate on the fragmentation of international law. Section 13.3 identifies opportunities for, and limitations of, well-known legal techniques for avoiding and resolving inter-regime conflicts. Section 13.4 then moves on to show how institutional cooperation between treaty bodies created under different legal regimes may lead to greater coherence between the climate change regime and other international legal regimes. It argues, however, that such a strategy also raises concerns, especially regarding its legitimacy. Section 13.5 provides concluding remarks, and identifies areas for further research.

13.2 The Fragmentation of International Law

13.2.1 What Is 'Fragmentation'?

'Fragmentation' means different things to different people. Indeed, the very use of the term has been the subject of a vigorous debate among international lawyers.[28] The discussions in the ILC and its 2006 report sparked a debate in international legal circles about the state of international law and governance, and about the potential threats and opportunities posed by the phenomenon of fragmentation.

The main reason why the term 'fragmentation' has caused so much controversy is its purported negative bias. Koskenniemi and Leino were among the first to point at a possible political agenda behind the use of the notion by several judges from the International Court of Justice (ICJ) at the turn of this century. They argued that the judges' "postmodern anxieties" concerning the unity of international law could be best explained as an attempt to raise the profile of the ICJ on the international plane at a time when an increasing number of other judicial bodies might undermine its relevance.[29] Likewise, the fears that some authors have expressed about the growing specialization in international law have been explained as a counter-reaction of general international lawyers afraid of becoming irrelevant within

[27] Notable exceptions are Wolfrum and Matz, *Conflicts in International Environmental Law*, supra, note 14, at 119–209; W. Bradnee Chambers, *Interlinkages and the Effectiveness of Multilateral Environmental Agreements* (Tokyo: United Nations University Press, 2008); Young, *Trading Fish, Saving Fish*, supra, note 24.

[28] The term also has also led to discussions in the international relations literature. See Frank Biermann et al., "The Fragmentation of Global Governance Architectures: A Framework for Analysis", 9 *Global Environmental Politics* (2009), 14, at 16–17.

[29] Koskenniemi and Leino, "Fragmentation of International Law?", supra, note 20, at 576–577.

their profession.[30] Yet, other terms such as 'diversity', 'pluralism', and 'polycentricity' have a positive subtext that would make them equally suitable to defend a certain position.[31] For the purposes of this chapter, the term 'fragmentation' refers to a landscape where various international legal instruments are overlapping in terms of substantive issue coverage. This definition is intended to be value-neutral, and is primarily aimed at describing the state of international law relevant for addressing climate change.

To further clarify some of the conceptual confusion, two typologies of 'fragmentation' can be distinguished. A first distinction can be made between *substantive* and *institutional* fragmentation. The ILC made this distinction when it decided not to examine "the competence of various institutions applying international legal rules and their hierarchical relations *inter se*" (i.e. institutional fragmentation), but instead focused on "the splitting up of the law into highly specialized "boxes" that claim relative autonomy from each other and from the general law" (i.e. substantive fragmentation).[32] Although the ILC study introduced this clear demarcation between substantive and institutional fragmentation, the two types are in fact inter-related. Abi-Saab describes this as a "law of legal physics": "To each level of normative density, there corresponds a level of institutional density necessary to sustain the norms."[33] This relation can also be seen in practice. For instance, the *Swordfish* dispute between the European Union and Chile is mostly seen as an example of institutional fragmentation, as the case was brought before the WTO dispute settlement mechanism and the International Tribunal for the Law of the Sea (ITLOS) simultaneously.[34] However, both dispute settlement mechanisms are inherently connected to substantive bodies of law, namely the various WTO Agreements and the United Nations Convention on the Law of the Sea (UNCLOS).[35]

---

[30] Mario Prost, "All Shouting the Same Slogans: International Law's Unities and the Politics of Fragmentation", 17 *Finnish Yearbook of International Law* (2006), 131, at 158.

[31] Anne-Charlotte Martineau, "The Rhetoric of Fragmentation: Fear and Faith in International Law", 22 *Leiden Journal of International Law* (2009), 1, at 27. For instance, "diversity" was contrasted with "cacophony" in a special issue of the Michigan Journal of International Law focusing on the advantages and drawbacks of the fragmentation of international law. See, for example, Bruno Simma, "Fragmentation in a Positive Light", 25 *Michigan Journal of International Law* (2004), 845, at 845. Pluralism is generally seen as a benign development by legal pluralists. See, for instance, Andreas Fischer-Lescano and Gunther Teubner, "Regime-Collisions: the Vain Search for Legal Unity in the Fragmentation of Global Law", 25 *Michigan Journal of International Law* (2004), 999. For a recent discussion of polycentricity in a positive light, see Elinor Ostrom, "Polycentric Systems for Coping with Collective Action and Global Environmental Change", 20 *Global Environmental Change* (2010), 550.

[32] ILC, Fragmentation of International Law, supra, note 20, para. 13.

[33] Georges Abi-Saab, "Fragmentation or Unification: Some Concluding Remarks", 31 *New York University Journal of International Law and Politics* (1999), 919, at 925.

[34] For a discussion of the case, see Marcos A. Orellana, "The Swordfish Dispute between the EU and Chile at the ITLOS and the WTO", 71 *Nordic Journal of International Law* (2002), 55.

[35] Tomer Broude, "Principles of Normative Integration and the Allocation of International Authority: The WTO, The Vienna Convention on the Law of Treaties, and the Rio Declaration", 6 *Loyola University Chicago International Law Review* (2008), 173, at 182–183.

Second, the fragmentation of international law may refer to the relationship between different interpretations of general international law, the relation between general international law and specialized regimes, or the relations among two or more overlapping specialized regimes.[36] An example of the first type is the *Tadić* case, in which the International Criminal Tribunal for the Former Yugoslavia came to a different judgment about the criterion applicable to assess when an armed military group can be said to be acting on behalf of a foreign power than the earlier decision by the ICJ in the *Nicaragua* case.[37] Under the second category, scholars have discussed, for instance, how the general law on state responsibility relates to non-compliance mechanisms used in international environmental law or other more specialized regimes that may conflict with, or complement the general rules.[38] The third type of fragmentation is exemplified by the various trade and environment disputes before the WTO dispute settlement body, and forms the focus of this chapter.

13.2.2 The Promises and Pitfalls of Fragmentation

While the very notion of 'fragmentation' may thus reveal assumptions about its consequences, the positive and negative implications of fragmentation have been discussed extensively in the literature. This section draws in particular on the claims about the promises and pitfalls of fragmentation that have been raised in the international law literature, although it will also refer to other studies that have discussed the advantages and drawbacks of fragmentation in the specific context of global climate governance.[39]

An ILC feasibility study on the fragmentation of international law conducted in 2000 indicated that the issue was one that should be looked at in terms mainly of "risks", "threats", or other negative connotations. In particular, it argued that fragmentation can be seen as detrimental, since "[d]oubts could be raised as to whether international law will be able to achieve one of its primary objectives, dispute avoidance and the stabilisation of international relations and, thus, achieve its genuine function of law. The credibility, reliability and, consequently, authority of international law would be impaired."[40] This rather general statement can be split up in various arguments against fragmentation.

One argument is that the growing body of international legal rules threatens the unity and coherence of international law, as various specialized rules are created

[36] ILC, Fragmentation of International Law, supra, note 20, para. 47.

[37] Prosecutor v. Dusko Tadic, Judgment, 15 July 1999, Case No. IT-94-1-A, A.Ch.

[38] Martti Koskenniemi, "Breach of Treaty or Non-Compliance: Reflections on the Enforcement of the Montreal Protocol", 3 *Yearbook of International Environmental Law* (1992), 123.

[39] See, for instance, Biermann et al., "The Fragmentation of Global Governance Architectures", supra, note 28; Robert O. Keohane and David G. Victor, "The Regime Complex for Climate Change", 9 *Perspectives on Politics* (2011), 7.

[40] Gerhard Hafner, "Risks Ensuing from Fragmentation of International Law", Official Records of the General Assembly, 55th session, Supplement No. 10 (A/55/10, 2000), Annex, 143 at 147.

which allow international judicial institutions to reach diverging decisions, in other words, the institutional fragmentation referred to above.[41] For instance, a dispute between Ireland and the United Kingdom regarding the construction of a plant reprocessing nuclear fuel led to three different legal procedures, all based on a different body of substantive law.[42]Another important drawback is that the fragmentation of international law can be used by a handful of powerful States to their advantage. These States can opt for a mechanism that best serves their interests, and can create new agreements if the old ones no longer serve their interests.[43] With regard to dispute settlement, this may lead to 'forum shopping': countries are likely to choose the forum that is most likely to deliver a positive outcome. This explains why in the *Swordfish* dispute, the EU initiated proceedings at the WTO, arguing that Chile had restricted the movement of goods. Conversely Chile, the state taking conservation measures with respect to swordfish, brought its case before ITLOS, alleging that the EU had violated the law of the sea.

Finally, a fragmented international legal system could lead to (some degree of) prioritization of certain fields of international law over others, for example, the dominance of international economic law over international environmental law or – less likely – vice versa.[44] Indeed, such prioritization may be inevitable, as "each legal regime will naturally assert itself as the proper forum in which to address the situation, claiming superior status for its particular descriptions and concerns."[45] Regimes, as Koskenniemi describes it, thus have a "structural bias" in favour of themselves.[46] This structural bias becomes important when one regime can be considered 'stronger' than others, because of the involvement of more powerful States, or because of stronger mechanisms to ensure compliance. This fear is often raised in the context of the trade and environment debate, where the WTO's dispute settlement system is considered to be stronger than the non-compliance mechanisms of most multilateral environmental agreements.[47]

---

[41] Pierre-Marie Dupuy, "The Danger of Fragmentation or Unification of the International Legal System and the International Court of Justice", 31 *New York University Journal of International Law and Politics* (1999), 791; Benedict Kingsbury, "Is the Proliferation of International Courts and Tribunals a Systemic Problem?", 31 *New York University Journal of International Law and Politics* (1999), 679.

[42] ILC, Fragmentation of International Law, supra, note 20, paras. 10 and 439–442.

[43] Benvenisti and Downs, "The Empire's New Clothes", supra, note 20, at 628.

[44] Craven, "Unity, Diversity and the Fragmentation of International Law", supra, note 20, at 5; ILC, *Fragmentation of International Law*, supra, note 20, para. 493.

[45] Khrebtukova, "A Call to Freedom", supra, note 20, at 56.

[46] Martti Koskenniemi, *From Apology to Utopia: The Structure of International Legal Argument,* 2nd ed. (Cambridge: Cambridge University Press, 2005), at 600–615. See also Martti Koskenniemi, "Hegemonic Regimes", in Margaret A. Young (ed.), *Regime Interaction in International Law: Facing Fragmentation* (Cambridge: Cambridge University Press, 2012), 305.

[47] For a comparison of the dispute settlement mechanisms of the WTO and multilateral environmental agreements, see Alexandra González-Calatayud and Gabrielle Marceau, "The Relationship between the Dispute-Settlement Mechanisms of MEAs and those of the WTO", 11 *Review of European Community and International Environmental Law* (2002), 275.

While plenty of arguments thus draw attention to the negative effects of fragmentation, the literature shows that it may also entail numerous advantages. Indeed, after initial fears were expressed about the phenomenon, international legal scholars quickly realized that fragmentation might also have positive effects. First, fragmentation is viewed as a positive indicator of increased diversity in legal norms and the expansion of international law to previously unregulated fields.[48] Over time, international law has come to cover important new issue areas of international relations such as international commerce, human rights, and the environment. As Koskenniemi and Leino aptly put it: "Special regimes and new organs are parts of an attempt to advance beyond the political present that in one way or another has been revealed unsatisfactory."[49] However, while the expansion to new areas could in principle be seen as a positive development, this does not necessarily mean that "more (international) law equals better (international) law."[50]

The increased specialization in international law is also arguably a way of accommodating diverging interests of States. As a result, governments view specialized regimes as better serving their interests and thus have stronger incentives to comply. As Hafner argues, a "less-than-global approach seems particularly necessary when different States clearly hold different beliefs about what basic values should be preserved by international regulation."[51] This argument has been reiterated in the context of international climate policy, where several observers have called for a 'minilateral' approach towards international decision-making on climate change.[52] Furthermore, some commentators have posited that fragmentation would not endanger the coherence of the wider body of international law, as it would lead to the global diffusion of the "best ideas".[53] Similarly, it has been argued that regulatory competition may allow for the development of different solutions in different regulatory contexts, of which the most effective will "survive" and be diffused to other regulatory contexts.[54]

In summary, while the use – or non-use – of the term 'fragmentation' may serve particular agendas, my modest claim is that it provides an accurate description of the current state of international affairs, where the emergence of different social rationalities at the global level has led to multiple international agreements that overlap in terms of their subject matter. Whether the phenomenon is beneficial or

---

[48] Anja Lindroos and Michael Mehling, "Dispelling the Chimera of 'Self-Contained Regimes': International Law and the WTO", 16 *European Journal of International Law* (2005), 857, at 859.

[49] Koskenniemi and Leino, "Fragmentation of International Law?", supra, note 20, at 578.

[50] Benvenisti and Downs, "The Empire's New Clothes", supra, note 20, at 602.

[51] Hafner, "Pros and Cons Ensuing from Fragmentation of International Law", supra, note 20, at 859.

[52] Moisés Naím, "Minilateralism. The Magic Number to Get Real International Action", *Foreign Policy* (2009), 135.

[53] Jonathan Charney, "The Impact on the International Legal System of the Growth of International Courts and Tribunals", 31 *New York University Journal of International Law and Politics* (1999), 697, at 700.

[54] Biermann et al., "The Fragmentation of Global Governance Architectures", supra, note 28, at 27.

malign is mainly in the eye of the beholder, and further depends on whether the term is used to describe the relationship between different specialized regimes or the relationship between such regimes and general international law. This chapter argues that the consequences of fragmentation do not necessarily depend on the existence of various overlapping agreements *per se*, but rather on how their interrelationships are managed. To this end, the following sections provide an overview of the opportunities for, and limitations of, different legal and political strategies for managing the fragmentation of international climate change law.

13.3 Managing Fragmentation Through Legal Techniques

An international lawyer's intuitive reaction to managing fragmentation is probably to resort to the rules provided by international law for dealing with norm conflicts. Indeed, the ILC report offers international lawyers a toolbox to address many of the challenges arising from the fragmentation of international law.[55] These tools include conflict avoidance techniques, such as treaty interpretation, as well as rules for deciding which treaty will prevail in case of a conflict, such as the maxims of *lex posterior* (i.e. the later treaty prevails) and *lex specialis* (i.e. the more specific treaty prevails). It is not my intention to review these various techniques here.[56] Instead, I will highlight some of the opportunities they provide for managing the fragmentation of international climate change law, as well as their inherent limitations.

13.3.1 Opportunities

13.3.1.1 Harmonious Treaty Interpretation

Treaty interpretation as a technique of avoiding a conflict between different climate-related treaties has been discussed in detail in the literature.[57] The ILC deemed Article 31.3(c) of the Vienna Convention on the Law of Treaties[58] a particularly promising avenue for avoiding conflicts,[59] and the provision has been the subject of

---

[55] ILC, Fragmentation of International Law, supra, note 20, para. 492.

[56] See, for instance, Joost Pauwelyn, *Conflict of Norms in Public International Law. How WTO Law Relates to other Rules of International Law* (Cambridge: Cambridge University Press, 2003); Christina Voigt, *Sustainable Development as a Principle of International Law. Resolving Conflicts between Climate Measures and WTO Law* (Leiden: Brill, 2008).

[57] See notably Voigt, ibid., at 265–292.

[58] Vienna Convention on the Law of Treaties, Vienna, 22 May 1969, in force 27 January 1980, 8 *International Legal Materials* (1989), 679.

[59] ILC, Fragmentation of International Law, supra, note 20, paras. 410–480.

an increasing number of analyses.[60] Article 31.3(c) provides that in the interpretation of treaties, "[t]here shall be taken into account, together with the context: … any relevant rules of international law applicable in the relations between the parties."[61] Like the other interpretation rules laid down in the Vienna Convention on the Law of Treaties, the provision is regarded to have the status of customary international law.[62] Moreover, to some authors, the interpretative guidance contained in this provision amounts to a "principle of systemic integration" that forms "a constitutional norm within the international legal system,"[63] or a "principle of mutual supportiveness."[64] Although there is a certain harmonizing appeal to the provision, no such principle has yet been explicitly recognized under general international law, and it still lacks an authoritative formulation.[65] Still, the notion finds some support in rules of treaty interpretation and also past case law, including decisions by the WTO dispute settlement system.[66] For instance, in the well-known *US-Shrimp* dispute, the WTO Appellate Body referred to the provision, indicating that it sought additional interpretive guidance from the general principles of international law.[67]

---

[60] Philippe Sands, "Treaty, Custom and the Cross-fertilization of International Law", 1 *Yale Human Rights and Development Law Journal* (1998), 85; Campbell McLachlan, "The Principle of Systemic Integration and Article 31(3)(c) of the Vienna Convention", 54 *International and Comparative Law Quarterly* (2005), 279; Duncan French, "Treaty Interpretation and the Incorporation of Extraneous Legal Rules", 55 *International Comparative Law Quarterly* (2006), 281; Broude, "Principles of Normative Integration and the Allocation of International Authority", supra, note 35; Anja Lindroos and Michael Mehling, "From Autonomy to Integration? International Law, Free Trade and the Environment", 77 *Nordic Journal of International Law* (2008), 253; Panos Merkouris, "Article 31(3)(c) of the VCLT and the Principle of Systemic Integration" (PhD thesis on file at the Queen Mary University of London, College of Law), 2010, available at: https://qmro.qmul.ac.uk/jspui/bitstream/123456789/477/1/MERKOURISArticle%2031(3)(c)2010.pdf (last accessed on 14 February 2012); Riccardo Pavoni, "Mutual Supportiveness as a Principle of Interpretation and Law-Making: A Watershed for the 'WTO-and-Competing-Regimes' Debate?", 21 *European Journal of International Law* (2010), 649; Mélanie Samson, "High Hopes, Scant Resources: A Word of Scepticism about the Anti-Fragmentation Function of Article 31(3)(c) of the Vienna Convention on the Law of Treaties", 24 *Leiden Journal of International Law* (2011), 701.

[61] Vienna Convention on the Law of Treaties, supra, note 58, Art. 31.3(c).

[62] Merkouris, "Article 31(3)(c) of the VCLT and the Principle of Systemic Integration", supra, note 60, at 8.

[63] McLachlan, "The Principle of Systemic Integration and Article 31(3)(c) of the Vienna Convention", supra, note 60, at 280.

[64] Pavoni, "Mutual Supportiveness as a Principle of Interpretation and Law-Making", supra, note 60, at 678.

[65] Lindroos and Mehling, "From Autonomy to Integration?", supra, note 60, at 268.

[66] McLachlan, "The Principle of Systemic Integration and Article 31(3)(c) of the Vienna Convention", supra, note 60, at 295–309. However, as Lindroos and Mehling observe, the case law is rather recent, and provides only a "weak basis for an actual principle of systemic integration". Lindroos and Mehling, "From Autonomy to Integration?", supra, note 60, at 268.

[67] *United States – Import Prohibition of Certain Shrimp and Shrimp Products*, Report of the Appellate Body, WTO Doc. WT/DS58/AB/R, 6 November 1998, para. 158.

The principle of systemic integration has been invoked as a possible tool to address (looming) conflicts between the UN climate regime and other legal regimes. For instance, in discussing a potential conflict between the Kyoto Protocol and biodiversity-related treaties, Pontecorvo argues that the principle confirms "a specific duty for Parties to interpret the provisions of the Kyoto Protocol relating to sinks potentially conflicting with pre-existing commitments under other treaties in such a way as to make them compatible with these commitments."[68] It has also been suggested that the climate treaties (and decisions adopted under them) could be of use in the interpretation of other ambiguous or indeterminate WTO norms in case of a climate-trade dispute.[69] For instance, the climate treaties – and in particular a possible future climate agreement – could inform the analysis of whether a climate-related trade measure is "*necessary* to protect human, animal or plant life or health"[70] or "*relating to* the conservation of exhaustible natural resources"[71] (emphasis added) under the general exceptions in Article XX of the General Agreement on Tariffs and Trade (GATT). A country adopting the measure could invoke the principle of systemic integration, and use its ratification of climate treaties in its defence of the non-commercial, environmental objectives of its measure. All other things being equal, participation in climate treaties would make the tests formulated in the exceptions of the GATT easier to meet.

It remains unclear whether Article 31.3(c) is indeed a "master-key"[72] in dealing with fragmentation. First, it remains unsettled what "taken into account" actually entails.[73] It is generally agreed that this phrase does not mean that the extraneous rules override the interpreted rules, but rather that their normative significance needs to be determined on a case-by-case basis.[74] So while the legal norms developed under the climate regime could inform a decision by the WTO dispute settlement bodies, they could not result in the setting aside of WTO norms.[75] Second, while the

---

[68] Concetta Maria Pontecorvo, "Interdependence between Global Environmental Regimes: The Kyoto Protocol on Climate Change and Forest Protection", 59 *Zeitschrift für ausländisches öffentliches Recht und Völkerrecht* (1999), 709, at 741.

[69] Harro van Asselt, Francesco Sindico and Michael A. Mehling, "Global Climate Change and the Fragmentation of International Law", supra, note 24, at 435–436; Navraj Singh Ghaleigh and David Rossati, "The Spectre of Carbon Border-Adjustment Measures", 2 *Climate Law* (2011), 63, at 71–72.

[70] Art. XX(b) of the General Agreement on Tariffs and Trade, 15 April 1994, in force 1 January 1995, Marrakesh Agreement Establishing the World Trade Organization, Annex 1A, The Legal Texts: The Results of the Uruguay Round of Multilateral Trade Negotiations, 33 *International Legal Materials* (1994), 1153.

[71] Ibid., Art. XX(g).

[72] McLachlan, "The Principle of Systemic Integration and Article 31(3)(c) of the Vienna Convention", supra, note 60, at 280–281.

[73] Sands, "Treaty, Custom and the Cross-fertilization of International Law", supra, note 60, at 103; ILC, Fragmentation of International Law, supra, note 20, para. 423.

[74] ILC, Fragmentation of International Law, supra, note 20, para. 473–474; Voigt, Sustainable Development as a Principle of International Law, supra, note 56, at 282.

[75] Voigt, Sustainable Development as a Principle of International Law, supra, note 56, at 284–286.

provision has been invoked by international adjudicatory bodies,[76] some bodies, including the dispute settlement mechanism of the WTO, have been rather reluctant to seek recourse to it,[77] and it is far from clear whether they would do so in the future. Most notably, in the *EC-Biotech Products* dispute, the WTO Panel rejected the argument to consider the Cartagena Protocol on Biosafety to the Convention on Biological Diversity[78] as one of the "relevant rules of international law" under Article 31.3(c), because the membership of the treaty was not identical to the WTO.[79] If this approach was adopted, this would significantly limit the scope of Article 31.3(c) for avoiding conflicts between international legal regimes related to climate change, as parallel memberships of multilateral treaties are rather limited.[80] Third, and more importantly, the extensive academic debate on systemic integration ignores the fact that other methods for interpreting treaties already provide ample opportunity to take into account other rules of international law, and that Article 31.3(c) of the VCLT has mainly "residual value."[81] For example, a teleological interpretation of the provisions of the WTO Agreements, taking into account their "context" and their "object and purpose" would already include the preambular language on the WTO's sustainable development objective,[82] and would likely allow for a balancing approach which would not be too different from the one envisaged under Article 31.3(c). Furthermore, treaty interpreters may adopt an "evolutionary approach" irrespective of their reference to Article 31.3(c) of the VCLT.[83]

The principle of systemic integration can thus be regarded as a strong integrative device in theory – but its theoretical strength is its weakness in judicial practice. Adjudicators will refrain from using it, as the resulting normative integration would also entail integration of authority – i.e. a direct influence on lawmaking.[84] As Broude explains, "to integrate (with) the norms of another system is to acknowledge the authority of that other system to produce pertinent norms" as well as "assert[ing]

[76] ILC, Fragmentation of International Law, supra, note 20, paras. 433–460.

[77] Lindroos and Mehling, "From Autonomy to Integration?", supra, note 60, at 270.

[78] Cartagena Protocol on Biosafety to the Convention on Biological Diversity, 29 January 2000, in force 11 September 2003, 39 *International Legal Materials* (2000), 1027.

[79] *European Community – Approval and Marketing of Biotech Products*, Reports of the Panel, WTO Doc. WT/DS291/R, WT/DS292/R, WT/DS293/R, 29 September 2006, at para. 7.71. (*Biotech*).

[80] For this reason, the Panel's approach has been criticized, for instance, by Margaret A. Young, "The WTO's Use of Relevant Rules of International Law: An Analysis of the Biotech Case", 56 *International and Comparative Law Quarterly* (2008), 907, at 914–918.

[81] Isabelle van Damme, *Treaty Interpretation by the WTO Appellate Body* (Oxford: Oxford University Press, 2009), at 375; Samson, "High Hopes, Scant Resources", supra, note 60, at 711–712.

[82] Vienna Convention on the Law of Treaties, supra, note 58, at Art. 31.1. See also Miguel A. Elizande Carranza, "MEAs with Trade Measures and the WTO: Aiming toward Sustainable Development", 15 *Buffalo Environmental Law Journal* (2007), 43, at 86–91.

[83] Voigt, Sustainable Development as a Principle of International Law, supra, note 56, at 275–276.

[84] Broude, "Principles of Normative Integration and the Allocation of International Authority", supra, note 35, at 200.

authority over them."[85] It is this integration of authority that dispute settlement bodies seek to avoid by using 'weaker' forms of integration, such as Article 31.1 of the VCLT.[86] Similarly, while a possible "principle of mutual supportiveness" has much theoretical appeal as an interpretative device,[87] it cannot be automatically inferred that it would be used in practice.

13.3.1.2 Conflict Clauses

In case a conflict between two different treaties arises, the starting point for its resolution is to examine whether a treaty contains any conflict clauses.[88] The purpose of such clauses is to clarify the relationship between treaties, and to prevent contradictions.

The climate treaties contain several provisions that regulate their relationships with other multilateral agreements and international organizations. For instance, the United Nations Framework Convention on Climate Change (UNFCCC)[89] and the Kyoto Protocol[90] delimit their scope by only covering "greenhouse gases not controlled by the Montreal Protocol."[91] While this provision shows awareness of the linkages between the problems of – and solutions to – climate change and ozone layer depletion, it does not in itself prevent or resolve conflicts between them.[92] It can also be argued that the Kyoto Protocol's reference to "relevant international environmental agreements" in Article 2.1(a)(ii) constitutes a conflict clause, with a view to ensuring that parties to the Protocol do not implement climate policies that frustrate the objectives of other environmental treaties. The provision requires that in implementing and elaborating climate change policies and measures, developed countries take into account their commitments under "relevant international environmental agreements."[93] However, the provision is unclear about which commitments in other agreements it refers to, and also merely states that such commitments should be "taken into account" by parties.[94] It is thus difficult to see how this formulation

[85] Ibid., at 186–187.

[86] Ibid., at 187.

[87] Pavoni, "Mutual Supportiveness as a Principle of Interpretation and Law-Making", supra, note 60.

[88] Pauwelyn, *Conflict of Norms in Public International Law*, supra, note 56, at 328.

[89] United Nations Framework Convention on Climate Change, New York, 9 May 1992, in force 21 March 1994, 31 *International Legal Materials* (1992), 849.

[90] Kyoto Protocol to the United Nations Framework Convention on Climate Change, Kyoto, 10 December 1997, in force 16 February 2005, 37 *International Legal Materials* (1998), 22.

[91] See, for instance, UNFCCC, supra, note 89, Art. 4.1(a); Ibid., Art. 5.1.

[92] Sebastian Oberthür, "Linkages between the Montreal and Kyoto Protocols: Enhancing Synergies between Protecting the Ozone Layer and the Global Climate", 1 *International Environmental Agreements: Politics, Law and Economics* (2001), 357.

[93] Kyoto Protocol, supra, note 90, Art. 2.1(a)(ii).

[94] Pontecorvo, "Interdependence between Global Environmental Regimes", supra, note 65, at 739–740.

could be construed in such a way that it would subordinate the commitments in the Kyoto Protocol to other international environmental agreements.[95] Article 3.5 of the UNFCCC and Article 2.3 of the Kyoto Protocol could also be seen as conflict clauses with respect to the WTO agreements. Article 3.5 of the UNFCCC provides, *inter alia*, that "[m]easures taken to combat climate change, including unilateral ones, should not constitute a mans of arbitrary or unjustifiable discrimination or a disguised restriction on international trade."[96] Under Article 2.3 of the Kyoto Protocol, developed countries must "strive to implement policies and measures… in such a way as to minimize adverse effects, including the adverse effects of climate change, effects on international trade, and social, environmental and economic impacts on other Parties…"[97] However, these provisions do not establish a clear hierarchy between the trade and climate change regimes, and do not – explicitly or implicitly – allow or prohibit climate-related trade measures.[98] In other words, they do not determine which treaty would prevail in case a conflict arises. Still, it is important to also note what the agreements do not explicitly state: they do not subordinate to the WTO Agreements, in contrast with, for instance, the Cartagena Protocol on Biosafety;[99] and they do not explicitly allow for trade measures against non-parties or non-compliers.[100] Furthermore, while not being explicit conflict clauses, the provisions may still provide interpretative guidance. For instance, it can be argued that Article 2.3 of the Kyoto Protocol directs Parties to adopt measures that minimize effects on international trade, except in cases where such effects are necessary to ensure the effectiveness of such measures.[101] Finally, the Kyoto Protocol contains a provision delimiting the scope of the climate treaties by delegating the negotiation of rules on emissions from international aviation and maritime transport to the International Civil Aviation Organization and the International Maritime Organization.[102]

---

[95] Frédéric Jacquemont and Alejandro Caparrós, "The Convention on Biological Diversity and the Climate Change Convention 10 Years After Rio: Towards a Synergy of the Two Regimes?", 11 *Review of European Community and International Environmental Law* (2002), 139, at 178.

[96] UNFCCC, supra, note 89, Art. 3.5.

[97] Kyoto Protocol, supra, note 90, Art. 2.3

[98] Voigt, *Sustainable Development as a Principle of International Law*, supra, note 56, at 299.

[99] Cartagena Protocol on Biosafety, supra, note 78, preamble (stating that "that this Protocol shall not be interpreted as implying a change in the rights and obligations of a Party under any existing international agreements"). For a discussion of the history of this clause, see Sabrina Safrin, "Treaties in Collision: The Biosafety Protocol and the World Trade Organization Agreements", 96 *American Journal of International Law* (2002), 606, at 614–618. .

[100] Olav Schram Stokke, "Trade Measures and Climate Compliance: Institutional Interplay Between WTO and the Marrakesh Accords", 4 *International Environmental Agreements: Politics, Law and Economics* (2004), 339, at 346.

[101] Voigt, *Sustainable Development as a Principle of International Law*, supra, note 56, at 298.

[102] Kyoto Protocol, supra, note 90, Art 2.2. See Sebastian Oberthür, "The Climate Change Regime: Interactions with ICAO, IMO, and the EU Burden-Sharing Agreement", in Sebastian Oberthür and Thomas Gehring (eds), *Institutional Interaction in Global Environmental Governance. Synergy and Conflict among International and EU Policies* (Cambridge: The MIT Press, 2006), 53, at 59–68.

Other climate-related treaties contain more clearly identifiable conflict clauses. For instance, the CBD gives priority to any existing agreement, "except where the exercise of those rights and obligations would cause a serious damage or threat to biological diversity."[103] This clause arguably serves to limit climate change mitigation activities that would cause a serious damage or threat to biodiversity.[104] However, it applies only to treaties *existing* at the time of the CBD's adoption in 1992, and is thus not applicable to the Kyoto Protocol, adopted in 1997. Furthermore, the phrase "serious damage or threat to biological diversity" is nowhere defined or elaborated upon, meaning that the practical application of the clause remains uncertain. It is therefore "doubtful that this clause can prevent or solve conflicts."[105] Article 311.3 UNCLOS is also a provision that claims priority over any other international agreement incompatible with it.[106] Consequently, if, for example, certain forms of geological carbon storage were inconsistent with UNCLOS, but were endorsed under the UNFCCC umbrella, this clause could be called upon to argue that UNCLOS prevails over the climate treaty.

There are various difficulties with the use of conflict clauses: their wording is often unclear and open to diverging interpretations (e.g. what would establish "a serious damage or threat to biological diversity"?); they are not dynamic enough to reflect new developments (e.g. changes in scientific insights); it is not always clear when a treaty comes into existence;[107] and chances are that such clauses may never be applied "in the absence of a single, unifying dispute settlement system."[108] Nevertheless, from a legal perspective, they provide the primary means for addressing the relationship between treaties. An opportunity for managing fragmentation thus lies in their drafting. Whenever a new treaty or amendment is negotiated – either within the UNFCCC context or outside of it – conflict clauses could be drafted in a way that fully considers the implications for other treaties, and preferably in an unambiguous manner.[109] Hence, making a list of all international legal instruments that may have an impact on the treaty under negotiation is sensible.[110] Under such a "stop and think approach" the impacts of a new treaty or a treaty amendment would

---

[103] Convention on Biological Diversity, supra, note 13, Art. 22.1.

[104] Wolfrum and Matz, *Conflicts in International Environmental Law*, supra, note 14, at 124.

[105] Ibid., at 125; ILC, Fragmentation of International Law, supra, note 20, para. 280; Malgosia Fitzmaurice and Olefumi A. Elias, *Contemporary Issues in the Law of Treaties* (Utrecht: Eleven Publishing, 2005), at 244–345.

[106] United Nations Convention on the Law of the Sea, Montego Bay, 10 December 1982, in force 16 November 1994, 21 *International Legal Materials* (1982), 1261, Art. 311.3.

[107] E.W. Vierdag, "The Time of the 'Conclusion' of a Multilateral Treaty: Article 30 of the Vienna Convention on the Law of Treaties and Related Provisions", 59 *British Yearbook of International Law* (1988), 75.

[108] Jacob Werksman, "Formal Linkages and Multilateral Environmental Agreements", 1999, available at: http://archive.unu.edu/inter-linkages/1999/docs/jake.PDF (last accessed on 14 February 2012).

[109] Wolfrum and Matz, *Conflicts in International Environmental Law*, supra, note 14, at 128.

[110] Wilfred Jenks, "The Conflict of Law-Making Treaties", 30 *British Yearbook of International Law* (1953), 401, at 452.

be carefully assessed, where appropriate, in cooperation with the relevant states, secretariats and international organizations.[111] This suggestion is certainly not new. In 1953, Jenks already noted the importance of consultations before and during the drafting of legal instruments.[112] However, there is as of yet still no standard procedure to assess the impacts of a new instrument on existing ones, or to consider how an instrument could contribute to the objectives of other treaties. An opportunity thus lies in introducing such a procedure in drafting new climate-related agreements.

13.3.2 Limitations

13.3.2.1 Definitions of 'Conflict'

Whether international law can play a role in resolving conflicts between the climate treaties and other agreements depends on whether a conflict – in the strict legal sense – actually exists. This may sound like a clear-cut exercise, but is everything but that. Indeed, the scholarly literature is divided on the issue, with some authors arguing for a 'narrow' definition, and others opting for a 'wide' definition.[113] The main challenge in defining conflicts is to capture the divergences between different legal regimes, while at the same time acknowledging that not all divergences need to be resolved through the establishment of a hierarchy between the relevant regimes.

According to the classical definition suggested by Jenks, a "conflict in the strict sense of direct incompatibility arises only where a party to the two treaties cannot simultaneously comply with its obligations under both treaties."[114] More recently, this test of 'impossible joint compliance' has become the subject of criticism.[115] In particular, critics argue that Jenks' focus on obligations only is too limited, and unjustifiably excludes incompatibilities between obligations and permissions. This would include cases of overlap in which a (future) climate change treaty permits a measure that restricts international trade, whilst a trade agreement contains a specific obligation not to restrict trade. Addressing this lacuna, Pauwelyn's treatise on the conflict of norms in international law includes a proposal to expand the definition to include conflicts involving permissive norms. Vranes similarly argues for a broader definition, which comprises "incompatibilities between permissions and obligations, permissions and prohibitions, and obligations and prohibitions," adding that there is a conflict if one of the norms "is necessarily or potentially violated."[116]

---

[111] Hicks, "Treaty Congestion in International Environmental Law", supra, note 25, at 1669–1673.

[112] Jenks, "The Conflict of Law-Making Treaties", supra, note 110, at 452.

[113] For an overview of the debate, see Erich Vranes, *Trade and the Environment. Fundamental Issues in International and WTO Law* (Oxford: Oxford University Press, 2009), at 10–38.

[114] Jenks, "The Conflict of Law-Making Treaties", supra, note 102, at 426.

[115] Pauwelyn, *Conflict of Norms in Public International Law,* supra, note 56, at 166–175 and Vranes, *Trade and the Environment*, supra, note 113, at 19–21.

[116] Vranes, *Trade and the Environment*, supra, note 113, at 38.

Jenks already acknowledged that his narrow definition might not cover all the divergences and inconsistencies between treaties that may have negative effects.[117] While the wider definitions proposed by Pauwelyn and Vranes ensure that certain obvious conflicts are not "defined away,"[118] even their construction of conflict may be insufficient to cover the various kinds of incompatibilities that may arise in international climate law (and, arguably, in international law more broadly). In this regard, Wolfrum and Matz identify several categories of conflicts in international environmental law that fall outside the aforementioned definitions.[119] These include, in the first place, conflicts between agreements resulting from their different objectives (e.g. trade liberalization versus environmental protection). The broader categorization also includes conflicts arising from the incorporation of different principles and approaches (e.g. a precautionary approach versus cost-effectiveness). Differing objectives and principles, however, do not necessarily need to lead to conflicts between two treaties, especially where they are phrased in unclear terms. It is especially in the instances where States have a wide margin of discretion that there may be a conflict in the implementation phase of the agreements.[120] In the case of international law on climate change, this is evidenced by the triggering of conflicts through decisions of treaty bodies, as will be discussed below.

It can thus be established that even wide legal definitions of 'conflict' seem to be insufficient to cover all potential climate-related conflicts. But here it should be asked: insufficient for what purpose? This is perhaps the most crucial question regarding the definition of conflicts, and it is emerging as a focus in the debate on the fragmentation of international law.[121] In this regard, it is useful to cite one of Vranes' main objections to a narrow definition: "The problematic consequence … is that conflicts maxims such as the *lex posterior* principle cannot come into play … ."[122] In other words, it is important to establish that there is a conflict if one wishes to decide which norm prevails. But this is based on an assumption that one of the norms *should* prevail, or that the existence of a conflict in a particular situation is undesirable. Such an assumption can be explained through the 'structural bias' of a specific regime: from a climate change perspective, climate-related norms should trump trade norms. From the trade perspective, trade norms should naturally prevail.[123] This assumption can be questioned if it is accepted that two legal regimes can pursue a similar objective, such as the pursuit of sustainable

[117] Jenks, "The Conflict of Law-Making Treaties", supra, note 110, at 426.

[118] Vranes, *Trade and the Environment*, supra, note 113, at 20.

[119] Wolfrum and Matz, *Conflicts in International Environmental Law*, supra, note 14, at 7–13.

[120] Ibid., at 11.

[121] In dealing with regime interactions, Dunoff highlights the lack of a "redemptive narrative". With this, he refers to the lack of overarching guidance that could help lawyers in deciding how to integrate regimes. See, Jeffrey Dunoff, "A New Approach to Regime Interaction", in Margaret A. Young (ed.), *Regime Interaction in International Law: Facing Fragmentation* (Cambridge: Cambridge University Press, 2012), 136, at 155.

[122] Vranes, *Trade and the Environment*, supra, note 113, at 19.

[123] Khrebtukova, "A Call to Freedom", supra, note 20, at 63.

development,[124] or if one seeks to identify an otherwise overarching 'narrative' that reconciles two regimes.[125] In other words, is it really desirable that a hierarchy between norms be established? This seems to be purpose of most conflict resolution techniques discussed by the ILC, such as the *lex specialis* and *lex posterior* rules, and conflict clauses. However, if one accepts that the climate regime and related regimes are actually pursuing common goals, this quest for normative hierarchy becomes rather futile.

13.3.2.2 Treaty Body Decisions

The climate regime, like many other international environmental regimes, is characterized by a form of lawmaking that departs from the traditional idea of treaty-based lawmaking. Lawmaking does not stop when the treaty text is agreed upon, but continues through the decision-making bodies constituted by those treaties, such as the Conference of the Parties (COP) established under the UNFCCC. The consequence of this innovative form of international lawmaking is that interactions between international legal regimes could well be 'triggered' by a decision by a treaty body rather than the treaty itself. In fact, the impacts of rule-development on forest carbon sinks under the Kyoto Protocol on issues discussed under the CBD shows that the potentially conflicting interaction has its origins in the decisions made by the UNFCCC COP,[126] as it is those decisions that allow for the implementation of Clean Development Mechanism projects that might result in adverse impacts on biodiversity.[127] Yet the debate on the fragmentation of international law is primarily concerned with conflicts stemming from treaties as such. This is in line with the aforementioned discussions on conflicts in international law, which have tended to focus primarily on treaties as the source of conflict. For instance, the classic definition by Jenks states that a "conflict in the strict sense of direct incompatibility arises only where a party to the two *treaties* cannot simultaneously comply with its obligations under *both treaties*" (emphasis added).[128] In addition, various legal techniques to avoid or resolve conflicts are based on, or linked to, the law of *treaties*.

---

[124] For instance, Voigt argues that sustainable development is enshrined in both the climate change and trade regimes. Voigt, *Sustainable Development as a Principle of International Law*, supra, note 56, at 89–144. Koskenniemi refers to sustainable development as one of the "regime hybrids… through which the experts representing the respective regimes may wage their struggle for influence". See Koskenniemi, "Hegemonic Regimes", supra, note 46, at 319–320.

[125] Dunoff, "A New Approach to Regime Interaction", supra, note 121. Koskenniemi is critical whether such a narrative in fact can be construed, citing the example of the legal scholarship on constitutionalization and global administrative law. Koskenniemi, "Hegemonic Regimes", supra, note 46, at 320–321.

[126] Van Asselt, "Managing the Fragmentation of International Environmental Law", supra, note 14.

[127] Van Asselt, "Integrating Biodiversity in the Climate Regime's Forest Rules: Options and Tradeoffs in Greening REDD Design", supra, note 14, at 141–143.

[128] Jenks, "The Conflict of Law-Making Treaties", supra, note 110, at 426.

This means that the question of how to deal with cases where the texts of two treaties are perfectly compatible, but subsequent rule-development under one of the treaties leads to a conflict has thus far been largely ignored.

Whether such situations are captured by the ongoing fragmentation debate depends on the legal status assigned to the decisions of the treaty bodies of international environmental agreements. In other words, to what extent do the decisions adopted by these bodies constitute international lawmaking in a traditional sense?[129] There is no straightforward answer, but it has been argued that while COP decisions are not devoid of normative substance, their legal force is intrinsically connected to the treaty obligation upon which they are based. As Wiersema concludes: "consensus-based COP activity … cannot be seen as giving rise to stand-alone legal or even political obligations" and COP decisions "hold little meaning but for their connection to the treaty."[130] However, even if it can be successfully argued that there are intricate linkages between COP decisions and underlying treaty provisions, this does not mean that the decisions themselves are covered by the law of treaties.[131]

The increasing relevance of decisions by treaty bodies in international environmental lawmaking hence limits the usefulness of the conventional conflict resolution techniques. This does not mean that any conflicts arising from such decisions cannot be dealt with, but rather points to the need to think about alternative means to manage them.

13.4 Managing Fragmentation Through Institutional Cooperation

While legal techniques hold some potential to manage the fragmentation of international climate change law, particularly in the case of normative conflicts, their limitations are also clear. This realization has directed attention towards less formal approaches to managing fragmentation. In particular, the question has been raised

---

[129] See, generally, Robin R. Churchill and Geir Ulfstein, "Autonomous Institutional Arrangements in Multilateral Environmental Agreements: A Little-Noticed Phenomenon in International Law", 94 *American Journal of International Law* (2000), 623; Jutta Brunnée, "COPing with Consent: Law Making under Multilateral Environmental Agreements", 15 *Leiden Journal of International Law* (2002), 1; Annecoos Wiersema, "The New International Law-makers? Conferences of the Parties to Multilateral Environmental Agreements", 31 *Michigan Journal of International Law* (2009), 231.

[130] Wiersema, "The New International Law-makers?", supra, note 129, at 245. See also Fitzmaurice and Elias, *Contemporary Issues in the Law of Treaties*, supra, note 105, at 262 (referring to the Kyoto Protocol provisions on flexible mechanisms as "enabling clauses" for subsequent decisions by the treaty bodies); and Brunnée, "COPing with Consent", supra, note 129, at 24 (referring to "enabling provisions").

[131] Wiersema, "The New International Law-makers?", supra, note 129, at 247.

to what extent treaty bodies, such as COPs and secretariats, could enhance coherence between different legal regimes. Again, I do not seek to provide an exhaustive overview of options for institutional cooperation, but rather focus on the most pertinent opportunities and limitations of this particular management strategy.

13.4.1 Opportunities

13.4.1.1 Bureaucracies

Bureaucracies, such as the UNFCCC Secretariat, are important actors in international environmental – and climate change – governance, yet they have largely flown under the radar of analysts. They can be regarded as key actors 'behind the scenes' in managing the fragmentation of international climate law. However, although their influence on the individual regimes they are tied with is becoming clearer,[132] their role in managing the relationships between different regimes remains under-researched.[133]

Whether there is a mandate for secretariats to engage in institutional cooperation is not the same question as whether these secretariats have the legal capacity to enter into external cooperation agreements in the first place. The latter question has been examined in-depth by Churchill and Ulfstein, who conclude that the institutions of multilateral environmental agreements "have implied powers to act on the external plane, including the capacity to enter into treaties when necessary to carry out their functions."[134] Specifically with respect to secretariats, Chambers argues that while the legal personality of secretariats may not be entirely clear, their power "would certainly include entering into agreements of collaboration with other [multilateral environmental agreements] where there is a clear overlap or interest."[135]

In terms of mandates, liaising with other secretariats is generally one of the tasks assigned to the secretariats of environmental treaties. This is the case, for instance, for the climate secretariat,[136] the ozone secretariat,[137] and the biodiversity

[132] On the role and influence of bureaucracies in global environmental governance, see the contributions in Frank Biermann and Bernd Siebenhüner (eds), *Managers of Global Change: The Influence of International Environmental Bureaucracies* (Cambridge: The MIT Press, 2009).

[133] There are some notable exceptions, such as Sikina Jinnah, "Overlap Management in the World Trade Organization: Secretariat Influence on Trade-Environment Politics", 10 *Global Environmental Politics* (2010), 64; Sikina Jinnah, "Marketing Linkages: Secretariat Governance of the Climate-Biodiversity Interface", 11 *Global Environmental Politics* (2011), 23.

[134] Churchill and Ulfstein, "Autonomous Institutional Arrangements in Multilateral Environmental Agreements", supra, note 129, at 649.

[135] Chambers, *Interlinkages and the Effectiveness of Multilateral Environmental Agreements*, supra, note 27, at 66.

[136] UNFCCC, supra, note 89, Art. 8.2(e); Kyoto Protocol, supra, note 90, Art. 14.2.

[137] Vienna Convention for the Protection of the Ozone Layer, supra, note 10, Art. 7.1(e).

secretariat.[138] Although cooperation is not explicitly included in the mandate of the WTO Secretariat, it has become active in enhancing the transparency of the WTO's activities related to climate change,[139] for instance, through the organization of side-events at COPs and the publication of reports on the linkages between trade and climate change.[140]

Institutional cooperation between the climate secretariat and other bureaucracies has remained largely limited to observership, mutual attendance at meetings, scientific cooperation, and information exchange. An interesting development in this regard has been the formation of the Joint Liaison Group, which comprises the secretariats of the CBD and the UNFCCC, subsequently also joined by the secretariat of the UN Convention to Combat Desertification (the third 'Rio Convention').[141] The mandate of the Joint Liaison Group is to "enhance coordination between the three conventions, including the exchange of relevant information" and "[t]o explore options for further cooperation between the three conventions, including the possibility of a joint work plan and/or a workshop."[142] By the end of 2011, the Joint Liaison Group had convened ten times, focusing on crosscutting issues such as research and monitoring, information exchange, technology transfer, capacity building, financial resources, education and public awareness, and adaptation to climate change. Its activities primarily consist of information exchange and coordination between the administrative bodies of the different regimes.[143] In 2004, the three secretariats drafted a joint paper identifying options for enhanced cooperation. Whereas some of the options identified in the paper (for instance, joint workshops or the sharing of information among secretariat staff) are relatively easy to implement, others (such as the harmonization of reporting) require much more preparation and consensus.[144]

There may be opportunities for enhancing the role of the secretariats in promoting synergies between different environmental regimes. For instance, the tool of memoranda of cooperation – widely used, for instance, by the biodiversity secretariat – has rarely been used by the climate secretariat. Such written agreements could formalize existing informal practices, thereby keeping the relationship with other regimes permanently on the agenda. However, it can be questioned

[138] Convention on Biological Diversity, supra, note 13, Art. 24.1(d).

[139] Jinnah, "Overlap Management in the World Trade Organization", supra, note 133, at 68.

[140] Tamiotti et al., *Trade and Climate Change,* supra, note 15.

[141] United Nations Convention to Combat Desertification in Those Countries Experiencing Serious Drought and/or Desertification, Paris, Particularly in Africa, 17 June 1994, in force 26 December 1996, 33 *International Legal Materials* (1994), 1328.

[142] Report of the Subsidiary Body on Scientific and Technological Advice on the Second Part of Its Fourteenth Session, Bonn, 24–27 July 2001. U.N. Doc. FCCC/SBSTA/2001/2, 18 September 2011, para. 42(d).

[143] Chambers, *Interlinkages and the Effectiveness of Multilateral Environmental Agreements*, supra, note 27, at 69.

[144] Options for Enhanced Cooperation Among the Three Rio Conventions, U.N. Doc. UNEP/CBD/SBSTTA/10/INF/9 Annex, 15 December 2004.

whether formalizing cooperation *per se* would result in synergies at the operational level. Indeed, one of the advantages of the secretariats' activities is that they avoid the cumbersome political decision-making processes of the COPs, and thereby provide a valuable informal and flexible way of integrating environmental regimes.

13.4.1.2 Decision-Making Bodies

While the bureaucracies of environmental treaties thus may play an important role in raising awareness of interactions and their consequences, they do not have any decision-making competencies. Nevertheless, also the decision-making bodies in environmental treaties are often guided to cooperate with other bodies. In this regard, the decision-making bodies of the climate regime are mandated to "[s]eek and utilize, where appropriate, the services and cooperation of, and information provided by, competent international organizations and intergovernmental and non-governmental bodies."[145] Other environmental conventions contain similar instructions. The Vienna Convention on Ozone Layer Depletion directs its parties (and parties to subsequent protocols adopted under the treaty) to cooperate with competent international bodies.[146] Likewise, the CBD COP is mandated to "[c]ontact, through the Secretariat, the executive bodies of conventions dealing with matters covered by [the CBD] with a view to establishing appropriate forms of cooperation with them."[147] This has formed the basis for the cooperation between the biodiversity secretariat and other secretariats mentioned above. Also in the area of the international trade regulation, the Agreement Establishing the WTO provides that the "General Council shall make appropriate arrangements for effective cooperation with other intergovernmental organizations that have responsibilities related to those of the WTO."[148] It thus seems clear that there is ample scope for decision-making bodies to cooperate with each other.

Not every decision-making body is equally active, however. In particular, the UNFCCC COP has been rather silent about its relationships with other international conventions. There has been only one COP decision on cooperation, which generally affirms the need for enhanced cooperation "with the aim of ensuring the environmental integrity of the [Rio Conventions] and promoting synergies under the common objective of sustainable development, in order to avoid duplication of efforts, strengthen joint efforts and use available resources more efficiently."[149]

In contrast, decision-making bodies of other international environmental regimes have sought to manage the overlap with the climate regime. For instance, the CBD

---

[145] UNFCCC, supra, note 89, Art. 7.2(l); Kyoto Protocol, supra, note 90, Art. 13.4(i).

[146] Vienna Convention on the Protection of the Ozone Layer, supra, note 10, Art. 2.2(d).

[147] Convention on Biological Diversity 2004, supra, note 13, Art. 23.4(h).

[148] Art. V.1 of the Agreement Establishing the World Trade Organization, 15 April 1994, in force 1 January 1995, 33 *International Legal Materials* (1994), 1144.

[149] Decision 13/CP.8, Cooperation with Other Conventions, U.N. Doc. FCCC/CP/2002/7/Add.1, 28 March 2003, preamble.

COP has adopted a wide range of decisions related to climate change and biodiversity, which have been instrumental in highlighting biodiversity concerns in UNFCCC decisions,[150] although they have not necessarily lead to stronger references to biodiversity protection in the climate regime's decisions. The parties to the Montreal Protocol have also been engaged in activities closely related to the climate regime, most notably by adopting a decision in 2007 that significantly accelerated the phasing out of the consumption and production of hydrochlorofluorocarbons, a potent greenhouse gas that also served as substitute for ozone depleting substances.[151] A similar decision to limit the use of another substitute with global warming potential, hydrofluorocarbons, has been proposed by some parties to the Montreal Protocol, but is still opposed by others.[152]

While institutional cooperation on climate-related overlaps between regimes thus mainly takes place unilaterally – initiated mainly by several proactive decision-making bodies outside the UNFCCC – enhanced cooperation could take place in a "more ambitious form of comprising joint planning of programmes or even the coordination of substantive decision-making or implementation activities."[153] There are examples of such enhanced cooperation in international environmental law, for instance, in biodiversity protection, fisheries management and chemical substances.[154] In the case of chemicals, it was even possible to hold a joint session of the decision-making bodies of three different multilateral environmental agreements. Although extending this type of institutional cooperation to the climate regime may sound attractive in theory, there are limitations to what is possible and desirable, as will be discussed in the next section.

13.4.2 Limitations

13.4.2.1 Unclear Mandates

Although institutional cooperation to manage linkages between the climate regime and other legal regimes is intensifying, the effects are as of yet uncertain. While institutional cooperation can create mutual awareness between regimes, and build capacity at various levels, it is often also plagued by rhetoric about the 'mutual supportiveness' of different treaties, and devoid of practical suggestions. Part of the

---

[150] Farhana Yamin and Joanna Depledge, *The International Climate Change Regime: A Guide to Rules, Institutions and Procedures* (Cambridge: Cambridge University Press, 2004), at 523–524.

[151] Decision XIX/6, Adjustments to the Montreal Protocol with Regard to Annex C, Group I, Substances (Hydrochlorofluorocarbons), U.N. Doc. UNEP/OzL.Pro.19/7, 21 September 2007.

[152] Oberthür, Dupont and Matsumoto, "Managing Policy Contradictions Between the Montreal and Kyoto Protocols", supra, note 12, at 128–129.

[153] Olav Schram Stokke, "The Interplay of International Regimes: Putting Effectiveness Theory to Work?" Fridtjof Nansen Institute (FNI) Report 10/2001, 2001, available at: http://www.fni.no/doc&pdf/FNI-R1401.pdf (last accessed on 2 March 2012), at 12.

[154] Scott, "International Environmental Governance", supra, note 24, at 202–208.

reason for this is that institutional cooperation is challenging because of unclear or restricted mandates.

Secretariats initiating cooperation with other bodies usually act upon a decision by the COP, thereby interpreting the mandate provided in such a decision. While it may seem "commonsensical that a secretariat would not engage in activities against the will of its member states,"[155] it is actually not always clear what this 'will of the parties' is. In other words, secretariats do not always have a clear legal authority regarding the extent of institutional cooperation.[156] This may either constrain or enable them. Parties often tend to interpret the secretariats' mandates restrictively, and secretariats will need to walk on eggshells when engaging in activities with other international actors. This is especially the case for the UNFCCC Secretariat, which has been said to be "living in a straitjacket" imposed by the parties.[157] However, other secretariats have taken a more proactive stance by adopting a wide interpretation of their mandate. The CBD secretariat, for instance, has made use of the limited space provided to it by the COP, partly due to a very active Executive Secretary.[158]

Cooperation between secretariats is even more difficult if their respective mandates differ in their scope. For instance, at its fifth meeting, the Joint Liaison Group argued for consistent guidance from the various COPs, indicating that it can only facilitate, but not guarantee such consistency.[159] Furthermore, at its ninth meeting, the Group noted that "there remains a disconnect between the roles and mandates given to the [Joint Liaison Group] by each convention with this disconnect resulting in limitations when considering the implementation of the requested activities."[160] Because of these limitations, the Joint Liaison Group acts primarily as a forum to facilitate information exchange, and to encourage harmonizing implementation of the Rio Conventions at the national level.[161]

13.4.2.2 Overstepping Regime Boundaries

The mandate for cooperation – and how it is interpreted – will, for an important part, depend on parties' willingness to construct linkages with other regimes. This brings

---

[155] Chambers, Interlinkages and the Effectiveness of Multilateral Environmental Agreements, supra, note 27, at 66.

[156] Ibid., at 70–71.

[157] Per-Olof Busch, "The Climate Secretariat: Making a Living in a Straitjacket", in Frank Biermann and Bernd Siebenhüner (eds), *Managers of Global Change: The Influence of International Environmental Bureaucracies* (Cambridge: The MIT Press, 2009), 225, at 225.

[158] Jinnah, "Marketing Linkages: Secretariat Governance of the Climate-Biodiversity Interface", supra, note 133.

[159] Report of the Fifth Meeting of the Joint Liaison Group. U.N. Doc. FCCC/SBSTA/2004/INF.9, 15 June 2004, para. 4(1).

[160] Report of the Meeting of the Joint Liaison Group of the Convention on Biological Diversity, the United Nations Convention to Combat Desertification, and the United Nations Framework Convention on Climate Change (New York, 14 May 2009), para. 11.

[161] Chambers, *Interlinkages and the Effectiveness of Multilateral Environmental Agreements*, supra, note 27, at 69.

us to one of the core challenges to enhancing institutional cooperation: the risk that states "may be unwillingly drawn into regimes that they are not party to or affiliated with, and implicitly become subject to obligations under those regimes, by virtue of cooperative arrangement."[162] It can be assumed that any effort by actors in one regime to influence the normative development in another will likely be limited by the extent to which memberships are congruent. For instance, while the United States is a party to the UNFCCC, it has not ratified the CBD. A broad mandate for the climate regime's treaty or administrative bodies to cooperate with the CBD could lead to the perception that state sovereignty is eroded by "importing" concepts or rules from the CBD.[163] Party submissions to the UNFCCC seem to confirm that this perception exists. Responding to the work of the Joint Liaison Group in 2004, the United States noted that the Rio Conventions "have a distinct legal character, mandate and membership."[164] Australia even argued that "[t]he CBD and the UNCCD do not have a legitimate role in greenhouse mitigation, which is clearly the work of the UNFCCC."[165] But even when membership is largely overlapping, there may be resistance to the idea of cooperation between bureaucracies. For instance, with respect to the WTO secretariat's role in managing the climate-trade overlap, Cossy and Marceau note that "the competences of the secretariats are limited (they do not normally include decision-making) and underlain by their obligation to remain neutral vis-à-vis the membership."[166]

More generally, cooperation between institutional arrangements of two different regimes gives rise to concerns about legitimacy and accountability.[167] If one adopts a more traditional legal perspective emphasizing the importance of state consent (and state sovereignty) in international lawmaking, it is difficult to see where the legitimacy of enhanced institutional cooperation comes from, particularly in the case of incongruent memberships. These concerns relate back to the 'structural bias' of each regime.[168] Can cooperation really take place in a fashion that gives equal weight to the norms of each regime? This may not be the case when 'stronger' and 'weaker' regimes are concerned. This could result in the prioritization of one regime over another, meaning that cooperation "may become dominated by procedures, principles and concepts that are prevalent within one regime at the expense of [others]."[169] Another matter is whether the norms of each regime *should* be given

---

[162] Scott, "International Environmental Governance", supra, note 24, at 212.

[163] Wolfrum and Matz, *Conflicts in International Environmental Law*, supra, note 14, at 163.

[164] Views on the Paper on Options for Enhanced Cooperation Among the Three Rio Conventions, Submissions from Parties, U.N. Doc. FCCC/SBSTA/2006/MISC.4, 23 March 2006, at 16.

[165] Ibid., at 5.

[166] Mireille Cossy and Gabrielle Marceau, "Institutional Challenges to Enhance Policy Co-ordination – How WTO Rules Could be Utilised to Meet Climate Objectives?", in Thomas Cottier, Olga Nartova and Sadeq Z. Bigdeli (eds), *International Trade Regulation and the Mitigation of Climate Change* (Cambridge: Cambridge University Press, 2009), 371, at 376.

[167] Scott, "International Environmental Governance", supra, note 24, at 211–215; Young, *Trading Fish, Saving Fish*, supra, note 24, at 281–287

[168] Koskenniemi, *From Apology to Utopia,* supra, note 46, at 600–615.

[169] Scott, "International Environmental Governance", supra, note 24, at 213.

equal weight. Young argues convincingly that bodies seeking to cooperate with other regimes should "scrutinise and review the 'sources' of external regimes."[170] Only in this way, she posits, can institutional cooperation be made accountable, and can the risk of 'managerialism' be avoided.

13.5 Concluding Remarks

The 'wicked' nature of the climate problem means that it is difficult, if not impossible, to govern the problem through a single legal regime. The argument here is that any effective response to the climate change problem will need to take into account the potential of other regimes to either mitigate or exacerbate the problem, while at the same time also considering the impacts of the climate regime on other legal regimes. International climate change law is thus inevitably fragmented. However, the consequence of such fragmentation does not have to be regulatory chaos, or the prioritization of one policy field over another – as has been feared by international lawyers participating in the general debate on the fragmentation of international law. Crucially, the implications of the fragmentation of international climate law depend on how it is *managed*.

With this in mind, this chapter has sought to illustrate the potential of well-known legal techniques to manage interactions between different international legal regimes. It has also addressed several less-well studied forms of institutional cooperation. With respect to legal techniques this chapter argued, first, that pursuing harmonious treaty interpretation, whereby treaty interpreters take into account extraneous rules, could avoid conflicts between climate-related treaties. It questioned, however, whether this necessarily needed to take place through a principle of 'systemic integration' or 'mutual supportiveness', which some scholars have suggested is embodied in Article 31.3(c) of the VCLT. Second, it indicated that in the course of international climate lawmaking, negotiators could take a step back, carefully considering the implications of the negotiations for other regimes and drafting provisions to regulate their inter-relationships. The chapter then moved on to point out that many of the tensions involving the climate regime cannot be adequately be captured by traditional legal definitions of conflict thereby limiting the usefulness of many techniques discussed in detail by the ILC. Furthermore, it questioned whether applying such techniques – leading to a normative hierarchy – is in fact desirable at all.

This chapter argued that informal institutional cooperation can complement the formal legal techniques for managing the fragmentation of international climate law. It showed how various secretariats as well as decision-making bodies in climate-related regimes have started to address overlapping issues, with a view to

[170] Young, *Trading Fish, Saving Fish*, supra, note 24, at 277.

avoiding conflicts and maximizing synergies. However, it is clear that there are also limitations as to what can be achieved through such means. Secretariats' mandates are not always clearly defined, and to avoid a rebuke by parties, secretariats will tend to stay away from intruding too much into the decision-making process through external cooperation with other institutions. This question is linked with more general concerns about the legitimacy and accountability of institutional cooperation. These concerns are to some extent based on traditional notions of state consent, but they point to the real risk that actors in one regime are sidelined through the use of norms borrowed from another.

Although this chapter has given a first indication of how the fragmentation of international climate law could be managed, further research into the (im)possibilities of other strategies could complement the existing body of knowledge. First, the focus of this chapter has been on the international level. It should, however, be clear that managing the relationship between different regimes, to an important extent, takes place at the national or subnational level – i.e. in the implementation phase of international agreements. While the coordination and integration of policies and laws has been the subject to attention of lawyers and political scientists at the domestic level, further research could shed light into the way in which such coordination could strengthen the coherence of international law. Vice versa, there has been little research on how cooperation at the international level could strengthen coherence in the implementation phase. Second, and related to the first point, there has been scant attention to the role of non-state actors, such as environmental organizations, the private sector, or public-private partnerships, in improving mutual coherence between different regimes. A third point relates to the legal form of international climate governance. While international legal instruments, including the climate treaties, other environmental treaties, and international trade law, still play a key role in steering behaviour, climate governance is characterized by the emergence of a wide array of non-state initiatives, and initiatives that could be rather regarded as soft law. The point here is that the role of legal techniques for managing the fragmentation of international climate law may further diminish if the role of international law in addressing climate change itself is further reduced. Indeed, the Vienna Convention on the Law of Treaties or conflict resolution principles such as *lex specialis* will not be applicable in case of interactions between hard law and soft law, since there will not be any norm conflict in the strict legal sense. The relationship between hard and soft law related to climate change, and the management of their relationship is therefore another appropriate area for further inquiry.

Chapter 14
No Need to Reinvent the Wheel for a Human Rights-Based Approach to Tackling Climate Change: The Contribution of International Biodiversity Law

Elisa Morgera

Abstract This chapter provides a systematic analysis of the ways in which international biodiversity law contributes to the fight against climate change by assessing and preventing the negative impacts on biodiversity and community livelihoods of measures to address climate change ('response measures'), and adopting the ecosystem approach to climate change mitigation and adaptation. In highlighting readily available legal avenues for ensuring the mutual supportiveness of the international biodiversity regime and the international climate change regime, the chapter argues that positive interaction between the two regimes can promote a human rights-based approach to the development of the international climate change regime and its implementation at the national level.

14.1 Introduction

Climate change is one of the main drivers of global biodiversity loss.[1] Consequently, the closely related challenges of biodiversity loss and climate change must be addressed "with equal priority" and in close coordination, if "tipping points in biodiversity loss" are to be avoided.[2] This objective is increasingly reflected in

---

Elisa Morgera holds an LL.M degree from University College London and a Ph.D. from the European University Institute, Florence. She is Lecturer in European Environmental Law and Director of the LL.M Programme in Global Environment and Climate Change Law, School of Law, University of Edinburgh, UK.

[1] CBD and United Nations Environment Programme (UNEP)-World Conservation Monitoring Centre (WCMC), *Global Biodiversity Outlook 3* (Montreal: CBD, 2010), available at: http://gbo3.cbd.int/ (last accessed on 10 April 2012), at 22 (hereinafter, GBO 3).

[2] Ibid., at 11 and 75.

E. Morgera (✉)
Old College, University of Edinburgh, South Bridge, Edinburgh, EH8 9YL, Scotland/UK
e-mail: elisa.morgera@ed.ac.uk

E.J. Hollo et al. (eds.), *Climate Change and the Law*, Ius Gentium: Comparative Perspectives on Law and Justice 21, DOI 10.1007/978-94-007-5440-9_14,
© Springer Science+Business Media Dordrecht 2013

international biodiversity law. This chapter thus proceeds from the argument that international biodiversity law has established close and important links with climate change law, making a multifaceted contribution to the fight against climate change.[3] Parties to the Convention on Biological Diversity (CBD)[4] and to the various other biodiversity-related conventions have, through normative activity of their governing bodies, sought to assess potential and actual threats that climate change and measures to respond to climate change ('response measures') pose to the conservation and sustainable use of biodiversity. They have also identified ways to prevent and address negative impacts of climate change and response measures on biodiversity through the mutually supportive interpretation and application of international climate change and biodiversity law.

This chapter provides a systematic analysis of the normative contribution of international biodiversity law to climate change law. This is particularly useful as guidance under the CBD has been developed in an obscure fashion,[5] with the result that these significant developments have escaped academic attention. Notably, these developments not only concern specifically climate change, but also include earlier and more general guidance providing innovative, environmentally holistic and people-centered approaches that can usefully apply for climate change-related purposes. These developments concern the assessment of the negative impacts of climate change response measures on biodiversity and community livelihoods, and the application of the ecosystem approach to climate change mitigation and adaptation. In analyzing them, the present contribution investigates readily available legal avenues to ensure mutual supportiveness between the international biodiversity regime and the international climate change regime, highlighting how a positive interaction between the two regimes can also support a human rights-based approach[6] to the development of the international climate change regime and its implementation at the national level.

---

[3] I preliminarily explored this argument in Elisa Morgera, "Far away, so close: A legal analysis of the increasing interactions between the Convention on Biological Diversity and climate change law", 2 Climate Law (2011), 85.

[4] Convention on Biological Diversity, Rio de Janeiro, 5 June 1992, in force 29 December 1993, 1760 *United Nations Treaty Series* (1993), 79, (hereinafter CBD).

[5] This is due to the fact that "CBD guidance on climate change and biodiversity is dispersed throughout a myriad of (generally long) COP decisions; and within these decisions, relevant passages are not always well organized or clearly separated by topic or addressee. Frequent qualifications and convoluted drafting further undermine the comprehensibility of COP decisions and of their legal implications under the CBD." Morgera, "Far away, so close", supra, note 3, at 86.

[6] In line with the hortatory reference in the Cancun Agreements that UNFCCC Parties "should in all climate change related actions, fully respect human rights", Decision 1/CP.16, Cancun Agreements: Outcome of the work of the Ad Hoc Working Group on Long-term Cooperative Action under the Convention, UN Doc. FCCC/CP/2010/7/Add.1, 15 March 2011 para. 8. A human rights-based approach has been described as "viewing certain human rights as essential precursors to achieving environmental protection" and focused on procedural rights, see Edward Cameron, "Human Rights and Climate Change: Moving from an Intrinsic to an Instrumental Approach", 38 *Georgia Journal of International and Comparative Law* (2009–2010), 673, at 699. In this chapter, however, a human rights-based approach is rather concerned both with procedural and substantive rights and aims to achieve both protection of human rights and the environment.

This chapter thus places itself in the context of the ongoing debate on the 'normative interplay' between the international biodiversity and climate change regimes, which are seen as overlapping and distinct, but not necessarily conflicting, systems of rules.[7] By outlining normative developments in international biodiversity law, this chapter aims to show that the abundant and timely normative activity of the CBD Conference of the Parties (COP), which not only embodies the consensus of 193 States but also the inputs of indigenous and local communities,[8] already provides useful and well-developed conceptual bridges not only between climate change law and biodiversity law, but also with human rights law.[9] It argues that normative activity under the CBD provides environmentally holistic and human rights-based standards that could fill gaps related to the protection of biodiversity and human rights in climate change law, both at the level of international law-making and national implementation.[10] The gaps in the climate change regime have already been identified, particularly in relation to human rights implications of the Clean Development Mechanism and reducing emissions from deforestation and

[7] Harro van Asselt, Francesco Sindico and Michael Mehling, "Global Climate Change and the Fragmentation of International Law", 30 *Law and Policy* (2008), 423; Margaret Young, "Climate Change and Regime Interaction", 5 *Carbon and Climate Law Review* (2011), 147; Harro van Asselt, *Managing the Fragmentation of International Environmental Law: Forests at the Intersection of the Climate and Biodiversity Regimes* (SSRN, 2010).

[8] Under the CBD Working Group on Article 8(j) (traditional knowledge), the fullest possible participation of indigenous and local communities is ensured in all Working Group meetings, including in contact groups, by welcoming community representatives as Friends of the Co-Chairs, Friends of the Bureau and Co-Chairs of contact groups; without prejudice to the applicable rules of procedure of the Conference of the Parties establishing that representatives duly nominated by parties are to conduct the business of CBD meetings so that any text proposal by indigenous and local communities' representatives must be supported by at least one party. Report of the Seventh Meeting of the Ad Hoc Open-ended Working Group on Article 8(j) and Related Provisions, UN Doc. UNEP/CBD/COP/11/7, 24 November 2011, para. 20.

[9] Human Rights Council, Resolutions on Human Rights and Climate Change: 7/23 of 2008; 10/4 of 2009; and 18/22 of 2011. See generally, Cameron, "Human Rights and Climate Change", supra, note 6; and Lavanya Rajamani, "The Increasing Currency and Relevance of Rights-based Perspectives in the International Negotiations on Climate Change", 22 *Journal of Environmental Law* (2010), 391.

[10] Young, "Climate Change and Regime Interaction", supra, note 7, at 152–153: although not at the level of adjudication, due to the absence of a compliance mechanism under the CBD. On the latter point, see Elisa Morgera and Elsa Tsioumani, "Yesterday, Today and Tomorrow: Looking Afresh at the Convention on Biological Diversity", 21 *Yearbook of International Environmental Law* (2011), 3, at 10–11 and 28. Other avenues could, however, be available: for instance, in the case of marine biodiversity, the UN Fish Stocks Agreement could provide access to international adjudication for disregarding the duty to protect biodiversity of species associated or dependent from fish stocks from climate change impacts. See, William Burns, "Potential Causes of Action for Climate Impacts under the United Nations Fish Stocks Agreement", in William Burns and Hari Osofsky (eds), *Adjudicating Climate Change* (Cambridge: Cambridge University Press, 2009), 14.

forest degradation in developing countries (REDD), as well as other measures on energy, biofuels and adaptation.[11]

Overall, this contribution aims to fill a gap in the current policy and academic debates on human rights and climate change.[12] Leaving aside the consideration of human rights in the context of the North–South divide in the ongoing UN climate change negotiations[13] and the potential of human rights-based litigation to contribute to the development or implementation of climate change law,[14] the present analysis offers significant insights on a human right-based approach to climate change law and policy at the international level as well as "within States."[15] The latter can be seen as the "most effective means of complying with positive obligations to protect individuals against the threats posed by climate change … in adaptation measures as well as climate-related development aid."[16] The chapter will further touch upon the relevance of the CBD normative activity in the context of a human rights-based approach to climate-related development assistance, as well as in relation to the responsibility of business entities to respect human rights in the context of climate change action.

---

[11] Naomi Roht-Arriaza, "'First, Do No Harm': Human Rights and Efforts to Combat Climate Change", 38 *Georgia Journal of International and Comparative Law* (2009–2010), 593, at 595; Ole Padersen, "The Janus Head of Human Rights and Climate Change: Adaptation and Mitigation", 80 *Nordic Journal of International Law* (2011), 403; Cameron, "Human Rights and Climate Change", supra, note 6, at 704–705, who emphasizes that response measures may "undermine" but not necessarily "violate" human rights.

[12] To the author's knowledge, none of the legal scholars writing on climate change and human rights has yet made an argument about the usefulness of the normative activity of the CBD. In addition to the sources cited elsewhere in this article, the author has also consulted: Ole Pedersen, "Climate Change and Human Rights: Amicable or Arrested Development?", 1 *Journal of Human Rights and the Environment* (2010), at 236; Amy Sinden, "Climate Change and Human Rights", 27 *Journal of Land, Resources and Environmental Law* (2007), 255; and Rebecca Tsosie, "Indigenous People and Environmental Justice: The Impact of Climate Change", 78 *University of Colorado Law Review* (2007), 1625, who briefly refers to the CBD, in ibid., at 1668.

[13] Rajamani, "The Increasing Currency and Relevance of Rights-based Perspectives", supra, note 9, at 395–398.

[14] Marilyn Averill, "Linking Climate Litigation and Human Rights", 18 *Review of European Community and International Environmental Law* (2009), 139; Eric A. Posner, "Climate Change and International Human Rights Litigation: A Critical Appraisal", 155 *University of Pennsylvania Law Review* (2007), 1925; Hari Osofsky, "The Inuit Petition as a Bridge: Beyond Dialectics of Climate Change and Indigenous Peoples' Rights", 31 *American Indian Law Review* (2007), 675; and Svitlana Kravchenko, "Right to Carbon or Right to Life: Human Rights Approaches to Climate Change", 9 *Vermont Journal of Environmental Law* (2008), 513.

[15] That is of a state vis-a-vis its citizens: Rajamani, "The Increasing Currency and Relevance of Rights-based Perspectives", supra, note 9, at 426.

[16] John von Doussa, Allison Corkery and Renee Chartres, "Human Rights and Climate Change", 14 *Australian International Law Journal* (2007), 161, at 161–162.

14.2 Systemic Interpretation and Mutual Supportiveness in the Context of the UNFCCC and CBD

Before proceeding to the systematic analysis of the multifaceted guidance provided by the CBD parties on climate change, it is necessary to clarify the overall relationship between the international climate change and biodiversity regimes. In doing so, the advantages of systemic interpretation and mutual supportiveness will be illustrated in order to better understand the interaction between the different legal instruments comprised in each legal regime.

At the treaty level, there is no insurmountable conflict between the international biodiversity and climate change regimes.[17] The United Nations Framework Convention on Climate Change (UNFCCC) makes reference to ecosystems in the context of its ultimate objective of stabilizing greenhouse gas concentrations and achieving international cooperation for the conservation of sinks and reservoirs.[18] It does not, however, link the application of the precautionary principle to potential environmental consequences or seek to prioritize mitigation measures based on their environmental impacts.[19] While the Kyoto Protocol[20] does not expressly provide incentives for meeting its legally binding emission reduction targets for developed countries "in a manner that minimizes negative impacts on biodiversity,"[21] it does require minimization of adverse environmental impacts by one Protocol party on another, particularly on developing States.[22] It also requires its governing body to assess the environmental impacts of measures taken pursuant to the Protocol,[23] and includes a clause calling upon parties to implement policies and measures taking into account commitments under relevant international agreements.[24]

---

[17] Van Asselt, *Managing the Fragmentation of International Environmental Law*, supra, note 7, at 17; on the basis of United Nations Framework Convention on Climate Change, New York, 9 May 1992, in force 21 March 1994, 31 *International Legal Materials* (1992), 849 (hereinafter, UNFCCC), Arts. 2, 4(1)(d), 1(1) and 4(8).

[18] Ibid., Arts. 2 and 4(1)(d).

[19] Meinhard Doelle, "Integration among Global Environmental Regimes: Lessons Learned from Climate Change Mitigation", in Aldo Chircop, Ted McDorman, Susan Rolston (eds), *The Future of Regime-Building in the Law of the Sea: Essays in Tribute to Douglas M. Johnston* (Leiden: Martinus Nijhoff, 2008), 63, at 75, based on UNFCCC, supra, note 17, Arts. 3(3) and 4.

[20] Kyoto Protocol to the United Nations Framework Convention on Climate Change, Kyoto, 10 December 1997, in force 16 February 2005, 37 *International Legal Materials* (1998), 22.

[21] Meinhard Doelle, "Linking the Kyoto Protocol and Other Multilateral Environmental Agreements: From Fragmentation to Integration?", 14 *Journal of Environmental Law and Practice* (2004), 75, at 83.

[22] Doelle, "Integration among Global Environmental Regimes", supra, note 19, at 76; and Van Asselt, Sindico and Mehling, "Global Climate Change and the Fragmentation of International Law", supra, note 7, at 18; based on Kyoto Protocol, supra, note 20, Art. 2(3).

[23] Kyoto Protocol, supra, note 20, Art. 13(4)(a); See comments by van Asselt, *Managing the Fragmentation of International Environmental Law*, supra, note 7, at 18.

[24] Kyoto Protocol, supra, note 20, Art. 2(a)(ii).

To compare, the CBD requires its Parties to cooperate through competent international organizations on matters of mutual interest for the conservation and sustainable use of biodiversity, which may well include climate-related issues.[25] In addition, on the basis of systemic interpretation,[26] the CBD can be read as calling on its Parties to: integrate biodiversity issues into climate change plans, programmes, and policies[27]; undertake environmental impact assessments of adaptation and mitigation projects that are likely to have significant adverse effects on biodiversity[28]; regulate climate-change-related processes and activities that have a significant adverse effect on biodiversity[29]; avoid or minimize adverse impacts from the use of biological resources for adaptation or mitigation purposes[30]; prevent the introduction of invasive alien species in the context of adaptation and mitigation measures[31]; bring about cooperation between national authorities and the private sector in ensuring the sustainable use of biodiversity for adaptation or mitigation purposes[32]; and provide incentives for the conservation and sustainable use of biodiversity components in the context of adaptation and mitigation activities.[33] Furthermore, the CBD can be interpreted as calling on Parties to respect and preserve the traditional knowledge and practices of indigenous and local communities when implementing mitigation and adaptation measures, involving those communities in climate-change-related decision-making and rewarding them for their intellectual contribution to mitigation and adaptation measures.[34] The latter notably offers a specific legal basis for the CBD to inject a right-based approach to the application of all the other above-outlined tools, thereby promoting synergies between biodiversity law, human rights and climate change law.

Furthermore, the CBD[35] gives "conditional priority" to its Parties' obligations arising from other treaties existing at the time of its conclusion only in the absence

---

[25] CBD, supra, note 4, Art. 5; Frédéric Jacquemont and Alejandro Caparrós, "The Convention on Biological Diversity and the Climate Change Convention 10 Years After Rio: Towards a Synergy of the Two Regimes?", 11 *Review of European Community and International Environmental Law* (2002), 169, at 179.

[26] Vienna Convention on the Law of Treaties, Vienna, 23 May 1969, in force 27 January 1980, 1513 *United Nations Treaty Series* 293 (1980), Art. 31(3)(c).

[27] CBD, supra, note 5, Art. 6(b).

[28] Ibid., Art. 14(1)(a).

[29] Ibid., Art. 8(l).

[30] Ibid., Art. 10(b).

[31] Ibid., Art. 8(h).

[32] Ibid., Art. 10(e).

[33] Ibid., Art. 11.

[34] Ibid., Art. 8(j). Note that this language is partly reflected in Decision 1/CP.16, supra, note 6, Appendix I, para. 2(c-d). For a discussion of the significant evolution in the interpretation of this provision by CBD Parties, see Elisa Morgera and Elsa Tsioumani, "The Evolution of Benefit-sharing: Linking Biodiversity and Community Livelihoods", 15 *Review of European Community and International Environmental Law* (2010), 150.

[35] CBD, supra, note 4, Art. 22(1).

of a serious threat or damage to biodiversity.[36] It thus leaves a wide margin of discretion to its Parties to determine the circumstances in which the CBD should take precedence over other international agreements.[37] In this light, the CBD can arguably be interpreted as authorizing its Parties to give precedence to their international obligations arising from the CBD in those specific instances where a serious threat of damage to biodiversity has been identified. In addition, this provision implicitly calls upon CBD Parties to be constantly alert to, and promptly identify, such a threat to biodiversity when it materializes.[38] Against this background, the normative activity of the CBD COP has periodically and progressively crystallized consensus on the identification of serious threats to biodiversity arising from climate change and from actions pursuant to the international climate change regime that warranted synergetic responses. By the end of 2010, climate change had evolved into a key cross-cutting component in the work of the CBD in two respects. As a threat to biodiversity through the negative impacts of climate change and response measures on biodiversity and the livelihoods of communities; and as a response that contributes to biodiversity conservation and sustainable use through climate change mitigation and adaptation measures with biodiversity co-benefits.[39] Thus, the impacts of climate change and of response measures that pose significant threats to biodiversity have been, and will continue to be, addressed comprehensively in normative work under the CBD.[40]

This, however, has not been reciprocated in the practice of the international climate change regime[41]: recent decisions on REDD, for instance, have only provided for a very general reference to relevant international instruments.[42] The coherence between the international biodiversity and climate change regimes thus appears to rest mostly on coherence between the decisions by their respective

---

[36] Riccardo Pavoni, "Mutual Supportiveness as a Principle of Interpretation and Law-Making: A Watershed for the WTO-and-Competing-Regimes Debate?", 21 *European Journal of International Law* (2010), 649, particularly, at 655.

[37] Jacquemont and Caparrós, "The Convention on Biological Diversity and the Climate Change Convention 10 Years After Rio", supra, note 25, at 178.

[38] Morgera, "Far away, so close", supra, note 3, at 89.

[39] I am grateful to Jaime Webbe, CBD Secretariat, for drawing my attention to this point, which I discussed in more detail in Morgera, "Far away, so close", supra, note 3.

[40] Ibid., at 113–115.

[41] The lack of cross-reference in decisions taken in the context of the international climate change regime to relevant decisions taken in the context of the CBD has been emphasized by van Asselt, *Managing the Fragmentation of International Environmental Law*, supra, note 7, at 36–37, referring specifically to decisions on forests, and Jamie Pittock, "A Pale Reflection of Political Reality: Integration of Global Climate, Wetland and Biodiversity Agreements", 1 *Climate Law* (2010), 343, at 355.

[42] Decision 1/CP.16, supra, note 6, Appendix I, para. 2(a, c–e). Note that the explicit reference to the CBD in Decision 2/CP.15, The Copenhagen Accord, UN Doc. FCCC/CP/2009/11/Add.1, 30 March 2010, para. 8, Annex, "[does] not reappear in subsequent COP decisions dealing with REDD" as highlighted by Annalisa Savaresi in her contribution to this volume.

treaty bodies.[43] This is particularly significant as both regimes evolve dynamically and continuously through COP decisions; several studies have been devoted to the legal nature and impacts of the climate change COP decisions,[44] and the few studies on the relevant CBD COP decisions indicate that the Convention on Biodiversity has been subject to a highly evolutionary interpretation by its Parties.[45] While CBD COP decisions, however, have been systematically taking into account normative developments occurring in the international climate change regime, the latter has not shown a reciprocal interest in parallel developments in the international biodiversity regime. Divergences in COP decisions under separate international regimes may represent "different ways of dealing with a problem" but can still "lead to mutually supportive outcomes,"[46] thereby paving the way for "fruitful interactions" between the two regimes.[47] As compatibility with COP decisions cannot be assured through the systemic interpretation approach reflected in the Vienna Convention on the Law of Treaties,[48] the emerging general principle of mutual supportiveness appears as a more appropriate legal avenue to promote coherence between the two regimes at the level of the normative work of their governing bodies. In addition to being more flexible than systemic interpretation concerning the instruments to which it can be applied, the added value of the principle of mutual supportiveness is that it goes beyond interpretation. This means that it not only calls on States, at the interpretative level, to avoid resolving tensions between competing international regimes through the subordination of one regime to the other; but that the principle of mutual supportiveness also has a law-making dimension. It calls upon States to exert good-faith efforts to negotiate and conclude instruments that clarify the relationship between competing regimes, particularly when interpretative reconciliation efforts have been exhausted.[49]

---

[43] Van Asselt, Sindico and Mehling, "Global Climate Change and the Fragmentation of International Law", supra, note 7, at 425.

[44] Jutta Brunnée, "COPing with Consent: Law-making under Multilateral Environmental Agreements", 15 *Leiden Journal of International Law* (2002), 1; Annecoos Wiersema, "The New International Law-Makers? Conferences of the Parties to Multilateral Environmental Agreements", 31 *Michigan Journal of International Law* (2009), 231.

[45] Morgera and Tsioumani, "Yesterday, Today and Tomorrow", supra, note 10. Strangely enough, none of the general studies on COP decisions has ever referred to the CBD as a case study: in addition to the sources cited supra, note 44, see also Malgosia Fitzmaurice, "Consent to Be Bound – Anything New Under the Sun?", 74 *Nordic Journal of International Law* (2005), 483; and Robert Churchill and Geir Ulfstein, "Autonomous Institutional Arrangements in Multilateral Environmental Agreements: A Little-Noticed Phenomenon in International Law", 94 *The American Journal of International Law* (2000), 623.

[46] Van Asselt, Sindico and Mehling, "Global Climate Change and the Fragmentation of International Law", supra, note 7, at 430.

[47] Young, "Climate Change and Regime Interaction", supra, note 10, at 147.

[48] Art. 31(3)(c); Van Asselt, Sindico and Mehling, "Global Climate Change and the Fragmentation of International Law", supra, note 7, at 430.

[49] Pavoni, "Mutual Supportiveness", supra, note 36, at 661–669.

Through the lens of mutual supportiveness, therefore, the following sections will discuss how the guidance from the CBD COP has sought to promote an environmentally holistic and human rights-based approach to the international law-making on climate change and national implementation, through guarantees for the conservation and sustainable use of biodiversity and the protection of indigenous and local communities' rights.[50] Although other biodiversity-related conventions have increasingly addressed climate change issues, notably the Convention on Migratory Species (CMS),[51] the Ramsar Convention on Wetlands of International Importance[52] and the World Heritage Convention,[53] these contributions appear less sophisticated or less comprehensive than those emerging from the CBD framework. Accordingly, this chapter will only draw on relevant normative benchmarks elaborated under other biodiversity-related treaties[54] when they provide value added to normative work under the CBD.

[50] Morgera, "Far away, so close", supra, note 3.

[51] Convention on the Conservation of Migratory Species of Wild Animals, Bonn, 23 June 1979, in force 1 November 1983, 1651 *United Nations Treaty Series* (1991), 333.

[52] Convention on Wetlands of International Importance, Ramsar, 2 February 1971, in force 21 December 1975, 996 *United Nations Treaty Series* (1976), 245.

[53] World Heritage Convention (Convention Concerning the Protection of the World Cultural and Natural Heritage), Paris, 16 November 1972, in force 7 August 1956, 1037 *United Nations Treaty Series* (1977), 151; and Policy Document on the Impacts of Climate Change on World Heritage Properties, WHC-07/16.GA/10, September 2008. Nonetheless the World Heritage Committee has been "reluctant to impose more than site-specific mitigation obligations on State Parties," basically "deferring to the general mitigation options contained in the UNFCCC": comments by Young, "Climate Change and Regime Interaction", supra, note 10, at 148–149 and 152. See also William Burns, "'Belt and Suspenders'? The World Heritage Convention's Role in Confronting Climate Change", 18 *Review of European Community and International Environmental Law* (2009), 148; and Anna Huggins, "Protecting World Heritage Sites from Adverse Impacts of Climate Change: Obligations for State Parties to the World Heritage Convention", 14 *Australian International Law Journal* (2007), 121.

[54] The other two biodiversity-related conventions have only begun to address climate change: the COP to the Convention on International Trade in Endangered Species (Convention on International Trade in Endangered Species of Wild Fauna and Flora, Washington DC, 3 March 1973, in force 1 July 1975, 993 *United Nations Treaty Series* (1976), 243) adopted in 2010 decisions on information-gathering related to climate change impacts on the Convention (Decisions 15.15 and 15.16); while a Ministerial Conference on Biodiversity, Food Security and Climate Change, held on 11 March 2011, in Bali, Indonesia, adopted the Bali Ministerial Declaration on the Role of the International Treaty on Plant Genetic Resources for Food and Agriculture on Biodiversity, Climate Change and Food Security. Note also that under this Treaty (Rome, 3 November 2001, in force 29 June 2004, 2400 *United Nations Treaty Series* (2006), 303) the multilateral benefit-sharing fund provides financial support for the development of strategic action plans to adapt plant genetic resources for food and agriculture to climate change, as well as financial support for the implementation of immediate action projects that in the second round prioritized climate change adaptation: accordingly, the Treaty's benefit-sharing fund is recognized as an adaptation-funding mechanism in the UNFCCC adaptation funding interface, available at: http://unfccc.int/adaptation/implementing_adaptation/adaptation_funding_interface/items/4638.php (last accessed on 10 April 2012). I am grateful to Elsa Tsioumani for drawing my attention to this development.

14.3 The Contribution of the International Biodiversity Regime: The Ecosystem Approach

The conceptual cornerstone of the interaction between the international climate change and biodiversity regimes is the ecosystem approach, which allows both regimes to integrate other environmental concerns beyond their specific objectives. While under the international climate change regime limited references are made to the ecosystem approach, the CBD COP has devoted significant time and energy to fleshing out this approach not only with a view to ensuring the balanced and coherent achievement of its three objectives,[55] but also to contributing to other areas of international law.[56] In doing so, CBD Parties have delved into key questions of relevance for both regimes, such as the role of precaution, the balance between cost-effectiveness and equity, and the need for procedural and substantive protection of indigenous and local communities.

In 2004, the CBD COP identified the ecosystem approach as a tool to facilitate climate change mitigation and adaptation while ensuring mutual supportiveness between the UNFCCC and the CBD.[57] The ecosystem approach as elaborated under the CBD entails a process aimed at integrating management of land, water and living resources, and promoting conservation and sustainable use in an equitable way, recognizing that human beings are an integral component of many ecosystems.[58] In a nutshell, the ecosystem approach focuses on the interconnectedness among species and between species and their habitats, on long-term timeframes and on the integrity of the structure and functions of genetic, species, population and ecosystem diversity for human wellbeing and ecosystem resilience.[59]

The ecosystem approach is thus tightly linked to precaution,[60] also included among the principles listed in the UNFCCC.[61] As aptly summed up by Burns, the precautionary approach entails taking into account the vulnerability of the environment, the limitations of science, the availability of alternatives, and the need for long-term, holistic environmental considerations, thus operating as a safeguard

---

[55] CBD, supra, note 4, Art. 1.

[56] Morgera and Tsioumani, "Yesterday, Today and Tomorrow", supra, note 10, at 38; Desiree McGraw, "The CBD: Key Characteristics and Implications for Development", 11 *Review of European Community and International Environmental Law* (2002), 17, at 24.

[57] CBD Decision 7/15, Biodiversity and Climate Change, UN Doc. UNEP/CBD/COP/7/21, 13 April 2004, para. 8.

[58] CBD Decision 5/6, Ecosystem approach, UN Doc. UNEP/CBD/COP/5/23, 22 June 2000, paras. 1–2.

[59] Arie Trouwborst, "The Precautionary Principle and the Ecosystem Approach in International Law: Differences, Similarities and Linkages", 18 *Review of European Community and International Environmental Law* (2009), 26, at 28.

[60] UNFCCC, supra, note 17, Art. 3(3). On the fact that the CBD is based on the ecosystem approach and the UNFCCC on the precautionary approach as a differentiating factor see Pittock, "A Pale Reflection of Political Reality", supra, note 41, at 349; based on Rudiger Wolfrum and Nele Matz, *Conflicts in International Environmental Law* (Berlin: Springer, 2003), at 119.

[61] UNFCCC, supra, note 17, Art. 3.3.

against asymmetric information and imperfect monitoring.[62] The precautionary approach can be implemented through adaptive management;[63] responding to changing circumstances and new knowledge, as well as generating new knowledge and reducing uncertainties, thereby allowing management to anticipate and cater for change as a result of an ongoing learning process.[64] As highlighted by Trouwborst, the precautionary and ecosystem approaches both embody responses to the failure of reactive and fragmented approaches to environmental protection: precaution is an integral component of the ecosystem approach, determining when action to prevent damage is necessary, that is, when there are reasonable grounds for concern that serious or irreversible harm to ecosystem integrity may occur.[65] Both approaches accordingly assign similar roles to scientific information, requiring continuous information-gathering and monitoring to feed back into decision-making, and mandate similar implementing measures that should be tailor-made and readily adaptable.[66] Trouwborst thus concludes that the ecosystem approach should be taken into account in the application of the precautionary principle, which addresses broader environmental protection issues than ecosystem integrity.[67]

The consideration of cost-effectiveness is also a common feature of the precautionary and ecosystem approaches. The ecosystem approach calls for assessing the costs and benefits of conserving, maintaining, using and restoring ecosystems and for taking into account the interests of all relevant stakeholders for equitably sharing the benefits according to national law.[68] This is particularly significant in light of the "prominent role afforded to cost-effectiveness in the climate regime,"[69] and the need to ensure that the economic and non-economic values of biodiversity and ecosystem services[70] are taken into account when planning and undertaking

---

[62] Burns, "Potential Causes of Action for Climate Impacts under the United Nations Fish Stocks Agreement", supra, note 10.

[63] CBD Decision 7/11, Ecosystem Approach, UN Doc. UNEP/CBD/COP/7/21, 13 April 2004, Annex I, Principle 6, Implementation Guideline 6.2.

[64] Ibid., Annotations to the Rationale of Principle 9.

[65] Trouwborst, "The Precautionary Principle and the Ecosystem Approach in International Law", supra, note 59, at 26 and 33–34.

[66] Ibid., at 36.

[67] Ibid., at 33–34.

[68] CBD Decision 7/11, supra, note 63, Annex I, para. 12(5).

[69] Van Asselt, Sindico and Mehling, "Global Climate Change and the Fragmentation of International Law", supra, note 7, at 428.

[70] *The Millennium Ecosystem Assessment, Ecosystems and Human Well-being: Synthesis* (Washington, DC: Island Press, 2005), also available at: www.maweb.org/en/index.aspx (last accessed on 10 April 2012) is a global scientific process commissioned by the UN Secretary-General to assess the consequences of ecosystem change on human well-being. The report is noteworthy for having facilitated far-reaching global endorsement of the term "ecosystem services" as the benefits people obtain from ecosystems, such as: food, water, timber, and fiber; regulating services that affect climate, floods, diseases, wastes, and water quality; cultural services that provide recreational, aesthetic, and spiritual benefits; and supporting services such as soil formation, photosynthesis, and nutrient cycling. For a discussion of legal implications, see Elisa Morgera, "The 2005 UN World Summit and the Environment: The Proverbial Half-Full Glass", 15 *Italian Yearbook of International Law* (2006), 53.

climate-change-related activities; and that incentives for such activities should be carefully designed to simultaneously consider cultural, social, economic, and biophysical factors, while avoiding market distortions.[71] The international reflection on the economic valuation of biodiversity, however, is only at incipient stages, although it is considered essential for mainstreaming biodiversity more effectively in other sectors and demonstrating the effectiveness of ecosystem protection and restoration towards climate change adaptation and mitigation.[72]

Much more clearly than the precautionary approach, the ecosystem approach entails a social process: interested communities must be involved through the development of efficient and effective structures and processes for decision-making and management.[73] From that perspective, a key emerging element of the ecosystem approach is benefit-sharing – the substantive dimension underpinning and reinforcing current efforts to ensure community involvement in decision-making and sustainable management of living resources. Benefit-sharing is thus the linchpin for addressing cost-effectiveness and equity concerns at the same time. It operates as a reward for the integration of the traditional knowledge of indigenous and local communities in planning and management, or as compensation for the costs and negative impacts of biodiversity conservation or sustainable management activities on indigenous and local communities.[74] According to the ecosystem approach, benefit-sharing is expected to target stakeholders responsible for the production and management of the benefits flowing from the multiple functions provided by biodiversity at the ecosystem level, including through capacity-building, especially at the level of local communities managing biodiversity in ecosystems, and local incentives for good management practices.[75] This is based on the understanding that where those who control land use do not receive benefits from maintaining natural ecosystems and processes, they are likely to initiate unsustainable practices for short-term gains.[76]

In line with the ecosystem approach, the CBD work programme on protected areas links the goal of promoting equity and benefit-sharing with the legal recognition and

---

[71] CBD Secretariat, *Connecting Biodiversity and Climate Change Mitigation and Adaptation: Report of the Second Ad Hoc Technical Expert Group on Biodiversity and Climate Change, Technical Series No. 41* (Montreal: CBD Secretariat, 2009), at 8–14.

[72] GBO 3, supra, note 1, at 83; Pavan Sukhdev, Heidi Wittmer, Christoph Schröter-Schlaack, Carsten Nesshöver, Joshua Bishop, Patrick ten Brink, Haripriya Gundimeda, Pushpam Kumar and Ben Simmons, *The Economics of Ecosystems and Biodiversity – Mainstreaming the Economics of Nature: A Synthesis of the Approach* (Malta: Progress Press, 2010); and CBD Decision 10/4, Third edition of the Global Biodiversity Outlook: implications for the future implementation of the Convention, UN Doc. UNEP/CBD/COP/10/27, 20 January 2011, para. 5; CBD Decision 10/2, The Strategic Plan for Biodiversity 2011–2020 and the Aichi Biodiversity Targets, UN Doc. UNEP/CBD/COP/10/27, 20 January 2011, paras. 7 and 17(e).

[73] CBD Decision 10/29, Marine and coastal biodiversity, UNEP/CBD/COP/10/27, 20 January 2011, para. 13(h) and Annex, para. d.

[74] Morgera and Tsioumani, "The Evolution of Benefit-sharing", supra, note 34, at 160.

[75] CBD Decision 5/6, Principles of the Ecosystem Approach, UN Doc. UNEP/CBD/COP/5/23, 22 June 2000, Annex B, Operational Guidance 2, para. 9.

[76] CBD Decision 7/11, supra, note 63, Annex I, annotations to rationale to Principle 4.

effective management of indigenous and local community conserved areas, using the social and economic benefits generated by protected areas for poverty reduction, and stresses the need for engaging indigenous and local communities and relevant stakeholders in participatory planning and governance.[77] Similarly, the CBD work programme on forest biodiversity explicitly refers to the fair and equitable sharing of the benefits from forest-related traditional knowledge,[78] emphasizing its link with community-based forest management[79] and the need to address socio-economic failures and distortions that lead to decisions that result in loss of forest biodiversity. To this end, the work programme makes reference to the use of forest planning and management, stakeholder analysis and mechanisms for transferring costs and benefits, providing market and other incentives for the use of sustainable practices, develop alternative sustainable income-generation programmes and facilitate self-sufficiency programmes of indigenous and local communities.[80]

The Addis Ababa Principles and Guidelines for the Sustainable Use of Biodiversity adopted under the CBD highlight that the involvement of local people facilitates compliance with legislation on the sustainable use of natural resources. They also underscore that management regimes are enhanced when training to identify income alternatives, or assistance in diversifying their management capacities is provided to communities.[81] Therefore, policies and regulations should ensure that indigenous communities and local stakeholders involved in the management of a resource for sustainable use receive an equitable share of any benefits derived, as well as additional benefits such as job opportunities for local people and support for co-management, or equal distribution of returns amongst locals and outside investors.

Overall, benefit-sharing in the context of the ecosystem approach implies that the State is expected to couple procedural guarantees for community participation in decision-making and management planning with substantive measures for the legal recognition of communities' sustainable practices, the provision of guidance and support to improve the environmental sustainability of community practices, and the proactive identification of opportunities for better/alternative livelihoods in these endeavours, with a view to facilitating understanding of, and compliance with, the law.

The underlying argument here is that, notwithstanding continued reluctance by some CBD Parties to use more explicit human rights language in CBD COP decisions,[82] the normative activity of the CBD COP has had far-reaching implications

---

[77] Ibid., Annex I, paras. 2.1.3–2.1.5.

[78] CBD Decision 6/22, Expanded Programme of Work on Forest Biological Diversity, UN Doc. UNEP/CBD/COP/6/20 27 May 2002, para. 13.

[79] Ibid., para. 19(h).

[80] CBD Decision 6/22, supra, note 78, Annex, activities (b) and (f) under Objective 1

[81] CBD, *Addis Ababa Principles and Guidelines for the Sustainable Use of Biodiversity* (Montreal: Secretariat of the Convention on Biological Diversity, 2004); CBD COP Decision 7/12, Sustainable Use (Article 10), UN Doc. UNEP/CBD/COP/7/21, 13 April 2004, Annex II, rationale to Principle 4.

[82] Morgera and Tsioumani, "Yesterday, Today and Tomorrow", supra, note 10, at 15–16 and 18–23.

for the protection of the rights of indigenous peoples and local communities in the context of the precautionary and ecosystem approaches, well in line with international human rights developments.[83] The following subsections will provide a coherent reading of the multiple sources of guidance by the CBD COP, designed to ensure environmentally holistic and human rights-based responses to climate change in a way that complements normative developments under the climate change regime.

14.3.1 Assessing and Reducing the Negative Impacts of Climate Change Response Measures on Biodiversity

The report of the CBD Expert Group on Climate Change in 2009 not only confirmed the reciprocal interactions between biodiversity loss and climate change, but also called attention more systematically to possible negative impacts of climate change response measures, depending on their design and implementation, on biodiversity.[84] Accordingly, the CBD COP has in a series of decisions spelt out guidance on carrying out appropriate assessments of response measures with a view to identifying environmentally holistic options and modalities for their design and implementation.

In more specific terms, the CBD COP has recommended undertaking environmental impact assessments and strategic assessments of renewable energy planning in mountain areas.[85] These assessments are to facilitate the consideration of all available options, with a view to avoiding the conversion or degradation of areas important for biodiversity. In so doing, CBD Parties are to consider traditional knowledge, including through the full involvement of indigenous peoples and local communities; they are also to consider the biodiversity components that are important for conservation and sustainable use; and they are to develop ecosystem- and species-vulnerability assessments.[86] Parties are also invited to consider the role of biodiversity and associated ecosystem services when enhancing the climate resilience of investments, projects, and programmes.[87] In addition, CBD Parties committed to assessing the impacts of climate change not only on biodiversity but also on the

---

[83] Notably, relevant human rights case law: Mauro Barelli, "The Interplay between Global and Regional Human Rights Systems in the Construction of the Indigenous Rights Regime", 32 *Human Rights Quarterly* (2010), 951, particularly at 971–972 and 975–978; and John Knox, "Climate Change and Human Rights Law", 50 *Virginia Journal of International Law* (2009–2010), 163, at 189–190.

[84] CBD Secretariat, *Connecting Biodiversity and Climate Change Mitigation and Adaptation: Report of the Second Ad Hoc Technical Expert Group on Biodiversity and Climate Change*, supra, note 71, 8–14.

[85] CBD Decision 10/30, Mountain biological diversity, UN Doc. UNEP/CBD/COP/10/27, 20 January 2011, para. 5.

[86] Ibid., para. 8(u)-(v).

[87] Ibid., para. 17.

biodiversity-based livelihoods of indigenous and local communities, with a view to identifying adaptation priorities. Particular attention is directed, in this respect, to livelihoods within ecosystems that have been identified as being particularly vulnerable to the negative impacts of climate change.[88]

Along similar lines, the CMS COP has called for the application of strategic environmental assessments to identify the appropriate construction sites of wind turbines, to avoid negative impacts on migratory species.[89] In turn, the Ramsar Convention urged using environmental impact assessments and strategic assessments before undertaking biofuel production and where avoidance of negative impacts is not feasible, to apply compensation and offsets including through wetland restoration.[90]

Earlier, more general guidance from the CBD COP provides further clarification on necessary procedural steps for a biodiversity-inclusive[91] and socio-cultural assessments that have great importance from an adaptation and mitigation perspective. These procedural[92] steps serve to assess the costs and benefits of conserving, maintaining, using and restoring ecosystems, take into account the interests of all relevant stakeholders and equitably share the benefits,[93] particularly when communities' traditional lands or protected areas are at stake.[94]

The most relevant tool developed by the CBD COP in that regard is the Akwé: Kon Guidelines for the conduct of cultural, environmental and social impact assessment on sacred sites and on lands and waters traditionally occupied or used by indigenous and local communities.[95] These guidelines illustrate how impact assessments can be used for identifying and weighting expected cultural, social and environmental costs and impacts of proposed climate change response measures that are proposed to take place on sacred sites and on lands and waters traditionally occupied or used by indigenous and local communities. In these circumstances, tangible benefits

---

[88] CBD Decision 10/33 Biodiversity and climate change, UN Doc. UNEP/CBD/COP/10/27, 20 January 2011, para. 8(b).

[89] CMS Resolution 7.5, Wind turbines and migratory species, Proceedings of the Seventh Meeting of the Conference of the Parties, March 2002.

[90] Ramsar Resolution X.25, Wetlands and biofuels, COP10 Conference Report, 2008, para. 15.

[91] CBD Decision 6/7, Identification, monitoring, indicators and assessments, UN Doc. UNEP/CBD/COP/6/20, 27 May 2002, Annex, Guidelines for incorporating biodiversity-related issues into environmental impact assessment legislation and/or process and in strategic environmental assessment.

[92] Svitlana Kravchenko, "Procedural Rights as a Crucial Tool to Combat Climate Change", 38 *Georgia Journal of International and Comparative Law* (2009–2010), 613.

[93] CBD Decision 7/11, supra, note 63, para. 12(5).

[94] CBD Decision 7/28, Programme of Work on Protected Areas, UN Doc. UNEP/CBD/COP/7/21, 13 April 2004, Annex, para. 2(1)(1).

[95] CBD Guidelines, Akwé: Kon Voluntary Guidelines for the Conduct of Cultural, Environmental and Social Impact Assessment regarding Developments Proposed to Take Place on, or which are Likely to Impact on, Sacred Sites and on Lands and Waters Traditionally Occupied or Used by Indigenous and Local Communities (Montreal: Secretariat of the Convention on Biological Diversity, 2004) in Article 8(j) and related provisions (CBD COP 7 Decision VII/16 F, Article 8(j), UN Doc. UNEP/CBD/COP/7/21, 13 April 2004), para. 56.

should accrue to such communities, such as payment for environmental services, job creation within safe and hazard-free working environments, viable revenue from the levying of appropriate fees, access to markets, and diversification of income-generating (economic) opportunities for small and medium-sized businesses.'[96] These more rounded assessments aim to achieve a multiplicity of goals, namely to support the full and effective participation and involvement of indigenous and local communities in all planning phases and properly take into account their cultural, environmental and social concerns and interests. In addition, these assessments are needed to take into account the knowledge, innovations and practices of these communities, with due regard to the ownership of and the need for the protection and safeguarding of traditional knowledge. Furthermore, they can contribute to promoting the use of appropriate technologies; identify and implement appropriate measures to prevent or mitigate any negative impacts of proposed developments; and take into consideration the interrelationships among cultural, environmental and social elements.[97] To these ends, the assessment needs to evaluate the likely impacts of a proposed development on the way of life of a particular group or community of people, their economic, social, cultural, civic and political rights, as well as their well-being, vitality and viability.[98] Assessments also need to provide a process whereby local and indigenous communities may have the option to accept or oppose a proposed development that may impact on their community; the conclusion of agreements on mutually agreed terms, between the proponent of the proposed development and the affected communities for the implementation of measures to prevent or mitigate any negative impacts of the proposed development; and of a review and appeals process.[99] Ultimately, against this framework, prior assessments of response measures having potential effects on lands and resources traditionally occupied by indigenous and local communities need to support the right of these communities to prior informed consent,[100] by taking into account their customary laws and procedures, through the use of appropriate language and process, the allocation of sufficient time and the provision of accurate, factual and legally correct information to them.[101]

[96] Ibid., para. 46.

[97] Ibid., para. 3.

[98] Ibid., para. 6.

[99] Ibid., para. 8.

[100] The understanding of "prior informed consent" proposed by the UN Special Rapporteur on indigenous peoples' rights is that prior informed consent does not provide indigenous people with a veto power when the State acts legitimately and faithfully in the public interest, but rather "establishes the need to frame consultation procedures in order to make every effort to build consensus on the part of all concerned" and that consensus-driven consultation processes should not only address measures to mitigate or compensate for adverse impacts of projects, but also explore and arrive at means of equitable benefit-sharing in a spirit of true partnership (Report of the Special Rapporteur on the situation of human rights and fundamental freedoms of indigenous peoples, UN Doc. A/HRC/12/34, 15 July 2009, paras. 48 and 53).

[101] CBD, Akwé: Kon Voluntary Guidelines, supra note 95, paras. 50 and 60.

Overall, undertaking cultural, social and environmental impact assessments with the full engagement of the relevant communities is an indispensable procedural step to ensure intra-generational equity in mitigation and adaptation.[102] Benefit-sharing in the context of these assessments provides incentives and rewards when community practices and knowledge contribute to biodiversity conservation and the fight against climate change. Benefit-sharing also promotes specific measures, such as payments for ecosystem services, diversification of income-generating opportunities, and other mitigation measures, to constructively address situations, and possibly prevent conflicts, when the interests of biodiversity protection and climate change response measures are in an irreconcilable conflict with the legitimate interests of communities, and the former need to prevail.[103]

14.3.2 Ecosystem Approach to Climate Change Mitigation

Systematic proofing of climate change mitigation policies for their impact on biodiversity and ecosystem services is considered essential to ensure that climate change itself is more effectively addressed; biodiversity conservation and, where necessary, restoration of ecosystems can be cost-effective interventions for mitigation purposes, with substantial co-benefits.[104] While the CBD Parties have just started consideration of international guidance on ecosystem restoration, it appears that this will be considered as the last-resort option, and not a substitute for conservation or sustainable use.[105] Conversely, the CBD COP has provided ample guidance on the conservation and sustainable use of ecosystems for mitigation purposes in relation to protected areas, inland waters, forests and biofuels. This guidance not only provides specific, technical adjustments to mitigation action to contribute – or at least avoid undermining – biodiversity conservation, but also includes guarantees for indigenous and local communities.

CBD Parties committed to identifying protected areas that are important for mitigation purposes, through carbon sequestration and maintenance of carbon stocks and to undertaking joint planning of protected-area networks and of mitigation measures, while recognizing that biodiversity conservation remains the primary objective of these areas.[106] The COP also invited Parties to evaluate and recognize

---

[102] Daniel Magraw and Lisa Hawke, "Sustainable Development", in Daniel Bodansky, Jutta Brunnée and Ellen Hey (eds.), *Oxford Handbook of International Environmental Law* (Oxford: Oxford University Press, 2007), 630.

[103] Morgera and Tsioumani, "The Evolution of Benefit-sharing", supra, note 34, at 165.

[104] CBD and UNEP-WCMC, *Global Biodiversity Outlook,* supra note 1, at 83.

[105] CBD Subsidiary Body for Scientific, Technical and Technological Advice, Recommendation 15/2, Ways and means to support ecosystem restoration, UN Doc. UNEP/CBD/COP/11/2, 7 December 2011.

[106] CBD Decision 10/31, Protected Areas, UN Doc. UNEP/CBD/COP/10/27, 20 January 2011, paras. 14(d) and (f), and 19(c).

the value and the benefits of comprehensive, effectively managed, and ecologically representative protected-area systems in climate change mitigation efforts.[107] Along similar lines, the CMS COP urged Parties to select sites for mitigation projects on the basis of environmental sensitivity and zoning maps signaling critical sites for migratory species.[108] In addition, CBD Parties undertook to ensure that any resettlement of indigenous communities as a consequence of the establishment or management of protected areas, including for mitigation purposes, will only take place with their prior informed consent that may be given according to national legislation and applicable international obligations.[109]

With regards to freshwaters, CBD Parties committed to ensuring that their climate change mitigation activities are designed and implemented while taking into account the needs and opportunities to sustain or enhance the services provided by inland water ecosystems and thereby contribute to the improvement of human well-being, as well as the mitigation capacities of wetlands[110] in the light of the interdependence of the carbon and water cycles.[111] In doing so, they are required to ensure opportunities for the active participation of indigenous and local communities in all stages of rapid assessments of biodiversity of inland waters traditionally occupied or used by these communities, consistent with the Akwé: Kon Voluntary Guidelines. This is coupled with the provision of support to these communities in re-establishing, developing and implementing traditional approaches and/or adaptive management approaches to conserve and sustain the use of the biodiversity of inland water ecosystems. CBD Parties are also to draw upon scientific, technical and technological knowledge of these communities, with their prior informed consent, in the implementation phase and promote the fair and equitable sharing of benefits gained from the use of inland water genetic resources and associated traditional knowledge.[112]

In the context of forest-based mitigation activities, CBD Parties undertook to promote forest biodiversity conservation and restoration in climate change mitigation measures and assess how the conservation and sustainable use of forest biodiversity can contribute to the international fight against climate change.[113] The COP specifically called upon the Parties to prioritize the use of native communities of

[107] Ibid., para. 14(a)-(c).

[108] CMS Resolution 11.19, Migratory species conservation in the light of climate change 2011, paras. 9–13, available at http://www.cms.int/bodies/COP/cop10/resolutions_adopted/resolutions.htm (meeting report unavailable at the time of writing).

[109] CBD Work Programme on Protected Areas, supra, note 94, para. 2.2.5.

[110] CBD Decision 10/28, Inland waters biodiversity, UN Doc. UNEP/CBD/COP/10/27, 20 January 2011, paras. 26(a)-(b) and 27.

[111] Ibid., para. 29.

[112] CBD Decision, 7/4, Biological diversity of inland water ecosystems, UN Doc. UNEP/CBD/COP/7/21, 13 April 2004, para. 24 and Annex, Revised programme of work on inland water biodiversity, para. 9.

[113] CBD Decision 6/22, supra, note 78, Objective 3.

tree species and limit the degradation and clearing of primary and secondary forests.[114] Parties were also encouraged, when designing, implementing, and monitoring afforestation, reforestation, and forest-restoration activities, to consider converting only low-biodiversity value or degraded lands, avoiding invasive alien species, and strategically locating afforestation activities within the landscape to enhance connectivity and increase the provision of ecosystem services within forest areas.[115] In that context, CBD Parties are generally expected to support the development of community-based approaches[116] and share benefits with indigenous and local communities.[117] In the context of these technical measures, CBD Parties called for the development of mechanisms to ensure that monetary and non-monetary costs and benefits of forest biodiversity management are equitably shared between stakeholders at all levels thorough, inter alia, the use of forest planning and management, the development of alternative sustainable income-generation programmes and the support of self-sufficiency programmes of indigenous and local communities.[118]

Along similar lines, Parties to the Ramsar Convention recommended that mitigation responses, including revegetation, forest management, afforestation and reforestation do not lead to serious damage to the ecological character of wetlands.[119] They also urged reducing the degradation and improving the management practices of peatlands for mitigation purposes.[120] CMS Parties, in turn, committed to conduct post-construction monitoring of energy and other mitigation projects as a standard requirement and ensure that such monitoring continues for the duration of plant operations. In addition, CMS Parties committed to ensure that energy and mitigation structures are operated in ways that minimize the mortality of migratory species, such as short-term shutdowns or higher turbine cut-in speeds with regards to wind farms for instance.[121]

CBD Parties then placed particular attention on sustainable biofuel production, recognizing the need to promote its positive impacts while minimizing the negative impacts on biodiversity and on the livelihoods of local and indigenous communities. To this end, the CBD COP called for the full and effective participation of these

[114] CBD Decision 10/36, Forest biodiversity, UN Doc. UNEP/CBD/COP/10/27, 20 January 2011, para. 8(o).

[115] Ibid., para. 8(p).

[116] Ibid., para. 34.

[117] CBD Decision 6/22, supra, note 78, Annex, activities (b) and (f) under Objective 1; see also UN Forum on Forests, Resolution on Forests for People, Livelihoods and Poverty Eradication, UN Doc. E/CN.18/2011/20, 2011.

[118] CBD Decision 6/22, supra, note 78, Annex, activities (b) and (f) under Objective 1.

[119] Ramsar Resolution VIII.3 Climate change and wetlands: impacts, adaptation and mitigation, COP8 Conference Report, 2002.

[120] Ramsar Resolution X.24: Climate change and wetlands, COP10 Conference Report, 2008, para. 32.

[121] CMS Resolution 11.19, supra, note 108, paras. 9–13.

communities in the implementation of activities relevant to the sustainable production and use of biofuels, and identified a series of international standards developed by the CBD in the context of precautionary and ecosystem-based approaches that governments should take into account.[122] In addition, the CBD COP called on Parties to assess and address direct and indirect land-use and water-use changes affecting areas of high value for biodiversity and areas of cultural, religious, and heritage interest and indigenous and local communities[123]; and put in place policies, supportive measures, environmentally sound technologies, and impact assessments to minimize negative impacts on broadly defined "biodiversity-related socio-economic conditions." These are understood by CBD Parties not only as concerns related to food and energy security, but also "the consideration of land tenure and resource rights, including water, where relevant for the CBD implementation, and in particular the implications for indigenous and local communities."[124] The COP further urged Parties to ensure that the sustainable agricultural practices of indigenous and local communities are respected, subject to national legislation, taking into account communities' customary laws where applicable.[125] In addition, CBD Parties urged governments to apply the precautionary approach to the release of synthetic life, cells, or genomes into the environment, acknowledging the parties' entitlement, in accordance with domestic legislation, to prevent such release.[126] Also the Ramsar Convention COP urged formulating appropriate land use policies for biofuels sustainable production, promote sustainable forest and agricultural practices that mitigate any adverse effects of biofuel production and consider the full range and value of ecosystem services and livelihoods provided by wetlands.[127]

Overall, all climate change mitigation measures relying on the use of biodiversity should ensure that such use is undertaken in a manner in which ecological processes, species and genetic variability remains above thresholds needed for long-term

[122] CBD Decision 9/2, Agricultural biodiversity: biofuels and biodiversity, UN Doc. UNEP/CBD/COP/9/29, 9 October 2008, paras. 1–3. Relevant guidelines were listed in the decision, namely: the Addis Ababa Principles and Guidelines on Sustainable Use, supra, note 81; the work programme on protected areas, supra, note 94; CBD Decision 5/16, the work programme on traditional knowledge, Article 8(j) and related provisions, UN Doc. UNEP/CBD/COP/5/23, 22 June 2000; the Akwé: Kon Voluntary Guidelines, supra, note 95; CBD Decision 6/9, The Global Strategy for Plant Conservation, UN Doc. UNEP/CBD/COP/6/20, 27 May 2002; the guiding principles on alien invasive species (CBD Decision 6/23, Alien species that threaten ecosystems, habitats or species, UN Doc. UNEP/CBD/COP/6/20, 27 May 2002); the application of sustainable forest management and best agricultural practices in relation to biodiversity; national biodiversity strategies and action plans; and relevant guidance developed under the Cartagena Protocol on Biosafety (Cartagena, 29 January 2000, in force 11 September 2003, 2226 *United Nations Treaty Series* 208 (2005)).

[123] CBD Decision 9/2, supra, note 122, paras. 6 and 8–10.

[124] CBD Decision 10/37, Biofuels and biodiversity, UN Doc. UNEP/CBD/COP/10/27, 20 January 2011, para. 2.

[125] Ibid., para. 4.

[126] Ibid., para. 16.

[127] Ramsar Resolution X.25, Wetlands and biofuels, COP10 report, 2008, paras. 16–19.

viability.[128] To that end, national legal frameworks should allow for timely and effective responses to unsustainable use and consideration of the customary law of indigenous and local communities, empowering communities through the recognition of their customary rights and effective opportunities for participating in relevant decision-making.[129] In addition, mitigation measures relying on biodiversity should avoid economic mechanisms and incentives having a negative impact on the sustainable use of biodiversity, and incorporate benefit-sharing systems targetting local and indigenous communities in order to support successful implementation.[130]

14.3.3 Ecosystem Approach to Climate Change Adaptation

Adaptive management is also key in the context of climate change adaptation.[131] According to the Addis Ababa Principles and Guidelines on the Sustainable Use of Biodiversity,[132] adaptive management should be based not only on science but also on local and traditional knowledge, which has led to sustainable use of biodiversity over long time-periods without detriment to the environment and is critical also for modern use systems.[133] Along these lines, the CBD COP elaborated more specific guidance on an ecosystem approach to climate change adaptation, focusing on protected areas, mountain, forests, inland waters and marine ecosystems, and *ex situ* conservation. Once again, technical guidance is coupled with procedural and substantive guarantees for indigenous and local communities.

First, CBD Parties committed to integrating climate change adaptation measures in protected areas planning, management strategies and in the design of protected area systems.[134] The COP further invited Parties to consider climate change adaptation in assessing the management effectiveness of protected areas, and in integrating protected areas into wider landscapes, seascapes and sectors, including through the use of connectivity measures and the restoration of degraded habitats and landscapes.[135] The CBD COP then underscored the need to enhance scientific knowledge, as well as traditional knowledge, to support the development of adaptive-management plans for

---

[128] In line with the definition of "sustainable use" at CBD, supra, note 4, Art. 2.

[129] CBD, Addis Ababa Guidelines for the Sustainable Use of Biodiversity, supra, note 81, paras. 8(a), Practical Principles 1–2 and 7.

[130] Ibid., Principles 3–4.

[131] Useful distinctions are drawn between hard and soft adaptation, as well as between first-generation and second-generation adaptation by Cameron, "Human Rights and Climate Change", supra, note 6, at 690.

[132] CBD, Addis Ababa Principles and Guidelines for the Sustainable use of Biodiversity, supra, note 81, Principle 4(a).

[133] Ibid., operational guidelines to Principle 4.

[134] CBD Work Programme on Protected Areas, supra, note 94, para. 1.4.5.

[135] CBD Decision 10/31, supra, note 106, para. 14(a).

protected areas, and evaluate and recognize the value and the benefits of comprehensive, effectively managed, and ecologically representative protected area systems in climate change adaptation.[136] CBD Parties are further called upon to recognize the role of areas conserved by indigenous peoples and local communities in strengthening ecosystem connectivity and resilience, with a view to supporting ecosystem services and biodiversity-based livelihoods in the face of climate change.[137]

Along similar lines, the World Heritage Convention Strategy to Assist State Parties to Implement Appropriate Management Responses[138] calls for the development of effective monitoring systems, the application of adaptive management responses, and the reduction of non-climatic stress factors on protected sites, including by integrating climate adaptation in site management plans and developing regional or transboundary plans to reduce the vulnerability of sites in larger landscape and seascape contexts.[139] The Strategy focuses on capacity building and financial assistance, improved knowledge sharing and inclusion of local communities and protected site users in climate change response measures.[140] The CMS COP, in turn, urged Parties to ensure that critical sites are sufficiently large to hold a variety of habitats, to strengthen physical and ecological connectivity between sites, and aiding species dispersal and colonization when distribution shifts. Following an assessment of the extent to which existing protected area systems address the needs of migratory species in terms of resilience to climate change, CMS Parties are further to consider the designation of seasonal protected areas where migratory species occur at critical stages of their lifecycle and would benefit from extra protection.[141]

The fragility of mountain ecosystems and species and their vulnerability to global climate change[142] has led the CBD COP to recommend preventing or mitigating the negative impacts of infrastructure projects and other human-induced disturbances on mountain biodiversity at all levels, paying particular attention to cumulative impacts, with a particular view to reducing the negative impacts of global climate change on mountain biodiversity.[143] The COP thus encouraged climate change adaptation by

---

[136] Ibid., para. 14(b)–(c).

[137] Ibid., para. 8(i)–(j).

[138] World Heritage Committee Decision 30 Com.7.1, Examination of the State of Conservation of World Heritage properties, UN Doc. WHC.06 /30. COM /19, 23 August 2006.

[139] Burns, "'Belt and Suspenders'", supra, note 53, at 157; and Huggins, "Protecting World Heritage Sites from Adverse Impacts of Climate Change", supra, note 53, at 126–127.

[140] See comments by Huggins, "Protecting World Heritage Sites from Adverse Impacts of Climate Change", supra, note 53, at 129.

[141] CMS Resolution 10.19, Migratory species conservation in the light of climate change, 2011, para. 8 and Resolution 10.3 (2011), The role of ecological networks in the conservation of migratory species, para. 9(i), both available at http://www.cms.int/bodies/COP/cop10/resolutions_adopted/resolutions.htm (meeting report unavailable at the time of writing).

[142] CBD Decision 7/27, Work Programme on Mountain Biodiversity, UN Doc. UNEP/CBD/COP/7/21, 13 April 2004, Annex, para. 8(c).

[143] CBD Decision 10/30, supra, note 85, para. 1.1.1–3 and 5.

conserving *in situ* and *ex situ* genetic resources and species currently and potentially under threat from climate change, reducing deforestation, restoring degraded mountain-forest ecosystems, favoring sustainable agricultural practices and conserving carbon in mountain soil.[144] In all these instances, CBD Parties are expected to promote the indigenous and local communities' techniques and technologies and community-based management systems, as well as support the use of mountain-related traditional knowledge, in particular concerning sustainable management of biodiversity, soil, water resources and slopes.[145]

Furthermore, CBD Parties committed to promoting the monitoring of climate change impacts on forest biodiversity and investigate the interface between forest components and the atmosphere; promote the maintenance and restoration of biodiversity in forests in order to enhance their capacity to resist to, recover from and adapt to climate change; and promote forest biodiversity conservation and restoration in climate change adaptation measures.[146]

Adaptation and the conservation of inland waters biodiversity have also been explored in detail. CBD Parties committed to encouraging the adoption of integrated river basin management strategies to maintain, restore or improve the quality and supply of inland water resources and the multiple functions and values of inland water ecosystems, including appropriate responses to combat, and prevent where possible, the negative impacts of climate change.[147] In addition, CBD Parties are to encourage the use of low-cost technology, non-structural and innovative approaches, and, through prior informed consent, traditional practices for inland water biodiversity assessment.[148] Parties to the Ramsar Convention also undertook to manage wetlands so as to increase their resilience to climate change and extreme climatic events,[149] promote the restoration of rivers, lakes, aquifer basins and wetlands, protect mountain wetlands and respect water allocations for wetland ecosystems.[150] According to the Ramsar Guidelines for establishing and strengthening local communities' and indigenous peoples' participation in the management of wetlands, these communities are to be ensured access to natural resources within the wetlands that are essential for their livelihoods, security and cultural heritage, coupling communities' long-term involvement through benefit-sharing and the maintenance of sustainable livelihoods.[151]

---

[144] Ibid., para. 5.

[145] CBD Decision 7/27, supra, note 142, Annex, paras. 1.3.7 and 1.3.2–1.3.4.

[146] CBD Decision 6/22, supra, note 78, Objective 3.

[147] CBD Decision 7/4, supra, note 112, objectives (b)-(c).

[148] Ibid., para. 2.2.2.

[149] Climate change and wetlands: impacts, adaptation and mitigation, Ramsar Resolution VIII.3, COP8 report, 2002, para. 14.

[150] Climate change and wetlands, Ramsar Resolution X.24, COP10 report, 2008, paras. 28–31.

[151] Guidelines for establishing and strengthening local communities' and indigenous peoples' participation in the management of wetlands, Ramsar Resolution VII.8, COP7 report, 1999.

In addition, the CBD COP encouraged Parties to maintain or restore the connectivity of inland water ecosystems with terrestrial and marine ecosystems for climate change adaptation purposes.[152] With specific regard to marine biodiversity, CBD Parties undertook to increase the resilience of coastal and marine ecosystems, particularly coral reefs and estuaries, and habitats such as tidal salt marshes, mangroves, and sea grasses, by *inter alia* establishing marine protected areas.[153] The CBD COP further called on Parties to incorporate emerging knowledge on ocean acidification into relevant (biodiversity, coastal management, and marine protected area) planning; and to incorporate climate change adaptation into development and disaster-reduction planning, particularly in coastal areas.[154]

Ex situ conservation measures have also been discussed with a view to contributing to climate change adaptation. CBD Parties are thus expected to take a precautionary approach when considering ex situ adaptation measures, such as species relocation, assisted migration and captive breeding, to avoid unintended ecological consequences, such as the spread of invasive alien species.[155] Parties are further encouraged to develop strategies for biodiversity conservation and sustainable use in areas that are becoming accessible to new uses as a consequence of climate change; to take specific measures for species that are particularly vulnerable to climate change, including migratory species; and to maintain genetic diversity in the face of climate change.[156] These measures are particularly significant for the protection of animal migratory species. Accordingly, the CMS COP specified that Parties should employ adaptive management and the ecosystem approach to address climate impacts and monitor the effectiveness of their migratory species conservation, develop a standardized methodology for evaluating the susceptibility of migratory species to climate change and prepare species-specific action plans for species considered most vulnerable to climate change. Parties are also to consider *ex situ* measures and assisted colonization as appropriate for migratory species most severely threatened by climate change; and implement monitoring regimes on the interaction between climate change and migratory species, including on impacts on local communities dependent on ecosystem services provided by these species with a view to sharing monitoring results regularly with range States.[157]

Overall, in providing indications on an ecosystem approach to mitigation and adaptation, the CBD COP pointed to the use of environmental and social impact assessments, the integration of traditional knowledge and community concerns in management plans, the legal recognition and active support of community-based

---

[152] CBD Decision 10/28, Inland waters biodiversity, UN Doc. UNEP/CBD/COP/10/27, 20 January 2011, paras. 10(l) and 26(c).

[153] CBD Decision 10/29, Marine and coastal biodiversity, UN Doc. UNEP/CBD/COP/10/27, 20 January 2011, para 8.

[154] Ibid., paras. 67 and 77.

[155] CBD Decision 10/33, supra, note 88, para. 8(e).

[156] Ibid., para. 8(f)–(g).

[157] CMS Resolution 10.19, supra note 141, paras. 4–7.

management arrangements, the setting-up of benefit-sharing mechanisms when revenue generated through conservation and sustainable use activities is accrued by the State or outside investors, the provision of livelihood-based mitigation and compensatory measures, the use of other incentives such as payments for ecosystem services, as well as the re-investment of benefits in the protection of traditional knowledge and traditional sustainable practices.[158] These tools can protect several human rights that may be negatively impacted by climate change: the right to life, adequate food, health, adequate housing, self-determination, access to safe drinking water and sanitation, and access to means of subsistence.[159]

14.4 The Contribution of International Biodiversity Law to a Human Rights-Based Approach to Tackling Climate Change

The burgeoning academic debate on human rights and climate change has shed much light on the need, benefits and conceptual challenges of developing a human rights-based approach to climate change. Accordingly, such an approach entails a conceptual framework for climate change policies focusing on the inclusion of marginalized populations; encourages accountability, participation and transparency in decision-making; and provides suitable outcomes by building the capacity of key stakeholders.[160] It thus emphasizes equity vis-a-vis right-holders, with the implication that States have to create 'specific channels' for the poor and marginalized on the basis of non-discrimination and substantive equality.[161] A human rights-based approach could also contribute to a determination of socially and culturally appropriate and 'acceptable levels of risks' in light of precaution in the climate change regime.[162] UNFCCC Parties that are also Parties to human rights treaties must, at a minimum, refer to them as benchmarks to address the climate change problem as a human rights concern and take procedural steps to integrate the relevant standards into policy-making with a view to identifying human rights that may be placed at risk by the impacts of climate change and taking protective action in that regard when devising mitigation and adaptation responses.[163]

---

[158] Morgera and Tsioumani, "The Evolution of Benefit-sharing", supra, note 34, at 167.

[159] Human Rights Council, Resolution 10/4, supra, note 9, at 1.

[160] Von Doussa, Corkery and Chartres, "Human Rights and Climate Change", supra, note 16, at 171.

[161] Ibid., at 174.

[162] Rajamani, "The Increasing Currency and Relevance of Rights-based Perspectives", supra, note 9, at 424.

[163] Ibid., at 412. Stephen Humphreys, *Climate Change and Human Rights: A Rough Guide* (Geneva: International Council on Human Rights, 2008).

What has not been explored yet in this debate, however, is that the normative developments under the CBD COP represent near-universal intergovernmental consensus on timely, comprehensive and sophisticated guidance that already adapts human rights considerations to the technicalities of the precautionary and ecosystem approaches, with inputs from indigenous and local community representatives. On the basis of the preceding analysis, in fact, a convincing argument can be put forward that the gaps related to a human rights-based approach in the context of the international climate change regime can be filled by the procedural and substantive steps that the CBD COP has spelt out to ensure the protection of the rights[164] and livelihoods[165] of local communities and indigenous peoples that are disproportionality affected by climate change.[166]

UNFCCC COP decisions could thus refer to relevant CBD guidance, thereby finding a way for human rights to be incorporated into the international climate regime at different levels, without the need to create new standards.[167] However, doing so would notably require buy-in, or at least acquiescence, by the United States as the only country that is a Party to the UNFCCC but not to the CBD. Beyond a strictly legal perspective, however, buy-in is also required from certain CBD Parties that fear that cross-referencing CBD guidelines in the context of the international climate change negotiations may influence the negotiating dynamics and bargaining power in the UNFCCC.[168] Even in the absence of cross-references between CBD and UNFCCC COP decisions, at the national level CBD Parties are required to comply with both sets of international obligations and guidance from both bodies. Nonetheless, the need for cross-reference to CBD guidance at the international level remains relevant in light of inherent limitations in ensuring normative coherence only at the national level.[169]

By focusing on local and indigenous communities, the CBD clearly "gives a human face" to these issues[170] and offers a bottom-up approach to building a true

[164] This is quite significant, given the silence of the Convention on human rights, as remarked by Dinah Shelton, "Fair Play, Fair Pay: Preserving Traditional Knowledge and Biological Resources", 5 *Yearbook of International Environmental Law* (1994), 76, at 80.

[165] The focus on livelihoods is also considered necessary in the context of the duty of cooperation in the climate change regime: Rajamani, "The Increasing Currency and Relevance of Rights-based Perspectives", supra, note 9, at 425.

[166] Von Doussa, Corkery and Chartres, "Human Rights and Climate Change", supra, note 16, at 167–168.

[167] As suggested by Roht-Arriaza, "'First, Do No Harm': Human Rights and Efforts to Combat Climate Change", supra, note 11, at 609–610.

[168] This political resistance emerges clearly in CBD negotiations of climate-related decisions. See, Asheline Appleton et al., "Analysis of SBSTTA 14", *The Earth Negotiations Bulletin* 9(514) (2010).

[169] Savaresi's contribution to this volume.

[170] The importance of this practical value of a human rights-based approach to climate change is stressed by von Doussa, Corkery and Chartres, "Human Rights and Climate Change", supra, note 16, at 171.

partnership with communities in preventing biodiversity loss and fighting climate change by proactively combining economic and non-economic benefits. Reliance on the relevant normative activity under the CBD not only allows to provide "much needed attention to individual welfare" in the context of the climate change regime,[171] but also a "community" dimension in the human rights-based approach to climate change mitigation and adaptation that may be otherwise easily under-emphasized.[172] Furthermore, the abundant normative activity under the CBD offers a pragmatic approach to ensure good governance and adaptive management for the conservation and sustainable use of biodiversity: ensuring benefit-sharing from the rational use of natural resources to resource-dependent communities may serve as an incentive for communities that in all events utilise and exercise control over resources.[173] This ultimately facilitates communities' compliance with applicable biodiversity and climate laws.

Although international human rights do not contain provisions on development aid,[174] the principle of common but differentiated responsibility underpinning the climate change[175] and biodiversity regimes[176] does. Thus, a human rights-based approach to addressing climate change could also imply a human rights-based approach to development cooperation,[177] as a facet of the application of the principle of common but differentiated responsibilities under the climate change and biodiversity regimes. This would entail informing appropriate levels of financing and appropriate choices of measures with poverty reduction concerns and bottom-up community empowerment in the development of climate policies in a locally grounded and culturally appropriate way.[178] Through this lens, the CBD can make an important contribution to the application of a human rights-based approach to climate responses not only between States and within States –that is, between governments and local and indigenous communities– but also between States and those subject to another State's

---

[171] Rajamani, "The Increasing Currency and Relevance of Rights-based Perspectives", supra, note 9, at 429.

[172] On a community-focused rather than individualistic human rights-based approach, see Francesco Francioni, "International Human Rights in an Environmental Horizon", 21 *European Journal of International Law* (2010), 41.

[173] See Gregory Maggio, "Recognizing the Vital Role of Local Communities in International Legal Instruments for Conserving Biodiversity", 16 *University of California Los Angeles Journal of Environmental Law and Policy* (1997–1998), 179, at 180 and 185.

[174] Possible reliance on article 2(1) of the International Covenant on Economic, Social and Cultural Rights to that end is explored by Daniel Bodansky, "Climate Change and Human Rights: Unpacking the Issues", 38 *Georgia Journal of International and Comparative Law* (2010), 511; and Knox, "Climate Change and Human Rights Law", supra, note 83, at 202 and 206–218.

[175] UNFCCC, supra, note 17, Art. 4(3–4). For a discussion of how a human rights approach does not preclude differential treatment, see Rajamani, "The Increasing Currency and Relevance of Rights-based Perspectives", supra, note 9, at 420–421.

[176] CBD, supra, note 4, Art. 20.

[177] Cameron, "Human Rights and Climate Change", supra, note 6, at 712–714.

[178] Von Doussa, Corkery and Chartres, "Human Rights and Climate Change", supra, note 16, at 176.

jurisdiction.'[179] In the latter case, this would be a reflection of the global nature of international environmental law since the functional exercise of national sovereignty aimed at conserving biodiversity and fighting climate change as a common concern of mankind not only is at the service of developing countries in light of the concept of common but differentiated responsibility, but also at the service of the well-being of individuals and groups within developing countries.[180] From that perspective, international biodiversity law serves to highlight the interactions between international, national and community customary law, as well as the relevance of international standards for non-State actors, notably the private sector.[181] Both dimensions have important implications for an even more ambitious human rights-based approach to climate change mitigation and adaptation.

Linking different levels of governance according to standards and procedures set out in community customs, national and international law may be necessary for the effective realisation of the goals of the international biodiversity and climate change regimes. A tool attempting to bridge inter-State legal developments with communities' needs, aspirations and livelihoods that is rapidly gaining currency under the CBD,[182] is the bio-cultural community protocol.[183] Supporting a bottom-up approach, a bio-cultural community protocol is a written document developed by a community, following a consultative process, to outline the core ecological, cultural and spiritual values and customary laws relating to the community's traditional knowledge and resources, based on which the community provides clear terms and conditions to regulate access to their knowledge and resources. The process leading to the

---

[179] The question is posed, although not replied to, by Rajamani, "The Increasing Currency and Relevance of Rights-based Perspectives", supra, note 9, at 428–429.

[180] Elisa Morgera, "Bilateralism at the Service of Community Interests? Non-judicial Enforcement of Global Public Goods in the context of Global Environmental Law" 23 European Journal of International Law (forthcoming, 2012).

[181] Ibid.

[182] These instruments are included in the Nagoya Protocol on Access to Genetic Resources and the Fair and Equitable Sharing of Benefits arising from their Utilization to the Convention on Biological Diversity, UN Doc. UNEP/CBD/COP/DEC/X/1, 29 October 2010, Arts.12 and 21; and also in recent recommendations on *sui generis* systems of protection of traditional knowledge and of customary sustainable use: Article 8(j) Working Group Recommendation 7/5, Development of elements of *sui generis* systems for the protection of traditional knowledge, innovations and practices and Recommendation 7/6, Article 10, with a focus on Article 10(c), as a major component of the programme of work on Article 8(j) and related provisions, both in UN Doc. UNEP/CBD/COP/11/7, 24 November 2011.

[183] United Nations Environment Programme (UNEP) and Natural Justice, "Biocultural Community Protocols: A community approach to ensuring the integrity of environmental law and policy", 2009, available at: http://www.unep.org/communityprotocols/index.asp (last accessed on 10 April 2012); Harry Jonas, Kabir Bavikatte and Holly Shrumm, "Community Protocols and Access and Benefit-Sharing", 12 *Asian Biotechnology and Development Review* (2010), 49. See also UNEP website on community protocol case studies, available at: http://www.unep.org/communityprotocols/casestudies.asp (last accessed on 10 April 2012); and the website of a coalition of different actors on community protocols, available at: http://www.community-protocols.org/ (last accessed on 10 April 2012).

bio-cultural protocol development allows a community to prepare in advance for negotiations of an arrangement with a public or private entity planning activities impacting on community livelihoods or utilising its traditional resources or knowledge, contributing thus to a more level-playing field among the parties. Furthermore, the development of bio-cultural community protocols allows a community to identify any question related to the governance of future benefit-sharing, thus preventing internal conflicts. Compliance with the provisions of these protocols, however, remains voluntary, unless it is secured through national legislation.[184] Nonetheless, biocultural community protocols can prove essential for a public or private entity planning adaptation or mitigation activities likely to negatively impact on community livelihoods or utilise traditional resources or knowledge. These protocols can significantly support public and private efforts to adopt an ecosystem and human rights-based approach to mitigation and adaptation in light of international standards and with respect for community customary rules and procedures.[185]

In addition, a human rights-based approach for mitigation and adaptation also needs to take into account the role of the private sector, which is increasingly prominent under the international climate change regime[186] and under international human rights law.[187] Significantly, normative activity under the CBD not only supports an environmentally holistic and right-based approach in the interactions between States and within States, but also between private entities and local and indigenous communities.[188] Guidelines adopted under the CBD that inform the ecosystem approach to adaptation and mitigation, such as the Addis Ababa Principles and Guidelines[189] and

[184] This draws on Morgera and Tsioumani, "The Evolution of Benefit-sharing", supra, note 34, at 157–158.

[185] In fact, biocultural community protocols are been pioneered in the context of REDD: see UNEP and Natural Justice, *Biocultural Community Protocols,* supra, note 183, chapter 4.

[186] Von Doussa, Corkery and Chartres, "Human Rights and Climate Change", supra, note 16, at 170; Amy Sinden, "An Emerging Human Right to Security from Climate Change: The Case of Gas Flaring in Nigeria", in William Burns and Hari Osofsky (eds.), *Adjudicating Climate Change* (Cambridge: Cambridge University Press, 2009), 173, at 190–191; Averill, "Linking Climate Litigation and Human Rights", supra, note 14 at 141; Knox, "Climate Change and Human Rights Law", supra, note 83, at 195–198.

[187] Report of the Special Representative of the Secretary-General on the issue of Human Rights and Transnational Corporations and Other Business Enterprises: Protect, Respect and Remedy: A Framework for Business and Human Rights, UN Doc. A/HRC/8/5, 7 April 2008; and Guiding Principles on Business and Human Rights to implement the UN Protect, Respect and Remedy Framework, UN Doc. A/HRC/17/31, 21 March 2011. The Framework and the Guiding Principles were adopted by the Human Rights Council by Resolutions 8/7, in UN Doc. A/HRC/8/52, 1 September 2008 and 17/4 (2011), in UN Doc. A/HRC/17/L.30 (advanced, undated version).

[188] Morgera and Tsioumani, "The Evolution of Benefit-sharing", supra, note 34, at 165–167.

[189] CBD, Addis Ababa Principles and Guidelines on the Sustainable Use of Biodiversity, supra, note 81, para. 1 clarifies that: "The principles provide a framework for advising Governments, resource managers, indigenous and local communities, the private sector and other stakeholders about how they can ensure that their use of the components of biodiversity will not lead to the long-term decline of biological diversity."

the Akwé: Kon Guidelines,[190] were drafted so as to specifically address also non-State actors, especially the private sector. In addition, these and other normative developments under the CBD have been increasingly integrated into international standard-setting on corporate environmental accountability[191] and in normative developments in the context of business responsibility to respect international human rights law.[192] Relevant CBD standards are thus also readily applicable to private entities responsible for carrying out climate change mitigation and adaptation activities. They could be influential in ensuring that also the private sector's contribution to the fight against climate change follows an ecosystem and human rights-based approach.

14.5 Conclusions

This chapter has sought to draw attention to the abundance of climate change- and human rights-related normative developments under the CBD and its great potential to fill key gaps in the international climate change regime and in implementation at the national level. Not only has the CBD COP "actively sought to manage the interactions between the two regimes", revealing itself as "instrumental in highlighting biodiversity concerns in UNFCCC decisions,"[193] but it has also made significant conceptual progress on the politically charged question related to environmentally holistic and human rights-based approaches to climate change mitigation and adaptation.[194] As a result, the normative activity undertaken by the CBD COP can con-

[190] Although they are directed to Parties and governments, as indicated by Akwé, Kon Voluntary Guidelines, supra, note 95, para. 1, the Guidelines are expected to provide a collaborative framework for Governments, indigenous and local communities, decision makers and managers of developments (para. 3).

[191] I am here referring to standard-setting led by intergovernmental organizations, not private standard-setting, discussed, for instance, in Roht-Arriaza, "'First, Do No Harm'", supra, note 11, at 607–609. See in particular 2012 Performance Standards of the International Finance Corporation, available at: http://www.ifc.org/ifcext/policyreview.nsf/Content/2012-Edition#PerformanceStandards (last accessed on 10 April 2012); and, "Final Statement by the UK National Contact Point for the OECD Guidelines for Multinational Enterprises", 25 September 2009, available at: http://www.berr.gov.uk/files/file53117.doc (last accessed on 10 April 2012), paras. 44–46. See generally, Elisa Morgera, *Corporate Accountability in International Environmental Law* (Oxford: Oxford University Press, 2009).

[192] For instance, Report of the Special Rapporteur on the situation of human rights and fundamental freedoms of indigenous people, UN Doc. A/HRC/15/37, 19 July 2010, Section III. For a discussion, Elisa Morgera, "From Corporate Social Responsibility to Accountability Mechanisms", in Pierre-Marie Dupuy and Jorge Vinuales (eds.), *Harnessing Foreign Investment to Promote Environmental Protection: Incentives and Safeguards* (Cambridge: Cambridge University Press, forthcoming, 2012).

[193] Van Asselt, *Managing the Fragmentation of International Environmental Law,* supra, note 7, at 36.

[194] Morgera and Tsioumani "Yesterday, Today and Tomorrow", supra, note 10, at 33.

tribute to ensuring coherence between the international climate change regime and international human rights instruments, linking international, national and local levels of governance and reaching into the relations between private entities and indigenous and local communities. Notably, international biodiversity law can provide both procedural[195] and substantive elements of a human rights-based approach to climate change.

It remains to be seen whether these multi-level normative developments under the CBD will be allowed to filter into UNFCCC COP decisions and national-level implementation, although practice under the international climate change regime so far has been disappointing. Given the urgency of constructing an effective international climate change regime,[196] however, reliance on the CBD guidance may save UNFCCC Parties precious negotiating time. Cross-reference to the CBD decisions can also provide a "social justice and development" dimension to the international climate change regime, thus facilitating "intersecting inequalities that contribute to vulnerability and allow for an exploration of a variety of approaches that offer redress and capacity-building to marginalized populations."[197] In addition, the CBD normative activity provides highly refined and intergovernmentally approved "methodologies for engaging the participation of, and consultation with, key stakeholders in the formulation of climate change and development strategies."[198]

In conclusion, this chapter represents an invitation not only to climate change lawyers, but also to human rights experts interested in climate change to engage with the normative activity of the governing bodies of international biodiversity-related conventions. In particular such an engagement would be useful to ascertain whether existing guidance under the CBD and related conventions covers all relevant vulnerable groups.[199] It would also be interesting to start a dialogue on the possible value added of supporting a human rights-based approach through the CBD COP decisions. For instance, an argument can be made that the CBD guidelines go beyond human rights instruments because they do not require an 'identifiable violation,'[200] but can rather be triggered by a threat of a negative impact, thereby injecting human rights with a preventive (and even precautionary) approach. In addition, the CBD guidelines can more easily reach across international borders, on

[195] *Contra* see Kravchenko, "Procedural Rights as a Crucial Tool to Combat Climate Change", supra, note 92, at 648, who argued that "a human rights approach helps to find solutions to problems for which environmental law does not have a response."

[196] Cameron, "Human Rights and Climate Change", supra, note 6, at 701.

[197] As appears needed to Cameron, ibid., at 709.

[198] Ibid.

[199] For instance, gender has only been recently addressed by the CBD COP. See CBD Decision 9/24, Gender Plan of Action, UN Doc. UNEP/CBD/COP/9/29, 9 October 2008; and CBD Decision 10/19, Gender mainstreaming, UN Doc. UNEP/CBD/COP/10/27, 20 January 2011. On gender and climate change, see Cameron, "Human Rights and Climate Change", supra, note 6, at 687.

[200] Which was considered a major barrier in applying human rights law to climate change response measures: Cameron, ibid, at 705.

the basis of the common concern of humankind,[201] whereas there are significant limitations to the extraterritorial application of human rights instruments.[202] Finally, the CBD can count on a virtually universal membership, whereas different UNFCCC parties are subject to different human rights instruments with varying membership.[203]

Finally, human rights experts, climate lawyers and biodiversity lawyers could engage in a certainly enriching debate on enforcement and compliance. Without explicit and operational links between the international law on climate change, biodiversity and human rights, state compliance with these interconnected obligations cannot be monitored and enforced.[204] Even if these links are established, however, monitoring compliance under the CBD would be very limited. The CBD does not have a compliance committee and does not use Parties' self-reporting or other types of monitoring to identify shortcomings in individual States' compliance.[205] In turn, while international human rights instruments have international tribunals and rapporteurs to hear and investigate complaints,[206] not all impacts on human rights arising from climate change response measure may trigger them[207] and not all human rights enforcement mechanisms are necessarily effective.[208] So, another question that merits discussion is whether the compliance mechanism under the international climate change regime has the potential to contribute to the respect of international biodiversity and human rights law between States, within States and possibly even in relations between the private sector and communities.

[201] Jutta Brunnee, "Common Areas, Common Heritage and Common Concern", in D. Bondansky, J. Brunnee and E. Hey (eds), *The Oxford Handbook of International Environmental Law* (2007), 550; Patricia Birnie, Alan Boyle and Catherine Redgwell, *International Law and the Environment* (Oxford: Oxford University Press, 2009), at 128–131.

[202] Cameron, "Human Rights and Climate Change", supra, note 6, at 706.

[203] Savaresi, in her contribution to this volume, underlines the "fragmented nature of States' obligations in the human rights field." On the limited relevance of customary international law on human rights for climate change-related purposes, see Knox, "Climate Change and Human Rights Law", supra, note 83, at 15; and generally Savaresi's contribution to this volume, at 140.

[204] Roht-Arriaza, "'First, Do No Harm'", supra, note 11, at 611.

[205] Morgera and Tsioumani, "Yesterday, Today and Tomorrow", supra, note 10, at 24–25, note possible indications of a change of practice in that regard.

[206] Kravchenko, "Procedural Rights as a Crucial Tool to Combat Climate Change", supra, note 92, at 616.

[207] Bodansky, "Climate Change and Human Rights: Unpacking the Issues", supra, note 174, at 517 and 519.

[208] Cameron, "Human Rights and Climate Change", supra, note 6, at 706.

Chapter 15
The Role of REDD in the Harmonisation of Overlapping International Obligations

Annalisa Savaresi

Abstract Since 2007, Parties to the United Nations Framework Convention on Climate Change (UNFCCC) have been negotiating "policy approaches and positive incentives on issues relating to reducing emissions from deforestation and forest degradation in developing countries; and the role of conservation, sustainable management of forests and enhancement of forest carbon stocks in developing countries," commonly referred to with the acronym REDD+. Albeit the negotiations on REDD+ remain *in fieri* at the time of writing, this article discusses its potential to complement international conventions and agreements dealing with biodiversity protection and human rights. The choice of these two areas relates both to their links with the subject matter of REDD+, and to the fact that Parties to the UNFCCC and international bureaucracies dealing with these matters have already taken some steps to address potential overlaps. Far from being merely a theoretical question, therefore, the issues discussed in this article have attracted ample attention as negotiations progress. This article gives an account of this ongoing debate, providing a snapshot of its evolution, as well as some predictions on its outcome.

Annalisa Savaresi is Researcher at the International Centre for Climate Governance, Venice, Italy and Ph.D. Candidate at the Faculty of Law, University of Copenhagen, Denmark.

A. Savaresi (✉)
School of Law, University of Edinburgh, Old College, Edinburgh, EH8 9YL
e-mail: annalisa.savaresi@gmail.com

E.J. Hollo et al. (eds.), *Climate Change and the Law*, Ius Gentium: Comparative Perspectives on Law and Justice 21, DOI 10.1007/978-94-007-5440-9_15,
© Springer Science+Business Media Dordrecht 2013

15.1 REDD+ and the International Climate Regime

Deforestation[1] and forest degradation[2] in the tropics account for a significant share of global carbon emissions.[3] Forest loss is caused by a complex combination of market drivers, and policy and governance failures that make it more attractive to fell trees than to keep them standing.[4] As a result, several tropical countries are rapidly depleting their forests and may lose them altogether by the end of the century.[5]

Forest loss raises concerns not only for its impact on carbon emissions, but also because of its effects on ecosystems and livelihoods. Remaining tropical forests provide precious ecosystem services the substitution of which may be impossible, especially with regard to their contribution to climate stability, watershed and soil protection, as well as biodiversity conservation. These matters are not just of concern for countries that presently harbour forests. Increased awareness of the dangers associated with forest loss has lead to a long strife to ensure that forest uses become more sustainable. However, so far states have failed to effectively address the global drivers of deforestation and, more specifically, the displacement of environmental pressure associated with the international trade in timber and agricultural products. No comprehensive, legally-binding international instrument for

[1] Whereas the term has not yet been defined for the purposes of REDD+, under the Kyoto Protocol "deforestation" means "direct human-induced conversion of forested land to non-forested land." Decision 16/CMP.1, Land Use, Land-use Change and Forestry, UN Doc. FCCC/KP/CMP/8/Add.3, 30 March 2006, annex, para. 1(d).

[2] Whereas the term has not yet been defined for the purposes of REDD+, the Intergovernmental Panel on Climate Change (IPCC) has suggested defining forest degradation as "direct human-induced long-term loss (persisting for X years or more) of at least Y per cent of forest carbon stocks (and forest values) since time (T) and not qualifying as deforestation." IPCC, *Definitions and methodological options to inventory emissions from direct human-induced degradation of forests and devegetation of other vegetation types* (Kanagawa, Japan: Institute for Global Environmental Strategies, 2003), at 16.

[3] According to the IPCC Fourth Assessment Report, forestry accounts for around 17% of global carbon emissions. See, Rajendra K. Pachauri and Andy Reisinger (eds.), *Climate Change 2007: Synthesis Report. Contribution of Working Groups I, II and III to the Fourth Assessment Report of the Intergovernmental Panel on Climate Change* (Cambridge: Cambridge University Press, 2007), at 36. According to more recent estimates, this number may be closer to 15%, taking into account emissions from peat lands (excluded from the IPCC estimate) as well as increased fossil fuel emissions and updated deforestation data, see Guido R. van der Werf et al., "CO2 Emissions from Forest Loss", 2 *Nature Geoscience* (2009), 737, at 737.

[4] See, for example, Kenneth M. Chomitz, *At loggerheads? Agricultural expansion, poverty reduction, and environment in the tropical forests* (Washington DC: The International Bank for Reconstruction and Development / The World Bank, 2007), at 1; and Gabrielle Kissinger, *Linking forests and food production in the REDD+ context* (Copenhagen: CGIAR Research Program on Climate Change, Agriculture and Food Security, 2011), at 12.

[5] UN Food and Agriculture Organization (FAO), *Global Forest Resources Assessment* (Rome: FAO, 2010), at 4.

protecting forest exists,[6] and forests are mainly regarded as natural resources under international law, which recognizes states' sovereign right to exploit their forest resources according to their own environmental policies.[7]

The numerous regional and multilateral initiatives created to promote sustainable forest management at the global level have achieved little in the way of concrete results.[8] And yet, forests arguably need to be managed in a way that is consistent with commitments embodied in international treaties, ratified by virtually all states in the world, concerning issues regarded as a 'common concern of humankind,' such as biodiversity protection and climate change.

Against this background, negotiations over enhancing international climate cooperation under the United Nations Framework Convention on Climate Change (UNFCCC)[9] have drawn unprecedented attention to the role of forests in mitigating climate change. Forests are both carbon sinks and emission sources under the Convention,[10] which specifically mentions that policies and measures to deal with

---

[6] For an analysis see, among others, David Humphreys, "The Elusive Quest for a Global Forests Convention", 14 *Review of European Community and International Environmental Law* (2005), 1; Ronnie D. Lipschutz, "Why Is There No International Forestry Law? An Examination of International Forestry Regulation, Both Public and Private", 19 *University of California Los Angeles Journal of Environmental Law and Policy* (2001), 153; Radoslav S. Dimitrov, "Hostage to Norms: States, Institutions and Global Forest Politics", 5 *Global Environmental Politics* (2005), 1; Jeremy Rayner, Alexander Buck and Pia Katila, *Embracing complexity: Meeting the challenges of international forest governance: A global assessment report prepared by the Global Forest Expert Panel on the International Forest Regime* (Vienna: International Union of Forest Research Organizations, 2010).

[7] See, Non-Legally Binding Authoritative Statement of Principles for a Global Consensus on the Management, Conservation and Sustainable Development of All Types of Forests (Forest Principles), Rio de Janeiro, 14 June 1992, UN Doc. A/CONF.151/26 Vol. III, 14 August 1992, Principle 1(a). The Principle provides that: "States have, in accordance with the Charter of the United Nations and the principles of international law, the sovereign right to exploit their own resources pursuant to their own environmental policies and have the responsibility to ensure that activities within their jurisdiction or control do not cause damage to the environment of other States or of areas beyond the limits of national jurisdiction." This principle recalls Principle 21 of the Stockholm Declaration on the Human Environment, Stockholm, 16 June 1972, UN Doc. A/CONF.48/14/Rev.1; and Principle 2 of the Rio Declaration, Rio de Janeiro, 14 June 1992, UN Doc. A/CONF.151/26 Vol. I. Principle 21 of the Stockholm Declaration may be regarded as a statement of customary international law, which appears also in Article 3 of the Convention on Biological Diversity and in the preamble of the United Nations Framework Convention on Climate Change. For an assessment on the status of the principle in international law, see Pierre M. Dupuy, "Soft Law and the International Law of the Environment", 12 *Michigan Journal of International Law* (1990–1991), 420, at 422.

[8] The most salient outcome of these processes was the adoption by the United Nations General Assembly of the Non-Legally Binding Instrument on Sustainable Forest Management of All Types of Forests, UNGA/Res/62/98, 17 December 2007. For a review, see for example Katharina Kunzmann, "The Non-legally Binding Instrument on Sustainable Management of all Types of Forests – Towards a Legal Regime for Sustainable Forest Management?", 9 *German Law Journal* (2008), 981.

[9] United Nations Framework Convention on Climate Change, New York, 9 May 1992, in force 21 March 1994, 31 *International Legal Materials* (1992), 849.

[10] Ibid., Art. 4.1 (c, d).

climate change should "be comprehensive, cover all relevant sources, sinks and reservoirs of greenhouse gases and adaptation, and comprise all economic sectors".[11] To date, this mandate has only been fulfilled partially with regard to forests, chiefly for reasons related to the approach undertaken under the Kyoto Protocol.[12] Avoided deforestation was excluded from the scope of project activities eligible under the Clean Development Mechanism (CDM), established under the Kyoto Protocol. This exclusion was motivated by methodological concerns that were initially perceived as insurmountable obstacles to incentivising avoided deforestation, as well as by reluctance to divert attention from efforts to reduce emissions from the combustion of fossil fuels.[13]

Addressing avoided deforestation as an option to take action on climate change mitigation gained new momentum in 2005.[14] The initial proposition was to "draw developing countries towards emission reductions" by addressing emissions from deforestation, either by including them in the Kyoto Protocol, or through an optional Protocol under the UNFCCC.[15] Negotiations on reducing emissions from deforestation and forest degradation in developing countries (REDD) have since progressed steadily and the related debate has proven to be a rare area of convergence amid chaotic efforts to find agreement on long-term cooperative action under the Convention. The main reason for the relative popularity of REDD+ is that the forest sector features peculiarities that make it particularly appealing "politically". At least in theory, REDD+ lends itself to bridge the divide between developed and developing country Parties, providing a testing field for large-scale climate change mitigation in the latter, with projected expeditious results.[16] In addition, these efforts may build upon extant international initiatives designed to tackle forest loss in the tropics, with collateral benefits for biodiversity and sustainable development.

Since 2007, the Ad Hoc Working Group on Long-term Cooperative Action under the Convention (AWG-LCA) has considered the establishment of a REDD+ mechanism in the framework of "nationally appropriate mitigation actions by developing country Parties."[17] Whereas negotiations initially centred on reducing emissions

---

[11] Ibid., Art. 3.3.

[12] Kyoto Protocol to the United Nations Framework Convention on Climate Change, Kyoto, 10 December 1997, in force 16 February 2005, 37 *International Legal Materials* (1998), 22.

[13] For an overview, see David Humphreys, "The Politics of Avoided Deforestation: Historical Context and Contemporary Issues", 10 *The International Forestry Review* (2008), 433; and Patrick Graichen, "Can Forestry Gain from Emissions Trading? Rules Governing Sinks Projects Under the UNFCCC and the EU Emissions Trading System", 14 *Review of European Community and International Environmental Law* (2005), 11.

[14] UNFCCC, Reducing emissions from deforestation in developing countries. Approaches to stimulate action: Submission from Parties, UN Doc. FCCC/CP/2005/MISC.1, 6 December 2005.

[15] Ibid., para. 8.

[16] For an overview of arguments, see, for example, Nicholas Stern, *The Economics of Climate Change: The Stern Review* (Cambridge et al.: Cambridge University Press, 2006), at 26. See also the comprehensive review in: Johan Eliasch, *Climate Change: Financing Global Forests. The Eliasch Review* (Oxford: Earthscan, 2008).

[17] Decision 1/CP.13, The Bali Action Plan, UN Doc. FCCC/CP/6/Add.1, 14 March 2008, para 1(b)(ii).

from deforestation and forest degradation in developing countries (hence the acronym REDD), the concept was subsequently expanded to include "the role of conservation, sustainable management of forests and enhancement of forest carbon stock in developing countries" commonly referred to with the acronym REDD+.[18] This wider scope relates to the realisation that a narrower focus may lead to perverse outcomes.[19] So while deforestation and forest degradation have remained the immediate priorities, the notion of REDD+ encompasses both, emission reductions, as well as measures to preserve and enhance forest carbon stocks.

Several proposals on REDD+ have been put forward, with significant differences in scope, reference levels for carbon crediting, and other design features.[20] In 2009, Parties called for the "immediate establishment" of a mechanism including REDD,[21] but concrete action in this direction has suffered from slow progress generally affecting negotiations under the AWG-LCA. The Cancun Agreements of 2010 sketched out some of the details of the REDD+ mechanism, effectively adopting a text that had been on the table already at the 2009 UN Climate Change Conference the year before, but that was not adopted as a result of the chaos that notoriously affected the closing hours of the Copenhagen Conference. The 17th session of the Conference of the Parties (COP 17) held in Durban, South Africa in 2011, has merely clarified some marginal details of this process, as REDD+ was arguably overshadowed by "larger issues".[22]

Against this background, the use of the term 'mechanism' in connection with REDD+ refers to work in progress. The timeline and ultimate outcome of the related process remain uncertain. The role of the mechanism within the broader framework of the UNFCCC is equally unsettled. Although it does not seem implausible to predict that the REDD+ mechanism will become one element of a post-2012 international architecture to address climate change, at the time of writing, its details are sketched out in a handful of COP decisions.[23]

---

[18] Ibid., para. 1(b)(iii).

[19] For an analysis, see Kathleen Lawlor et al., *Expanding the Scope of International Terrestrial Carbon Options: Implications of REDD+ and Beyond* (Durham: Nicholas Institute for Environmental Policy Solutions, Duke University, 2010).

[20] For a review, see Charlie Parker et al., *The Little REDD+ Book: a guide to governmental and non-governmental proposals for reducing emissions from deforestation and degradation* (Oxford: Global Canopy Programme, 2009).

[21] Decision 2/CP.15, Copenhagen Accord, UN Doc. FCCC/CP/2009/11/Add.1, 30 March 2010, para. 6.

[22] Bruce Cabarle, head of WWF's Climate and Forests initiative, as quoted in Yana Marull, "Little headway in Durban on deforestation", 15 December 2011, available at: http://www.iol.co.za/scitech/science/environment/little-headway-on-deforestation-experts-1.1199192?showComments=true (last accessed on 10 March 2012).

[23] Decision 2/CP.13, Reducing Emissions from Deforestation in Developing Countries: Approaches to Stimulate Action, UN Doc. FCCC/CP/6/Add.1, 14 March 2008; Decision 4/CP.15, Methodological Guidance for Activities Relating to Reducing Emissions from Deforestation and Forest Degradation and the Role of Conservation, Sustainable Management of Forests and Enhancement of Forest Carbon Stocks in Developing Countries, UN. Doc. FCCC/CP/2009/11/Add.1, 30 March 2010; Decision 2/CP.15, supra, note 22; Decision 1/CP.16, Cancun Agreements: Outcome of the Work of

The most salient document in this connection is Decision 1/CP.16, which encourages developing country Parties to contribute to climate change mitigation through REDD+ activities.[24] The envisioned process is set to start with the development of national strategies or action plans, national forest reference emission levels and/or forest reference levels, and a robust and transparent national forest monitoring systems.[25] The ultimate result is expected to be the establishment of a system of incentives for developing countries to maintain and sustainably manage their forest carbon stocks.

In the meantime, demonstration activities have been undertaken[26] and numerous multilateral and bilateral processes scaled up actions and finance to support "REDD+ readiness" or, in other words, countries' capacity to carry out REDD+ activities and handle REDD+ financing effectively and equitably.[27] The most important processes established to carry out demonstration activities are the UN Collaborative Programme on REDD+ (UN-REDD Programme) and the Forest Carbon Partnership Facility (FCPF). Both processes are expected to contribute to the first two phases of REDD+ identified in the Cancun Agreements, that is readiness planning and implementation of pilot REDD+ initiatives.[28] In addition, the FCPF is also testing a pilot programme

---

the Ad Hoc Working Group on Long-Term Cooperative Action under the Convention, UN Doc. FCCC/CP/7/Add.1, 15 March 2011; Decision 12/CP.17, Guidance on systems for providing information on how safeguards are addressed and respected and modalities relating to forest reference emission levels and forest reference levels as referred to in decision 1/CP.16, UN Doc. FCCC/CP/2011/9/Add.2, 15 March 2012, Appendix I; and Decision 2/CP.17, Outcome of the work of the Ad Hoc Working Group on Long-term Cooperative Action under the Convention, UN Doc. FCCC/CP/2011/9/Add.1, 15 March 2012.

[24] Decision 1/CP.16, supra, note 24, paras. 68–79.

[25] Ibid, para. 71.

[26] Decision 2/CP.13, supra, note 24, Annex, para. 2 encouraged Parties to "explore a range of actions, identify options and undertake efforts, including demonstration activities, to address the drivers of deforestation relevant to their national circumstances, with a view to reducing emissions from deforestation and forest degradation and thus enhancing forest carbon stocks due to sustainable management of forests."

[27] For an overview see The Voluntary REDD+ Database provided by the REDD+ Partnership at: http://reddplusdatabase.org (last accessed on 07 March 2012). The most well-known bilateral activities undertaken to date are those between Norway and Brazil, Indonesia, Guyana and Mexico respectively. More detailed information see Norwegian Ministry of the Environment, The Government of Norway's International Climate and Forest Initiative at: http://www.regjeringen.no/en/dep/md/Selected-topics/climate/the-government-of-norways-international-.html?id=548491 (last accessed on 30 April 2012). For a review, see *NORAD, Real-Time Evaluation of Norway's International Climate and Forest Initiative Contributions to a Global REDD+ Regime 2007–2010* (Oslo: Norwegian Agency for Development Cooperation, 2011).

[28] In this connection, see for example Benoit Bosquet and Andre Rodrigues Aquino, "Forest Carbon Partnership Facility: Demonstrating Activities that Reduce Emissions from Deforestation and Forest Degradation", 2010, available at: http://www.forestcarbonpartnership.org/fcp/sites/forestcarbonpartnership.org/files/Documents/PDF/English_54462_WorldBank_FCPF_Brochure.pdf (last accessed on 07 March 2012), at 17; and the UN Food and Agriculture Organization (FAO), "UNDP and UNEP, UN Collaborative Programme on Reducing Emissions from Deforestation and Forest Degradation in Developing Countries (UN-REDD) Framework Document", 2008, available at: www.unredd.net/index.php?option=com_docman&task=doc_download&gid=4&Itemid=53 (last accessed on 7 March 2012), at 7.

for the implementation of "results-based actions", which is expected to take place in phase three.[29]

The Cancun Agreements also emphasise the potential of REDD+ to complement or be consistent with the "relevant international conventions and agreements" and to enhance "other social and environmental benefits", such as biodiversity conservation, 'poverty alleviation' as well as improved forest governance.[30] However, the generation of co-benefits (or multiple benefits, as they are also called)[31] does not necessarily correspond with carbon sequestration.[32] So, while REDD+ may bring about a triple-win solution that addresses climate change, sustainable development and biodiversity conservation simultaneously, the extent of these co-benefits depends on the design and implementation of the REDD+ mechanism.

After the initial optimism, increased awareness of challenges associated with ensuring environmental integrity and sustainable development through climate change mitigation activities[33] has revealed that REDD+ is neither going to be quick, nor easy.[34] Similar concerns had already surfaced in connection with the CDM.[35]

---

[29] World Bank, "Charter establishing the Forest Carbon Partnership Facility", 2010, available at: http://www.forestcarbonpartnership.org/fcp/sites/forestcarbonpartnership.org/files/Documents/PDF/Sep2010/ FCPF_Charter-August_2010_clean.pdf (last accessed on 07 March 2012), at 10.

[30] Decision 1/CP.16, supra, note 24, Appendix I, para. 2 (a,e).

[31] For the use of this terminology, see for example UN-REDD Programme, "Multiple Benefits-Issues and Options for REDD", 2009, available at: http://www.unredd.net/index.php?option=com_docman&task=doc_download&gid=472&Itemid=53 (last accessed on 7 March 2012).

[32] For an review of arguments, see for example Jaboury Ghazoul et al., "REDD: a reckoning of environment and development implications", 25 *Trends in Ecology and Evolution* (2010), 396; Oliver Springate-Baginski and Eva Wollenberg (eds), *REDD, forest governance and rural livelihoods* (Bogor: CIFOR, 2010); and Thomas Sikor et al., "REDD-plus, Forest People's Rights and Nested Climate Governance", 20 *Global Environmental Change* (2010), 423.

[33] See, for example, Christof Arens, Hanna Wang-Helmreich and Timon Wehnert "Mitigation Versus Sustainable Development? Why NAMAs Shouldn't repeat the CDM's Mistakes", 17 *Joint Implementation Quarterly* (2011), 6.

[34] For an overview, see, for example, Frieds of the Earth (FOE), *REDD Myths: A Critical Review of Proposed Mechanisms to Reduce Emissions from Deforestation and Degradation in Developing Countries* (Amsterdam: FOE, 2008); Larry Lohmann, *Chronicle of a Disaster Foretold REDD-with-Carbon-Trading* (Sturminster Newton, UK: The Corner House, 2008); Tom Griffiths and Francesco Martone, *Seeing REDD? Forests, Climate Change Mitigation and the Rights of Indigenous Peoples and Local Communities* (Moreton in Marsh: Forest Peoples Programme, 2009); Francesco Martone, *The emergence of the REDD Hydra* (Washington DC: Rights and Resources Initiative, 2010); Terence Sunderland, "'Win-win' is too simplistic a description for REDD+ – and possibly wrong", 20 April 2011, available at: http://blog.cifor.org/2585/win-win-is-too-simplistic-a-description-for-redd-%E2%80%93-and-possibly-wrong/ (last accessed on 17 February 2012).

[35] See for example Christina Voigt, "Is the Clean Development Mechanism Sustainable?", 8 *Sustainable Development Law and Policy* (2007–2008), 15, at 18; Diana Liverman and Emily Boyd, "The CDM, ethics and development", in Karen Holm Jensen and Jørgen Fenhann (eds), *A reformed CDM – including new mechanisms for sustainable development* (Roskilde: UNEP Risø Center, 2008), 47, at 55; and Gary Cox, "The Clean Development Mechanism as a Vehicle for Technology Transfer and Sustainable Development – Myth or Reality?", 179 *Law, Environment and Development Journal* (2010), 179, at 194.

The CDM has generally been criticized for not adopting standards to assess projects' contribution to sustainable development.[36] The assessment of CDM projects' contribution to sustainable development is presently left to the discretion of the countries hosting projects. The decision to delegate such assessment to host countries has solved the intractable problem of reaching agreement on international sustainable development indicators for CDM projects. It has nevertheless left national authorities in the difficult position of acting as watchdogs for the 'integrity' of the CDM, while they share a common interest with project developers in maximizing the number of Certified Emission Reductions awarded to a given CDM project activity. This situation has created perverse incentives for national authorities to overlook sustainability concerns, in what has been defined as a "race to the bottom in sustainable development standards."[37] CDM guidelines used at the national level in this connection are reportedly weak, and their observance is not scrupulously monitored and verified over the life of the project.[38] The result is that no CDM project has seemingly ever failed validation due to sustainable development and social equity requirements,[39] whereas host countries do not appear to have favoured projects with high sustainable development benefits.[40]

Afforestation and reforestation projects pose specific challenges under the CDM. Whereas the implementation of these activities was generally meant to "contribute"[41] to the conservation of biodiversity and sustainable use of natural resources, this requirement has not been strictly enforced and some registered CDM projects have reportedly had negative effects on the livelihoods of local communities and on the conservation of biodiversity.[42]

Lessons learnt from the CDM seem to suggest that ensuring the provision of co-benefits and complementarity with the aims and objectives of other relevant international conventions and agreements[43] through REDD+ arguably requires the drafting of international standards, together with specific monitoring and verification

---

[36] See for example, Lena Ruthner et al., *Study on the Integrity of the Clean Development Mechanism (CDM)* (London: AEA, 2011), at 16, available at: http://ec.europa.eu/clima/policies/ets/linking/docs/final_report_en.pdf.

[37] Aaron Cosbey et al., *Realizing the Development Dividend: Making the CDM Work for Developing Countries (Phase I Report)* (London: International Institute for Sustainable Development, 2005), at 43.

[38] Ibid.

[39] Tanguy du Monceau and Arnaud Brohé, *Sustainable Development and Social Equity, Study on the Integrity of the Clean Development Mechanism* (London: AEA, 2011), at 19, available at http://ec.europa.eu/clima/policies/ets/linking/docs/sustainable_development_en.pdf

[40] Micheal Gillenwater and Stephen Seres, *The Clean Development Mechanism. A Review of the First International Offset Program* (Arlington: Pew Center on Global Climate Change, 2011), at 30.

[41] Decision 16/CMP.1, supra, note 2, para. 1(e).

[42] For a discussion, see for example Cox, "The Clean Development Mechanism as a Vehicle for Technology Transfer and Sustainable Development ", supra, note 36, at 193.

[43] Decision 2/CP.13, Preamble.

requirements.[44] As a result, REDD+ has subsumed what have been described as "long-standing, perhaps intractable, development policy challenges" within its remit, in the attempt to create favourable conditions to incentivise climate change mitigation in the forest sector in developing countries.[45] These issues have played an increasingly prominent role in the negotiations on REDD+ through the debate on so-called 'safeguards'. The notion of 'safeguards' has become standard terminology to refer to "policies and measures that aim to address both direct and indirect impacts on communities and ecosystems, by identifying, analysing, and ultimately working to manage risks and opportunities."[46] As this article argues, the discourse on safeguards may be regarded as an attempt to address overlaps between international regimes.

15.2 REDD+ and the Fragmentation of International Law

Because of its intricate subject matter and nature the climate change regime is particularly likely to overlap with other international regimes.[47] While such overlaps unavoidably lead to some complications, they arguably also provide a potential avenue to improve coordination and communication.[48] This potential is particularly evident in connection with REDD+. As the forest sector lies at the intersection of economic, environmental, and social policies, the potential for overlaps with other areas of the law is great. It has emerged quite clearly that REDD+ is particularly likely to overlap with instruments concerning the protection of biodiversity and human rights.

Overlaps between international regimes can be addressed through conflict avoidance, resolution of conflicts through the application of interpretative principles and institutional cooperation and coordination.[49] Lack of actual treaty rules on REDD+ has however limited the scope to deploy these techniques.

---

[44] For an overview, see Annalisa Savaresi, "Reducing emissions from deforestation in developing countries under the UNFCCC. Caveats and opportunities for biodiversity", 21 *Yearbook of International Environmental Law* (2011). Advance access edition: http://yielaw.oxfordjournals.org/content/early/2011/11/10/yiel.yvr004.extract (last accessed on 9 March 2012).

[45] Simon West, "'Command Without Control': Are Market Mechanisms Capable of Delivering Ecological Integrity to REDD?", 6 *Law, Environment and Development Journal* (2010), 298, at 301.

[46] For this definition, see Deborah Murphy, *Safeguards and Multiple Benefits in a REDD+ Mechanism* (Winnipeg, CA: IISD, 2011), at 1.

[47] For an overview, see for example Harro van Asselt, Francesco Sindico and Michael A. Mehling, "Global Climate Change and the Fragmentation of International Law", 30 *Law & Policy* (2008), 423, at 424; and Cinnamon Pinon Carlarne, "Global Climate Governance: Only a Fragmented System of International Law Away?", 30 *Law & Policy* (2008), 450, at 450.

[48] Carlarne, "Global Climate Governance", supra, note 48, at 452.

[49] International Law Commission (ILC), *Fragmentation of International Law: Difficulties Arising from the Diversification and Expansion of International Law. Report of the Study Group of the International Law Commission*, UN Doc. A/CN.4/L.682, 13 April 2006, at 13.

As rules on REDD+ are still in the process of being drafted, the best way to address overlaps and avoid perverse outcomes may be that of addressing potential conflicts *ex ante*. The UNFCCC does not include a clause specifically addressing conflicts with other international instruments. In the context or REDD+, overlaps have been chiefly addressed through safeguards, which specifically mention that REDD+ activities "complement or are consistent with" the objectives of national forest programmes and relevant international conventions and agreements.[50] Safeguards for REDD+ may be viewed as an attempt to pre-empt conflict with other international law objectives and rules. However, it is important to note here that such efforts are being carried out by means of COP decisions, rather than by the inclusion of specific conflict clauses in the text of the UNFCCC.[51]

The question as to whether decisions by international treaty bodies have enough legal strength to amount to legal obligations has received considerable scholarly attention.[52] Technically speaking treaty based institutions, such as COPs, are mere diplomatic conferences or "coalitions of the willing".[53] *Stricto jure*, the rules for treaty interpretation embodied in the Vienna Convention on the Law of Treaties (VCLT)[54] cannot be deployed in connection with COP decisions, as they are not treaties.[55] COP decisions may nevertheless be regarded as part of the normative environment of the provisions set out in the UNFCCC and its Kyoto Protocol and they "influence" the substantive obligations of the Parties in numerous ways.[56] Harmonization of rules with their 'normative environment' is instrumental to the orderly functioning of the (international) legal system.[57] Even though the law of the treaties may not *de jure* be

[50] Decision 1/CP.16, supra, note 24, Appendix I, para. 2(a).

[51] Harro van Asselt, "Managing the Fragmentation of International Environmental Law: Forests at the Intersection of the Climate and Biodiversity Regimes", *New York University Journal of International Law and Politics* (2012, forthcoming), at 27. On the notion of conflict clause, see ILC, supra, note 50, paras. 268–271.

[52] For an analysis, see for example Robin R. Churchill and Geir Ulfstein, "Autonomous Institutional Arrangements in Multilateral Environmental Agreements: A Little-Noticed Phenomenon in International Law", 94 *The American Journal of International Law* (2000), 623; Jutta Brunnée, "COPing with Consent: Lawmaking Under Multilateral Environmental Agreements", 15 *Leiden Journal of International Law* (2002), 1; Annecoos Wiersema, "The New International Law-Makers? Conferences of the Parties to Multilateral Environmental Agreements", 31 *Michigan Journal of International Law* (2009), 231.

[53] Patricia Birnie, Alan Boyle and Catherine Redgwell, *International Law and the Environment* (Oxford: Oxford University Press, 2009), at 11.

[54] See ILC, *Fragmentation of International Law*, supra, note 50, Appendix, at 250. The Vienna Convention codifies rules applicable to treaties concluded after its entry into force. The Convention's articles on interpretation (Arts. 31–3) are regarded as customary international law and have been deployed as an aid to interpret all treaties. Vienna Convention on the Law of Treaties, Vienna, 23 May 1969, in force 27 January 1980, 8 *International Legal Materials* (1969), 679.

[55] In this sense, see Van Asselt, "Managing the Fragmentation of International Environmental Law", supra, note 52, at 32–33.

[56] See Wiersema, "The New International Law-Makers?", supra, note 52, at 245; and Birnie, Boyle and Redgwell, International law and the environment, supra, note 53, at 19.

[57] International Law Commission, supra, note 50, at 37 and 120.

deployed as a hermeneutic tool in this connection,[58] overlaps between rules emanating from decisions by treaty bodies should arguably be addressed through systemic integration.[59] This approach seems consistent with the role of COP decisions within the legal framework that has developed under the UNFCCC and its Kyoto Protocol, whereby the COP is tasked with keeping under regular review the implementation of the Convention and any related legal instruments, and make the decisions necessary to promote their effective implementation.[60]

These doctrinal considerations are particularly important in connection with the discourse on REDD+ safeguards. The debate on safeguards under the AGW-LCA has so far avoided taking a precise stand on their legal nature. Safeguards included in the Cancun Agreements certainly do not feature the characteristics of conflict clauses typically included in treaties.[61] The fact that the safeguards are embedded in COP decisions raises a series of questions concerning their legal force. However, the inclusion of information on implementation of safeguards in Parties' national communications potentially opens the way to their review in the framework of the UNFCCC.[62]

COP 17 left it to the Subsidiary Body for Scientific and Technological Advice (SBSTA) to shed further light on the details.[63] Demonstration activities have provided a 'learning by doing' approach to REDD+ expected to influence the drafting of rules under the UNFCCC.[64] In this process, overlaps with instruments concerning the protection of biodiversity and human rights have been addressed through the adoption of specific safeguards and standards, as the following sections explain in greater detail.

15.3 Biodiversity Safeguards for REDD+

The conservation of forest carbon stocks holds great potential for synergy with the conservation of biodiversity, although the two objectives do not necessarily coincide. In fact, a focus on maximising carbon sequestration may have negative

[58] See van Asselt, "Managing the Fragmentation of International Environmental Law", supra, note 52, at 33; van Asselt, Sindico and Mehling, "Global Climate Change and the Fragmentation of International Law", supra, note 48, at 430.

[59] On systemic integration see ILC, *Fragmentation of International Law*, supra, note 50, paras. 37–43 and 410–480.

[60] UNFCCC, supra, note 10, Art. 7.2. See also ibid., Art. 10.2, where the COP is described as the "supreme body" of the Convention.

[61] See van Asselt, "Managing the Fragmentation of International Environmental Law", supra, note 52, at 33.

[62] Decision 12/CP.17, supra, note 24, para 4.

[63] Ibid., para 5.

[64] FCPF, "Information Memorandum", 2008, available at: http://www.forestcarbonpartnership.org/fcp/sites/forestcarbonpartnership.org/files/Documents/PDF/FCPF_Info_Memo_06-13-08.pdf (last accessed on 7 March 2012), at 3.

impacts on biodiversity.[65] For example, the plantation of invasive species may provide rapid carbon sequestration at the expense of other biodiversity rich ecosystems.[66]

As negotiations on REDD+ progressed under the UNFCCC, numerous observer organizations underscored opportunities to combine biodiversity conservation with climate change mitigation.[67] This issue had already arisen in connection with afforestation and reforestation projects under the CDM, where, however, the provision of co-benefits was not subjected to specific criteria and verification, leading to some perverse outcomes.[68]

At the time of writing, the extent to which the UNFCCC Parties may be willing to pursue synergies between biodiversity conservation and climate change mitigation remains unclear. The Convention on Biological Diversity[69] (CBD) and the UNFCCC view regard forests from different perspectives. While the CBD is concerned with forests as habitats and as components of biodiversity, they are chiefly regarded as carbon sinks and sources under the UNFCCC. Despite these different approaches to forests, both conventions address forest management to a certain degree. When implementing REDD+ activities, countries that are Parties to both conventions are likely to be faced with overlapping obligations.

The CBD and the UNFCCC are equally binding upon Parties and the principle *pacta sunt servanda* requires Parties to fulfil their commitments under both treaties

---

[65] For an overview, see Secretariat of the Convention on Biological Diversity, *Connecting Biodiversity and Climate Change Mitigation and Adaptation: Report of the Second Ad Hoc Technical Expert Group on Biodiversity and Climate Change. Technical Series No. 41* (Montreal: CBD Technical Series, 2009).

[66] For an overview, see e.g. Erik Nelson et al., "Efficiency of incentives to jointly increase carbon sequestration and species conservation on a landscape", 105 *Proceedings of the National Academy of Sciences* (2008), 9471.

[67] See, among others, Till Pistorius et al., *Greening REDD+. Challenges and opportunities for forest biodiversity conservation* (Freiburg, Germany: University of Freiburg,2010); Celia A. Harvey, Barney Dickson and Cyril Kormos, "Opportunities for achieving biodiversity conservation through REDD", 3 *Conservation Letters* (2010), 53; Jonah Busch et al., "Biodiversity Co-benefits of Reducing Emissions from Deforestation Under Alternative Reference Levels and Levels of Finance", 4 *Conservation Letters* (2011), 101; Barney Dickson and Matea Osti, *What are the ecosystem-derived benefits of REDD+ and why do they matter? Multiple Benefits Series 1. Prepared on behalf of the UN-REDD Programme* (Cambridge: United Nations Environment Programme World Conservation Monitoring Centre, 2010); Lera Miles, Emily Dunning and Nathalie Doswald, *Safeguarding and enhancing the ecosystem derived benefits of REDD+. Multiple Benefits Series 2. Prepared on behalf of the UN-REDD Programme* (Cambridge: United Nations Environment Programme World Conservation Monitoring Centre, 2010); and Cordula Epple, Nathalie Doswald, Barney Dickson, *Potential links between monitoring for multiple benefits of REDD+ and the monitoring requirements of the Rio Conventions. Multiple Benefits Series 9. Prepared on behalf of the UN-REDD Programme.* (Cambridge: United Nations Environment Programme World Conservation Monitoring Centre, 2010).

[68] For a discussion, see for example Cox, "The Clean Development Mechanism as a Vehicle for Technology Transfer and Sustainable Development", supra, note 36, at 193.

[69] Convention on Biological Diversity (CBD), Rio de Janeiro, 2 June 1992, in force 29 December 1993, 79 *United Nations Treaty Series* (1992), 1760.

in good faith.[70] Both conventions deal with global environmental problems and establish regimes of almost universal application, which prohibit Parties from making specific reservations to their provisions.[71] Furthermore, the objectives of the two conventions are not mutually exclusive, and provide several areas for mutually supportive action. Arguably, the CBD and the UNFCCC may also be regarded as 'common interest treaties,' as their negotiations took place in parallel and both were adopted at the UN Conference on Environment and Development in 1992.[72] The existence of common interests is confirmed, *inter alia*, by the establishment of a Joint Liaison Group, as an informal forum for exchanging information, exploring opportunities for synergistic activities, and increasing co-ordination between the Rio conventions.[73]

Overlapping obligations under the CBD and the UNFCCC should arguably be viewed as an integrated whole, and Parties to both treaties should adopt a harmonizing approach to their respective obligations. The same should apply to decisions by their treaty bodies. Also here, Parties are faced with some 'potential for synergy.'[74] As the next section shows, treaty bodies established under the CBD and the UNFCCC have already taken some action to address this potential.

15.3.1 Biodiversity Safeguards Under the UNFCCC

The indicative guidance initially adopted by the UNFCCC COP on REDD+ demonstration activities specifically mentions that they should be "consistent with sustainable forest management, noting, inter alia, the relevant provisions" of the UN Forum on Forests, the United Nations Convention to Combat Desertification[75]

[70] VCLT, supra, note 54, Art. 26.

[71] CBD, supra, note 68, Art. 34; UNFCCC, supra, note 10, Art. 24.

[72] See Concetta M. Pontecorvo, "Interdependence Between Global Environmental Regimes: The Kyoto Protocol on Climate Change and Forest Protection", 59 *Heidelberg Journal of International Law* (1999), 705, at 742; and Frédéric Jacquemont and Alejandro Caparrós, "The Convention on Biological Diversity and the Climate Change Convention 10 Years After Rio: Towards a Synergy of the Two Regimes?", 11 *Review of European Community and International Environmental Law* (2002), 169, at 178.

[73] The Joint Liaison Group was established to enhance co-ordination between the three conventions, including the exchange of relevant information, and to explore options for further co-operation between the conventions, including the possibility of a joint work plan and/or a workshop. See Report of the Subsidiary Body for Scientific and Technological Advice on its Fourteenth Session Bonn, UN Doc. FCCC/SBSTA/2001/2, 18 September 2001. The establishment of the group was later endorsed by Decision 13/CP.8, Cooperation with Other Conventions, UN Doc. FCCC/CP/2002/7/Add.1, 28 March 2003. Since then, co-operation with the CBD has been discussed under the UNFCCC under the agenda item "cooperation with relevant international organizations."

[74] van Asselt, "Managing the Fragmentation of International Environmental Law", supra, note 52, at 4.

[75] Convention to Combat Desertification in Countries Experiencing Serious Drought and/or Desertification, Particularly in Africa (UNCCD), Paris, 17 June 1994, in force 24 June 1998, 33 *International Legal Materials* (1994), 1016.

and the CBD.[76] This remarkable reference to other international instruments and processes dealing with forests may be regarded as a sign of the Parties' willingness to take into consideration guidance provided in the context of other international instruments and processes. This reference is especially notable given the fact that not all Parties to the UNFCCC are signatories to the CBD or the UNCCD.

However, expressions like these do not appear in subsequent decisions by the UNFCCC COP dealing with REDD+.[77] The Cancun Agreements, instead, generically assert that REDD+ activities should be "consistent with the conservation of natural forests and biological diversity" and "are not used for the conversion of natural forests, but are instead used to incentivise the protection and conservation of natural forests and their ecosystem services, and to enhance other social and environmental benefits."[78] Given the hortatory language deployed in the decision, the implications of this safeguard depend entirely on Parties' interpretation and enforcement of the inherent requirements.

The Cancun Agreements also requested the SBSTA to develop "a system for providing information" on how the safeguards are being addressed and respected throughout the implementation of the activities, "while respecting sovereignty."[79] The SBSTA has invited Parties and accredited observers to submit their observations in this regard.[80] Submissions by Parties indicate some support for exploiting synergies between the CBD and the UNFCCC. Australia, for example, has suggested that safeguards information systems should "include application of existing international frameworks such as the Food and Agriculture Organisation or the Convention on Biological Diversity."[81] The EU submission specifically makes reference to guidance provided in the context of the CBD,[82] whereas submissions by El Salvador[83] and Japan[84] recommend using information provided under existing international instruments, including the CBD.

---

[76] Decision 2/CP.13, supra, note 24 Annex, para. 8.

[77] The preamble of Decision 2/CP.15, supra, note 25, for example, merely r*ecognizes* the importance of promoting sustainable management of forests and co-benefits that may complement the aims and objectives of national forest programmes and relevant international conventions and agreements.

[78] Decision 1/CP.16, supra, note 25, Appendix I, para. 2(e).

[79] Ibid., para. 71(d).

[80] Decision 4/CP.15, supra, note 25, para. 4.

[81] UNFCCC, Views on methodological guidance for activities relating to reducing emissions from deforestation and forest degradation and the role of conservation, sustainable management of forests and enhancement of forest carbon stocks in developing countries. Submissions from Parties, UN Doc. FCCC/SBSTA/2011/MISC.7, 19 October 2011, Paper no. 1: Australia, at 6.

[82] Ibid. Paper no. 8: Poland and the European Commission on behalf of the European Union and its member States, supported by Albania, Croatia, Iceland, the Former Yugoslav Republic of Macedonia, Montenegro and Serbia, at 59.

[83] Ibid. Paper no. 7B: El Salvador on behalf of Dominican Republic, El Salvador, Honduras and Panama, at 38–42.

[84] Ibid. Paper no. 12: Japan, at 82.

None of these suggestions was endorsed by COP 17, where the Parties adopted a decision that makes only general reference to the pursuit of environmental integrity and the need to avoid perverse incentives.[85] If, however, the experience of the CDM offers an apt model for comparison, concerted international action is necessary to establish a level playing field, ensuring the pursuit of co-benefits as well as compliance with extant international obligations. At COP 17, Parties have arguably missed an opportunity to take more decisive action in this direction.[86]

Pending the outcome of the UNFCCC negotiations, REDD+ readiness processes are the main source of guidance for national lawmakers. Both the FCPF and the UN-REDD Programme require partner countries to consider ecosystem benefits, although they do not impose any mandatory requirements in this connection.[87] The related standards are still in the process of being drafted. There is, however, already a considerable gap between the two processes. While the UN-REDD Programme has made considerable efforts in incorporating suggestions provided in the literature, and in guidance elaborated in the context of the CBD,[88] the FCPF has been criticised for its lax approach towards enforcement of World Bank's safeguard policies,[89] and its scarce consideration for guidance provided in the context of the CBD.[90]

While the UN-REDD Programme is proving to be a more proactive and ambitious process than the FCPF, the extent to which these processes will succeed in ensuring the pursuit of synergies remains to be ascertained. In the meantime, more comprehensive and detailed guidance on issues relevant for REDD+ has been elaborated in the framework of the CBD.

15.3.2 The CBD and Biodiversity Safeguards

Already in 2008, the CBD COP issued a decision calling on Parties, non-Party governments, and international organizations to ensure that REDD+ activities support the aims and implementation of the CBD, provide benefits for forest biodiversity, and

[85] Decision 12/CP.17, supra, note 25, preamble.

[86] Michelle Kovacevic, "Durban talks both good and bad for REDD+", 14 December 2011, available at: http://blog.cifor.org/6507/durban-talks-both-good-and-bad-for-redd-says-expert/ (last accessed on 17 February 2012).

[87] See Miles, Dunning and Doswald, *Safeguarding and enhancing the ecosystem derived benefits of REDD+,* supra, note 66, at 7.

[88] Ibid., at 2 and 11.

[89] Kate Dooley et al., *Smoke and mirrors. A critical assessment of the Forest Carbon Partnership Facility* (Brussels and Moreton in Marsh: FERN not an acronym and Forest Peoples Programme, 2011). Compare also the ample coverage provided by the NGOs REDD Monitor, 2012, available at: http://www.redd-monitor.org/tag/forest-carbon-partnership-facility/ (last accessed on 17 February 2012) and by the Forest Peoples Programme, 2012, available at: http://www.forestpeoples.org/topics/climate-forests (last accessed on 17 February 2012).

[90] Baastel and NORDECO, First Program Evaluation for the Forest Carbon Partnership Facility (FCPF) Evaluation Report (Gatineau and Copenhagen: Baastel and Nordeco, 2011), at 49.

involve biodiversity experts in REDD+ program design.[91] In 2010, CBD COP 10 took further steps in this direction, by adopting a decision providing "guidance on ways to conserve, sustainably use and restore biodiversity and ecosystem services while contributing to climate change mitigation and adaptation."[92] Decision X.33 includes a wide range of recommendations on REDD+ and its impact on biodiversity.[93]

This detailed guidance is further strengthened through the provision of arrangements to enhance collaboration with the UNFCCC and other international bodies.[94] In addition, the CBD Secretariat has recently submitted views on methodological guidance for REDD+ activities to the UNFCCC SBSTA.[95]

The submission by the CBD Secretariat to the UNFCCC contains summaries of four expert workshops[96] aimed at supporting CBD Parties' efforts to address REDD+ in a way that contributes to the implementation of the CBD programme of work on forest biodiversity.[97] The workshops reviewed the approach to safeguards undertaken in the framework of the UN-REDD Programme and the FCPF. The summaries of the workshops emphasise that some progress has been made towards the incorporation of biodiversity concerns in REDD+ readiness activities. However, there are risks inherent in the proliferation of terms and approaches to REDD+ safeguards,[98] and some gaps remain to be addressed.[99] The summaries also underscore the need for improved coordination and communication between the Secretariats of the Rio Conventions.[100]

---

[91] CBD COP Decision IX/5, Forest biodiversity, UN Doc. UNEP/CBD/COP/9/29, 9 October 2008, para. 2a.

[92] CBD COP Decision X/33, Biodiversity and climate change, UN Doc. UNEP/CBD/COP/DEC/X/33, 29 October 2010, para. 8.

[93] For a comprehensive review of the decisions adopted at CBD COP10, see Elisa Morgera, "Faraway, So Close: A Legal Analysis of the Increasing Interactions between the Convention on Biological Diversity and Climate Change Law", 2 *Climate Law* (2011), 85.

[94] CDB COP Decision X/33, supra, note 91, paras. 9–10.

[95] Decision 4/CP.15, supra, note 25, para. 4.

[96] The workshops were organised pursuant to CBD COP Decision IX/16, Biodiversity and climate change, UNEP/CBD/COP/DEC/IX/16, 9 October 2008; CBD COP Decision IX/5, supra, note 90; and CBD COP Decision X/33, supra, note 91.

[97] CBD COP Decision VI/22 on Forest Biological Diversity, UN Doc. UNEP/CBD/COP/6/22, 22 May 2002.

[98] CBD, Asia-Pacific Regional Consultation and Capacity-Building Workshop on Reducing Emissions From Deforestation And Forest Degradation in Developing Countries (REDD-Plus), Including on Relevant Biodiversity Safeguards, UN Doc. UNEP/CBD/WS/CB/REDD/APAC/1/2, 18 March 2011, para. 8(c).

[99] CBD, Africa Regional Consultation and Capacity-Building Workshop on Reducing Emissions from Deforestation and Forest Degradation in Developing Countries (REDD-Plus), Including on Relevant Biodiversity Safeguards, UN Doc. UNEP/CBD/WS/CB/REDD/AFR/1/2, 23 September 2011, at 19. For a similar assessment, see also CBD, Latin America – Caribbean Regional Consultation and Capacity-Building Workshop on Reducing Emissions From Deforestation And Forest Degradation in Developing Countries (REDD-Plus), Including on Relevant Biodiversity Safeguards, I. Co-Chairs Summary UNEP/CBD/WS/CB/REDD/LAC/1/2, 8 July 2011, para. 29.

[100] Ibid., para. 46.

Decision X/33 and the CBD Secretariat's submission to the UNFCCC SBSTA may be viewed as an attempt to "influence the design of REDD" and to address overlaps between biodiversity protection and climate change mitigation in the forest sector.[101] Arguably, the CBD COP has taken a "holistic" and proactive stance on the matter,[102] which is fitted with the all-encompassing objective of the related convention and its earlier efforts to promote enhanced concerted action under the Rio Conventions,[103] as it is shown in the next section.

15.3.3 Biodiversity Safeguards for REDD: Some Preliminary Conclusions

Although Parties to the UNFCCC have undertaken some action to address biodiversity concerns associated with REDD+, the main obstacle to further integration of guidance supplied in the framework of the CBD is that the Parties have tended to interpret the conventions' mandates restrictively. While the CBD has been rather vocal in requesting enhanced concerted action under the two conventions, these calls have so far remained largely unanswered by the UNFCCC COP, save for a few erratic exceptions.[104] Some states have expressed the view that the CBD does not have a legitimate role in climate change mitigation, which remains a concern to be addressed under the UNFCCC.[105] Such intransigent approach seems counter-intuitive, given that only two States (Andorra and the United States) are not parties to both conventions. This approach is apparent also in the limited role played to date by the Joint Liaison Group,[106] perhaps confirming lack of political will to strengthen institutional cooperation under the Rio Conventions.[107]

The CBD Secretariat has recently promoted the drafting of Proposed Terms of Reference and Modus Operandi for the Joint Liaison Group.[108] If approved by the

---

[101] van Asselt, "Managing the Fragmentation of International Environmental Law", supra, note 51, at 14.

[102] Morgera, "Faraway, So Close", supra, note 92, at 95.

[103] For analysis, see ibid., at 91–92.

[104] See Decision 13/CP.8 Cooperation with Other Conventions, UN Doc. FCCC/CP/2002/7/Add.1, 28 March 2003, preamble.

[105] See van Asselt, "Managing the Fragmentation of International Environmental Law", supra, note 51, at 41, quoting UNFCCC, Views on the Paper on Options for Enhanced Cooperation among the Three Rio Conventions, Submissions from Parties, UN Doc. FCCC/SBSTA/MISC.4 16, 23 March 2006, submission by Australia, para. 5.

[106] See supra note 74.

[107] For an overview on co-operative action so far, see UNFCCC, Summary of Cooperative Activities with United Nations Entities and Intergovernmental Organizations to Contribute to the Work under the Convention, UN Doc. FCCC/SBSTA/INF.3, 11 May 2011.

[108] CBD, "Proposed Terms of Reference and Modus Operandi for the Joint Liaison Group between the Three Rio Conventions", 2011, available at: http://www.cbd.int/cooperation/doc/jlg-modus-operandi-en.pdf (last accessed on 20 March 2012).

COPs of all three Rio Conventions, the mandate of the Joint Liaison Group will be defined by a set of guiding principles. The principles emphasise that synergies and coordination between the Rio Conventions can be best implemented at the national level,[109] seemingly demonstrating Parties' reluctance to undertake more coordinated action, and their preference towards leaving it to states to interpret their international commitments in a synergistic fashion, if they so desire.[110] This cautious approach may be further evinced from proposed Principle 3, according to which the Joint Liaison Group recognizes the distinct and specific objectives of each convention and their different Parties, as well as the individual mandates and independent status of their treaty bodies and secretariats.

The very fact that guidance from the CBD COP has not percolated through the UNFCCC texts may be read as a signal of this continued restraint. The text of the Cancun Agreements provides scope for limited optimism, as it asserts that REDD+ activities should "complement" or be "consistent with the objectives of national forest programmes and relevant international conventions and agreements".[111] As mentioned earlier, whereas some Party submissions to the UNFCCC SBSTA supported exploiting synergies between the CBD and the UNFCCC, little has so far been done to take action in this direction and COP17 may be regarded as a missed opportunity.[112]

Rebus sic stantibus, it remains up to individual countries to interpret their obligations under the CBD and the UNFCCC, as well as guidance by their treaty bodies, in an integrated fashion. In this regard, guidance included in the decisions of the CBD treaty bodies should be interpreted and implemented in good faith by the Parties to that convention. As the next section shows, these considerations largely apply also to human rights.

15.4 REDD+ and Human Rights

REDD+ activities raise specific human rights concerns. Millions of people in developing countries live near or in forests, or depend on forest resources for their livelihood. For this reason, ensuring that REDD+ activities do not affect the human rights of indigenous peoples and forest-dependent communities is a major challenge.[113]

---

[109] Ibid., Principle 1.

[110] Ibid.

[111] Ibid., Appendix I, para. 2(a).

[112] Kovacevic, "Durban talks both good and bad for REDD+", supra, note 83.

[113] For an overview of the numerous implications that REDD+ may have in this connection, see, for example: Report of the Office of the United Nations High Commissioner for Human Rights on the relationship between climate change and human rights, UN Doc. A/HRC/10/61, 15 January 2009, para. 94; Simone Lovera, *The Hottest REDD Issues: Rights, Equity, Development, Deforestation and Governance by Indigenous Peoples and Local Communities* (Gland: IUCN, 2008); David Humphreys, *Climate Change and Human Rights: A Rough Guide* (Versoix: International Council on Human Rights Policy, 2008), at 33; and David J. Kelly, "The Case for Social Safeguards in a Post-2012 Agreement on REDD", 6 *Law, Environment and Development Journal* (2010), 61.

Similarly to biodiversity, so far human rights concerns have been addressed in the context of safeguards that Parties should promote and support when undertaking REDD+ activities.[114] The present section therefore investigates how 'social' safeguards may address synergies and overlaps with international obligations for the protection of human rights. The term 'human rights' is deployed here to refer to the rights of individuals and groups that are recognised as such in international law.[115] States' obligations concerning the protection of human rights depend on their commitments under the relevant international treaties. Virtually all developing country Parties to the UNFCCC eligible to participate to REDD+ have ratified the two 'foundational' human rights treaties, the International Covenant on Civil and Political Rights[116] and the International Covenant on Economic, Social and Cultural Rights.[117] Several of them have also ratified regional human rights treaties.[118]

Climate change is likely to undermine the realization of a range of human rights.[119] The question of whether climate change is in itself a violation of human rights exceeds the scope of this article.[120] A report by the Office of the High Commissioner for Human Rights has summarised difficulties in framing the relationship between climate change and human rights in these terms.[121]

---

[114] Decision 1/CP.16, supra, note 25, section IIIc and Annex I.

[115] For this use of the term, see Thomas Buergenthal, "Human Rights", in *Max Planck Encyclopedia of Public International Law* (Oxford: Oxford University Press, 2007), at 2.

[116] International Covenant on Civil and Political Rights, New York, 16 December 1966, in force 23 March 1976, 999 *United Nations Treaty Series* (1983), 171. As of 17 March 2012, the convention had 167 Parties.

[117] International Covenant on Economic, Social and Cultural Rights, New York, 16 December 1966, in force 3 January 1976, 993 *United Nations Treaty Series* (1983), 3. As of 17 March 2012, the convention had 160 Parties.

[118] Client Earth and World Resources Institute, "Lessons from International and Regional Instruments. A Submission to SBSTA", 2011, available at: http://www.clientearth.org/climate-and-forests/climate-forests-publications/reddsafeguards-sbsta-submission-1549 (last accessed on 9 March 2012), at 24–25.

[119] In this connection, see for example International Law Association, The Hague Conference, The legal principles relating to climate change, 2010, at 35.

[120] So far the most salient attempt to argue that this is the case has been the so–called Inuit Petition before the Inter-American Commission on Human Rights. See: Organization of American States, "Inter-American Commission on Human Rights, Petition Seeking Relief from Violations Resulting from Global Warming Caused by Acts and Omissions of the United States", 2005, available at: http://www.inuitcircumpolar.com/files/up- loads/icc-files/FINALPetitionICC.pdf (last accessed on 20 March 2012). For a review see for example Joanna Harrington, "Climate Change, Human Rights and the Right to Be Cold", 18 *Fordham Environmental Law Review* (2007), 513; Amy Sinden, "Climate Change and Human Rights", 27 *Journal of Land Resources and Environmental Law* (2007), 255; Hari M. Osofsky, "The Inuit Petition as a Bridge? Beyond the Dialectics of Climate Change and Indigenous Peoples' Rights", 31 *American Indian Law Review* (2007), 675.

[121] Report of the Office of the United Nations High Commissioner for Human Rights on the relationship between climate change and human rights, UN Doc. A/HRC/10/61, 15 January 2009, at 70: "Qualifying the effects of climate change as human rights violations poses a series of difficulties. First, it is virtually impossible to disentangle the complex causal relationships linking historical greenhouse gas emissions of a particular country with a specific climate change-related effect,

For the purposes of the present article, it suffices to underscore how the relationship between human rights protection and climate change action is composite. On the one hand, human rights may be viewed as constraints to action to address climate change, for instance, with regard to the impacts of some climate change mitigation activities and initiatives on the rights of vulnerable subjects, as seen in the case of the CDM. On the other hand, human rights may support and guide the drafting of rules concerning action to mitigate and adapt to climate change. In both connections, there are areas of potential friction and synergies with action to carry out climate change mitigation and adaptation.

These questions have attracted growing attention[122] and have ultimately gained recognition in the Cancun Agreements, which emphasize that "Parties should, in all climate change related actions, fully respect human rights".[123] The preamble to the agreements also "takes note" of the UN Human Right Council Resolution 10/4, Human Rights and Climate Change,[124] which includes a list of rights "most" affected by climate change[125] and calls for all relevant human rights special procedures to "give consideration to the issue of climate change within their respective

let alone with the range of direct and indirect implications for human rights. Second, global warming is often one of several contributing factors to climate change related effects, such as hurricanes, environmental degradation and water stress. Accordingly, it is often impossible to establish the extent to which a concrete climate change-related event with implications for human rights is attributable to global warming. Third, adverse effects of global warming are often projections about future impacts, whereas human rights violations are normally established after the harm has occurred." For a review of the report, see for example Marilyn Averill, "Linking Climate Litigation and Human Rights", 18 *Review of European Community and International Environmental Law* (2009), 139; Ron Dudai, "Climate Change and Human Rights Practice. Observations on and around the Report of the Office of the High Commissioner for Human Rights on the Relationship Between Climate Change and Human Rights", 1 *Journal of Human Rights Practice* (2009), 294; and Marc Limon, "Human Rights Obligations and Accountability in the Face of Climate Change", 38 *Georgia Journal of International and Comparative Law* (2009–2010), 543.

122 Several non-governmental organizations have addressed the issue. See, for example, Oxfam, *Climate Wrongs and Human Rights: Putting People at the Heart of Climate-Change Policy. Oxfam Briefing Paper 117* (Oxford: Oxfam, 2009); Greenpeace, *Human Rights and the Climate Crisis: Acting Today to Prevent Tragedy Tomorrow* (Amsterdam: Greenpeace, 2009); Earthjustice and Center for International Environmental Law (CIEL), *Global Warming and Human Rights* (Washinghton DC: Earthjustice and CIEL, 2009); and CIEL, *Climate Change and Human Rights: Practical Steps for Implementation* (Washinghton DC: CIEL, 2009).

123 See Decision 1/CP.16, supra, note 24, para. 8.

124 Ibid., Preamble, para. 7.

125 In this connection, UN Human Rights Council Resolution 10/4 notes that: "climate change-related impacts have a range of implications, both direct and indirect, for the effective enjoyment of human rights including, inter alia, the right to life, the right to adequate food, the right to the highest attainable standard of health, the right to adequate housing, the right to self-determination and human rights obligations related to access to safe drinking water and sanitation, and recalling that in no case may a people be deprived of its own means of subsistence." UN Human Rights Council, Resolution 10/4, 41st meeting, UN Doc. A/HRC/10/L., 25 March 2009, preamble.

mandates."[126] The scope for institutional cooperation in this connection is however limited, due to the fragmented nature of States' obligations in the human rights field. And while there are some potential overlaps and synergies between the climate change regime and human rights, the Cancun Agreements have left open the question of how states will take human rights impacts into account in construing, developing, and operationalising their commitments to combat climate change.

With regard to REDD+, it is clear that all Parties must comply with their extant international obligations, including human rights obligations, whenever they undertake REDD+ activities.[127] These obligations are especially relevant to the treatment of subjects that live near or in forests, or depend on forest resources.[128] More controversially, it may be argued that also actors, such as states, but also corporations, international and non governmental organizations, providing funding for REDD+ and purchasing REDD-generated offsets may indirectly or directly be liable for the human rights impact of these activities.[129]

The impact of REDD+ on human rights greatly depends on how governments choose to implement the related activities. International rules on REDD+ should not engender frictions with the protection of human rights. These concerns are particularly conspicuous with regard to human rights most closely related to access to land and forest resources, as well as procedural rights concerning participation to the design and implementation of REDD+ activities.

Although it is not realistic to expect that REDD+ may address all human rights concerns related to access and use of forests resources, it is paramount to ensure that at the very least it does not provide perverse incentives to carry out human rights violations. And while states carrying out REDD+ activities must comply with their human rights obligations, the need to provide some internationally coordinated guidance on the issue has become increasingly apparent, as the next section illustrates.

15.4.1 Social Safeguards Under the UNFCCC

Concerns regarding the social impact of REDD+ have been increasingly, albeit not comprehensively, addressed in COP decisions and in the framework of processes dealing with the implementation of REDD+. Adhesion to REDD+ may not become

[126] Ibid., para. 3. In 2011 the Human Rights Council adopted another resolution, requesting the Office of the High Commissioner for Human Rights to convene a seminar "on addressing the adverse impacts of climate change on the full enjoyment of human rights, with a view to (…) forging stronger interface and cooperation between the human rights and climate change communities." Resolution A/HRC/18/L.26/Rev1 on Human Rights and Climate Change, 30 September 2011, para. 2.

[127] Decision 12/CP.17, supra, note 25, Appendix I, para. 2.

[128] For an overview, see, for example, Griffiths and Martone, *Seeing REDD?*, supra, note 36.

[129] This argument is made by David Takacs, "Forest Carbon Offsets and International Law: A Deep Equity Legal Analysis", 22 *Georgetown International Environmental Law Review* (2010), 521, at 572–573.

an instrument to impose upon states obligations contained in treaties they have not ratified. In this regard, the inclusion of stringent rules on participation of indigenous peoples and local communities in COP decisions may be regarded as infringements upon national sovereignty.[130] It is however theoretically possible that, for example, only Parties that have ratified specific human rights instruments be given access to REDD+ funding. As it shall be seen, this approach has seemingly been supported in the framework of the UN-REDD programme. The UNFCCC COP, instead, has undertaken a more cautious stance to the matter.

The Cancun Agreements include an all-encompassing reference to human rights, according to which "Parties should, in all climate change related actions, fully respect human rights".[131] The Agreements further mention the need to engage a broad range of stakeholders at the global, regional, national and local levels, adding that "effective participation of women and indigenous peoples are important for effective action on all aspects of climate change".[132] More specifically, the Agreements request that, when developing and implementing their REDD+ national strategies or action plans, developing country Parties address, amongst others, "land tenure issues, forest governance issues, gender considerations and the safeguards ensuring the full and effective participation of relevant stakeholders, inter alia indigenous peoples and local communities".[133]

Safeguards included in the Cancun Agreements do not provide human rights as such, but address some human rights concerns raised by REDD+ and may be regarded as *sui generis* conflict avoidance devices. As mentioned earlier, the Cancun Agreements generally require that REDD+ activities "complement or are consistent with relevant international conventions and agreements".[134] In addition, the Agreements include a list of specific 'social safeguards' which REDD+ activities should promote and support. The list includes: respect for the knowledge and rights of indigenous peoples and members of local communities, by taking into account relevant international obligations, national circumstances and laws; as well as the full and effective participation of relevant stakeholders, in particular indigenous peoples and local communities.[135]

Some safeguards clearly encompass elements included in a number of international and regional human rights instruments ratified by several Parties to the UNFCCC. As mentioned earlier, developing country Parties undertaking REDD+ activities will be expected to periodically provide a summary of information on how safeguards are being addressed and respected, to be included in their periodic

---

[130] For this opinion, see for example Daniel Zarin et al., *Reducing Emissions from Deforestation and Degradation (REDD): An options assessment. Report prepared for the Government of Norway* (Washington DC: Meridian Institute, 2009), at 25.

[131] Decision 1/CP.16, supra, note 25, para. 8.

[132] Ibid., para. 7.

[133] Ibid., para. 72.

[134] Ibid., Appendix I, para. 2(a).

[135] Ibid., Appendix I, para. 2(c,d). Emphasis added.

national communications pursuant to the UNFCCC.[136] The systems for providing information on how the safeguards are addressed and respected should "provide transparent and consistent information that is accessible by all relevant stakeholders and updated on a regular basis".[137]

In this regard, observers' submissions to the SBSTA in preparation for COP 17 pointed to potential synergies with action undertaken towards implementing Parties' obligations under other international law instruments and, in particular, human rights instruments.[138] Some submissions even included lists of relevant international human right instruments and provisions,[139] and underscored the potential to integrate related reporting obligations.[140] The problem with this proposal is, however, that States' obligations clearly depend on those human rights instruments that they have ratified.[141] Moreover, only a handful of Party submission to the SBSTA mentioned human rights.[142]

At COP 17, Parties were unwilling to endorse suggestions that information systems for safeguards embody and reinforce the guidance and rules of existing environmental and human rights treaties. Indeed, the decision that was eventually adopted does not make any reference to human rights, human rights treaties or related reporting obligations.[143] Concerns regarding overlaps with human rights obligations have, however, been addressed in greater detail in the context of initiatives concerning REDD+ readiness.

The FCPF and the UN-REDD Programme have sought to coordinate their action concerning the social impact of REDD+, but so far their efforts have achieved limited results. There has instead been an inherent duplication of efforts between the two processes, which is motivated by the diverging human rights

[136] Decision 12/CP.17, supra, note 25, Appendix I, paras. 3–4.

[137] Ibid., para. 2(b).

[138] See for example Client Earth and World Resources Institute, Lessons from International and Regional Instruments, supra, note 117, at 2.

[139] Ibid., at 8–13.

[140] Ibid., at 7–20.

[141] The existence of customary international law concerning human rights is contentious. It may be argued that the practice of a state in relation to its own citizens is a matter of domestic jurisdiction, which is in principle without significance for the establishment of a customary rule. See for example Hugh Thirlway, "The Sources of International Law," in Malcolm D. Evans (ed.), *International Law* (Oxford University Press, 2010), 95, at 104. However, the argument has also been made that general *opinio iuris* as well as *consuetudo* exist to maintain that at least some human rights 'are today crystallized in the realm of customary international law'. See, for example, International Law Association, *The Hague Conference Report, Rights of Indigenous Peoples* (International Law Association, 2010), found at http://www.ila-hq.org/en/committees/index.cfm/cid/1024, at 43.

[142] In the submission by Switzerland, for example, it is suggested that "information systems for safeguards embody and reinforce the guidance and rules of existing environmental and human rights treaties, particularly UNDRIP and FLEGT, when relevant." UNFCCC, Submissions from Parties: Switzerland, supra, note 79, para 100. For a synopsis of submissions, see Gaia Larsen, Daniela Rey, and Florence Daviet, "Map of SBSTA Submissions: REDD+ Safeguard Information System" (World Resource Institute, Washington, 2012).

[143] Decision 12/CP.17, supra note 25.

agendas of the institutions supporting these two processes the World Band and the UN. While in fact the World Bank has been reluctant to engage with human rights,[144] the UN-REDD Programme is subject to UN policies on the matter.[145] The discrepancy between approaches undertaken by the two institutions is therefore more evident in connection with human rights, than with biodiversity safeguards.

The UN-REDD Programme is in the process of adopting a set of Social and Environmental Principles and Criteria, drawing upon the guidance provided by the Cancun Agreements, as well as existing "knowledge and literature on safeguards."[146] The principles and criteria are meant to reflect the UN-REDD Programme's "responsibility to apply a human-rights based approach to its programming and uphold UN conventions, treaties and declarations."[147] They are furthermore meant to help countries meet their commitments under a number of international agreements and, when applicable, the decisions taken by their treaty bodies. The UN-REDD Programme is also developing specific Guidelines for Seeking the Free, Prior, and Informed Consent (FPIC) of Indigenous Peoples and other Forest Dependent Communities,[148] outlining a normative, policy and operational framework' for partner countries.[149] Finally, the UN-REDD Programme is in the process of establishing a mechanism to address grievances from affected individuals and communities, as well as reports of non-compliance with its guidance and policies.[150]

---

[144] In this connection, at the World Bank website specifies that: "The World Bank needs to undertake analytic work to examine how human rights fit within the constitutional framework and what positive contribution they could make to the development process." The World Bank, "Human Rights", 2011, available at: http://go.worldbank.org/72L95K8TN0 (last accessed on 17 February 2012).

[145] The mainstreaming of human rights within the UN system has been at the centre of a series of UN reform efforts since 1997, when the former Secretary-General designated human rights as a crosscutting theme to the work of the Organization. At the request of the Secretary-General, an interagency plan of action on strengthening human rights related UN action at country level has been developed and adopted. See the dedicated portal: http://hrbaportal.org/?page_id=929 (last accessed on 17 February 2012).

[146] UN-REDD Programme, "Social and Environmental Principles and Criteria. Version 1", 2011, available at: http://www.un-redd.org/Multiple_Benefits/SEPC_BeRT/tabid/991/Default.aspx (last accessed on 08 March 2012), para 2.

[147] Ibid. This approach is outlined in the UN Common Understanding on the Human Rights-Based Approach to Development Cooperation, 2003, according to which "all programmes of development co-operation, policies and technical assistance should further the realisation of human rights as laid down in the Universal Declaration of Human Rights and other international human rights instruments", available at: http://www.undg.org/archive_docs/6959-The_Human_Rights_Based_Approach_to_Development_Cooperation_Towards_a_Common_Understanding_among_UN.pdf (last accessed on 17 February 2012).

[148] UN-REDD Programme, "Guidelines on Free, Prior and Informed Consent Draft for Comment", 2011, available at: http://www.unredd.net/index.php?option=com_docman&task=doc_download&gid=6369&Itemid=53 (last accessed on 8 March 2012).

[149] Ibid., para. 4.

[150] Ibid., para. 17.

The FCPF approach has been far more restrained. Operations of the FCPF are subject to the World Bank's operational policies. Activities affecting indigenous peoples must comply with the World Bank Operational Policy 4.10 on indigenous peoples,[151] which aims to "ensure that the development process fully respects the dignity, human rights, economies, and cultures of indigenous peoples,"[152] and calls for the recipient country to engage in "a process of free, prior, and informed *consultation*."[153] The terminology deployed here requires mere 'consultation' rather than 'consent.' There is no textual reference to FPIC, although guidance subsequently adopted by the World Bank's International Finance Corporation explicitly includes FPIC requirements.[154] Further discrepancies with the approach undertaken by the UN-REDD Programme are evident also where the policy mentions "measures to avoid potentially adverse effects on the indigenous peoples' communities and, when avoidance is not feasible, minimize, mitigate, or compensate for such effects."[155]

These aspects of World Bank's operational policy have attracted much criticism. For example, the UN Permanent Forum on Indigenous Issues has emphasised that "displacement and exclusion of indigenous peoples from their forests, which may be triggered by projects funded by the Partnership Facility, should be avoided at all costs" and also that "the choice not to participate in REDD+ or in the projects supported by the Partnership Facility should be respected." [156] Also the FCPF's decision not to deploy World Bank's policies in full in the REDD+ readiness phase has been stigmatised.[157]

The duplication of human rights standards under the two processes could well lead to a situation where the same activities in the same countries may be subject to different standards, depending on which entity is providing the funding.[158] In this regard, COP17 has specified that REDD+ activities should be consistent with safeguards "regardless of the source or type of financing."[159] However, while both FCPF and the UN-REDD Programme remain subjected to the safeguards included in the Cancun Agreements, their diverging approaches towards the social impact of REDD+ clearly provides an element for fragmentation and potential incongruence.

---

[151] World Bank, "Operational Policy 4.10 – Indigenous Peoples", 2 May 2011, available at: www.worldbank.org/safeguards (last accessed on 17 February 2012).

[152] Ibid., at 1.

[153] Ibid., para 1. Emphasis added.

[154] IFC, Performance Standard 7, Indigenous Peoples (World Bank, 2012), at 11–12, available at http://www1.ifc.org/wps/wcm/connect/1ee7038049a79139b845faa8c6a8312a/PS7_English_2012.pdf?MOD=AJPERES.

[155] Ibid.

[156] See United Nations Permanent Forum on Indigenous Issues Report on the Seventh Session, UN Doc. E/2008/43. E/C.19/2008/13, 21 April 2008, para. 7.

[157] Dooley et al., *Smoke and Mirrors,* supra, note 88, at 7 and 12.

[158] At the time of writing, while 37 countries have adhered to the FCPF, only 14 receive direct support from the UN-REDD Programme. Only eight countries currently receive funds from both, although this number is likely to grow, as the number of REDD+ Partner countries increases.

[159] Decision 2/CP.17, supra, note 25, para. 63.

15.4.2 Social Safeguards for REDD+: Some Preliminary Conclusions

As seen in connection of biodiversity safeguards, the fact that states must comply with their extant obligations, and that most countries eligible to host REDD+ activities are also Parties to a range of human rights agreements, makes it desirable that the REDD+ mechanism be designed in such a way as to be in line with widely endorsed human rights. This proposition may, however, be hard to put into practice because not all state Parties eligible to carry out REDD+ activities are Parties to the same human rights treaties. As recalled earlier, REDD+ may not become an instrument to impose upon states obligations contained in treaties they have not ratified. It has for example been argued that "stringent international rules on participation of indigenous peoples and local communities in the international climate regime may infringe on national sovereignty," as "issues of human rights are addressed through other international instruments, and therefore need not be dealt with under the UNFCCC."[160]

Contrary to what was observed with regard to biodiversity safeguards, furthermore, there is no obvious institutional interlocutor that may provide guidance on overlaps between processes dealing with human rights, as a result of the fragmented nature of states' commitments under different treaties.

So far the UNFCCC COP has provided only some cautious guidance on the issue and COP 17 may be regarded as a missed opportunity to bring REDD+ activities in line with international human rights commitments.[161] In the meantime, the FCPF and UN-REDD Programme have adopted their own separate standards on the matter.

While the UNFCCC COP may even decide not to adopt specific rules concerning compliance with safeguards, it remains up to the FCPF and UN-REDD Programme to adequately implement and monitor the enforcement of the standards they have elaborated. This seems a challenging endeavour for the two organisations, and reliance upon other international bodies and procedures may provide some scope to avoid duplicating efforts. As COP 17 has recently confirmed, however, there does not presently seem to be sufficient political will to undertake such a course of action. Although the FCPF and the UN-REDD Programme have no formal links to the UNFCCC system, they presently remain the main sources of reference for countries seeking to address human rights concerns associated with REDD+. Fragmented guidance provided by these two institutions is in urgent need of increased and enhanced coordination to ensure the establishment of a level playing field and the pursuit of co-benefits. Some reflections on the means for the establishment of such level playing field are made in the following section.

---

[160] Zarin et al., *Reducing Emissions from Deforestation and Degradation (REDD): An options assessment*, supra, note 129, at 25.

[161] Kate Dooley and Kate Horner, "FW Special Report. Durban aimed to save the market not the climate", 2012, available at: http://www.fern.org/node/5106 (last accessed on 13 March 2012), at 1.

15.5 Conclusions: REDD+ as a Vehicle for Coordination

Establishing a REDD+ mechanism under the UNFCCC has turned out to be an extremely complex and challenging undertaking. Awareness of shortcomings in the CDM has complicated the rule-making process for REDD+. The normative framework for REDD+ that has emerged so far is highly fragmented and made of instruments of mixed legal nature, which are likely to lead to overlaps and even conflict with one another.

Although a single integrated legal instrument dealing with REDD+ would be the most coherent and structurally appealing option, such an outcome has not materialized yet. In fact, little law on REDD+ exists. In the meantime, REDD-readiness activities have been developed in a fragmented and largely uncoordinated fashion, under the oversight of multiple institutions. Fragmentation is therefore looming. When and if Parties to the UNFCCC will draft operational rules for REDD+, they will be confronted with pitfalls inherent in overlaps with extant processes and international instruments, most saliently those concerning the protection of biodiversity and human rights.

As mentioned earlier, fragmentation may be addressed through conflict avoidance, the application of interpretative principles and institutional cooperation. The fragmented nature of states' commitments limits the possibility to address overlaps by means of institutional cooperation. Whereas this potential is evident in connection with biodiversity, the issue is more complex in relation to human rights. In both cases, however, lack political will has so far resulted in substantial inaction.

Since states undertaking REDD+ activities must comply with their extant obligations, and most REDD+ eligible countries are Parties to the same international agreements, it seems desirable to design REDD+ in line with widely ratified treaties concerning the protection of biodiversity and human rights. It would seem more efficient that conditions such as these become part of the normative apparatus supporting REDD+ at the international level, thus providing rules that are certain and equal for all parties. This outcome, however, presently remains distant.

Within this context, the adoption of safeguards may be regarded as a pragmatic effort to address overlaps with states' obligations and pursue synergies with other international instruments, albeit in an unconventional fashion. The vague and hortatory safeguards adopted by the UNFCCC COP thus far, however, seem ill-suited to avoid perverse outcomes. In the meantime, the two key processes in charge of facilitating REDD-readiness have each developed their own sets of standards. As they strive towards a common end, these processes risk functioning as a vehicle of further fragmentation.

Systemic integration presently remains the main tool to address overlaps between REDD+ and other international law regimes. And while REDD+ may not become a means to impose upon States obligations contained in treaties they have not ratified, it should not become a justification to overlook obligations that they have already undertaken. Given the large number of Parties to the instruments analysed here, and the fragmented nature of the legal landscape relevant for REDD+, ensuring that these simple tenets are upheld is going to be a difficult task.

Ultimately, national lawmakers retain the tools to interpret their obligations in a 'synergic' fashion. However, if the experience of the CDM offers an apt model for comparison, there are inherent limits to autonomous management and good-faith pragmatism. Lack of internationally coordinated standards may lead to perverse outcomes and concerted international action is arguably necessary to ensure the pursuit of REDD's co-benefits.

The evolving debate on REDD+ safeguards is precisely the result of a growing awareness on the need to complement economic incentives for carbon sequestration with measures to protect other ecosystems services, as well as their providers and users. In this regard, REDD+ constitutes an opportunity to adopt a set of rules about acceptable forest uses, which also streamline commitments undertaken in connection with the protection of biodiversity and human rights.

The realisation of multiple potential benefits of REDD+ is, however, faced with great challenges. The questions being discussed in connection with the establishment of the REDD+ mechanism are not new and have hampered official development assistance and forest conservation efforts for decades. Any instrument designed to reform the *status quo* is going to face considerable and potentially even insurmountable challenges.

In this process, perfect should not become the enemy of good. REDD+ should instead capitalise upon synergies that may facilitate the pursuit of triple-win outcomes associated with the coincidence of international law objectives on climate change mitigation, biodiversity protection and human rights. So far, Parties to the UNFCCC have taken opportunities for synergies into account only marginally. While it is arguably unwise to further complicate the already troubled negotiation process under the UNFCCC with detailed requirements on REDD+ co-benefits, the rewards of 'getting it right' stretch beyond climate change mitigation and include the conservation of some of the world's richest terrestrial ecosystems and the promotion of sustainable development. REDD+ potentially lends itself to this purpose and could provide the leverage to finally overcome obstacles that have hindered international cooperation this far. Only time, however, will tell whether REDD+ will manage to achieve this complex set of objectives.

Chapter 16
Climate Change and Trade: At the Intersection of Two International Legal Regimes

Kati Kulovesi

Abstract This chapter examines substantive and institutional linkages between the United Nations Framework Convention on Climate Change and the World Trade Organization. It focuses on identifying potentially sensitive areas in their relationship, including sustainability requirements targeting processes and production methods, as well as measures targeting carbon leakage and competitiveness concerns. It also discusses institutional and doctrinal challenges related to fragmentation of international law and highlights problems that could arise if a climate change related dispute was considered by the WTO dispute settlement system. The chapter concludes that the trade and climate regimes are increasingly relevant for each other and that they are not necessarily rivals – both could benefit from identifying and promoting unexploited synergies between the two regimes. However, closer cooperation and institutional coordination may be needed in the future in order to avoid mutually unhelpful institutional and legal clashes.

16.1 Introduction

Efforts are currently taking place under the United Nations Framework Convention on Climate Change (UNFCCC)[1] to strengthen international climate change cooperation. The number of countries implementing climate change mitigation policies continues to rise and climate change law expands. Ultimately, the battle against

---

Kati Kulovesi is Adjucnt Professor in Climate Law at the Department of Law, University of Eastern Finland. She holds a Ph.D. from the London School of Economics and Political Science.

[1] United Nations Framework Convention on Climate Change, 9 May 1992, New York, in force 21 March 1994, 31 *International Legal Materials* (1992), 849 (UNFCCC).

K. Kulovesi (✉)
Department of Law, University of Eastern Finland, PO Box 111, 80101 Joensuu, Finland
e-mail: kati.kulovesi@uef.fi

E.J. Hollo et al. (eds.), *Climate Change and the Law*, Ius Gentium: Comparative Perspectives on Law and Justice 21, DOI 10.1007/978-94-007-5440-9_16,
© Springer Science+Business Media Dordrecht 2013

climate change necessitates a fundamental transformation to a global low-carbon economy in the coming decades. This objective has, without a doubt, important economic implications. As Newell and Patterson indicate, "[i]n responses to climate change, we have the first instance of societies seeking a dramatic transformation of the entire global economy."[2] Many climate policies will have repercussions on the trade realm and are thus relevant from the point of view of the World Trade Organization (WTO) law. As a result, the territory shared between the UNFCCC and the WTO legal regimes is expanding. While both of these prominent international legal regimes have evolved significantly over the past 20 years, they have done so in a relatively comfortable insulation from each other. Their linkages, synergies and tensions are, however, becoming increasingly apparent.

Against this background, this chapter focuses on substantive and institutional linkages between the UNFCCC and WTO. It proceeds from the argument that a shift is taking place in attitudes towards the relationship between climate change and trade, bringing the two legal regimes closer together. It first studies substantive linkages between the UNFCCC and WTO regimes. It then analyses institutional issues, most notably potential role of the WTO dispute settlement system system in solving conflicts between climate change and trade. The chapter concludes that in terms of substance, the territory shared between the UNFCCC and WTO is already considerable and can be expected to expand in the future. In institutional terms, however, links between the two regimes remain weak. One of the key challenges is that WTO law and the WTO dispute settlement system are likely to dominate in disputes concerning linkages between trade and climate change. Addressing substantive fragmentation of international law through the WTO dispute settlement system is, however, far from ideal solution.[3] To avoid damaging conflicts, more attention to substantive synergies and institutional cooperation between the UNFCCC and WTO will be necessary in the future.

16.2 Climate Change and Trade: Shifting Attitudes

The relationship between climate change and trade remains subject to a rich debate. One of the most profound questions is whether it will be possible to reconcile trade and economic growth with the objective of avoiding dangerous anthropogenic climate change.[4] Not surprisingly, there are divergent views on this fundamental issue.[5]

[2] Peter Newell and Matthew Patterson, *Climate Change Capitalism: Global Warming and the Transformation of the Global Economy* (Cambridge et al.: Cambridge University Press, 2010), at 1.

[3] For detailed analysis, see Kati Kulovesi, *The WTO Dispute Settlement System: Challenges of the Environment, Legitimacy and Fragmentation* (The Netherlands: Kluwer Law International, 2011), at 261–267.

[4] UNFCCC, supra, note 1, Art. 2.

[5] For an useful overview of the spectrum of political views in the climate change debate, see Anthony Giddens, *The Politics of Climate Change* (Cambridge, UK and Malden, MA, USA: Polity Press, 2009), at 49 et seq.

For some, climate change strengthens the case against capitalism and the market economy. The 2010 World People's Conference on Climate Change and the Rights of Mother Earth in Bolivia highlighted capitalism as the cause of climate change, arguing that:

> The capitalist system has imposed on us a logic of competition, progress and limitless growth. This regime of production and consumption seeks profit without limits, separating human beings from nature and imposing a logic of domination upon nature, transforming everything into commodities: water, earth, the human genome, ancestral cultures, biodiversity, justice, ethics, the rights of peoples, and life itself.[6]

At the other extreme, a shrinking but vocal group continues to deny that scientific evidence on anthropogenic climate change is strong enough to warrant action. According to Carter's recent book, "to say that human-caused global warming is proven to be a dangerous problem is untrue, and to introduce futile policies aimed at 'stopping climate change' is both vainglorious and hugely expensive."[7] The skeptical environmentalist Lomborg argues, in turn, that reducing greenhouse gas emissions is one of the least helpful ways of serving humanity or the environment as, in his view, a focus on global warming could make future generations worse off.[8] Much of the early debate about climate change tended to be polarized around these extremes.

Given the mounting scientific evidence of climate change and its impacts, concerted efforts have taken place to frame climate change mitigation both as an economic necessity and opportunity. As a result, an important shift seems to be gradually taking place in attitudes concerning the relationship between climate change and trade. The 2006 Stern Review constituted a milestone by making the economic case for prompt action to mitigate climate change:

> if we don't act, the overall costs and risks of climate change will be equivalent to losing at least 5% of global GDP [Gross Domestic Product] each year, now and forever. If a wider range of risks and impacts is taken into account, the estimates of damage could rise to 20% of GDP or more In contrast, the costs of action – reducing greenhouse gas emissions to avoid the worst impacts of climate change – can be limited to around 1% of global GDP each year.[9]

Shortly afterwards, the European Commission presented its proposal for the Climate and Energy Package as the "climate change opportunity" of the European Union (EU).[10] It argued that the challenge of adapting to the demands of a low-carbon

---

[6] People's Agreement of Cochabamba, adopted by the World People's Conference on Climate Change and the Rights of Mother Earth, 24 April 2010, available at: http://pwccc.wordpress.com/2010/04/24/peoples-agreement/ (last accessed on 6 March 2012).

[7] Robert M. Carter, *Climate: The Counter Consensus* (UK: Stacey International, 2010), at 218.

[8] Björn Lomborg, Cool It: *The Sceptical Environmentalist's Guide to Global Warming* (New York: Alferd A. Knop, 2007), at 8–9.

[9] Nicholas Stern, *The Economics of Climate Change: The Stern Review* (Cambridge et al.: Cambridge University Press, 2007), at xv.

[10] Commission Communication: 20 20 by 2020: Europe's Climate Change Opportunity, COM(2008)30.

economy can be met and "it also opens the door to new opportunities. There is a real potential to make climate-friendly policies a major driver for growth and jobs in Europe. Europe can show that necessary change can go hand in hand with the process of securing a competitive and prosperous economy."[11]

Subsequently, 'green economy' and 'green growth' have become popular notions in a world trying to come to grips with the financial and economic crisis, and preparing to mark the 20th anniversary of the 1992 Rio Conference on the Environment and Development. Transition to a green economy is depicted as a move that will give those who succeed significant advantage over competitors. According to Friedman and Mandelbaum: "There is every reason to believe… that clean energy will become the successor to information technology as the next major cutting-edge industry on which the economic fortunes of the richest countries will depend."[12] They lament that the US "does not have in place the rules, standards, regulations and price signals – the market ecosystem – to stimulate thousands of green innovators in thousand of green garages to devise the break-through technologies that will give us multiple sources of abundant, cheap, reliable, carbon-free energy."[13] While the US has retaken the top position in investment in clean energy,[14] President Barack Obama recently urged Congress to "double-down" on the clean energy industry, indicating he would not "cede the wind or solar or battery industry to China or Germany because we refuse to make the same commitment here."[15]

It seems, then, that the trend is towards what Newell and Paterson call climate capitalism: "a model which squares capitalism's need for continual economic growth with substantial shifts from carbon-based industrial development."[16] As a result, climate change is increasingly penetrating international economic reality. From the legal perspective, these developments render the relationship between the UNFCCC and WTO legal regimes increasingly important.

Both the UNFCCC and WTO have gone through important progress during the past two decades. With its 195 Parties, the UNFCCC is now virtually universal in scope. It has given birth to a complex and detailed legal regime, which continues to evolve through the annual sessions of the Conference of the Parties (COP).[17] Around the same time that the UNFCCC entered into force in 1994, international trade negotiators concluded the Uruguay Round, marking a watershed in the evolution of the international trade regime. The WTO was established to administer the regime

---

[11] Ibid., at 3.

[12] Thomas L. Friedman and Michael Mandelbaum, *That Used to Be US: What Went Wrong with America – and How Can It Come Back?"* (USA: Little Brown, 2011), at 196.

[13] Ibid., at 197.

[14] 16 *Bridges Weekly Trade News Digest*, 25 January 2012.

[15] Ibid.

[16] Newell and Paterson, *Climate Capitalism*, supra, note 2, at 1.

[17] For a general overview, see Farhana Yamin and Joanna Depledge, *The International Climate Change Regime. A Guide to Rules, Institutions and Procedures* (Cambridge, UK: Cambridge University Press, 2004).

and the substantive scope of international trade law expanded. Attention began to shift towards non-tariff trade barriers,[18] including intellectual property, technical barriers to trade as well as sanitary and phytosanitary measures. The Uruguay Round also lead to the establishment of a strong dispute settlement mechanism. In contrast to previous practice under the General Agreement on Tariffs and Trade (GATT),[19] which required a consensus by all Parties to adopt a dispute settlement report, under the new WTO dispute settlement system, the adoption of reports can only be prevented by a consensus. A permanent Appellate Body was also established and the WTO dispute settlement system has a compulsory and exclusive jurisdiction in the field of WTO law. It is also competent to authorize trade sanctions against non-compliant WTO Members. Largely due to these reforms, international trade law currently stands out as one of the strongest areas of international law. Given that the WTO dispute settlement system would be the likeliest forum for settling a dispute related to trade and climate change, much of the discussion about the relationship between climate change and trade also tends to be dominated by the perspective of WTO law.

16.3 Substantive Linkages Between the UNFCCC and WTO Legal Regimes

Scholarly analysis concerning the relationship between the international trade and climate change regimes often alludes to the possibility of conflicts between the two regimes. This is linked to the debate concerning fragmentation of international law. In 2006, the International Law Commission (ILC) finalised a report dedicated to "difficulties arising from the diversification and expansion of international law."[20] According to the ILC, the essential concern about fragmentation is "the rise of specialized rules and rule-systems that have no clear relationship to each other."[21] There are often valid reasons for treating topics such as trade and climate change separately. As the ILC explains, "new types of specialized law do not emerge accidentally but seek to respond to new technical and functional requirements."[22] The downside is that: "Each rule-complex or regime comes with

[18] Thomas Cottier, "From Progressive Liberalization to Progressive Regulation in WTO Law", 9 *Journal of International Economic Law* (2006), 779, at 783.

[19] General Agreement on Tariffs and Trade 1994, 15 April 1994, Marrakesh Agreement Establishing the World Trade Organization, Annex 1A, The Legal Texts: The Results of the Uruguay Round of Multilateral Trade Negotiations 17 (1999), 33 *International Legal Materials* (1994), 1153.

[20] International Law Commission, *Fragmentation of International Law: Difficulties Arising from the Diversification and Expansion of International Law*. Report of the Study Group of the International Law Commission on the Fragmentation of International Law. Finalized by Martti Koskenniemi, UN Doc. A/CN.4/L.682, 13 April 2006.

[21] Ibid., at 245.

[22] Ibid., at 14.

its own principles, its own expertise and 'ethos,' not necessarily identical to the ethos of the neighbouring specialization. 'Trade law' and 'environmental law,' for example, have highly specific objectives and rely on principles that may often point in different directions."[23]

Reflecting the trend of fragmentation and specialization, the UNFCCC and WTO regimes have evolved in parallel but largely in isolation from each other. International climate change and trade negotiations are frequented mostly by different delegates and experts. There are no formal mechanisms to coordinate the two processes and ensure that their outcomes are mutually compatible. Also in the domestic sphere, trade and climate issues are mostly dealt with by different ministries and government experts. This means that the UNFCCC world remains relatively unknown to WTO experts, and vice versa. However, in recent years, calls have increasingly been made to enhance the mutual supportiveness of the two regimes.[24] My intention in this section is to examine substantive links between climate policies related to achieving the UNFCCC's ultimate objective of avoiding dangerous anthropogenic climate change and WTO rules.

16.3.1 Trade Measures and Other Climate Policies

The 2007 Fourth Assessment Report (AR4) by the Intergovernmental Panel on Climate Change (IPCC) had an important influence on international climate policy. It painted a grim picture of rapidly increasing greenhouse gas concentrations and already observable impacts of climate change.[25] However, it also drew attention to "substantial economic potential" to mitigate global greenhouse gas emissions in the coming decades.[26] In this regard, the IPCC identified the following as the most promising climate change mitigation policies:[27]

---

[23] Ibid., at 14.

[24] See, for example, Ludvine Tamiotti et al., *Trade and Climate Change: A Report by the United Nations Environment Programme and the World Trade Organization* (Geneva: UNEP and WTO, 2009); and Tracey Epps and Andrew Green, *Reconciling Trade and Climate: How the WTO Can Help Address Climate Change* (*Cheltenham*, UK and Northampton, MA, USA: Edward Elgar, 2010).

[25] IPCC, "Summary for Policymakers," in: Rajendra K. Pachauri and Andy Reisinger (eds.), *Climate Change 2007: Synthesis Report. Contribution of Working Groups I, II and III to the Fourth Assessment Report of the Intergovernmental Panel on Climate Change* (Geneva: Intergovernmental Panel on Climate Change, 2007).

[26] IPCC, "Summary for Policymakers," in: Bert Metz et al. (eds.), *Climate Change 2007. Mitigation of Climate Change. Working Group III Contribution to the Fourth Assessment Report of the Intergovernmental Panel on Climate Change* (Cambridge and New York: Cambridge University Press, 2007), at 9.

[27] I have used these two lists as examples also in Kulovesi, *The WTO Dispute Settlement System*, supra, note 3, at 232–233.

- regulations and standards (which provide "some certainty about emission levels" and "may be preferable to other instruments")[28];
- taxes and charges (which set a price for carbon "but cannot guarantee a particular level of emissions")[29];
- tradable permits (which "will establish a price for carbon")[30];
- financial incentives such as subsidies and tax credits (which generally come at higher economic cost but "are often critical to overcome barriers")[31];
- voluntary agreements between the government and industry (the majority of which "has not achieved significant emissions reductions beyond business as usual")[32];
- information instruments (however, "their impact on emissions has not been measured yet")[33];
- research, development and deployment (to "stimulate technological advances")[34]; and
- voluntary actions by corporations, local and regional authorities, NGOs etc. (which, on their own, generally have limited impact on the emissions).[35]

The IPCC AR4 makes no mention of the WTO or international trade law, illustrating the insulation of the international climate change and trade communities from each other. However, WTO scholars have identified a number of potential conflicts between climate policies and WTO law, including the GATT, General Agreement on Trade in Services (GATS),[36] Agreement on Technical Barriers to Trade (TBT Agreement)[37] and the Agreement on Subsidies and Countervailing Measures (SCM).[38] From the point of view of WTO law, some of the most relevant potential climate policies include:

- trade bans or punitive tariffs on certain products or on products originating from countries that are not participating in climate change mitigation;[39]

---

[28] IPCC Working Group III, "Summary for Policymakers," supra, note 26, at 19.

[29] Ibid.

[30] Ibid.

[31] Ibid.

[32] Ibid.

[33] Ibid.

[34] Ibid.

[35] Ibid.

[36] General Agreement on Trade in Services, 15 April 1994, Marrakesh Agreement Establishing the World Trade Organization, Annex 1B, The Legal Texts: The Results of the Uruguay Round of Multilateral Trade Negotiations 284 (1999), 33 *International Legal Materials* (1994) 1167.

[37] Agreement on Technical Barriers to Trade, 15 April 1994, 1867 *United Nations Treaty Series*, 493.

[38] Agreement on Subsidies and Countervailing Measures, 15 April 1994, Marrakesh Agreement Establishing the World Trade Organization, Annex 1A, The Legal Texts: The Results of the Uruguay Round of Multilateral Trade Negotiations 275 (1999), 1867 *United Nations Treaty Series* 14.

[39] The World Bank, *International Trade and Climate Change. Economic, Legal and Institutional Perspectives* (Washington DC: World Bank, 2008), at 37.

- product standards and regulations, including energy efficiency and other sustainability requirements;
- border tax adjustments, including taxing imported products based on their carbon content and other similar requirements on imported products or importers, such as the requirement to purchase emission allowances[40];
- using the system for Generalised Trade Preferences to encourage mitigation by developing countries[41];
- various climate change related subsidies[42]; and
- compulsory licensing and other measures to relax intellectual property rights for climate-friendly technologies.[43]

Concerning the compatibility of the climate policies and measures included in the list with WTO rules, legal analysis would depend largely on the detailed design of the measure. In more general terms, it is possible to imagine a conflict whereby measures based on specific provisions of the UNFCCC, Kyoto Protocol[44] or a possible new climate treaty adopted in 2015 for the post-2020 period are challenged under WTO law. However, it is useful to keep in mind that neither the UNFCCC nor the Kyoto Protocol contains trade sanctions nor has their introduction been contemplated in the ongoing negotiations.[45] In this sense, it has been argued that the provisions of the Kyoto Protocol "do not conflict directly with the WTO regime"[46] and a straightforward conflict with WTO rules therefore appears as unlikely. It is, however, possible to imagine a WTO dispute involving such climate policies or measures that have not been not clearly prescribed under the UNFCCC regime but that are closely related to the implementation of its ultimate objective in Article 2 of the Convention to avoid dangerous anthropogenic climate change. This scenario looks, in fact, fairly plausible.

From the point of view of WTO law, one of the key challenges in the relationship between climate change and trade is the tendency under the UNFCCC regime to

---

[40] For discussion, see for example, Tamiotti et al., *Trade and Climate Change*, supra, note 24, at 98–110; Epps and Green, *Reconciling Trade and Climate*, supra, note 24, at 122–141.

[41] Epps and Green, *Reconciling Trade and Climate*, supra, note 25, at 180–188; Michael McKenzie, "Climate Change and the Generalized System of Preferences," 11 *Journal of International Economic Law* (2008), 679.

[42] Tamiotti et al., *Trade and Climate Change*, supra, note 24, at 110–117; Epps and Green, *Reconciling Trade and Climate*, supra, note 24, at 103–121; Magnus Lodefalk and Mark Sotery, "Climate Measures and WTO Rules on Subsidies," 39 *Journal of World Trade* (2005), 23.

[43] I have used these examples also in Kulovesi, *The WTO Dispute Settlement System,* supra, note 24, at 233–234.

[44] Kyoto Protocol to the United Nations Framework Convention on Climate Change, Kyoto, 10 December 1997, in force 16 February 2005, 37 *International Legal Materials* (1998), 22.

[45] See, however, Epps and Green, *Reconciling Trade and Climate*, supra, note 24, at 56–60.

[46] Matthieu Wemaere and Charlotte Streck, "Legal Ownership and Nature of Kyoto Units and EU Allowances," in David Freestone and Charlotte Streck (eds) *Legal Aspects of Implementing the Kyoto Protocol Mechanisms. Making Kyoto Work* (Oxford et al.: Oxford University Press, 2005) 35, at 46.

avoid prescribing detailed climate policies and measures. This trend is reflected in the Kyoto Protocol, otherwise based on 'top down' legally binding emission reduction targets for developed countries. The recent shift under the UNFCCC towards informal, 'bottom up' mitigation pledges means even less clarity in terms of mitigation commitments.

While Article 3.1 of the Kyoto Protocol contains a clear obligation for Annex I countries to reduce greenhouse gas emissions, it does not contain any binding details on how this should be achieved. It merely indicates that:

> The Parties included in Annex I shall, individually or jointly, ensure that their aggregate anthropogenic carbon dioxide equivalent emissions of the greenhouse gases listed in Annex A do not exceed their quantified emission limitation and reduction commitments inscribed in Annex B…with a view to reducing their overall emissions of such gases by at least 5% below 1990 levels in the commitment period 2008 to 2012.

According to Article 2.1 of the Protocol, in achieving its emissions target, each Annex I country "shall implement and/or further elaborate policies and measures in accordance with its national circumstances." Article 2.1 also contains a non-exhaustive and non-binding list of policies and measures that its implementation could entail.[47] The Kyoto Protocol thus leaves ample discretion for each Annex I country in terms of the climate policies and measures that it will implement in order to comply with its legally binding emission reduction target.[48]

Some WTO scholars have argued that climate polices would be easier to justify under WTO rules if they were specifically prescribed by the Kyoto Protocol.[49] While the argument does have its merits from the perspective of WTO law, it sounds rather unrealistic when taking into consideration the political realities and evolution of the UNFCCC regime. The vagueness of the UNFCCC and the Kyoto Protocol in terms of policies and measures for their implementation is not an accident. On the contrary, countries have been firm in international climate negotiations on the need to minimise external constraints on domestic policy choices, particularly in such sensitive sectors as energy, transport, industry, agriculture and forestry.[50] In addition, countries are not identical in terms of their emissions profiles and mitigation potential. International climate negotiators have therefore chosen to defer to

---

[47] Kyoto Protocol, supra, note 44, Art. 2.1. The policies and measures listed in Article 2.1 are: Enhancement of energy efficiency; protection and enhancement of carbon sinks; promotion of sustainable forms of agriculture; taking measures related to renewable energy and carbon dioxide sequestration; addressing market imperfections (such as tax and duty exemptions and subsidies in greenhouse gas emitting sectors); encouraging appropriate reforms to promote policies and measures that limit or reduce emissions in relevant sectors; addressing emissions in the transport sector; and addressing methane emissions.

[48] UNFCCC, supra, note 1, Art. 4.2(e) and the Kyoto Protocol, supra, note 44, Art. 2.4 also contain some provisions on the possible coordination of policies and measures, but these issues have been highly controversial and there have been no formal attempts for coordination. See, Yamin and Depledge, *The International Climate Change Regime,* supra, note 17, at 113–117.

[49] Andrew Green, "Climate Change, Regulatory Policy and the WTO: How Constraining Are Trade Rules?," 8 *Journal of International Economic Law* 8 (2005), 143, at 187.

[50] Yamin and Depledge, *The International Climate Change Regime,* supra, note 17, at 115.

countries' national decisions on which economic sectors to involve in climate change mitigation and how much each sector will contribute to such efforts.

As indicated above, the UNFCCC regime has began to shift towards an even more flexible approach to climate change mitigation than the 'top down' legal structure of the Kyoto Protocol. The first commitment period under the Kyoto Protocol expires at the end of 2012. While general agreement on a second commitment period was reached at the 2011 UN Climate Change Conference in Durban, negotiations on key details are pending. Furthermore, countries like Japan and Russia have already announced that they will not participate in a second commitment period and Canada has withdrawn from the Protocol completely. As is widely known, the US never ratified the Protocol, which does not introduce targets for major emerging economies, such as China, India, Brazil and South Africa. From 2013 onwards, mitigation by several key countries is therefore likely be based on voluntary mitigation pledges, most of which were originally made in the context of the 2009 UN Climate Change Conference in Copenhagen. Following COP 16 in Cancun, the respective pledges by developed and developing countries have been 'anchored' in two information documents.[51] Their international legal status is unclear as is their relevance under WTO law. This introduces an unavoidable element of uncertainty into the relationship between the UNFCCC and WTO regimes.

The basic treaty provision that addresses the relationship between climate change mitigation and international trade is Article 3.5 of the UNFCCC, which provides that:

> The Parties should cooperate to promote a supportive and open international economic system that would lead to sustainable economic growth and development in all Parties, particularly developing country Parties, thus enabling them better to address the problems of climate change. Measures taken to combat climate change, including unilateral ones, should not constitute a means of arbitrary or unjustifiable discrimination or disguised restriction on international trade.

The last sentence of Article 3.5 echoes language used in Article XX of the GATT. Also the Kyoto Protocol gives some guidance on the relationship between climate change mitigation and other policy objectives. According to its Article 2.3, Annex I parties "shall strive to implement" their policies and measures "in such a way as to minimize adverse effects, including adverse effects of climate change, on international trade, and social, environmental and economic impacts on other Parties," especially in developing countries. In other words, international trade has been listed in Article 2.3 as one of the several areas potentially affected by the implementation of the Kyoto Protocol. Adverse effects on other Parties should be minimised, including "adverse impacts of climate change."

---

[51] UNFCCC, Compilation of economy-wide emission reduction targets to be implemented by Parties included in Annex I to the Convention, Revised Note by the Secretariat, UN Doc. FCCC/SB/2011/INF.7 June 2011; UNFCCC, Compilation of information on nationally appropriate mitigation actions to be implemented by Parties not included in Annex I to the Convention, Note by the Secretariat, UN Doc. FCCC/AWGLCA/2011/INF.1, 18 March 2010.

In the ongoing long-term negotiations under the UNFCCC, oil producing countries and emerging economies have proposed clearly prohibiting unilateral trade measures to address climate change.[52] The issue remains controversial, however, and these proposals have not lead to the adoption of more specific language on the relationship between trade and climate change.[53] At the 2011 UN Climate Change Conference in Durban, climate negotiators did, however, agree to establish a work programme and a forum on response measures. This aspect of the UNFCCC regime will focus on the negative and positive impact of measures taken to mitigate climate change.[54] It seems reasonable to expect that trade measures will be among the issues considered under the new initiative. This means that there is a new process under the UNFCCC where the relationship between trade and climate change could be considered.

16.3.2 Climate-Related Regulations and Standards

In practice, interaction between the WTO and UNFCCC regimes will focus on trade aspects of climate change policies and measures designed to reduce greenhouse gas emissions. As discussed above, the IPCC AR4 shows that climate mitigation policies and measures could take a variety of forms. Along with various other climate policies, technical regulations and standards related to energy efficiency have increased in recent years.[55] In principle, such requirements can apply to either products themselves or production methods. Their key objectives include reducing greenhouse gas emissions and energy consumption from either the use or production of products.

Under WTO law, such measures are mainly regulated under the GATT and the TBT Agreement, which covers both mandatory technical regulations[56] and voluntary standards.[57] For technical regulations, the TBT Agreement requires that they do not discriminate between domestic and imported 'like' products[58] and create unnecessary obstacles to international trade, in other words, that they are not more trade-restrictive than necessary to fulfil a legitimate objective, such as protection of human health or

[52] Kati Kulovesi, Sabrina Shaw and Stanley W. Burghiel, "Trade and Environment: Old Wine in New Bottles?", in Pamela S. Chasek and Lynn M. Wagner (eds), *The Roads from Rio: Lessons Learned from Twenty Years of Multilateral Environmental Negotiations* (New York and London: Routledge, 2012), 174.

[53] Ibid.

[54] Decision 8/CP.17, Forum and work programme on the impact of the implementation of response measures, UN Doc. FCCC/CP/2011/9/Add.2, 15 March 2012.

[55] For examples, see Tamiotti et al., *Trade and Climate Change*, supra, note 24, at 118–120.

[56] The definition of "regulation" is found in the TBT Agreement, supra, note 38, Annex I, para. 1.

[57] Ibid., Annex I, para. 2 contains definition of a "standard."

[58] Ibid., Art. 2.1 provides that imported products "shall be accorded treatment no less favourable than accorded to like products of national origin and to like products originating in any other country."

the environment.[59] Regulations must also be based on international standards where they exist unless these are ineffective or inappropriate.[60] Examples of climate policies already considered by the TBT Committee include fuel standards for cars, eco-design requirements for energy-using products, energy-efficiency programmes for consumer products and emission limit values for diesel engines.[61]

EU climate change law, for instance, includes several examples of regulations and standards to mitigate climate change. Given that voluntary agreements with the automobile industry failed to produce the desired outcome, the EU will enforce binding targets for carbon dioxide from passenger cars from 2012.[62] The EU has also adopted various eco-design requirements, including a Regulation to phase-out inefficient light bulbs from the market – a policy estimated to reduce carbon dioxide emissions by 32 million tonnes by 2020.[63] This is an area where synergies have been identified with climate change mitigation and the WTO regime: the World Bank indicates that developing countries such as China and India have emerged as major players in the fluorescent lamps market and that liberalising trade in fluorescent lamps could promote energy-efficient lighting.[64] Also trade in energy products, including renewable energy, could raise questions under WTO law, including the GATT and GATS.[65]

16.3.2.1 The Case of Biofuels

Trade-issues surrounding biofuels are a topical example of complex linkages between the WTO and climate change related policies.[66] With rising concerns over climate change and national energy security, biofuels are becoming increasingly popular.[67] Several concerns have, however, been identified in relation to biofuels production.[68]

---

[59] Ibid., Art. 2.2.

[60] Ibid., Art. 2.4.

[61] WTO, "Activities of the WTO and the Challenge of Climate Change", available at: http://www.wto.org/english/tratop_e/envir_e/climate_challenge_e.htm (last accessed 8 March 2012).

[62] European Parliament and Council Regulation (EC) No 443/2009 setting emission performance standards for new passenger cars as part of the Community's integrated approach to reduce CO_2 emissions from light-duty vehicles, OJ 2009 L 140/1.

[63] Commission Regulation (EC) No 244/2009 implementing Directive 2005/32/EC of the European Parliament and of the Council with regard to ecodesign requirements for non-directional household lamps, OJ 2009 L 76/3.

[64] The World Bank, *International Trade and Climate Change*, supra, note 39, at 67–68.

[65] Christina Voigt, "WTO Law and International Emissions Trading: Is There Potential for Conflict," 2 *Carbon and Climate Law Review* (2008), 52, at 55–57; Robert Howse, "World Trade Law and Renewable Energy: The Case of Non-Tariff Measures", 2009, available at: http://www.unctad.org/trade_env/test1/publications/UNCTAD_DITC_TED_2008_5.pdf (last accessed on 8 March 2012).

[66] I have discussed these issues similarly in Kulovesi, *The WTO Dispute Settlement System*, supra, note 3, at 247–251.

[67] For an overview of biofuels as a trade issue, see Doaa Abdel Motaal, "The Biofuels Landscape: Is There a Role for the WTO?," 42 *Journal of World Trade* 42 (2008), 61.

[68] See in general, Elisa Morgera, Kati Kulovesi and Ambra Gobena, (eds), *Case Studies on Bioenergy Policy and Law: Options for Sustainability* (Rome: Food and Agriculture Organization of the United Nations, 2009).

Depending on where and how they are produced, biofuels can have only a limited impact on greenhouse gas emissions. They are also associated with other environmental concerns, such as deforestation and loss of biodiversity. The relationship between biofuels and food security has also received ample attention. The key concern is that especially in developing countries, agricultural land will be used for biofuels production for export markets rather than for feeding local populations.

In 2007, the EU adopted a 10% target for renewable energy in the transport sector by 2020.[69] After a lengthy debate on the sustainability of biofules, the Directive on the promotion of the use of energy from renewable sources came to include sustainability criteria applicable to both domestically produced and imported biofuels.[70] Only biofuels complying with the criteria will be counted towards the 10% target. The Directive requires that greenhouse gas emission savings from biofuels must be at least 35% until 2017, and 50% from 2017 onwards.[71] It also contains the requirement that raw material for biofuels counted against the 10% target cannot originate from land with high biodiversity value and lays down detailed criteria for determining what constitutes such land.[72] Furthermore, the EU sustainability criteria exclude biofuels originating from peatland[73] or land with high carbon stock in 2008 where the land has subsequently lost this status.[74] The Directive also includes provisions on verification of compliance with the sustainability criteria, including that the Commission must endeavour to conclude bilateral or multilateral agreements with third countries on sustainability criteria that corresponds with the requirements set out in the Directive.[75]

From the point of view of WTO law, the EU's sustainability criteria for biofuels are interesting in that they seek to impact land use in foreign countries, touching upon the long-standing debate over processes and production methods (PPMs). The key question is whether two goods can be distinguished based on greenhouse gas emissions or other environmental criteria associated with their production process but not affecting the physical characteristics of the product.[76] Under Article

[69] We have analyzed these in detail in Kati Kulovesi, Elisa Morgera and Miquel Muñoz, "Environmental Integration and Multifaceted International Dimensions of EU Law: Unpacking the 2009 Climate and Energy Package", 48 *Common Market Law Review* (2011), 829.

[70] European Parliament and Council Directive 2009/28/EC on the promotion of the use of energy from renewable sources and subsequently repealing Directives 2001/77/EC and 2003/30/EC, OJ 2009 L 140/16.

[71] Ibid., Art. 17.2.

[72] Ibid., Art. 17.3.

[73] Ibid., Art. 17.5.

[74] Ibid., Art. 17.4.

[75] Ibid., Art. 18.4.

[76] There has been extensive scholarly debate on this issue for the past 20 years. For discussion in the climate change context, see Richard G. Tarasofsky, "Heating Up International Trade Law: Challenges and Opportunities Posed by Efforts to Combat Climate Change," 2 *Carbon and Climate Law Review* (2008), 7, at 8–10. For an overview of legal arguments in the PPMs debate, see: Gabrielle Marceau and Joel P. Trachtman, "The Technical Barriers to Trade Agreement, the Sanitary and Phytosanitary Measures Agreement and the General Agreement on Tariffs and Trade. A Map of World Trade Organization Law of Domestic Regulation of Goods," 36 *Journal of World Trade* 36 (2002), 856.

III:4 of the GATT, imported products may not be treated less favourably than domestic 'like' products. According to the WTO Appellate Body, the key criteria for analysing the 'likeness' of products takes into consideration their physical characteristics, end-uses, consumer preferences and tariff classification.[77] The Appellate Body also accepted that health risks were relevant for determining whether asbestos and other products with similar end uses were 'like.'[78] In the context of climate change, it has been argued that differences in consumer preferences could be used to justify differences in regulatory treatment of climate-friendly and non-friendly products, including biofuels.[79]

In cases where a violation of the GATT is found, the measure could still be justifiable under the general exceptions in Article XX of the GATT. Under Article XX(b) of the GATT, WTO Members can justify measures that are "necessary" to protect human, animal or plant life or health. Under Article XX(g) of the GATT, they can adopt measures relating to conservation of exhaustible natural resources.[80] According to a two-tired analysis developed by the Appellate Body, a measure must also comply with the chapeau of Article XX. The chapeau requires that the measure does not constitute "a means of arbitrary or unjustifiable discrimination" or "disguised restriction on international trade." The TBT Agreement is interesting in that it goes beyond the non-discrimination requirement in Article III:4 of the GATT. This means that under the TBT Agreement, regulations may not create unnecessary obstacles to international trade, in other words, they may not be more restrictive than necessary to achieve a legitimate objective, such as protecting human health or safety, or the environment.

Some scholars have subsequently argued that the EU's sustainability criteria for biofuels probably violate the GATT.[81] According to Mitchell and Tran, for instance, such "biofuels that differ only on the basis of the emissions-related sustainability criteria are probably not like products, because the emissions they generate are arguably a physical characteristic of the final product."[82] However, in their view,

---

[77] Report of the Appellate Body Report in *European Community – Measures Affecting Asbestos and Asbestos-Containing Products,* WT/DS135/AB/R, 12 March 2001, paras. 113–142.

[78] For discussion: Robert Howse and Elisabeth Tuerk, "The WTO Impact on Internal Regulations – A Case Study of the Canada-EC Asbestos Dispute," in George A. Bermann and Petros C. Mavroidis, (eds), *Trade and Human Health and Safety. Columbia Studies in WTO Law and Policy* (New York: Cambridge University Press, 2006), 77.

[79] For discussion of consumer preferences in the context of emission trading, see Voigt, "WTO Law and International Emissions Trading," supra, note 65, at 54.

[80] For discussion on how Article XX of the GATT might be applied in the context of climate change: Green, "Climate Change, Regulatory Policy and the WTO," supra, note 50, at 175–179 and 183–187; Voigt, "WTO Law and International Emissions Trading," supra, note 66, at 59–63; and Tarasofsky, "Heating Up International Trade Law," supra, note 76, at 9–10.

[81] Andrew Mitchell and Christopher Tran, "The Consistency of the EU Renewable Energy Directive with the WTO Agreements", Georgetown Law Faculty Working Papers, October 2009, available at: http://scholarship.law.georgetown.edu/fwps_papers/119 (last accessed on 8 March 2012).

[82] Ibid., at 3.

such biofuels "that differ only on the basis of the land-related sustainability criteria are probably like products, because the land from which they are derived does not affect the physical characteristics of the final product."[83] They conclude that the less favourable treatment of biofuels not meeting the land-related sustainability criteria "is likely to result in inconsistency with the EC"s substantive obligations under GATT."[84] Scott, in turn, has indicated that: "Those familiar with the contours of WTO law will perceive in the text of the renewable energy directive efforts to align the scope and application of the sustainability criteria with the multiple requirements of WTO law… Yet, while the EU's sustainability criteria have clearly been designed with WTO law in mind, still they *may* be vulnerable to challenge in a number of respects."[85] This is because the criteria addresses PPMs and seeks to protect the environment outside the EU.[86] It remains to be seen whether controversies related to biofuels eventually end up in the WTO, or whether they will be resolved outside the WTO. Under the UNFCCC, however, there have been no proposals to address specific questions concerning biofuels or create internationally-agreed sustainability criteria.[87]

Biofuels also raise a number of other trade-related questions, including the custom classification of biofuels,[88] their technical specifications, as well as tariff reductions.[89] Also various governmental measures to support the production and use of biofuels are relevant from the point of view of WTO law: tax exemptions, regulatory exemptions, subsidies, government procurement preferences and so on. These examples illustrate how the territory shared by the WTO and UNFCCC regimes is growing – and that it can be expected to expand even further as countries across the world strive to achieve a transition towards a highly energy efficient low-carbon economy.

16.3.3 Introducing a Price for Carbon

Creating a price for greenhouse gas emissions is commonly viewed as one of the most efficient ways to mitigate climate change.[90] As seen above, the IPCC AR4 indicated that emissions trading and a carbon tax are the key tools to achieve

---

[83] Ibid.

[84] Ibid., at 12.

[85] Joanne Scott, "The Multi-Level Governance of Climate Change", Centre for Law and Governance, University College London Working Paper, 009/10, at 58–59, available at: http://www.ucl.ac.uk/laws/clge/wp-series/ucl_clge_009_10.pdf (last accessed 9 March 2012).

[86] Ibid., at 60.

[87] Outside the UNFCCC and the WTO, however, there have been various sustainability certification initiatives for biofuels by governments and non-governmental actors alike. For an overview, see ibid., 59–66.

[88] Motaal, "The Biofuels Landscape," supra, note 67, at 76–78.

[89] Ibid., at 78–83.

[90] See for example, Stern, *The Economics of Climate Change*, supra, note 9, at 354 et seq; IPCC Working Group III, "Summary for Policymakers," supra, note 26, at 19.

this objective.[91] One of the main concerns for countries introducing a price for carbon emissions is that all countries are not participating in climate change mitigation efforts in equal terms. The use of carbon border adjustments to address carbon leakage and competitiveness concerns has been widely discussed. This section discusses the relationship between WTO rules and national emissions trading schemes and, more specifically, plans to introduce measures to address energy-intensive imports.

In order to introduce a price for greenhouse gas emissions, emissions trading schemes are being planned and implemented in various countries.[92] The most prominent example is the EU Emissions Trading Scheme (EU ETS), launched in 2005. It currently involves some 11,500 installations and represents around 40% of the total greenhouse gas emissions of the EU.[93] The second phase of the EU ETS, taking place in 2008–2012, has been designed to comply with the emission reduction targets under the Kyoto Protocol. In its third phase, taking place from 2013 to 2020, the EU ETS will cover new industries and greenhouse gases, an annually declining EU-wide emissions cap will be introduced and the auctioning of allowances will be increased.[94] In 2012, the ETS expanded to cover emissions from the vast majority of flights landing and taking off from EU airports, including foreign airlines.[95] As it will be explained below, the inclusion of foreign airlines in the ETS has been highly controversial, illustrating the political sensitivity of the topic discussed in this section. While several countries are taking steps towards introducing a price for greenhouse gas emissions, the world is still far away from a global carbon price or even trading among the major emitters or most polluting economic sectors. Questions concerning carbon leakage and competitiveness are therefore close to the surface in countries that have introduced a price for carbon or are contemplating doing so.

Border tax adjustments have been described as "a controversial area of overlap between international trade rules and climate policy."[96] Their rationale is to offset the negative environmental and competitiveness effects caused by national climate policies, most notably, the introduction of a price for greenhouse gas emissions. The idea of taking measures against imports first gained ground in Europe after the US decision in 2001 not to join the Kyoto Protocol. The question was whether the EU should level the playing field by imposing a carbon tax on imports from the US.[97]

---

[91] IPCC Working Group III, "Summary for Policymakers," supra, note 26, at 19.

[92] For more details, see chapters in Part V of this book, including those on Australia and Japan.

[93] European Parliament and Council Directive 2003/87/EC establishing a scheme for greenhouse gas emission allowance trading within the Community, OJ 2003 L 275/32.

[94] European Parliament and Council Directive 2009/29/EC amending Directive 2003/87/EC so as to improve and extend the greenhouse gas emission allowance trading system of the Community, OJ 2009 L 140/63.

[95] European Parliament and Council Directive 2008/101/EC amending Directive 2003/87/EC so as to include aviation activities in the scheme for greenhouse gas emission allowance trading within the Community, OJ 2008 L 8/3.

[96] Epps and Green, *Reconciling Trade and Climate*, supra, note 24, at 122.

[97] Frank Biermann and Rainer Brohm, "Implementing the Kyoto Protocol without the United States: The Strategic Role of Energy Tax Adjustments at the Border," 4 *Climate Policy* (2005), 289.

According to the former European Trade Commissioner Mandelson, however, taxing imports from countries that have not ratified the Kyoto Protocol is,

> highly problematic under current WTO rules and almost impossible to implement in practice.... Not participating in the Kyoto process is not illegal. Nor is it a subsidy under WTO rules. How would we choose what goods to target? China has ratified the Kyoto but has no Kyoto targets because of its developing country status. The US has not, but states like California have ambitious climate change policies.[98]

With the new Obama Administration taking office in 2009, the US re-engaged the negotiations under the UNFCCC and for a while, planned a federal cap-and-trade scheme for greenhouse gas emissions that would have also included imports of energy intensive products. The European Commission also raised the idea of a 'carbon equalization system' when proposing revisions to the ETS for the third trading period in 2013–2020. The rationale is that if other developed countries and major emitters of greenhouse gases are not participating in an international climate agreement,

> ...this could lead to an increase in greenhouse gas emissions in third countries where industry would not be subject to comparable carbon constraints ("carbon leakage"), and at the same time could put certain energy-intensive sectors and sub-sectors in the Community which are subject to international competition at an economic disadvantage. This could undermine the environmental integrity and benefit of actions by the Community.[99]

Plans to launch a federal emissions trading scheme in the US have subsequently been frozen. Also the European Commission has taken a cautious stance on the inclusion of imports in the ETS. Regardless, the question of border carbon adjustments has anything but disappeared from the academic and policy discussion.[100] From the point of view of WTO law, one of the key problems is that measures targeting greenhouse gas emissions from the manufacturing of imported products bring to the fore controversial themes from the classic trade-environment debate: Are trade measures targeting PPMs sometimes justifiable under WTO law, as it would seem in the light of the *Shrimp-Turtle* decisions,[101] and under what conditions?

---

[98] EU Trade Commissioner Peter Mandelson, "How Trade Can Be Part of the Climate Change Solution, "18 December 2006, available at: http://trade.ec.europa.eu/doclib/docs/2006/December/tradoc_131802.pdf (last accessed on 31 March 2012).

[99] European Parliament and Council Directive 2009/29/EC amending Directive 2003/87/EC so as to improve and extend the greenhouse gas emission allowance trading system of the Community, OJ 2009 L 140/63, para. 25 of the chapeau.

[100] See, for example, Susanne Droege, "Do Border Measures Have a Role in Climate Policy?" 11 *Climate Policy* (2011), 1185; Ludvine Tamiotti, "The Legal Interface between Carbon Border Measures and Trade Rules", 11 *Climate Policy* (2011), 1202; Stéphanie Monjon and Philippe Quirion, "A Border Adjustment for the EU ETS: Reconciling WTO Rules and Capacity to Tackle Carbon Leakage," 11 *Climate Policy* (2011), 1212.

[101] Report of the Appellate Body in *United States – Import Prohibition of Certain Shrimp and Shrimp Products*, WT/DS58/AB/R, 12 October 1998; and *Report of the Appellate Body in United States – Import Prohibition of Certain Shrimp and Shrimp Products, Recourse to Article 21.5 of the DSU by Malaysia*, WT/DS58/AB/RW, 22 October 2001.

In general, taxes on imported products are subject to the national treatment principle, enshrined in Article III:2 of the GATT. Furthermore, imposing discriminatory taxes only on imports from certain WTO Members would probably violate the Most Favoured Nation principle in Article I of the GATT.[102] However, measures violating these provisions could sometimes be justifiable under Article XX of the GATT. Also some other legal questions could arise. Could, for instance, the free allocation of allowances to certain sectors be seen as a subsidy under the SCM?

It is interesting to note that the principle of common but differentiated responsibilities and respective capabilities in Article 3 of the UNFCCC could pose some challenges from the point of view trade measures targeting imports from developing countries. Most developing countries would be likely to argue that targeting their imports circumvents some of the two key principles that have been guiding international climate change cooperation, namely the principle of common but differentiated responsibilities and the leadership role of industrialised countries.[103] Similar arguments have been made in the context of the inclusion of aviation emissions in the EU ETS.[104]

Overall, the debate concerning measures to address carbon leakage remains inconclusive. Much would seem to depend on the detailed design and application of the measures. Some have argued that measures to address carbon leakage could be designed in a way that is compatible with WTO law.[105] Others are more sceptical[106] and some have also warned that in response to such measures, developing countries could start imposing tariffs on products from developed countries based on criteria such as *per capita* greenhouse gas emissions.[107] It is interesting to note that closely related questions concerning, for example the exercise of extraterritorial jurisdiction and unilateralism, have recently surfaced in the context of the inclusion of emissions from foreign airlines in the ETS from 2012 onwards.[108] In the aviation context, the

---

[102] Tarasofsky, "Heating Up International Trade Law," supra, note 76, at 8.

[103] David Stanway, "China says 'carbon tariff' proposals breach trade rules," Reuters, 3 July 2009, available at: http://www.reuters.com/article/2009/07/03/us-china-climate-idUSTRE5620FV20090703 (last accessed on 31 March 2012).

[104] Joanne Scott and Lavanya Rajamani, "EU Climate Change Unilateralism: International Aviation in the European Union Emissions Trading Scheme," 23 *European Journal of International Law* (2012), 469.

[105] For discussion, see Voigt, "WTO Law and International Emissions Trading," supra, note 66, at 59–63; Epps and Green, *Reconciling Trade and Climate*, supra, note 24, at 122.

[106] Jason E. Bordoff, "International Trade Law and the Economics of Climate Policy. Evaluating the Legality and Effectiveness of Proposals to Address Competitiveness and Leakage Concerns", June 2008, available at: http://www.brookings.edu/events/2008/~/media/Files/events/2008/0609_climate_trade/2008_bordoff.pdf, (last accessed on 31 March 2012).

[107] Rachel Brewster, "The Problem with Carbon Tariffs: They Aren't Fair," *The Christian Science Monitor*, 20 April 2009, available at http://www.csmonitor.com/Commentary/Opinion/2009/0420/p09s01-coop.html (last accessed on 31 March 2012).

[108] I have analyzed this in detail in Kati Kulovesi, "Make Your Own Special Song even if Nobody Else Sings Along: International Aviation Emissions and the EU Emissions Trading Scheme", 2 *Climate Law* (2011), 535.

WTO's role has thus far been marginal as air traffic is mostly excluded from the GATS. It has thus been argued that the WTO is not relevant to the heated international dispute on aviation emissions.[109] There have, however, been attempts to argue that the GATT could apply to the case.[110]

16.3.4 Promoting Clean Energy: Pending Disputes at the WTO

For many years, the debate on trade and climate change remained rather abstract. However, as the proliferation of climate policies continues and climate change law expands, concrete linkages between the UNFCCC and the WTO are becoming more apparent. Interestingly, this reality is already reflected in the WTO dispute settlement system where some disputes related to renewable energy technologies have found their way. The most advanced is the *Canada-Certain Measures Affecting the Renewable Energy Generation Sector* case between Japan and Canada where a panel was composed in the autumn of 2011.[111] Another dispute, *China-Measures concerning Wind Farm Equipment* between the US and China reached the consultation stage before China ended the disputed measures.[112] The US has, however, also considered bringing a similar dispute to the WTO concerning China's subsidies to its solar panel industry.[113]

In the renewable energy dispute between Canada and Japan, the key question relates to a Canadian feed-in tariff, which, as such, is a popular way to promote the generation of renewable energy. Under its feed-in tariff programme, Ontario guarantees electricity purchase prices, grid access, and long-term contracts to renewable energy producers thus limiting their risks and supporting needed investments.[114] However, to receive such support, renewable energy producers must ensure that a

---

[109] See, for example, Eckhard Pache, "On the Compatibility with International Legal Provisions of Including Greenhouse Gas Emissions from International Aviation in the EU Emission Allowance Trading Scheme as a Result of the Proposed Changes to the EU Emission Allowance Trading Directive", Legal Opinion Commissioned by the German Federal Ministry for the Environment, Nature Conservation and Nuclear Safety, 2008, available at: www.bmu.de/files/pdfs/allgemein/application/pdf/aviation emission trading.pdf (last accessed on 31 March 2012), at 5–6.

[110] This argument by India is briefly mentioned, for example, in Lavanya Rajamani, *European Union, Climate Action Hero?* IndianExpress.com, 3 August 2011, available at: www.indianexpress.com/news/european-union-climate-action-hero/826290/1 (last accessed on 31 March 2012).

[111] *Canada-Certain Measures Affecting the Renewable Energy Generation Sector*, WTDS412/1, 16 September 2010.

[112] *China-Measures concerning Wind Farm Equipment*, Request for Consultations, WT/DS419/1, 22 December 2010.

[113] Marie Wilke, "US vs China: Renewable Energy Competition Hits the WTO", ICTSD China Programme, April 2011, available at: http://ictsd.org/i/news/bioresreview/103556/ (last accessed on 8 March 2012).

[114] "Canada-Japan Renewable Energy Spat Arrives at WTO," 11 *Bridges TradeBioRes*, 27 June 2011.

certain percentage of the goods and services used for setting up the facility comes from Ontario.[115] This can be as high as 60%.[116] Japan alleges that the measure violates the national treatment provisions of the GATT and the Agreement on Trade Related Investment Measures.[117] It also claims that the local content requirement makes the feed-in tariff programme a "prohibited subsidy," under the terms of the SCM Agreement.[118] Also the EU initiated WTO consultations against Canada concerning the same measure in August 2011. It indicates that:

> Exports from the EU into Canada in wind power and photovoltaic power generation equipment are significant, ranging from 300 to 600 million € in 2007–2009. These figures could be higher should the local content requirements be removed from the legislation in question. The EU is also increasingly concerned by such measures taken by other trading partners.[119]

In the wind farm equipment dispute between China and the US, the disputed measure related to grants, funds, or awards to Chinese enterprises manufacturing wind power equipment.[120] According to the US, the support appeared to be contingent on the use of domestic over imported goods, thus violating Article 3 of the SCM Agreement.[121] China initially responded that its measures were helping to save energy and protect the environment.[122] It has been argued that two important legal questions could have emerged in the dispute, namely whether the environmental exceptions under Article XX of the GATT extend to the SCM Agreement, and if so, could the local content requirement be defended under Article XX.[123] However, China has subsequently ended the disputed wind power equipment subsidies.[124]

Attention in the US is now increasingly turning towards Chinese subsidies to its solar industry.[125] The price of solar panels has dropped significantly – by more than 30% in 2011 – due to cheap production in China.[126] While this is good news for those installing solar panels, many US solar panel manufacturers have gone bankrupt.[127]

---

[115] Ibid.

[116] Ibid.

[117] Ibid.

[118] Ibid.

[119] "The EU requests WTO consultations with Canada over Ontario's renewable energy policy", Press Release, 11 August 2011, available at: http://trade.ec.europa.eu/doclib/press/index.cfm?id=732 (last accessed on 8 March 2012).

[120] *China-Measures concerning Wind Farm Equipment*, supra, note 112, at 1.

[121] Ibid.

[122] Wilke, "US vs China", supra, note 113.

[123] Ibid.

[124] "China Ends Wind Power Equipment Subsidies Challenged by the United States in WTO Dispute" Office of the United States Trade Representative Press Release, June 2011, available at: http://www.ustr.gov/about-us/press-office/press-releases/2011/June/china-ends-wind-power-equipment-subsidies-challenged (last accessed on 8 March 2012).

[125] James Melik, "US Turns Up Heat on China Solar Subsidies", BBC News, 22 February 2012, available at: http://www.bbc.co.uk/news/business-17094881 (last accessed on 8 March 2012).

[126] Ibid.

[127] Ibid.

On the other hand, new jobs continue to be created in the US for people installing solar panels.[128] Demand for solar panels is driven by a US government support scheme to encourage their installation – prompting, in turn, an announcement from China that it is looking into the US government support for renewable energy.[129]

The pending WTO cases and the US-China solar panel controversy illustrate that a new era may be dawning in the relationship between trade and climate change as the transition towards a low-carbon economy takes off. As argued above, climate change is increasingly framed as an economic opportunity and transition to cleaner energy is depicted as a question of competitiveness. To achieve the necessary transformation, governments are seeking to support clean energy technologies and other industries relevant for the green economy. While some WTO lawyers have drawn attention to questions concerning the desirability of climate-related subsidies in general,[130] one of the most evident legal challenges is that governments tend to design their support schemes in such a way that seeks to aid the domestic industries to gain an edge in greener technologies. As the pending disputes at the WTO show, the situation is therefore problematic from the perspective of international trade law. From the climate policy perspective, however, support for cleaner technologies is commonly seen as a necessity. According to the IPCC AR4, financial incentives, such as subsidies and tax credits, "are often critical to overcome barriers."[131] All this highlights the increasingly relevant interaction between the WTO and UNFCCC legal regimes – are their objectives and rules mutually supportive of the necessary but highly ambitious transition to a low-carbon economy? Based on the disputes currently pending at the WTO, questions concerning government support for green technologies may well end up being more significant in practice than the much-debated question of climate-motivated trade measures.

16.4 Institutional Linkages: Role of the WTO Dispute Settlement System

As we have seen, a growing number of substantive linkages between trade and climate change can be identified. Linkages between the two legal regimes are not, however, reciprocated at the institutional level. Still, as Epps and Green point out, institutional questions are critical: "the institutional framework… will determine or at least strongly influence who decides which policy (whether it be a climate policy or trade policy) is permissible."[132] One of the key legal questions in this regard is

[128] Ibid.

[129] Ibid.

[130] Epps and Green, *Reconciling Trade and Climate,* supra, note 24, at 103–121.

[131] IPCC Working Group, "Summary for Policymakers", supra, note 26, at 19.

[132] Epps and Green, *Reconciling Trade and Climate,* supra, note 24, at 10.

what would happen in case of a legal dispute surfaced involving a conflict between the UNFCCC and the WTO regimes. Institutionally, the WTO system appears much stronger than the combined force of the UNFCCC and the Kyoto Protocol. As explained above, the Understanding on the Rules and Procedures Governing the Settlement of Disputes (DSU)[133] created a quasi-judicial dispute settlement system[134] with a compulsory and exclusive jurisdiction on WTO law binding on all WTO Members. In contrast, the UNFCCC and the Kyoto Protocol reflect the general trend under international environmental law towards compliance assessment and facilitation, and hence do not contain provisions on legally binding dispute settlement.[135] The WTO dispute settlement system would therefore be the likeliest forum to settle a controversy involving the UNFCCC regime and WTO rules.

This has some important implications. Epps and Green argue that "WTO rules and the interpretation of those rules by panels and the WTO Appellate Body determine whose values prevail"[136] While this is true, I have argued elsewhere that this could be damaging both for the relationship between the WTO and UNFCCC, and for the legitimacy of the WTO dispute settlement system.[137] It is interesting to note, however, that the new Forum on Response Measures under the UNFCCC could, in principle, also address the question of trade measures implemented to address climate change.[138] While its institutional features do not match the compulsory nature of the WTO dispute settlement system and the Forum's status under the UNFCCC remains politically highly sensitive, it nevertheless provides an opportunity to consider the trade-climate change linkage under the UNFCCC. In theory at least, the Forum could recommend to the climate COP either a decision on climate-related trade measures in general or with respect to a concrete case. Reaching consensus on such politically sensitive issues under the UNFCCC remains, however, highly unlikely.

---

[133] Understanding on Rules and Procedures Governing the Settlement of Disputes, Marrakesh Agreement Establishing the World Trade Organization, Annex 2, The Legal Texts: The Results of the Uruguay Round of Multilateral Trade Negotiations 354 (1999), 33 *International Legal Materials* (1994), 1226.

[134] For an explanation of why the WTO dispute settlement system is commonly described as "quasi-judicial," see Georges Abi-Saab, "The WTO Dispute Settlement and General International Law," in Rufus Yerxa and Bruce Wilson, (eds), *Key Issues in WTO Dispute Settlement. The First Ten Years* (Cambridge et al.: Cambridge University Press, 2005), 7, at 9–10.

[135] UNFCCC, supra, note 1, Art. 14 and the Kyoto Protocol, supra, note 44, Art. 19 provide for optional recourse to the International Court of Justice or arbitration, and a mandatory recourse to non-binding conciliation. Article 13 of the UNFCCC also foresees the establishment of a multilateral consultative process "for the resolution of questions regarding the implementation of the Convention." The text has been negotiated apart from two paragraphs on representation but the issue has not been resolved "due to lack of interest in view of more pressing in developing compliance procedures under the Protocol." Yamin and Depledge, *The International Climate Change Regime*, supra, note 17, at 384–385.

[136] Epps and Green, *Reconciling Trade and Climate* supra, note 25, at 35.

[137] Kulovesi, *The WTO Dispute Settlement System,* supra, note 3, at 254–257 and 266–267.

[138] Decision 8/CP.17, supra, note 54.

The consideration of a dispute involving the WTO and the UNFCCC through the WTO dispute settlement system involves some important legal challenges. Most notably, the status of international environmental law, including the UNFCCC, the Kyoto Protocol and a possible new climate treaty, in WTO dispute settlement proceedings involves important uncertainties.[139] In theory, non-WTO norms of international law could play a role in the WTO dispute settlement system in three different ways: through direct application; as a source of interpretative material; or as factual evidence.[140] Scholars disagree as to whether the WTO dispute settlement system may directly apply non-WTO norms.[141] What may perhaps be seen as the prevailing view on the relationship between the WTO system and other norms of international law can be summarised as follows:

> WTO adjudicating bodies cannot formally interpret other treaties and customs and thus cannot apply or enforce other treaties or customs or determine the legal consequences of rights and obligations that WTO Members may have under other treaties or by custom; these may be examined only when necessary for the interpretation of WTO law and/or as a factual determination.[142]

There are, however, other interpretations. Pauwelyn argues that unless an international treaty explicitly contracts out of general international law, general international law automatically applies to the regime created and fills gaps left by the treaty.[143] Since the WTO Agreement contains no such "contracting out" provision, Pauwelyn argues that it is unnecessary for the DSU to explicitly refer to general international law as a source of law: the WTO system is automatically part of general international law.[144] He also argues that the expression "cannot add or diminish rights and obligations" in Article 3.2 of the DSU does not limit the competence of the WTO dispute settlement system in terms of applicable law.[145] Instead, it constrains the interpretative powers of the WTO dispute settlement system by setting out the limits of the judicial function.[146] What follows is that the WTO dispute settlement system can apply but not enforce non-WTO rules.[147] In light of the scholarly debate and existing WTO dispute settlement practice it does not seem

---

[139] For a detailed discussion, see Kulovesi, *The WTO Dispute Settlement System*, supra, note 3, at 156–178.

[140] Ibid., at 135–149.

[141] Compare Joel P. Trachtman, "The Domain of the WTO Dispute Resolution," *Harvard International Law Journal*, Spring (1999), at 333–377 and Gabrielle Marceau, "WTO Dispute Settlement and Human Rights," 14 *European Journal of International Law* (2002), 753; with Joost Pauwelyn, "The Role of Public International Law in the WTO," 95 *American Journal of International Law* (2005), 535.

[142] Marceau, "WTO Dispute Settlement and Human Rights," supra, note 140, at 753.

[143] Joost Pauwelyn, "How to Win a World Trade Organization Dispute Based on Non-World Trade Organization Law: Questions of Jurisdiction and Merits," 37 *Journal of World Trade* (2003), 997 at 1001–1002.

[144] Ibid.

[145] Pauwelyn, "The Role of Public International Law in the WTO," supra, note 141, at 561.

[146] Pauwelyn, "How to Win a WTO Dispute Based on Non-WTO Law," supra, note 143, at 1003.

[147] Pauwelyn, "The Role of Public International Law in the WTO," supra, note 141, at 566.

possible to conclude with certainty whether the UNFCCC, the Kyoto Protocol or its possible successor treaty could sometimes be directly applied during WTO dispute settlement proceedings.

While the direct application of non-WTO norms by the WTO dispute settlement bodies remains controversial, it is widely accepted that non-WTO rules of international law can play a role in WTO disputes through interpretation. This is in conformity with the customary rules of treaty interpretation and Article 31.3(c) of the Vienna Convention on the Law of Treaties (VCLT) providing that: "There shall be taken into account, together with the context… any relevant rules of international law applicable in the relations between the parties."[148] For Marceau, this provisions serves to attain a degree of coherence in international law and helps to remedy some of the problems arising out of the limited substantive applicability of non-WTO law in the dispute settlement system.[149] Also van Asselt has highlighted the provision and the principle of systemic integration that the VCLT arguably embodies as potential tools to manage fragmentation of international law.[150] Indeed, on the face of it, recourse to Article 31.3(c) of the VCLT to ensure the mutual supportiveness of the WTO and UNFCCC regimes seems like an attractive solution. There are, however, some important caveats: While it is clear that relevant rules of international law must be taken into account in the interpretation of WTO law, it is far less clear what constitutes such "'relevant rule of international law applicable in the relations between the parties." Are they only such rules that are binding on all WTO Member States?[151] This would mean that the practical relevance of non-WTO norms is very limited: "the more WTO members we have, the less relevant rules we can refer to. Because there are more WTO members, there will be less 'other rules' that are binding on all WTO members."[152]

Or, are relevant rules such rules that are binding on the parties to a particular dispute? This would seem like a sensible solution and one that could help to promote the coherence and unity of international law. Unfortunately, the answer to this question also remains open. In the *Biotech* dispute, the panel did not consider the Cartagena Protocol on Biosafety, closely related to the dispute's subject matter, as a relevant rule of international law. This was largely expected, given that none of the three complainants were Parties to the Protocol.[153] However, the panel also left open

---

[148] Convention on the Law of Treaties, Vienna, 22 May 1969, in force 27 January 1980, 8 *International Legal Materials* (1989), 679.

[149] Marceau, "WTO Dispute Settlement and Human Rights," supra, note 141, at 785–786. See also Gabrielle Marceau, "A Call for Coherence in International Law: Praises for the Prohibition Against 'Clinical Isolation' in WTO Dispute Settlement System," 33 *Journal of World Trade* (1999), 87, at 108.

[150] Harro van Asselt, "Fragmentation of International Climate Law" in Chapter 13 of the present volume.

[151] For discussion see, Marceau, "WTO Dispute Settlement and Human Rights," supra, note 141, at 780–783.

[152] Joost Pauwelyn, "Speech Delivered at the Fourth Annual WTO Conference," in Mads Andenas and Frederico Ortino (eds), *WTO Law and Process* (United Kingdom: British Institute for International and Comparative Law, 2005), 494, at 496.

[153] WTO Panel Report, *European Communities – Measures Affecting the Marketing and Approval of Biotech Products*, WT/DS291/R, WT/DS292/R, WT/DS293/R, 29 September 2006, paras. 7.71–7.75 (Hereafter: Biotech Panel Report).

the possibility that only such agreements could be considered as 'relevant rules' of international law to which all WTO Members are parties. In other words, the *Biotech* panel ruled that since the case was not one where relevant rules of international law were applicable between all parties to the dispute but not between all WTO Members, it did not need to decide whether, in such a situation, it would be entitled to take the relevant rules of international law into account.[154]

While the Kyoto Protocol has more Parties (192) than there are WTO Members (157), membership in these two international legal agreements is not completely overlapping. For instance, the US is a Member of the WTO but will never ratify the Kyoto Protocol and Canada, also a WTO Member, recently withdrew from the Protocol. Hence, in light of the *Biotech* decision, the possibility remains that the Kyoto Protocol is never relevant for the interpretation of WTO law within the meaning of the VCLT. It is also conceivable that some key countries from the point of view of climate change mitigation choose to remain outside of the possible new post-2020 climate treaty. According to the ILC, however, a better approach would be to permit references to another treaty in cases where the parties to the dispute are also parties to the other treaty – otherwise, the coherence of the WTO regime comes at the expense of the coherence of the multilateral treaty system as a whole.[155] It is easy to agree with the ILC here.

The *Biotech* panel emphasized that legal norms could also be considered, not in the legal sense but in the same way as dictionaries:

> Such rules would not be considered because they are legal rules, but rather because they may provide evidence of the ordinary meaning of terms in the same way that dictionaries do. They would be considered for their informative character. It follows that when a treaty interpreter does not consider another rule of international law to be informative, he or she need not rely on it.[156]

While the Appellate Body had considered at least the Convention on Biological Diversity and possibly also other environmental instruments in this sense in the *Shrimp-Turtle* dispute,[157] the *Biotech* panel decided that the Cartagena Protocol was not relevant in a dispute that related to its very subject matter of transboundary movement of living modified organisms.[158] What this shows is that the WTO dispute settlement system has broad discretionary powers when it comes to considering MEAs as factual evidence. In other words, this interpretative practice could also serve to promote coherence in the international legal system, also addressing the tricky issue of non-Parties.[159] In light of the *Biotech* panel decision, however, the

[154] Ibid., para. 7.71.

[155] International Law Commission, *Fragmentation of International Law*, supra, note 20, at 238.

[156] Panel Report, *Biotech*, supra, note 153, para. 7.91.

[157] *Shrimp-Turtle* Appellate Body Report, supra, note 101, para 130.

[158] Panel report – *Biotech*, supra, note 53, para. 7.95.

[159] See Kulovesi, *The WTO Dispute Settlement System,* supra, note 3, at 176–177 for detailed analysis.

extreme possibility cannot be entirely ruled out that the Kyoto Protocol would be deemed completely irrelevant in a WTO dispute involving policies and measures designed to implement the Kyoto Protocol.[160] Overall, the situation seems far from satisfactory: the legal relevance of the Kyoto Protocol or its successor in the WTO dispute settlement proceedings remains unclear with the options ranging from its direct application to complete ignorance. Such legal and institutional discrepancies are hardly conducive to fruitful cooperation between the trade and climate change regimes. Finally, as described by van Asslet in the present volume, there are also other ways in which institutional cooperation between the trade and climate regimes could be enhanced.[161]

16.5 Conclusions: Towards Increasing Linkages and Enhanced Cooperation?

Over the past 20 years, the international trade and climate change regimes have evolved in relatively comfortable isolation from each other. Different delegates and experts tend to attend the respective negotiations and few, if any, people can claim to have comprehensive understanding of both regimes. In this sense, concerns over fragmentation of international law are highly relevant and valid in the relationship between climate change and trade. However, substantive links between these two prominent international regimes are becoming increasingly apparent as their shared territory expands due to rapid evolution of climate change law and policy.

The question concerning unilateral trade measures and border carbon adjustments has received ample attention over the years. However, the evolution of climate change law shows that it is by no means the only relevant question in the relationship between the climate change and trade regimes –and perhaps not even the most important one. The recent surge of disputes concerning clean energy technologies at the WTO illustrates that the transformation to a low carbon or green economy may well shift the focus from the GATT and associated legal issues towards other WTO Agreements, including the SCM Agreement and also the TBT Agreement due to the growing importance of energy efficiency standards and similar measures. Enhanced efforts may be needed in the future to ensure and improve the mutual supportiveness of the two regimes.

In institutional terms, the relative institutional strength of the WTO and its dispute settlement system adds a layer of complexity to the relationship between the international trade and climate change regimes. Here, the biggest concern is the status and relevance of international climate treaties in the WTO dispute settlement

[160] Ibid., at 254–256

[161] van Asselt, "Fragmentation of International Climate Law", supra, note 150.

system. While the *Shrimp-Turtle* case made several advances in the relationship between trade and environment, it also left some crucial questions unanswered concerning the relationship between WTO law and international environmental law. Most critically, it did not explain the legal relevance of the various environmental instruments to which the WTO Appellate Body referred in its decision. It may well be that they were merely used as factual evidence to define the ordinary meaning of the wording of the GATT.[162] The subsequent *Biotech* panel avoided accepting the relevance of the Cartagena Protocol on Biosafety in the WTO proceedings, including as factual evidence. The *Biotech* decision was not appealed and the Appellate Body, which has traditionally been more open to considering international environmental law than the panels, never had a chance to consider the question. Regardless, the *Biotech* decision goes to show that the WTO dispute settlement system holds ample discretion to either consider or ignore MEA sas factual evidence. In this light, it is not clear what role the UNFCCC, Kyoto Protocol or a new, post-2020 climate treaty would play in WTO dispute settlement proceedings. This means that a considerable degree of uncertainty remains in the relationship between the UNFCCC and WTO legal regimes. However, after two decades of relative insulation, attention is gradually turning, as it should be, towards mutual supportiveness and ways in which trade and climate regimes could promote each other's objectives. For the transformation of low carbon economy, this is an essential step.

[162] For detailed analysis, see Kulovesi, *The WTO Dispute Settlement,* supra, note 3, at 173–175.

Chapter 17
Climate Law and Geoengineering

Ralph Bodle

Abstract Geoengineering describes a range of techniques that are proposed to counteract some of the negative impacts of climate change at a global scale, without actually reducing emissions. This chapter provides an overview of geoengineering techniques and the existing international law applicable to them. Geoengineering techniques are not as such prohibited and are hardly addressed by international law. They pose fundamental challenges to a potentially emerging area of international climate law. The main challenge for policy makers is deciding whether and how to get involved without providing an incentive or excuse for stepping away from reducing emissions. A key component is to clearly separate scientific input and political decision-making.

17.1 Introduction: What Is Geoengineering?

Geoengineering is a generic term describing a range of techniques that are proposed to counteract some of the negative impacts of climate change without actually reducing emissions. The main proposals involve techniques to cool the earth at a global scale by reducing incoming solar radiation or by removing carbon dioxide from the atmosphere.[1] A recent study suggested defining climate-related geoengineering as a "deliberate

---

Ralph Bodle is Senior Fellow at Ecologic Institute, Berlin. He holds an LLM from UCL College, London, and a Ph.D. in law from Humboldt University, Berlin. The author has been a member of the German delegation in the UN climate negotiations since 2008 and also participated in other international negotiations. The views presented in this chapter are entirely his own and do not necessarily represent the views of Germany or the EU. The author would like to thank Gesa Homann, Simone Schiele and Elizabeth Tedsen for their input to previous studies that are the basis of this article.

[1] See below for details on particular techniques.

R. Bodle (✉)
Ecologic Institute, Pfalzburger Str. 43/44, 10717 Berlin, Germany
e-mail: ralph.bodle@ecologic.eu

E.J. Hollo et al. (eds.), *Climate Change and the Law*, Ius Gentium: Comparative Perspectives on Law and Justice 21, DOI 10.1007/978-94-007-5440-9_17,
© Springer Science+Business Media Dordrecht 2013

intervention in the planetary environment of a nature and scale intended to counteract anthropogenic climate change and/or its impacts through, inter alia, solar radiation management or removing greenhouse gases from the atmosphere."[2]

All proponents of geoengineering stress that it is no substitute for reducing emissions, and that these proposals are primarily considered as an additional option, complementing other efforts to limit the magnitude of human-induced climate change.[3]

Most geoengineering techniques are at a conceptual or modeling stage, although there have been a number of field experiments on ocean fertilisation and initial attempts at aerosol injection (see below). Currently, further geoengineering research is being funded by e.g. the European Union,[4] the United Kingdom and the United States.[5] Apart from the fast growing scientific literature on particular techniques, several overview studies have addressed geoengineering.[6] Two coordinated Parliamentary hearings at the US House of Representatives[7] and the UK House of Commons[8] also contributed to raising public awareness and starting a more political debate. In 2012, the Secretariat of the Convention on Biological Diversity (CBD) published two overview studies in preparation for further work in the CBD SBSTTA.[9]

---

[2] Phillip Williamson et al., "Impacts of climate-related geoengineering on biological diversity", UNEP/CBD/SBSTTA/16/INF/28, 28 March 2012, at 9, available at www.cbd.int/SBSTTA 9.

[3] Ibid., at 3–4

[4] See http://implicc.zmaw.de/.

[5] See United States Government Accountability Office, *Climate Change: A Coordinated Strategy Could Focus Federal Geoengineering Research and Inform Governance Efforts*, Report to the Chairman, Committee on Science and Technology, House of Representatives, GAO-10-903 (Washington DC: United States Government Accountability Office, 2010).

[6] Selected main studies are: Royal Society (UK), *Geoengineering the Climate: Science, Governance and Uncertainty* (London: The Royal Society, 2009); United States Government Accountability Office: Center for Science, Technology, and Engineering, *Technology Assessment: Climate engineering: Technical status, future directions, and potential responses,* Report to Congressional Requester, GAO-11-71, (Washington DC: United States Government Accountability Office, 2011); Bipartisan Policy Center, Task Force On Climate Remediation Research, *Geoengineering: A National Strategic Plan for Research on the Potential Effectiveness, Feasibility, and Consequences of Climate Remediation Technologies* (Washington DC: Bipartisan Policy Center, 2011); German Federal Environment Office (Umweltbundesamt), *Geo-engineering – Effective climate protection or megalomania?* (Dessau-Rosslau: Umweltbundesamt, 2011); W. Rickels et al., *Large-Scale Intentional Interventions into the Climate System? Assessing the Climate Engineering Debate*, Scoping report conducted on behalf of the German Federal Ministry of Education and Research (BMBF) (Kiel: Kiel Earth Institute 2011).

[7] Bart Gordon, *Engineering the Climate: Research Needs and Strategies for International Coordination.* Staff Report, Committee on Science and Technology, US House of Representatives (Washington DC: Committee on Science and Technology, 2010).

[8] House of Commons Science and Technology Committee (UK), *The Regulation of Geoengineering: Fifth Report of Session 2009–10.* (London: The Stationery Office, 2010).

[9] Williamson et al., "Climate-Related Geoengineering", supra, note 2; Bodle et al., "Regulatory Framework for Climate-related Geoengineering Relevant to the Convention on Biological Diversity", UNEP/CBD/SBSTTA/16/INF, 2 April 2012, available at www.cbd.int/SBSTTA.

Concerns about geoengineering have been raised by civil society organizations[10] and debates are on-going within the scientific community.[11] Geoengineering will be part of the IPCC's Fifth Assessment Report.[12]

17.2 Geoengineering Techniques and Current International Law

Except for recent efforts under treaty regimes by the London Convention and London Protocol and the CBD,[13] international law has not addressed geoengineering as such. Most of the relevant treaties were adopted before geoengineering was an issue. However, many treaties, actual and potential customary rules and general principles of law, as well as other regulatory instruments and mechanisms could be interpreted so as to apply to all or some geoengineering concepts.[14] As a detailed analysis of all legal implications of each geoengineering technique would be beyond the scope of this chapter, the following section outlines the applicable rules of international law and the main legal problems and uncertainties.[15]

Geoengineering is a generic and general term comprising several different concepts.[16] Proposals for deliberate climate engineering emerged as early as the late nineteenth century, although the term geoengineering in the sense of counteracting

[10] See for instance ETC Group, "Geopiracy: The Case against Geoengineering", ETC Group Communique 103 (2010), available at www.etcgroup.org/upload/publication/pdf_file/ETC_geopiracy_4web.pdf.

[11] Cf. Williamson et al., "Climate-Related Geoengineering", supra, note 2; Mark New et al., "Four Degrees and Beyond: The Potential for a Global Temperature Increase of Four Degrees and its Implications", 369 *Philosophical Transactions of the Royal Society* (2010), 4; Christine Bertram, "Ocean Iron Fertilisation in the Context of the Kyoto Protocol and the Post-Kyoto Process" (Kiel Inst. for the World Econ., Working Paper No. 1523 2009), at 3; Geoff Brumfiel, "Controversial Research: Good Science Bad Science", 484 *Nature*, (2012), at 432.

[12] IPCC, "Scope, Content and Process for the Preparation of the Synthesis Report (SYR) of the IPCC Fifth Assessment Report (AR5)", IPCC-XXXII/Doc. 4 (2010), at 3, available at http://www.ipcc.ch/meetings/session32/syr_final_scoping_document.pdf. Previous IPCC reports briefly mentioned geoengineering, see Williamson et al., "Climate-Related Geoengineering", supra, note 2, at 7.

[13] See below.

[14] United Nations Convention on Biological Diversity, Rio de Janeiro, 5 June 1992, in force 29 December 1993, International Legal Material (2012), para. 12.

[15] For more detailed studies see e.g. Rickels et al., *Large-Scale Intentional Interventions*, supra, note 6; Bodle et al., "Regulatory Framework", supra, note 9; Ralph Bodle, "Geoengineering and International Law: The Search for Common Legal Ground", 46 *Tulsa Law Review* (2010), 305; Rex J. Zedalis, "Climate Change and the National Academy of Sciences' Idea of Geoengineering: One American Academic's Perspective on First Considering the Text of Existing International Agreements", 19 *European Energy and Environmental Law Review* (2010), 18.

[16] The following short description follows the most recent overview of the different geoengineering techniques and their potential impacts, Williamson et al., "Climate-Related Geoengineering", supra, note 2, at 3–4.

human-induced changes in the climate first arose in the 1970s and entered mainstream debate in the 1990s.[17] Geoengineering is the most common term of reference, but others such as "climate remediation" or "climate engineering" are alternatively applied to the same concept.[18]

Geoengineering techniques are usually subdivided into the two overarching categories of either solar radiation management (SRM) or carbon dioxide removal (CDR).

17.2.1 Solar Radiation Management Techniques

SRM techniques aim to change the earth's energy balance by reducing the incidence and subsequent absorption of short-wave solar radiation.[19] These techniques do not address greenhouse gas emissions or their concentration in the atmosphere.

Injecting of aerosols into the stratosphere aims to increase the planetary albedo, thereby reducing incoming solar radiation. Similar to the effects observed after the eruption of the Mount Pinatubo in 1991, the aerosol particles would filter incoming solar radiation and thus reduce overall global temperature. Although a wide range of potential aerosols are being discussed, the focus has been on the use of sulphate aerosols. A fleet of aircraft could introduce hydrogen sulphide (H_2S) or sulphur dioxide (SO_2) into the stratosphere as gases, where they are expected to oxidize into sulphate particles.[20]

Injecting H_2S or SO_2 into the stratosphere is at present not prohibited or significantly restricted by the main international treaties governing the emission of those substances. The LRTAP Convention[21] as such does not contain provisions that are specific enough to prohibit or significantly restrict the introduction of SO_2 into the stratosphere. The three protocols relating to sulphur establish reporting obligations for parties, but would restrict the introduction of SO_2 into the stratosphere only to the extent that it would lead to exceeding a party's emission ceiling. Similarly, the Ozone Convention[22] does not contain a sufficiently specific obligation requiring parties to ban or significantly restrict the introduction of such substances.[23]

[17] Cf. the historical overviews in Williamson et al., "Climate-Related Geoengineering", supra, note 2, at 7–8; J.R. Fleming, *Fixing the Sky: The Checkered History of Weather and Climate Control* (New York: Columbia University Press, 2010).

[18] *See e.g.* Bipartisan Policy Center, *National Strategic Plan*, supra, note 6; Gordon, *Engineering the Climate*, supra, note 7.

[19] German Federal Environment Office, "Effective Climate Protection or Megalomania?", supra, note 6, at 9; Williamson et al., "Climate-Related Geoengineering", supra, note 2, at 4.

[20] Royal Society, *Science, Governance and Uncertainty*, supra, note 6, at 29–32.

[21] Convention on Long-Range Transboundary Air Pollution, 13 December 1979, in force March 3 1983, *United Nations Economic Commission for Europe.*

[22] Vienna Convention for the Protection of the Ozone Layer, Vienna, 22 March 1985, in force 1 January 1989, *Audiovisual Library of International Law* (2008).

[23] For a similar assessment see John Virgoe, "International Governance of a Possible Geoengineering Intervention to Combat Climate Change", 95 *Climatic Change* (2009), 103, at 111.

The Montreal Protocol[24] does not cover H_2S and SO_2. Even if did, e.g. following an amendment, it would regulate their import, export, production and consumption, but not their use or injection.

Cloud brightening, also referred to as cloud-albedo enhancement or cloud seeding, describes a geoengineering technique by which clouds are increased and whitened over parts of the ocean, therefore reflecting more short-wave solar radiation back to space.[25] This would be achieved be releasing a "suitable hydrophilic powder"[26] from ships into the troposphere, particularly over ocean areas. Generating fine particles of sea-salt derived from ocean water is the most prominently discussed technique.[27]

It is unclear whether and to what extent cloud brightening is addressed by the relevant rules of international law. The Ozone Convention, even if it applied, does not impose significant restrictions. The LP does not prohibit cloud brightening provided that sea water vapor is used and the activity does not constitute dumping. The rules of UNCLOS provide for user conflict resolution on a case by case basis in the EEZ, while for the high seas it is arguable but not clear that cloud brightening would fall under the UNCLOS provisions against marine pollution.

Desert reflectors would cover desert surfaces with highly reflective materials so as to increase solar radiation reflection. Deserts are considered most suitable as they are largely uninhabited, flat surfaces with limited vegetation, and because they have high levels of incident solar radiation.[28] Desert reflectors do not appear to be in conflict with international law, although large-scale land use changes could indirectly be restricted by international law requiring the protection of biodiversity, ecosystems and habitats.

Incoming solar radiation could also be reduced by placing *installations in outer space*. Some proposals include installations in the near-earth orbits or further away from the earth, such as mirrors or other reflective material.[29] The lack of experience as well as the likely effort and costs required make it unlikely that geoengineering in outer space will be carried out in the near future.[30] Moreover, there are a number of uncertainties on their intended and unintended impacts on the climate system. Space law would apply,

---

[24] Montreal Protocol on Substances that Deplete the Ozone Layer, Montreal, 16 September 1987, in force 1 January 1989, *United Nations Environmental Programme* (2009).

[25] Williamson et al., "Climate-Related Geoengineering", supra, note 2, at 3; See also Royal Society , *Science, Governance and Uncertainty*, supra, note 6, at 26 and United States Government Accountability Office, *Technology Assessment*, supra, note 6, at 35.

[26] Royal Society, *Science, Governance and Uncertainty,* supra, note 6, at 27.

[27] Ibid., at 27. See also Bipartisan Policy Center, *National Strategic Plan*, supra, note 6, at 10.

[28] Royal Society, *Science, Governance and Uncertainty*, supra, note 6, at 26.

[29] Overview of all proposals in Royal Society, *Science, Governance and Uncertainty*, supra, note 6, at 32 et seqq., United States Government Accountability Office, *Technology Assessment*, supra, note 6, at 36 et seqq.

[30] Royal Society, *Science, Governance and Uncertainty*, supra, note 6, at32; Kelsi Bracmort, Richard Lattanzio, and Emily C Barbour, *Geoengineering Governance and Technology Policy. US Congressional Research Service Report*, US Congressional Research Service Reports (Washington, DC: Congressional Research Service, Library of Congress, 2010)

but there is no rule that explicitly prohibits space-based geoengineering as such, and the general obligations provide few restrictions regarding impacts on earth.

17.2.2 Carbon Dioxide Removal Techniques

CDR includes techniques that are intended to remove CO_2 from the atmosphere and therefore reduce one of the main contributors to climate change. CDR techniques involve two steps: (i) removal of CO_2 from the atmosphere and (ii) subsequent long-term storage of the captured CO_2 in order to take it out of circulation for a climatically relevant period.[31] Several techniques are being discussed for each step.[32]

Carbon capture and storage (CCS) at the point of emission is frequently excluded from the definition of geoengineering,[33] although it poses the same storage problems as storing CO_2 that is captured from the atmosphere.[34] Without prejudice to this debate, regulatory aspects regarding CCS can be relevant to those geoengineering techniques that require the storage of CO_2 as a second step.

There is no international legal regime that specifically addresses CCS on land. However, it is addressed by the climate regime in the rules on national greenhouse gas inventories, and has recently been included into the Kyoto Protocol's CDM.[35] CO_2 storage in the ocean and in sub-surface geological formations in the seabed potentially falls within the scope of UNCLOS, London Convention and London Protocol and OSPAR. Generally, in contrast to UNCLOS, the London Protocol prohibits CCS.[36] However, the parties to the LP have expressly permitted sub-seabed CO_2 storage under certain conditions, while disposal of CO_2 in the water column or on the seabed is prohibited.[37] The same will be true under recent amendments to OSPAR for CO_2 storage from vessels, once they enter into force. However, it is not clear whether the rules for CCS under the London Convention and London Protocol would apply to CO_2 captured after release into the atmosphere.[38]

---

[31] Williamson et al., "Climate-Related Geoengineering" supra, note 2, at 44.

[32] Cf. German Federal Environment Office, "Effective Climate Protection or Megalomania?", supra, note 6, at 18.

[33] It was, for instance, expressly excluded from the CBD's working definition of geoengineering in Decision X/33, para. 8 (w).

[34] Gordon, *Engineering the climate*, supra, note 7, at 21; Royal Society, *Science, Governance and Uncertainty*, supra, note 6, at 20. For differences see ibid. at 14.

[35] Decision 10/CMP.7, Clean Development Mechanism included under article 12, of the Kyoto Protocol, UN Doc FCCC/KP/CMP/2011/10/Add.2., 2006. See also decision 2/CMP.5, para. 29.

[36] Elizabeth Wilson, Timothy Johnson, and David Keith, "Regulating the Ultimate Sink: Managing the Risks of Geologic CO_2 Storage", 37 *Environmental Science and Technology* (2003), 3476, at 3479.

[37] Following an amendment that entered into force in 2007. A further amendment of 2009 regarding the sharing of sub-seabed geological formations for CCS projects is not yet in force, resolution LP.3(4) on the amendment to Article 6 of the London Protocol, 30 October 2009, IMO, Report of the Secretary-General on the status of the 1996 Protocol to the London Convention 1972, LC 33/2/1, at 2.

[38] Bodle et al., "Regulatory Framework", supra, note 9, para. 97.

Ocean fertilization involves enhancing the supply of nutrients to the marine environment with the aim of increasing the uptake of CO_2 in the oceans through biological processes and the subsequent long-term storage of a portion of the additional organic carbon in the deep sea.[39] Several field experiments on iron-based ocean fertilization were carried out over the last 20 years,[40] including the LOHAFEX experiment of January 2009 that not only sparked political controversy and public attention[41] but also intensified efforts to address geoengineering at an international regulatory level. To date ocean fertilisation is the geoengineering technique subject to the most detailed regulatory efforts. It was addressed by the United Nations General Assembly[42] and UNESCO's IOC.[43] Ocean fertilization experiments are now regulated under the London Convention and London Protocol's provisions on dumping and additional non-binding guidance including a risk assessment framework.[44] The CBD has referred to and incorporated this work in its own decisions, which extended the application of the guidance beyond the smaller number of Parties to the London Convention and London Protocol.[45] In 2010, the London Convention and London Protocol agreed to continue its work towards providing a more comprehensive "control and regulatory mechanism" for ocean fertilisation.[46]

---

[39] Williamson et al., "Climate-Related Geoengineering", supra, note 2, at 49, noting that enhanced downwelling, without necessarily increasing marine primary production, has also been proposed.

[40] Williamson et al., "Climate-Related Geoengineering", supra, note 2, at 49; Secretariat of the Convention on Biological Diversity, *Scientific Synthesis of the Impacts of Ocean Fertilisation on Marine Biodiversity.* Technical Series No. 45. CBD, (Montreal: Secretariat of the Convention on Biological Diversity, 2009); Wallace D.W.R, Law C.S., Boyd P.W., Collos Y., Croot P., Denman K., Lam P.J., Riebesell U., Takeda S. and Williamson P., *Ocean Fertilisation: A Scientific Summary for Policy Makers* (Paris: IOC/UNESCO, 2010) (IOC/BRO/2010/2).

[41] See Press release by the German Federal Ministry for the Environment, nature Conservation and Nuclear Safety, 1 January 2009 available at http://www.etcgroup.org/en/node/712.

[42] See UN GA Res. 62/215, para. 97–98, 14 March 2008; Res. 63/111, paras. 115–116, 12 February 2009; Res. 64/71, paras. 132–133, 12 March 2010; Res. 65/37, para. 149, 17 March 2011.

[43] Bodle et al., "Regulatory Framework, supra", note 9, at 6.

[44] Resolution LC-LP.1 (2008), para. 1. For views on the legal implications of the London Convention and London Protocol statements and decisions as well as the LOHAFEX experiment carrying out ocean fertilisation in 2009 see: David Freestone and Rosemary Rayfuse, "Contribution to the Theme Section 'Implications of Large-scale Iron Fertilization of the Oceans' Ocean Iron Fertilization and International Law", *Marine Ecology Progress Series 364* (2008), 227.; Harold Ginzky, "Ocean Fertilization as Climate Change Mitigation –Consideration Under International Law", 7 *Journal for European Environmental & Planning Law* (2010), 57; Bodle et al., "Regulatory Framework", supra, note 9, paras. 92 et seqq.

[45] Bodle et al., "Regulatory Framework, supra", note 9, at 4.

[46] IMO note to UNFCCC COP16, "Resolution LC-LP.2(2010), para. 5", 29 November 2010, available at http://www.imo.org/OurWork/Environment/PollutionPrevention/AirPollution/Documents/COP%2016%20Submissions/IMO%20note%20on%20LC-LP%20matters.pdf

Ocean liming describes techniques for enhancing ocean alkalinity by adding alkaline minerals or their dissolution products in order to chemically enhance fixing of atmospheric carbon dioxide and their marine storage. Current proposals cover a range of alkaline minerals and dissolution products that could be added through direct ocean releases, pipelines to the sea, or indirectly through discharges into river systems draining to the ocean.[47] Ocean liming could have the added positive benefit of offsetting acidification caused by climate change. Negative impacts to the marine environment and biodiversity are not well understood but may result from local spatial and temporal pH spikes and from extreme alkalinity levels.[48] Indirect impacts would result from mining, processing and transporting the required volumes of minerals. Ocean liming is not directly addressed under current international law regimes. However, the technique may be subject to provisions governing protection of the marine environment and ocean dumping under the London Convention and London Protocol, UNCLOS, and the OSPAR Convention. It has also been argued that potentially adverse impacts of ocean liming could be considered as contrary to the objective of the London Convention and London Protocol and therefore justify an interpretation bringing it into its scope. On the other hand, the purpose of combating climate change could equally be deemed environmentally beneficial, particularly considering the added benefit of countering ocean acidification.[49]

Biomass storage in the ocean involves deposition of crop residues or other terrestrial vegetative material into deep ocean waters. The deep ocean conditions would severely slow decomposition of the organic materials and thereby store carbon dioxide contained in the biomass for possibly thousands of years.[50] Potential impacts of ocean biomass storage are poorly understood due to limited understanding of deep sea ecosystems. Impacts would depend upon the type and permeability of biomass packaging used as well as energy consumption required for transport, burying, and processing.[51] Ocean sequestration of biomass is not directly addressed under current international law and does not appear to be clearly prohibited under UNCLOS, London Convention and London Protocol[52] or OSPAR. International law also does not appear to prohibit the production of biomass materials to be used specifically for geoengineering purposes.

CO_2 captured in *biomass* could also be sequestered by converting it to so-called *biochar*, which is then applied to soil, where it would gradually decompose over a long period of time. The storage is intended to prevent the release of CO_2 into the

---

[47] Williamson et al., "Climate-Related Geoengineering", supra, note 2, at 52.

[48] Williamson et al., "Climate-Related Geoengineering", supra, note 2, at 53; German Federal Environment Office, "Effective Climate Protection or Megalomania?", supra, note 6, at 28.

[49] Ginzky, "Ocean Fertilization", supra,note 44, at 64, in respect to ocean fertilisation under the London Convention and London Protocol; Rickels et al., *Large-Scale Intentional Interventions*, supra, note 6, at 6. Similar arguments could be made under OSPAR.

[50] German Federal Environment Office, "Effective Climate Protection or Megalomania?", supra, note 6, at 29.

[51] Royal Society, *Science, Governance and Uncertainty*, supra, note 6, at 11; Williamson et al., "Climate-Related Geoengineering", supra, note 2, at 58.

[52] Williamson et al., "Climate-Related Geoengineering", supra, note 2, at 58.

atmosphere during natural processes of decomposition of dead vegetation.[53] From a legal perspective, international law does not specifically address or prohibit the production of biomass, of biochar, or the application of biochar on soil. However, in order to produce and apply the necessary amount of biomass and biochar, these techniques could entail considerable large-scale land use changes. These could indirectly be restricted by international law requiring the protection of biodiversity, ecosystems and habitats.

Under the climate regime, the rules on the calculation of sinks such as LULUCF provide an incentive for states to generate sinks, even for parties to the UNFCCC without quantified reduction obligations under the Kyoto Protocol.[54] Although this does not amount to permitting geoengineering by biomass and biochar, it is conceivable to imagine moves towards crediting certain types of LULUCF under the Kyoto Protocol's flexible mechanisms or in future new market-based mechanisms.[55]

Enhanced weathering is a technique that accelerates the slow natural reaction of silicate rocks with CO_2 by spreading finely-ground silicate minerals such as olivine over agricultural soils.[56] Similar to geoengineering by biomass and biochar, enhanced weathering mainly has land-use change impacts. And similar to ocean liming, this technique would require considerable efforts in the extracting, processing and transporting of the minerals to the soil. In addition to these indirect impacts, the potential direct impacts on land include effects on soil structure and fertility and increased soil albedo. In addition, the scale required in order to be effective could potentially also result in impacts on rivers, coastal seas and the open ocean.[57] The legal framework is similar to that applying to biomass and biochar.

Air capture of CO_2 ("artificial trees") comprises a range of industrial processes aimed at extracting CO_2 directly from ambient air,[58] which subsequently has to be stored. This technique is relatively technically advanced and is well understood, although it is said to be "decades away from large-scale commercialization".[59] While impacts are likely to be low, the energy required will push costs and might significantly increase its climate footprint.[60] International law does not address

---

[53] On the process see German Federal Environment Office, "Effective Climate Protection or Megalomania?", supra, note 6, at 22; Williamson et al., "Climate-Related Geoengineering", supra, note 2, at 57.

[54] Cf. Article 4(1)(a) UNFCCC and Articles 3(3), 3(4), 3(7) and 4 KP as well as the overview of LULUCF rules available at http://unfccc.int/methods_and_science/lulucf, last accessed 2 May 2012.

[55] Virgoe, "International Governance of a Possible Geoengineering Intervention", supra, note 23; Bertram, "Ocean Iron Fertilisation in the Context of the Kyoto Protocol", supra, note 11.

[56] Williamson et al., "Climate-Related Geoengineering", supra, note 2, at 52–55.

[57] Williamson et al., "Climate-Related Geoengineering", supra, note 2, at 46–47.

[58] Williamson et al., "Climate-Related Geoengineering", supra, note 2, at 67; Royal Society, *Science, Governance and Uncertainty*, supra, note 6, at 16.

[59] United States Government Accountability Office, *Technology Assessment*, supra, note 6, at vi.

[60] United States Government Accountability Office, *Technology Assessment*, supra, note 6, at 23; Royal Society, Science, Governance and Uncertainty, supra, note 6, at 15–16. Williamson et al., "Climate-Related Geoengineering", supra, note 2, at 68 mentions potential risks of pollution from producing and handling the required chemicals.

geoengineering by artificial trees. It could become relevant once this technique reaches a certain scale that would for instance have potential transboundary impacts. Air capture installations might potentially be addressed as sinks by the UNFCCC regime and process, although the current rules do not apply to this type of CO_2 removal.

17.2.3 The Climate Regime

Neither the UNFCCC nor the Kyoto Protocol prohibit geoengineering as such. Although the objective of the climate regime according to Article 2 UNFCCC is to stabilise greenhouse gas concentrations in the atmosphere, this "ultimate" aim of stabilising greenhouse gas concentrations does not necessarily mean that the UNFCCC or the Kyoto Protocol prohibit other measures intended to prevent global warming.[61]

The precautionary principle embodied in Article 3.3 UNFCCC is binding, but its wording allows for interpreting it as not precluding geoengineering (see below on the precautionary principle). The few other provisions in the UNFCCC that could apply to geoengineering, such as Article 3.1, 3.3, 4.1 and 4.2(a) UNFCCC, are general in wording and normative content. The duty in Article 4.1(f) requires parties to take climate change considerations into account in national policies, but does not provide a particular direction regarding geoengineering. The provision only to some extent requires an environmental impact assessment, and would require that geoengineering qualifies as mitigation or adaptation. Article 4.2(a) UNFCCC requires developed countries to take measures on mitigation by limiting their emissions and by protecting and enhancing their sinks and reservoirs. These obligations do not restrict geoengineering measures.

Apart from CCS, the Kyoto Protocol does not address or prohibit geoengineering. There is a thematic overlap with land use change and sinks, as the Kyoto Protocol provides incentives to generate sinks from land-use and forestry projects (see above).

17.2.4 General Rules

Besides international rules provided by specific treaties or regimes, some cross-cutting international rules and principles are relevant to all geoengineering concepts and represent common legal ground because they are customary law[62] or treaty rules with near universal application. These are the duty to respect the environment, the

[61] Bodle et al., "Regulatory Framework", supra, note 9, 100.

[62] *Ius cogens* and obligatons *erga omnes* do not have practical relevance for geoengineering at this stage, Bodle et al., "Regulatory Framework", supra, note 9, para. 27 fn. 16.

precautionary principle in Art. 3.3 UNFCCC, the general obligation to carry out an environmental impact assessment and the rules on state responsibility.

Other general principles or concepts such as sustainable development or intergenerational equity also play a role in the considerations and debate on geoengineering. However, from a legal point of view such concepts are not universally recognized as legal obligations on states, or their content is too open to provide commonly accepted legal ground of international law relevant to geoengineering.

17.2.4.1 Duty to Respect the Environment

In several decisions over the last years, the ICJ held that all states are under a general obligation to ensure that activities within their jurisdiction or control respect the environment of other States or of areas beyond national jurisdiction or control.[63] The obligation has evolved from previous cases such as the *Trail Smelter* arbitration[64] and the *Corfu Channel* case,[65] as well as from several references in key international documents and treaties such as principle 2 of the Rio Declaration,[66] Article 3 CBD[67] and the preamble to the UNFCCC.[68] It is customary law and in its recent formulation encompasses the environment in general as well as areas beyond national jurisdiction.

This general duty to prevent transboundary harm applies to all geoengineering activities unless special rules apply. Generally, this obligation does not mean that any environmental impact is for that reason prohibited. It is common ground that the obligation to respect the environment requires a due diligence standard and that the problem of which diligence is "due" depends on the particular case.[69] Its general nature and the scarce state practice and case law therefore raise problems in

[63] *Legality of the Threat or Use of Nuclear Weapons,* Advisory Opinion – General Assembly, *ICJ Reports*, 8 July 1996, at 22, para. 29; *Gabcikovo-Nagymaros Project* (Hungary v. *Slovakia),* Judgment, 25 September 1997, *ICJ Reports* (1997), at 7, para. 53; *Case concerning pulp mills on the river Uruguay (Argentia v. Uruguay),* Judgment, 20 April 2010, para. 193. Note that the ICJ's formulation is "activities within their jurisdiction *and* control".

[64] Trail Smelter Arbitration (USA/Canada), Judgement, 16 April 1938 and 11 March 1941, Reports of International Arbitral Awards (RIAA), Bd. III, 1905 ff., 1963–1965.

[65] Corfu Channel (United Kingdom v. Albania), Judgment, 9 April 1949, ICJ Reports (1949), at 35.

[66] 31 ILM 876 (1992); cf. Principle 21 of the preceding 1972 Declaration of the UN Conference on the Human Environment (Stockholm Declaration), 11 ILM 1416 (1972).

[67] United Nations Convention on Biological Diversity, Rio de Janeiro, 5 June 1992, in force 29 December 1993, 31 International Legal Material (1992), 818.

[68] United Nations Framework Convention on Climate Change, Rio de Janeiro, 9 May 1992, in force 21 March 1994, 31 International Legal Materials (1992), 849.

[69] Patricia W Birnie, Alan E Boyle, and Catherine Redgwell, *International Law and the Environment*, 3rd ed., (Oxford: Oxford University Press, 2009), at 147; Bodle, "Geoengineering and International Law", supra, note 15, at 307; Rickels et al., *Large Scale Intentional Interventions,* supra, note 6, at 99; Bodle et al., "Regulatory Framework", supra, note 9, para. 44.

determining in advance whether a specific activity would be prohibited or allowed. There are legal and factual difficulties relating to causation and the standard of conduct required. For instance, while attributing a geoengineering activity to a state is likely to be feasible,[70] it might be difficult to show which precise effects resulted from the particular geoengineering activity and what harm they caused. For instance, a breach would require a causal link between the particular geoengineering activity and changes in precipitation patterns, as well as between those changes and specific environmental harm.[71] In view of the extent of the potential damage, reversing the burden of proof is being discussed on the basis of the precautionary principle.[72]

The obligation not to cause transboundary harm is at the center of an initiative by the Pacific island nation of Palau to request an advisory opinion from the ICJ on obligations of states relating to climate change.[73] If the proposal for a request is adopted in the General Assembly, it remains to be seen the extent to which the ICJ will be prepared and able to clarify more specific guidance.

The main problem with using the obligation to respect the environment in order to protect the climate is that the obligation is mainly retrospective. In respect of geoengineering, it would be difficult to argue that a state is in breach of this obligation before the geoengineering activity has already taken place. International law provides only very limited means to obtain advanced provisional measures in order to stop activities that could be in breach of international obligations.[74]

17.2.4.2 Precautionary Principle

The precautionary principle (or: "approach") is frequently underlying arguments in favour of and against geoengineering. However, there is no uniform formulation or usage for the precautionary principle and its legal status in customary international law has not yet been clearly established, although it has been invoked several times.[75]

Article 3.3 UNFCCC is of particular relevance for geoengineering. Under the heading "principles", it incorporates the precautionary principle in the operative part of the fundamental treaty on climate change which has near universal participation,

---

[70] See below on state responsibility.

[71] Bodle, "Geoengineering and International Law", supra, note 15, at 306–307.

[72] See below on the precautionary principle.

[73] UN Press Conference, "Press Conference on Request for International Court of Justice Advisory Opinion on Climate Change", 3 February 2012, available at http://www.un.org/News/briefings/docs/2012/120203_ICJ.doc.htm (last accessed 1 May 2012).

[74] Bodle, "Geoengineering and International Law", supra, note 15, at 308, with references to ICJ case law.

[75] Bodle et al., "Regulatory Framework", supra, note 9, para. 58, with further references. Some reject the term precautionary "*principle*" and prefer the term "*approach*", see the overview in Birnie et al., *International Law and the Environment,* supra, note 69, at 154–155. I use the term "precautionary principle" for ease of reference and without prejudice to these concerns.

including the US.[76] The first sentence requires that Parties "should" actively take precautionary measures. Proponents of geoengineering could argue that CDR techniques are measures addressing the cause of climate change and that SRM techniques mitigate its adverse effects. The second sentence provides that "lack of full scientific certainty should not be used as a reason for postponing such measures," e.g. geoengineering measures considered on the basis of the first sentence, provided there are threats of serious or irreversible damage. Proponents of geoengineering could argue that such threats exist, in view of the expected impacts of climate change, the current state of play in reducing emissions, and the short remaining time period during which emission trends need to be reversed. By this rationale, the lack of full scientific certainty about geoengineering would not be a reason for postponing it.[77] However, Article 3.3 UNFCCC could not be read as actually *requiring* geoengineering measures.[78] In addition, Article 4.1(f) UNFCCC could provide a safeguard against unfettered geoengineering, but geoengineering would have to qualify as a mitigation or adaptation measure within the scope of this provision.

The precautionary principle can also support the other side of the argument: The scientific uncertainty about the effectiveness and potential risks of geoengineering is a reason to refrain from or slow down potentially harmful activities such as geoengineering and follow the less risky action of implementing emission reductions. This view can be supported by the fact that Article 3.3 UNFCCC was drafted with regards to scientific uncertainty about climate change, not geoengineering.[79]

More generally, it is also argued that one of the precautionary principle's legal implications could be to change the burden of proof.[80] Sectoral applications of the precautionary principle under specific regimes such as UNCLOS may adopt such or similar legal implications.[81] However, there is insufficient evidence that international law generally requires a state to prove that activities within its jurisdiction or control are environmentally safe.[82] In the *Pulp mills on the river Uruguay* case, the ICJ accepted that a precautionary approach "may be relevant" in the interpretation and application of the treaty in question, but rejected that it operates as a reversal of the burden of proof.[83]

---

[76] Currently 194 parties, http://unfccc.int/parties_and_observers/parties/items/2352.php. The US is one of the major emitters and potential geoengineering states but not party to the Kyoto Protocol.

[77] More detailed argument in Bodle, "Geoengineering and International Law", supra, note 15, at 310.

[78] On the precautionary approach in this regard see Birnie et al., *International Law and the Environment,* supra, note 69, at 162–164.

[79] Bodle et al., "Regulatory Framework", supra, note 9, para. 61.

[80] For instance, Daniel Bodansky, "Governing Climate Engineering: Scenarios for Analysis", Discussion Paper 2011-47, Harvard Project on Climate Agreements, Belfer Center for Science and International Affairs (Harvard Kennedy School, 2011), at 15.

[81] See for instance ITLOS case No.17, "Responsibilities and obligations of States sponsoring persons and entities with respect to activities in the Area (Request for Advisory Opinion submitted to the Seabed Disputes Chamber)", para. 125–135.

[82] Birnie et al., *International Law and the Environment*, supra, note 69, at 158.

[83] Case concerning Pulp Mills on the River Uruguay (Argentina/Uruguay), Judgment, 20 April 2010, ICJ Reports, para. 164.

The precautionary principle embodies the core arguments both for and against geoengineering.[84] It does not resolve the conflict between the objectives of avoiding the effects of global climate change *vis a vis* avoiding the risks of geoengineering – in particular as there are shades of grey between these two objectives. The opposite view[85] appears to result in a cost-benefit analysis across the board, which would be at odds with many environmental legal rules that are not open to such "net" approaches. On the other hand, it has been argued that if the precautionary principle is applied in isolation, there is a risk of perpetuating the scientific uncertainty that gives rise to its application in the first place.[86]

At the very least, the precautionary principle or approach provides interpretative guidance and procedural safeguards for dealing with scientific uncertainty. Yet it does not provide a sufficient legal tool for making essentially political decisions about conflicting objectives and managing risks.[87]

17.2.4.3 Environmental Impact Assessment

The ICJ has recently recognized "a requirement under general international law to undertake an environmental impact assessment where there is a risk that the proposed industrial activity may have a significant adverse impact in a transboundary context, in particular, on a shared resource."[88] The general obligation recognised by the ICJ complements similar obligations in several treaties and instruments relevant to geoengineering, for instance in Art. 14 CBD or Art. 206 UNCLOS. The London Convention and London Protocol's rules on ocean fertilization are complemented by additional non-binding guidance including a risk assessment framework, which provides detailed steps for completion of an environmental assessment, including risk management and monitoring.[89]

The general obligation recognised by the ICJ is a significant development and would apply to all states carrying out geoengineering activities in cases where no specific duty applies. The ICJ expressly stated that the obligation also involves continuous monitoring of the activity's effect on the environment. However, the ICJ judgment in the *Pulp Mills* case refers to particular industrial activities and does not necessarily establish a general requirement for a strategic impact assessment.

[84] Bodle et al., "Regulatory Framework", supra, note 9, para. 63.

[85] Cf. Rickels et al., *Large-Scale Intentional Interventions,* supra, note 6, at 101–103.

[86] Rickels et al., *Large-Scale Intentional Interventions,* supra, note 6, at 102.

[87] See also Birnie et al., *International Law and the Environment*, supra, note 69, at 161.

[88] Case Concerning Pulp Mills on the River Uruguay (Argentina/Uruguay), Judgement, 20 April 2010, ICJ Reports, paras. 204–206.

[89] Resolution LC-LP.2(2010) on the assessment framework for scientific research involving ocean fertilisation, adopted on 14 October 2010. For the Assessment framework see the draft elaborated by the Scientific Group of the London Protocol and the Scientific Group of the London Protocol, LC/SG/32/15, Annex 2.

There is not sufficient evidence to assume a customary obligation to carry out a strategic impact assessment of proposed geoengineering policies, plans or programmes into potential geoengineering policy development.

17.2.4.4 State Responsibility

The customary rules on state responsibility[90] deal with the consequences of breaches of international law. The International Law Commission's Articles on Responsibility of States for Internationally Wrongful Acts of 2001 ("Articles on State Responsibility")[91] for the most part reflect customary law, although some concepts may not be universally accepted. In the absence of specific rules, they govern the general conditions under which a state is responsible for wrongful geoengineering actions or omissions, and the resulting legal consequences. However, the rules on state responsibility do not define the requirements of the obligation which is said to have been breached. They do not define the requirements or conditions under which geoengineering is permitted or not.

The rules specify "circumstances precluding wrongfulness" by which states can avoid responsibility.[92] They include "necessity" as "the only way for the State to safeguard an essential interest against a grave and imminent peril". This could be of particular interest in the geoengineering debate, as a state might argue that it is severely affected by climate change and on this basis invoke distress or necessity as a legal defence. On the other hand, the defence could arguably be excluded for states who contributed to climate change and thus to the state of necessity.[93]

17.2.4.5 ENMOD Convention

The ENMOD Convention addresses environmental modification techniques having widespread, long-lasting or severe effects. The definition provided in the treaty would cover geoengineering techniques.[94] However, the ENMOD Convention's applicability to geoengineering is limited by its material scope, its limited number of parties and the lack of practice to draw from.[95] In terms of material scope, the ENMOD Convention only applies in armed conflict. The prohibited activity is to engage in "military or any other hostile use" of certain large-scale environmental

---

[90] Annex to UNGA Res. A/RES/56/83 of 12.12.2001, ("Articles on State Responsibility"). The rules relevant to this chapter are customary law.

[91] Ibid.

[92] Ibid., Article 25

[93] Ibid., Article 25(2)(b)

[94] Article II and an interpretative understanding which clarify that its scope covers inducing changes in climate patterns, which would arguably apply to at least some geoengineering concepts.

[95] Bodle, "Geoengineering and International Law", supra, note 15, at 312–313.

modification techniques. It does not prohibit geoengineering in peacetime nor does it expressly permit it. Although it may be tempting for a state to unilaterally regard a particular geoengineering activity as "hostile" and therefore prohibited, this should be determined in accordance with the laws of armed conflict in order not to erode the crucial distinction between the law applying in peacetime and the law of armed conflict.[96]

17.2.4.6 CBD Decision X/33

At CBD COP10 in 2010, the parties went beyond previous decisions addressing ocean fertilization and adopted a decision addressing geoengineering *in general*.[97] Although it is not binding in form or language, CBD COP decision X/33, para 8(w) appears to be the only all-encompassing governance measure at this level to date: The chapeau of para 8 "invites Parties and other Governments, according to national circumstances and priorities," to consider the guidance given by this decision, which includes:

- "Ensure, in line and consistent with decision IX/16 C, on ocean fertilization and biodiversity and climate change, in the absence of science based, global, transparent and effective control and regulatory mechanisms for geoengineering, and in accordance with the precautionary approach and Article 14 of the Convention, that no climate-related geoengineering activities that may affect biodiversity take place, until there is an adequate scientific basis on which to justify such activities and appropriate consideration of the associated risks for the environment and biodiversity and associated social, economic and cultural impacts, with the exception of small scale scientific research studies that would be conducted in a controlled setting in accordance with Article 3 of the Convention, and only if they are justified by the need to gather specific scientific data and are subject to a thorough prior assessment of the potential impacts on the environment;"

Leaving aside the on-going debate on semi-legal and *de facto* implications of COP decisions within treaty regimes, the decision under a treaty with near universal membership[98] sends a political signal that would be difficult to ignore in practice.

The main implications of the decision can be summarised as follows:[99] The decision provides a tentative definition that explicitly excludes CCS, but is broad enough to cover all geoengineering techniques currently discussed. The core of the

[96] Bodle, "Geoengineering and International Law", supra, note 15, at 312.

[97] CBD, "Decision X/33, UNEP/CBD/COP/DEC/X/33", 29 October 2010, available at www.cbd.int/doc/decisions/COP-10/cop-10-dec-33-en.pdf/, (last accessed 2 May 2012).

[98] The decision was adopted by consensus of the CBD's 193 parties. The US is a signatory but not a party to the CBD.

[99] For a detailed analysis see Bodle, "Geoengineering and International Law", supra, note 15, at 313–318; Masahiro Sugiyama, and Taishi Sugiyama, "Interpretation of CBD COP10 Decision on Geoengineering", SERC Discussion Paper 10013 (Tokyo, Japan: Socio-Economic Research Center, 2010).

operative part of paragraph 8(w) is the guidance that no climate-related geoengineering activities that may affect biodiversity take place. Although the language and grammar are not entirely clear, the intended restriction of geoengineering appears to be subject to three conditions[100]: First, it is a transitional measure that applies "in the absence of science based, global, transparent and effective control and regulatory mechanisms for geoengineering". Second, the restriction applies "until there is an adequate scientific basis on which to justify" geoengineering activities, which includes a comprehensive risk assessment. Third, it exempts small-scale scientific research studies, provided that they are conducted in a controlled setting, justified by the need to gather specific scientific data and subject to a thorough prior assessment of the potential impacts on the environment. The decision leaves it to parties to determine whether the conditions for the second and third condition are met.

In accordance with the mandate in decision X/33, the CBD Secretariat has prepared two studies, one on the impacts of geoengineering and one on gaps in the international regulatory framework. The two studies[101] were submitted to CBD SBSTTA 16 for consideration.

17.3 Challenges to Climate Law

Geoengineering poses fundamental challenges to climate policy in general and climate law in particular. To some extent these may resemble those of other high-risk or controversial technologies such as genetic modified organisms, nuclear power and perhaps nanotechnology. Yet geoengineering is different in that it is presented as a plan B to mitigation, as an unasked-for fallback option that is not desirable as such but is pursued in order to at least find out whether it is viable. Specific aspects of geoengineering cause particular challenges, which include e.g. the broad range of concepts, most of which become high-risk only when deployed at large scale; the difficulty of seeking more knowledge on geoengineering without endorsing it or causing a lock-in effect; the focus of climate law on its traditional categories of mitigation and adaptation; adding political legitimacy and responsibility to a largely science-driven debate; and the potential of the geoengineering debate to obstruct the climate change negotiations and depart from emission reductions.

17.3.1 Inadequacy of Existing Law

The geoengineering debate has taken international law somewhat by surprise. The main legal studies so far show an emerging consensus that -details aside- existing international law hardly addresses the potential impacts of geoengineering or related

[100] Bodle et al., "Regulatory Framework", supra, note 9, paras. 79–81.

[101] Williamson et al., "Climate-Related Geoengineering", supra, note 2; Bodle et al., "Regulatory Framework", supra, note 9.

key questions. Most international law was developed before geoengineering was a significant issue and does not currently contain explicit references to geoengineering techniques. Geoengineering is currently not as such prohibited by international law. The application of specific rules and restrictions on geoengineering would often depend on specific actual or potential impacts. Whether such impacts would actually occur is difficult to assess or predict at this stage.[102]

In the absence of clear specific rules, and in view of the potential implications and impacts of some geoengineering techniques, it is tempting to seek legal guidance from cross-cutting general rules and principles of international law. While it is legitimate and necessary to explore and develop the meaning of general rules by applying them to specific cases, there is a risk of reading one's own desired normative content into such rules. Overburdening general rules could be detrimental to their acceptance and legal value.

The precautionary principle does not help in resolving the problem of determining the "lesser evil", i.e. choosing between the potential impacts of geoengineering and the impacts of climate change that are inevitable or assumed to happen in the absence of geoengineering. From a specific climate law perspective, Article 3.3 UNFCCC is ambiguous: Depending on how we assess the risk posed by geoengineering (i) in relation to a scenario with substantial mitigation as well as (ii) in relation to a scenario of unmitigated climate change, the precautionary principle embodies the core arguments both for and against geoengineering.[103]

At a more specific legal level, geoengineering leads to the problem of whether legal rules protecting the environment allow for a "net" approach to environmental harm by taking into account the harm avoided by the activity in question. Most treaties do not appear to provide for consideration of the overall "net" effects on the broader environment in comparison to harm avoided.[104]

17.3.2 Governance

The London Convention and London Protocol as well as the CBD have developed rules specifically on geoengineering in general and on particular techniques. However, the CBD decision on geoengineering does not mean that the question of whether and how address geoengineering is resolved. The existing rules and guidance are unlikely to be able to contain the risks posed by geoengineering or be able to avoid related political conflicts.[105] From a governance perspective, the existing legal hooks are not strong enough to carry the political weight of geoengineering.

[102] Bodle et al., "Regulatory Framework", supra, note 9, para. 182 and 186.

[103] Bodle et al., "Regulatory Framework", supra, note 9, para. 63.

[104] Bodle et al., "Regulatory Framework", supra, note 9, para. 196.

[105] Bodle, "Geoengineering and International Law", supra, note 15, at 321. On the social, economic and cultural considerations regarding geoengineering have significant inter- and intra-generational equity issues see Williamson et al., "Climate-Related Geoengineering", supra, note 2, at 63.

Yet it is not self-evident that a governance framework for geoengineering is needed at the *international* level. Geoengineering techniques are still at an early stage which might not justify or muster the political will and effort to develop an international framework. In addition, as it is likely that at least some geoengineering concepts could be tested and deployed by a single state, a state capable of doing so might prefer to address geoengineering in its domestic jurisdiction only, and be reluctant to wait for or subject itself to international agreement. However, even for those states there are compelling reasons why it is in the national interest to participate in an international governance framework:[106] First, the mere potential for transboundary impacts of geoengineering is likely to have serious foreign policy implications. The main risk is political tension *regardless* of whether any impacts can be proven to be caused by the geoengineering activities in question. It is in the national interest of any state pursuing geoengineering research to avoid the political costs of such tensions as well as preventing others from engaging in unilateral and uncoordinated geoengineering.[107] Second, international governance could provide legitimacy to one's own policy. A polarised debate, perhaps similar to instances regarding climate change, would make it difficult for a state to adopt and implement any policy on geoengineering. Third, depending on the particular geoengineering concept, at some stages research activities might need to be coordinated at the international level in order to ensure that data can be correctly attributed to particular experiments and to ensure validity of results.

Geoengineering presents a shift in climate law from providing incentives and international obligations to *do* something (reduce emissions) to providing incentives and international obligations *not to do* something (unfettered geoengineering). The efforts under the LC/LP and the CBD are first steps towards articulating common ground amongst states on how to address geoengineering. If geoengineering is to be further addressed at the international level, virtually all treaties impose some procedural obligations on geoengineering activities falling within their scope of application.[108] Yet governance of geoengineering in all likelihood also requires institutions.[109] Assuming a need for governance, under which regimes or in which fora should governance be exercised? The mandate of many international regimes and institutions would allow them to address geoengineering or some aspects of it.

---

[106] For detailed argument and further references see Ralph Bodle, "International governance of geoengineering: Rationale, functions and forum", in: William C.G. Burns and A. Strauss, (eds.), Climate Change Geoengineering: Legal, Political and Philosophical Perspectives (Cambridge: Cambridge University Press, forthcoming).

[107] Cf. Dan Bodansky, "May we engineer the climate?", 33 *Climatic Change* (1996), 309, at 310; see also Albert C. Lin, "Geoengineering Governance", 8 *Issues in Legal Scholarship 3* (2009); Scott Barrett, "Geoengineering's Governance" Written Statement Prepared for the U.S. House of Representatives Committee on Science and Technology Hearing on "Geoengineering III: Domestic and International Research Governance" (2010), available at http://science.house.gov/publications/Testimony.aspx?TID=15386.

[108] Bodle et al., "Regulatory Framework", supra, note 9, para. 193.

[109] Bodle et al., "Regulatory Framework", supra, note 9, para. 160.

This could lead to different treaties or institutions potentially competing for addressing geoengineering with overlapping or inconsistent rules or guidance.[110]

The climate regime seems to be an obvious candidate for addressing geoengineering. The regime has a strong institutional structure and a scientific underpinning linked to work of the IPCC. Accordingly, there have been suggestions outside the climate negotiations to address geoengineering under the UNFCCC, for instance by a new protocol.[111] However, the UNFCCC and Kyoto Protocol have not addressed geoengineering concepts or governance (see above Sect. 17.2). There were but few instances where geoengineering was mentioned: At one point the Executive Secretary of the UNFCCC warned that carbon dioxide removal techniques might have to be developed due to the slow process of the negotiations.[112] A planned Joint IPCC Expert Meeting of several Working Groups on geoengineering sparked a submission by Bolivia to the UNFCCC demanding that the meeting's agenda be changed.[113] Geoengineering was also included in a 2012 submission by the group of least developed countries containing a list of themes to be addressed at the regular research dialogue.[114]

There are good reasons why the climate regime should continue to focus on its already highly complex body of rules and on-going negotiations on a future regime. At this stage all options for introducing geoengineering could seriously jeopardize the current climate negotiations and make geoengineering part of the trade-offs that are part of them.[115] If other fora begin or continue to address geoengineering, the need for co-ordination and consistency with climate objectives and law should be assessed. It may be that the existing formal and informal channels between treaty regimes and international fora are sufficient.

17.3.3 Definition

The broad range of techniques discussed as "geoengineering" present a challenge at a more technical level. There is no universal and uniform use of the term

---

[110] Bodle, "Geoengineering and International Law", supra, note 15, at 321.

[111] Barrett, "Geoengineering's Governance", supra, note 107, at 10–11; Karen N. Scott, "Marine Geoengineering: A New Challenge for the Law of the Sea", 18th Annual Australia New Zealand Society of International Law (ANZSIL) Conference (Canberra, Australia: 2010, 2009).

[112] Fiona Harvey "Global warming crisis may mean world has to suck greenhouse gases from air", The Guardian, 5 June 2011, available at www.guardian.co.uk.

[113] "Bolivian Submission to Joint Workshop of Experts on Geoengineering", available at http://unfccc.int/files/meetings/ad_hoc_working_groups/lca/application/pdf/bolivian_submission_on_geoingeneering.pdf.

[114] Submission on Specific Research Themes by Republic of The Gambia on behalf of the Group of Least Developed Countries (LDCs), FCCC/SBSTA/2012/MISC.2, 30 March 2012, at 8, available at http://unfccc.int.

[115] For details see Bodle, "International governance of geoengineering: Rationale, functions and forum", supra, note 106.

"geoengineering".[116] For some techniques there is no consensus as to whether they should be labeled and discussed as "geoengineering". For instance, afforestation and reforestation in order to enhance carbon sinks are considered a type of CDR by some, but not others.[117] The same goes for carbon capture and storage (CCS).[118] There are plausible scientific and technical reasons for such distinctions, for instance drawing a line between mitigation meaning reducing the *generation* of greenhouse gas emissions, and geoengineering meaning reducing pre-existing atmospheric *concentrations*. However, potentially negative implications of being classified as "geoengineering", in particular for regulatory purposes, play a role as well. For instance, classifying forestry techniques as geoengineering might affect programmes such as REDD+, and CCS is a technology that has recently been introduced with special rules into the Kyoto Protocol's clean development mechanism.

Due to the broad range of geoengineering techniques, any overarching definition for regulatory purposes is unlikely to be sufficiently comprehensive to capture all relevant techniques while being sufficiently precise to exclude uncontroversial techniques or scale of activities. In a regulatory context, a definition would have to be complemented by further details on determining and measuring unspecific terms such as scale. This could be achieved for instance, by complementing the definition with a positive list that expressly mentions specific techniques or activities which are considered to be geoengineering. Such a list could be combined with an amendment procedure allowing for updating it to new developments. Another option is a process or institution providing further guidance on a case by case basis.

17.3.4 Addressing Research

Another key question is how to address further research. Proponents of further research argue that it is needed in order to obtain reliable information about feasibility and risks. However, this would at some stage require real-world field experiments that would have to be gradually scaled up in order to know the impacts of a particular technique and whether it is effective. Apart from the difficulty of drawing the line

---

[116] See Overview of some definitions in Williamson et al., "Climate-Related Geoengineering", supra, note 2, at 75.

[117] *See e.g.* National Research Council, Committee on Science, Engineering, and Public Policy (U.S.). Panel on Policy Implications of Greenhouse Warming, *Policy Implications of Greenhouse Warming: Mitigation, Adaptation, and the Science Base* (National Academies Press, 1992); Rickels et al., *Large-Scale Intentional Interventions*, supra, note 6; as opposed to Royal Society, *Science, Governance and Uncertainty*, supra, note 6; also German Federal Environment Office, "Effective Climate Protection or Megalomania?", supra, note 6, at 18 and 23; Williamson et al., "Climate-Related Geoengineering", supra, note 2, at 14.

[118] German Federal Environment Office, "Effective Climate Protection or Megalomania?", supra, note 6, at includes it as a geoengineering technique, while e.g. the Royal Society does not, Royal Society, *Science, Governance and Uncertainty*, supra, note 6, at 6.

between research and deployment, most existing rules of international law do not make this distinction. To what extent should international law endorse research activities even if they could cause severe impacts, on the grounds that this is the only way to know for sure that a geoengineering technique causes such impacts?

Scientists have drafted a set of five basic principles that should guide and govern geoengineering research. These five "Oxford principles"[119] include geoengineering to be regulated as a public good, public participation in decision making, disclosure of geoengineering research and open publication of results, independent assessment of impacts and having in place clear governance arrangements before deployment. The principles appear inadequate to address the challenges presented here, as they seem to resemble common sense and expected practice regarding any research involving potential risks to the environment.[120] The fifth principle, that governing arrangements be made clear prior to any actual use of the technologies, implicitly entrenches the distinction between research and "actual use" without a clear rationale for it. The Solar Radiation Management Governance Initiative has taken these ideas further and provided a more detailed assessment of governance needs for different research activities.[121]

At the heart of this challenge is the question of what constitutes research and could be a reason for privileging it. If the objective of governance is to address risks and potential impacts of an activity, then activities involving the same risks and potential impacts should be treated the same regardless of whether an activity is carried out as "science" or as "deployment". On the other hand, some argue that following certain procedures and implementing safeguards is what constitutes research, and that therefore such activities should be treated differently. In contrast to the former view, this latter understanding appears to include plausible conditions for privileging geoengineering research to some extent.

17.3.5 Relation to the Mitigation and Adaptation Categories

International climate law, in particular the few rules under the UNCCC and Kyoto Protocol that could be of relevance to geoengineering, are based on the traditional approach of distinguishing mitigation and adaptation (see above). These categories

---

[119] S. Rayner et al., *Climate Geoengineering Governance:* Memorandum on draft principles for the conduct of geoengineering research. House of Commons Science and Technology Committee inquiry into The Regulation of Geoengineering. (2009), available at http://www.sbs. ox.ac. uk/ centres/ insist /Documents/regulation-of-geoengineering.pdf.

[120] The same goes for the five similar principles recommended as the outcome of the Asilomar conference in March 2010, International Conference on Climate Intervention Technologies, Asilomar Conference Center, March 22–26, 2010, Pacific Grove, USA, http://www.climate.org/ resources/climate-archives/conferences/asilomar.html.

[121] Solar Radiation Management Governance Initiative (SRMGI), *Solar Radiation Management: The Governance of Research* (Environmental Defense Fund, The Royal Society and TWAS, 2011).

can have legal implications. However, geoengineering does not easily fit into these categories. While all geoengineering techniques are intended to counteract climate change and its effects, they do not address emission reductions, and basically they do not address how to adapt to a changed climate. This strict view might have advantages, as it might avoid interpretations of UNFCCC and Kyoto Protocol rules with surprising or undesirable implications. Nevertheless, several geoengineering approaches that may be considered as geoengineering can also be considered as climate change mitigation or adaptation, or both, for example, some ecosystem restoration activities.[122] The Kyoto Protocol has recently opened the traditional approach to some extent by allowing CCS into the CDM, although CCS does not reduce the *production* of emissions. Against this background, if geoengineering were to move forward, there could be pressure to credit certain geoengineering activities that do not fit easily into the mitigation category. In addition, treating geoengineering as mitigation or adaptation could for instance have implications for funding institutions and their eligibility criteria.

17.4 Conclusions and Suggestions: Laws for Geoengineering?

Geoengineering is hardly addressed by international law in general and by climate law in particular. The small legal hooks are unlikely to hold the heavy political weight of geoengineering.[123] The main challenge for policy makers is deciding whether and how to get involved. Although the debate about geoengineering is still largely driven by scientists, it is gaining attention at the policy interface. For instance, in a resolution adopted in preparation for the Rio + 20 Summit in 2012, the European Parliament has expressed its "opposition to proposals for large scale geo-engineering".[124] Apart from the CBD decision on geoengineering, its potential effects will be part of the IPCC's fifth assessment report, including the possible role, options, risks, and status of geoengineering as a response option.[125]

Good quality information on many aspects of geoengineering is still very limited, and there are knowledge gaps regarding the effectiveness of many of the geoengineering techniques and their environmental impacts as well as their social

[122] Williamson et al., "Climate-Related Geoengineering", supra, note 2, at 6.

[123] Bodle, "International governance of geoengineering: Rationale, functions and forum", supra, note 106.

[124] European Parliament resolution of 29 September 2011 on developing a common EU position ahead of the United Nations Conference on Sustainable Development (Rio + 20), P7_TA(2011)0430, para. 90.

[125] IPCC, "Scope, Content and Process for the Preparation of the Synthesis Report (SYR) of the IPCC Fifth Assessment Report (AR5)", 14 October 2010, available at www.ipcc.ch/meetings/session32/doc04_p32_cont_process_SYR.pdf.

and economic implications.[126] This situation seems to inevitably call for more research in order to make informed decisions. However, more research could be a slippery slope towards a lock-in towards geoengineering. It would also be a political choice with implications for instance for other research areas where less research funding could be available, or for climate policy in general. The "moral hazard" describes the risk that pursuing geoengineering could provide an incentive or excuse for stepping away from reducing emissions. At the more technical level of international climate law, introducing geoengineering would add another layer to the already over-complex climate negotiations. For instance, states might push for crediting some geoengineering techniques (see above).

On the other hand, geoengineering could still be a storm in a teacup. The technical and economic feasibility might be confounded at early stages rather than after years of research and gradually scaled-up experiments. In addition, for better or worse, geoengineering might not be politically viable.

The main challenge is to consider governance arrangements that are commensurate with these parameters. In addition to the rationale for international governance proposed above, I have outlined key aspects and functions elsewhere.[127] The study for the CBD on regulatory aspects has identified further aspects for designing a future governance framework.[128] A key component is to clearly separate scientific input and political decision-making. In order to keep the focus on the transition to a low-carbon economy, international climate law and governance need to decide when and how *political* decisions are to be made about geoengineering.

[126] Williamson et al., "Climate-Related Geoengineering", supra, note 2, at 9.

[127] Bodle, "International governance of geoengineering: Rationale, functions and forum", supra, note 106.

[128] Bodle et al., "Regulatory Framework", supra, note 9, para. 197.

Part V
Comparative Climate Law

Chapter 18
Climate Law in the United States: Facing Structural and Procedural Limitations

Michael Mehling and David John Frenkil

Abstract Just 5 years ago, the official position of the White House on the issue of climate change was that there was no such position. President George W. Bush and his administration declined to address whether climate change was even occurring, much less how to mitigate the causes of a phenomenon that had potentially contributed to billion-dollar disasters, thousands of fatalities during Hurricanes Rita and Katrina, and a significant number of displaced U.S. citizens.

By the time that President Bush departed the White House in January 2009, his administration had at least acknowledged the existence of climate change. Also, during his second and final term as President, the United States Supreme Court in 2007 issued a decision, *Massachusetts v. EPA*, which mandated the U.S. Environmental Protection Agency (EPA) to regulate greenhouse gas (GHG) emissions under the federal Clean Air Act.

But still, the U.S. Federal Government did not take an active stance on mitigating GHG emissions until President Bush's successor, Barack H. Obama, entered office. Armed with the *Massachusetts v. EPA* decision, the Obama Administration has since issued a comprehensive array of regulations through the EPA aimed at mitigating GHG emissions from stationary sources such as electric power plants and increasing fuel efficiency standards for the automotive sector. At the international level, President

---

Michael Mehling is President of the Ecologic Institute, Washington, DC, and an Adjunct Professor at Georgetown University; David John Frenkil is an energy attorney and publisher of EfficiencyLaw.com, an online journal covering electric power and energy development.

M. Mehling (✉)
Ecologic Institute, 1630 Connecticut Ave. NW, Suite 300, 20009
Washington, DC, USA
e-mail: michael.mehling@ecologic-institute.us

D.J. Frenkil
1015 33rd St NW, Apt 410, 20007 Washington, DC, USA
e-mail: frenkil@gmail.com

E.J. Hollo et al. (eds.), *Climate Change and the Law*, Ius Gentium: Comparative Perspectives on Law and Justice 21, DOI 10.1007/978-94-007-5440-9_18,

© Springer Science+Business Media Dordrecht 2013

Obama has also affirmed the U.S. commitment to reduce greenhouse gas emissions by 17% below 2005 levels until 2020, as pledged internationally under the 2009 Copenhagen Accord.

This drastic shift in U.S. climate policy over a relatively short period provides a glimpse into the existing structural and procedural workings of the decision-making process behind climate and energy policy in the United States. In theory, this system has the potential to drive substantive change, but, in practice, climate change is occurring much faster than U.S. policy can keep up.

18.1 Current Energy Landscape in the United States

As the world's largest economy, the United States is also one of the largest consumers of fossil energy sources and has above-average per capita emissions of GHGs. Over the past decade, emissions have mostly continued to grow, largely as a result of an expanding transportation sector and a traditionally heavy reliance on coal in the energy sector.

Perhaps the most remarkable trend in recent years, however, has been a dramatically falling price of natural gas relative to coal due to the discovery of substantial natural gas reserves in domestic shale formations and higher compliance costs associated with power production from coal resulting from more stringent environmental regulations. Together, these two factors have led to a shift away from coal in electricity generation, and hence reduced the carbon intensity of U.S. energy consumption.

The recent shift to natural gas for electricity production and the economic downturn have led to a reversal in emissions growth. Nevertheless, the U.S. has retained its reputation on the international stage as a laggard in the realm of climate policy due to a federal strategy that has been based on a fragmented array of voluntary measures and incentives, executive regulation, and issue-specific legislation. While these efforts add up to more GHG mitigation than observers will often give credit, there are important structural and procedural limitations impeding a more coherent and ambitious national climate policy. Such barriers are highlighted throughout the remainder of this chapter.

Politics and public opinion also pose significant obstacles to legislative progress on climate change in the United States, although these are far more difficult to capture in a straightforward analysis. Domestically, public perceptions regarding climate change and relevant policy responses have proven to be volatile. While nationwide surveys suggest that a majority of Americans consider global warming a serious or very serious problem, fewer believe that climate change should be a priority for government action.[1] Meanwhile, during the 2010 midterm elections and 2012 presidential campaign, global warming became an increasingly partisan issue,

[1] See the results of the 2012 National Survey of American Public Opinion on Climate Change (NSAPOCC), summarized in Chris Borick and Barry Rabe, *Public Views on Climate Policy Options: Spring 2012 NSAPOCC Findings* (Washington, DC: Brookings Institution, 2012).

as candidates for the Republican Party nomination repeatedly challenged the accuracy of the science behind climate change in order to gain favor with conservative voters. Powerful interest groups committed to portraying GHG emission reduction efforts as unnecessary, costly and economically harmful have invested substantial resources to influence public opinion and ultimately exerted a noticeable effect on the political process. A comprehensive analysis of the legal and institutional realities of climate policy in the U.S. would therefore also have to consider aspects such as rules on campaign finance, the electoral system, and the role of the media; yet that would exceed the scope of this chapter, which instead focuses on the separation of powers and legislative process.

18.2 Separation of Powers in the United States

The U.S. Constitution establishes the separation of powers between the three branches of government – legislative, executive and judicial. While the system creates an arrangement of "checks and balances" that was designed to ensure that too much power would never be vested in one individual or body within the Federal Government, in practice, the system yields policy outcomes that are oftentimes confusing and incoherent, and sometimes indeed non-existent.

18.2.1 Overview of the Federal Government

The nuances of the system, and its associated rules, can be very difficult to understand, especially for outsiders. While details of each system are outside the scope of this article, in general, it is important for readers to understand the following.

In the legislative branch, Congress is comprised of the "lower" chamber – the U.S. House of Representatives – and the "upper" chamber, the U.S. Senate. There are 435 members in the House, where congressional seats are re-apportioned every 10 years by population and based on the decennial census, and 100 members in the Senate, where each of the 50 states are afforded two seats.

Prior to coming to a vote, legislation is considered by committee. In the House, a bill typically goes before multiple committees covering various issue areas; whereas in the Senate, only one committee has jurisdiction over a bill. In order to pass the House, all that is required is a simple majority of 50% plus one vote. In the Senate, however, voting rules call for a "super-majority" of 60 votes to close the debate on a bill and proceed to a substantive vote. Additional rules cover different scenarios, but in general, once a bill is passed by the House and the Senate, a "conference committee" comprising members of both chambers works through the legislation in order to present a coordinated piece of legislation to the U.S. President, who then proceeds to sign the bill into law.

Once a bill becomes the law, the Executive Branch – headed by the U.S. President – has the authority to implement the laws. The Executive Branch is comprised of numerous federal agencies, such as the Departments of State, Defense, Energy, Interior, Transportation, and the Environmental Protection Agency (EPA). Each of these federal agencies has the authority to promulgate additional rules and regulations, and most have procedures for administrative law judges to resolve disputes associated with such regulation.

Finally, the Judicial Branch consists of a hierarchy of court systems. At the lowest level of the U.S. federal judicial system are the U.S. District Courts, which are located in every state and province in the United States. Decisions made by U.S. District Courts can be appealed to the U.S. Court of Appeals (also referred to as a "Circuit Court"), which is organized on a regional basis. There are 11 Circuit Courts plus the U.S. Court of Appeals for the Federal Circuit. Decisions of any federal Circuit Court can be appealed to the U.S. Supreme Court. Decisions made by this Court are considered the supreme law of the land and take precedence over any related decision at the lower levels of the judiciary system.

18.2.2 Legislative Branch

Given the scope and complexity of climate change, a formal act of Congress as the federal legislature is generally considered the most suitable vehicle for a comprehensive policy response. While efforts to pass relevant legislation date back more than a decade, the U.S., unlike a majority of industrialized nations, currently lacks a coherent legislative climate regime. Even after winning both houses of Congress and the White House in the 2008 elections, Democrats failed to pass legislation in the 111th Congress (2009–2010) that would address climate change.

While it never became law, the U.S. House of Representatives did pass a comprehensive climate bill in June 2009, the first legislation to successfully cross this threshold in U.S. history. Entitled the "Clean Energy and Security Act of 2009" (ACES), this bill passed the House by a narrow margin of 219–212 votes.[2] Once in force, it would have placed limits on GHG emissions from a large section of the U.S. economy, and would have introduced a combined energy efficiency and renewable electricity standard. Specifically, it called on electric utilities to meet 20% of their electricity demand through renewable energy sources and improved energy efficiency by 2020, established energy-saving standards for new buildings and appliances, and mandated CO_2 emission reductions from major domestic sources of 17% by 2020, 42% by 2030, and 83% by 2050 over 2005 levels. An economy-wide emissions trading

[2] H.R. 2454, "American Clean Energy and Security Act of 2009", 111th Congress, 1st Session, 26 June 2009, available at http://energycommerce.house.gov/Press_111/20090701/hr2454_house.pdf (last accessed on 10 June 2012).

system would have been phased in starting in 2012 so as to help achieve these objectives at reduced cost. ACES never entered into force, however, because lawmakers in the U.S. Senate could not gain the 60 votes necessary to close the debate on the bill and proceed to a substantive vote.

18.2.3 Executive Branch

Although Congress has been the focal point of media attention in U.S. federal energy policy, not all policy is required to pass the Congress. As noted earlier in section 18.2.1, the Executive Branch has the authority to implement certain types of regulation, even without seeking Congressional approval.

Notably, the President has directed the EPA to regulate carbon dioxide and other GHGs under the existing Clean Air Act (CAA). Established on 2 December 1970 as an independent agency of the U.S. government, the EPA implements federal legislation on a broad range of environmental subject matters, frequently adopting regulations and supervising their implementation by federal and state authorities. On the issue of climate change, however, the EPA had not become significantly involved until a landmark decision of the Supreme Court in the case of *Massachusetts v. EPA*, discussed below, declared that GHGs are pollutants and hence fall within the jurisdiction of the EPA.

On 7 December 2009, less than a year after President Obama had taken office, the EPA formally adopted an "Endangerment" and "Cause and Contribute" Finding under Section 202 of the CAA stating that anthropogenic climate change threatens the environment and public health, a prerequisite for the adoption of rules to limit greenhouse gas emissions from mobile and stationary sources.[3]

Responding to this mandate, the administration announced a plan to integrate federal fuel economy standards under the Energy Policy and Conservation Act, called CAFE standards, with federal vehicle emissions standards under the Clean Air Act, increasing these to an average of 35.5 miles per gallon by 2016. Overall, this would translate into an emissions limit of 250 g of CO_2 per mile by 2016. A joint final rule was adopted by the EPA and the National Highway Safety Administration (NHTSA) on 1 April 2010, and its requirements will apply beginning 1 October 2012.

For stationary sources, a comprehensive rule on "Mandatory Reporting of Greenhouse Gases" issued by the EPA adopted 30 October 2009 requires more than 10,000 facilities throughout the US – accounting for nearly 85% of US GHG

[3] U.S. Environmental Protection Agency, "Endangerment and Cause or Contribute Findings for Greenhouse Gases Under Section 202(a) of the Clean Air Act, Final Rule", 74 Federal Register 66496, 15 December 2009, available at: http://www.epa.gov/climatechange/endangerment/downloads/Federal_Register-EPA-HQ-OAR-2009-0171-Dec.15-09.pdf (last accessed on 15 June 2012).

emissions – to report their emissions of eight categories of GHGs on an annual basis.[4] In force since 1 January 2010, this rule requires certification of the emissions inventory and imposes penalties for failure to report. While it does not impose reduction targets in itself, this rule yields vital information for the design and implementation of further mitigation measures.

Since 1 January 2011, new or substantially modified emitters have also been subject to a permitting requirement under the CAA, with a Tailoring Rule adopted on 13 May 2010 ensuring that these requirements apply only to the largest stationary sources of GHGs.[5] While this system of operating permits allows the EPA to elaborate guidance for implementing state authorities on how to define best available control technologies (BACTs) for each covered source, it does not provide the means to specify GHG emission standards. Rather, the CAA provides additional pathways through which to regulate GHG emissions from stationary sources, including, most importantly, national ambient air quality standards (NAAQS) and performance standards for new and existing stationary sources (NSPS), with the latter considered more suitable for non-toxic greenhouse gases.

Exercising this prerogative, the EPA signed decrees on 23 December 2010 requiring the Agency to propose new source performance standards and emission guidelines for greenhouse gas emissions from refineries and electric generating units fired with natural gas, oil and coal, two sectors that, combined, account for approximately 40% of all U.S. emissions. Following several delays, the EPA finally proposed to set a nationwide standard for emissions of carbon dioxide from new fossil fuel electric generating units on 27 March 2012.[6] Under this proposed NSPS, new fossil fuel electric generating units (EGUs) would be subject to a maximum CO_2 emissions rate of 1,000 lb/MWh, which essentially precludes new coal-fired power plants unless these are equipped with abatement technologies such as carbon capture and sequestration

[4] U.S. Environmental Protection Agency, "Mandatory Reporting of Greenhouse Gases: Final Rule", 22 September 2009, available at http://www.epa.gov/climatechange/emissions/downloads/FinalMandatoryGHGReportingRule.pdf (last accessed on 10 June 2012). Under the rule, suppliers of fossil fuels or industrial greenhouse gases, manufacturers of vehicles and engines, and facilities that emit 25,000 metric tons or more per year of GHG emissions are required to submit annual reports to the EPA; the gases covered by the rule are carbon dioxide (CO_2), methane (CH_4), nitrous oxide (N_2O), hydrofluorocarbons (HFC), perfluorocarbons (PFC), sulfur hexafluoride (SF_6), and other fluorinated gases including nitrogen trifluoride (NF_3) and hydrofluorinated ethers (HFE).

[5] U.S. Environmental Protection Agency, "Prevention of Significant Deterioration and Title V Greenhouse Gas Tailoring Rule", Federal Register, 75 Federal Register (2010), 31513–31608; Under the Tailoring Rule, permitting focuses on the largest industrial sources, starting with new or substantially modified facilities that already are subject to permitting requirements for conventional pollutants and have the potential to emit 75,000 tons per year of carbon dioxide equivalent (CO_{2e}) or more, and later adding all sources that emit at least 100,000 tons of GHG per year. Sources emitting less than 50,000 tons of GHGs per year will not be required to obtain permits for the time being.

[6] U.S. Environmental Protection Agency, "Standards of Performance for Greenhouse Gas Emissions for New Stationary Sources: Electric Utility Generating Units", 27 March 2012, 40 CFR Part 60, available at http://epa.gov/carbonpollutionstandard/pdfs/20120327proposal.pdf (last accessed on 12 June 2012).

(CCS). Still, judicial proceedings filed against these requirements by several states and the private sector as well as efforts by Republican legislators in Congress to limit the authorities of the EPA regarding their implementation have introduced a measure of uncertainty with respect to the forthcoming regulations.

18.2.4 Judicial Branch

As discussed above, the U.S. Supreme Court decision in *Massachusetts v. EPA* provided the EPA with authority to regulate GHGs, irrespective of Congressional action on climate change.[7] The case originated when a group of 19 private organizations filed a rulemaking petition[8] on 20 October 1999 asking the EPA to regulate "greenhouse gas emissions from new motor vehicles under § 202 of the Clean Air Act".[9] On 8 September 2003, the EPA entered an order denying the rulemaking petition, explaining that the CAA does not provide authority to the EPA to issue regulations addressing climate change[10] and such regulations would be unwise.[11] In response, environmental organizations and state and local governments sought review of the EPA's decision in the United States Court of Appeals for the District of Columbia Circuit. However, the Court denied the petition for review because two of the three judges on the panel agreed that "the EPA Administrator properly exercised his discretion under § 202(a)(1) in denying the petition for rule making."[12]

On appeal, the U.S. Supreme Court issued a landmark decision when it ruled on 2 April 2007 that the EPA has the authority to regulate GHG emissions. The Court said that, due to the rise in sea levels, Massachusetts had standing to challenge the EPA's denial of the rulemaking petition.[13] Thus, the Court remanded the petition to the EPA, holding that, "[u]nder the clear terms of the Clean Air Act, EPA can avoid taking further action only if it determines that greenhouse gases do not contribute to climate change or if it provides some reasonable explanation as to why it cannot or will not exercise its discretion to determine whether they do."[14]

---

[7] *Massachusetts v. Environmental Protection Agency et al.*, 549 U.S. 497 (2007), available at www.supremecourtus.gov/opinions/06pdf/05-1120.pdf (last accessed on 10 June 2012).

[8] The Administrative Procedures Act provides for petitions for rulemaking in order for the public to express its desire for new regulations, deregulations or modifications to regulations already in effect.

[9] *Mass. v. EPA*, supra, note 7, at 510. Section 202 of the Clean Air Act requires the EPA to prescribe standards applicable to the emission of "any air pollutant" from any class of new motor vehicle which, in the EPA Administrator's judgment, has caused or contributed to air pollution reasonably anticipated to endanger public health or welfare.

[10] U.S. Environmental Protection Agency, "Notice of Denial of Petition for Rulemaking", 68 Federal Register 52922, 23 September 2003.

[11] 68 Federal Register 52922, at 52929–52931.

[12] *Mass. v. EPA* 415 F.3d 50, 58 (2005).

[13] *Mass. v. EPA*, supra, note 7, at 526.

[14] Ibid. at 533.

During the 2008 Presidential campaign season, the EPA under the Bush Administration responded to the Supreme Court's order for remand by delaying a decision until after the election.[15] In December 2009, under the Obama administration, the EPA responded to the *Massachusetts v. EPA* ruling with the so-called "Endangerment" and "Cause and Contribute" findings discussed above.[16] While this rulemaking did not establish a regulation by which the EPA would regulate GHGs, they set the stage for the agency's action that followed.

18.3 Federalism and the Tug and Pull Between Federal and State Governments

Under the U.S. Constitution, both the national ("federal") and state governments are granted certain exclusive powers, while they share other powers in common. This system of "federalism" – the sharing of power between the federal and state governments – is unlike the centralized system in many foreign countries where national governments maintain far more comprehensive power over matters of state, including energy policy.

State governments typically mirror the legislative, executive and judicial systems discussed above at the federal level. Thus, federalism creates a balkanized system of governance in which certain authorities overlap between different levels of government, characterized by divergent interests and a diverse constituent base. This can create benefits in some areas of policy, such as the highly coordinated tax regime that provides revenue for federal, state and local government coffers. However, it can also lead to costly and inefficient systems such as the electric transmission grid, which was developed on a local and regional basis, making it difficult at the national level to, for example, structure consistent standards for the reliable delivery of electricity or policies to incentivize the development of renewable energy.

As Table 18.1 illustrates, decision-making authority for energy policy is shared between different levels of government and even quasi-governmental entities, such as the North American Electric Reliability Corporation (NERC). NERC has the authority to regulate on certain matters of national energy policy, even though it is technically a private corporation. Congress has provided NERC with statutory responsibility to regulate the electric transmission grid's users, owners, and operators through the adoption and enforcement of standards in order to ensure reliability of the delivery of electricity.

Also, as noted in Table 18.1, states maintain a significant level of authority over energy policy within their jurisdiction. However, the divergent ways in which states choose to govern complicates matters even further for participants in the U.S. energy

[15] U.S. Environmental Protection Agency, Advanced Notice of Proposed Rulemaking, "Regulating Greenhouse Gas Emissions under the Clean Air Act", 73 Federal Register 44354, 31 July 2008.

[16] EPA, "Endangerment and Cause or Contribute Findings", supra, note 3.

Table 18.1 Balkanization of U.S. climate and energy policy decision-making

	Federal	State	Local	Regional
Electricity				
Rates	X			
Transmission siting	X	X		X
Tax incentives	X			
Environmental safety	X	X		
Reliability	X	X		X
Nuclear	X			
Hydro	X	X		
Renewable/energy efficiency incentives	X	X	X	
Gas pipelines				
Rates	X			
Pipeline siting	X	X	X	
Tax incentives	X			
Pipeline safety	X	X		
Oil and gas exploration and production				
Off-shore leases	X			
On-shore leases	X	X		
Tax incentives	X			
Coal				
Mine siting	X	X		
Mine health/safety	X	X		
Transportation				
Road construction	X	X	X	
Fuel efficiency	X	(CA only)		
Safety	X	X		
Water				
Rates		X	X	
Pipelines		X	X	
Legal rights	X	X		
Misc.				
Energy R&D funding	X	X	X	

markets. This lack of uniformity is demonstrated in the ways that many states have sought to address climate change and accelerate the deployment of renewable energy sources.

18.3.1 California

California has traditionally been a frontrunner in the arena of energy policy, with pioneering measures on the promotion of renewable energy sources, energy efficiency, and mitigation of transport emissions. On 27 September 2006, the state adopted legislation with the intention of cutting state GHG emissions to

1990 levels by 2020. Known as the Global Warming Solutions Act of 2006,[17] this legislation directs the California Air Resources Board (CARB) to establish a system for GHG emissions reporting and to monitor and enforce compliance. Although it does not mandate specific measures to reduce GHG emissions, the legislation authorizes CARB to adopt market-based compliance mechanisms such as emissions trading.

In 2010, CARB approved the design of a cap-and-trade system,[18] paving the way for its operation starting 2013. Major sources have been required to report their GHG emissions since 1 January 2012, and a first trial auction of allowances (California Carbon Allowances, or CCAs) took place in August 2012. Electricity production as well as imports and large industrial facilities emitting more than 25,000 metric tonnes of carbon dioxide per year will be covered by this system. Although a flurry of lawsuits at both the state and federal levels has delayed implementation of the program, the climate regime in California provides a useful model for the rest of the country.

18.3.2 State Renewable Portfolio Standards

At the time of writing, well over half of all U.S. state governments had adopted some form or renewable portfolio standard (RPS), which requires regulated utilities and electricity retailers to acquire a minimum percentage of the energy they sell in a given year from renewable energy resources. By contrast, the U.S. Federal Government has not been able to establish a similar policy, despite repeated efforts in Congress to do so over the past few years. Under an RPS, the amount of energy obtained from renewable resources is proven either by records submitted to the applicable regulatory state or regional agency (depending on the RPS regime within each state) or by renewable energy certificates (RECs) issued by a certifying organization. The RECs represent the production of electric energy by generators using eligible renewable resources.

In many cases, utilities demonstrate compliance with applicable RPS standards through the submission of RECs obtained directly by the utility from the certifying organization for utility-owned renewable generation, RECs obtained from the seller of energy to the utility, or through purchase of the REC from available markets. In most states, but not all, RECs generally reflect a MWh of energy produced. They accrue monetary value through purchase and sale either in bilateral arrangements or through sales in markets. While the term REC is typically referred to as a general concept, participants in the REC markets must navigate quite heterogenous

[17] Assembly Bill 32, California Global Warming Solutions Act of 2006, 17 October 2006, adopted as Division 25.5 of the Health and Safety Code, available at http://www.leginfo.ca.gov/cgi-bin/postquery?bill_number=ab_32&sess=cur&house=b&author=nunez (last accessed on 10 June 2012).

[18] See California Air Resources Board, "Proposed Regulation to Implement the California Cap-and-Trade Program", 28 October 2010, available at http://www.arb.ca.gov/regact/2010/capandtrade10/capv1appa.pdf (last accessed on 10 April 2012).

treatment among each of the U.S. states. Inconsistencies in REC programs include issues as basic as the definition of the certificates and even the term "renewable." Also, states diverge *inter alia* in their treatment of rules related to REC ownership, whether the certificate may be bundled with the commodity electricity, and limits on the price at which RECs may be traded.[19]

This non-standardization of RPS obligations, rules and regulations is particularly problematic in the utility industry – the very participants targeted by the myriad state RPS programs. As the utility industry continues to consolidate – for instance, over a 4 month period in late 2010 and early 2011, four separate mergers were announced between major utilities – the customer bases increasingly span different states and regions. Also, ratepayers may be far removed from the prime locations for siting renewable energy facilities, which means that the electrons generated from renewable resources must often be carried across multiple states and regions in order to reach the consumer. This system, resulting from the model of federalism established under the U.S. Constitution in the eighteenth century, clearly impedes the successful commercialization and proliferation of twenty-first century renewable energy technologies.

18.3.3 Regional Cooperation on Climate Change

While the division of powers between the federal and state levels has manifestly hampered progress on certain issues such as sustainable energy, it has also afforded states and regions scope for action where the Federal Government has been unable to act. In particular, sub-federal cooperation between states has resulted in two formal initiatives covering multiple jurisdictions and partly even involving participants and observers from foreign countries: the Regional Greenhouse Gas Initiative (RGGI) in the Northeast and Mid-Atlantic, and the Western Climate Initiative (WCI) on the West Coast.[20] Although both initiatives differ in coverage and scope, they share the objective of reducing GHG emissions and harnessing market instruments to that end.

Not only do such efforts provide an important fallback given the absence of comprehensive federal legislation, they also may improve the case for progress in Washington, D.C. by pioneering new instruments and technologies, and creating pressure at the federal level to pre-empt a regulatory patchwork and ensure uniform conditions throughout the US economy.

---

[19] For more on the non-standardization of RPS regimes among U.S. state governments, see David John Frenkil and David P. Yaffe, "Renewable Energy Certificates: A Patchwork Approach to Deploying Clean Technologies", 5 *Oxford Journal of World Energy Law & Business* (2012), 1–12.

[20] More recently, North America 2050: A Partnership for Progress (NA2050) was established as an informal network to facilitate state and provincial efforts related to the design, promotion and implementation of policies that reduce GHG emissions and create economic opportunities, see http://na2050.org (last accessed on 24 June 2012). Its creation is in part a reflection of the challenges faced in many parts of the U.S. to establish mandatory efforts in the area of climate policy.

18.3.3.1 Regional Greenhouse Gas Initiative (RGGI)

Operational since 1 January 2009, the Regional Greenhouse Gas Initiative (RGGI) is a cooperative effort by nine US states in the Northeast and Mid-Atlantic to limit GHG emissions from the electricity sector. Connecticut, Delaware, Maine, Maryland, Massachusetts, New Hampshire, New Jersey, New York, Rhode Island, and Vermont are all signatories to a Memorandum of Understanding (MoU)[21] released on 20 December 2005 that sets out common objectives and design elements of a regional "CO_2 Budget Trading Program." Under the MoU, these ten states – at the time – committed to stabilize CO_2 from their electric utilities from 2009 to 2014, followed by a 10% reduction between 2015 and 2019;[22] overall emissions are broken down into state allocations specified in the MoU.[23]

Each participating state was required to adopt the necessary rules for implementation of the Budget Trading Program. In order to ensure consistency across states, a Staff Working Group (SWG) consisting of state officials issued a Draft Model Rule on 15 August 2006, providing a template for state legislation.[24] As a result, the individual trading programs in each of the participating states are linked through CO_2 allowance reciprocity. Regulated entities are able to use a CO_2 allowance issued by any of the participating states to demonstrate compliance with the state program governing their facility. Overall, thus, the individual state programs function as a single regional compliance market for carbon emissions, and form the first mandatory, market-based GHG emissions reduction program in the U.S.

While the Model Rule creates a uniform framework for the Budget Trading Program, it also leaves states with flexibility regarding various design features, including allowance allocation. Under the MoU, participating states agreed to allocate a minimum of 25% of allowances to support consumer benefit programs, with the remaining 75% left for states to decide on. Auctions are conducted in regular intervals on an electronic platform, pursuant to a uniform auctioning format. In the first auction, held on 25 September 2008, 12.5 million allowances were sold to 59 bidders at a clearing price of US$3.07 per allowance. Because of the modest initial target, reduced electricity demand due to the recession and a significant shift from coal to natural gas for electricity generation, the market has

[21] Regional Greenhouse Gas Initiative, "Memorandum of Understanding", 20 December 2005, available at http://www.rggi.org/docs/mou_12_20_05.pdf (last accessed on 20 June 2012).

[22] For the period between 2009 and 2014, total CO_2 emissions for power producers in all ten states are limited to 188 million short tons (a short ton is a unit of weight equivalent to 2000 lb, or 907.4 kg); thereafter, the cap will decrease by 2.5% each year until 2018, reaching 169 million short tons by 2018/2019, or 90% of the initial cap. Covered are all fossil-fuel-fired electric generating units serving a generator of 25 MW or larger.

[23] The largest allocation – over 64 million short tons per year – goes to the State of New York, which has the largest population and economy in the region; conversely, the smallest allocation – just over 1.2 million short tons – goes to Vermont, a small state with one nuclear plant powering most of its area.

[24] Regional Greenhouse Gas Initiative, "Model Rule", 15 August 2006, available at http://www.rggi.org/docs/model_rule_8_15_06.pdf (last accessed on 20 June 2012).

been oversupplied with allowances and the price has since fallen to the minimum clearing price allowed at auction.

Despite these low price levels, however, a recent analysis of the economic impacts of RGGI commissioned by a group of nonprofit foundations and conducted by an independent research organization concluded that RGGI has had a positive macro-economic impact while helping reduce emissions in participating states, mainly through investments in energy efficiency measures and renewable energy deployment.[25] Going forward, it is likely that a comprehensive review of the RGGI program mandated by the MoU will result in proposals on how to address the excess supply of allowances and resulting price weakness.[26] Already, several participating states have agreed to retire a large part of unsold allowances, which indicates a clear interest in strengthening the market.

But political support is not equally strong in all participating states: effective 1 January 2012, New Jersey withdrew from RGGI, citing concerns about its effectiveness and economic impacts. Environmental and economic effects of the program will therefore remain under close scrutiny. In June 2012, RGGI released the first 3-year Compliance Summary Report, reviewing performance during the control period from 1 January 2009 to 31 December 2011. It found that 206 of the 211 power plants subject to RGGI compliance requirements met their obligations, and average annual CO_2 emissions for the same period were 23% lower than during the preceding 3-year period, 2006–2008, and 33% below the annual pollution cap of 188 million short tons.[27]

18.3.3.2 Western Climate Initiative (WCI)

On 26 February 2007, the Western Climate Initiative (WCI) was launched to develop regional strategies to address climate change.[28] It initially brought together Arizona, California, Montana, New Mexico, Oregon, Utah, and Washington, and the Canadian

[25] Specifically, the analysis suggests that the first 3-year control period added 1.6 billion USD in net present value (NPV) to the region, with capital flows into economic goods and services as well as ratepayer savings from energy efficiency improvements clearly outweighing net revenue losses in the energy sector, Paul J. Hibbard et al., *The Economic Impacts of the Regional Greenhouse Gas Initiative on Ten Northeast and Mid-Atlantic States: Review of the Use of RGGI Auction Proceeds from the First Three-Year Compliance Period* (Boston et al.: Analysis Group, 2011).

[26] As specified in the MoU, the program review should start in 2012 and include, among other things, an evaluation of program success, program impacts, additional reductions, leakage effects, and offsets. Recommendations ensuing from the review are expected for later in 2012.

[27] Regional Greenhouse Gas Initiative, "97% of RGGI Units Meet First Compliance Period Obligations", Press Release, 4 June 2012, available at http://www.rggi.org/docs/PR060412_Compliance.pdf (last accessed on 24 June 2012).

[28] For the original agreement signed by the Governors of Arizona, California, New Mexico, Oregon, and Washington, see Western Climate Initiative, "Western Regional Climate Action Initiative", 27 February 2007, available at http://www.westernclimateinitiative.org/ewebeditpro/items/O104F12775.pdf (last accessed on 20 June 2012).

provinces of British Columbia, Manitoba, Ontario, and Quebec; these states and provinces committed to a regional goal of lowering GHG emissions by 15% below 2005 levels by 2020.

In July 2010, the WCI released a detailed program design for the emission trading system serving as the central compliance instrument in the program.[29] Pursuant to this design, the system will only cover emissions from large downstream emitters, notably electricity, industrial processes, and industrial and commercial sources. From 1 January 2015, however, coverage is set to extend to upstream emissions from fuel combustion for transportation purposes and at residential, commercial, and industrial facilities, to the extent that these are not already covered.

Once implemented, coverage could extend to nearly 90% of the emissions from participating states. Initially, at least 10% of the allowances will be auctioned, rising to a minimum of 25% by 2020. With a view to ensuring a consistent and strong price signal, the first 5% of allowances auctioned by each partner will have a minimum price. If part of the allowances is not purchased at or above the minimum price, a fraction will be retired. Additionally, no more than 49% of emissions reductions may be achieved through offsets.

Due to the economic downturn and political changes at the state level, however, six states withdrew from the WCI in 2011, leaving only California and the Canadian provinces of British Columbia, Manitoba, Ontario, and Quebec as active WCI members. Only two partners of the WCI are currently working to adopt emissions trading systems, Quebec and California. Again, this illustrates the current political challenges even in progressive states to pursue climate policy efforts.

18.3.4 Local Cooperation on Climate Change

On 16 February 2005, the date when the Kyoto Protocol entered into force, the mayor of Seattle, Gregory J. Nickels, launched the US Mayors Climate Protection Agreement.[30] Its objective was to encourage at least 141 US cities to adopt the reduction objective agreed to for the US under the Kyoto Protocol prior to its withdrawal: a GHG emissions reduction of 7% below 1990 emissions levels by the 2008–2012 period. Specifically, participating cities committed to strive to meet or beat the Kyoto Protocol targets in their own communities, urge their state governments and the federal government to enact necessary policies, and urge the US Congress to pass bipartisan legislation to establish a federal emission trading system.

---

[29] See Western Climate Initiative, "Design for the WCI Regional Program", 27 July 2010, available at: http:// westernclimateinitiative.org/component/remository/func-download/282/chk,9785c3fccdd14a166e4daac6467df394/no_html,1 (last accessed on 10 April 2012).

[30] US Mayors Climate Protection Agreement, as endorsed by the 73rd Annual US Conference of Mayors meeting, Chicago, 16 February 2005, available at http://www.usmayors.org/climateprotection/documents/mcpAgreement.pdf (last accessed on 24 June 2012).

By 1 June 2012, 1,054 mayors, representing in excess of 88 million citizens, had signed the Climate Protection Agreement.[31] In 2007, the US Conference of Mayors launched the Mayors Climate Protection Center to administer and track the agreement. While it appears that few signatories to the Agreement will achieve the Kyoto Protocol reduction target by 2012, the agreement has prompted several cities to launch policy initiatives aimed at reducing municipal GHG emissions, including energy efficiency improvements to city buildings and transportation fleets, expansion of public transportation networks, renewable energy mandates, new building codes with efficiency requirements for residential and commercial structures, urban development plans that discourage vehicle use, and tax incentives and grants for community groups that take additional steps to reduce their GHG footprints.

18.4 Conclusion

Despite a new administration that elevated climate change and energy sustainability to one of the U.S. Federal Government's central areas of concern, preoccupation with the cost of climate policy, intensified by the economic downturn, ultimately prevented passage of comprehensive climate legislation in the U.S. Congress. Since 2008, it has become perhaps the most divisive issue between the Democratic and Republican parties, with ideology and political opportunism replacing the sober, fact-based approach climate change calls for. As a result, the U.S. response to climate change remains highly fragmented, drawing on federal, regional and local efforts to achieve the official U.S. target of a 17% GHG reduction below 2005 levels by 2020.

If the presidential campaign of 2012 is any indication, climate change will remain highly politicized in the foreseeable future. At worst, this will continue to hinder the emergence of a coherent policy to address GHG emissions as well as future energy needs of the country. The pace and scope of the domestic debate has direct implications for any international engagement by the U.S., with uncertainties at the national level directly translating into the international negotiations on a future climate regime. Without strong commitments on the part of the United States, other regions will find it more difficult to justify support for strong domestic and international action. Future developments in the U.S. may thus be the single most important condition for meaningful progress in the global struggle to address climate change and growing energy demand.

[31] U.S. Conference of Mayors Climate Protection Agreement, "List of Participating Mayors", available at http://www.usmayors.org/climateprotection/list.asp (last accessed on 20 June 2012).

Chapter 19
Canada and the Kyoto Protocol: An Aesop Fable

Jane Matthews Glenn and José Otero

Abstract This chapter attempts to explain why Canada, a country whose citizens like to think of themselves as progressive in matters of both international cooperation and environmental protection, has waivered in the performance of its commitments under the Kyoto Protocol and ultimately withdrawn entirely from it. We suggest that Canada's inconsistency reflects not so much a change of heart of Canadians as a change of political leadership at the federal level. For much of the period, Canada was governed by a left-of-centre government which enthusiastically supported the Protocol. However, it did so in a way that alienated most of the provinces, particularly in the west, and tied the hands of future federal governments. And in 2006 – just a year after Kyoto came into force – the pro-Kyoto government was replaced in Ottawa by a right-of-centre anti-Kyoto one, which set about dismantling previous climate change initiatives and replacing them with their own, less stringent ones. Paradoxically, Canada's emissions rose steadily in the first period and fluctuated downward in the latter. We suggest this is due more to the actions taken at the provincial level than to actions or inactions at the federal level.

Jane Matthews Glenn, Professor Emeritus, Faculty of Law and School of Urban Planning, and Associate Member, McGill School of Environment, McGill University; José Otero, Ph.D. Candidate, School of Urban Planning, McGill University, Montreal, Canada; Attorney (US: District of Columbia).

J. Matthews Glenn (✉)
Faculty of Law, McGill University, 3644 rue Peel,
H3A 1W9 Montreal, QC, Canada
e-mail: jane.glenn@mcgill.ca

J. Otero, Ph.D.
School of Urban Planning, McGill University,
Suite 400, 815 Sherbrooke Ouest, H3A 0C2 Montreal, QC, Canada
e-mail: jose.otero@mail.mcgill.ca

E.J. Hollo et al. (eds.), *Climate Change and the Law*, Ius Gentium: Comparative Perspectives on Law and Justice 21, DOI 10.1007/978-94-007-5440-9_19,

© Springer Science+Business Media Dordrecht 2013

19.1 Introduction

An Aesop fable tells of a man who first blew on his hands to warm them up and later blew on his porridge to cool it down, earning him the disapproval and distrust of his companion who exclaimed, "I will have nought to do with a man who can blow hot and cold with the same breath".[1] Canada is in the same position as the man in Aesop's fable, as it has blown first hot and later cold on the Kyoto Protocol, earning itself the disapproval and distrust of most in the international community and the shame of many Canadians.

19.1.1 Canada's Aesop Fable

In the years leading up to Kyoto and afterwards, Canada blew hot on the idea of international cooperation on climate change in general and the Kyoto Protocol in particular.[2] In 1986, it ratified the 1985 Vienna Convention on the Protection of the Ozone Layer, the first nation to do so, and the following year it hosted the meeting at which the Montreal Protocol on Substances that Deplete the Ozone Layer was signed. In 1988, Canada sponsored the first major international conference on climate change, the Toronto Conference on the Changing Atmosphere, at which leaders of industrialized countries agreed to a voluntary reduction in greenhouse gas (GHG) emissions of 20% by 2005 (the "Toronto target"). In 1992, Canada was a self-described "key player"[3] at the Rio Conference on Environment and Development in obtaining agreement on the United Nations Framework Convention on Climate Change (UNFCCC), which it signed and ratified the same year. In 1998, Canada signed the Kyoto Protocol, agreeing to reduce its GHG emissions to 6% below 1990 levels by 2012. And in 2005, it hosted the first meeting of the parties to Kyoto, in Montreal, after the Protocol entered into force earlier that year.

Later, however, Canada began to blow cold on the Kyoto Protocol. In 2006, it started to talk in terms of a "made-in-Canada" approach to climate policy. In 2009, Canada insisted on a substantially lower Kyoto target, 17% below 2005 levels by 2020, at the Copenhagen Summit. In 2011, it opposed an extension of the Kyoto Protocol beyond 2012 at the Durban Summit. And immediately thereafter, Canada withdrew entirely from Kyoto.

---

[1] Aesop, "The Man and the Satyr" in *Aesop's Fables* (translated with introduction and notes by Laura Gibbs, Oxford: Oxford University Press, 2002).

[2] Although its activities in this regard might be described, perhaps unkindly, as "all talk; no action": e.g. Nic Rivers & Mark Jaccard, "Talking without Walking: Canada's Ineffective Climate Effort" in Burkard Eberlein & G. Bruce Doern, *Governing the Energy Challenge: Canada and Germany in a Multi-level Regional and Global Context* (Toronto: U. of Toronto Press, 2009), at 285.

[3] Canada, *The Rio Earth Summit: Summary of the United Nations Conference on Environment and Development*, Doc. BP-317E (Ottawa: Govt of Canada, Depository Services Program, 1992), at A.1, available at http://dsp-psd.pwgsc.gc.ca/Collection-R/LoPBdP/BP/bp317-e.htm#Change.

Why this blowing hot and cold? After all, most Canadians – ordinary citizens as well as most political leaders – like to think of themselves as progressive in matters of both international cooperation and environmental protection. In our view, Canada's inconsistent rhetoric on Kyoto reflects a change in political leadership rather than a change of heart on the part of Canadians. And the change in leadership reflects the vagaries of applying "first-past-the-post" voting in the context of a multi-party system – particularly where there is a reconfiguration of parties across the political spectrum, one which sees the votes on the left, favourable to Kyoto, fragmenting among several parties and those on the right, opposed to Kyoto, coalescing around a single party – rather than a change in the political leanings of the majority of Canadians.[4]

In short, it was first Canada's left-of-centre government in office in Ottawa for much of the Kyoto period (first in majority, later in minority) which blew hot on Kyoto, and it is now Canada's right-of-centre government (first in minority, now in majority) which is blowing cold. It is somewhat ironic, therefore, that Canada's greenhouse gas emission rose significantly during the long period of left-of-centre government, despite pro-Kyoto rhetoric from Ottawa, and have plateaued and even decreased modestly during the more recent right-of-centre period, with its strongly anti-Kyoto rhetoric.

19.1.2 Setting the Scene

The constitutional backdrop for Canada's Aesop fable is the fact that authority over international relations, including treaty-making, rests with the federal government, and more particularly with the executive branch (Cabinet) rather than the legislative branch of government. This is well-accepted and said to flow from British constitutional conventions applicable throughout the British Commonwealth and to be reflected in the Constitution Act, 1867.[5] Treaties thus made bind Canada internationally but have no direct legal effect domestically (although the courts generally try to harmonize international engagements and domestic law as much as possible).[6] If some change in domestic law is required to implement a treaty, the legislation must

---

[4] For example, the present Conservative government received only 39% of votes in the last election although it holds the majority of seats in the House of Commons. For a more detailed discussion, see Jane Matthews Glenn & José Otero, "Addressing Climate Change in Canada: '(Un)cooperative Federalism'?", 32 *Finnish Journal of Environmental Law* (2012), 82. See generally Dennis Pilon, *The Politics of Voting: Reforming Canada's Electoral System* (Toronto: Emond-Montgomery, 2007).

[5] *Constitution Act, 1867* [formerly *British North America Act*] (U.K.), 30 & 31 Victoria, c. 3 (reprinted in Revised Statutes of Canada [R.S.C.] 1985, App. II, No. 5), s. 132; see e.g. Gerald L. Morris, "The Treaty-Making Power: A Canadian Dilemma", 45 *Canadian Bar Review* (1967), 478, at 482 f. (Canadian statutes and court decisions, both federal and provincial, are readily available on the Canadian Legal Information Institute's open website, http//www.canlii.org.)

[6] See e.g. René Provost, "Judging in Splendid Isolation", 56 *American Journal of Comparative Law* (2008), 125.

be enacted by the legislature having jurisdiction over the matter under the general division of powers clauses in the constitution.[7] In other words, the fact that treaty-making is a matter of federal jurisdiction does not entitle the federal legislature (Parliament) to encroach on provincial law-making powers to implement the treaty.[8]

This limitation is particularly important in regard to climate change because the legislative authority to implement international environmental agreements such as the Kyoto Protocol is divided between the federal and provincial governments, with the provincial governments having the lion's share of the authority. As the Supreme Court of Canada observed in 1997, "the environment is not, as such, a subject matter of legislation under the Constitution Act, 1867.... [the Act] has not assigned the matter of 'environment' sui generis to either the provinces or Parliament. Rather, it is a diffuse subject that cuts across many different areas of constitutional responsibility, some federal, some provincial."[9] The same can be said of "climate change".

Federal regulatory powers over GHG emissions are the subject of some debate, but are generally accepted to be grounded in its enumerated power to make laws in relation to "Criminal Law", on the one hand, and in its residual power to make laws for "the Peace, Order and good Government" (POGG) of Canada, on the other. The former is the more important of the two, as the Supreme Court interprets the scope of the criminal power liberally and the POGG power relatively narrowly. Other possible sources of federal power are its enumerated powers over the "Regulation of Trade and Commerce" (supporting a national emissions trading program) and "the raising of Money by any Mode or System of Taxation" (justifying a national carbon tax).[10]

In contrast, provincial regulatory powers over GHG emissions are well-recognized as being grounded in several of its enumerated powers, including "the Management and Sale of Public Lands belonging to the Province" (an important power as much of

---

[7] See especially *Constitution Act,* ss. 91 (federal legislative powers) and 92 (provincial powers).

[8] *Attorney-General for Canada v. Attorney-General for Ontario (Labour Conventions)* [1937] Appeal Cases [A.C.] 326 (Judicial Committee of the Privy Council) at 347–348 (setting out general rule); but see Stewart Elgie, "Carbon Emissions Trading and the Constitution" in Thomas J. Courchene & John R. Allan, eds., *Carbon Pricing and Environmental Federalism*, Canada: The State of the Federation Series (Montreal: McGill-Queen's Press, 2010), 161, at 167f (arguing that the Supreme Court of Canada should recognize an unlimited federal treaty-*implementing* power).

[9] *R. v. Hydro-Québec* [1997] 3 S.C.R. 213 (S.C.C.) [*Hydro-Québec*] at para. 112, citing *Friends of the Oldman River Society v. Canada (Minister of Transport)* [1992] 1 S.C.R. 3 (S.C.C.) at 63–64.

[10] See *Constitution Act,* opening words of s. 91 (POGG), ss. 91(2) (trade and commerce power), 91(3) (taxation power) and 91(27) (criminal law power). See also *Hydro-Québec* (criminal law power); *R. v. Crown-Zellerbach Canada Ltd.* [1988] 1 S.C.R. 401 (S.C.C.) (POGG); and *Reference re Securities Act* [2012] 3 S.C.R. 837 (S.C.C.) (trade and commerce). See generally e.g. Nigel D. Bankes & Alastair R. Lucas, "*Kyoto*, Constitutional Law and Alberta's Proposals", 42 *Alberta Law Review* (2004), 355; Peter Hogg, "Constitutional Authority over Greenhouse Gas Emissions", 46 *Alberta Law Review* (2008–2009), 507; Shi-Ling Hsu & Robin Elliot, "Regulating Greenhouse Gases in Canada: Constitutional and Policy Dimensions", 54 *McGill Law Journal* (2009), 463; and Nathalie J. Chalifour, "The Constitutional Authority to Levy Carbon Taxes" in Courchene & Allan, *Carbon Pricing and Environmental Federalism*, supra, note 8, at 177. See also the discussion in Matthews Glenn & Otero, "Addressing Climate Change in Canada", supra, note 4.

rural Canada is publicly owned, with an overwhelming amount of it being provincial rather than federal land) and "Property and Civil Rights in the Province" (giving the provinces authority over privately owned land and resources, as well as over private law generally and the regulation of commercial activities in particular), as well as in a 1982 addition to the constitution affirming provincial jurisdiction over non-renewable natural resources in the province, including the "exploration for" and the "development, conservation and management of" such resources.[11]

The economic foreground for the fable is the importance of the energy-intensive, trade-related sector to the economy of Canada in general and some provinces in particular. Revenues from the bituminous or tar sands[12] and other non-renewable resources, for example, represent about a fifth of Alberta's budget, and that amount is bound to grow.[13] But the energy-rich provinces, notably Alberta and Saskatchewan, are also those with the highest rates of GHG emissions by far, both in absolute terms and in terms of emissions per unit of economic output and per capita.[14] Climate policy in Canada thus takes a back-seat to economic development and the promotion and protection of trade, and political parties ignore this at their peril. The Liberal Party did so when it adopted a "National Energy Program" (NEP) in 1980 in response to the world-wide oil crisis of the 1970s. The NEP was intended to ensure stable supplies and low prices throughout the country, more equal sharing of oil revenues across the regions, and a larger federal share of oil profits at the expense of industry and provincial governments. It included strong pricing controls and new federal fuel taxes to achieve this. The program was fiercely opposed by the oil-rich western provinces, which viewed it as an unfair attempt to favour eastern interests at their expense. A 1981 remark of the then mayor of Alberta's major oil city and

[11] *Constitution Act*, ss. 92(5) (public lands), 92(13) (property and civil rights), 92A (added by *Constitution Act, 1982* (being Schedule B to the *Canada Act* (U.K.), 1982, c 11), s. 50). Other provincial enumerated powers include those over "Direct Taxation within the Province in order to the raising of a Revenue for a Provincial Purpose" (and indirect as well as direct taxation in respect of non-renewable natural resources) (s. 92(2) & 92A), most "Local Works and Undertakings" (s. 92(10)) (reinforcing provincial authority over large industrial emitters such as oil and gas installations), and "Generally all Matters of merely a local or private Nature in the Province" (s. 92(16)).

[12] The appropriate terminology is a matter of some debate. The most accurate is probably "bituminous sands", and this is what is used in Quebec's French language press ("*les sables bitumineuses*"): see e.g. Robin Rowland, "Accuracy is the Best Neutrality; It's All about the Bitumen", *Northwest Coast Energy News*, 28 Sept. 2011. Two other terms, easier to say, are "oil sands" and "tar sands". The former is preferred by the oil industry and the federal and Alberta governments (said to be because it sounds cleaner) and is now used almost exclusively in Canada's English language press; the latter is preferred by environmentalists and other opponents. "Tar sands" also seems to be used more often outside Canada (e.g. the U.K. and the U.S.) and we have opted to use it for this reason.

[13] The tar sands, for example, place Canada as second only to Saudi Arabia in terms of proven global crude oil reserves. See generally Alastair R. Lucas, "Mythology, Fantasy and Federalism: Canadian Climate Change Policy and Law", 20 *Pacific McGeorge Global Business & Development Law Journal* (2007), 41.

[14] See generally Environment Canada, *National Inventory Report 1990–2009: Greenhouse Gas Sources and Sinks in Canada* (May 2011), available at http://unfccc.int/national_reports/annex_i_ghg_inventories/national_inventories_submissions/items/5888.php.

future Premier of the province, "Let the eastern bastards freeze in the dark", captured the extent of this opposition and set the tone of future energy discourse. The NEP was abolished by a Conservative government in 1985 and replaced with a more market-oriented, province-friendly "Western Accord" with British Columbia, Alberta and Saskatchewan. But the political damage has proved lasting, as the present-day Conservative Party grew out of a western alienation party spawned by the conflict and the Liberal Party remains hugely unpopular in the West.[15]

19.2 Blowing Hot: The pro-Kyoto Political Left

Canada had a pro-Kyoto Liberal government in power in Ottawa when the Kyoto Protocol was negotiated and adopted in 1997, when Canada signed it in 1998, when it ratified it in 2002, and when the Protocol came into effect in 2005. This was a majority government until 2004, with the Liberals holding an absolute majority of the seats in the House of Commons and thus being in full control of the legislative and executive branches of government. In 2004, they were reduced to minority status, with only a plurality of the seats in the House, and governed with the general support of two other parties on the political left, the socialist New Democratic Party (NDP) and the separatist Bloc Québécois. All three parties support the Kyoto Protocol.

In 2006, however, shortly after Kyoto came into effect, the Conservatives replaced the Liberals as the governing party, albeit also with only a plurality of seats and thus as a minority government. This meant that the political left no longer had direct control over executive decisions and had only limited ability to influence the legislative agenda. Their only real leverage was the threat of defeating the government in a vote of no-confidence and thus triggering a new election, but this was not a practical threat where the electorate was weary of election, as was (and still is) the case in Canada. And in fact, a further election held in 2008 produced substantially the same minority results. Nevertheless, the political left continued to press for respect of the Kyoto Protocol during this period.

19.2.1 Kyoto Adoption

As we have seen, when Canada signed the Kyoto protocol in April 1998, it agreed to reduce its greenhouse gas (GHG) emissions to 6% below 1990 levels by 2012. It is not clear why the Canadian government agreed to such a stringent target.

[15] See e.g. Patrick James, "The Canadian National Energy Program and Its Aftermath: A Game-Theoretic Analysis", 16 *Canadian Public Policy / Analyse de Politiques* (1990), 174. See also Parliament of Canada, "Regional Representation: 1867 to Date" available at http://www.parl.gc.ca/parlinfo/compilations/houseofcommons/regionalrepresentation.aspx?menu=hoc-representation.

Agreeing to this target has been described as "a triumph of moral voluntarism over rational analysis"[16] and there may be some truth in this, as many of the Annex B countries that Canada looks to as models agreed to higher emissions reduction targets (e.g. most European countries, 8%; the United States, 7%). This might have encouraged or embarrassed Canada into following suit, disregarding the fact that sister countries merely agreed to cap their increases (e.g. New Zealand, 0%; Australia, +8%).[17] Canada's ambitious target has also been said to take "no account of our economic growth, population growth, cold temperature, vast distances and fossil fuel production", and there is most certainly much truth in this. Canada's population had already grown by 9% over 1990 levels when Canada signed the Protocol in 1998, and is now over 20% higher;[18] its continental climate of cold winters and hot summers contributes to higher heating and cooling demands; its large size and dispersed centres of population generate greater transportation demands; and fossil fuel production, notably from unconventional sources (e.g. tar sands, coalbed methane), is an important motor of the economy that contributes disproportionately to the country's GHG emissions.[19] And finally, adopting a 6% target for GHG emissions reduction has been said to have "contravened our federal system, because Ottawa broke a fragile federal-provincial consensus" hammered out in extremis in a series of federal-provincial discussions described as "heated".[20] This consensus saw the provinces agreeing to a maximum reduction target of 3% (with only Quebec favouring more stringent cuts). Canada's agreement to a target twice as high at Kyoto was seen by many as a federal (read Liberal) betrayal on a par with the National Energy Program.[21]

[16] Thomas J. Courchene, "Climate Change, Competitiveness and Environmental Federalism: The Case for a Carbon Tax", Background Document for an Address to the *Canada 2020 Speakers' Series* (Ottawa: National Press Club, 2008), at 3.

[17] The United States subsequently failed to ratify the Protocol; Hungary, Japan and Poland joined Canada in agreeing to 6%; Russia, a country similar to Canada in its geography, climate, natural resources and population distribution, was among those capping increases (0%): see UNFCCC, "Countries included in Annex B to the Kyoto Protocol and their emissions targets", available at http://unfccc.int/kyoto_protocol/items/3145.php.

[18] Being 27.7 million in 1990, 30.3 in 1998 and 33.4 in 2011: Statistics Canada, "Estimated population of Canada, 1605 to present", available at http://www.statcan.gc.ca/pub/98-187-x/4151287-eng.htm; The Sustainability Report, "Canada's Population", available at http://www.sustreport.org/signals/canpop_ttl.html.

[19] See Matthews Glenn & Otero, "Addressing Climate Change in Canada: '(Un)cooperative Federalism'?", supra, note 4.

[20] Jeffey Simpson, Mark Jaccard & Nic Rivers, *Hot Air: Meeting Canada's Climate Change Challenge* (Toronto: McClelland Stewart, 2007) at 249–250, as quoted in Courchene, "Climate Change, Competitiveness and Environmental Federalism", supra, note 16, at 4–5; Pamela J. Robinson, *Canadian Municipal Response to Climate Change: A Framework for Analyzing Barriers*, Ph.D. dissertation (Toronto: University of Toronto, Dept of Geography, 2000), available at http://www.collectionscanada.gc.ca/obj/s4/f2/dsk1/tape4/PQDD_0018/NQ53743.pdf, at 3.

[21] See generally Douglas Macdonald & Heather A. Smith, "Promises Made, Promises Broken: Questioning Canada's Commitment to Climate Change", 55 *International Journal* (1999–2000), 107.

The stage was thus set for political opposition to the Protocol which delayed its ratification for more than four years, until December 2002. As part of the negotiations leading to ratification, the Liberal majority government in power during this period promised not to impose new climate taxes and to allow emitters to purchase carbon credits at a backstop price of $15 per tonne as an alternative compliance mechanism. These promises now limit in practice the federal government's political margin of manoeuvre in the actions it takes in regard to climate change, whether these actions are taken in implementation of Kyoto or not.

19.2.2 Kyoto Implementation

The Liberals announced various plans to implement the Kyoto Protocol while they formed the government, first as a majority government and then (from June 2004 to January 2006) as a minority one. And the parties on the political left also promoted implementation of Kyoto when they were in opposition to a Conservative minority government (from 2006), but to limited avail.

19.2.2.1 As Government

The Liberal government proposed at least three different climate plans while in office. Two of these were adopted when it was in the majority but prior to the coming into force of Kyoto. The first, entitled "Action Plan 2000 on Climate Change", was adopted in 2000.[22] It was a 16-page skeletal outline, thin on specifics, of what the federal government would like to see included in an eventual federal-provincial "National Climate Change Business Plan". The second, entitled "Climate Change Plan for Canada", was adopted in 2002.[23] It was a more robust 68 pages and privileged voluntary compliance through such measures as negotiated agreements with major emitters, voluntary reductions by the automobile industry and a much-advertised "one tonne challenge" for individual households.

The third plan, entitled "Moving Forward on Climate Change: A Plan for Honouring Our Kyoto Commitment", was adopted in 2005, thus after the coming into force of Kyoto but when the Liberal government was in a minority position.[24] It built upon the earlier plans but with added detail. In particular, it proposed a cap-and-trade system similar in outline to all subsequent proposals, that is to say: an intensity-based

[22] Canada, *Action Plan 2000 on Climate Change* (Ottawa: Govt of Canada, 2000), available at http://dsp-psd.pwgsc.gc.ca/Collection/M22-135-2000E.pdf.

[23] Canada, *Climate Change Plan for Canada* (Ottawa: Govt of Canada, 2002), available at http://dsp-psd.pwgsc.gc.ca/Collection/En56-183-2002E.pdf.

[24] Canada, *Moving Forward on Climate Change: A Plan for Honouring Our Kyoto Commitment* (Ottawa: Govt of Canada, 2005) (better known as "Project Green").

emissions reduction target for major emitters (15% of their avoidable emissions – i.e. those that are controllable, not inherent in production) which could be met either by in-house reduction (with sale of any surplus reductions to other emitters or to a federal "Climate Fund") or by purchasing off-set credits from others. That the proposal is for intensity-based targets for major emitters is interesting to note because, as we shall see, the Liberals reacted vigorously to a similar proposal by the Conservatives when they took office.

Some initial implementation measures were also taken by the government. The aforementioned Climate Fund to facilitate the purchase of GHG credits nationally and internationally was set up and funded.[25] A Greenhouse Gas Technology Investment Fund Act was enacted but not declared in force and thus not funded.[26] And most importantly, the federal Canadian Environmental Protection Act was amended to add six GHGs to the "List of Toxic Substances" in Schedule 1 of the Act.[27] This is important because the Supreme Court of Canada has upheld the Act's provisions governing the release of "toxic substances" into the environment as constitutionally valid under the federal criminal law power.[28]

The early Liberal voluntary measures were largely ineffectual, and the later ones were only beginning to take effect when the Conservatives replaced the Liberals as government in February 2006 – one year, almost to the day, after the coming into force of Kyoto.

19.2.2.2 In Opposition

The minority position of the new government meant that the parties on the political left could, if they worked together, use their control of Parliament and the legislative process to try to ensure the implementation of Kyoto. They began to do so almost immediately. It is difficult to isolate the specific trigger for the opposition parties' actions, but it might have been the admission of the Conservative environment

---

[25] *Canada Emissions Reduction Act*, enacted in *Budget Implementation Act, 2005*, S.C. 2005, c. 30, s. 87 ($1 billion funding authorized).

[26] Enacted in *Budget Implementation Act, 2005*, s. 96.

[27] *Canadian Environmental Protection Act*, S.C. 1999, c. 33; "Order Adding Toxic Substances to Schedule 1 to the Canadian Environmental Protection Act, 1999", SOR/2005-345, *Canada Gazette: Part II*, Vol. 139, No. 24, 30 Nov. 2005, available at http://canadagazette.gc.ca/archives/p2/2005/2005-11-30/html/sor-dors345-eng.html.

[28] *Hydro-Québec*, supra, note 9. The case concerned the discharge of polychlorinated biphenyls (PCBs) into a watercourse; the Supreme Court was divided in its decision, with five judges holding the provisions valid under the criminal law power, but four judges arguing that they were too regulatory and not prohibitive enough to be characterized as criminal law. A change in the composition of the Court (whose judges are appointed by the Governor General on the advice of the Prime Minister) or a broad climate change regulation might sway the Court to reverse, limit or distinguish its decision in *Hydro-Québec*.

minister, made on 15 May 2006 while attending a Kyoto meeting in Bonn, that Canada would not meet its Kyoto target as this was "unachievable".[29]

The Bloc Québécois was the first to react. The day following the Bonn announcement, it introduced a motion calling on the government to publish "an effective and equitable plan for complying with the Kyoto protocol" by 15 October 2006, to establish a plan "that includes a system of emission objectives for large emitters along with an exchange of emission rights", to implement "an equitable plan that would call on large emitters to make a contribution proportional to their emissions", and to establish bilateral agreements with Quebec (and other provinces that might want one). The motion was adopted by the House of Commons on the same day, with its binding effect predicated on the idea that the Prime Minister "must now be consistent with the remarks he made when he was Leader of the Opposition and he criticized the attitude of the Liberal government of the day because it did not respect the votes of the House of Commons. Stephen Harper must now respect the position of parliamentarians on the Kyoto protocol."[30]

The Liberal Party was next to respond, submitting a Kyoto Protocol Implementation Act (KPIA) (Bill C-288) to the House of Commons for first reading on 17 May 2006 (i.e. the day after the Bloc motion). As the opposition parties together had enough votes in both the House of Commons and the Senate to adopt the bill, it made it safely through all legislative hurdles and received Royal Assent to bring it into effect on 22 June 2007.[31] Because the Act did not emanate from the government, it was introduced as a "private member's bill" (i.e., a bill sponsored by a member who is not a Cabinet minister). The content of such bills is rather limited as they may not involve the expenditure of public funds or impose taxes and may be treated as a motion of non-confidence by the government if it wishes, in which case their adoption would precipitate the fall of the government and a new election. This meant that the bill was necessarily long on rhetoric but short on specifics: it could not include any spending or taxation provisions, such as carbon taxes, nor could it prescribe any other specific mitigation measures, such as cap-and-trade. Instead, it merely required regular government reporting on measures taken and progress made in meeting its international obligations, with public opinion as the implicit enforcement mechanism.

The Kyoto Protocol Implementation Act thus simply requires the government to prepare annual "Climate Change Plans" setting out the measures to be taken "to ensure that Canada meets its obligations under … the Kyoto Protocol"; the

---

[29] What seems to have been particularly galling about this admission was that it was made at (but seemingly not in) an international Kyoto meeting of which the Minister was the chair (as President of the Conference of Parties, a position she owed to the success of the (Liberal) Montreal meeting the previous year), and it represented the first major embarrassment for Canada (at least on climate policy) on the international stage: see "[Environment Minister] Ambrose Feels Heat Heading UN Climate-change Meeting", CBC News, 15 May 2006, available at http://www.cbc.ca/news/canada/story/2006/05/15/kyoto.html.

[30] Bloc Québécois Parliamentary Wing, Press Release, "Parliament Calls on the Harper Government to Respect Canada's Commitment to Kyoto", [no date] available at http://www.blocquebecois.org/document.aspx?doc=1033e1d8-82a2-441d-ac96-2e2f4390a892 .

[31] *Kyoto Protocol Implementation Act*, S.C. 2007, c. 30.

Plans are to be tabled in both Houses of Parliament and submitted to the relevant Standing Committee as well as to the National Round Table on the Environment and the Economy (NRTEE)[32] for analysis and comment. As well, the government is required "to ensure" that Canada "fully meets" its Kyoto obligation by making, amending or repealing the necessary regulations.

The Conservative government's compliance with the obligations under this Act has been token at best. It has taken the necessary steps to comply with its formal requirements of the Act by issuing annual "Climate Change Plans for the Purposes of the Kyoto Protocol Implementation Act", tabling them in Parliament, submitting them for analysis and comment, and publishing the comments. But the Plans make it clear that the government has no intention of complying with Kyoto's substantive GHG reduction obligation, as it has now become too onerous, exacting too high a price in terms of jobs and economic growth.[33]

For this reason, Friends of the Earth, a Canadian not-for-profit environmental organization, brought suit against the federal government after the first Plan was tabled in 2007, seeking a declaration that the government had breached its statutory duties to prepare a plan consistent with Canada's obligations under Kyoto and to adopt the necessary regulations to implement it. However, the trial judge found that the issues were not justiciable – although not so much on the basis of principle as for reasons of statutory interpretation – and dismissed the action in 2008; the appeal court adopted the reasons of the trial judge and dismissed an appeal in late 2009; and the Supreme Court of Canada refused leave to appeal in 2010.[34]

Finally, the New Democratic Party was the third to react, tabling a private member's bill, the Climate Change Accountability Act (Bill C-377), in Parliament on 31 October 2006. It received all three readings in the House of Commons and first reading in Senate by mid-2008, but further progress was desultory, presumably in light of the adoption of the similarly intended Kyoto Protocol Implementation Act. However, it was reintroduced (as Bill C-311) in early 2009 after the trial decision in *Friends of the Earth* denying the KPIA's justiciability. Although the two Acts were generally similar, the NDP's Bill C-377/311 called for plans consistent with "a

[32] The Conservative government has now abolished the NRTEE (which employs about 30 people) as a cost-saving measure, saying "It was created before the Internet, when there were few such sources of domestic, independent research and analysis on sustainable development": Mike de Souza, "[Environment Minister] Kent Defends Killing Advisory Panel: Says Environmental Advice Free on Net", The Montreal Gazette, 31 March 2012.

[33] The annual Plans and analyses of the National Round Table on the Environment and the Economy are available online at http://www.climatechange.gc.ca/default.asp?lang=En&n=4044AEA7-1. The economic analysis draws on an earlier report, Environment Canada, *The Cost of Bill C-288 to Canadian Families and Business* (Environment Canada, 2006), available at http://www.ec.gc.ca/Publications/CA074D5B-18B6-4C41-AAEC-4A3C82FCCAA6%5CTheCostOfBillC288ToCanadianFamiliesAndBusiness.pdf with which other, independent economists have differed: e.g. Jaccard, M. & N. Rivers, *Estimating the Effect of the Canadian Government's 2006–2007 Greenhouse Gas Policies*, Working Paper (Toronto: C.D. Howe Institute, 2007).

[34] *Friends of the Earth v. Canada (Governor in Council)* (2008) 299 Dominion L. Reports [D.L.R.] (4th) 583 (Fed. Ct., Trial Div.); affirmed (2009) 313 D.L.R. (4th) 767 (Fed. C.A.); leave to appeal refused 2010 CanLII 14720 (S.C.C.) *[Friends of the Earth]*.

responsible contribution by Canada to the UNFCCC's ultimate objectives of preventing dangerous anthropogenic interference with the climate system and with Parliament's strong commitment to the Kyoto Protocol" rather than with Canada's obligations under the Protocol itself; it also called for GHG emission reductions of 25% below 1990 levels by 2020. Bill C-311 was adopted by the opposition-controlled House of Commons in May 2010, but was defeated in Senate in October of the same year (without debate on the substance, in breach of a long-standing Parliamentary convention to the contrary) in a snap vote called in the absence of enough key supporters for the vote to be lost.[35]

The hope of continued pressuring for the implementation of Kyoto ended when the Conservatives formed a majority government after an election in May 2011.

19.3 Blowing Cold: The anti-Kyoto Political Right

The 2006 election of a Conservative minority government with Stephen Harper as Prime Minister marked a turning point in the federal government's position on Kyoto. The Conservatives in general, and Harper in particular, are wary of climate change regulation for economic, political and ideological reasons. Economically, they believe that the sort of aggressive measures needed to meet Kyoto targets would exact too high a price in terms of jobs and economic growth, particularly since the United States did not ratify the Protocol. Politically, they need to retain the support of their electoral base in western Canada and particularly Alberta (Harper himself represents a riding in Calgary, Canada's oil industry hub), where oil and gas business interests dominate. And ideologically, they are opposed in principle to big government and to public sector regulation of the private sector. This wariness is summed up in a 2002 fundraising letter, in which Harper famously described the Kyoto regime as being "essentially a socialist scheme to suck money out of wealth-producing nations".[36]

Soon after coming to power, therefore, the Conservatives set about dismantling previous climate change initiatives and replacing them with their own.[37] They did so first by avoiding mention of Canada's obligations under Kyoto as much as possible, and then by rejecting them.

---

[35] See *Climate Change Accountability Act*, Bill C-377, 39th Parl., 1st & 2nd Sess. 2006–2008; Bill C-311, 40th Parl., 3rd Sess., 2010, available at http://www.parl.gc.ca/LEGISInfo/BillDetails.aspx?billId=3073285& Language=E&Mode=1 and http://www.parl.gc.ca/LegisInfo/BillDetails.aspx?Language= E&Mode=1&billId=4328110; Parliament of Canada, "The Senate Today: Making Canada's laws", available at http://www.parl.gc.ca/About/Senate/Today/laws-e.html.

[36] "Harper Letter Called Kyoto 'Socialist Scheme'", The [Toronto] Star, 30 Jan. 2007.

[37] See generally Meinhard Doelle & Dennis Mahony, "A Shift in the Legal Climate: The Emergence of Climate Change as a Dominant Legal Issue across Canada" in Stanley D. Berger & Dianne Saxe, eds., *Environmental Law: The Year in Review 2007* (Aurora, Ont.: Canada Law Book, 2008), at 7f, available at http://law.dal.ca/Files/MEL_Institute/Reports/Year_in_Review_Climate_Change_Update_Bali1_doelle.pdf.

19.3.1 Kyoto Avoidance

A first example of avoidance by the Conservative government was the proposed Canada's Clean Air Act (Bill C-30), which was introduced in the House of Commons on 19 October 2006 and which subsumed the question of GHG emissions into the larger one of air pollution. An important proposal in the Bill was to amend the Canadian Environmental Protection Act[38] to include a new section about air pollutants in general and GHGs in particular, including emissions-trading programs.[39] Opponents argued that no amendment to the environmental legislation was necessary as the Liberals had already listed GHGs as "toxic substances" under it.[40] They feared that removing GHGs from the list and recasting them as a separate concern grouped with air pollutants under a new section of a revised Act could reopen debate about the constitutionality of any resulting GHG regulation: was it a valid exercise of the federal criminal law power, or not?[41] Opponents also worried that the recasting would reopen scientific debate about the nature of GHGs and, in particular, whether they could be defined as "toxic" within the framework of the present Act.[42] Finally, and perhaps most significantly, opponents objected that the proposed legislation ignored the Kyoto Protocol and Canada's obligations under it. In the result, the Bill was substantially modified in the opposition-dominated Legislative Committee, and the government then withdrew its support. The Bill languished and ultimately "died on the order paper", in Canadian parliamentary parlance, when Parliament was dissolved and an election called in 2008. It was not reintroduced.

Avoidance was also evident in the "Notice of Intent", published two days after the introduction of the Canada's Clean Air Act, explaining the Act and setting out an agenda to accomplish the necessary reduction in air pollution.[43] This agenda anticipated all subsequent Conservative GHG emissions agendas. It favoured a "sector-by-sector" approach and focused on regulations dealing with emissions in the transportation, consumer and commercial products, and key industrial sectors. Regulation of the first two sectors was to be introduced rapidly, with the general aim

---

[38] *Canadian Environmental Protection Act,* supra, note 27.

[39] *Canada's Clean Air Act* [changed to *Canada's Clean Air and Climate Change Act* in Committee], Bill C-30, 39th Parliament, 1st Session, 2006, Part 1, available at http://www.parl.gc.ca/LegisInfo/BillDetails.aspx? Language=E&Mode=1&billId=2397040.

[40] See the discussion of the *Canadian Environmental Protection Act* supra, the text at note 27.

[41] See the discussion of the *Constitution Act*, supra, the text preceding note 10.

[42] S. 64 of the *Canadian Environmental Protection Act* requires that toxic substances have "an immediate or long-term harmful effect on the environment or its biological diversity" or be "a danger to the environment on which life depends" or "a danger in Canada to human life or health".

[43] "Notice of Intent to Develop and Implement Regulations and Other Measures to Reduce Air Emissions", *Canada Gazette, Part I*, Vol. 140, No. 42, 21 Oct. 2006. The Notice identifies the reduction in air emissions as "a matter of national concern", presumably to assert federal jurisdiction under its constitutional power to make laws for the "Peace, Order and Good Government" of Canada. See *R. Hydro-Québec* supra, the discussion following note 9.

of aligning Canadian standards with those of the United States. Regulation of the key industrial sector was to proceed more cautiously and to take place over a longer term. This sector was identified as including fossil-fuel fired electricity generation, upstream oil and gas, and downstream petroleum, and as accounting for 47% of Canada's GHG emissions (and 52% of its air pollution).[44] The eventual regulations were to set "realistic emissions targets" for this sector over the short (2010–2015), medium (2020–2025) and long (2050) terms. For air pollutants, the emissions targets were to be established using a fixed cap approach. For GHGs, on the other hand, targets were to be set using an emissions intensity approach, "one that will yield a better outcome for the Canadian environment than under the [Liberal] plan previously proposed on July 16, 2005, and show real progress on the environment here in Canada."[45] The intensity-based approach was to be the only approach in the short-term period; in the medium term, the government was to set intensity targets "that are ambitious enough to lead to absolute reductions in emissions and thus support the establishment of a fixed cap on emissions during this period"; in the long term, the government was said to be "committed to achieving an absolute reduction in GHG emissions between 45 and 65% from 2003 levels by 2050". Compliance options included "self-supporting market mechanisms that are not reliant on taxpayer dollars" such as an industry-led emissions trading system, opt-in mechanisms for unregulated industries to voluntarily assume emissions targets, incentives for industry investment in technology (e.g. carbon capture and storage), credit for early action, and a verifiable domestic offset system.[46]

Finally, avoidance is also evident in the content of a Conservative government climate change plan, entitled "Turning the Corner: An Action Plan to Reduce GHG Emissions and Air Pollution", which was released in April 2007.[47] The Plan follows the lead of the "Notice of Intent" in ignoring Canada's Kyoto GHG emissions target, and generally echoes the Notice's sector-by-sector regulatory approach.

[44] Notice of Intent, item no. 5. Other key industrial sectors mentioned are base metal smelters, iron and steel, cement, forest products, and chemicals.

[45] An intensity-based approach ties emissions reduction targets to production: the target is met if the emissions per value of product output fall, even if rising production makes overall emissions rise. See generally Nic Rivers & Mark Jaccard, "Intensity-Based Climate Change Policies in Canada" 36 *Canadian Public Policy/Analyse de politiques* (2010), 409. Intensity-based emissions came under question in *Pembina Institute for Appropriate Development v. Canada (Attorney General)* (2008) 80 Administrative Law Review (4th) 74 (Federal Court, Trial Div.) (environmental panel reviewing major tar sands project required to give reasons why intensity-based target "would be effective to reduce the [annual] greenhouse gas emissions, equivalent to 800,000 passenger vehicles, to a level of insignificance"); for panel response see http://www.imperialoil.ca/Canada-English/Files/ThisIs/EUB_Kearl_Addendum_May_008-06.pdf; for additional proceedings see *Imperial Oil Resources Ventures Limited v. Canada (Minister of Fisheries and Oceans)*, 2008 FC 382 and 2008 FC 598.

[46] Notice of Intent, item no. 9. The Notice stresses that the government "will not purchase credits or otherwise participate in the emissions trading market".

[47] Canada, *Turning the Corner: An Action Plan to Reduce GHG Emissions and Air Pollution* (Ottawa: Govt of Canada, 2007).

The Plan sets a long-term commitment to reduce GHG emissions 20% below 2006 levels by 2020 (and 60–70% by 2050),[48] and outlines a variety of programs to reach this target. Most programs are incentive-based (e.g. development of renewable energy sources, improved freight technology, more energy-efficient building construction) and many will yield only minimal reductions (e.g. phasing out incandescent light bulbs, offering tax credits for public transport passes, scrapping older vehicles). However, the heart of any climate change program in Canada has to be the energy sector and, like the Notice of Intent, "Turning the Corner" anticipates a special regime for industrial GHG emitters, including coal-fired electricity plants and tar sands projects, the basic outline of which was set out in a separate document.[49]

Under this special regime, existing industrial facilities are to reduce emissions intensity 18% below 2006 levels by 2010 and a further 2% annually until 2020, at which time they are apparently to move from an intensity-based to a fixed cap regime; new facilities (2004 or later) are to have 3 years of grace and then to be subject to a 2% intensity cap thereafter; still newer facilities (2012 or later) would be subject to additional carbon capture and storage requirements. Emitters could comply with these targets either by their own abatement actions or through several flexible alternatives recalling those mentioned in the Notice of Intent: a one-time credit for verified "early action" (from 1992 to 2006) to reduce emissions; payment of $15 per excess tonne to a climate change technology fund (for up to 70% of target in the first year and declining amounts thereafter until the program is phased out in 2018); purchase of domestic credits (either surplus credits generated by other regulated emitters or "off-set credits" generated by activities not covered by the regulation); purchase of some types of international credits generated by Kyoto's "Clean Development Mechanism" (limited to 10% of total target); and carbon capture and storage (if and when this technology becomes available).[50] This program is eventually to be implemented under the Canadian Environmental Protection Act, which the Conservatives now agree needs no amendment and which provides for the use of economic instruments and market-based approaches, including the development of "tradeable units" systems.[51]

There has been some limited implementation of this Plan, such as the adoption in 2010 of Passenger Automobile and Light Truck Greenhouse Gas Emission Regulations to establish standards and procedures "that are aligned with the federal

[48] The change in base years between the Notice of Intent (2003) and Turning the Corner (2006) makes comparison of their commitments difficult.

[49] Environment Canada, *Turning the Corner: A Regulatory Framework for Industrial Greenhouse Gas Emissions* (Environment Canada, 2008).

[50] See generally Grant Boyle, "A Review of Emerging GHG Emissions Trading in North America: Fragmentation or Progress?", 46 *Alberta Law Review* (2008), 173.

[51] *Canadian Environmental Protection Act*, supra, note 27, s. 322 (adoption of guidelines by Minister) & s. 326 (adoption of regulations by Cabinet). See also Hogg, "Constitutional Authority Over Greenhouse Gas Emissions supra", note 10 at 514–515 (concluding that provision of alternative means of compliance represents valid exercise of federal criminal law power).

requirements of the United States".[52] In the energy sector, however, oft-promised regulations of coal-fired electricity plants and the all-important tar sands facilities remain elusive: adoption of the promised regulations of emissions from new coal-fired electricity plants is now long overdue[53]; and regulations of tar sands facilities are once again on hold pending further industry consultation.[54]

The slow pace of implementation undoubtedly reflects the fact that implementation depends mainly on the United States, as it is a basic tenet of the present government that the Canadian economy is so integrated into the American that "it makes absolutely no sense" for Canada to adopt different climate policies from those of the United States.[55] And climate change does not appear high on the agenda of either country, as the recession and its aftermath have occupied centre stage in both.

19.3.2 Kyoto Rejection

The Conservative government has long been on record as rejecting Canada's obligations under the Kyoto Protocol. It rejects Canada's 6% emissions reduction target agreed to under Kyoto; it questions the use of international credits (particularly generated under Kyoto) to meet this target, as it regards them as not environmentally credible[56]; and it objects to the fact that not all major emitting countries are required to take action under Kyoto. It is therefore antipathetic to extension of the Kyoto Protocol beyond its 2012 term.

This has been made plain domestically in most, if not all, of the government's climate change documents: in 2006, in its proposed Canada's Clean Air Act (Bill C-30) and accompanying "Notice of Intent"; in 2007, in its climate change plan, "Turning the Corner"; and thereafter, in the annual plans filed under the Kyoto Protocol Implementation Act.

[52] *Passenger Automobile and Light Truck Greenhouse Gas Emission Regulations*, SOR/2010-201, s. 2 (includes possibility of limited trading of "emission credits" and recognition of "early action credits"). See also *Marine Spark-Ignition Engine, Vessel and Off-Road Recreational Vehicles Regulations*, SOR/2011-10; and *Renewable Fuels Regulations*, SOR/2010-189 (includes possibility of some trading of "distillation compliance units").

[53] "Reduction of Carbon Dioxide Emissions from Coal-Fired Generation of Electricity Regulations" *Canada Gazette Part I*, Vol. 145, No. 35, 27 August 2011, p. 2779 at 2842 (draft regulations published for comment).

[54] Simon Dyer, "Faulty Premise Underlies Budget 2012 'Streamlining' of Environmental Review Process", *iPolitics*, 4 April 2012, available at http://www.pembina.org/op-ed/2327.

[55] Canada, "Canada's Continental Action", 10 January 2011, available at http://www.climatechange.gc.ca/default.asp?lang=En&n=A4F03CA6-1.

[56] Particularly excess "assigned amount units" (AAUs) acquired from other countries in economic decline, which it labels "hot air" credits: Environment Canada, "*The Cost of Bill C-288 to Canadian Families and Business*", 2007, supra, note 33, 11, 22–23.

It has also been made plain internationally at every UNFCCC Conference of Parties (or Kyoto summits) held since the Conservatives took office in 2006. We have already mentioned the environment minister's comments at Bonn in 2006. At Bali in 2007, the government resisted the idea of preparing for post-Kyoto commitments generally, and particularly to the suggestion that industrialized countries and rapidly industrializing countries should not be subject to the same rules. At Poznan in 2008, the Conservative government continued to resist suggestions that any new agreement should require industrialized countries to reduce GHG emissions by 25–40% below 1990 levels by 2020. At Copenhagen in 2009, it again resisted such stringent GHG targets, supporting instead the more flexible approach reflected in the Copenhagen Accord which left each country free to choose its own "quantified economy-wide emissions targets for 2020".[57] The Conservative choice was a target of 17% below 2005 levels by 2020, "to be aligned with the final economy-wide emissions target of the United States in enacted legislation". The formal letter of submission stated that it was made "with the expectation that" other Annex 1 Parties and "major Non-Annex 1 Parties" would also submit emissions targets and mitigation actions in a timely fashion.[58] By way of contrast, most other countries used 1990 as the base year and most European countries agreed to reduction targets of 20% or better, whereas Canada's Copenhagen target is equivalent to 2.5% above 1990 levels.[59] Finally, at Cancun in 2010 and Durban in 2011, the Conservative government continued to resist both firmer and deeper targets and differential treatment of major emitters (i.e. between industrialized and rapidly developing countries) in any post-Kyoto agreement. In other words, at every international meeting Conservative action weakened rather than enhanced the emerging consensus, earning Canada's numerous "fossil awards" from environmental groups.

Conservative rejection of Kyoto was made manifest shortly after the Durban summit, when the government announced Canada's withdrawal from the Protocol. This withdrawal is to become effective as of 15 December 2012.[60] A motion for

---

[57] Copenhagen Accord, para. 4, in UNFCCC, "Report of the Conference of the Parties on its Fifteenth Session, Held in Copenhagen from 7 to 19 December 2009", Doc. FCCC/CP/2009/Add.1 30 March 2010, available at http://unfccc.int/resource/docs/2009/cop15/eng/11a01.pdf.

[58] Canada's submission letter, 29 January 2010, available at http://unfccc.int/files/meetings/cop_15/copenhagen_accord/application/pdf/canadacphaccord_app1.pdf. In the accompanying government announcement, "with the expectation that" became "is contingent on" all major emitters signing on: Environment Canada, "News Release: Canada Lists Emissions Target under the Copenhagen Accord", 1 February 2010, available at http://www.ec.gc.ca/default.asp?lang=En&n=714D9AAE-1&news=EAF552A3-D287-4AC0-ACB8-A6FEA697ACD6.

[59] UNFCCC, "Appendix I – Quantified Economy-Wide Emissions Targets for 2020", available at http://unfccc.int/meetings/copenhagen_dec_2009/items/5264.php; Council of Canadians, "Prentice Pledges Emissions Cuts 2.5% Above 1990 Levels" available at http://www.canadians.org/campaignblog/?p=2793. The Conservatives' 2007 target, 20% below 2006 levels by 2020, was equivalent to 3% *below* 1990 levels (mainly because of the difference between 2006 and 2007 emissions levels): Council of Canadians.

[60] UN Secretariat, Reference: C.N.796.2011.TREATIES-1 (Depositary Notification), "Kyoto Protocol to the United Nations Framework Convention on Climate Change – Canada: Withdrawal", available at http://treaties.un.org/doc/Publication/CN/2011/CN.796.2011-Eng.pdf.

judicial review of Canada's withdrawal was filed in Federal Court on 15 January 2012 and the case has now been set down for hearing on 1 June 2012. A main argument is that the withdrawal contravenes Canada's obligations under the Kyoto Protocol Implementation Act.[61] And Canada's Commissioner of the Environment and Sustainable Development is said to be seeking legal advice about the on-going effect of his reporting obligations under the Act.[62]

19.4 Conclusions

Canada admittedly faces a number of constraints in trying to reduce its GHG emissions, as we have seen. These include physical factors over which the country has little or no control such as geography, northern climate, growing population and so on; they also include economic factors over which the country has some control, such as its economic dependence on natural resources, including oil and gas. Other constraints are legal in nature, and include the constitutional division of powers which divides jurisdiction over different aspects of climate policy between the federal and provincial governments. Still others are political, including the concentration of power in the Prime Minister's office which is implicit in a Parliamentary (or "Westminster") system of government.[63] But the most important constraint is arguably a lack of political will, on the part of all parties of the political spectrum, to tackle in any meaningful way the question of GHG emanating from the tar sands. Fossil fuel production in general, and from the tar sands in particular, is crucially important to the economy of the western provinces: Canada's economic, and therefore political, centre of gravity is shifting westward.

These constraints undoubtedly account for Canada's disappointing record in meeting its Kyoto GHG emissions target. Its most recent "National Inventory Report", filed with the UNFCCC on 11 April 2012, notes that Canada's emissions were 17% above the 1990 total in 2010. But this the picture is slowly changing. Steady annual increases in GHG emissions for the first 15 years after 1990 (the base year for Kyoto targets) are now giving place to more fluctuating downward rates and Canada's emission rate is now on a par with its rate a decade ago. The decrease in

---

[61] Équipe Kyoto, Press Release, "Court challenge to Canada's denunciation of the Kyoto Protocol at the Federal Court of Canada: Équipe Kyoto Speeds Up Its Work since the Announcement of the Hearing Date of June 1, 2012", 24 March 2012, available at http://www.polymemes.com/polymemes/press%20release%2024-03-12.pdf.

[62] Jason Fekete, "Even out of Kyoto, Canada's Environment Commissioner may still have legal mandate to report on accord", National Post, 13 Dec. 2011, available at http://news.nationalpost.com/2011/12/13/even-if-canada-is-out-of-kyoto-government-may-still-have-legal-mandate-to-follow-accord-environment-commissioner/.

[63] For a more detailed discussion, see Matthews Glenn & Otero, "Addressing Climate Change in Canada: '(Un)cooperative Federalism'?", supra, note 4.

emissions intensity (per GDP and per capita) is even more striking.[64] Some of this is due to the effects of the recession, of course. But not all, as one must not overlook the role played by the provinces.

For the provinces have an important role to play. They have the constitutional authority to address climate change, and are largely responsible for such progress as Canada has made. Most take the threat of climate change seriously and have adopted GHG reduction targets that are as aggressive, if not more so, as Canada's commitments under Kyoto. And they have adopted a variety of approaches to try to meet these targets. A first approach is clean energy, and a number of provinces are phasing out coal-fired electricity plants. Ontario is a good example of this approach, and is in the process of replacing coal with increased nuclear capacity, natural gas thermal generation and, most interestingly, renewable energy. Its "feed-in tariff" (or FIT) program is North America's first comprehensive, long-term guaranteed pricing structure for renewable electricity production.[65] A second approach is carbon taxes, and Quebec's (introduced in 2007) and British Columbia's (2008) are Canada's best examples.[66] A third approach is cap-and-trade and, perhaps surprisingly, Alberta was the first Canadian jurisdiction to adopt framework legislation (2003)[67] and is the only one to have a fully operating system in place today, but other provinces are beginning to develop cap-and-trade systems as well, particularly in the context of several North American sub-national associations of states and provinces of which the "Western Climate Initiative" is the best-known example.[68]

This provincial activity reminds us that Canada is a confederation made up of provinces with extensive constitutional powers and budgets to match. They are concerned about climate change and recognize that they have a role to play. This makes it less important, overall, whether the federal government blows hot or blow cold about Kyoto.

---

[64] Environment Canada, *National Inventory Report 1990–2010: Greenhouse gas sources and sinks in Canada* (April 2012), Pt. 1, at 17–19, available at http://unfccc.int/national_reports/annex_i_ghg_inventories/national_inventories_submissions/items/6598.php.

[65] It is being rolled out under the *Green Energy and Green Economy Act*, Statutes of Ontario [S.O.] 2009, c. 12.

[66] Quebec: *Regulation respecting the annual duty payable to the Green Fund*, Revised Regulations of Quebec [R.R.Q.], c. R-6.01, adopted under Ch. VI.3 ("Financing of Measures to Reduce Greenhouse Gas Emissions and Fight Climate Change") of the *Act respecting the Régie de l'énergie*, Revised Statutes of Quebec [R.S.Q.], c. R-6.01, c. 46, s. 48; B.C.: *Carbon Tax Act*, Statutes of British Columbia [S.B.C.] 2008, c. 40; *Carbon Tax Regulation*, B.C Reg. 125/2008 as am.

[67] *Climate Change and Emissions Management Act*, Statutes of Alberta [S.A.] 2003, c. C-16.7.

[68] Western Climate Initiative, "The WCI Cap & Trade Program", available at http://www.westernclimateinitiative.org/the-wci-cap-and-trade-program.

Chapter 20
Climate Law and Policy in the European Union: Accidental Success or Deliberate Leadership?

Michael Mehling, Kati Kulovesi, and Javier de Cendra

Abstract Internationally, the European Union has become known as a forerunner in the adoption and implementation of ambitious climate policies, pioneering new regulatory instruments and voluntarily committing itself to ambitious targets both with regard to greenhouse gas mitigation and energy sustainability. Underlying Europe's perceived leadership in the struggle against climate change is a proliferation of internal laws and regulations, including the largest environmental permit trading scheme in history. It would be erroneous, however, to assume that the emergence of this comprehensive policy landscape has followed an easy or straightforward path; more aptly, the development of climate and energy policy in the European Union has often been preceded by arduous negotiations and significant controversy. Going forward, what are the prospects for continued climate leadership in the European Union, and what role will the legal and institutional framework play?

Michael Mehling is President of the Ecologic Institute, Washington DC, and Adjunct Professor at Georgetown University; Kati Kulovesi is Adjunct Professor at the Department of Law, University of Eastern Finland; Javier de Cendra is Senior Research Associate at the University College London (UCL) Energy Institute and the Faculty of Laws, and visiting professor at University of Leuven (KU Leuven).

M. Mehling (✉)
Ecologic Institute, 1630 Connecticut Ave. NW, Suite 300, 20009
Washington, DC, USA
e-mail: michael.mehling@ecologic-institute.us

K. Kulovesi
Department of Law, University of Eastern Finland, Joensuu, Finland
e-mail: kati.kulovesi@uef.fi

J. de Cendra
University College London, Central House, 14 Upper Woburn Place London, WC1H 0HY
email: j.decendra@ucl.ac.uk

E.J. Hollo et al. (eds.), *Climate Change and the Law*, Ius Gentium: Comparative Perspectives on Law and Justice 21, DOI 10.1007/978-94-007-5440-9_20,
© Springer Science+Business Media Dordrecht 2013

20.1 Introduction

In stark contrast to the greenhouse gas (GHG) emissions pathways of most other regions, whether developed or developing, the European Union (EU) has seen emissions fall by 15.4% in the two decades between 1990 and 2010.[1] Unsurprisingly, therefore, Europe is often described as a climate leader[2] and the region that is "doing more than any other part of the world to address global climate change and to share the burdens associated with it."[3] Yet to what extent is this success due to a deliberate policy strategy, and if it is, what factors have allowed the EU to act with greater apparent success than other jurisdictions? Tracing the evolution of Europe's domestic and international climate strategy, this chapter hopes to identify and explain some of the underlying dynamics that have resulted in the EU's perceived leadership.

It is worth noting that supranational cooperation on energy issues dates back to the very beginning of the European integration project, where it played an important role in efforts to promote peaceful cooperation following World War II. Both the establishment of the European Coal and Steel Community (ECSC) in 1951 and of the European Atomic Energy Community (EURATOM) in 1957 by a small group of West European states[4] reflected the priorities of the time: peace and economic integration, two objectives that were further promoted in 1957 with the creation of the European Economic Community (EEC).

As the European Commission itself has observed, "as far back as 1954, energy was regarded as one of the motors of European integration since it is at the heart of economic activity and social welfare and because it is a key factor in Community solidarity."[5] At the time, of course, climate change had not yet been identified as a relevant policy issue. It was not until the 1980s, when mounting scientific evidence led to a growing demand for political responses, that the transboundary nature of climate change and the desire to maintain uniform policy requirements across Europe provided a compelling argument for Community action. In 1986, the

[1] European Environment Agency (EEA), Annual European Union Greenhouse Gas Inventory 1990–2010 and Inventory Report 2012 (Copenhagen: EEA, 2012).

[2] Such leadership can manifest itself in many ways, and the European Union has been designated a structural leader by virtue of its membership, combined population and economic power, but also as a directional and instrumental leader for setting an example through domestic policies and building dynamic coalitions within the international community, see Joyeeta Gupta and Lasse Ringius, "The EU's Climate Leadership: Reconciling Ambition and Reality", 2 *International Environmental Agreements: Politics, Law and Economics* (2001), 281–299.

[3] Peter G. Harris, "Europe and the Politics and Foreign Policy of Global Climate Change", in Peter G. Harris (ed.), *Europe and Global Climate Change: Politics, Foreign Policy and Regional Cooperation* (Cheltenham: Edward Elgar, 2007), at 31.

[4] These states were Belgium, France, Germany, Italy, Luxembourg and the Netherlands.

[5] European Commission, Report from the Commission to the Council on Civil Protection, Tourism and Energy, SEC (1996) 496 final of 3 April 1996, 2.

European Parliament adopted its first resolution on climate change, while the Council passed a – legally non-binding – resolution to improve the energy efficiency of final energy demand by 20% until 1995.[6]

At this early stage, however, such measures were not yet considered part of a coherent climate strategy. Incited by an environmental movement that had successfully campaigned against conventional pollutants in the previous decade and was now beginning to embrace climate change for its agenda, the international community adopted the United Nations Framework Convention on Climate Change (UNFCCC) in 1992, prompting the European Commission to propose a strategy to limit carbon dioxide emissions and improve energy efficiency. More concrete measures followed at a rapid pace, including a greenhouse gas monitoring mechanism in 1993[7] and various measures on the promotion of energy efficiency and energy labeling. As will be discussed below, this marks one of many instances in which domestic European action was preceded by an international commitment.

Despite these efforts, European emissions would have grown considerably throughout the early 1990s without a series of historical events: the "wallfall profits" of industrial collapse in Germany after reunification, the British "dash for gas" following the defeat of coal miners in a series of strikes throughout the 1980s and subsequent introduction of affordable North Sea gas, and finally a broader economic slowdown throughout the region.[8] Progress at the international level once again inspired more earnest action in the European Union: under the Kyoto Protocol, a supplementary agreement to the UNFCCC negotiated in 1997, Europe bound itself to mandatory greenhouse gas mitigation commitments by accepting the highest reduction target among major industrialized countries. In a unique arrangement, the efforts of each Member State were internally distributed through a burden-sharing agreement, which accounted for domestic circumstances such as the expectation for economic growth, the prevailing energy mix, and the structure of the industrial sector.[9] A result of intense political negotiations, this arrangement helped Europe accommodate very different factual circumstances and levels of ambition while maintaining a common position at international negotiations. Its ramifications for the domestic and international climate efforts of the EU are outlined below.

[6] Marc Pallemaerts et al., *Climate Change and Sustainable Energy Policies in Europe and the United States* (Brussels and Washington, DC: IEEP/NRDC, 2006), at 22.

[7] Council, Decision 93/389/EC of 24 June 1993 for a Monitoring Mechanism of Community CO_2 and other Greenhouse Gas Emissions, OJ 1993 L167/31.

[8] Jørgen Henningsen, *EU Energy and Climate Policy – Two Years On*, EPC Issue Paper No. 55 (Brussels: EPC, 2008), at 9.

[9] Council, Decision 2002/358/EC of 25 April 2002 concerning the Approval, on Behalf of the European Community, of the Kyoto Protocol to the United Nations Framework Convention on Climate Change and the Joint Fulfilment of Commitments Thereunder, OJ 2002 L130/1. Commitments range from a reduction of 28 set for Luxembourg or 21 for Germany and Denmark, to an increase of no more than 27 for Portugal and 25 for Greece.

20.2 Towards a Domestic EU Climate Strategy

Responding to this newly entered commitment, the Commission launched a European Climate Change Programme (ECCP) in early 2000 with the overall objective of identifying and developing "all those elements of a European Climate Change strategy that are necessary for the implementation of the Kyoto Protocol."[10] In accordance with the legislative roadmap set out by this program, the Commission proceeded to draft a number of measures on energy labeling and ecodesign requirements, energy services, renewable energy sources, energy taxation, research funding, and emissions trading, all of which were subsequently adopted by the Council and the European Parliament. By this time, climate change had clearly become part of a comprehensive regulatory strategy at the European level. Yet even this proliferation of individual measures was unable to reverse emission trends in Europe: by 2005, the statistical agency Eurostat reported that "both greenhouse gas emissions and energy consumption" had increased since 2000, and that several Member States were "moving away from their agreed targets."[11] It would be erroneous, therefore, to assume that the emission reductions achieved since were purely due to historical events such as the expansion of the European Union: hard policy choices were instrumental to reverse the foregoing trend.

Accordingly, a more ambitious second phase of the European Climate Change Programme was elaborated in 2005 to address the remaining shortcomings of the preceding phase.[12] Among other things, the Commission acknowledged that a successful policy framework would have to bridge the existing divide between environmental policy and energy and transport policy. In other words, the policy responses to the climate challenge – which had largely evolved within the institutional purview of two different Council formations and separate departments at the European Commission – would have to become more streamlined and ultimately integrated.

In 2007, the Commission responded with a landmark document setting out its proposal for an integrated climate and energy policy.[13] Described as a "watershed" in European energy and climate policy,[14] this document represented the first cooperative effort by two Commissioners – the Commissioner for Energy Policy and the Commissioner for the Environment – to formulate a joint and comprehensive

---

[10] European Commission, EU Policies and Measures to reduce Greenhouse Gas Emissions – Towards a European Climate Change Programme (ECCP), COM (2000) 88 final, at 8.

[11] Eurostat, Measuring Progress towards a More Sustainable Europe: Sustainable Development Indicators for the European Union (Luxembourg: Office for Official Publications of the European Communities, 2005).

[12] European Commission, Winning the Battle Against Global Climate Change, COM (2005) 35 of 9 February 2005.

[13] European Commission, An Energy Policy for Europe, COM (2007) 1 of 10 January 2007.

[14] Sebastian Oberthür and Claire Roche Kelly, "EU Leadership in International Climate Policy: Achievements and Challenges", 43 *The International Spectator* (2008), 35–50, at 41.

response to the challenges of climate change and energy sustainability.[15] According to the Commission, its ambitious objectives would ensure that Europe becomes "a highly energy efficient and low CO_2 energy economy, catalyzing a new industrial revolution, accelerating the change to low carbon growth and, over a period of years, dramatically increasing the amount of local, low emission energy that we produce and use."[16] With strong endorsement from the Presidency at the time, all heads of state and government agreed to this policy strategy.

Rendering this strategy operational would prove more difficult, however: in early 2008, the European Commission announced a set of legislative proposals to revise existing policies and implement the new targets and objectives. Measures in this package included a sweeping reform of the emissions trading system, domestic emission targets for sectors not covered by carbon market, revised legislation on the promotion of renewable energy including biofuels in the transport sector, and a legal framework for Carbon Capture and Storage (CCS) activities. Overall, these proposals shifted more responsibility to the European level, and while initial reactions from the Member States suggested that the measures had found an adequate balance between ambition and flexibility, arduous negotiations and political concessions were needed to reach a compromise on the final package, which was adopted nearly a year later. Disagreement among Member States on the various details of the package heralded internal divisions that would effectively prevent the European Union from exerting leadership at the momentous climate summit in Copenhagen in December 2009, where some observers felt it had indeed become "marginalized."[17]

What seems nevertheless clear is that a policy creation phase has given way to a policy implementation phase, whereby both the EU and its Member States have to cope with the considerable challenge of ensuring that existing policies deliver real and significant emission reductions on the ground. For instance, there is abundant evidence that, for all the grandeur of its legal framework, the EU ETS is failing to make a substantial difference in emissions, and moreover is being bogged down by a myriad challenges that threaten to undermine it even further. For instance, the tensions that are building between, on the one hand, declarations and actions by some member states to unilaterally increase the stringency of the scheme in order to ensure its environmental effectiveness, and on the other hand the efforts of the Commission to avoid that those actions undo the level of harmonization already achieved ‹ tensions which might end up being adjudicated by the Court of Justice of

[15] Specifically, the document outlined a new set of mitigation targets to be achieved by 2020: a firm and independent commitment to reduce greenhouse gas emissions by 20% relative to 1990 levels, and a commitment to reduce emissions by up to 30% if an international climate protection regime sets comparable ambitious targets for other countries; a mandatory target of 20% for the share of renewable energy in the overall energy mix; and an objective to reduce primary energy consumption by 20% compared to projections for 2020 through improved efficiency.

[16] European Commission, An Energy Policy for Europe, supra, note 14, at 5.

[17] Joseph Curtin, *The Copenhagen Conference: How Should the EU Respond?* (Dublin: Institute of International and European Affairs, 2010), at 1; see also below, Sect. 20.3.

the European Union (CJEU) ‹ suggest the real risk that the whole scheme falls into environmental irrelevance. Moreover, it must be pointed out that, despite the ambition of the EU strategy, given the limits of what the EU can do in terms of implementation and enforcement, it falls upon member states to put in place and enforce domestic legal frameworks that can deliver such ambition. And given the increasing evidence about the very different degrees in ambition and regulatory strategies pursued by different member states, only time will tell whether the EU ambitions will be realized.

Moreover, it must be constantly kept in mind that until energy policy is not completely integrated with climate change policy, it is impossible to make radical progress in climate mitigation ‹ given the contribution of the energy sector to total emissions. The 2007 Climate Change and Energy Package mentioned above did not really achieve such integration, chiefly because it does not impose far-reaching consequences in relation to all the issues that require unanimous decision making in the Council pursuant to Article 192.2 TFEU, namely Member State's choice between different energy sources and the general structure of their energy supply. However, this does not mean that the EU is not making progress on this front; the suite of measures being adopted to fully implement the Third Energy Package, together with the adoption of two highly significant roadmaps, namely the Energy 2050 Roadmap and the Roadmap for moving to a competitive low carbon economy, are proving both impetus and direction to the integration of climate change and energy at the EU level. All these developments are underlined by the new Treaty title on Energy Policy, which explicitly refers to a ŒUnion policy on energy[1] within which the goals of completing of the internal market for energy, enhancing energy security, and mitigating climate change, are becoming deeply intermingled."[18]

In many ways, thus, the European Union has reached a crossroads in terms of its domestic climate and energy agenda: on the positive side, Europe has been able to

[18] See, generally, Javier de Cendra de Larragán, "EU Climate and Energy Law: Challenges for Member States", in Marjan Peeters, Mark Stallworthy and Javier de Cendra de Larragán (eds), Climate Law in EU Member States: Towards National Legislation for Climate Protection (Cheltenham: Edward Elgar, forthcoming 2012); Javier de Cendra de Larragán, "Achieving Deep Integration Between the Climate-Change and Energy Agendas: Some Reflections on the EU Approach", 3 Journal of Energy & Environmental Law (2012), 240; the Third Energy Package is composed by the following legal instruments: Parliament and Council Directive 2009/72/EC of 13 July 2009 concerning Common Rules for the Internal Market in Electricity and Repealing Directive 2003/54/EC, OJ 2009 L211/55; Directive 2009/73/EC of 13 July 2009 concerning Common Rules for the Internal Market in Natural Gas and Repealing Directive 2003/55/EC, OJ 2009 L211/94; Parliament and Council Regulation (EC) No 713/2009 of 13 July 2009 Establishing an Agency for the Cooperation of Energy Regulators, OJ 2009 L211/1; Parliament and Council Regulation (EC) No 714/2009 of 13 July 2009 on Conditions for Access to the Network for Cross-border Exchanges in Electricity and Repealing Regulation (EC) No 1228/2003, OJ 2009 L211/15; Parliament and Council Regulation (EC) No 715/2009 of 13 July 2009 on Conditions for Access to the Natural Gas Transmission Networks and Repealing Regulation (EC) No 1775/2005, OJ 2009 L211/36; for the roadmaps, see European Commission, Energy Roadmap 2050, COM(2011)885, and European Commission, A Roadmap for Moving to a Competitive Low Carbon Economy, COM(2011)112.

create a solid policy framework to draw upon, and has also learned important lessons from earlier mistakes; yet at the same time, new challenges will join a number of unresolved issues going forward. Clearly, with the enlargement of the European Union, action on climate and energy needs to accommodate a growing array of actors and interests. While per capita energy use and greenhouse gas emissions have remained surprisingly harmonious across Europe, Member States are currently on very different emission trajectories, and generally advocate different priorities as regards their energy mix and energy security.

20.3 Climate Change in the EU's External Relations

Likewise, at the international level, the EU has strived to play a global leadership role in the battle against climate change. The origins of this goal can be traced to the late 1980s when various European institutions embraced the idea that the EU should seek to play a strong role internationally with respect to climate change and other global environmental challenges. For example, the Rhodes European Council in December 1988 signaled the determination of "the Community and its Member States to play a leading role in the action needed to protect the world's environment" and "strive for an effective international response, particularly to such global problems as depletion of the ozone layer, the greenhouse effect and the ever-growing threats to the natural environment."[19] The goal of playing a global leadership role in the fight against climate change has recently been given a legal formulation in the Lisbon Treaty as one of the objectives of EU environmental policy.[20] This arguably reinforces the EU's commitment to fighting climate change in its external relations.[21] It is useful to bear in mind, however, that the EU's global climate leadership aspirations have always been strongly influenced by internal factors.[22] For example, the rejection of the Treaty Establishing a Constitution for Europe in 2005[23] led European leaders to "desperately cast around for issues that they hoped would be

[19] European Council, "Conclusions of the Presidency, Annex I: Declaration on the Environment," Rhodes 2–3 December 1988.

[20] According to Article 191(1) of the Treaty on the Functioning of the European Union, one of the objectives of the EU's environmental policy is to contribute to: preserving, protecting and improving the quality of the environment; protecting human health, prudent and rational utilization of natural resources; and promoting measures at international level to deal with regional or worldwide environmental problems, *and in particular combating climate change*. Emphasis added.

[21] Maria Lee, "The Environmental Implications of the Lisbon Treaty," 10 *Environmental Law Review* (2008), 131, at 133.

[22] Oberthür and Roche Kelly, "EU Leadership in International Climate Policy," *supra*, note 15, at 35.

[23] The Constitutional Treaty would have replaced the EU founding treaties by a single text, given legal force to the Charter of Fundamental Rights, and strengthened majority voting. While a number of Member States ratified it, Dutch and French voters rejected it in a referendum in May 2005. The Treaty of Lisbon was subsequently created to replace the Constitutional Treaty.

better received by citizens."[24] As a consequence, climate change became a central theme in the European integration process, and the European institutions saw it as a window of opportunity to advance both internal and external EU policy.[25]

The means employed by the EU to assert its global climate change leadership range from using its international influence to strengthen the multilateral framework for climate change mitigation under the UNFCCC to "leadership by example" and unilateral measures. While the EU has consistently sought to be a positive force in the UNFCCC negotiations, it has seldom been able to broker major breakthroughs.[26] In the aftermath of the 2009 Copenhagen conference, for example, the EU and multilateralism were identified as the biggest failures.[27] According to the *Financial Times*, "Europe is the big loser from Copenhagen. Climate has been the one issue where Europe has led the world. In the end the continent was too weak to succeed when it counted."[28] At the recent Durban conference, however, the EU played a highly visible role, confronting India in order to make the mandate of the new *Ad Hoc* Working Group under the Durban Platform for Enhanced Action more ambitious.[29]

Concerning "leadership by example," the EU has often sought to use its internal legislation to inspire developments abroad. In this regard, the ETS and the 2009 climate and energy package are the EU's flagships. The EU attempted to adopt a far-reaching climate and energy package as far back as 1992 for showcasing at the UN Conference on the Environment and Development in Rio de Janeiro.[30] The proposed CO_2 tax met with strong opposition from the main industry lobbies and from Member

[24] Andrew Jordan, David Huitema and Harro van Asselt, "Climate Change Policy in the European Union: An Introduction," in Andrew Jordan et al. (eds), *Climate Change Policy in the European Union: Confronting Dilemmas of Mitigation and Adaptation?* (Cambridge, UK et al.: Cambridge University Press, 2010), 3, at 11.

[25] Oberthür and Roche Kelly, "EU Leadership in International Climate Policy," *supra*, note 15, at 43.

[26] For a comprehensive overview of the EU's role in the UNFCCC negotiations, see Kati Kulovesi, "Climate Change in the EU External Relations: Please Follow My Example (Or I Might Force You To)," in Elisa Morgera (ed.), *The External Environmental Policy of the European Union: EU and International Law Perspectives* (Cambridge University Press, forthcoming 2012).

[27] See for example, "Winners and losers in Copenhagen", *The Economist* Blog, 21 December 2009, available at http://www.economist.com/blogs/charlemagne/2009/12/winners_and_losers_in_copenhag (last accessed on 20 June 2012).

[28] Thomas Kleine-Brockhoff, "Lessons of a Memorably Chaotic Global Gathering", *The Financial Times*, 21 December 2009.

[29] For discussion, see Kati Kulovesi, "A New Chapter in the UN Climate Change Negotiations? First Steps under the Durban Platform for Enhanced Action," 3 Climate Law (2012), 181 Lavanya Rajamani, *Deconstructing Durban,* IndianExpress.com, 15 December 2011, available at: http://www.indianexpress.com/news/deconstructing-durban/887892/ (last accessed on 21 June 2012).

[30] Jorgen Wettestad, "The Complicated Development of EU Climate Policy," in Joyeeta Gupta and Michael Grubb (eds.), *Climate Change and European Leadership: A Sustainable Role for Europe?* (Kluwer Academic Publishers, 2000) 25, at 27.

States reluctant to relinquish fiscal power to Brussels, however.[31] The EU's first attempt to lead by example through internal climate legislation therefore became considerably less ambitious than originally envisaged. Subsequently, however, the EU was among the first developed countries to ratify the Kyoto Protocol and it also successfully launched the ETS in 2005, thereby introducing a carbon price to energy-intensive economic sectors. In December 2008, the EU managed to reach agreement on the climate and energy package ahead of the Copenhagen conference, hoping that the comprehensive package of legal and policy instruments would inspire other countries to follow the EU's footsteps.[32] Through the package, the EU took measures to, *inter* alia, expand the ETS, promote renewable energy and energy efficiency, and integrate climate change considerations into various non-environmental sectors and policies.[33] However, other countries have been slow to follow the EU's example and introduce equivalent national climate change legislation. A further challenge for the EU's leadership by example is that since the Copenhagen conference, the EU has been struggling internally to strengthen its leadership by increasing its unilateral emissions target from 20 to 30% reduction from 1990 levels by 2020, which would be more in line with the requirements of climate science.

Given the relative lack of success of its cooperative international efforts to strengthen global climate change cooperation, the EU has recently taken certain unilateral steps to force the direction of international climate policy.[34] It has included international aviation emissions in the ETS and banned credits from controversial industrial gas projects under the Kyoto Protocol's Clean Development Mechanism (CDM). It has also introduced sustainability criteria for biofuels and considered the possibility of trade measures against imports of energy-intensive products from countries lacking effective climate policies.

Some of these unilateral steps have been highly controversial. Most notably, the inclusion of aviation emissions into the ETS is subject to an escalating international row.[35] The main point of controversy is that the ETS will apply to foreign airlines to the extent they operate flights to and from EU airports.[36] Because of this,

---

[31] Denny Ellerman, Frank Convery and Christian de Perthius, *Pricing Carbon: The European Union Emissions Trading Scheme* (Cambridge: Cambridge University Press, 2010), at 16.

[32] For a comprehensive overview of the package, its negotiating history and international dimensions, see Kati Kulovesi, Elisa Morgera and Miquel Muñoz, "Environmental Integration and Multifaceted International Dimensions of EU Law: Unpacking the 2009 Climate and Energy Package," 48 *Common Market Law Review* (2011), 829.

[33] Ibid.

[34] See similarly, Joanne Scott, "The Multi-Level Governance of Climate Change," 5 *Carbon and Climate Law Review* (2011), 25, at 27–28.

[35] For discussion, see Kati Kulovesi, "'Make Your Own Special Song, Even If Nobody Else Sings Along': International Aviation Emissions and the EU Emissions Trading Scheme", 2 *Climate Law* (2011), 535; and Kati Kulovesi, "Addressing Sectoral Emissions outside the UNFCCC: What Roles for Multilateralism, Minilateralism and Unilateralism?" 21 *Review of European Community and International Environmental Law* (forthcoming, 2012).

[36] See Directive 2008/101/EC of the European Parliament and of the Council of 19 November 2008 amending Directive 2003/87/EC so as to include aviation activities in the scheme of greenhouse gas emissions allowance trading within the Community, OJ L 8, at 3, 13 January 2009.

the EU has been accused of, *inter alia*, using unilateral trade measures and exercising extraterritorial jurisdiction in violation of international law,[37] and failing to adequately reflect the principle of common but differentiated responsibilities and respective capabilities (CBDRRC) in the design of its aviation scheme.[38] The EU aviation scheme has also been subject to legal action. In 2009, American Airlines, Continental Airlines, United Airlines and the Air Transport Association of America (ATA) launched a complaint against the scheme through UK courts. This led to a request for a preliminary ruling from the Court of Justice of the European Union concerning the validity of the relevant Directive 2008/101/EC in light of its alleged incompatibility with certain rules and principles of international law.[39] In December 2011, the Court affirmed the validity of Directive 2008/101/EC, finding its provisions to be compatible with international law.[40] The outcome failed, however, to satisfy key foreign countries whose airlines will be affected by the scheme. The US considered the "European Union Emissions Trading Prohibition Act of 2011," which passed the House of Representatives in October 2011 and would have prohibited US-based airlines from participating in the ETS if a counterpart bill passed the Senate.[41] A bill with somewhat less stringent language is expected to be adopted by the full Congress in 2012.[42] Meanwhile, China has prohibited its airlines from participating in the ETS and increasing fares or imposing other charges related to the scheme, and India has instructed its airlines not to participate in the scheme.[43]

In accordance with Directive 2008/101/EC, airlines failing to surrender the required number of emission allowances will incur an excess-emissions penalty of €100 for each tonne of CO_2 equivalent emitted for which the airline has not surrendered

---

[37] See Kulovesi, "International Aviation Emissions in the EU Emissions Trading Scheme", supra, note 36, for an overview of legal arguments in this regard. See also Lavanya Rajamani, *European Union: Climate Action Hero?* IndianExpress.com, 3 August 2011, available at: http://www.indianexpress.com/news/european-union-climate-action-hero/826290/1 (last accessed on 21 May 2012).

[38] Joanne Scott and Lavanya Rajamani, "EU Climate Change Unilateralism: International Aviation in the European Emissions Trading Scheme", 11 *European Journal of International Law* (2012), 339.

[39] Case C-366/10. Reference for a preliminary ruling from High Court of Justice Queen's Bench Division (Administrative Court) (United Kingdom) made on 22 July 2010 – The Air Transport Association of America, American Airlines, Inc., Continental Airlines, Inc., United Airlines, Inc. v. The Secretary of State for Energy and Climate Change, OJ C 260, at 9, 25 September 2010. For detailed legal analysis of the arguments and the October 2011 advisory opinion by Advocate General Juliane Kokott, see Kulovesi, "International Aviation Emissions in the EU Emissions Trading Scheme", supra, note 36, at 544 et seq.

[40] Case C-366/10, supra, note 40.

[41] H.R. 2594, European Union Emissions Trading Prohibition Act of 2011.

[42] "US Congress to Oppose EU Law on Aircraft Emissions", *Carbon Market Europe*, 3 February 2012.

[43] BBC News, "China 'Bans' Airlines from Joining EU Carbon Scheme", 6 February 2012, available at: http://www.bbc.co.uk/news/business-16901106 (last accessed on 2 March 2012); "India Confirms Boycott of EU Aviation Emissions Rule", *Bridges Weekly*, 29 March 2012, available at: http://ictsd.org/i/trade-and-sustainable-development-agenda/129985/ (last accessed on 30 April 2012).

allowances.[44] Such a payment will not release the airline from the obligation to surrender the missing allowances. Ultimately, a failure to comply with the Directive may lead to a decision by the European Commission that the airline in question is banned from operating in the EU.[45] This is the legal consequence that airlines from the US, China and India could ultimately face if choosing to comply with legislation in their native countries rather than the provisions of Directive 2008/101/EC applicable in the EU. In light of this, it is clear that the aviation row holds potential to become a complex legal dispute between overlapping jurisdictions.[46] It also serves to highlight that the external dimension is highly relevant when considering the development of climate change law and policy in the EU.

20.4 Explaining European Climate Leadership

Looking back at what has been anything but a straightforward development trajectory for Europe's climate strategy, it is nonetheless evident that the leadership role assumed by the EU has not been accidental, but the consequence of deliberate policy decisions. Yet what prompted the EU to embark on this often lonesome and challenging path, to the extent of inscribing an explicit leadership mandate in the Lisbon Treaty and adopting unilateral measures at the risk of antagonizing important economic partners? Much analysis has been devoted to the motivations behind European climate leadership, and various explanations have been suggested at different points in time. Mostly, these relate to the *raison d'être* of the European Union as an institution, the scope of its powers and activities, and its role in the international political arena. For Europe, energy policy has been at the heart of its development into a Community: coming out of a period of devastating conflict in the first half of the twentieth century, and challenged with a comparatively low resource base as well as rapidly growing energy demand, the small group of European nations forming what has become today's European Union chose energy as one of the initial focus areas for economic and political integration. Recognizing the role of energy as a strategic backbone for any industrialized society, Europe has ever since – albeit not always successfully – sought to achieve greater integration of energy markets and policies.

A few decades into the integration process, high population densities and widespread degradation of air, water, and soil as a result of intensive industrialization led to growing public awareness of environmental threats and engendered an active green movement, promoting environmental concerns onto the European political agenda. Early action on energy efficiency and pricing – for instance through excise taxes on mineral oil – is yielding compounded effects many decades later, with

[44] Directive 2008/101/EC, supra, note 37, Art. 16.3.

[45] Ibid., Arts. 16.5–16.12.

[46] This argument has also been made in Kulovesi, "Addressing Sectoral Emissions", supra, note 36.

adaptive responses in infrastructure, housing, transport and other sectors reverberating throughout the economy and creating stakeholders for new and different economic opportunities. Over time, therefore, the tolerance for more restrictive or costly access to commodities such as energy has increased noticeably vis-à-vis countries with less stringent regulations and pricing systems.

In many ways, therefore, European leadership in climate and energy policy is an extension of two inherent trends in the larger process of European integration. When several setbacks threatened this process, including the rejection of a European Constitution in 2005, climate and energy policies were also seen as a unifying and sufficiently urgent agenda for the EU. Internationally, moreover, the high profile of climate negotiations has been described as a useful vehicle to promote the multilateral and rule-based approach to global governance espoused by the European Union, thereby enhancing its role as a global actor and its ability to build coalitions.[47] As seen in the preceding sections, significant climate policy developments in the EU were often preceded by acceptance of strong international commitments, an active executive body – the European Commission – setting the agenda, and competitive leadership by the rotating presidencies of the European Council. All this has been conditional on a deeply seated belief in multilateralism and the value of supranational integration.

But just as such very concrete circumstances and objectives may explain the European motivation to seek a position of leadership in climate negotiations, more abstract considerations, such as institutional structures and dynamics, have also been cited as a vital precondition. In particular, scholars have argued that the design of European multi-level governance creates numerous leadership points where competitive leadership has been initiated, opening avenues by which advocates of climate change action have been able to inject their priorities and concerns into the policy debate. But they also highlight an active network of environmental citizen groups and internal division among the traditional opponents of stringent climate policies, industry and commerce, and see them as fostering a unique and fertile political landscape.[48] Indeed, climate change is one of the few policy issues around which European citizens have been largely united over a number of years. Surveys of public opinion consistently show that two thirds or more of Europeans consider climate change one of the most serious problems facing humanity.[49] Whilst such abstract public support for action on climate change should not mask the fact that European publics often remain deeply skeptical of certain mitigation policies, and even strongly opposed to the prospects of having to face the economic consequences

[47] Oberthür and Roche Kelly, "EU Leadership in International Climate Policy," *supra*, note 15, at 43.

[48] Miranda Schreurs and Yves Tiberghien, "Multi-Level Reinforcement: Explaining European Union Leadership in Climate Change Mitigation", 7 *Global Environmental Politics* (2007), 19–46, at 25.

[49] European Commission, *Europeans' Attitudes Towards Climate Change*, Special Eurobarometer Report (Brussels: European Commission, 2008), at 6, 9.

that inevitably accompany them, it is nevertheless undeniable that climate policies are therefore seen as "a good political 'spin'", evidencing the capacity of the Community "to deliver effective policies to address the public's concerns in an area where citizens believe the Union can – and should – play a strong role."[50] And ultimately, that might be the most important factor in the EU's ability to sustain its climate leadership going forward.

20.5 Conclusion

In retrospect, a number of factors have motivated European leadership on climate policy. German reunification and the ensuing collapse of heavy industry in its eastern states are unique historical developments that cannot be easily replicated elsewhere. While such events are by no means the only – or even the main – factor behind the European success in reducing greenhouse gas emissions, their consequences can still be felt today: with much of the European power generation fleet built after the war and currently nearing the end of its useful life, the need to steer future generation capacities in a more sustainable direction poses itself with far greater urgency in the European Union than in other regions, where the electricity infrastructure has emerged over a longer and more gradual period, and the risk of technology lock-in hence is lower.

Likewise, unique geographical conditions have important implications for the EU's emissions profile, with historically evolved settlement patterns in Europe favoring greater efficiency in transport and urban housing. It is important, however, to point out that climate leadership in Europe is not merely a *fait accompli*. Quite the contrary: rising public concern about climate change impacts, emerging divisions within traditionally opposed stakeholders from industry and trade, and the desire to reduce dependency on energy imports and thereby increase energy security have been important motivators of policy action in Europe as much as elsewhere in the world. Arguably, however, the translation of political will into legislative action has greatly benefited from favorable institutional conditions, such as the simple majority needed to adopt climate legislation in the Council. And the importance of such structural and procedural factors cannot be overemphasized.

Ultimately, the prospects for continued European leadership on climate change prompts challenging questions about the ability of its political system to adequately tackle complex, long-term challenges. As the ongoing divisions over monetary and fiscal policy in Europe clearly illustrate, unity is not always a given in Brussels' hallways of power. If it is true that, as one group of authors puts it, climate governance involves "making difficult choices between alternative options that are supported by different groups of actors who often have incommensurate

[50] Henningsen, *EU Energy and Climate Policy*, supra, note 9, at 7.

values",[51] continued leadership will depend on whether decision makers are able to agree on these choices. In a Europe of currently 27 Member States, the formulation of a "consensus among many voices"[52] has invariably become more challenging, and it is difficult to predict what the future role of the Union will be both in the medium and the long term.

[51] Andrew Jordan et al., "Governing the European Union: Policy Choices and Governance Dilemmas", in Andrew Jordan et al. (eds), *Climate Change Policy in the European Union: Confronting the Dilemmas of Mitigation and Adaptation?* (Cambridge: Cambridge University Press, 2010), at 29.

[52] Nuno Lacasta, Suraje Dessai and Eva Powroslo, "Consensus Among Many Voices: Articulating the European Union's Position on Climate Change", 32 *Golden Gate University Law Review* (2002), 351.

Chapter 21
Climate Law in Germany

Felix Ekardt

Abstract Despite its impressive quantity current climate protection law is not suited to solve the climate problem – neither on a global level through public international law nor in the EU or Germany. In Germany, not only the absolute emission levels raise concerns. Relative development, too, is much worse than is often assumed. German climate law is characterized by a variety of rules, although a substantial part (more or less) implements EU law. The – internationally often copied – German Renewable Energy Sources Act (EEG) contains a fixed tariff for renewable electricity similar to a subsidy. In addition to that and to a number of energy efficiency rules, there are a number of legal rules that directly flank the regulatory, financial, and informational regulations on efficiency, sufficiency, and renewable energies. It remains true, however, that renewable energies and energy efficiency do not per se reduce greenhouse gas emissions or replace fossil fuels; in fact there may also be shifts in emissions and fuel transfers to other countries and/or increases in overall energy consumption. These rebound and shifting effects are a common major barrier to effective climate policy, including energy efficiency policy. A completely new cap and trade approach on the EU level (combined with border adjustments) might be the best way to solve these problems.

Felix Ekardt is a Lawyer, Sociologist and Philosopher. He is a professor of environmental law and philosophy of law at the University of Rostock and Director of the Research Unit Sustainability and Climate Policy in Leipzig.

F. Ekardt (✉)
Forschungsstelle Nachhaltigkeit und Klimapolitik,
Könneritzstrasse 41, Leipzig 04229, Germany
e-mail: felix.ekardt@uni-rostock.de

E.J. Hollo et al. (eds.), *Climate Change and the Law*, Ius Gentium: Comparative Perspectives on Law and Justice 21, DOI 10.1007/978-94-007-5440-9_21,
© Springer Science+Business Media Dordrecht 2013

21.1 Germany – A Leader in Climate Protection?

Anthropogenic climate change is at its heart a consequence of the release of various greenhouse gases mainly from fossil fuels (related to electricity, heat, fuel and material usage) and land use.[1] Therefore, policies which attempt to combat climate change aim at potentially far-reaching changes in those sectors. Scientific and economic research – which on a global level is bundled in the Intergovernmental Panel on Climate Change (IPCC) – develops statements about necessary reduction targets; those are needed to evaluate the political and legal call for action. They state that, in order to avoid resource wars, huge migration flows, an endangered food and water supply, natural disasters, substantial economic damage and millions of deaths, global emissions' reductions of about 80%, and in the industrialised world of up to 95%, are needed by 2050 on the basis of 1990. One reason for this specific reduction statement for industrialised states is that currently, on a global level, per capita emissions are very unequally distributed: The annual per-capita emissions of an average German still add up to 20–30 times the amount of a person in Sub-Saharan Africa and two and a half times the amount of a Chinese.

Despite its impressive proliferation in recent years, climate protection law is currently not suited to solve these problems – neither on a global level through public international law nor in the EU or Germany. Notwithstanding any details, this is evident from the results of previous attempts. Even though Europe and Germany often claim to be a "leader in climate protection", one German still emits several times the greenhouse gas volume of developing country counterparts; this inequality is even larger with respect to those greenhouse gases already accumulated in the atmosphere. This is all the more noteworthy given that residents of developing countries will be disproportionately affected by climate change. A fortiori, future generations are expected to be greatly injured by climate change without having caused it at all. Total global emissions have increased by more than 40% since 1990.

In Germany, concerns arise not only the absolute emission levels. Relative development, too, is much worse than is often assumed. If (a) the industrial collapse of Eastern Europe in 1990, (b) the relocation of production facilities to developing countries, and (c) the financial crisis since 2008 are eliminated from calculations, emissions in Germany since 1990 have not (starting at a high level) fallen but risen. For the financial crisis will hardly result in a lasting drop in production, including permanent greenhouse gas reductions; and relocation of production only shifts greenhouse gas emissions from one country to another, such as from Germany to China or Malaysia. Therefore, Germany is not the imagined leader in climate

[1] On all topics, questions and arguments of this contribution see in more detail Felix Ekardt, Theorie der Nachhaltigkeit: Rechtliche, ethische und politische Zugänge – am Beispiel von Klimawandel, Ressourcenknappheit und Welthandel (Baden-Baden: Nomos, 2011); Felix Ekardt, Bettina Hennig and Herwig Unnerstall (ed.), Erneuerbare Energien: Ambivalenzen, Governance, Rechtsfragen (Marburg: Metropolis, 2012).

protection – debates, normative standards and technical innovations are impressive, but at the end of the day it comes down to the actual emissions budget.

21.2 Fundamentals of German Climate Policy and Nuclear Power Phase-Out

German climate protection law is characterized by a variety of rules, although a substantial part (more or less) implements EU law. However, important parts are independent from EU law, since EU law mainly sets detailed provisions for emissions trading. In all other areas of climate protection, essentially only framework provisions exist on a European Union level.

As at the EU level, energy and climate protection law in Germany is regularly advanced in "packages", e.g. in the federal government's Integrated Energy and Climate Programme of 2007 (IECP).[2] This programme – which is also referred to as the "Meseberg decisions" – was worked out at the federal government's retreat in Meseberg in 2007 and later that year adopted by the Cabinet. Such packages are regularly comprised of a multitude of individual actions concerning existing laws. Another major energy package was adopted in the summer of 2011 after the nuclear catastrophe at Fukushima. Programmatically important is also the quite comprehensive federal government's energy concept which was established in the summer of 2010. It defines the general goal of energy and climate policy to reduce greenhouse gas emissions in the energy sector (i.e. not all greenhouse gases are covered) by 35–40%. In addition, there are sub-goals like the expansion of renewable energies in different subject areas, e.g. in the energy sector to 35% by 2020.

Thus, the various instruments – regulatory law, economic instruments, informational instruments, rules of competition, financial support, etc. – are subject to constant development. There are also bans on technology: It has been widely publicized that, as a result of Fukushima – and after several twists – Germany decided to gradually phase out nuclear power generation. Similarly, so far, carbon capture and storage (CCS), i.e. coal-fired power plants without emissions, has not been legally permitted in Germany. This does not mean, however, that the issue was removed from the political agenda, since the EU's CCS Directive must still be implemented – the respective deadline has already expired. Overall, the debate in Germany is of course often narrowed considerably. It is centred on electricity, compared to heat and fuel. And, with respect to electricity, there is a clear focus on nuclear power, neglecting the removal of fossil fuels.

The strategy of German climate policy is to strengthen renewable energies and energy efficiency. There is no final estimation as to the national and international

[2] For details of all programmes see http://www.bmu.de/klimaschutz/downloads/doc/40514.php (last accessed on 15 February 2012).

long-term need of additional greenhouse-gas-free coal-fired power plants. However, the dominant perception is that radical climate gas reductions like EU-wide minus 95% (excluding effects from production shifts) can be achieved by "purely technical" means. German politics avoid the question whether perhaps sufficiency, i.e. voluntary or forced absolute reduction of resource consumption and climate gas emissions (if necessary by renunciation), rather than only more efficiency – in the sense of more economic use e.g. of energy in relation to a definite result – is necessary. Yet EU emissions trading also includes absolute reduction targets; admittedly weak ones and without a ban on shifting emissions (or the production of goods) to other countries.

21.3 Subsistence for Renewable Energies

The advantages of renewable energies such as (in principle) climate neutrality, creation of new jobs, replacement of finite resources, economic innovation, security of supply independent from unstable regions and resource conflicts, etc. are obvious. This huge potential, however, cannot hide the fact that renewable energy sources often cannot yet compete in the market without some form of assistance. On the one hand, this is due to the partly developing technology, on the other hand, to the fact that conventional fossil fuels such as petroleum, coal, uranium, and natural gas can be offered at supposedly more favourable prices because energy prices do not fully reflect the external costs, such as anthropogenic climate change or the risks of nuclear energy. Accordingly, legal frameworks that support renewable energies are obvious. Currently, those are designed differently in different Member States within a more general European framework. The European framework under the Renewable Energy Directive is known to define only pan-European and national development targets to be achieved in a given period.

The – internationally often copied – German Renewable Energy Sources Act (EEG) contains a fixed tariff for renewable electricity similar to a subsidy.[3] In addition, it contains some incentives for the coupled generation of electricity and heat from renewable energy sources. According to Section 1 paragraph 2 of the EEG, the share of renewable energies in electricity supply shall reach at least 35% by 2020 and then gradually be increased to 80% by the year 2050. Hereto, the EEG provides anyone who generates electricity from renewable energy sources and feeds it into the grid system for general supply with a claim against the respective grid system operator for the connection of her installation to the operator's grid system, the purchase and transmission of this electricity, as a priority, respectively, and the

[3] Gesetz für den Vorrang erneuerbarer Energien (EEG), available at: http://www.erneuerbare-energien.de/files/pdfs/allgemein/application/pdf/eeg_2012_bf.pdf (last accessed on 25 February 2012).

payment of a statutory minimum tariff (EEG Sections 5 paragraph 1, 8 paragraph 1, and 16 paragraph 1). The EEG minimum tariff is significantly higher than current market prices – e.g. on major trading centres such as stock markets or in bilateral supply contracts – and, in accordance with EEG Section 21 paragraph 2, it is guaranteed for a period of 20 years from the time when the EEG-generator first produces electricity. The respective tariff is determined mainly by the energy sources used. In addition, the time of first energy production as well as, partly, the installation's capacity and location, and other criteria are used. The economic burden on grid system operators which results from the payment of EEG-tariffs is ultimately apportioned via the EEG-surcharge mechanism to the majority of electricity consumers in Germany. First, a grid system operator is obliged to accept electricity, which is fed into the operator's grid system. Second, a grid system operator (unless already being an upstream transmission system operator) shall immediately deliver the electricity to an upstream transmission system operator who in turn is bound by EEG Section 8 paragraph 1 with respect to the grid system operator. According to this scheme, all the electricity which is paid for under EEG tariffs ultimately gets to upstream transmission system operators. The latest major reform once again increased the number of EEG rules and led to a partial revision of the support framework for solar radiation, offshore wind energy, biomass, and direct selling. Still, solar energy, particularly, remains an infinite source of (opposing) demands for new reforms – and that, at least in the long run, will come as a detriment of legal certainty and planning security.

The expansion of renewable energy in the electricity sector as such is perhaps the biggest (and only real) success story of recent German climate policy. Nevertheless, further discussion is necessary. A feed-in tariff system is not always perfectly in harmony with emissions trading (which will be introduced infra). While the latter requires a reduction of overall greenhouse gas emissions, e.g. for the EU, the former results in climate protection preferably by switching to renewable electricity (instead of e.g. increased energy efficiency) within the latter's "greenhouse gas cap." Economists, in particular, assume that this renders the expansion of renewable energies via the EEG meaningless and yet unnecessarily expensive. However, closer investigation reveals that this apparent paradox is very limited; for inter alia feed-in regulations also result in innovations, which is why they need be included in any effective climate protection regime.

It remains true, however, that subsistence for renewable energies does not per se reduce greenhouse gas emissions or replace fossil fuels; in fact there may as well be shifts in emissions and fuel transfers to other countries or increases in overall energy consumption. These rebound and shifting effects are a common major barrier to effective climate policy. For instance, the generation of energy from biomass is highly ambivalent. Often, its greenhouse gas balance is no improvement compared to fossil fuels. Such ambivalence of renewable energies cannot be solved with "sustainability criteria" as recently introduced by the EU for bio-energy imports. Such regulations again potentially fail due to rebound and shifting effects, as well as enforcement problems. I will briefly return to this aspect at the end of this paper.

The Renewable Energies Heat Act (EEWärmeG)[4] which came into force on 01/01/2009 aims to support of renewable energies in the heating market in order to (i) reduce greenhouse gas emissions; (ii) optimise the security of supply by decreasing the dependence on foreign supply; (iii) permanently immunise energy prices against oil and gas price shocks; (iv) respond to the steadily declining availability of fossil fuels; and (v) gain economic and innovative benefits by becoming active "on time." Therefore, the EEWärmeG defines the target to increase the share of renewable energies in the production of heat by the year 2020 to 14% (cf. EEWärmeG Section 1 paragraph 2). Regarding only the replacement of fossil fuels it is to be welcomed that hereto renewable heat under EEWärmeG Section 5 must originate either from solar energy (15%), biogas (30%), certain solid or liquid biofuels (50%) or geothermal and environmental heat (50%). From a climate policy perspective, however, it need be criticised that this only applies to new buildings. With respect to old buildings, there is only an incentive programme[5] with investment subsidies. Moreover, the target for new buildings is far too low. Furthermore, there seems to be a massive enforcement problem resulting not least from the number of individual acts that must be controlled.[6]

In the area of fuel, there is also a law setting biofuel quotas which – as with heat and electricity – again is based only on a very general European law foundation.

21.4 Energy Efficiency and Sufficiency

Perhaps energy-efficient building renovations, i.e. measures in the area of heat which save resources and protect the climate by increasing efficiency (and using renewable energy), offer the greatest potential in one single area for climate protection in OECD countries; after all, buildings are responsible for more than a third of Germany's greenhouse gas emissions. At the same time, climate protection through the refurbishment of buildings is economically viable not only because of long-term results from climate change but also because of mid-term energy prices and energy security issues. Old buildings are of special importance. This is due to the large number of old buildings and the likely gradually dropping demand for new housing

---

[4] Gesetz zur Förderung Erneuerbarer Energien im Wärmebereich (Erneuerbare-Energien-Wärmegesetz), available at http://www.erneuerbare-energien.de/files/pdfs/allgemein/application/pdf/ee_waermeg.pdf (last accessed on 25 February 2012).

[5] Cf. Richtlinien zur Förderung von Maßnahmen zur Nutzung erneuerbarer Energien im Wärmemarkt, 20 February 2009, http://www.bafa.de/bafa/de/energie/erneuerbare_energien/index.html

[6] On many topics of energy efficiency law see also Thomas Schomerus et al., *Rechtliche Konzepte für eine effizientere Energienutzung* (Berlin: Erich Schmidt Verlag, 2008); Martin Winkler, *Klimaschutzrecht* (Münster: LIT Verlag, 2005); Cimin Keyhanian, *Rechtliche Instrumente der Energieeinsparung* (Baden-Baden: Nomos, 2008).

as a result of demographic development in Germany. Furthermore, old buildings require significantly more heat than new ones.

Consequently, since coming into effect on 8 July 2010, the revised EU Energy Performance of Buildings Directive (EPBD) defines (at least in the mid- and longer term) quite relevant energy efficiency standards for new buildings and significant alterations.[7] This includes methods for calculating the efficiency of buildings and a commitment to provide high standards for the refurbishment of old buildings. By the end of 2020, the zero-energy house standard applies to all new buildings. Beyond that, the EPBD is of course primarily informational; Member States shall comprehensively report on the steps taken (and they shall introduce any regulations as well as targets at all), and citizens will receive energy certificates as an orientation for all buildings. In Germany, these European requirements are specified by the Energy Conservation Act and the Energy Saving Ordinance (EnEV),[8] which try to boost energy efficiency in the building sector. In two steps, the EnEV requires a considerable increase in energy efficiency for new buildings as well as for old buildings where there are significant alterations, though only under much lower standards (EnEV Section 9). As with the EEWärmeG, the problem remains that with respect to new buildings existing potential is not fully used and that regulation regarding old buildings is incomplete.

In the EU and in Germany, a variety of energy efficiency regulations for technical equipment exist in addition to those concerning buildings. For some time there has been a German federal regime for some types of equipment and for motor vehicles – the Energy Consumption Labelling Act – as well as the Energy Consumption Labelling Ordinance and the Energy-using Products Act.[9] In addition, there is a Combined Heat and Power Act which attempts to promote the combined generation of heat and electricity through (weak) incentives. Furthermore, following the British example, on 29 September 2000, the German Energy Agency was founded by the Federal Ministry of Economics and the Bank for Reconstruction. Its task at the federal level is to take care of improvements in energy efficiency in households, businesses, and public administrations, as well as of the use of renewable energy sources, and to provide information.

In addition, Germany is part of the EU emissions trading system; its greenhouse gas reduction targets include incentives to the participating industries for greater energy efficiency, but theoretically also for sufficiency. The Emissions Trading Directive is transformed into German law through a German Greenhouse Gas Emissions Trading Act, a Project Mechanisms Act, an Allocation Act, and an Allocation Ordinance. The German Emissions Trading Authority is responsible for

---

[7] On details see Schomerus et al., supra, note 6, at 127 et seq.

[8] Verordnung über energiesparenden Wärmeschutz und energiesparende Anlagentechnik bei Gebäuden (Energieeinsparverordnung), available at: http://www.enev-online.org/enev_2009_volltext/index.htm (last accessed on 25 February 2012).

[9] For an overview of existing German energy law see Wilfried Erbguth and Sabine Schlacke, *Umweltrecht* (Baden-Baden: Nomos, 3rd edition 2010).

all administrative activities concerning emissions trading. Of course, from 2013, the importance of purely national standards will clearly be reduced for emissions trading due to the then more intense level of European regulation regarding reduction targets, auction duties, etc. However, the reduction targets even for emissions trading are certainly insufficient to achieve existing climate targets. Moreover, they do not avoid the effects of shifting into other countries.

In addition to this EU ETS, there is a German "environmental tax", which – just like emissions trading – aims at providing incentives for efficiency and sufficiency through an additional burden on prices. Hereto, the German electricity and fuel tax under the Electricity Tax Act and the Energy Tax Act[10] surcharges fuel and electricity primarily to consumers; the manufacturing sector, however, is partly exempt through a reduced tax rate (StromStG Section 9a), because it is assumed to be covered in particular by the (rather modest) EU ETS. The German environmental tax, however – paralleling EU ETS frictions – is currently so low that there is only a limited effect on ingrained behaviours (such as picking up rolls at the nearest bakery with your own car). Furthermore, as with the EU ETS, the lack of long-term tax rate increases and a strong reduction for the manufacturing sector show adverse results. Another concern is the favourable treatment of coal and nuclear power over natural gas; only natural gas is levied with an additional tax on fuels under the energy tax act (in addition to the environmental tax). Moreover, with regard to the regional and sectoral approach what has been said about the EU ETS applies analogously.

The tax reduction for the manufacturing industry (StromStG Section 9a) leads to another issue: currently, a variety of regulations in the industrialised countries even subsidise a non-sustainable behaviour. In Germany, this effect results from explicit subsidies, e.g. for German coal mining, as well as tax reductions. In addition to the manufacturing sector e.g. the company car privilege (which encourages individual transport and large cars), the distance flat expense (which supports transportation and production energy consumption as well as land use, cf. Income Tax Act Section 9 paragraph 1 number 4), the tax exemption for aviation gasoline (which favours a (due to altitudes) particularly climate-damaging use of fossil fuels, cf. Energy Tax Act Section 27), etc. Many other benefits can be found for example in the field of conventional agriculture which often proves little sustainable regarding biodiversity, climate, and energy. Moreover, there are indirect subsidies for various activities by not charging their external costs like damages on climate, forests, etc. In November 2008, the Federal Environment Agency calculated 42 billion Euros of environmentally harmful subsidies in Germany every year through direct payments or tax benefits. This was only referring to federal measures; states and municipalities are barely taken into account.

---

[10] Energiesteuergesetz (EnergieStG), available at: http://www.gesetze-im-internet.de/energiestg/index.html (last accessed on 25 February 2012).

21.5 Planning Law and Energy Law

In addition, there are a number of legal rules that directly flank the regulatory, financial, and informational regulations on efficiency, sufficiency, and renewable energies. Of great importance is the Energy Management Act. It compliments the EU energy law directives, demands a liberalised energy market, and provides, inter alia, priority access for renewable electricity.

In the field of renewable energies, municipalities may also apply planning-specific legal instrumentalities. In addition to the call for further individual and business action, municipalities are a frequent addressee of a "climate change policy from below." In sum, despite many discourses, approaches, and lots of thinking, the results in this area, again, are too small to achieve existing climate targets; at the same time, the issue inevitably hangs in the air whether under a situation of global economic competition a global problem can be approached locally.

From an administrative law perspective, municipalities mainly have opportunities but no obligations regarding climate protection – given the often limited motivation of administration, politics, and citizenship this leads to predictable outcomes. The classical control mechanism available to municipalities for a variety of objectives is land use planning. In addition, the law of street use (StVO Section 45) offers ways to direct individual transport and thus to reduce it: through measures such as parking management, noise control, reduction of road construction or the establishment of traffic-free zones.

Development plans decide on the admissibility of construction projects and the design of buildings. If a municipality wants to ensure a sustainable energy supply in its area, it can require specific energy supply or corresponding construction measures (e.g. the installation of solar panels). This of course leads to the general question whether, e.g. climate protection, can be a permissible objective of land use planning. Since its amendment in 2004,[11] Building Code Section 1 paragraph 5 and 6 explicitly states that land use planning contributes to environmental protection also considering its responsibility for climate protection in general. This visibly manifests the intention of the legislature to promote climate protection more firmly on the local level. In this sense, the legislative materials clarify that this in particular also includes "global" and not only regional or local climate protection. In this regard, however, legal practice stumbles upon Building Code Section 1 paragraph 3 (and Building Code Section 9 paragraph 1) which states that land use planning must have an "urban" reference. Therefore, it is partly assumed that determinations must not be made only on the basis of general energy considerations, to save energy or to protect the global climate, but that they rather require a justification on the basis of urban characteristics and the local situation. However, this is doubtful for two reasons. First, this view is incompatible with Section 1 paragraph 5 sentence 2 of the

[11] Gesetz zur Anpassung des Baugesetzbuches an EU-Richtlinien, 20 July 2004 (BGBl I p. 2424).

2004 Building Code ("general climate protection"). Second, in the end, local climate protection has "never" the potential to specifically prevent a local e.g. flood – because climate change is a global problem. But if that is the case, anyway, and yet "general climate protection" is a target, a local reference cannot be required; for it would be rendered meaningless.

Municipal building planning is particularly important for the provision of land necessary for electricity and heat installations. While, for instance, photovoltaic systems are preferably erected on roofs and thus mostly in town, wind power or biomass installations are primarily site-variable outer space projects because unlike geothermal and hydroelectric power plants they are not linked to geographical or geological land characteristics. Power generation plants using renewable energy sources are therefore often subject to licensing requirements under construction and immission control laws. For example, this has been outlined elsewhere regarding bio-energy, including existing reliefs. The legislature, however, has learned from the negative experience with wind turbines and therefore established in Building Code Section 35 paragraph 3 the planning law option to regionally and nationally direct the spread of Building Code Section 35 paragraph 1 numbers 2–6 projects, thus including biomass facilities. Consequently, the local practice should control the creation of such installation and thus any ambiguities through allocation plans. Determinations in allocation plans favouring renewable energy sources can also be used for inner regions, i.e. the inner city urban areas inside of towns and villages (Building Code Sections 30 and 34).

21.6 Power Lines and Energy Storage

Renewable energy sources for electricity and heat can often be produced locally, but they are not equally available all the time. Against this background, it is undisputed that especially for electricity many new lines must be built and power storage technologies must be improved. However, the details to what extent one or the other should happen are heavily debated.[12]

Basically, under German law there is a system of obligations and incentives for the development of networks. According to Energy Management Act Section 11 paragraph 1, operators of energy supply systems shall "operate a safe, reliable, and efficient energy supply system and, as needed, develop it to the extent it is economically reasonable." This expansion obligation is oriented on the (long-term) demand and is subject to economic reasonableness. For transmission system operators, the

[12] On this topic see Felix Ekardt and Justus Wulff, Energiespeicherung und Energieleitungsbau als Governance- und Rechtsproblem, 115 Jahrbuch des Umwelt- und Technikrechts (2012), forthcoming.

general development obligation is specified in Energy Management Act Section 12 paragraph 3 which states that they shall permanently ensure the ability of the system to meet the demand for transmission of electricity and contribute to the security of supply, in particular through adequate transmission capacity and reliability of the network. In Germany, according to Energy Management Act Sections 11 paragraph 1 and 12 paragraph 2 the cost of expansion is regulated under an incentive regulation ordinance. The concept of regulating incentives with its differentiated ways of apportioning network expansion costs to electricity customers offers a monetary incentive to network operators to take advantage of cost savings and thus reduce inefficiency. However, so far, these long established basic rules could only modestly promote network expansion.

In terms of a broader approach, therefore, the German energy policy 2011 includes another reform. The legislative package of the summer of 2011 includes measures, as the Grid Development Acceleration Act (NABEG) which in essence amends the Energy Act in order to reduce the duration of planning and licensing processes and to ensure greater acceptance of network expansion among the people. Hereto, a future federal technical planning is envisaged, which shall be conducted by the Federal Network Agency in coordination with the states concerned. A result of this planning shall be a federal network plan which will identify the nationwide necessary route corridors and reserve them for the construction of highest voltage transmission lines. However, it remains to be seen whether this will stimulate a rapid network expansion. Regarding the creation of energy storage, so far, there have been only sporadic incentives.

21.7 Climate Change Legislation at the State Level

There has long been a debate in Germany whether, in addition to classical instruments of, e.g. regulation, information, subsistence, land use planning, charges, and market certificates, explicit climate target systems would be useful as a kind of climate protection framework. At the federal level, however, such a concept could not yet gather a majority. Nevertheless, at the state level, such a regulation is currently sought after in North Rhine-Westphalia and Baden-Wuerttemberg.

Generally, state policy is related to climate protection in multiple ways. The lack of a clear result of climate policy on a public international, European and federal law level which would be suited to achieve the given climate objectives raises the question whether regional units such as the states need to give impetus or even fill this existing gap. In any case, state approaches to climate change are valuable experiments for higher regulatory levels. Even if in the foreseeable future an effective global and European climate policy should evolve, e.g. through certificate markets or additional charges, it still remains dependent on certain supplements including measures of state land use planning. Thus, in a federal state like Germany, this results in a strong call for the single states.

21.8 Structural Deficits of German Climate Policy

Why is the overall effect of all the legal climate protection instruments that were introduced above on reducing greenhouse gases so little? In short, the answer is (a) that the instruments' targets are not strict enough, (b) that there are enforcement problems, and (c) that mere "technical solutions" without any behavioural changes are insufficient to achieve absolute emission reductions because of rebound effects and (d) shifting/displacement effects. More specifically: Renewable energy sources are clearly important for an effective climate policy. The same applies to energy efficiency: food, clothing, building heat, consumer electronics; energy lurks in all things. And light bulbs or cars could often be many times more efficient in their production and operation. However, efficiency alone is not enough to permanently satisfy the growing European and global hunger for electricity, heat, and fuel with wind energy, geothermal energy, solar energy, and hydropower. For instance, some metals, from which solar panels are built, will soon be in short supply. Therefore, absolute energy consumption must be limited.

However, this cannot be achieved by simply making any car or any device slightly more efficient, while at the same time cars are getting bigger and more numerous because Germany, Europe, and the world are getting richer. And who wanted to tell the Chinese that they cannot live like us? Neither can energy-efficient homes per se solve our climate and energy problem, if their living space is getting bigger and we use the money saved from heating to acquire flights to remote vacation locations. And even if energy were infinite: building materials for cars and airplanes are clearly not. And nor is energy, at any rate.

Therefore, absolute emissions limits are necessary. Prescriptions and prohibitions which, e.g., require more efficient products or houses cannot achieve this goal: Growing prosperity partly consumes those efficiency gains (rebound effect). Moreover, the vacation example shows that energy consumption and greenhouse gas emissions are often easily shifted to other countries, other resources or other activities in response to regulatory prescriptions and prohibitions (shifting/displacement effect). Regarding climate and energy supply, it is no good to save energy in this country but let cell phones and cars be produced in East Asia. Even fewer cars would not help, if they were replaced by more flights.

Especially the controversial bio-energy leads to shifting and displacement effects. For it conserves fossil fuels but consumes potential food, water, and soil in a starving world. Moreover, land use and especially conventional fertilisers – such as for bio-energy plants – are in themselves a climate problem. Therefore, as mentioned supra, in terms of climate protection large-scale industrial bio-energy is often not better than the use of fossil fuels; especially when large areas of grassland are destroyed, e.g. through the destruction of Amazon rainforest for the production of Western animal feed and bio-energy export crops. Neither can this issue be solved by bans such as the EU's sustainability criteria ("no bio-energy plants in the rain forest"); though the EU might currently try. However, the enforcement of such provisions in the proverbial Amazon region is doubtful. Moreover, new displacement effects are looming: energy plant crop growers might fulfil these requirements in order to continue selling their bio-energy into the EU – and instead grow animal feed and cosmetics raw materials for the West at the same area which perhaps have been grown out of the rain forest before.

Ultimately, rebound effects and shifting/relocating effects can only be solved by a climate policy that provides for an overall cap on energy and land consumption.[13] But this can only be achieved through a charge on fossil fuels and land use instead of single products' regulations. If we do not want to create new shifting effects, this need be established at the highest possible level. Therefore, a global charge or an entirely new global emissions trading is advisable, as described in my first article in this volume. For the moment, however, this is likely to remain visionary.

But there is a real politically feasible alternative which could make the EU, for the first time, become a real and not primarily rhetorical "climate leader." Hereto, existing EU ETS had to be expanded to a primary energy emissions trading. Unlike previous approaches, this would cover all emissions – at least if land use is also included in the ETS or (if the enforcement were too difficult) it is levied with higher charges. The control of the few existing primary energy companies would be simple and much less bureaucratic than complex detailed rules like bans or regulations on a variety of products. The effects of shifts to other countries could be avoided by allowing all non-European States to participate in the system. In case of States reject the offer, border adjustments for exports and imports are introduced in relation to those countries. This avoids the effects of shifts and creates a pressure to commit to a worldwide charge. At the same time, the ETS could then (unlike now) be linked to slowly and gradually increasing reduction targets. For competitive disadvantages in comparison to other States are meant to be avoided by the border adjustment. Furthermore, for the same reasons, a full auction of emissions certificates could then easily be introduced.

Energy companies and farmers would pass the slowly rising cost of the new primary energy ETS to consumers. Electricity, heat, and fuel from fossil sources would thus gradually become more expensive. Efficiency and renewable energy sources would be more attractive. But there would also be absolute energy savings since the charge on fossil fuels would persistently increase. And rebound and shifting effects would be eliminated, because fossil fuels and land use were covered in all areas of life. A number of other energy and climate protection schemes, such as the regulatory regimes for thermal insulation, could in turn be abolished.

The ETS revenues could compensate the socially weaker in the EU and especially the developing countries for higher energy prices and those climate change

[13] On details of the following ideas see note 1 and Felix Ekardt and Antonia von Hövel, "Distributive Justice, Competitiveness, and Transnational Climate Protection: 'One Human – One Emission Right'", 2 *Carbon & Climate Law Review* (2009), 102; Felix Ekardt and Andrea Schmeichel, "Border Adjustments, WTO Law, and Climate Protection", 6 *Critical Issues in Environmental Taxation* (2009), 737.

damages that have partly already occurred, until the transition to renewable energies is complete. At the same time, the world would realise that the fossil trail of the West cannot be repeated. However, a more effective climate protection law requires an interaction of political and legal standards. On the part of the citizens these factors require a process of learning and ability to learn. Whether this can be started in time, likely remains an open question.

Chapter 22
Climate Law in the United Kingdom

Colin T. Reid

Abstract There is no single legislative source for the United Kingdom's legal response to climate change. Initial measures to tax large energy users, enable participation in the European Union Emissions Trading Scheme and encourage renewable electricity generation have subsequently been joined by the Climate Change Acts operating at UK and Scottish levels. These Acts set demanding targets for reductions in greenhouse gas emissions and break new and uncertain legal ground in making these legally binding on Ministers. The targets are supported by detailed reporting mechanisms, to Parliament and the public, that are intended to be the main route to enforcement and by provisions enabling many detailed changes to the law to allow progress towards the targets. The separate legislation in Scotland highlights the difficulties that dealing with pervasive issues, especially those with an European Union and international dimension, pose for by sub-national governments with distinct political ambitions but limited jurisdiction.

22.1 Introduction

The Climate Change Acts[1] passed by the United Kingdom and Scottish Parliaments in 2008 and 2009 are the most obvious feature of the response to climate change in the United Kingdom, but this legislation came several years after the first substantial legal measures endeavouring to achieve a reduction in greenhouse gas emissions. The Acts are innovative in establishing legally bind-

[1] Climate Change Act 2008; Climate Change (Scotland) Act 2009.

Colin T. Reid is Professor of Environmental Law at the University of Dundee.

C.T. Reid (✉)
Dundee Law School, University of Dundee,
Nethergate, Dundee, DD1 4HN, Scotland, UK
e-mail: c.t.reid@dundee.ac.uk

E.J. Hollo et al. (eds.), *Climate Change and the Law*, Ius Gentium: Comparative Perspectives on Law and Justice 21, DOI 10.1007/978-94-007-5440-9_22,
© Springer Science+Business Media Dordrecht 2013

ing targets for emissions reductions, but the full legal impact of these provisions is uncertain. More important in practical terms are the substantial framework that the Acts create for ensuring that progress towards the targets is reported, to Parliament and the public, and the policy measures that contribute to their achievement. The role of the devolved administrations in Scotland, Wales and Northern Ireland, each with distinct legal powers and policy goals, adds further complexity to the picture.

In 1989, Mrs. Thatcher's speech at the United Nations[2] placed the United Kingdom at the forefront of raising the issue of climate change. The following years produced significant policy documents,[3] and significant reductions in emissions, although these were the result much less of environmental concerns than of the economic tides that led to the decline of energy-intensive heavy industry and the 'dash for gas' as the newly privatised electricity industry built new gas-powered generating stations, able to out-compete the ageing coal-fired plants.[4] Electricity generation was the focus of the strongest initial measures, aimed at encouraging the use of renewable sources, whilst the United Kingdom was a pioneer in emissions taxation and trading, introducing a domestic scheme in advance of the European Union (EU) Emissions Trading Scheme (ETS).

The growing international pressure on climate change during the first decade of this century, supported by the Stern Review's[5] conclusions on the economic case for action, led to the proposal for a Climate Change Bill which was thoroughly scrutinised in the UK Parliament[6] before its enactment in 2008. A separate Climate Change (Scotland) Act followed in 2009, called for both by the division of legal responsibilities following the devolution settlement at the very end of last century and by the political importance of the issue for Scotland. The comparatively high profile of climate issues in Scotland is driven largely by the existing wealth of fossil fuel resources, the exploitation of which is a major industrial activity, and the outstanding potential for renewable energy generation: wind, wave and tidal, in addition to some hydro-electric potential beyond that already utilised.[7] Such factors

[2] Margaret Thatcher, Prime Minister, speech at the United Nations General Assembly on 8 November 1989, available at: http://www.margaretthatcher.org/document/107817 (last accessed on 1 May 2012).

[3] Climate Change – The UK Programme: United Kingdom's First Report under the Framework Convention on Climate Change (Cm 2427, 1994).

[4] Emissions figures going back to 1990 are available through the UK Emissions Statistics, Department of Energy and Climate Change, "UK Emissions Statistics", 29 March 2012, available at: http://www.decc.gov.uk/en/content/cms/statistics/climate_stats/gg_emissions/uk_emissions/uk_emissions.aspx (last accessed on 1 May 2012).

[5] Nicholas Stern, "The Economics of Climate Change: The Stern Review", 2007, available at: http://www.hm-treasury.gov.uk/stern_review_report.htm (last accessed on 8 May 2012).

[6] See, for example the First Report of the Joint Committee on the Draft Climate Change Bill (2006–2007 HL 170-I/HC 542-I), which sets out the policy background in paras. 7–14.

[7] Garrad Hassan et al., "Scotland's Renewable Resource 2001", 28 June 2005, available at: http://www.scotland.gov.uk/Publications/2003/09/18270/27258 (last accessed on 8 May 2012).

have made responding to climate change, and the economic opportunities this presents, an important area of policy and achieving a 'greener Scotland'[8] was one of the strategic objectives set out by the Scottish National Party government that took office after the Scottish election in 2007.[9] In Northern Ireland there has been much less support for policy initiatives in this area, whilst in Wales the focus has been on a more all-encompassing approach to sustainability.

22.2 Tax, Trading and Turbines

Substantial legal measures were introduced in relation to energy generation and use prior to the more general climate change legislation.[10] These fall into three categories: taxation, emissions trading and measures to promote the use of renewable sources for electricity generation. Nevertheless, these are just the main elements among a large number of ever-changing other initiatives, including incentives for improved insulation of homes and the fitting of carbon-efficient heating systems,[11] higher building standards[12] and support for private and public sector bodies to review and reduce their emissions.[13] The changing regulation of the energy industry has also had a significant impact on how such issues are addressed.[14] The overwhelming impression is of the sheer volume and complexity of the financial measures seeking to encourage reduced emissions and to ensure that the various schemes and the EU ETS work together.

---

[8] Defined as "improv[ing] Scotland's natural and built environment and the sustainable use and enjoyment of it"; *Scottish Government, Principles and Priorities: The Government's Programme for Scotland* (2007).

[9] Initially as a minority government but winning an overall majority in the 2011 election, an outcome that was a surprise in view of a system of proportional representation that was thought unlikely ever to return a single party majority.

[10] For a useful chronology and summary, see Alex Bowen and James Rydge, "Climate-Change Policy in the United Kingdom", OECD Economics Department Working Papers, No. 886, 2011, available at: http://dx.doi.org/10.1787/5kg6qdx6b5q6-en (last accessed on 8 May 2012), at 16–18.

[11] This area has been marked by a continuous revision of often fairly short-lived schemes. A current overview is available at: http://www.direct.gov.uk/en/Environmentandgreenerliving/Energyandwatersaving/index.htm (last accessed on 8 May 2012).

[12] See, for example, Scottish Government, "Progress Report on the Low Carbon Building Standards Strategy for Scotland", 2011, available at: http://www.scotland.gov.uk/Resource/Doc/217736/0113638.pdf (last accessed on 8 May 2012).

[13] For example through the Carbon Trust; see http://www.carbontrust.co.uk (last accessed on 8 May 2012).

[14] See Department of Energy and Climate Change, "Ofgem Final Report", 2011, available at: http://www.decc.gov.uk/assets/decc/11/meeting-energy-demand/energy-markets/2151-ofgem-review-final-report.pdf (last accessed on 8 May 2012), part 1.

The most significant tax measure[15] is the Climate Change Levy, introduced in 2001.[16] This is a tax on energy use in industry, commerce and the public sector, charged on the supplier.[17] The detailed exemptions and reliefs include exemptions for fuel supplied to approved combined heat and power plants and more significantly for those who enter a Climate Change Agreement.[18] These are agreements reached with operators in certain energy-intensive industries, either on a sectoral or individual basis, whereby a discount on the climate change levy[19] is granted in exchange for agreeing to and meeting energy efficiency or carbon saving targets. A further tax measure is the CRC Energy Efficiency Scheme, initially introduced as predominantly a trading scheme for operators at a level below that captured by the EU ETS, but now with revenues being directed to the Treasury rather than being returned to participants on the basis of their performance.[20] A feature of the scheme is a published league table of performance by participants.[21]

In relation to trading, the UK introduced a voluntary scheme in advance of the EU ETS coming into effect, the first national scheme to apply across all sectors of industry. This scheme began with the first auction of allowances in 2002 and although closed to new participants in 2006 some trading can still continue. Giving industry and government experience of a trading approach was as much part of the purpose of the scheme as achieving direct emissions reductions and some success was achieved on both counts, although it was considered that some of the emission targets were undemanding.[22] Trading is an element of the CRC Energy Efficiency Scheme, but for large users of energy it is now the EU ETS that provides the regulatory framework. The potentially distinctive UK contribution is the proposal for a

---

[15] The duty on petrol is a further relevant tax, but plans to increase this at a rate higher than inflation have proved very vulnerable to public opposition at times of high prices and economic gloom (*Budget 2011*, (HM Treasury, 2011; 2010–2011 HC 836), para. 2.131). Even the already postponed increase in line with inflation was deferred from January to August 2012 (*Autumn Statement 2011* (HM Treasury, 2011; Cm 8231), para. 1.132).

[16] Finance Act 2000, s.30 and Scheds 6 & 7.

[17] For an overview see HM Revenue and Customs, "A general guide to Climate Change Levy", November 2011, Notice CCL1, available at: http://customs.hmrc.gov.uk/channelsPortalWebApp/channelsPortalWebApp.portal?_nfpb=true&_pageLabel=pageExcise_ShowContent&id=HMCE_CL_000290&propertyType=document#downloadopt (last accessed on 8 May 2012).

[18] An overview is available at: http://www.decc.gov.uk/en/content/cms/emissions/ccas/ccas.aspx (last accessed on 8 May 2012).

[19] Over time this has varied between 65 and 80%.

[20] This scheme has been undergoing considerable change since its inception as the Carbon Reduction Commitment and its precise future shape remains unclear; see: http://www.decc.gov.uk/en/content/cms/emissions/crc_efficiency/crc_efficiency.aspx (last accessed on 8 May 2012).

[21] The 2010/2011 CRC Performance League Table is available at: http://crc.environment-agency.gov.uk/pplt/web/plt/public/2010-11/CRCPerformanceLeagueTable20102011 (last accessed on 8 May 2012).

[22] National Audit Office, *The UK Emissions Trading Scheme: A New Way to Combat Climate Change* (2003–2004 HC 517).

carbon price floor starting in 2013, imposing additional charges if the price of EU allowances is below the set level (as at the time of writing in January 2012). This has proved a controversial proposal.[23]

The third major area of activity has been in relation to the promotion of renewable sources of energy, particularly for electricity generation. Central to this has been the Renewables Obligation, a requirement on electricity generators to produce a certain proportion of their supply from renewable sources, a proportion that increased over the years.[24] The operation of the scheme, which operates separately in Scotland and Northern Ireland,[25] has more recently been used to give specific encouragement to particular generating methods by varying the amount of credit given to electricity from different sources, especially since the early response has been dominated by on-shore wind turbines, as the most mature, readily accessible and cost-effective technology. The scheme again involves an element of trading for generators who cannot provide from their own activity a sufficient number of certificates.[26] As the background to the recent litigation over the withdrawal of some support for small-scale solar generation has shown,[27] in a field that is developing rapidly in technology and in the commercial and customer response to opportunities it presents, it has proved difficult to find the level and duration of support that provides effective incentives for low-carbon generation without incurring disproportionate costs for the state or distorting the market too much in undesired ways.

22.3 Climate Change Acts

The host of detailed measures noted above (and others) are not directly affected by the Climate Change Acts passed by the UK and Scottish Parliaments which focus on the issue of national targets. The two Acts, the Climate Change Act 2008 ("2008 Act") and the Climate Change (Scotland) Act 2009 ("Scottish Act") share many features but also have some significant differences in their mechanisms and in the targets set. At the core of both Acts is the setting of targets for reductions in greenhouse gas emissions,[28] targets which are expressed in the form of a legal duty on

---

[23] Energy and Climate Change Select Committee, *The EU Emissions Trading System* (10th Report of 2010–2012 HC 1476).

[24] Rising from 3% in 2002 to over 15% for 2015–2016 (the Northern Ireland figure is lower at just over 6%).

[25] Renewables Obligation Order 2009, SI 2009/785; Renewables Obligation (Scotland) Order 2009, SSI 2009/140; Renewables Obligation Order (Northern Ireland) 2009, SR 2009/154.

[26] Ofgem, Renewables Obligation: Guidance for licensed electricity suppliers (GB and NI) (63/11, 2011).

[27] *Secretary of State for Energy and Climate Change v. Friends of the Earth* [2012] EWCA Civ 28.

[28] The Acts cover six gases: carbon dioxide, methane and nitrous oxide with baselines in 1990 and hydrofluorocarbons, perfluorocarbons and sulphur hexafluoride with baselines in 1995; 2008 Act, ss.24–25; Scottish Act, ss.10–11.

Ministers to achieve the specified reductions. These targets are set at an 80% reduction from the 1990 baseline by 2050,[29] with interim targets for 2020 of 34% for the UK and 42% for Scotland.[30] The more demanding target for Scotland must be reviewed and made no lower if the EU increases its commitment by adopting measures to reduce emissions by 2020 by at least 30%.[31] The targets are based on the net level of emissions produced, taking account of the removal of emissions from the atmosphere, by land-use, forestry, etc., as well as any international credits permitted.[32]

There are no specific sanctions provided for a failure to meet these targets and their precise legal status is unclear. The imposition of explicit duties on Ministers to achieve a specific target is an innovation in British law and there are arguments that these provisions should be interpreted in line with previous target-setting legislation which merely requires Ministers to make reasonable endeavours to see that they are achieved. On the other hand the distinctive and absolute way in which the duties are expressed[33] can be viewed as demanding that the duties are given full legal force, so that any failure to achieve them does amount to a breach of duty, although who might be able to enforce this, and how, remains open to argument.[34] In any event, the embedding of such long-term goals in legislation, determining policy priorities for decades to come, is remarkable and has led to suggestions that the climate change legislation should be seen as having "constitutional significance".[35]

The key targets for 2020 and 2050 are supported by shorter-term targets, set well in advance, but there is no legal obligation on Ministers to ensure that these are met. In the 2008 Act these operate on the basis of 5-year budgets,[36] whereas for Scotland the targets are set on an annual basis.[37] Both Acts specify the schedule for setting the targets and parameters within which they must be determined, aiming to ensure that

---

[29] 2008 Act, s.1; Scottish Act, s.1; for some gases the baseline is 1995 (see previous note).

[30] 2008 Act, s.5(1)(a), as amended by Climate Change Act 2008 (2020 Target, Credit Limit and Definitions) Order 2009, SI 2009/1258, art.2; Scottish Act, s.2.

[31] Scottish Act, s.2(9)–(14).

[32] 2008 Act, ss.11, 26–31; Scottish Act, ss.13–23. In both cases there is provision to limit the amount of credits under international trading schemes that can be used in calculating the net emissions.

[33] "It is the duty of the Secretary of State to ensure…" (2008 Act, s.1(1)); "The Scottish Ministers must ensure…" (Scottish Act s.1(1)).

[34] Colin T. Reid, "A New Sort of Duty? The Significance of 'Outcome' Duties in the Climate Change and Child Poverty Acts", *Public Law* (2012), 748.

[35] Lord Rooker in the House of Lords during the passage of the 2008 Act; HL Deb vol.696 col.1209 (27 November 2007); Aileen McHarg, "Climate Change Constitutionalism? Lessons from the United Kingdom", 2 *Climate Law* (2011), 469.

[36] 2008 Act, s.4; Carbon Budgets Order 2009, SI 2009/1259; Carbon Budget Order 2011, SI 2011/1603. Limited amounts can be carried forward and back between budgets; 2009 Act, s.17.

[37] Scottish Act, s.3; Climate Change (Annual Targets) (Scotland) Order 2010, SSI 2010/359; Climate Change (Annual Targets) (Scotland) Order 2011, SSI 2011/353.

the legal targets for 2020 and 2050 are achieved.[38] The setting of the initial series of targets for Scotland was disputed, with the relevant parliamentary committee and the Parliament itself rejecting the initial proposals as being insufficiently demanding during the first few years and considerable adjustments being required before the targets were finally approved.[39]

The main mechanism for securing that all the targets are met, and the only one for the targets other than the legally binding ones set for 2020 and 2050, is a detailed scheme of reporting obligations. Under the UK Act, the Secretary of State must prepare proposals and policies to meet the carbon budgets,[40] and report to Parliament each year on the level of emissions.[41] At the end of each budgetary period, a report must be prepared on the total net emissions, providing an explanation of the reasons if the budget has not been met and proposals to compensate in future budget periods for any excess emissions.[42] The Scottish Ministers must report to Parliament their proposals and policies for meeting the annual targets[43] and each year must report on the level of emissions. If a target has not been met, then a report setting out proposals and policies to compensate in future years for the excess emissions must be made. Further reports are required in relation to the 2020 and 2050 targets.[44] Additional reports in Scotland are required on how the exercise of ministerial functions relating to electricity generation has affected net emissions[45] and on the impact of budget proposals on emissions.[46] This latter analysis is recognised as being in part experimental and endeavours to take account of direct, indirect and induced emissions.[47]

These reporting obligations are supported by a major role for the independent Committee on Climate Change.[48] This body serves a range of advisory functions, most notably being required to advise on the setting of targets and to make its own

[38] 2008 Act, ss.4–10; Scottish Act, ss.3–7.

[39] Climate Change (Annual Targets) (Scotland) Order 2010, SSI 2010/359. See the Official Report of the Transport, Infrastructure and Climate Change Committee for 18 May and 5 October 2010 and of the Parliament for 27 May and 7 October 2010.

[40] 2008 Act, ss.13–14.

[41] Ibid., s.16.

[42] Ibid., ss.16, 18–20. A final statement is required after 2050.

[43] The first report was published in March 2011: Low Carbon Scotland: Meeting the Emissions Reduction Targets 2010–2022: The Report on Proposals and Policies (Scottish Government, 2011).

[44] Scottish Act, ss.33–43.

[45] Ibid., s.38.

[46] Ibid., s.94.

[47] I.e. those generated by direct recipients of government funding (21% of the total in 2009), those arising from supplying goods and services to such recipients (51%) and those resulting from the spending of employees engaged in the previous activities (28%); *Carbon assessment of the 2010–2011 Draft Budget* (Scottish Government, 2009).

[48] 2008 Act, s.32 and Sched.1. The Scottish Act makes provision for a separate Scottish Committee, but at present the one body operates under both Acts; Scottish Act, ss.24–25.

annual reports on progress towards the targets,[49] and what further efforts might be necessary to ensure that they are met.[50] The Ministers must respond to these reports and provide reasons for departing from the advice given on some issues.[51] The various reports produced by the Committee, not just the national progress reports but further papers on adaptation and particular sectors,[52] provide a wealth of well-researched and closely argued material on the issue of climate change, providing thorough scrutiny of the governments' progress and keeping the issue in the public eye and on the political agenda, so that it is impossible for the obligations set under the Acts to fall out of sight, even as other political concerns and goals emerge as competing priorities.

The main structures created by the Climate Change Acts are supported by a number of other measures. The UK Act requires reports on the impact of and adaptation to climate change, again with advice from the Committee on Climate Change, and confers powers to require such reports from a wide range of public authorities.[53] Further reports are required on improving the efficiency and sustainability of government buildings[54] and Ministers are bound to ensure that information on greenhouse gas emissions are included in annual company reports, or to explain why they have decided not to.[55] The Act also provides powers to introduce trading schemes relating to greenhouse gas emissions[56] and charging schemes for single use carrier bags.[57]

The Scottish Act contains a wider range of provisions to assist in turning the Act's targets from goals into achievements. One of the most significant of these is the duty imposed on all public authorities to act in the way 'best calculated' to contribute to the delivery of the climate change targets and to deliver the adaptation strategy, as well as to act in the way they consider most sustainable. These duties are supported by the issue of guidance, the imposition of reporting requirements and the establishment of

[49] The latest Reports are not wholly encouraging, concluding for the UK that "A step change in the pace of emissions reduction is still required." Committee on Climate Change, *Meeting Carbon Budgets – 3rd Progress Report to Parliament* (2011), at 39. For Scotland, the 2020 target is likely to be missed without a tightening of the EU ETS cap or further measures beyond existing proposals; Committee on Climate Change, *Reducing Emissions in Scotland: 1st Progress Report* (2012), at 35.

[50] 2008 Act, ss.33–36; Scottish Act, ss.24–32

[51] 2008 Act, s.37; Scottish Act, s.29.

[52] Reports by the Committee on Climate Change are available at: http://www.theccc.org.uk/reports (last accessed on 8 May 2012).

[53] 2008 Act ss.56–70.

[54] Ibid., s.86.

[55] Ibid., s.85

[56] 2008 Act, ss.44–55; these provisions provide the authority for the CRC Energy Efficiency Scheme (see above; CRC Energy Efficiency Scheme Order 2010, SI 2010/768).

[57] 2008 Act, s.77; a scheme has been introduced in Wales (Single Use Carrier Bags Charge (Wales) Regulations 2010, SI 2010/2880) and separate legislation for this exists in Northern Ireland (Single Use Carrier Bags Act (Northern Ireland) 2011).

monitoring mechanisms with powers of investigation.[58] A high level of energy performance is also demanded of government buildings.[59] Ministers must publish a plan for the promotion of energy efficiency, including the use of renewable sources of energy, and are also required to promote the use of renewable heat and to require the assessment of energy performance of buildings.[60] Other requirements include the preparation of adaptation programmes in response to the reports on the impact and risks of climate change prepared by the UK government.[61] A public engagement strategy is also required,[62] reflecting the importance of engaging wider society in the changes needed to meet the ambitious targets that have been set.

Net emissions can be considerably affected by land use and the Scottish Act requires Ministers to produce a land use strategy,[63] setting out objectives for sustainable land use that contribute to achieving the emission reduction targets set in the Act, the objectives set out in the adaptation programme and sustainable development. The legal duties of the Forestry Commission (the body responsible for regulating forestry and directly in charge of large areas of woodland) may be altered in order to assist in the reduction of net emissions or for other purposes in relation to climate change[64] and there is also power to alter the dates when muirburn (the burning of heather and grass on moorland to rejuvenate the vegetation) can lawfully take place to ensure that this land management tool can be used in a way that is beneficial in terms of net emissions.[65] Further measures adjust the rules of land law to enable obligations aimed at reducing greenhouse gas emissions to become "real burdens", adding them to the limited class of obligations that automatically bind successive owners of the land.[66] The installation of insulation is included within the range of maintenance activities which can go ahead in tenement buildings even without all owners' consent,[67] and planning rules are relaxed to permit micro-generation equipment to be installed without the need for express planning permission.[68] Discounts will also be available on local taxes where premises have undergone energy efficiency improvements.[69]

---

[58] Scottish Act, ss.44–52; *Public Bodies Climate Change Duties: Putting Them Into Practice – Guidance Required by Part 4 of the Climate Change (Scotland) Act 2009* (Scottish Government, 2011).

[59] Scottish Act, s.75.

[60] Ibid., ss.60–64.

[61] Ibid., ss.53–56; *Scotland's Climate Change Adaptation Framework* (Scottish Government, 2009). The UK reports are required under the Climate Change Act 2008, s.56.

[62] Scottish Act 2009, s.91; *Low Carbon Scotland: Public Engagement Strategy* (Scottish Government, 2010).

[63] Scottish Act, s.57; *Getting the best from our land – A land use strategy for Scotland* (Scottish Government, 2011).

[64] Scottish Act, s.59.

[65] Ibid., s.58.

[66] Ibid., s.68, adding s.46A to the Title Conditions (Scotland) Act 2003.

[67] Scottish Act, s.69, amending the Tenements (Scotland) Act 2004.

[68] Ibid., ss.70–71, requiring the making of regulations under the Town and Country Planning (Scotland) Act 1997, ss.30–31.

[69] Scottish Act, ss.65–67.

Waste is a further issue addressed in the Act, encouraging the reduction of waste and the promotion of reuse and recycling which in turn should reduce emissions.[70] These measures all depend on the introduction of further regulations which may never be made if the voluntary measures which are already evident in some of these areas show adequate progress. The matters covered include requiring specified people to prepare and comply with plans for the prevention, reduction, management, recycling, use and disposal of waste, for example, construction site waste management plans, and to report on the waste produced.[71] Further regulations may require the provision of recycling facilities at certain sites, such as supermarkets and events, for instance, sporting or cultural events, and the use of certain percentages of recycled materials, as well as setting targets for the reduction of packaging, establishing deposit and return schemes and charging for the supply of carrier bags.[72]

The listing of these detailed requirements serves to emphasise the fact that the legislative response to climate change is far-reaching and complex. The national targets and reporting requirements under the Climate Change Acts may attract the most attention, but they are only part of the overall picture. Much of the hard work of effecting emission reductions is done through the energy-related measures noted earlier, most of which pre-date the Climate Change Acts, rather than under the Acts themselves. Moreover, a host of other provisions, including minor adjustments to the legal regimes from land law to waste is also needed if a comprehensive policy is to be delivered. It remains to be seen how effective the statutory targets and reporting requirements are in achieving the significant reductions in emissions that are required, and in forcing greater efforts if[73] it appears that the targets might be missed, whilst there will doubtless be many other detailed measures needed to ensure that other legal frameworks contribute to, or at least do not obstruct, the steps required to meet the targets.

22.4 Impact of Devolution

For the UK, the complexities of tackling such a pervasive issue as climate change have been exacerbated by the impact of devolution. The Scotland Act 1998 created a Scottish Parliament and Government[74] that have control over all matters that have

---

[70] A comprehensive approach to waste has been put forward in *Scotland's Zero Waste Plan* (Scottish Government, 2010).

[71] Waste Information (Scotland) Regulations 2010, SSI 2010 No.435.

[72] Scottish Act, ss.78–90.

[73] Perhaps not "if" but "when"; see supra, note 49.

[74] The Scotland Act 1998 refers to the "Scottish Executive" but when the Scottish National Party won power in 2007, it started to refer to the "Scottish Government" and this usage is now virtually universal and is granted legal recognition in the Scotland Act 2012, s.12.

not been reserved to the UK.[75] There are devolved assemblies and governments in Wales and Northern Ireland, but the details of their structures and the extent of powers devolved are different in each case,[76] and in all cases the original settlement has been or is being significantly changed.[77] No new arrangements have been made for England, so that the UK Parliament and Government on some matters act for the UK as a whole, on others for England only[78] and where the extent of devolved powers do not coincide, for England and varying combinations of the other constituent nations. Ultimately, though, the UK Parliament retains its authority to legislate on any matter, even those within devolved competence, as well as to amend, or even repeal, the legislation creating the devolved structures.

The division of powers leaves the devolved administrations in a position where they have significant constraints on their ability to respond to climate change. In the first place, it is the UK authorities alone which operate on the international field, including any matters relating to the EU. Therefore the devolved administrations have no direct role in negotiating the successor to or prolongation of the Kyoto Protocol, nor in any matter of EU policy. Yet such issues will have a massive impact on what is happening, particularly for Scotland where if the major sources of greenhouse gases under the EU ETS achieve only a limited reduction in emissions because the EU target remains modest, the burden of achieving the statutory target of 42% reductions by 2020 will fall even more heavily on those areas of activity outwith that trading scheme.[79]

A second constraint is imposed by the extent of the powers reserved to the UK authorities. Many of the potential levers for implementing climate policy are in the hands of the UK, not the devolved, authorities. The tax and trading measures described above are matters for the UK government alone. Even for Scotland, which has the most devolved power, energy is a reserved matter – although significant

[75] See generally, Alan Page, Colin Reid and Andrea Ross, *A Guide to the Scotland Act 1998* (Edinburgh: Butterworths, 1999); Chris M.G. Himsworth and Colin R. Munro, *The Scotland Act 1998*, 2nd edition (Edinburgh: W. Green, 2000); Jean McFadden and Mark Lazarowicz, *The Scottish Parliament: An Introduction*, 4th edition (London: Bloomsbury Professional, 2010).

[76] Government of Wales Act 1998, as significantly amended by Government of Wales Act 2006; Northern Ireland Acts 1998, 2000, 2006 and 2009, Northern Ireland (St. Andrews Agreement Act) 2006.

[77] See previous note and the Scotland Act 2012, to say nothing of the contested plans for a referendum on Scottish independence; *Your Scotland – Your Referendum – A Consultation Document* (Scottish Government, 2012).

[78] The anomaly that in the UK Parliament Members of Parliament from outside England can still vote on matters affecting England alone, even when for their own constituencies the matter is the responsibility of the devolved authorities, generates considerable debate, and is known as the "West Lothian question" after the constituency of Tam Dalyell MP who raised it tenaciously during the debates on the earlier attempts at devolution legislation in the late 1970s.

[79] Committee on Climate Change, *Reducing Emissions in Scotland: 1st Progress Report* (2012), at 35.

powers in this area have been transferred to the Scottish Ministers through "executive devolution,"[80] – limiting the scope for action on this central issue. Other less obvious constraints stand in the way of possible measures to reduce emissions. Transport is a large contributor to greenhouse gas emissions and reducing speed limits is one measure that could be introduced to reduce these, but this is also a reserved matter outwith Scotland's control,[81] although the Scotland Act 2012 does transfer to Scotland control over some speed limits.[82] Further reservations in the devolution legislation may constrain other initiatives that seek to take effect by influencing consumer and commercial behaviour, e.g. the reservation of 'consumer affairs', including advertising, and 'competition'.[83]

Nevertheless, as shown by the account of the Climate Change (Scotland) Act 2009 given above, there is plenty of scope within devolved competence for legislative initiative and the development of distinct policies. Indeed, there have been marked differences between the devolved administrations in their response to climate change, driven by their different factual and economic situations and political aims. In Northern Ireland, which is disproportionately reliant on comparatively high-carbon electricity generation, the authorities have been opposed to taking strong climate change measures,[84] and the current target is for an emissions reduction of only 25% from 1990 levels by 2025.[85] By contrast, Scotland has set very ambitious targets – requiring a 42% reduction by 2020 as opposed to the 34% reduction for the UK as a whole[86] – and sees the combination of the existing industrial expertise in the energy industry and Scotland's exceptional potential for renewable energy gen-

[80] Under the Scotland Act 1998, s.63 the UK Government can authorise Scottish Ministers to act on their behalf in exercising powers within reserved areas. For example, the Scottish Ministers decide on approval for new electricity generating stations and transmission lines and on the renewables obligation. Nevertheless an amendment to the devolution legislation was needed even to give the Scottish Environment Protection Agency the competence to impose conditions in relation to energy efficiency which is a required element in authorisations under the Pollution Prevention and Control regime introduced by EU legislation; Scotland Act 1998 (Transfer of Functions to the Scottish Ministers, etc.) Order 2008, SI 2008/1776; Pollution Prevention and Control (Scotland) Amendment Regulations 2009, SSI 2009/336.

[81] Scotland Act 1998, Sched.5 Part II Section E1.

[82] Scotland Act 2012, ss.21–22.

[83] Scotland Act 1998, Sched.5 Part II Sections C3 and C7.

[84] One Minister blocked an advertising campaign encouraging energy conservation; *BBC News*, "Quit Call Over Blocked Green Act", 9 February 2009, available at: http://news.bbc.co.uk/1/hi/northern_ireland/7878399.stm (last accessed on 8 May 2012).

[85] Cross-Departmental Working Group on Greenhouse Gas Emissions, *Northern Ireland Greenhouse Gas Emissions Reduction Action Plan* (2011), available at: http://www.doeni.gov.uk/northern_ireland_action_plan_on_greenhouse_gas_emissions_reductions.pdf (last accessed on 8 May 2012). The Committee on Climate Change has reported on the potential for greater reductions and for specific climate change legislation; *The Appropriateness of a Northern Ireland Climate Change Act* (2011).

[86] See supra, note 30.

eration as offering both environmental and economic gains on a substantial scale.[87] In Wales, which until recently has had more limited legislative competence, again there is a strong commitment to emissions reduction,[88] but the focus has been on a more holistic approach to sustainability rather than concentrating specifically on climate change.[89]

22.5 Conclusion

The introduction of legally binding targets for reductions in greenhouse gas emissions is the highlight of the UK's legal response to climate change. Yet their precise status is unclear and the emphasis will be on reporting requirements and political and public pressure rather than the law in order to ensure that the targets are met. At the same time, much of the work of achieving these targets is achieved through energy legislation which does not directly derive from the Climate Change Acts, whilst a host of other lesser measures contribute to the overall goal. That fragmentation is exacerbated by the effects of devolution, which both enables the development and implementation of distinct policies, but can obstruct the adoption of a holistic approach. As the decade continues, the emergence from economic recession and the approach of the 2020 target date will put increasing pressure on meeting the targets, and it is then that we shall see how powerful and effective the legislation really is.

[87] See, for example, the *10 Energy Pledges* announced in 2009 as part of the *Greener Deal for Scotland*, available at: http://www.scotland.gov.uk/Topics/Business-Industry/Energy/Action/economic-recovery (last accessed on 8 May 2012).

[88] *Energy Wales: A Low Carbon Transition* (2012), available at http://wales.gov.uk/docs/desh/publications/120314energywalesen.pdf (last accessed on 24 May 2012). The aim is for a 40% reduction from 1990 levels by 2020 from areas within devolved responsibility (by contrast, the Scottish target is for all emissions in Scotland). The Welsh Government, "Our Targets", 17 January 2012, available at: http://wales.gov.uk/topics/environmentcountryside/climatechange/emissions/targets/?lang=en (last accessed on 8 May 2012).

[89] *One Wales: One Planet – A Welsh Government Discussion Paper – Sustainable Development Bill* (2011), also available at: http://wales.gov.uk/docs/desh/publications/111201susdevdiscussionen.pdf (last accessed on 8 May 2012).

Chapter 23
Climate Law and Policy in Russia: A Peasant Needs Thunder to Cross Himself and Wonder

Yulia Yamineva*

Abstract Being one of the largest emitters of greenhouse gases and having an enormous carbon storage capacity in its forests, Russia plays a significant role in addressing global climate change. Yet, for a long time, its domestic climate policy remained under-developed and lagged behind other countries. The presidential term of Dmitry Medvedev and his modernisation agenda brought about the necessary transformation. The Climate Doctrine adopted in 2009 acknowledges the anthropogenic nature of climate change, setting principles and goals for mitigation and adaptation policies. The adoption of the Doctrine coincided with the development of a comprehensive framework for energy efficiency and energy conservation which, if fully implemented, will lead to significant reductions in greenhouse gas emissions. Adaptation policies should also be urgently formulated, as according to the Russian Federal Service for Hydrometeorology and Environmental Monitoring, climate change, alongside some benefits, will bring more droughts, floods and other extreme events as well as negative consequences for infrastructure, agriculture and other sectors of the economy.

A Russian proverb: Пока гром не грянет, мужик не перекрестится.

*Yulia Yamineva holds a Ph.D. from the University of Cambridge. She is currently a postdoctoral researcher at the Department of Law, University of Eastern Finland. She followed the UNFCCC negotiations with the International Institute for Sustainable Development – Reporting Services in 2007–2009 and was a consultant based at the UNFCCC Secretariat in 2010–2011.23.1

Y. Yamineva (✉)
Department of Law, University of Eastern Finland,
PO Box 111, 8010 Joensuu, Finland
e-mail: yulia.yamineva@uef.fi

E.J. Hollo et al. (eds.), *Climate Change and the Law*, Ius Gentium: Comparative Perspectives on Law and Justice 21, DOI 10.1007/978-94-007-5440-9_23,
© Springer Science+Business Media Dordrecht 2013

Introduction

Frankly, after the colossal drought and wildfires in Russia last summer even some of the sceptics in our country have realised that climate processes are extremely complex. We must think about what should be done next.

Dmitry Medvedev, President of the Russian Federation, speech following the EU-Russia summit in Brussels, 7 December 2010

Russia is an important player in international climate policy. Despite the decline of its economy and industrial activities since the collapse of the Soviet Union, Russia remains one of the top world emitters of carbon dioxide (CO_2). The country's latest strategy for economic development outlines ambitious plans for becoming one of the world's top five economies in terms of Gross Domestic Product (GDP) by 2020.[1] This suggests further economic growth and increase in greenhouse gas emissions. Russia also has vast forest reserves, serving as an important carbon sink for the planet. Furthermore, impacts of climate change are apparent in Russia and should be relevant in the context of Russian strategic interests in the development of the Arctic, on the one hand, and potential release of methane, a powerful greenhouse gas, from thawing Siberian permafrost, on the other.

Despite its importance for the global battle against climate change, Russian domestic climate change policy has been lagging behind many other countries. Although Russia ratified the Kyoto Protocol in 2004, it delayed the development of a legal and institutional framework for carbon market mechanisms and approved the first Joint Implementation (JI) projects only in 2010. For a long time, government-level attitude towards anthropogenic climate change and related policies was marked by distrust, scepticism or mere ignorance. In recent years, the situation has started to change as evidenced by the adoption of the Climate Doctrine in 2009[2] and formulation of energy efficiency and renewable energy policies.

This chapter provides an overview of the evolution of climate change law and policy in Russia, starting with the country's role under the United Nations Framework Convention on Climate Change (UNFCCC)[3] and the Kyoto Protocol.[4] It then describes the main characteristics of Russian domestic climate policy up to 2009. The chapter elaborates on the Climate Doctrine adopted in 2009, arguing that the document marks a turning point for the evolution of a climate policy framework in Russia. It then proceeds to outline mitigation measures, in particular in relation to energy efficiency and renewables, and the Kyoto Protocol's flexibility mechanisms.

---

[1] Directive No. 1662-p of the Government of the Russian Federation. Concept for Long-Term Social and Economic Development of the Russian Federation up to 2020, 17 November 2008.

[2] Directive No. 861-рп of the President of the Russian Federation. Climate Doctrine of the Russian Federation, 17 December 2009.

[3] United Nations Framework Convention on Climate Change, New York, 9 May 1992, in force 21 March 1994, 31 *International Legal Materials* (1992), 849.

[4] Kyoto Protocol to the United Nations Framework Convention on Climate Change, Kyoto, 10 December 1997, in force 16 February 2005, 37 *International Legal Materials* (1998), 22.

The chapter concludes with a summary of the current status of Russia's policies to adapt to the impacts of climate change.

23.2 Russia and the UN Climate Change Regime

The Russian Federation ratified the UNFCCC in 1994. It is listed in Annex I of the Convention, meaning that it has committed to taking action to limit its greenhouse gas emissions and report relevant policies to the Conference of the Parties (COP), among other responsibilities. However, like many other Central and Eastern European countries, Russia has subsequently been given the status of a country undergoing the process of transition to a market economy – "economy in transition." This means that it is allowed "a certain degree of flexibility" in the implementation of its commitments.[5] Russia is also exempt from financial commitments to support developing country Parties in mitigation of, and adaptation to, climate change.

During the negotiations for the Kyoto Protocol, Russia played an active role, at least when it came to the textual negotiations, putting forward a number of formal proposals and in-room suggestions on various issues.[6] Korppoo, Karas and Grubb have characterised Russia's position as defensive, noting that Russia advocated for lower level of ambition for economies in transition and opposed also the consideration of forests as carbon sinks.[7] Russia's ratification of the Protocol on 18 November 2004 was a significant milestone, effectively allowing the Protocol to enter into force after the US decided in 2001 not to ratify the treaty. However, concerns over climate change apparently played only a limited role in the Russian government's decision to adopt the Kyoto Protocol. Allegedly, the Protocol's endorsement became a token in the diplomatic game between Russia and the European Union (EU) concerning Russia's accession to the World Trade Organisation.[8] This explanation appears plausible given the heated internal debates in Russian political circles concerning the costs of joining the legally-binding climate treaty. One of the fierce opponents of the Kyoto Protocol was Andrey Illarionov, President Vladimir Putin's economic policy adviser, who went as far as to characterise the Kyoto Protocol as an "economic Auschwitz" for Russia and "an assault on science, economic growth and

---

[5] UNFCCC, supra, note 3, Art. 4.8.

[6] Joanna Depledge, "Tracing the Origins of the Kyoto Protocol: an Article-by-Article Textual History", UNFCCC Technical Paper, UN. Doc. FCCC/TP/2000/2, 25 November 2000.

[7] Anna Korppoo, Jacky Karas and Michael Grubb (eds), *Russia and the Kyoto Protocol: Opportunities and Challenges* (London: The Royal Institute of International Affairs, Brookings, 2006), at 7.

[8] See, for instance, Deutsche Welle, "Russia Will Join WTO and Sign Kyoto Protocol", available at: http://www.dw.de/dw/article/0,,1209875,00.html (last accessed on 23 March 2012).

human freedoms."[9] The scientific community represented by the Russian Academy of Sciences also cautioned strongly against joining the agreement, casting doubts over its scientific basis and potential to reduce greenhouse gas emissions, also stressing economic risks and the Protocol's discriminatory nature towards Russia.[10] Despite these internal disagreements, once the decision was made at the political level, the legal procedure for the ratification of the Protocol advanced at a surprisingly fast speed – in mere weeks Russia deposited its instrument of ratification with the UN.[11]

In accordance with the Protocol's Annex B, Russia committed itself to not exceeding its emissions during the baseline year of 1990. This unambitious target was due to Russia's special status as an economy in transition and it was designed to allow unhindered economic growth following the post-Soviet economic collapse. At the time of the adoption of Kyoto Protocol in 1997, Russia's emissions were about 37% lower than in 1990.[12] Under the most pessimistic scenario at that time, Russia's emissions were forecasted to exceed 1990 levels by 4% in 2010.[13] This placed Russia, together with Ukraine, in a highly beneficial position as holders of excess carbon credits – Assigned Amount Units (AAUs) – which can be sold to other industrialised countries under the Protocol's emissions trading scheme created under Article 17. These excess units became known as "hot air." In addition, Article 6 of the Protocol also made it possible for Russia to take part in the Joint Implementation mechanism, potentially leading to investments in various sectors of the economy.

23.3 Evolution of Russia's Domestic Climate Change Policy

In spite of the rising importance of climate change in the international arena, it has remained a peripheral issue on Russia's domestic policy agenda. Until recently, Russian officials, if ever mentioning climate change, described it as primarily beneficial for the country. Their statements were, however, mostly speculative and rarely backed by analytical studies. Vladimir Putin, President of the Russian

[9] Kirill Sukhotskiy, "Interview with Andrey Illarionov", *BBC Russia*, 20 February 2004, available at: http://news.bbc.co.uk/hi/russian/russia/newsid_3507000/3507913.stm (last accessed on 6 March 2012). See also Gerg Walters, "Illarionov Makes His Case on Kyoto", *The Moscow Times*, 18 December 2003, available at: http://www.themoscowtimes.com/news/article/illarionov-makes-his-case-on-kyoto/233950.html (last accessed on 23 February 2012).

[10] Izvestiya, "Kyoto Protocol Does Not Respond to Russian Interests", 18 May 2004, available at: http://www.izvestia.ru/news/290059 (last accessed on 23 February 2012).

[11] Tatyana Avdeeva, "Russia and the Kyoto Protocol: Challenges Ahead", 14 *Review of European Community and International Environmental Law* (2004), 293, at 293.

[12] Third National Communication of the Russian Federation to the UNFCCC, 2002, available at: http://unfccc.int/resource/docs/natc/rusncr3.pdf (last accessed on 23 February 2012), at 8.

[13] Second National Communication of the Russian Federation to the UNFCCC, 1998, available at: http://unfccc.int/resource/docs/natc/rusncr2.pdf (last accessed on 23 February 2012), at 11.

Federation in 2000–2008 and Prime-Minister in 2008–2012, on several occasions expressed doubt over the anthropogenic nature of current climate change,[14] while making comments on potential advantages of global warming for the exploration of the Arctic.[15] International efforts to address climate change were often viewed with suspicion and scepticism. As Korppoo notes the Kyoto Protocol, for example, was not seen as an environmental agreement as such but more as an instrument for wealth redistribution.[16]

The sceptical attitude of Russian politicians, often bordering poor understanding of the climate change issue, can be partly explained by the lack of a unified position on climate change among scientists in Russia. On the one hand, scientists of the Russian Federal Service for Hydrometeorology and Environmental Monitoring (Roshydromet) have been contributing to the work of the Intergovernmental Panel on Climate Change (IPCC), with Professor Yuri Israel, its high-level official, serving as a member of the IPCC Bureau for a long time. As part of the Panel, they officially endorsed its findings, including those confirming the anthropogenic nature of the current climate change. On the other hand, it has not been rare to hear the same scientists contradicting conclusions of the IPCC when addressing the domestic audience. For instance, Professor Israel, in an interview to a major Russian newspaper in May 2010, stated that: "During centuries the temperature on the planet has risen and fallen again. The reason behind such cycles is still unclear."[17] Also Wilson, in her research on the framing of climate change in the Russian media in 2000–2007, reports that it was characteristic of scientists in their interviews to explain current climate change as part of a natural cycle whereby the climate undergoes periods of warming and cooling.[18] Not all scientists in Russia seem to share the same understanding, however. The same governmental agency of Roshydromet has published several assessment reports, referring to the anthropogenic causes of climate change and alarming of its negative consequences for the country, such as: Strategic Forecast of Climate Change in the Russian Federation for the Period of

---

[14] See, for instance, the records of his meeting with scientists in Yakutiya, 23 August 2010, available at: http://premier.gov.ru/pda/visits/ru/11848/events/11882/ (last accessed on 23 March 2012).

[15] See, for instance, RIA Novosti, "Putin: Russia Intends to Implement Projects in the Arctic Not Harming the Environment", 22 September 2011, available at: http://ria.ru/arctic_news/20110922/441898381.html (last accessed on 23 March 2012).

[16] Anna Korppoo, "Russia and the Post-2012 Climate Regime: Foreign Rather Than Environmental Policy", Finnish Institute of International Affairs Briefing Paper No. 23, 24 November 2008, available at: http://www.upi-fiia.fi/assets/events/UPI_Briefing_Paper_23_2008.pdf (last accessed on 23 February 2012), at 8.

[17] Yuri Medvedev, "Climategate: Interview with Yuri Izrael", *Rossiyskaya Gazeta*, 14 May 2010, available at: http://www.rg.ru/2010/05/14/izrael-nauka.html (last accessed on 23 February 2012). Author's translation from Russian.

[18] Elana Wilson Rowe, "Who is to Blame? Agency, Causality, Responsibility and the Role of Experts in Russian Framings of Global Climate Change", 61 *Europe-Asia Studies* (2009), 593, at 600–602.

2010–2015 and Its Impacts on the Russian Economy (2005),[19] Assessment Report of Climate Change and Its Consequences in Russian Federation (2008)[20] and Assessment of Macroeconomic Impacts of Climate Change in Russian Federation until 2030 and beyond (2011).[21]

This range of opinions on climate change among Russian scientists must be placed in a wider context since in reality research activities in the country have significantly shrunk in the last 20 years due to financial constraints. Roshydromet experts are themselves highly critical of the state of climate science in Russia, bluntly stating that: "Beginning from the 1990s, Russian climate science has been primarily relying on the achievements of the previous decades. Today, this resource is almost exhausted and perspectives of its replenishment are, at the very least, modest. It can be said that by the beginning of the twenty-first century Russia has lost its leading position in global climate science."[22]

Overall, up to around 2009 the achievements of climate policy in Russia were modest. As required by the Kyoto Protocol, the country established a national system for estimating emissions and their reductions and a national registry of carbon units in 2006. The government also adopted several regulations giving the green light to the Protocol's flexible mechanisms but those were never implemented. More importantly, Russia lacked a unified and consistent official position on the issue of climate change and mitigation and adaptation policies. It would not be unfair to say that before 2009 there was simply no climate policy in Russia.

The situation started to change during the presidential term of Dmitry Medvedev who placed modernisation at the heart of his policy programme. Undoubtedly, accelerating UN negotiations on a post-2012 agreement also affected that change, as did so realisation of potential financial gains from implementing Kyoto Protocol emission reduction projects in Russia.

In this regard, an important transformation in the development of policy and legislation took place with the adoption of Climate Doctrine in late 2009. The Doctrine is a political document that sets out a unified state stance on the issue of

---

[19] Federal Service for Hydrometeorology and Environmental Monitoring (Roshydromet), "Strategic Forecast of Climate Change in the Russian Federation for the Period of 2010–2015 and Its Impacts on the Russian Economy", 2005, available at: http://www.meteo.ru/publish/obzor/klim_r.pdf (last accessed on 23 March 2012).

[20] Federal Service for Hydrometeorology and Environmental Monitoring (Roshydromet), "Assessment Report of Climate Change and Its Consequences in Russian Federation, General Summary", 2008, available at: http://climate2008.igce.ru/v2008/pdf/resume_ob_eng.pdf (last accessed on 23 March 2012).

[21] Federal Service for Hydrometeorology and Environmental Monitoring (Roshydromet), "Assessment of Macroeconomic Impacts of Climate Change in Russian Federation until 2030 and beyond", 2011, available at: http://voeikovmgo.ru/download/publikacii/2011/Mokryk.pdf (last accessed on 23 March 2012).

[22] Vladimir Katsov, Valentin Meleshko and Sergey Chicherin, "Climate Change and National Security in the Russian Federation", 1–2 *Law and Security* (2007), at 29. Author's translation from Russian.

climate change, and identified the goals, principles and necessary actions for climate policy. For the first time at such a high political level, the document recognised the anthropogenic nature of global climate change and acknowledged that the problem goes beyond a scientific phenomenon and demands economic, environmental and social responses. The Doctrine sets the fundamentals and strategic direction for a comprehensive domestic climate policy and the country's position on the international arena. According to the document, climate policy in Russia builds on the following principles:

- Russia's interests in climate change have a global nature and are not limited to its national boundaries;
- National interests are given a priority in the development and implementation of climate policy;
- Clarity and openness of information, and public dialogue;
- The need for actions even under scientific uncertainty about future climate change and its impacts, including readiness for responsible and constructive participation in international initiatives;
- Comprehensive consideration of both positive and negative impacts of climate change;
- Precautionary principle in developing and implementing policy responses to negative impacts of climate change; and
- Flexibility of climate policy to allow for regular and timely update according to new knowledge and changes in the international framework for climate change and national policies of other countries.

The Doctrine also sets the concrete tasks for: strengthening and developing of information and scientific knowledge on climate change; developing and implementing of short- and long-term adaptation and mitigation policies; and participating in international initiatives on climate change and related issues. Importantly, the document recognises that effective climate change mitigation policy, mainly through energy efficiency measures, can serve as a catalyst for the technological modernisation of the Russian economy, strengthening its position in the world economic community and increasing its competitiveness.

The Climate Doctrine set the principles and directions for developing a comprehensive policy framework for addressing climate change in Russia. It took, however, more than a year to develop the Implementation Plan for the Climate Doctrine, which was finally adopted in April 2011.[23] The Implementation Plan requests a number of ministries and agencies to develop climate-related policies for various sectors of the economy, including energy, agriculture, transportation, infrastructure, scientific research and monitoring, public health, forestry and others. In particular, the Implementation Plan requests: the Ministry for Economic Development to make

[23] Directive No. 730-p of the Government of the Russian Federation. Comprehensive Implementation Plan of the Climate Doctrine of the Russian Federation for the Period up to 2020, 25 April 2011.

adjustments in the programmes for social and economic development in order to take into account climate risks and mitigation and adaptation measures; Roshydromet to work towards the establishment of an integrated centre for monitoring, assessment and prediction of climate change and dangerous natural phenomena; Ministry for Regional Development to conduct regional climate vulnerability assessments; and the Ministry for Public Health and Social Development to develop assessment methodologies for climate change impacts on public health and related adaptation scenarios. The ministries and agencies in question are requested to report annually on this work to the Ministry of Natural Resources for it to prepare an annual report on the implementation of the Climate Doctrine to the Russian government. It has to be said that although comprehensive, the Implementation Plan has no specific financial and human resources support, which weakens its significance and potential impact.

In institutional terms, there is no one governmental body in Russia dealing with the entire set of climate change issues or coordinating the development and implementation of climate policies across different ministries. According to the aforementioned Implementation Plan for the Climate Doctrine, the Ministry of Natural Resources is becoming the focal point for climate policy implementation, although without a coordinating role. Historically, Roshydromet, which is part of the Ministry of Natural Resources, has already been playing an important role in domestic climate policy as a provider of monitoring data and scientific information on climate change. The agency has also been actively participating in the UNFCCC negotiations, frequently representing Russia, alongside with the Ministry for Foreign Affairs. In addition, before the 2009 UN Climate Change Conference in Copenhagen, President Medvedev appointed the retired head of Roshydromet, Alexander Bedritsky, as a Special Adviser to the President and Special Representative of the President on Climate Issues. Bedritsky's responsibilities and role are not clearly defined but so far he has formally been the head of the Russian delegation at COPs and represented Russia at high-level meetings.

23.4 Mitigation of Greenhouse Gas Emissions in Russia

Sharp economic downfall following the collapse of the Soviet Union in 1991 caused a dramatic decrease in Russia's greenhouse gas emissions. Figure 23.1 below illustrates country's annual emissions from 1990 to 2009, based on the UNFCCC greenhouse inventory data.[24]

According to the UNFCCC data, Russia's total emissions in 2008, without land use, land-use change, and forestry, constituted around 66.6% of 1990 emissions. This percentage amounts to 2,243,477.72 Gg CO_2 equivalent, making Russia the

[24] UNFCCC, "Greenhouse Gas Inventory Data from UNFCCC", 2012, available at: http://unfccc.int/ghg_data/ghg_data_unfccc/items/4146.php (last accessed on 17 January 2012).

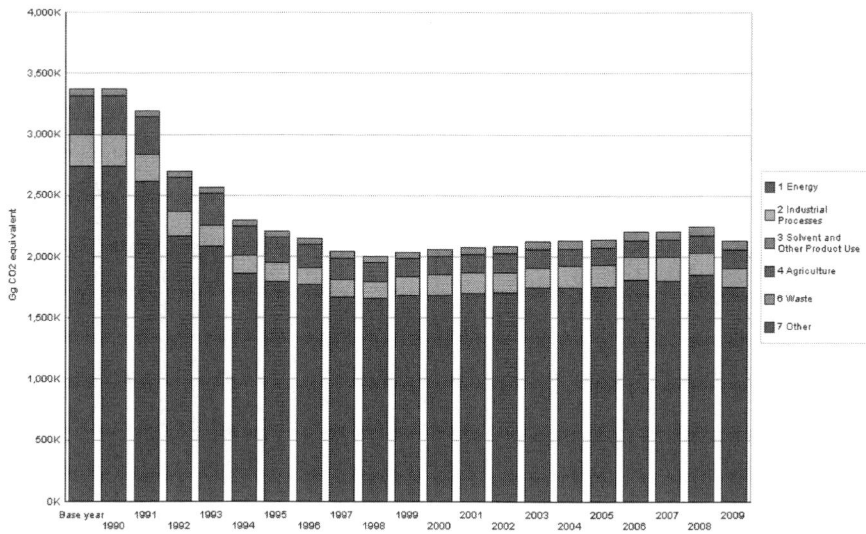

Fig. 23.1 Annual greenhouse gas emissions of the Russian Federation (1990–2009) (Source: UNFCCC greenhouse gas inventory data)

world's third largest emitter of greenhouse gases. The energy sector was the domineering source comprising about 82.4% of the country's total emissions in 2008.

Over time, emissions in Russia decreased until the late 1990s when they started showing a slow but steady growth, increasing by more than 10% from 1999 to 2008. A recent world financial and economic crisis affected energy consumption and industrial processes in the country negatively, and so the recent data from the UNFCCC shows a decline in greenhouse gas emissions in 2009.

Future emissions trends will be influenced by a number of factors, including macroeconomic situation in the world and in Russia itself, GDP growth, sectoral policies and mitigation actions. Russia's fifth national communication suggested three scenarios of how greenhouse gas emissions will develop in the energy sector[25] up to 2030: a scenario based on moderate growth, most close to business-as-usual; innovative scenario which supposes full deployment of technologies for higher energy efficiency; and a scenario with measures additional to those under the second scenario aiming at limiting and reducing greenhouse gas emissions. None of these emissions scenarios suggest reaching the 1990 emissions levels before 2020, and only under the moderate growth scenario, emissions in 2030 will exceed those in 1990.[26]

---

[25] It is assumed that emissions trends in other sectors of the economy will be similar. Fifth National Communication of the Russian Federation to the UNFCCC, 2010, available at: http://unfccc.int/resource/docs/natc/rus_nc5_resubmit.pdf (last accessed on 23 February 2012), at 92.

[26] Ibid., at 90–94.

When it comes to mitigation measures, the Climate Doctrine of 2009 defines their scope as related to: increasing energy efficiency in all sectors of the economy; developing renewables and alternative sources of energy; reducing market failures and implementing financial and tax measures to promote greenhouse gas emission reductions; and protection and enhancement of sinks and reservoirs of greenhouse gases, including sustainable forest management, afforestation and reforestation on a sustainable basis.

Hitherto, action has focused on increasing energy efficiency and, to a lesser extent, developing renewables. These goals are in line with the Concept for Long-Term Social and Economic Development of the Russian Federation up to 2020, which calls for transformation of the current economic model based on the export of natural resources towards socially oriented development built on innovations and new technologies.[27] In this regard, noting that energy intensity of the Russian economy is significantly higher than in countries in the Organization for Economic Co-operation and Development, the Concept sets a priority for the energy sector to decrease energy intensity of GDP by 40% in 2020 compared to 2007.[28] The document also states that Russia should take a leading role in the development of renewables as well as deploy environmentally sound technologies for energy production on an industrial scale.

Increasing energy efficiency has indeed become a policy priority for the Russian government in the last years. Estimates show that the energy efficiency potential of the Russian economy is simply massive and stands at about 45%, being equal to the annual primary energy consumption of France.[29] The government itself acknowledges that the energy intensity of GDP in Russia is 2.5 times higher than the world average and 2.5–3.5 higher than in developed countries.[30]

Decreasing the energy intensity of the Russian economy as a long-term policy goal is further detailed in the Russia's Energy Strategy for the period up to 2030 adopted by the government in November 2009.[31] The Strategy suggests the following actions to achieve the energy efficiency goal: creating a favourable economic environment through legislative and institutional measures, and stimulating entrepreneurial activities; adopting a set of standards and norms; supporting strategic initiatives such as formulating federal, regional and municipal programmes, and developing new technologies and pilot projects. As a result of implementing the strategy, it is expected that in 2030, the energy intensity of GDP will decrease by no less than 2.3 times, and that greenhouse gas emissions will remain at 100–105% of the 1990 levels.

---

[27] Directive No. 1662-p, supra, note 1.

[28] Ibid.

[29] World Bank, "Energy Efficiency in Russia: Untapped Reserves", World Bank Working Paper, 2008, available at: www.ifc.org/ifcext/rsefp.nsf/AttachmentsByTitle/FINAL_EE_report_Engl.pdf/$FILE/Final_EE_report_engl.pdf (last accessed on 23 February 2012), at 6.

[30] Directive No. 2446-p of the Government of the Russian Federation. Energy Conservation and Increasing Energy Efficiency for the period up to 2020, State Programme of the Russian Federation, 27 December 2010.

[31] Directive No. 1715-p of the Government of the Russian Federation. Energy Strategy of Russia for the period up to 2030, 13 November 2009.

In parallel to the Energy Strategy, the Federal Assembly – Russian parliament – also passed a law concerning energy conservation and increasing energy efficiency,[32] and in response, the government adopted a plan of actions to achieve the goals set by the federal statute.[33] A year later, the Ministry for Energy developed a state programme on energy conservation and increasing energy efficiency up to 2020[34] outlining actions in various sectors of the economy necessary to achieve a 40% decrease of GDP energy intensity.

In sum, in 2008–2010, Russia has developed a comprehensive policy and legislative framework for decreasing the energy intensity of its economy. Given the major contribution of the energy sector to country's greenhouse gas emissions, this framework, if fully implemented, will lead to significant emission reductions.

As far as renewables are concerned, their development, except for large hydropower, has so far enjoyed limited attention from policy-makers in Russia, possibly due to the historical reliance on abundant oil and gas resources, and lack of political will, domestic experience, investments and technologies. The total share of renewables is currently at the minuscular figure of less than 1%.[35] This is despite the truly gigantic potential of renewables for energy generation: according to Russian experts, realising just the economic potential of renewables can allow for the 30% contribution to electricity production while using a technological potential brings a figure 25 times higher than that.[36] The Energy Strategy itself recognises the colossal potential of renewables, in particular solar and wind energy, referring to estimates of energy generation at 4.5 billion tons per year, which is four times higher than the current total energy consumption in Russia. Notwithstanding, the targets for renewables set by the government are modest compared to those of industrialised countries and emerging economies[37]: Russia aims to increase the share of renewables, excluding large hydropower, from 0.5 to 4.5% by 2020.[38]

---

[32] Federal Law of the Russian Federation No. 261-ФЗ. Energy Conservation and Increasing of Energy Efficiency and Introducing Amendments to Specific Legislative Acts of the Russian Federation, 23 November 2009.

[33] Directive No. 1830-p of the Government of the Russian Federation. Action Plan for the Implementation of the Federal Law on Energy Conservation and Increasing of Energy Efficiency and on Introducing Amendments to Specific Legislative Acts of the Russian Federation, 1 December 2009.

[34] Directive No. 2446-p, supra, note 30.

[35] This figure excludes the contribution of large hydropower stations, which is around 18 % of total electricity generation. See Yuri Fedorov, Georgiy Safonov and Adil Bagirov, *Low-Carbon Economy in Russia: Trends, Problems, Opportunities* (Moscow: National Carbon Sequestration Foundation, 2009), at 16.

[36] Pavel Bezrukikh (ed), *Reference Book on Renewable Energy Resources in Russia and Local Fuel Types* (Moscow: Institute for Energy Strategy, 2007), at 67; Fedorov, Safonov and Bagirov, *Low-Carbon Economy in Russia*, supra, note 35, at 16.

[37] For example, the EU aims to derive 20 % of its energy from renewable sources by 2020 and China 15 % by 2020.

[38] Directive No. 1-p of the Government of the Russian Federation. Main Directions for the State Policy in the Area of Increasing Electrical Energy Efficiency on the Basis of the Use of Renewables for the Period up to 2020, 8 January 2009.

The question remains how these policies on energy efficiency and renewables will affect the country's emissions of greenhouse gases in the long run. Russian experts estimate that even without these additional measures, greenhouse emissions will still be in the range of 72–80% of the 1990 levels in 2020 depending on the rate of GDP growth.[39] Full implementation of the energy efficiency and renewables policy goals and modernisation of the economy will stabilise emissions at about 60–65% if GDP growth is at 4.5% a year, and at around 70–75% assuming GDP growth of 6–7%. These estimates are, however, based primarily on emission trends in the energy sector; and little analytical work exists so far for other sectors such as industry, waste, agriculture, and forests.

At COP 15 in Copenhagen, Russian President Medvedev underlined that Russia is willing to commit to reducing emissions by 25% below 1990 levels by 2020 if other countries join a future legally-binding agreement.[40] In its subsequent submission to the UNFCCC as part of the Cancun Agreements, Russia communicated a more modest target of a 15–25% reduction, conditional upon the appropriate accounting of Russia's forestry sector and participation of all major emitters in a legally-binding agreement.[41] From the beginning of the post-2012 negotiations under the *Ad Hoc* Working Group on Long-term Cooperative Action, the Russian position has been that a new legal instrument, binding on all major emitters amongst developed and developing countries, should be developed.[42] In this context, Russia, first of all, referred to the US, which is not a Party to the Kyoto Protocol, and to major developing countries, such as China and India, implying that without their commitments to emission reductions, any intergovernmental arrangement would be meaningless as it would not stop further warming of the planet. For this reason, at the 2011 UN Climate Change Conference in Durban, Russia refused to participate in a second commitment period under the Kyoto Protocol. At the same time, Russia communicated in Durban its intention to start taking measures to achieve its target of a 15–25% reduction by 2020 in parallel with the post-Durban negotiations on a new climate agreement.[43]

---

[39] Fedorov, Safonov and Bagirov, *Low-Carbon Economy in Russia*, supra, note 35, at 23.

[40] Dmitry Medvedev, "Address by the President of the Russian Federation Dmitry Medvedev to COP 15", 18 December 2009, available at: http://kremlin.ru/news/6384 (last accessed on 23 February 2012).

[41] Compilation of Economy-Wide Emission Reduction Targets to Be Implemented by Parties Included in Annex I to the Convention. Revised Note by the Secretariat, UN Doc. FCCC/SB/2011/INF.1/Rev.1, 7 June 2011.

[42] See, for instance, submission of the Russian Federation to UNFCCC contained in Additional Views on Which the Chair May Draw in Preparing Text to Facilitate Negotiations among Parties. Submissions from Parties. Addendum, UN Doc. FCCC/AWGLCA/2010/MISC.2/Add.1, 17 May 2010.

[43] Address by Alexander Bedritskiy to COP 17, 8 December 2011, available at: http://meteorf.ru/default_doc.aspx?RgmFolderID=a4e36ec1-c49d-461c-8b4f-167d20cb27d8&RgmDocID=0c47f6a9-671c-48f3-a69e-f67d7ddb0ac3 (last accessed on 23 March 2012).

23.5 Russia and the Kyoto Protocol's Flexibility Mechanisms

According to the Kyoto Protocol, Russia is eligible to participate in Joint Implementation under Article 6 and emissions trading under Article 17. Developing a domestic institutional and legal framework for the Protocol's flexibility mechanisms, however, proved to be an arduous process characterised by long delays and lack of transparency, resulting in over-bureaucratic procedures and low confidence among investors in the Russian market. Despite potential financial gains, it took the government exceptionally long to finalize domestic regulations for Joint Implementation projects in 2007. Those rules, however, were never put to practice and not a single project was endorsed, causing much frustration among interested businesses. Finally, in 2009, the government revoked the 2007 regulations and introduced a new legal and institutional framework. Projects were to be approved on the basis of a tender with a certain ceiling for emissions reductions for each round. The Ministry for Economic Development was appointed as the national authority for the final endorsement of proposed Joint Implementation projects with Sberbank, the largest bank in Russia with the state as the main shareholder, conducting all operations and playing a significant role in the selection of projects. The first round of tenders ended in mid-2010 resulting in the approval of 15 projects which would lead to emissions reductions of 29.99 million tons of CO_2 equivalent and €1.6 billion in total investments, 19.2% of which would be "Kyoto" money.[44] One more tendering round was conducted in the second half of 2010, bringing the total potential emissions reductions from Joint Implementation projects in Russia to approximately 60 million tons of CO_2 equivalent. Taking into account the first experiences with Joint Implementation projects, in September 2011, the government replaced the 2009 rules with a new set of regulations, which are conceptually similar but aim at making the procedures more efficient. The rules allocate 300 million of emission reduction quotas in total for Joint Implementation projects. Importantly, project proponents are now required to re-invest the money received through the scheme into activities beneficial for the environment and society. It remains to be seen whether the new rules will prove to be viable. According to the Head of Sberbank's department in charge for Kyoto Protocol's mechanisms Vsevolod Gavrilov, implementation of Joint Implementation projects in the time remaining until the end of the first commitment period can generate around €6–8 billion of new direct investments.[45]

With regards to emissions trading, there has not been much progress in developing a regulatory framework. Although the Russian surplus of AAUs is projected to be very high – in the order of gigatons, demand is limited since potential buyer

[44] Sberbank, "Outcomes of the first round of tender to select joint implementation projects", 23 July 2010, available at: http://www.sbrf.ru/tula/ru/about/concurs/archive/2010/index.php?id114=11006763 (last accessed on 23 March 2012).

[45] Alexey Shapovalov, "Interview with Vladimir Gavrilov", Kommersant-Online, 20 July 2011, available at: http://www.kommersant.ru/doc-y/1682278 (last accessed on 23 March 2012).

nations have indicated that they would not purchase "hot air" to comply with their emission reduction targets. In 2000, Russia proposed a presumably more attractive Green Investment Scheme for investors, whereby it would voluntarily re-invest the money generated through selling its excess quotas into emission reduction projects.[46] Since then, the Green Investment Schemes have been successfully implemented in several Eastern European countries, but in Russia the Schemes have stumbled upon the lack of a regulatory framework. Although in 2011, according to Sberbank, it was in the process of preparing pilot deals,[47] overall development of a Green Investment Scheme does not seem to be of interest for the Russian government in the short term. At the same time, Russia strongly holds to its unused AAUs in the first commitment period and has indicated on several occasions that those should be transferred to the subsequent periods.[48] It is a widespread approach among those involved in the formulation of climate policy in Russia that as a result of the massive economic downfall and decline in emissions, "an environmental service was provided [by Russia] to the humankind, in the full accordance with the Kyoto Protocol and Marrakesh Accords."[49]

23.6 Climate Change Adaptation in Russia

When it comes to adaptation responses to the negative consequences of climate change, the picture is bleak as there are virtually no specific adaptation measures in place. With the Implementation Plan for the Climate Doctrine requesting ministries to develop relevant policies, Russia is currently at the very beginning of developing a comprehensive adaptation framework covering a range of industries and applying a region-specific approach. The lack of financial and human resources support for the implementation of the Climate Doctrine, however, reflects a low sense of urgency over adaptation to climate change.

At the same time, assessments of climate change impacts across Russia by Roshydromet contain several sober warnings. In 2008, Roshydromet published a comprehensive report, which gave a worrying picture of negative consequences of

---

[46] On Green Investment Schemes, see, for instance, Alexander Averchenkov, *Economy and Climate: Russia's Participation in Solving a Global Environmental Problem*, (Moscow: Institute for Sustainable Development/Centre for Environmental Policy in Russia, 2009), at 33–51.

[47] Sberbank, "Presentation at the meeting of the Presidential Commission for Modernisation and Technology Development of the Russian Economy", 27 June 2011, available at: http://i-russia.ru/sessions/25.html (last accessed on 23 March 2012).

[48] See, for instance, Ecolife, "Interview with Alexander Bedritsky", 27 December 2011, available at: http://www.ecolife.ru/intervju/4380/ (last accessed on 23 March 2012). Given that to the date, Russia has not taken commitments for the second term of the Protocol, the issue of carryover of surplus AAUs becomes irrelevant unless Russia changes its position in the future.

[49] Averchenkov, *Economy and Climate*, supra, note 46, at 62.

climate change for infrastructure, agriculture and other sectors of the economy.[50] In September 2011, Roshydromet followed with a report on the assessment of macroeconomic impacts of climate change until 2030 and beyond, which analyses major climate change impacts on economic sectors and regional development as well as adaptation policies and options.[51] While noting strong regional variations, the report identifies both positive and negative impacts of climate change.[52] On the beneficial side, global warming implies for Russia the increase in habitable territories, less energy usage for district heating and opening of new transport routes as well as access to natural resources in the Arctic. This however comes alongside with species displacement, increase of droughts in some regions and of floods in other regions, and permafrost degradation leading to damages for infrastructure and communications in the northern regions of Russia. Governmental experts conclude that "many – and most likely, the majority – of such [climate] changes will have a negative impact on economy development and public health."[53] Referring to the transitionary period of the Russian economy and high concentration of population in certain areas, Roshydromet further warns of high vulnerability to impacts of extreme weather and climate events.[54] Ignoring the need for adaptation policies will lead to huge social and economic losses, with just one example of the forest fires in the unusually hot summer of 2010 leading to about 500 billions of roubles in damages, or 1.2% of the Russian GDP.[55] The conclusion by Roshydromet is clear: urgent action on adaptation to current and future impacts of climate change in Russia is required and the "climate factor" must be accounted for in developing programmes and projects to modernise industry and services in Russia.[56]

23.7 Conclusions

Having been behind many industrialised and large developing countries in formulating a domestic climate policy, Russia is now slowly catching up. Its main policy statement on climate change and related mitigation and adaptation responses, known as the Climate Doctrine, was adopted in 2009. Together with the Implementation Plan followed in 2011, the Doctrine lays out a road towards a comprehensive legislative and policy framework to address climate change in Russia. In parallel,

---

[50] Roshydromet, Assessment Report of Climate Change and Its Consequences in Russian Federation, supra, note 20.

[51] Roshydromet, Assessment of Macroeconomic Impacts of Climate Change in Russian Federation until 2030 and beyond, supra, note 21.

[52] Ibid., at 9.

[53] Ibid., at 174.

[54] Ibid., at 9.

[55] Ibid., at 10.

[56] Ibid., at 176.

decreasing the energy intensity of the Russian economy became a long-term policy priority leading to the adoption of a number of policy and legislative acts on energy conservation and energy efficiency. As the energy sector is the domineering source of greenhouse gas emissions in Russia, these steps will potentially result in significant emissions reductions in the future.

Despite a relatively optimistic outlook, climate change still remains a peripheral issue in the current policy agenda in Russia. The country is far from being in a position to take up more ambitious mitigation targets or aspire for a leadership role in environmental sustainability and low-carbon development. Current policy goals for developing renewable sources of energy are modest. The recent strive for a less energy intensive economy, although a win-win solution for the climate, has been driven by concerns over economic development rather than by a sense of urgency to reduce emissions of greenhouse gases. Despite warnings by Russian scientists that most of climate change impacts will be adverse and lead to increased costs for the economy, the development of adaptation frameworks has been delayed and lacks budgetary support.

Finally, climate policy measures in Russia have been formulated in the challenging context of the world financial and economic crisis. This constraint, coupled with the return of Vladimir Putin – who is little enthused about pro-active climate policies – to the presidential post in 2012, will undoubtedly impact the effectiveness of existing and future mitigation and adaptation policies. Yet, an ambitious policy programme for modernising the economy and bringing about innovations opens a window of opportunity for no-regret climate change policies.

Chapter 24
Australia: From "No Regrets" to a Clean Energy Future?

Sharon Mascher and David Hodgkinson

Abstract In September 2011 Australia passed the Clean Energy Act 2011 (Cth), a piece of legislation that will for the first time introduce a carbon price into the Australian economy. The passage of this Act marks a momentous step forward for Australia, a country that until now has been dominated by a domestic climate change policy of 'no-regrets'. This Chapter explores the evolution of climate change policy in Australia from the late 1980s through to the passage of the Clean Energy Act 2011 (Cth).

Sharon Masher is a Professor in the Faculty of Law, University of Calgary and on Honorary Research Fellow in the Faculty of Law, University of Western Australia. David Hodgkinson is an Associate Professor in the Faculty of Law, University of Western Australia and Special Counsel with Clayton Utz, a national Australian law firm.

S. Mascher (✉)
Faculty of Law, University of Calgary, 2500 University Drive,
NW, Calgary AB, Canada
e-mail: smascher@ucalgary.ca

D. Hodgkinson
Faculty of Law, University of Western Australia,
35 Stirling Highway, WA, Crawley, Australia
e-mail: david.hodgkinson@uwa.edu.au

E.J. Hollo et al. (eds.), *Climate Change and the Law*, Ius Gentium: Comparative Perspectives on Law and Justice 21, DOI 10.1007/978-94-007-5440-9_24,
© Springer Science+Business Media Dordrecht 2013

24.1 Introduction

> Australia is said to be more vulnerable to the effects of climate change than other developed countries.[1] The reasons for this are two-fold. First, the country's already hot, dry climate makes it highly vulnerable to predicted changes in the climate system.[2] Second, Australia's terms of trade and geographical location leave it exposed indirectly to the impacts of climate change on its trading partners and regional neighbors. In particular, climate change related impacts on countries such as China, India and Indonesia could result in a decline in demand for Australia's mineral and energy resources and agricultural products.[3] In addition, as Australia is physically situated in a region of developing countries which are in both highly vulnerable to climate change and in a weaker position to adapt, climate change related impacts such as human displacement due to rising sea-levels and geopolitical and food security issues will be magnified in the region.[4] As a result, it is in Australia's national interest that effective action be taken to mitigate the effects of climate change.

At the same time, Australia has a very emission intensive economy. While the country's contribution to overall global greenhouse gas emissions is small, accounting for only 1.5% of global emissions in 2005,[5] its per capita emissions are the highest in the Organization for Economic Cooperation and Development (OECD) and amongst the highest in the world.[6] The country's high emission levels are largely a consequence of its ready access to low-cost fossil fuel reserves, around which a very energy-intensive economy has developed.[7] Indeed, Australia derives more than 40% of its total primary energy supply from brown and black coal, with coal accounting for 84% of its total electricity generation in 2007–2008.[8] With primary energy consumption on an upward trajectory,[9] absent a change in climate change governance Australia's greenhouse gas emissions will also continue to rise. Not surprisingly, given its abundance of fossil fuel reserves, Australia is also net energy exporter. Indeed, energy exports accounted for 33% of Australia's total exports in 2008–2009,

---

[1] Ross Garnaut, *Garnaut Climate Change Review* (Melbourne: Cambridge University Press, 2008), at xix.

[2] Intergovernmental Panel on Climate Change, "Technical Summary" in: M.L. Parry et al. (eds), Climate Change 2007, Impacts, Adaptation and Vulnerability: Contribution of Working Group II to the Fourth Assessment Report of the Intergovernmental Panel on Climate Change (Cambridge: Cambridge University Press, 2007), 25, at 50.

[3] Garnaut, *Garnaut Climate Change Review*, supra, note 1, at 145.

[4] Ibid., at 145–150.

[5] Ibid., at 65, Table 3.2.

[6] Ibid., at 153; Prime Ministerial Task Group on Emissions Trading, *Report of the Task Group on Emissions Trading* (Canberra: Commonwealth of Australia, 2007), at 20–22.

[7] Department of Prime Minister and Cabinet, *Energy White Paper, Securing Australia's Energy Future* (Canberra: Commonwealth of Australia, 2004), at 1.

[8] Garnaut, *Garnaut Climate Change Review*, supra, note 1, at 158; Australian Bureau of Agriculture and Resource Economics, *Energy in Australia* 2010 (Canberra: Department of Resources, Energy and Tourism, 2010), at 12–13 and 21.

[9] Clara Cuevas-Cubria and Damien Riwoe, *Australian Energy: National and State Projections to 2029–2030* (Canberra: Australian Bureau of Agriculture and Resource Economics, 2006), at 27.

at a value of AUD $78 billion.[10] Of its energy resources, coal is by far the country's largest export earner,[11] with the export value of this single resource increasing from $11 billion in 2000–2001 to $43 billion in 2010–2011.[12]

Australia is therefore both highly vulnerable to the effects of climate change and economically privileged by virtue of its consumption and export of emission intensive fossil fuels. For these reasons, successive Australian governments have struggled with the competing tensions associated with protecting the country's economic interests and moving beyond "no-regrets" measures to achieve climate change mitigation objectives. The result has been decades of intensive debate around Australia's domestic climate policy. With the passage of Clean Energy Act 2011 (Cth)[13] on 8 November 2011 and the mandated introduction of a carbon price into the economy from 1 July 2012, Australia's domestic climate policy took a momentous step forward. The tensions, however, remain.

This chapter will explore the evolution of climate policy in Australia from the late 1980s through to the passage of the Clean Energy Act 2011 (Cth). The discussion is organized into three parts. The first part examines the era of "no-regrets", an approach that dominated Australian domestic climate policy for over two decades. The second part explores the post-2007 era, and the attempts to move from a policy of "no-regrets" through the introduction of emissions trading legislation. The third part of this chapter focuses on the Clean Energy Act 2011 (Cth), describing the central features of the legislation and exploring whether it will afford an effective mechanism to transition Australia to a clean energy future.

24.2 The Era of "No-Regrets" in Domestic Climate Change Policy

Climate change policy has been on the agenda of successive Australian governments for over two decades. While pro-active in some respects, the dominant policy in the years preceding 2007 can best be described as one of voluntary "no-regrets" measures constrained by overriding concerns for the economy.

---

[10] Australian Bureau of Agriculture and Resource Economics, *Energy in Australia* 2010, supra, note 8, at 2.

[11] Ibid., at 2.

[12] Department of Resources, Energy and Tourism, *Draft Energy White Paper 2011 – Strengthening the Foundation for Australia's Energy Future* (Canberra: Commonwealth of Australia, 2011), at 82.

[13] The Clean Energy Act 2011 (Cth) is the central piece of legislation in a legislative package, which also includes: the Clean Energy (Consequential Amendments) Act 2011 (Cth); the Climate Change Authority Act 2011 (Cth); and, the Clean Energy Regulator Act 2011 (Cth). For information about each of these pieces of legislation, available at: http://www.climatechange.gov.au/government/clean-energy-future/legislation.aspx (last accessed on 22 February 2012).

24.2.1 The Origins of 'No-Regrets'

When the climate change issue began attracting international attention in the late 1980s, the Australian Government, led by Australian Labor Party (Labor) Prime Minister Bob Hawke, took a leadership role, strongly supporting international action.[14] Accepting that developed countries should take the lead, and implicitly recognizing Australia's obligation to act, in 1990 the Hawke Labor Government adopted a domestic Interim Planning Target to stabilize greenhouse gas emission at 1988 levels by 2000 and to reduce these emissions by 20% by the year 2005.[15] While one of the most stringent national targets through to the mid-1990s,[16] this commitment was subject to the caveat that the measures taken would not "have net adverse economic impacts nationally or on Australia's trade competitiveness, in the absence of similar action by major ghg producers".[17] Failing from the start to consider the benefits of taking the lead in climate change mitigation and focusing instead on the costs, the caveat meant that early action came in the form of a 'no-regrets' strategy, directed at those activities where the economic benefits outweighed the costs.

The 'no-regrets' approach was further entrenched when the Australian Government, now led by the Labor Prime Minister Paul Keating, released its National Greenhouse Response Strategy (NGRS). While the Strategy contained measures directed towards achieving the qualified Interim Planning Target, they were largely voluntary and designed to cause minimal disturbance to the community as a whole or to any single industry sector or particular geographical region.[18] Criticized for "prioritization of economic and industry concerns over environmental ones", this approach laid the foundation for the country's subsequent approach to climate change policy.[19]

During this time the Keating Labor Government also became increasingly concerned that taking "similar action" to that of other developed countries would cost the Australian economy more. As a result, the Australian Government began to emphasize at the international level the principle of "common but differentiated responsibilities", the need to share equitably the burden of taking action, and the need to take account of the special needs of fossil-fuel dependent economies.[20]

---

[14] Roslyn Taplin, "International Cooperation on Climate Change and Australia's Role", 26 *Australian Geographer* (1995), 16, at 16; Matt McDonald, "Fair Weather Friend? Ethics and Australia's Approach to Global Climate Change", 51 *Australian Journal of Politics and History* (2005), 216, at 221.

[15] Matt McDonald, "Fair Weather Friend? Ethics and Australia's Approach to Global Climate Change", 51 *Australian Journal of Politics and History* (2005), 216, at 221–222.

[16] Ibid., at 221.

[17] Ian Rowlands, "Explaining National Climate Change Policies", 5 *Global Environmental Change* (1995), 235, at 245; see also, Paul Kay, *Australia and Greenhouse Policy – A Chronology 1997–1999,* Background Paper 4 (1997), available at: http://www.aph.gov.au/About_Parliament/Parliamentary_Departments/Parliamentary_Library/Publications_Archive/Background_Papers/bp9798/98bp04 (last accessed on 22 February, 2012).

[18] Commonwealth of Australia, *National Greenhouse Response Strategy* (Canberra: Australian Government Public Service, 1992), at 12.

[19] McDonald, "Fair Weather Friend?", supra, note 15, at 222–223.

[20] Ibid., at 223.

Paradoxically, while arguing that a differentiated approach was required to accommodate its resource rich and emission intensive economy, Australia joined countries such as the United States in calling for a greater commitment to mitigation from developing countries.

24.2.2 The Howard Government Era – "No-Regrets" Entrenched

In March 1996, Prime Minister John Howard was elected as leader of a Liberal-National Coalition Government. Holding power from 1996 through to late 2007, the Howard Liberal Government played a central role in determining Australia's domestic and international position on climate change during the crucial time surrounding the Kyoto Protocol negotiations and the decade that followed.

Relying on economic modeling to demonstrate the disproportionate costs Australia's resource intensive economy would incur to reduce its ghg emissions,[21] the Howard Liberal Government entered the Kyoto Protocol negotiations strongly opposed to uniform emission reduction targets and asserting that Australia was entitled to the benefit of the principle of common but differentiated responsibilities.[22] Threatening to block consensus at COP 3 in Kyoto, the Australian Government finally won late night concessions on its targets and on allowances for changes in land use. As a result, Australia received a Kyoto target of an 8% increase in emissions in 2008–2012 above 1990 levels[23] and a concession in the "Australia clause", which authorized Kyoto signatories to include net carbon emissions from land clearing as part of their targets.[24] With Australia's 1990 unusually high land clearance rates of 675,000 ha having already fallen substantially as a result of restrictions on clearing imposed by the States, the resulting compromise meant that Australia's commitment was even less onerous than its otherwise generous Kyoto target suggested. Despite this, and despite characterizing the Kyoto outcome as a "win for the environment and a win for Australian jobs",[25] the Howard Liberal Government later refused to ratify the Kyoto Protocol on the basis that it would unfairly hurt the Australian economy, heavily reli-

[21] Brian Fisher, *International Climate Change Policy Economic Implications for Australia* (Canberra: Australian Bureau of Agricultural and Resource Economics, 1997); Rosemary Lyster, "Common but Differentiated? Australia's Response to Global Climate Change", *16 Georgetown International Environmental Law Review* (2003–2004), 561, at 564.

[22] Ibid., at 563–564.

[23] Kyoto Protocol to the United Nations Framework Convention on Climate Change, Kyoto, 10 December 1997, in force 16 February 2005, 37 *International Legal Materials* (1998), 22, Art. 3(a) and Annex B.

[24] Ibid., Art. 3.7.

[25] Prime Minister John Howard, AM Radio Program, 19 December 1997.

ant as it is on coal for both domestic energy and export income, while countries like India, China and the United States were not bound by targets.[26]

Having adopted this stance internationally, the Howard Liberal Government maintained the policy of 'no-regrets' at home. While responsible for creating the world's first government agency focused solely on ghg emissions, the Australian Greenhouse Office, and for passing legislation which put in place a mandatory renewable energy target, the Howard Liberal Government's "no-regrets" policy agenda became the subject of growing criticism[27] as Australia's direct emissions continued to rise. While still on track to meet its Kyoto target, estimates suggested Australia's direct emissions would increase by approximately 33% between 1990 and 2010[28] and structural changes required to transition to a low carbon economy were not taking place in Australia.

Dissatisfied with the Commonwealth Government's 'no-regrets' stance, the Australian States and Territory governments began to put in place a variety of state-based mitigation measures. In 2003, the State of New South Wales took the significant step of introducing one of the world's first mandatory emissions trading schemes[29] and the following year the State and Territory governments formed a National Emissions Trading Taskforce (NETT) to develop a model for a national emissions trading scheme. The NETT released a discussion paper in 2006 on the design of a national emission trading scheme which "invited" the Commonwealth government to join but also contemplated the possibility of the States and Territories pursuing a national emissions trading scheme "in the absence of Commonwealth support."[30] With the release of the Stern Review, Al Gore's Inconvenient Truth and experience of severe drought in the eastern States of Australia, the public also became increasingly concerned about the adequacy of the Australian Government's response to climate change.[31]

---

[26] Prime Minister John Howard, Hansard: Commonwealth Parliamentary Debates, 5 June 2002, at 3163; Prime Minister John Howard, Hansard: Commonwealth Parliamentary Debates, 26 May 2004.

[27] Lyster, "Common but Differentiated?", supra, note 21, at 573–577.

[28] Andrew Macintosh, The National Greenhouse Accounts and Land Clearing: Do the numbers stack up? Research Paper No. 38 (2007), at 3.

[29] Information about the New South Wales Greenhouse Gas Reduction Scheme, a baseline and credit scheme aimed at reducing the ghg emissions associated with the production and use of electricity is available at: http://www.greenhousegas.nsw.gov.au/ (last accessed on 22 February 2012).

[30] National Emissions Trading Taskforce, "Possible Design for a National Greenhouse Gas Emissions Trading Scheme", 2006, available at: http://www.climatechange.gov.au/en/government/initiatives/cprs/~/media/publications/cprs/nett-discussion-paper.ashx (last accessed on 22 February, 2012), at ii and 13.

[31] Andrew Macintosh, "The Garnaut Review's Targets and Trajectories: A Critique", 26 *Environmental and Planning Law Journal* (2009), 88, at 88.

Bowing to the pressure, on 10 December 2006 then Prime Minister Howard announced the establishment of a joint government-business Task Group on Emissions Trading. An important emphasis however remained the protection of the economy, with the Task Group's terms of reference indicating that in assessing Australia's further contribution to reducing greenhouse gas emissions, the competitive advantages Australia enjoyed through the possession of large reserves of fossil fuels and uranium must be preserved.[32] When the final report by the Prime Ministerial Task Group on Emissions Trading recommended the adoption of an emissions trading scheme, it seemed at the time to mark a "critical turning point in the climate debate in Australia".[33]

While still opposed to the ratification of the Kyoto Protocol, on 17 July 2007 the Howard Government committed to establishing a national emissions trading scheme in Australia by 2011.[34] With the official opposition strongly supporting both the ratification of the Kyoto Protocol and the introduction of an emissions trading scheme, Australia seemed poised to finally move beyond a 'no-regrets' approach as it headed into a federal election in November 2007.

24.3 The Post-2007 Era – Beyond "No-Regrets"?

On 24 November 2007, Australia elected Prime Minister Kevin Rudd and his Labor Government on a platform that included the promise to effect change in domestic climate change governance. Describing climate change as "the great moral challenge of our generation",[35] the promise of the Rudd Government was action on climate change. This promise had the support of a high proportion of Australians, even if it meant paying higher prices.[36] Indeed, Ross Garnaut, commissioned by the Rudd Labor Government and the State and Territory governments to conduct an independent review on impacts of climate change on the Australian economy, concluded in 2008 that there was "a much stronger base of support for reform and change on this issue than on any other big question of structural change in recent decades, including trade, tax and public business ownership reform."[37]

[32] Prime Ministerial Task Group on Emissions Trading, *Report of the Task Group*, supra, note 6, at 8–9.

[33] Warwick McKibbin, "The Prime Ministerial Task Group on Emissions Trading", 14 *Agenda* (2007), 13, at 13.

[34] Prime Minister John Howard, Speech Transcript, "Address to the Melbourne Press Club", 2007, available at: www.pm.gov.au/media/Speech/2007/Speech24445.cfm (last accessed on 22 February 2012).

[35] Hon. Kevin Rudd MP, Opening Remarks to the National Climate Change Summit, Parliament House, Canberra, 31 March 2007.

[36] Garnaut, *Garnaut Climate Change Review*, supra, note 1, at xviii.

[37] Ibid., at xviii–xvix.

Signaling its intention to move quickly on its climate change commitment, the new Rudd Government ratified the Kyoto Protocol and set about putting in place the central piece of its climate change strategy, the Carbon Pollution Reduction Scheme (CPRS). The fate of the CPRS Bill (together with several associated Bills), introduced into Parliament on 14 May 2009, exposed the ongoing struggle in Australia between protecting economic interests, often associated with emission intensive activities, and moving beyond "no-regrets" measures to meet climate change mitigation objectives.

24.3.1 The CPRS in Broad Overview

In very broad overview, the CPRS proposed a market-based cap and trade approach to put a price on carbon.[38] Including all six greenhouse gas listed in Annex A of the Kyoto Protocol, the CPRS covered the stationary energy, transport, fugitive emissions, industrial processes and waste sectors. Of the covered sectors, the CPRS obligations applied to operators of facilities within these sectors with annual direct emissions of greater than 25,000 tonnes or more of carbon dioxide equivalent (CO_2-e). The transport sector was to be captured by applying obligations to upstream fuel suppliers resulting from the combustion of the fuel supplied. The Scheme was anticipated to capture approximately 1,000 liable entities, totaling 75% of Australia's total greenhouse gas emissions.

The CPRS required each liable entity in a covered sector to acquire and surrender a permit for each tonne CO2-e emitted. Permits up to the limit of the annual cap were to be allocated by a scheme Regulator by way of auction or free allocation to eligible emissions intensive, trade-exposed (EITE) entities, until such time as it was not longer warranted, and to strongly affected industries (being coal-fired electricity generators) on a limited transitional basis. It was also to be possible for liable entities to purchase international permits and permits generated by reforestation projects that "opted into" the Scheme. As a transitional measure, the CPRS also included a "safety valve", allowing liable entities to purchase permits for a fixed charge. Failure to surrender the requisite permits attracted both a penalty and an obligation to make good the following financial year.

24.3.2 Progress of the CPRS Bill in the Australian Parliament

Following its initial introduction and passage by the House of Representatives, on 13 August 2009 the CPRS Bill was rejected by the Senate, with all non-government Senators voting against it. That all non-government Senators voted against

[38] The text of the original Carbon Pollution Reduction Scheme Bill 2009 (Cth), together with Explanatory memoranda, is available at: http://parlinfo.aph.gov.au/parlInfo/search/display/display.w3p;query=Id:legislation/billhome/R4127 (last accessed on 22 February, 2012).

the CPRS Bill reflects the significance of the division that existed across the political spectrum in relation to key elements of the Bill. The Greens, on the one hand, called for tougher 2020 emission reduction targets and less assistance for large EITE industries. The Liberals, on the other hand, sought increased levels of protection for EITE industries, concerned to ensure that protection compared favorably to jurisdictions such as the United States. The Nationals, indicated they would not support the CPRS Bill until after the Copenhagen Conference on Climate Change.

The CPRS Bill was re-introduced into the Australian Parliament on 22 October 2009 and again passed by the House of Representatives before moving into the Senate on 17 November 2009. To secure the passage of the Bill through the Senate, the Rudd Government turned to the opposition Liberal party to negotiate. On 24 November 2009, the Government released a package of amendments to the CPRS Bill, which it said represented "the culmination of over a month of detailed negotiations between the Government and the Opposition, and over a decade of policy development" and delivered a "deal to the Opposition" aimed at passing the CPRS.[39] Instead, a tumultuous debate on the issue of climate change followed which exposed strong divisions within the Liberal party and culminated in the election of a new Opposition leader. On 2 December 2009, despite the changes put forward in the 24 November package, the Senate again defeated the CPRS Bills by 42 votes to 30. That same day the Liberal party withdrew its support for an emissions trading scheme and announced it would also not implement a carbon tax.[40]

The CPRS Bill was again reintroduced into Parliament on 2 February 2010 and passed the House of Representatives on 11 February 2010. However, rather than putting the legislation to a third vote in the Senate, on the 27 April 2010 Prime Minister Rudd announced the implementation of CPRS would be delayed until after the end of the current commitment period of the Kyoto Protocol.

24.3.3 Obstacles to the Passage of the CPRS Bill

A closer look at two of the most significant obstacles to achieving legislative consensus on the CPRS Bill – the overall emission reduction commitments and the treatment of EITE industries –highlight the ongoing tensions between protecting

---

[39] Senator Penny Wong, Minister for Climate Change, Energy Efficiency and Water, *A Carbon Pollution Reduction Scheme in the National Interest*, Media Release, 29 November, 2009, available at: http://www.climatechange.gov.au/minister/previous/wong/2009/media-releases/November/mr20091124.aspx (last accessed on 22 February, 2012).

[40] Senator Penny Wong, Minister for Climate Change, Energy Efficiency and Water, *New Opposition Policy – No ETS and No Carbon Tax*, Media Release, 3 December, 2009, available at: http://www.climatechange.gov.au/en/minister/previous/wong/2009/media-releases/December/mr20091203.aspx (last accessed on 22 February, 2012).

Australia's emission intensive economy and taking effective domestic action to mitigate climate change.

24.3.3.1 Overall Emission Reduction Commitments

Rising levels of greenhouse gas emissions associated with industrial and agricultural activities have sustained rising living standards over the past two centuries and the transformation of existing production and consumption patterns to reduce emissions dramatically requires change that reaches deep into current practices.[41] This is undoubtedly the case for the emission intensive economy of Australia, with many years of 'no-regrets' policies demonstrating the need for significant reform in order to bring about structural change to the economy. Nevertheless, a prevailing concern in designing the CPRS, and particularly in setting the overall emission reduction commitments, was to structure a transition which balanced the need to protect the Australian economy with the objective of achieving strong mitigation outcomes.

The original CPRS proposal included a reduction target aimed at reducing greenhouse gas emissions to 60% below 2000 levels by 2050 and a reduction target of 5–15% below 2000 levels by 2020. These targets were said by the Government to be acceptable in that they were expected to impose, in aggregate, a modest cost to the economy while at the same time providing a "credible and constructive contribution to achieving a long-term global solution capable of protecting the planet and promoting our national interest."[42] However, having accepted that "Australia's national interest was best served by a comprehensive global agreement to stabilize atmospheric concentrations of greenhouse gases at around 450 ppm of CO_2-e or lower",[43] the Rudd Government proposed to set post-2020 reduction targets "so as to ensure it plays its full role in achieving the agreed goal" should such an agreement emerge.[44] After sustained criticism that the existing targets were inadequate, the CPRS Bill was amended to contemplate a reduction of 25% below 2000 levels by 2020 conditioned, however, on Australia becoming a party to a comprehensive international agreement capable of stabilizing atmospheric concentrations of greenhouse gases at around 450 ppm CO_2-e or lower.[45]

---

[41] James Meadowcroft, *Climate Change Governance, Policy Research Working Paper No. 4941* (Washington, DC: World Bank, 2009), at 4, available at: http://papers.ssrn.com/sol3/papers.cfm?abstract_id=1407959 (last accessed on 22 February 2012).

[42] Commonwealth Government, *Carbon Pollution Reduction Scheme: Australia's Low Pollution Future* (Canberra: Commonwealth Government, 2008), at 4-16–4-17.

[43] Ibid., at 4-1.

[44] Commonwealth of Australia, *Commentary: Exposure Draft Carbon Pollution Reduction Scheme Bill 2009* (Canberra: Commonwealth Government, 2009), at 87–88.

[45] Carbon Pollution Reduction Scheme Bill (Cth) 2009, cl 3(4)(a).

While the addition of the conditional 25% reduction target by 2020 was viewed as a more credible target and enough to garner the support of some,[46] others viewed both the 2020 and 2050 targets contained in the CPRS Bill as inadequate.[47] However, rather than setting aggressive unconditional targets directed at achieving deep structural changes, the Rudd Government was not prepared to move beyond a modest unconditional 5% reduction target by 2020 absent global action.

24.3.3.2 Assistance Packages for EITE

The CPRS proposed three different types of industry assistance: assistance to new and existing energy intensive and trade exposed (EITE) industry; compensation for loss in asset value to coal-fired electricity generators; and transitional assistance to coal mines with high fugitive emissions. The most contentious form of assistance was that proposed for EITE industries through the allocation of free permits.

The justification for providing assistance to EITE industry was two fold: first to avoid the risk of carbon leakage; and second to "smooth the transition for individual firms, rather than just have them take a hit on their profit."[48] In attempting to determine an appropriate level of assistance, the Rudd Government was again confronted with the extent to which the most emission intensive sectors of the Australia economy were be buffered from the impacts of a carbon price. This necessarily included decisions relating to how to spread the patterns of risk and opportunity across the economy. As Garnaut put it, assisting this type of industry presents "a truly dreadful problem" for policy makers as "it undermines attempts to limit national ghg emissions or increases the adjustment burden elsewhere in the economy".[49]

As the CPRS Bill progressed, the level of assistance available to EITE industry changed substantially. The resulting EITE assistance package was the subject of two main types of criticism. The first argued that EITE should receive more assistance. This position was based on the assertion that trade-exposed industry would otherwise be unable to compete internationally or would be driven offshore inviting resulting risk of carbon leakage or, alternatively, required transitional assistance to adjust to the carbon constrained economy.[50] The second argued that EITE industry should

[46] See for example: Climate Institute, "How will the CPRS carnival end?", 2009, available at: http://www.climateinstitute.org.au/index.php?option=com_content&view=article&id=571:how-will-the-cprs-carnival-end-&catid=112:blogs&Itemid=49 (last accessed on 22 February 2012).

[47] Michael Power, "Emissions Trading in Australia: Markets, Law and Justice Under the CPRS", 27 *Environmental Planning and Law Journal* (2010), at 131.

[48] Standing Committee on Economics, Exposure Draft of the Legislation to Implement the Carbon Pollution Reduction Scheme (Canberra: Senate Printing Unit, 2009), at 44.

[49] Garnaut, *Garnaut Climate Change Review*, supra, note 1, at 316.

[50] Standing Committee on Economics, *Exposure Draft*, supra, note 48, at 42–48; Senate Select Committee on Fuel and Energy, *Interim Report: The CPRS: Economic Cost without Environmental Benefit* (Canberra: Senate Printing Unit, 2009), at 151–152; Select Committee on Climate Policy, *Report* (Canberra: Senate Printing Unit, 2009), at 77–78 and 81–86.

receive less assistance. This position was based on the lack of evidence to support the risk of carbon leakage,[51] the fact a significant proportion of the assistance would go to a handful of very large companies,[52] the consequential shifting or burden away from these polluters to the rest of the economy,[53] and, of course, the fact that the assistance "mutes the incentives" for EITE industry to reduce their ghg pollution.[54]

Rather than adopt a vigorous approach to climate change governance, which "cannot avoid disturbing power economic and political interests"[55] the CRPS sought to accommodate them with the result that the assistance available to EITE industries increased at each step in the development of the CPRS. Ultimately the treatment of EITE industries, together with the veracity of the emission reduction commitments, both polarized and dominated the debate and undermined both the validity and acceptability of the overall CPRS Scheme.[56]

24.4 Clean Energy Act 2011 (Cth)

24.4.1 "Carbon Pricing is a Reform We Need to Make to Keep Our Economy Competitive, to Protect Our Environment and to Do the Right Thing for Our Children"[57]

24.4.1.1 Background

On 24 June 2010, for reasons largely unrelated to climate change legislation, Julia Gillard replaced Kevin Rudd as Labor leader and Prime Minister of Australia. Prime Minister Gillard subsequently called a federal election and, during the election

---

[51] Standing Committee on Economics, *Exposure Draft,* supra, note 50, at 42–44 and 49; Select Committee on Climate Policy, *Report,* supra, note 50, at 78–79; and, Power, "Emission Trading in Australia", supra, note 47, at 156.

[52] RiskMetrics Group, *Research Note: The Impact of Industry Assistance Measures under the CPRS* (2009), at 6–7, available at: http://www.acfonline.org.au/sites/default/files/resources/RiskMetrics_CPRS_Industry_Assistance_May09.pdf (last accessed on 22 February, 2012).

[53] John Daley and Tristan Edis, *Restructuring the Australian Economy to Emit Less Carbon: Main Report* (Victoria: Grattan Institute, 2010), at 11 and 14; and, Power, "Emissions Trading in Australia", supra, note 47.

[54] Daley and Edis, *Restructuring the Australian Economy*, supra, note 53, at 12–13.

[55] Meadowcroft, *Climate Change Governance*, supra, note 41, at 35.

[56] Tim Flannery and Nick Rowley, "Comment: Carbon Omissions", *The Monthly* (2009), available at: http://www.themonthly.com.au/Tim-Flannery-Nick-Rowley (last accessed on 22 February, 2012).

[57] Prime Minister of Australia, The Hon Julia Gillard MP, "Securing a clean energy future for Australia", 10 July 2011, available at: http://www.pm.gov.au/press-office/securing-clean-energy-future-australia (last accessed on 24 February 2012).

campaign, made it clear that an emissions trading scheme would be introduced in (or, rather, delayed until the end of) 2012. The Labor Party was able to form government after the election, but only with the support of a number of independent members of the Commonwealth Parliament.

In September 2010 the establishment of a Multi-Party Climate Change Committee (the CCC) was announced by the Prime Minister, the aim of the CCC being to "consult, negotiate, and report to the Cabinet … on agreed options for the implementation of a carbon price in Australia."[58] Membership included senior members of the Government, including the Prime Minister, together with two members of the Australian Greens and two independent members of parliament (those whose support had enabled the Prime Minister to form government). The CCC was advised by Ross Garnaut, amongst others. It did not include members from the opposition parties; they had declined to participate.

The rationale – largely economic – for the introduction of a carbon price by the government was put succinctly by the federal Minister for Climate Change and Energy Efficiency when he said that the costs of carbon pollution had not been borne by its producers but by society as a whole and that this "must now change".[59] He argued that such costs need to be considered when companies and individuals make decisions about what to produce, what to invest in and what to consume. This means that the true cost of carbon pollution needs to be attached to its production and use, that is carbon emissions need to have a price … a carbon price will create the incentive for large emitters to reduce pollution, and stimulate investment in low emissions technologies and processes. It will provide greater certainty for business investment.[60]

The CCC established terms of reference, commissioned a number of studies, and published a set of principles to guide the development of a price on carbon: environmental effectiveness; economic efficiency; budget neutrality; competitiveness of Australian industries; energy security; investment certainty; fairness; flexibility; administrative simplicity; clear accountabilities; and support for Australia's international objectives and obligations.[61] In February 2011, the Prime Minister announced the outline of a "broad architecture" of the Government's plan "to cut pollution, tackle climate change and deliver the economic reform Australia needs to move to

[58] Australian Government, Department of Climate Change and Energy Efficiency, "Carbon pricing", available at: http://www.climatechange.gov.au/government/reduce/carbon-pricing.aspx (last accessed on 24 February 2012).

[59] The Hon Greg Combet AM MP, Minister for Climate Change and Energy Efficiency, "Address to the AIGN/BCA Carbon Pricing Forum", 23 March 2011, available at: http://www.climatechange.gov.au/minister/greg-combet/2011/major-speeches/March/sp20110323.aspx (last accessed on 25 February 2012).

[60] Ibid.

[61] Multi-Party Climate Change Committee, "MPCC Agreed Principles to Guide Development of a Carbon Price Mechanism", 24 February 2011, available at: http://www.climatechange.gov.au/government/initiatives/~/media/publications/mpccc/mpccc-carbon-price-mechanism.pdf (last accessed on 20 March 2012).

a clean energy future".[62] The Government proposed a "carbon price mechanism" (the CPM) which would start on 1 July 2012 with a fixed price period of between 3 and 5 years, with a transition to an emissions trading scheme after that period.[63] The commencement of this CPM, of course, was subject to the government's ability to negotiate agreement with a majority of both houses of Parliament (negotiation made more difficult given the close election result) and to pass the legislation.

24.4.1.2 Starting Price, Flexible Price, Price Floor and Ceiling, Energy Intensive Trade Exposed Industries

On 10 July 2011 the Government announced further details of and refinements to the CPM, the result in part of additional negotiations with the CCC which released its Clean Energy Agreement.[64] The CPM was to commence with a carbon price of AUD 23 per tonne and with a 3-year "fixed price" period (although the price of a permit – or a "carbon unit" under the CPM – would increase each year by 2.5%).[65] A "flexible price" period and an open "cap-and-trade" emissions trading scheme would operate from 1 July 2015 onwards, with a price "floor" and "ceiling" for the first 3 years after that date.[66] Details announced regarding EITE industries were broadly similar to those set out under the CPRS.

24.4.2 Legislation to "Encourage the Use of Clean Energy"[67]: Main Design Features

24.4.2.1 Introduction

Australia's climate change legislation passed by the Senate in 2011 – the main piece of which is the Clean Energy Act 2011 (Cth) (the Act), legislation "to encourage the use of clean energy", and for other purposes[68] – introduces a price on carbon by way

---

[62] Prime Minister, Minister for Climate Change and Energy Efficiency, "Climate Change Framework Announced", 24 February 2011, available at: http://www.climatechange.gov.au/minister/greg-combet/2011/media-releases/February/mr20110224.aspx (last accessed on 25 February 2012).

[63] Ibid.

[64] Australian Government, Multi-Party Climate Change Committee, Clean Energy Agreement, 10 July 2011, available at: http://www.climatechange.gov.au/government/initiatives/~/media/publications/mpccc/mpccc_cleanenergy_agreement-pdf.pdf (last accessed on 25 February, 2012).

[65] Australian Government, *Securing a Clean Energy Future: The Australian Government's Climate Change Plan* (Canberra: Commonwealth of Australia, 2011), at xiii.

[66] Ibid.

[67] Clean Energy Act 2001 (Cth), supra, note 13, Long Title.

[68] Ibid.

of a "carbon price mechanism" (again, a CPM) which commences on 1 July 2012.[69] Liable entities under the CPM (those corporations generating over 25,000 tonnes of CO_2e emissions each year[70]) must purchase and surrender carbon units for each tonne of carbon pollution they emit.

24.4.2.2 Objects

The objects of the Act include giving effect to Australia's obligations under the United Nations Framework Convention on Climate Change (the UNFCCC)[71] and its Kyoto Protocol[72]; supporting "the development of an effective global response to climate change;" taking action directed towards meeting Australia's target of reducing its net greenhouse gas emissions to 80% below 2000 levels by 2050 in "a flexible and cost-effective way": and putting a price on GHG emissions such that investment in clean energy is encouraged, jobs and competitiveness in the economy is supported; and economic growth is supported while pollution is reduced.[73]

24.4.2.3 Sectors Covered

The CPM will cover emissions from about 500 "liable entities" across the stationary energy, industrial processing, waste and resources sectors, covering approximately 60% of Australia's emissions.[74]

24.4.2.4 Carbon Units

The price of carbon units will be fixed in the fixed price period. In the flexible price period, carbon units will be freely tradable. In both periods, units (howsoever described) from offset projects both domestic and international may be used, although with some restrictions. In both the fixed and flexible price periods, liable entities under the CPM must acquire and surrender carbon units that are equal to their annual emissions from activities covered by the CPM.[75]

---

[69] Ibid., Section 4.

[70] Ibid., Part 3, Division 2.

[71] United Nations Framework Convention on Climate Change, New York, 9 May 1992, in force 21 March 1994, 31 *International Legal Materials* (1992), 849.

[72] Kyoto Protocol, supra, note 23.

[73] Ibid., Section 3.

[74] Australian Government, Securing a Clean Energy Future: The Australian Government's Climate Change Plan, supra, note 65, at xii, xiii.

[75] Clean Energy Act 2001 (Cth), supra, note 13, Part 4, Division 2.

International carbon units, including Certified Emission Reductions (CERs) from Clean Development Mechanism (CDM) projects and Emission Reduction Units (ERUs) from Joint Implementation (JI) projects under the Kyoto Protocol can be used to meet CPM liabilities up to 50% of the relevant entity's carbon unit surrender obligation. Other permitted international carbon units include removal units issued by a Kyoto Protocol state on the basis of land use, land-use change and forestry (LULUCF) activities and other international units permitted by Government regulation.[76]

24.4.2.5 EITE Assistance Units

EITE industries (together with coal-fired power generators[77]) will receive assistance in the form of free carbon units.

24.4.2.6 Liability Transfer

Liability under the CPM can be transferred from one corporate facility to another member of the corporate group or another person who has financial control of the facility. Corporate members of unincorporated joint ventures may make application to transfer emissions liability to joint venture participants in proportion to their interest in the facility.[78]

24.4.2.7 Agriculture, the Land Sector and the Carbon Farming Initiative

The CPM excludes the agricultural sector. As a result, farmers, forestry operators and other land managers will not be liable entities under the CPM. However, under the Carbon Credits (Carbon Farming Initiative) Act 2011 (Cth) (the CFI), which sets up a scheme for the issue of Australian carbon credit units (ACCUs) in relation to certain eligible offsets projects,[79] farmers and other entities can generate credits from a sector not covered by the CPM which can be used by liable entities to meet obligations under the CPM.

Although the CFI can work independently of the carbon price mechanism, "compliance" ACCUs can be used under the CPM to meet up to 5% of compliance obligations in the 3 years, fixed price period,[80] with no restrictions as to use after that period.

[76] Ibid., Part 6, Division 1.

[77] Ibid., Part 8.

[78] Ibid., Part 3, Division 6.

[79] Carbon Credits (Carbon Farming Initiative) Act 2011 (Cth), Parts 2 and 3.

[80] Clean Energy Act 2001 (Cth), supra, note 13, Section 125(7).

24.4.3 *An Effective Mechanism to Transition Australia to a Clean Energy Future?*

The objects of the Act include putting a price on greenhouse gas emissions such that "investment in clean energy" is simply encouraged, and taking action which is merely "directed towards meeting" Australia's long-term target of reducing its net greenhouse gas emissions to 80% below 2000 levels by 2050 – a target just less than 40 years away.[81] While the mechanisms which the Act puts in place *may* encourage the *use* of clean energy (as the Act's long title suggests), it seems clear that transitioning Australia to a clean energy future is not one of its objectives.

This becomes clearer when one considers what the Act does not do. For example, it would not appear (given its relatively narrow scope and its generous levels of industry compensation) to deliver 'least cost' emissions reductions, and it does not remove subsidies for fossil fuel use.

The Act is also concerned to give effect to Australia's obligations under the UNFCCC and its Kyoto Protocol.[82] However, the first commitment period under the Kyoto Protocol ends in 2012, with no formal targets negotiated beyond a decision on a second commitment period to begin on 1 January 2013 and end either in 2017 or 2020.[83]

COP-17 at Durban also launched a Platform for Enhanced Action, a non-binding agreement "to develop a protocol, another legal instrument or an agreed outcome with legal force" under the UNFCCC.[84] Any such protocol, legal instrument or "agreed outcome with legal force" is to be concluded by 2015, with "pledges" from developed and developing state parties to reduce emissions, and ostensibly to come into effect and be implemented from 2020.[85] These parties would also, of course, need to ratify such agreement. The Durban Platform is, however, simply an agreement to reach agreement. Additionally, then, it is also possible to argue that giving effect to Australia's obligations under the UNFCCC and Kyoto Protocol, such as they are, does not assist Australia – or the world – to transition to a clean energy future.

24.5 Conclusions

With the passage of the Clean Energy Act 2011 (Cth), Australia has finally stepped beyond the 'no-regrets' approach that for decades has dominated it domestic climate change mitigation policy. However, it seems unlikely that this Act, alone, will

[81] Ibid., Section 3.

[82] Ibid.

[83] See Decision 1/CMP.7, Outcome of the work of the Ad Hoc Working Group on Further Commitments for Annex I Parties under the Kyoto Protocol at its sixteenth session, FCCC/KP/CMP/2011/10/Add.1, 15 March 2011, para. 1.

[84] Decision 1/CP.17, Establishment of an Ad Hoc Working Group on the Durban Platform for Enhanced Action, UN. Doc. FCCC/CP/2011/9/Add.1, 15 March 2012, para. 2.

[85] Ibid., para. 4.

provide a mechanism to transition the country to a clean energy future. Rather, in putting a price on carbon, this legislation seeks only to "encourage" investment in clean energy while at the same time supporting jobs and competitiveness in the economy. Until these objectives are seen as one and the same, the tensions between adopting effective measures to mitigate climate change in Australia and protecting the country's economic interests remain an obstacle to a full transition to a clean energy economy.

Chapter 25
Climate Law and Policy in Japan

Hitomi Kimura

Abstract Japanese climate law and policy have developed rapidly since the 1990s after the ratification of the United Nations Framework Convention on Climate Change and the Kyoto Protocol. They are characterized by the staged introduction of policies and measures, as well as by a step-by-step approach, with a review in 3-year intervals. The regulatory approach combines a framework law, the Law Concerning the Promotion of the Measures to Cope with Global Warming, and specific laws, as well as the proactive use of voluntary approaches such as Keidanren's Voluntary Action Plan. Unique policies have also been introduced, including the Japanese Voluntary Emissions Trading Scheme, a domestic offset system and bilateral offset mechanisms. In comparison to the European Union, the Japanese approach to climate law and policy has been passive. It was impossible for the Japanese Diet to pass the 2010 Basic Bill to Cope with Global Warming pending adoption of a legally-binding and comprehensive international climate treaty. Furthermore, Japanese decision not to participate in the second commitment period under the Kyoto Protocol spells out a gloomy future for a quick passage of the Bill. Japan will comply with its 6% emission reduction target under the Kyoto Protocol until the end of 2012, take reluctant policies and measures based on Japan's voluntary mitigation pledge under the Cancun Agreements. The voluntary target of reducing emissions by 25% below 1990 levels is most likely to be decreased without any binding international commitment after 2012, and Japan will then consider whether to join the legally-binding comprehensive treaty expected after 2020.

Hitomi Kimura is Assistant Professor at the Otsuma Women's University.

H. Kimura (✉)
Department of Social Information Studies, Otsuma Women's University,
Karakida 2-7-1, Tokyo, 206-8540, Japan
e-mail: kimura@otsuma.ac.jp

E.J. Hollo et al. (eds.), *Climate Change and the Law*, Ius Gentium: Comparative Perspectives on Law and Justice 21, DOI 10.1007/978-94-007-5440-9_25,
© Springer Science+Business Media Dordrecht 2013

25.1 Characteristics of Japanese Climate Policy

Japanese climate law and policy has rapidly developed since the 1990s after the ratification of the United Nations Framework Convention on Climate Change (UNFCCC)[1] and the Kyoto Protocol.[2] Their evolution has been characterized by: a staged introduction of policies and measures; a step-by-step approach, with a review in 3-year intervals; a regulatory approach combining a framework law, the Law Concerning the Promotion of the Measures to Cope with Global Warming, and specific laws; as well as proactive use of voluntary approaches, such as Keidanren's Voluntary Action Plan, strongly influenced by the industrial sector; and a relatively reluctant use of economic instruments, including emissions trading.[3] As a compromise, unique climate policy and measures were introduced, including the Japanese Voluntary Emissions Trading Scheme (JVETS), a domestic offsetting system and bilateral offset mechanisms.

25.2 Evolution of Climate Change Law and Policy in Japan

25.2.1 *Ratification of the UNFCCC and the Kyoto Protocol*

The evolution of Japanese climate policy began with the formulation of the Action Program to Arrest Global Warming in preparation for the adoption of the UNFCCC in 1992.[4] Japan ratified the UNFCCC in 1994, the same year as the Convention entered into force. In Japan, environmental treaties are not self-executing as domestic laws, but it is necessary to enact and implement domestic laws to comply with the mandate and achieve the objective of the treaty.[5] Global warming was included in the 1994 Basic Environment Act and 1995 Basic Environmental Plan domestically, but these policies consist of various policies from different ministries.[6] With no strong legislation and policy in place, the biggest industrial group "Keidanren"

---

[1] United Nations Framework Convention on Climate Change, New York, 9 May 1992, in force 21 March 1994, 31 *International Legal Materials* (1992), 849.

[2] Kyoto Protocol to the United Nations Framework Convention on Climate Change, Kyoto, 10 December 1997, in force 16 February 2005, 37 *International Legal Materials* (1998), 22.

[3] Tadashi Otsuka, *Global Warming Law and Policy* (Tokyo: Showadou, 2004).

[4] Japanese Ministry of the Environment, "Action Program to Arrest Global Warming," Tokyo, 23 October 1990, available at: http://www.env.go.jp/hourei/syousai.php?id=03000015 (last accessed on 1 May 2012).

[5] Otsuka, *Global Warming Law and Policy*, supra, note 3.

[6] Tadashi Otsuka, *Environmental Law* (Tokyo: Yuhikaku, 2010), at 157. Basic Environmental Act, Tokyo, 11 November 1993, in force 11 November 1993, Roppou Zensho (2012), 2393.

set the target to limit its carbon dioxide emissions below the 1990 level by 2010 from the industrial and energy sectors, including power and oil as part of their Voluntary Action Plan in 1997. The Global Warming Prevention Headquarters were also established in 1997, consisting of the Prime Minister as well as ministers from the Ministry of Environment (MoE), Ministry of Economy, Trade and Industry (METI) and other related ministries.

In 1998, 1 year after the adoption of the Kyoto Protocol, energy conservation was given a central role in the Guideline of Measures to Global Warming,[7] which substituted the Action Program to Arrest Global Warming. Huge plants were required to improve their energy efficiency by 1% annually through the amendment of Energy Conservation Law, enacted to cope with the oil shock in the 1970s, and companies were required to achieve the best available energy efficiency level in the market through the Top Runner Approach. The Law Concerning the Promotion of the Measures to Cope with Global Warming[8] was also enacted in 1998 as an umbrella law to cope with global warming.

Adoption of the Marrakesh Accord at the seventh session of the Conference of the Parties (COP) in 2001 gave impetus for the adoption of the New Guideline of Measures to Global Warming (New Guideline).[9] The New Guideline lists policy packages, targeting each sector, with a total of 115 policies and measures, but there were no fundamental changes in comparison to the old guidelines. Soon after Japan's ratification of the Kyoto Protocol, the Law Concerning the Promotion of the Measures to Cope with Global Warming was also amended along with the Energy Conservation Law to enforce action in household and office buildings. In the context of the first review of climate policy in 2004, the New Guideline was revised. The MoE and METI shared the view that further action was necessary to achieve Japan's 6% emission reduction target under the Kyoto Protocol, but their proposals for specific measures differed from each other. While the METI insisted on compliance without any tax increases and enforcement of the existing policies and measures, such as energy conservation, improvement of transparency of Keidanren's Voluntary Action Plan and promoting the use of the Kyoto Protocol's flexibility mechanisms, the MoE insisted on the introduction of additional measures, including an environmental tax, mandatory reporting scheme for greenhouse gas emissions, emissions trading and the use of the flexibility mechanisms.

[7] "Guideline of Measures to Global Warming," Tokyo, 19 March 2002, adopted by the Global Warming Prevention Headquarters, Japanese Ministry of the Environment, available at: http://www.env.go.jp/earth/ondanka/taiko/all.pdf (last accessed on 1 May 2012).

[8] Law Concerning the Promotion of the Measures to Cope with Global Warming, Tokyo, 9 October 1998, in force 8 April 1999, Roppou Zensho (2012), 2485.

[9] New Guideline of Measures to Global Warming. 19 March 2002, adopted by the Global Warming Prevention Headquarters, Japanese Ministry of the Environment, available at: http://www.env.go.jp/council/16pol-ear/y161-07/mat01.pdf (last accessed on 1 May 2012).

25.2.2 Implementation of the Kyoto Protocol

With the entry into force of the Kyoto Protocol in February 2005, the Cabinet approved the Kyoto Target Achievement Plan (KTAP). To implement this plan, action and follow-up of Keidanren's Voluntary Action Plan was given a central role. The 2005 amendment to the Energy Conservation Law required companies to submit an emission reduction plan and regular reports in the industrial sector, enforced energy conservation in the transport sector and expanded Top Runner products from 9 to 21 in the household sector.

In 2005, the MoE also introduced the JVETS. Its voluntary nature was due to strong opposition against a mandatory emissions trading scheme by the industrial sector and the METI. The JVETS is characterized by voluntary targets. However, the targets are absolute rather than intensity-based, and they are also binding with penalties once a firm has agreed to participate in the scheme. Subsidies were made available to firms to assist them in achieving emissions reductions, but could not be used after April 2009, reflecting criticism by other countries that this was a subsidy and distortion of the market. Tradable units under the JVETS include: excess units accumulated under Keidanren's Voluntary Action Plan; units tradable under the JVETS (Japanese Emission Allowances, JPA and Japanese Certified Emission Reductions, j-CERs); and credits from a domestic offsetting system that is similar to the Clean Development Mechanism (CDM) under the Kyoto Protocol. Accordingly, small and medium enterprises (SMEs) not covered by the JVETS would be granted credits for emission reductions achieved through projects undertaken voluntarily, under rules similar to those in operation for the CDM. Large companies participating in the JVETS will be able to purchase emission credits generated from these projects and use them for their own compliance. In case of non-compliance, companies have to refund any subsidies received, and the names of corporations that fail to meet their targets will be made public. A scheme for calculation, reporting and publication of greenhouse gas emissions was also introduced to support the ETS.[10]

The introduction of JVEST was a revolutionary event in Japanese climate policy. However, its effectiveness is limited because many major emitters did not join the scheme, the emission reduction targets do not require deep cuts, and the penalties are not severe. In general, a voluntary emissions trading scheme attracts participants that can easily achieve the pledged targets. Although the number of

[10] Hitomi Kimura and Andreas Tuerk, *Emerging Japanese Emissions Trading Schemes and Prospects for Linking* (Cambridge: Climate Strategies, 2008), available at: http://www.climatestrategies.org/component/reports/category/33/81.html (last accessed on 1 May 2012). Japanese Ministry of the Environment, "Approach to Japanese Emissions Trading Scheme: Interim Report, Executive Summary", 2008, available at: http://www.env.go.jp/en/headline/file_view.php?serial=233&hou_id=788 (last accessed on 1 May 2012).

participants in the JVETS has been increasing, participation in the scheme is not sufficient to support a truly effective and efficient market. During Phase I (from April 2005 to March 2006) of the JVETS, the scheme included 31 target participants and seven trading participants. During Phase II (from April 2006 to March 2007) this increased to 61 target participants and 12 trading participants. During Phase III (from April 2007 to March 2008), the JVETS included 61 target participants and 25 trading participants. The number of transaction is small with only 24 transactions during Phase I and 51 during Phase II. The JVETS, with its absolute targets, does not include the firms from the most energy intensive sectors, such as steel and power, although such firms do participate in the Keidanren Voluntary Action Plan which has intensity-based targets.[11]

Following the formal adoption of the Marrakesh Accord and precise rules for the Kyoto Protocol's flexibility mechanisms at the first session of the Conference of the Parties serving as the meeting of the Parties to the Kyoto Protocol (COP/MOP), New Energy and Industrial Technology Development Organization (NEDO) under METI was appointed as an institution to procure Kyoto credits. As part of the second review of the Japanese climate policy in 2006, the two Committees under MOE and METI were combined to consider additional measures. The 2006 amendment to the Law Concerning the Promotion of the Measures to Cope with Global Warming was designed to assure the legally safe platform for emissions trading by recognizing asset value for credits.[12]

In 2008, the Tokyo Metropolitan Government (TMG) passed a law to establish a regional mandatory emissions trading scheme, starting in April 2010.[13] In 2008, the TMG also joined the International Carbon Action Partnership (ICAP), a group of representatives of trading schemes who try to ensure sufficient harmonization and compatibility to support direct bilateral links between the schemes. This action emphasized the TMG's positive position on linkage. The initiatives by the TMG are expected both to drive the introduction of a mandatory emissions trading at the national level and to push linkage considerations forward.[14]

Japan's desire to demonstrate political leadership at the Summit of the Group of Eight (G-8) held in 2008 in Toyako, Hokkaido, was the critical factor in bringing climate change to the top of political agendas in Japan.[15] At the end of the G-8 Summit, Japan was successful in mentioning "the goal of achieving at least 50% reduction of global emissions by 2050", setting "ambitious economy-wide mid-term

[11] Kimura and Tuerk, *Emerging Japanese Emissions Trading Schemes*, supra, note 10.

[12] Tadashi Otsuka, "EU ETS and issues for Japan. In Special issue of the entry into force of the Kyoto Protocol and action against global warming", 1296 *Jyurist* (2008), at 30.

[13] Tokyo Metropolitan Government, "Amendment of Tokyo Metropolitan Ordinance on Environmental Preservation", 2012, available at: http://www.kankyo.metro.tokyo.jp/joureikaisei2008/index.htm (last accessed on 1 May 2012).

[14] Kimura and Tuerk, *Emerging Japanese Emissions Trading Schemes,* supra, note 11.

[15] Ibid.

goals in order to achieve absolute emissions reductions…, reflecting comparable efforts among all developed economies", and that "sectoral approaches can be useful tools to improve energy efficiency and reduce GHG emissions."[16] Hosting the G-8 Summit pushed the development of Japanese climate law and policy in the same vein as other international conferences held in Japan have triggered the progress of Japanese environmental policy, including COP 3 of the UNFCCC in Kyoto in 1997 and COP 10 of the Convention on Biological Diversity in Nagoya in 2010. In order to achieve Japan's 6% emission reduction target of the Kyoto Protocol, the Kyoto Target Achievement Plan (KTAP) and the Law Concerning the Promotion of the Measures to Cope with Global Warming and Energy Conservation Law were amended. To achieve long-term emission reductions, the Action Plan for Building Low Carbon Society, setting out a 60–80% greenhouse gas emission reduction target by 2050, was developed.

25.2.3 *Turning Points for the Mid- and Long-Term Japanese Climate Policy*

Political shift from the long-ruling Liberal Democratic Party to the Democratic Party of Japan in September 2009 constituted a turning point for Japanese climate policy. The 2010 Basic Bill to Cope with Global Warming decided by the Cabinet in 2010 included targets to reduce emissions by 25% by 2020, and 80% by 2050 below the 1990 level.[17] It also included the objective of increasing the portion of renewable energy to 10% of the primary energy supply by 2020.[18] The 25% mid-term emission reduction target was submitted to the UNFCCC by Japan as voluntary pledge under the Copenhagen Accord and was "premised on the establishment of a fair and effective international framework in which all major economies participate and on agreement by those economies on ambitious targets."[19]

[16] Group of Eight Summit, "G8 Hokkaido Toyako Summit Leaders Declaration of 8 July 2008," available at: http://www.mofa.go.jp/policy/economy/summit/2008/doc/doc080714__en.html%2 (last accessed on 1 May 2012).

[17] Japanese Cabinet, "Basic Bill to Cope with Global Warming", 8 October 2000, available at: http://www.env.go.jp/info/hoan/index2.html#176 (last accessed on 1 May 2012).

[18] Ibid.

[19] Japanese Ministry of Foreign Affairs, "Submission of the information on Japan's willingness to be associated with the Copenhagen Accord and its quantified economy-wide emissions target for 2020", Tokyo, 26 January 2010, available at: http://www.mofa.go.jp/announce/announce/2010/1/PDF/012601e.pdf (last accessed on 1 May 2012).

25.2.3.1 Possible Change of Nuclear-Dependent to Renewable-Based Climate and Energy Policy After Fukushima Nuclear Accident

On 11 March 2011, a strong earthquake with a magnitude of 9 hit Tohoku area of Japan. The Tohoku Earthquake resulted in a huge Tsunami, recorded at maximum of 38.9 m, according to a survey by University of Tokyo, and left around 20,000 dead or missing. The nightmare continued by the subsequent accident at the Fukushima nuclear power station, which exploded and melted down due to station black out triggered by the Tsunami. Even in Tokyo, located more than 400 km from the source, many people suffered from the strong earthquake and its aftershocks, traffic paralysis and planned blackouts due to the halt of nuclear power plants, and lack of food and water in the supermarkets.

The Fukushima Nuclear Accident triggered a nation-wide discussion about Japanese energy policy, more specifically on how to shift from nuclear energy, amounting to 23.8% of Japan's total electricity generation, to renewable energy which currently only accounts for 9%.[20] The Japanese climate policy has been dependent on nuclear energy, which emits less carbon dioxide than oil-based thermal power generation. The 2003 Basic Energy Plan included the objective to increase the ratio of nuclear energy to 53% of total power generation by constructing 14 new nuclear power plants by 2030.[21] The 2005 Framework for Nuclear Energy Policy, based on the 1955 Atomic Energy Basic Act, sought to maintain the current 30–40% level or more of nuclear energy in total electricity generation even after 2030 in order to prevent global warming and secure a safe energy supply.[22] The 2006 New National Energy Strategy also sought to increase the proportion of nuclear energy to 30–40% in order to cope with the hike of oil price.[23]

The promotion of renewable energy started through the 1997 Law Concerning Special Measures to Promote the Use of New Energy (New Energy Law).[24] New energy includes both renewable and natural energy, being a wider concept of

---

[20] Organization for Economic Cooperation and Development, *OECD Environmental Performance Reviews: JAPAN* (Paris: OECD Publishing, 2010), at 132.

[21] Japanese Ministry of Trade, Economy and Industry, "Basic Energy Plan", Tokyo, October 2003, available at: http://www.meti.go.jp/report/downloadfiles/g31006b1j.pdf (last accessed on 1 May 2012).

[22] Atomic Energy Basic Act, Tokyo, 1 January 1956, in force 19 December 1957, Roppou Zensho (2012), 6226.

[23] Japanese Ministry of Trade, Economy and Industry, "New National Energy Strategy", Tokyo, 31 May 2006, available at http://www.meti.go.jp/committee/materials/downloadfiles/g60710a18j.pdf (last accessed on 1 May 2012).

[24] Law Concerning Special Measures to Promote the Use of New Energy, Tokyo, 18 April 2003, in force 2 June 1997, Kankyo Roppou (2011), 576.

energy alternatives to oil, originally inspired by experiences during the two oil shocks in the 1970s. The 2007 revision of the 2002 Renewable Portfolio Standards (RPS) Law increased the target of mandatory use of new energy in 2014 to 1.63%.[25] The introduction of a feed-in-tariff in 2009 allows owners of solar panels to sell the surplus of solar generation at 48 Yen per kilowatt hours during 10 years. The 2010 Basic Bill for Prevention of Global Warming, which intends to increase the ratio of renewable energy to 10% of the primary energy supply by 2020, was decided by the Cabinet but has not yet been passed by the Parliament as of May 2012.

Soon after the Fukushima accident, the former Prime Minister Kan announced that the 2003 Basic Energy Plan would be withdrawn and that the goal was to secure 20% of total power generation from renewable energy, and phase out nuclear energy. In August 2011, his Diet passed the Act on Special Measures concerning the Procurement of Renewable Electric Energy by Operators of Electric Utilities, which requires power companies to purchase all the electricity generated by the individual company from five renewable energies, namely solar, wind, water, geothermal and biomass, at a certain price to be fixed by the METI.[26] The succeeding Prime Minister Noda also mentioned the policy to decommission an end-of-life nuclear reactor and phase out the establishment of new power plants. Based on the policy drafted in late 2011 by the Energy and Environment Committee, various committees under related ministries will discuss the electricity mix in spring of 2012 and compile as strategy in the summer of 2012. In addition, the introduction of Tax on the Action against Global Warming from October 2012 will also be expected to reduce the use of fossil fuels.[27]

25.2.3.2 Japan's Role in the UN Climate Negotiations

Unfortunately, Japan has not really been a leader in the international climate change negotiations under the UN, and has often been given the infamous "Fossil of the Day Awards" as a member of the Umbrella Group, which also includes, *inter alia*, Australia, Canada, New Zealand, Norway, Russia and the United States. Japanese domestic climate policy has also not been strong enough to provide international leadership, except for energy conservation based on the Top Runner approach and energy efficient technologies. Furthermore, Japanese climate policy is influenced by

[25] Renewable Portfolio Standards (RPS) Law, Tokyo, 7 June 2002, in force 1 April 1993, Kankyo Roppou (2012), 591.

[26] Act on Special Measures concerning the Procurement of Renewable Electric Energy by Operators of Electric Utilities, Tokyo, 30 August 2011, in force 1 July 2012, Roppou Zensho (2012), 454.

[27] Japanese Ministry of Finance, "Outline of Tax Reform Guideline", Tokyo, 30 March 2012, passed by the Diet, available at: http://www.mof.go.jp/tax_policy/tax_reform/outline/fy2012/24taikougai you.html (last accessed on 1 May 2012).

the strong industry lobby and diverging views on climate policies by the MoE and METI. These factors have sometimes confused the Japanese position in the international negotiations, for instance, concerning sectoral approaches. Against this background, it was exceptional that in 2010, Japan provided leadership for international climate policy by setting the ambitious 25% emission target by 2020 from 1990 levels, which is higher than, for instance, the European Union's (EU) unilateral target of 20%. As of May 2012, however, the Basic Bill for Prevention of Global Warming has not been adopted. Instead, Prime Minister Noda has mentioned the possibility of decreasing the 25% target, reflecting the increasing reliance on thermal power plants after the 2011 Fukushima Nuclear Accident.

Overall, Japan tends to do its best to achieve its international obligations, once ratified, through transposition of multilateral treaties into domestic law, sound compliance and implementation. In general, however, the Japanese approach to climate law and policy has been passive, in comparison to active actors, such as the EU, which seeks to provide international examples through aggressive and ambitious internal legislation. In general, it is rare that Japanese domestic law contributes to the development of environmental treaties or international standardization of environmental standards.[28] Therefore, it is highly unlikely for the Japanese Diet to pass the 2010 Basic Bill to Cope with Global Warming pending the adoption of legally-binding and comprehensive international climate treaty.

Furthermore, Japanese decision at COP/MOP 7 in Durban not to undertake a mandatory emission reduction target under the Kyoto Protocol's second commitment period spells out a gloomy future for a quick passage of this Bill given that in the post-2012 period, Japan only has a voluntary international emission reduction pledge of 25%. Japan will certainly comply with its 6% Kyoto target until 2012 and then take reluctant domestic policies and measures based on the voluntary mitigation pledge under the Cancun Agreement. However, Japan's voluntary target of reducing emissions by 25% below 1990 levels is most likely to be decreased. Japan will then consider whether to join a legally-binding, comprehensive climate treaty, expected after 2020. In general, Japanese climate diplomacy has tended to focus on UN-based multilateral forums, seeking a careful balance between positions by the United States and the EU.[29] However, COP 17 shows that the Japanese position towards climate negotiation is shifting from the UN-based multilateral diplomacy towards unilateral or bilateral diplomacy. Having played a crucial role in the adoption of the Kyoto Protocol, Japan should, however, return to promoting multilateral climate diplomacy to combat climate change (Table 25.1).

---

[28] Hiroshi Isozaki, Relation between domestic environmental law and international environmental law: its reciprocal relation and mechanism of domestic implementation (Tokyo: Nihon Hyoronsha, 2009), at 17.

[29] Sebastian Oberthür and Hermann Ott, *The Kyoto Protocol: International Climate Policy for the Twenty-First Century* (Berlin: Springer, 1999), at 77–79.

Table 25.1 History of international and Japanese Climate Policy

	International climate policy	Japanese climate policy
1990		Action Program to Arrest Global Warming enacted
1992	UNFCCC adopted	
1993		Basic Environmental Law enacted
1994	UNFCCC entered into force	Japan ratifies the UNFCCC
		First Environmental Plan enacted
1995	COP 1	
1997	Kyoto Protocol adopted at COP 3, Kyoto	Global Warming Prevention Headquarters established
1998		Energy Conservation Law amended
		Guideline of Measures to Global Warming adopted
		Law Concerning the Promotion of the Measures to Cope with Global Warming enacted
2001	The Marrakesh Accords adopted at COP 7, Marrakesh	
2002		Guideline of Measures to Global Warming adopted
		Law Concerning the Promotion of the Measures to Cope with Global Warming amended
		Japan ratifies the Kyoto Protocol
2004		First Review of Climate Policy
2005	Kyoto Protocol entered into force	Kyoto Target Achievement Plan (KTAP) decided by the Cabinet
	G-8 Summit, Gleneagles	JVETS introduced
	First G-20 Dialogue, London	GHG calculation, reporting, publication scheme introduced
	Marrakesh Accord adopted by COP/MOP 1, Montreal	
2006		NEDO Law decided by the Cabinet
		Energy Conservation Law amended
		Second Review of Climate Policy
2007	UN High-Level Event on Climate Change	Strategy for an Environmental Nation in the Twenty-First Century developed
	Major Economies Meeting (MEM), Washington, DC	
	Fourth Assessment Report by the Intergovernmental Panel on Climate Change	
	COP 13 and COP/MOP 3, Bali	
2008	MEM, Washington, DC	Cool Earth Partnership declared
	Fourth G-20 Dialogue, Chiba	KTAP revised
	G-8 Summit, Japan	Action Plan for Building Low-Carbon Society developed

(continued)

Table 25.1 (continued)

	International climate policy	Japanese climate policy
2009	Major Economies Forum (MEF), Washington, DC	Sophisticated Methods of Energy Supply Structures enacted
	G-8/MEF Summit, L'aquila	Law Concerning Promotion of the Development and Introduction of Alternative Energy amended
	COP 15 and COP/MOP 5, Copenhagen	
2010	COP 16 and COP/MOP 6, Cancun	2010 Basic Bill to Cope with Global Warming decided by the Cabinet
2011	COP 17 and COP/MOP 7, Durban	Fukushima Nuclear Accident
		Act on Special Measures concerning the Procurement of Renewable Electric Energy by Operators of Electric Utilities passed by the Diet
		Japan decided not to participate in a second commitment period under the Kyoto Protocol after 2012
2012	COP 18 and COP/MOP 8, Doha	Tax on the Action against Global Warming introduced

Chapter 26
Sustainable Development and Climate Policy and Law in China

Christopher Tung

Abstract China has taken significant steps to advance sustainable development and transition to a low carbon economy. Since 1994, a national sustainable development strategy has underpinned the creation of policies and law that directly and indirectly impact the environment and climate change. While the policy and legislative process has been broadly successful, this chapter considers some of the problems found in practice and how certain decisions by Chinese regulators may compromise China's ability to enhance the development and implementation of effective sustainable development and climate change policies and law. This chapter concludes with recommendations on: (a) how these problems might be avoided; and (b) how sustainable development objectives and principles may be strengthened in the implementation of Chinese government policies and laws directly or indirectly impacting climate change and low carbon economy objectives.

26.1 Introduction

In 1994, the White Paper on China's Population, Environment and Development in the Twenty-First Century (China's Agenda 21 or A21) was approved by the State Council.[1] In the same year, the State Council issued a notice calling on Chinese government institutions at all levels to consider A21 as an overarching strategic guideline for the formulation of economic and social development plans, environmental

---

Christopher Tung is a partner in the Hong Kong office of global law firm K&L Gates.

[1] "China's Agenda 21", available at http://www.acca21.org.cn/ca21pa.html (last accessed on 15 June 2012).

C. Tung (✉)
K&L Gates, 44/F Edinburgh Tower, 15 Queens Road Central, Hongkong, China
e-mail: christopher-tung@klgates.com

E.J. Hollo et al. (eds.), *Climate Change and the Law*, Ius Gentium: Comparative Perspectives on Law and Justice 21, DOI 10.1007/978-94-007-5440-9_26,
© Springer Science+Business Media Dordrecht 2013

protection and day-to-day management. While A21 does not itself have the force of law, Chapter 3 of A21 provides for the incorporation of sustainable development objectives and principles into Chinese legislation.[2] Sustainable development objectives effectively require Chinese policy makers and regulators to reconcile and balance competing economic, social and environmental considerations in their decision making, so as to produce a sustainable outcome or solution for the long term. Relevant sustainable development law principles recognized in Chinese policy and law include: equity and eradication of poverty; common but differentiated responsibility; and precaution. With China's ambitious aspirations to transition sustainably to a low carbon economy, this chapter considers by reference to three examples discussed below, whether actions and decision making in China's climate policy and laws (both generally and those specifically aimed at supporting low carbon economic development) have been consistent with and compatible with these sustainable development objectives and principles.

26.2 Climate Policy and Law Consistent with Sustainable Development?

Whether climate policy and law are consistent with sustainable development is not a straightforward question. However, where there is an overarching commitment to develop and implement policies and laws that are consistent with sustainable development principles as provided for by A21, it is a question that should be continuously posed and at the forefront of the minds of policy makers, legislators and regulators. In theory and rhetoric, there is substantial reference in China to the application of sustainable development principles. In practice and substance, however, Chinese government actors may fall short of adhering properly to sustainable development principles. Conversely, it is possible to characterize non-optimum actions and solutions as a result of weighing up competing sustainable development considerations of a particular issue or problem. Seen in this light, what may at first brush appear to be a decision or action that is incompatible or divergent from sustainable development principles, may in fact be justified as a compromise between competing economic, social and environmental considerations. However, to avoid or minimise the risk of being perceived as or actually failing to properly observe sustainable development principles in matters affecting climate policy or law, regulators should take extra care to articulate their positions and decisions, especially in areas where there are diverse and competing interests. Three of these areas are discussed below.

---

[2] Ibid., at Chapter 3, para.3.1. For further reading see Christopher Tung, *Carbon Law and Practice in China*, Chapter 22, in: David Freestone and Charlotte Streck (eds.), *Legal Aspects of Carbon Trading Kyoto, Copenhagen and Beyond* (Oxford: Oxford University Press, 2009).

26.3 The Flash Points: Aviation, Trade and Power

Three key sectors are briefly considered in this chapter, aviation, trade and power. There are other important sectors including agriculture, building and construction, real estate and other forms of transportation, which are not dealt with here.

26.3.1 Aviation: Arguing for Exclusion from the EU ETS?

China has a large and fast growing aviation industry. It is mostly state controlled. The principal Chinese government ministry responsible for aviation is the Civil Aviation Administration of China or CAAC. The National Development and Reform Commission or NDRC is the lead ministry responsible for climate change and low carbon economy matters. A senior representative of the NDRC co-ordinates the National Leading Committee on Climate Change (NLCCC), which is chaired by Premier Wen Jiabao. The NLCCC is responsible for overseeing and reviewing key national strategies, guidelines and measures related to climate change, and to co-ordinate and resolve climate issues between government agencies. The members of the NLCCC are representatives of 20 government ministries including the CAAC. Despite the existence of the NLCCC and participation of the CAAC, the CAAC has until very recently had a low profile in climate change matters. This changed on 6 February 2012 when the CAAC issued a directive to prohibit Chinese airlines from participating in the European Union Emissions Trading Scheme (EU ETS).[3] In the directive, the CAAC accused the European Union (EU) of unilaterally imposing emissions trading obligations on Chinese airlines flying into and out of the EU in violation of the United Nations Framework Convention on Climate Change and the Chicago Convention but did not identify the provisions of those conventions allegedly contravened by the EU. The CAAC also has no apparent legal basis to ban Chinese airlines from complying with the EU ETS under Chinese law (nor to compensate Chinese airlines in the event of them being sanctioned by the European Commission for failing to comply with the EU ETS). In a decision rendered by the European Court of Justice (ECJ) dated 21 December 2011,[4] the court had already rejected the challenge of the Air Transportation Association of America and a number of US airlines on the basis that inclusion of international aviation in the EU ETS is contrary to international law and to the Chicago Convention, the Kyoto Protocol and the Open Skies Agreement. There is

---

[3] "CAAC Directive", 6 February 2012, available (in Chinese) at, http://www.caac.gov.cn/A1/201202/t20120206_45737.html (last accessed on 15 June 2012).

[4] "ECJ Judgment", 21 December 2011, available at http://curia.europa.eu/juris/document/document.jsf?text=&docid=117193&pageIndex=0&doclang=EN&mode=req&dir=&occ=first&part=1&cid=356247 (last accessed on 18 June 2012).

no indication that China might put forward arguments against the inclusion of Chinese airlines in the EU ETS that would produce a different result before the ECJ. As such, Chinese airlines which fail to comply with the EU ETS are liable to being banned from flying into and out of the EU. In the circumstances, rather than contesting the validity of the inclusion of Chinese airlines under international law, a better solution for China would be to demonstrate that its own national measures to reduce aviation emissions qualify for exemption or partial exemption of its national airlines from the EU ETS.[5] Such a course would also be consistent with the sustainability objectives of A21 as a balanced approach and low carbon policy by taking positive action to reduce aviation emissions. A confrontational stance would on the other hand put the CAAC and Chinese airlines in a difficult position under EU and Chinese law and contrary to those sustainability objectives.

26.3.2 Trade: Border Tax Adjustments and Carbon Intensive Goods

Little needs to be said about China's massive industrial production and export prowess. This success has a lot to do with China's early "Open Door" policy and foreign investment laws, as well as plentiful and cheap manpower. This export success and the belief that Chinese products and production processes are carbon intensive has lead to threats by foreign governments of the imposition of border tax adjustments.[6] While the clamor for border tax adjustments or equivalent measures has subsided recently, arguably because China has made significant commitments to reducing its carbon intensity,[7] it is still a contentious issue in light of the continuing concerns over the carbon intensity of goods produced and exported by China. In this regard, although policy such as the 12th National Five Year Plan (12FYP) and laws such as the Cleaner Production Promotion Law 2002 and Circular Economy Promotion Law 2008 encourage and provide support for clean, green and low carbon production, it is plain that these national policies and laws have had limited success in reducing the use of materials and curbing production processes that remain carbon

[5] Amended Council Directive 2003/87/EC on Establishing a Scheme for Greenhouse Gas Emission Allowance Trading within the Community, which included aviation activities within the EU ETS, OJ 2003 L 275/32.

[6] Border tax adjustments are used to put foreign and domestic producers on relatively equal footing in both international and domestic markets by offering domestic producers a tax rebate upon export and imposing charges on imports. A detailed discussion of border tax adjustments is beyond the scope of this chapter but for further reading see: Matthew Genasci, "Border Tax Adjustments and Emissions Trading: The Implications of International Trade Law for Policy Design", 1 *Carbon & Climate Law Review* (2008), 33.

[7] On 26 November 2009, China's Premier Wen Jiabao announced a national emissions intensity reduction target of 40–45% by 2020, adopting 2005 as the base year. In the current 12th National Five Year Plan (2011–2015), a 17% national emissions intensity reduction target has been allocated.

intensive. A rigorous application and enforcement of these policies and laws at provincial and local levels together with financial and tax incentives would help to reduce if not eliminate carbon intensive production and products. Further, a more clearly articulated commitment to this objective would give China a stronger hand to argue against the imposition of border tax adjustments, and also produce outcomes more aligned with A21 and sustainable development objectives in the Cleaner Production Promotion Law and the Circular Economy Promotion Law.[8]

26.3.3 Power: The CDM and Low Carbon Energy, Not Enough Emissions Reductions?

China's general and broad policy support for renewable energy and low carbon development is not in serious doubt. However, the devil, as always, is in the details. The power sector is probably the most intractable sector that confronts Chinese policy makers and legislators in their efforts to make real and substantial improvements in reducing emissions and delivering clean energy. A major lynchpin continues to be the issue of electricity pricing, which is strictly controlled by the Chinese central government. So while China successfully introduced laws to support the development of CDM projects and renewable energy projects,[9] including the introduction of feed-in-tariffs for a range of renewable energy and emissions abatement projects, these feed-in-tariffs and other financial and tax incentives have not made the investment in renewable energy and CDM projects as attractive to investors, both domestic and foreign, as they might have been. Chinese renewable energy and CDM project development has relied heavily on investment by Chinese state owned power companies (compelled by policy rather than economic reasons to invest), instead of the private sector.

Chinese CDM projects have attracted a large number of foreign buyers for the Certified Emissions Reductions (CERs) they generate. However, this interest dropped off significantly from 2011, as it became clear that the EU ETS would not accept CERs generated by projects in non-Least Developed Countries for compliance in Phase III of the EU ETS.[10] It is still open to the EU and China to agree to the inclusion of Chinese CDM projects, but no such agreement has emerged to date. This situation leaves the fate of further CDM project development in China uncertain and reduces much needed domestic and foreign investment in emissions

[8] Cleaner Production Promotion Law, 29 June 2002, at Article 1; Circular Economy Promotion Law, 29 August 2008, at Article 1.

[9] The Measures on Administration and Operation of Clean Development Mechanism Projects 2004 (CDM Measures) and Renewable Energy Law 2005 (RE Law). Note Article 2 of the CDM Measures and Article 1 of the RE Law in relation to sustainable development.

[10] Council Directive 2009/29/EC of 23 April 2009 amending Directive 2003/87/EC So as to Improve and Extend the Greenhouse Gas Emission Allowance Trading Scheme of the Community.

abatement projects. Absent viable future CDM projects, there is some urgency to establish a domestic scheme for offset projects that would produce domestic carbon credits for compliance and trading within a Chinese carbon market.

In addition, there have been questions over the sustainability of CDM projects approved by the NDRC which have affected the credibility and bankability of Chinese CDM projects.[11] Concerns over the sustainability of Chinese hydro, wind, landfill and HFC-22 projects have continued to taint the domestic CDM project approval process, due largely to the absence of clear guidance on sustainable development criteria for CDM projects. Reform to the domestic CDM approval process should introduce sustainability guidelines to assist project developers and screen out unsustainable projects.

Further issues that have arisen which require the careful attention of Chinese government decision makers concern recent preliminary rulings of the US Department of Commerce against Chinese solar power cell manufacturers in relation to unlawful subsidies provided by local and provincial governments to these manufacturers and the dumping of solar power cells by these manufacturers in the US market.[12] While final rulings may be less unfavourable to Chinese solar cell manufacturers, these actions demonstrate the importance of transparency in any financial assistance granted by Chinese government agencies to manufacturers and the need to monitor the pricing of exports to reduce the risk of similar actions against Chinese clean energy equipment manufacturers in the US or the EU in the future.

26.4 Conclusion

China's sustainable development strategy, policies and laws set a robust standard for the application of sustainable development objectives and principles into policy and legislative decision making. While sustainable development principles are dynamic and have to be implemented according to national circumstances and interests, there are nevertheless areas where the implementation of policies and laws in relation to climate and low carbon issues requires more attention from policymakers and regulators. From the above analysis, a further embedding of sustainable development objectives and principles in law and a deeper understanding of the inter-relationships and cross-impacts between issues in different industry sectors would help to produce more sustainable outcomes, and inherently more effective and equitable decision making in those industry sectors. In order to achieve this aim and to reinforce A21 and Chinese laws which contain sustainable development objectives, the

---

[11] See Christopher Tung, "The CDM and a Low Carbon Economy", in: Michael Mehling, Amy Merril and Karl Upston-Hooper (eds.), *Improving the Clean Development Mechanism: Options and Challenges Post-2012* (Berlin: Lexxion, 2011), Section C, Chapter 3.

[12] International Trade Administration, "Fact Sheet", available at, http://ia.ita.doc.gov/download/factsheets/factsheet-prc-solar-cells-ad-prelim-20120517.pdf and http://ia.ita.doc.gov/download/factsheets/factsheet-prc-solar-cells-adcvd-prelim-20120320.pdf (last accessed 18 June 2012).

Chinese government should actively consider the introduction of an overarching sustainable development law to:

- give direct legal effect to A21 (and to update A21 to address developments in climate change and low carbon economy actions outlined in the 12FYP);
- outline sustainable development principles recognized under Chinese law;
- require A21 and sustainable development principles to be observed in all government planning and decision making;
- articulate sustainable development objectives, guidelines and criteria, to provide an objective point of reference for both government and private sector actors;
- provide for administrative sanctions for the failure to comply with obligations under the sustainable development law.

Chapter 27
India's Evolving Climate Change Strategy

Patodia Rastogi

Abstract India, along with most other developing countries, has viewed climate change as an environmental concern that first and foremost must be addressed by the industrialized west. As a developing country with a massive population living in poverty, India's priority lies with the development challenges it faces. As a result, domestic action on climate change has been minimal and to the extent that it existed, it was primarily viewed as a "co-benefit" of another policy. Only in 2008, primarily due to increasing pressure from the international community and India's growing status as a major economy, that a dramatic shift was seen in India's approach to addressing climate change. The release of the National Action Plan on Climate Change, a comprehensive framework policy where climate change was the central focus, marked this change. Since then, India has built on the National Plan and undertaken various initiatives that point towards its commitment on this issue. This chapter explores the evolution of domestic climate policy-making in India – from the period where climate was considered purely a "first world problem" to one where India is now proactively engaging at all levels to address climate change.

27.1 Introduction

India's current status as a developing country as well as a major emerging economy poses it with unique challenges in relation to climate change. India has a population of over a billion people, second only to China's, and is expected to become the

Patodia Rastogi is an independent consultant formerly the International Fellow at the Center for Climate and Energy Solutions, VA, USA formerly the Pew Center on Global Climate Change. She can be contacted at npatodia@yahoo.com.

P. Rastogi (✉)
Ecologic Institute Washington, 1630 Connecticut Ave. NW,
Suite 300, Washington, DC 20009, USA
e-mail: npatodia@yahoo.com

E.J. Hollo et al. (eds.), *Climate Change and the Law*, Ius Gentium: Comparative Perspectives on Law and Justice 21, DOI 10.1007/978-94-007-5440-9_27,
© Springer Science+Business Media Dordrecht 2013

world's most populous country by 2025.[1] With almost 42% of its population living in poverty[2] and about 350 million people with no access to electricity,[3] India's primary challenge remains poverty eradication and raising the standards of its people. Compounding these challenges is the potential impact of climate change. Approximately half a million people are dependent for their livelihood on the glacier-fed Himalayan region which is one of the main freshwater sources for the Gangetic basin. Studies have shown that in the longer-term, melting glaciers and the resulting water stress could have a crippling effect on that region.[4] India's agriculture-dependent economy could be affected by lower crop yields, another impact of climate change. Hundreds of millions of people live in coastal India and climate change impacts like sea level rise could have a devastating effect on them.[5]

In spite of the developmental challenges India faces, its economy has been experiencing unprecedented growth. It is currently the fourth largest economy in the world[6] and to meet its economic and development goals, the Indian government has targeted economic growth rates of about 8–10% a year for the next two decades.[7] The Government of India estimates that to meet these goals a three- to four-fold increase in primary energy supply and a five- to six-fold increase in electricity generation capacity will need to occur.[8] Coal and oil, both fossil fuels, account for about 75% of India's energy consumption and approximately 70% of electricity is generated from coal-fired power plants.[9] It is expected that due to the high demand for energy, coal will remain the mainstay of the Indian economy for the foreseeable future.

Even though India contributes only about 5% to global greenhouse (GHG) emissions, it still is ranked fourth in terms of absolute emissions. Furthermore, its emissions are projected to experience one of the highest growth rates in the next few

---

[1] US Census Bureau, "China's Population to Peak at 1.4 Billion Around 2026", 15 December 2009, available at: http://www.census.gov/newsroom/releases/archives/international_population/cb09-191.html (last accessed on 30 April 2012).

[2] World Bank, "New Global Poverty Estimates – What It Means for India", available at http://go.worldbank.org/51QB3OCFU0 (last accessed on 30 April 2012).

[3] World Bank, "India's Power Sector", available at http://go.worldbank.org/HQBS4S8190 (last accessed on 30 April 2012).

[4] Elizabeth L. Malone, *Changing Glaciers and Hydrology in Asia: Addressing Vulnerabilities to Glacier Melt Impacts* (Washington, DC: USAID, 2010) at 3.

[5] UNFCCC, "Indian Minister of Environment Jairam Ramesh: Press Conference at COP 16", 7 December 2010, available at http://unfccc.int/resource/webcast/player/app/play.php?id_episode=2988 (last accessed on 23 April 2012).

[6] On a GDP PPP basis for the year 2011. IMF, "World Economic Outlook Database 2012", available at http://www.imf.org/external/pubs/ft/weo/2010/02/weodata/index.aspx (last accessed on 22 April 2012).

[7] Government of India, *Integrated Energy Policy Report of the Expert Committee* (New Delhi: Planning Commission, 2006).

[8] Ibid.

[9] Energy Information Administration, "Country Analysis Briefs: India", 21 November 2011, available at http://www.eia.doe.gov/cabs/India/Full.html (last accessed on April 22 2012).

decades – about 47% between now and 2020. In terms of per capita emissions, India however ranks extremely low – only about a tenth of the United States' per capita emissions and a third of the world average.[10]

Striking the right balance between these two conflicting factors – on the one hand tackling the country's development challenges and on the other being recognized as a major emerging economy – is the basis on which India has developed its recent approach to climate change. Historically, India's stance on climate change has been driven primarily by its categorization within the United Nations Framework Convention on Climate Change (UNFCCC) as a Non Annex 1 party (developing country party) where it has no obligation to reduce its GHG emissions. More recently though, its emergence on the world stage as a major economy cast a spotlight on the country and it has felt an increasing pressure to undertake domestic action to address climate change. This chapter explores the evolution of climate policy-making in India. The following section elaborates on India's historical stance of the climate issue being a "first world problem" and highlights relevant domestic legislation that was developed in this context. The third section discusses the drivers that led to the launch of the National Action Plan on Climate Change (NAPCC) and further elaborates on the relevant Missions of the NAPCC. It also describes policies and measures that have been developed since the NAPCC and concludes with a brief overview of engagement at the sub-national level on climate change.

27.2 Until the Mid 2000s: India's Traditional Approach to Climate Change

As a Non Annex 1 party to the UNFCCC and the Kyoto Protocol, India has no legal obligation to reduce its GHG emissions. Instead India has long argued that the historical responsibility[11] to deal with climate change lies with the developed countries as they are the creators of the problem and that they must take on a leadership role in addressing climate change. The "overriding priority" of India (and other developing countries) is to address development and poverty eradication and any action to reduce emissions should be voluntary. This principle, commonly known as the principle of common but differentiated responsibility, is one that India has long been an advocate of. The firewall created through this principle between developed

---

[10] Based on author's calculations. Data is from International Energy Agency, *CO2 Emissions from Fuel Combustion 2010 Highlights* (Paris: OECD/IEA, 2010); International Energy Agency, *World Energy Outlook 2010* (Paris: OECD/IEA, 2010) and United States Environmental Protection Agency (USEPA), *Global Anthropogenic Non-CO2 Greenhouse Gas Emissions: 1990–2020* (Washington, DC: USEPA, 2006). Data does not include emissions from land-use, land-use change and forestry.

[11] Mukund G. Rajan, *Global Environmental Politics: India and the North–South Politics of Global Environmental Issues* (New Delhi: Oxford University Press, 1997).

and developing countries shielded India from undertaking any meaningful action to reduce emissions.

India's domestic approach to climate change largely mirrored its multilateral stance. Domestic policies per se did not exist and climate change was addressed only in a piecemeal manner, mainly through energy or forestry policies in which climate was never the central focus. The main thrust of these policies was social and economic development. Climate change was mentioned, if at all, only as a "co-benefit" of another policy.

27.2.1 Five Year Plans

Many policies developed during this period that had a climate "co-benefit" originated in India's Five Year Plans. Five Year Plans, much like in other countries, lay the foundation for economic planning and development for the country. In India's case, its Five Year Plans are developed by its Planning Commission. Environmental concerns were addressed for the first time in the Sixth Five Year Plan (1980–1985)[12] and one of the primary objectives of the Ninth Five Year Plan (1997–2002) was to "ensure environmental sustainability of the developmental process through social mobilization and participation of people".[13] Climate change was given due importance for the first time as late as the Eleventh Five Year Plan (2007–2012). Most of the objectives in the Eleventh Five Year Plan were driven by the need to build energy resources to meet the growing energy demand and ensuring India's economic growth. Objectives included reducing energy intensity by 20% by 2016–2017, building additional capacity for nuclear power and renewable energy, and the Plan also set a specific objective of increasing forest cover by 5%.[14]

27.2.2 Energy Conservation Act (2001)

The Energy Conservation Act of 2001[15] laid the foundation for promoting energy efficiency by establishing the Bureau of Energy Efficiency (BEE) with the primary objective of reducing energy intensity. The Act empowered the government to set

---

[12] Planning Commission, *Eighth Five Year Plan Volume 2*, available at http://planningcommission.nic.in/plans/planrel/fiveyr/8th/vol2/8v2ch4.htm (last accessed on April 28 2012).

[13] Planning Commission, *Ninth Five Year Plan Volume 1*, available at http://planningcommission.nic.in/plans/planrel/fiveyr/9th/vol1/v1c1-1.htm (last accessed on April 28 2012).

[14] Planning Commission, *Eleventh Five Year Plan 2007–2012 Volume 1*, available at http://planningcommission.nic.in/plans/planrel/fiveyr/welcome.html (last accessed on April 28 2012).

[15] Ministry of Law, Justice and Company Affairs, *Energy Conservation Act*, 29 September 2001, available at http://www.powermin.nic.in/acts_notification/pdf/ecact2001.pdf (last accessed on 30 April 2012).

energy consumption standards for equipment and appliances, specify energy conservation building codes for commercial buildings and set energy consumption norms and standards for consumers. It also established a compliance mechanism through energy audits to be conducted by accredited auditors.

27.2.3 Integrated Energy Policy (2006)

The Integrated Energy Policy (2006)[16] was an effort by the Planning Commission to address the country's growing energy demands. The Integrated Energy Policy provides a comprehensive framework for energy policy in India. Its vision is to "to reliably meet the demand for energy services of all sectors including the lifeline energy needs of vulnerable households in all parts of the country with safe, clean and convenient energy at the least-cost."[17] The policy states that India should pursue all available fuel options and forms of energy and that coal will remain the mainstay of the Indian economy till 2031–2032. In terms of addressing climate change, the Integrated Energy Policy makes recommendations that include power sector reforms, ramping up mass transit, increasing nuclear power and renewables, and highlighting energy efficiency in all sectors.

27.2.4 Other Relevant Legislation

The Electricity Act 2003[18] was a landmark legislation for power sector reforms in India. It encouraged the development of renewable energy by mandating that State Electricity Regulatory Commissions (SERCs) promote renewable energy by allowing connectivity and sale of electricity to any interested person and specified that a certain percentage of the electricity consumption should be from renewables. Building on this Act, the National Electricity Policy (2005) called for increased participation by the private sector to exploit non-conventional energy resources and allowed for renewable energy procurement through a competitive bidding process. It also called for a more level playing field between non-conventional technologies and conventional ones by allowing for differential tariffs.[19] To further encourage

---

[16] Planning Commission, Integrated Energy Policy. 2006.

[17] Press Information Bureau, "Integrated Energy Policy", 26 December 2008, available at http://www.pib.nic.in/newsite/erelease.aspx?relid=46172 (last accessed on 25 April 2012).

[18] Ministry of Law and Justice, *The Electricity Act 2003*, 26 May 2003, available at http://www.powermin.nic.in/acts_notification/electricity_act2003/pdf/The%20Electricity%20Act_2003.pdf (last accessed on 17 April 2012).

[19] Ministry of Power, *National Electricity Policy*, 12 February 2005, available at http://www.powermin.nic.in/whats_new/national_electricity_policy.htm (last accessed on 10 April 2012).

renewable energy development, the National Tariff Policy (2006)[20] stipulated the SERCs to purchase a minimum percentage of power from renewable sources.

Most of the legislative and policy initiatives described in this section were driven primarily by India's development imperatives – a need to exploit all potential energy resources to ensure that there is an adequate energy supply and to continue to spur economic growth. It was only in the late 2000s, due to various factors most notably the growing international pressure to address climate that led India to develop a climate-specific policy framework.

27.3 Late 2000s Onwards: India's Growing Status as a Major Emerging Economy

The twenty-first century saw the arrival of India on the world stage as a global power. With the Indian economy growing at a phenomenal rate of 8% a year, India was no longer viewed solely as a "developing country". In fact, during the 2007 financial crisis both India and China were considered critical partners in the G20 talks that helped avert a global meltdown due to their growing economies. This acknowledgement of India's status as a major emerging economy brought with it more responsibility. To be recognized as a legitimate global player the world needed to see India as a country willing to shoulder responsibilities. This expectation was reflected within the multilateral climate negotiations where India began to feel international pressure to be more proactive and take action to reduce its own GHG emissions.[21] The shift in India's stance on taking action domestically and launching the National Action Plan on Climate Change (NAPCC)[22] can be attributed to a large extent to this growing international pressure. Facing the fear of isolation[23] and continued international pressure, prior to the much-hyped Copenhagen Summit in 2009, India further announced that it would voluntarily reduce its emissions intensity between 20 and 25% below 2005 levels by 2020.[24]

---

[20] Ministry of Power, *National Tariff Policy*, 6 January 2006, available at http://www.powermin.nic.in/whats_new/pdf/Tariff_Policy.pdf (last accessed on 25 April 2012).

[21] Shyam Saran "Climate Change Negotiations: The Challenge for Indian Diplomacy", Speech delivered at Vivekanand International Foundation, 19 March 2010, available at http://www.vifindia.org/node/299 (last accessed on 29 April 2010).

[22] Government of India, *National Action Plan on Climate Change*, Prime Minister's Council on Climate Change, New Delhi, July 2008, available at http://pmindia.nic.in/Pg01-52.pdf (last accessed on 22 April 2012).

[23] Aaron Atteridge et al. "Climate Policy in India: What shapes International, National and State Policy?" 41(1) *Ambio* (2012) 68 at 71.

[24] Excluding agriculture emissions. UNFCCC, "Letter including India's Domestic Mitigation Actions", 30 January 2010, available at: http://unfccc.int/files/meetings/cop_15/copenhagen_accord/application/pdf/indiacphaccord_app2.pdf (last accessed on 30 April 2012).

While international diplomatic pressure played a critical role that led to India's more proactive stance, one cannot dismiss the importance that domestic priorities played in this context as well. A realization by the Indian government that it is extremely vulnerable to climate change impacts meant that it was in India's own interest to take action both domestically and internationally. With its accelerating energy demands and India importing more than 70% of its oil, it made sense for India to prioritize energy security and focus on energy efficiency and renewables. This is reflected in specific missions of the NAPCC. Energy access, another domestic concern, could put India's development at risk. The Indian government's focus on developing renewable energy helps it to achieve this goal.

Another driver worth noting is the immense opportunity that the clean energy technology sector provides for the twenty-first century. With the race for fossil fuels becoming increasingly competitive, many countries are looking at building their own domestic capacities to shore up energy supplies. Potential growth in this sector and the economic opportunity that it opens up in the global marketplace has driven great interest in this sector. Projections of global investment in clean energy technologies are far from trivial – estimates show that from now until 2020, cumulative global investment totals for clean power generation technologies could reach nearly USD 2.3 trillion assuming strong action on climate change.[25] Countries that are first-time movers in the clean energy technology space are more likely to become market leaders and be able to exploit the potential of this sector. Global competition has already begun in this sector with China being the leader in solar power manufacturing and wind generation. Some European countries, Denmark and Germany specifically, have taken steps to enhance their renewable technology capacities. India's Solar Mission is an acknowledgement of the opportunity the clean technology space offers. Success in this Mission will enable India to become a frontrunner in solar manufacturing.[26] By establishing aggressive renewable policies through its climate programs, India has begun laying the path towards becoming a leader in the technology choices of the future.

27.3.1 National Action Plan on Climate Change

The National Action Plan on Climate Change (NAPCC) released by Prime Minister Manmohan Singh in 2008 marked a watershed moment in India's domestic engagement on the climate issue. Consisting of eight national missions that run till 2017, the

[25] Center for Climate and Energy Solutions, *Clean Energy Markets: Jobs and Opportunities* (Arlington: 2011), available at http://www.c2es.org/docUploads/clean-energy-markets-update2011_0.pdf (last accessed on 1 May 2012).

[26] Farookh Abdullah, "A Renewable Future for Mankind: Challenges and Prospects", 13 January 2011, *Making It Magazine.net*, available at: http://www.makingitmagazine.net/?p=2849 (last accessed on 26 April 2012).

NAPCC is a package of existing and planned initiatives, policies and programs focused both on adaptation and mitigation. While the Plan emphasized India's development objectives, it established for the first time, a concrete framework to address climate specifically in the domestic context.

The eight Missions comprehensively address mitigation, adaptation as well as research and development. The missions are the National Solar Mission, National Mission on Sustainable Habitat, National Mission for Sustaining the Himalayan Ecosystem, National Water Mission, the National Mission on Enhanced Energy Efficiency, National Mission for a Green India, National Mission for Sustainable Agriculture and National Mission on Strategic Knowledge for Climate Change. In early 2012, the Government of India announced its plans to establish an additional mission on clean coal technologies.[27] These Missions are to be institutionalized through inter-sectoral groups, consisting of members from the relevant ministries, civil society, industry and academia. Detailed plans that include targets, timelines and objectives are to be developed by these groups and submitted to the Prime Minister's Council on Climate Change.

Some of the missions build on existing legislation or policies, like the National Mission on Sustainable Habitat or the Green India Mission, while others like the Solar Mission, chart a new and ambitious path by the government to access a previously untapped energy source. Considering the significant progress made by the National Mission on Enhanced Energy Efficiency (NMEEE) and the National Solar Mission, this section provides an overview of the achievements specifically of these two Missions.

27.3.1.1 National Mission on Enhanced Energy Efficiency – Perform Achieve and Trade Scheme

Under the NAPCC, The National Mission on Enhanced Energy Efficiency was mandated to implement four new initiatives: the Perform, Achieve and Trade Scheme; Market Transformation for Energy Efficiency; Energy Efficiency Financing Platform and; Framework for Energy Efficient Economic Development.[28] While the latter three are still in the early stages of implementation, the Perform, Achieve and Trade scheme (PAT Scheme) was recently launched by the Government of India.[29] The PAT scheme is an innovative market mechanism for

[27] Hindu Bureau, "National Mission on Clean Coal Technologies on the Cards", 27 February 2012, *The Hindu Business Line*, available at http://www.thehindubusinessline.com/industry-andeconomy/economy/article2938979.ece?ref=wl_industry-and-economy (last accessed on 30 April 2012).

[28] Ministry of Power, *National Mission on Enhanced Energy Efficiency Draft Mission Document: Implementation Framework*, December 2008, available at http://www.indiaenvironmentportal.org.in/files/National%20Mission%20for%20Enhanced%20Energy.pdf (last accessed on 1 May 2012).

[29] Ministry of Power "Notification (Energy Conservation Rules 2012)", 30 March 2012, available at http://220.156.189.23/schemes/documents/nmeee/pat/PAT_Rules_English.PDF (last accessed on 1 May 2012).

trading energy efficiency certificates in energy-intensive sectors. After many delays mostly due to industry pushback and legal holdups, the PAT scheme was finally rolled-out in April 2012 making India the first developing country to implement a market-based mechanism.

The PAT scheme is being implemented in three phases – the first phase covers 478[30] facilities from eight energy-intensive sectors, namely aluminum, cement, chlor-alkali, fertilizer, iron and steel, pulp and paper, textiles and thermal power plants. The government expects the scheme to deliver reductions of about 100 million tons of CO_2 annually by the end of its first phase. The Energy Conservation Act 2001, the legal framework on which the PAT scheme is based, was amended by the Parliament in 2010 to allow for the establishment of the scheme.

The PAT scheme is a baseline-credit scheme that allows facilities to trade certificates to meet their compliance requirements and simultaneously reduce costs. Each facility has a specific energy consumption target (a reduction in energy consumption from the facility's baseline) with less energy efficient facilities having a greater reduction target compared to more efficient ones. A facility's baseline is based on its historic specific energy consumption over the period 2007–2010. Facilities that make greater reductions than their target will be receive "Energy Saving Certificates" (EsCerts) which can be traded with other facilities that have difficulty meeting their target or bank them for use in a subsequent phase. Those facilities that are unable to meet their targets must buy EsCerts or pay a penalty. One Energy Saving Certificate is equivalent to 1 ton of oil equivalent, an energy consumption measure rather than a carbon reduction measure.

The first phase extends over a 3-year period (2012–2015) and covered facilities are expected to meet their target by the end of the first phase. Monitoring and verification will be conducted by auditors at the end of the first phase and Energy Saving Certificates will be issued ex-post.

While details of the subsequent phases of the PAT scheme are still being fleshed out, early signs have hinted at the possibility of broadening the scheme to include other energy-intensive sectors like petroleum refineries, petrochemicals, chemicals etc. and further tightening the targets.

27.3.1.2 Solar Mission

Prime Minister Manmohan Singh has emphasized the importance of the Indian economy to gradually shift away from a fossil fuels-based economy to one more dependent on non-fossil fuels and renewable sources of energy.[31] This principle is

[30] Ministry of Power "Notification (Rules)", 30 March 2012, available at http://220.156.189.23/schemes/documents/nmeee/pat/PAT_Notification_English.pdf.

[31] Manmohan Singh, "Release of the National Action Plan on Climate Change", Prime Minister's Speech, 30 June 2008, available at http://pmindia.gov.in/speech-details.php?nodeid=667 (last accessed on 1 May 2012).

reflected in the Solar Mission,[32] one of the most ambitious Missions launched by the government. The Mission sets a goal of generating 20 GW of grid-connected solar power plants by 2022, a several thousand-fold increase from current levels. While extremely ambitious, the Government of India has already demonstrated some early success in getting this Mission off the ground.

The Government of India decided to award solar projects to the private sector for the first phase of this Mission (till 2013) through a process of reverse auction rather than a feed-in tariff, primarily due to great interest shown by the Indian solar industry. The first reverse auction was conducted in 2010 wherein 150 MW of solar PV and 470 MW of concentrated solar thermal power were auctioned. Response to this first auction was overwhelming – bid applications totaled 5,126 MW, about eight times more than the maximum allotted capacity of 620 MW and price quotes received were on average 25–32% below the central government's declared tariffs.[33] The government selected the projects based on the criteria of a maximum discount offered on the declared tariff. Due to the extremely low bids, there was some concern raised about the viability of some projects selected in the first auction and as a result about half the bids were discarded. Learning from this experience, guidelines were revised for the second round of auctioning held in 2011. This round is expected to award 350 MW of solar PV and concentrated solar thermal power projects to eligible project developers. Bids received have been significantly lower than even the first round, with the lowest one being 38% below the average price in the first round.

27.3.2 *Other Relevant Initiatives*

While the National Action Plan on Climate Change was primarily developed to address international concerns, it helped elevate the issue of climate change domestically. Apart from building on the work of the Missions, the Indian government has made further advancements to strengthen its policies in the climate and clean energy sectors, both at the national and at the state level.

27.3.2.1 Coal Levy

India showed its commitment to addressing climate change by being one of the first developing countries to implement a levy on coal. What is most noteworthy of this effort is the fact that the revenues solely go towards a National Clean Energy Fund.

---

[32] Ministry of New and Renewable Energy, *Jawaharlal Nehru National Solar Mission*, available at http://india.gov.in/allimpfrms/alldocs/15657.pdf (last accessed on 1 May 2012) and; Ministry of New and Renewable Energy, *Resolution Jawaharlal Nehru National Solar Mission*, 11 January 2010, available at http://www.mnre.gov.in/solar-mission/jnnsm/resolution-2/ (last accessed on 26 April 2012).

[33] Ranjit Deshmukh, Ashwin Gambhir and Girish Sant "India's Solar Mission: Procurements and Auctions" 46 *Economic and Political Weekly* (2011) 22, at 24–25.

First introduced in India's budget in February 2010 and implemented in July 2010, a levy of 50 rupees a ton (approximately USD 1 a ton) was imposed on domestic and imported coal, lignite and peat.[34] The levy is expected to generate about USD 500 million a year. The National Clean Energy Fund is to support research and development for clean energy technologies primarily including critical renewable energy infrastructure projects like silicon manufacturing, advanced solar manufacturing, geothermal energy, hydrogen and fuel cells, and clean fossil energy (Carbon Capture and Storage, coal gasification etc.)[35]

27.3.2.2 Renewable Energy Certificate Mechanism

In March 2011, the Government of India launched its Renewable Energy Certificate Mechanism, another example of the Indian government's use of innovative ways to encourage development of renewable energy.

Under the Electricity Act 2003, State Electricity Regulatory Commissions (SERCs) are required to specify Renewable Purchase Obligations (RPOs), a requirement for distribution companies to purchase a certain percentage of electricity from renewable sources. The Central Electricity Regulatory Commission (CERC) set regulations[36] for the implementation of the Renewable Energy Certificate (REC) mechanism in 2010. The mechanism provides flexibility to states that do not have adequate renewable sources to meet their Renewable Purchase Obligations through a trading mechanism. The mechanism allows a renewable energy generator to either sell its renewable energy at a feed-in tariff previously determined or sell its electricity and "renewable" attributes separately. The "renewable" electricity attributes can be exchanged in the form of a Renewable Energy Certificate (REC). One REC is equivalent to 1 mega-watt hour (MWh) of renewable electricity fed into the grid. RECs are differentiated into two types – solar and non-solar – both with floor and ceiling prices determined by the CERC. Currently the Indian REC market has conducted trades worth about USD 5 million since its launch. It is expected that monthly sales will be valued at over USD 20 million by the end of 2012.[37]

---

[34] Ministry of Finance "Levy of Clean Energy Cess", 24 June 2010, available at http://www.cbec.gov.in/excise/cx-circulars/cx-circulars-10/circ-cec01-2k10.htm (last accessed on 25 April 2012).

[35] Ministry of Finance "Guidelines for Appraisal and Approval of Projects/Schemes Eligible for Financing under the National Clean Energy Fund", 18 April 2011, available at http://finmin.nic.in/the_ministry/dept_expenditure/plan_finance2/Guidelines_proj_NCEF.pdf (last accessed on 1 May 2012).

[36] Central Electricity Regulatory Commission, "Notification (Central Electricity Regulatory Commission Regulations 2010", 14 January 2010, available at https://www.recregistryindia.in/pdf/REC_Regulation/2(a)CERC_Regulation_on_Renewable_Energy_Certificates_REC.pdf (last accessed on 23 April 2012) and; ABPS Infra "Report on the Conceptual Framework for Renewable Energy Certificate Mechanism for India", June 2009, available at http://mnre.gov.in/file-manager/UserFiles/MNRE_REC_Report.pdf (last accessed on 27 April 2012).

[37] PTI, "Energy Credits Trading likely to touch 100 crores by year-end", *Hindu Business Line*, January 4 2012, available at http://www.thehindubusinessline.com/industry-and-economy/economy/article2774310.ece (last accessed on 27 April 2012).

27.3.2.3 Expert Group on Low Carbon Strategies

Recognizing the importance of transitioning to a low carbon economy, the Indian government set up an "Expert Group on Low Carbon Strategies for Inclusive Growth" consisting of experts from various ministries, civil society and the private sector.[38] The Expert Group is based in the Planning Commission. Recommendations from the Expert Group's work are expected to feed into India's Twelfth Five Year Plan (2013–2017). The group submitted its interim report in May 2011. The report, while interim, gives a flavor of the range of actions that will be required for India to transition to a low carbon economy. These include actions to increase investment in renewable technologies, reduce losses from transmission and distribution from the power sector; adopt super-critical technologies in coal based thermal power generation; get iron, steel and cement sectors to adopt best-available technology and; to increase the share of rail in overall freight transport.

27.3.3 Engagement at the State level

Soon after the launch of the NAPCC in August 2009, Prime Minister Manmohan Singh addressed state environment ministers and urged them to develop State Action Plans on Climate Change (SAPCC). Following this, the Ministry of Environment and Forests provided additional guidance by providing states with a common framework for developing these plans. The State Plans are to be prepared under the auspices of the NAPCC and the intention is to have a top-down approach where national and state actions are "harmonized".[39]

Responses by states have been mixed with some states making climate change a priority while others have yet to submit their action plans.[40] As can be expected, most of these sub-national plans are driven by each state's assessment of their vulnerabilities and opportunities, as well as their own development agenda.

As a coastal state where most of its population lives below the poverty line, the state of Orissa's plan gives greater importance to adaptation and focuses on the agriculture sector and coastal disasters.[41] Due to the abundance of coal in the state,

---

[38] Planning Commission, *Interim Report of the Expert Group on Low Carbon Strategies for Inclusive Growth*, May 2011, available at http://planningcommission.nic.in/reports/genrep/Inter_Exp.pdf (last accessed on 27 April 2012).

[39] Ministry of Environment and Forests "Towards a common framework for preparation of State Level Strategy and Action Plans on Climate Change", National Consultation Workshop, 19 August 2010, available at http://moef.nic.in/downloads/others/Experts-SAPCC-Preeti.pdf (last accessed on 26 April 2012).

[40] Draft reports submitted by states to the MoEF can be found at http://moef.nic.in/modules/others/?f=sapcc-2012. As of April 2011, 10 states were listed on this website. However, draft plans of some of the states can be found on other websites and/or in form of power point presentations.

[41] Government of Orissa, "Orissa Climate Change Action Plan 2010–2015", available at http://moef.nic.in/downloads/public-information/Orissa-SAPCC.pdf (last accessed on 28 April 2012).

the plan makes clear that most of the additional power will be generated from coal but prioritizes clean coal as well. Karnataka, another coastal state, emphasizes actions in the agriculture, water and energy[42] sectors according equal importance to both adaptation and mitigation. The desert state Rajasthan's action plan[43] sketches out action for the short-, medium- and long-term and includes specific time frames and targets for implementation. Its plan focuses on its regional concerns – desertification and land degradation, and human health. It also emphasizes the opportunities that abound in the state for exploiting the use of renewables, specifically solar and biomass. With a large population whose livelihood is dependent on agriculture and forestry, Madhya Pradesh's plan focuses more on strategies that ensure that the state is "climate-resilient" one.[44] The state of Gujarat is one that has been extremely active in implementing "on the ground" climate action, and as one of the most industrialized states in the country it comes as no surprise that its focus is on mitigation opportunities. It was not only one of the first states to establish its own climate change department[45] but also has its own Solar Policy and Wind Energy Policy and is the leader in developing Clean Development Mechanism (CDM) projects in the country.[46] While specific timeframes or targets seem to be lacking in most of the plans, some states have included budgets for implementing their action plans. For example, the state of Orissa estimates an ambitious budget of USD 3,200 million approximately over 5 years.

27.4 Conclusion

As a country that has routinely resisted taking action on climate change, in the last few years India has taken some substantial strides in undertaking ambitious domestic climate policy. International pressure and domestic concerns have played a critical role in catalyzing action. Action at the national level has trickled down to all levels, with involvement from the states, private sector as well as civil society.

---

[42] EMPRI and TERI "Karnataka State Action Plan on Climate Change Prepared for the Government of Karnataka", 17 September 2011, available at http://www.empri.kar.nic.in/Karnataka%20SAPCC%20draft%20-%20EMPRI,%20TERI%202011-09-17.pdf (last accessed on April 23 2012).

[43] Government of Rajasthan "Rajasthan State Action Plan on Climate Change" available at http://210.212.96.131/rpcb/ReportsAndPaper/ClimateChange_15_12_2011.PDF (last accessed on 25 April 2012).

[44] Government of Madhya Pradesh, "Madhya Pradesh State Action Plan on Climate Change", February 2012, available at http://moef.nic.in/downloads/public-information/MP-SAPCC.pdf (last accessed on 23 April 2012).

[45] Business Standard, "Gujarat to set up Asia's First Climate Change Department", Business Standard, 25 February 2009, available at http://www.business-standard.com/india/news/gujarat-to-setasias-first-dept-for-climate-change/350044/ (last accessed on 1 May 2012).

[46] Government of Gujarat "Climate Change Action and Adaptation", available at http://moef.nic.in/downloads/others/States-SAPCC-gujarat.pdf (last accessed on 1 May 2012).

However, India should remain cautious that its impressive list of climate undertakings is not bogged down by bureaucratic hurdles and is implemented in a timely manner. As India heads into the next few decades of critical economic growth, it will continue to grapple with the competing tensions of tackling economic development and climate change. To ensure that neither priority is compromised, it is essential that the Indian government continues to find that balance where the economy and the environment are not at loggerheads with each other. Instead, by exploring innovative ways to engage the private sector in tackling climate change, continuing a dialogue with civil society, and laying the foundation for a low carbon and climate-resilient economy, India can ensure that its economic growth is in sync with the environment.

Chapter 28
Climate Change Responses in South Africa

Michael Kidd and Ed Couzens

South Africa is a significant emitter of greenhouse gases. Despite a long history of policy development, there is insufficient legislation addressing climate change (the chapter briefly canvassing what legislation there is) – and numerous policy imperatives which might undermine the effectiveness of recent policy innovations. Policy documents – especially the recent *White Paper on the National Climate Change Response* (2011) – are considered in some detail. Also considered is the recent *White Paper on South Africa's Foreign Policy* (2011), which provides important insight into South Africa's intentions in respect of international commitments and both national and regional growth. As a counterweight to the environmental aspirations of the White Paper, South Africa's energy policy is then considered and it is concluded that while South Africa continues on the path it is presently treading it is going to be all but impossible to reconcile the goals of strong economic growth and poverty alleviation with environmental protection generally, and South Africa's international commitments in the climate change issue-area specifically.

Michael Kidd is Professor of Law at the University of KwaZulu-Natal, Pietermaritzburg, South Africa. He holds a B Com, LLB, LLM and PhD all from the University of Natal; Ed Couzens holds the degrees of PhD from the University of KwaZulu-Natal; LLM Environmental Law from the Universities of Natal and Nottingham; and BA Hons LLB from the University of the Witwatersrand. He is currently an Associate Professor in the School of Law, University of KwaZulu-Natal, Durban. He is an admitted attorney of the High Court, RSA. He has been since 2006 a co-editor of the University of Eastern Finland/United Nationa Environment Programme Review of International Environmental Law-making and Diplomacy; and has been since 2001 an assistant editor on the South African Journal of Environmental Law and Policy.

M. Kidd (✉) • E. Couzens
College of Law and Management Studies,
University of KwaZulu-Natal, Scottsville 3209, South Africa
e-mail: Kidd@ukzn.ac.za; Couzense@ukzn.ac.za

E.J. Hollo et al. (eds.), *Climate Change and the Law*, Ius Gentium: Comparative Perspectives on Law and Justice 21, DOI 10.1007/978-94-007-5440-9_28,

© Springer Science+Business Media Dordrecht 2013

28.1 Introduction[1]

South Africa is a globally significant greenhouse gas (GHG) emitter, and is the highest emitter in Africa. According to United Nations statistics for 2007, South Africa emitted 433.53 million tonnes of carbon dioxide (CO_2) in that year, placing the country 13th amongst all states and ahead of countries such as Australia and France. The next highest emitter in Africa was Egypt with 184 million tonnes. South Africa fared somewhat better in per capita figures, placing 47th internationally with a figure of 8.82 tonnes per capita, second in Africa behind Libya.[2] South Africa has also become a significant player in the international climate change negotiations regime, and hosted the latest Conference of the Parties to the United Nations Framework Convention on Climate Change (UNFCCC) in Durban in 2011.[3]

These facts, coupled with the fact that South Africa has a relatively sophisticated system of environmental laws, suggests that it ought to have a climate change legal regime in place to address its significant climate impact; but this chapter will indicate that this is not the case. Although the country published a climate change response policy in 2011, following on a near decade-long process of policy development, there is little (even indirect) legislation addressing climate change. The policy does envisage legislative innovation, but its effectiveness looks likely to be undermined by simultaneous policy development in other branches of government confirming continued commitment to fossil-fuel based sources of energy and indeed increased generation of energy from fossil fuels.

28.2 South Africa's Greenhouse Gas Emissions Profile

According to the 2011 *White Paper on the National Climate Change Response*,[4] South Africa's 2000 Greenhouse Gas Inventory[5] (the latest available) shows that the main source of South Africa's energy emissions is electricity generation, which

[1] Portions of this chapter are derived from Michael Kidd, *Environmental Law*, 2nd edition (Cape Town: Juta, 2011), Chapter 10.

[2] United Nations Statistics Division, "Environmental Indicators", available at http://unstats.un.org/unsd/environment/air_co2_emissions.htm (last accessed 2 March 2012). There are different data available from different sources (see, for example, sources cited in Kidd, supra note 2). The data differ according to the year of the statistics' derivation and the differences are not dramatic.

[3] The 17th Conference of the Parties to the United Nations Convention on Climate Change and the 7th Meeting of the Parties to the Kyoto Protocol. See, generally, "Meetings", available at: http://unfccc.int/meetings/items/6240.php and, specifically, "COP17/CMP7", available at: http://www.cop17-cmp7durban.com/ (both last accessed on 1 March 2012).

[4] Department of Environmental Affairs, *White Paper on the National Climate Change Response* (2011), at §6.2, Gen N 757 in *GG* 34695 of 19 October 2011.

[5] Department of Environmental Affairs and Tourism, *Greenhouse Gas Inventory South Africa: 1990–2000* (2009), available at: http://www.pmg.org.za/files/docs/090812greenhouseinventory.pdf (last accessed 1 March 2012).

constituted about half of the energy emissions and just under 40% of total emissions.[6] Other significant sources of emissions are transportation and energy used in industry (just under 10% each) and industrial process emissions (about 14% of total emissions). Emissions from agriculture and land-use change "constitute only around 5% of emissions, compared to an average of 44% in developing countries as a whole".[7] To put these figures into perspective, "average energy use emissions for developing countries constituted 49% of total emissions, whereas South Africa's energy use emissions constituted just under 80% of total emissions".[8]

In recognition of the significant contribution to climate change for which South Africa is responsible, the government in late 2009 announced that it would implement mitigation actions collectively resulting in 34 and 42% deviations below its "Business As Usual" emissions growth trajectory by 2020 and 2025, respectively.[9] Although this announcement was made conditional on certain aspects of the international regime under the UNFCCC coming to fruition, these commitments have subsequently been declared as domestic government policy in the White Paper.

28.3 Current Climate Change Legislation in South Africa

South Africa has few legislative provisions directly addressing climate change issues, although there are some specific legislative provisions that can be used for that purpose and a legislative framework which arguably requires more pro-action by government in this regard.

At the apex of South Africa's legal system is the Constitution, including a Bill of Rights which "applies to all law, and binds the legislature, the executive, the judiciary and all organs of state".[10] The Constitution includes the so-called "environmental right"[11] which provides that:

Everyone has the right-

(a) to an environment that is not harmful to their health or well-being; and
(b) to have the environment protected, for the benefit of present and future generations, through reasonable legislative and other measures that:

1. prevent pollution and ecological degradation;
2. promote conservation; and
3. secure ecologically sustainable development and use of natural resources while promoting justifiable economic and social development.

[6] cf. Ibid. at 17, where the higher figure of 47.6% is given.

[7] White Paper, supra, note 4.

[8] Ibid.

[9] Ibid at §6.1.

[10] Section 8(1) of the Constitution of the Republic of South Africa, 1996.

[11] Section 24.

The national environmental management principles, in the National Environmental Management Act,[12] are intended to give effect to the Constitutional right. The principles have at their core the concept of people being placed at the centre of environmental management and sustainable development.[13] Section 24 of the Constitution together with the national environmental management principles require, at a general normative level, the South African government to address climate change and its possible impacts on South Africa, by means of legislation and other reasonable measures.

As for more specific legislation, the National Environmental Management: Air Quality Act[14] has potential for addressing GHG emissions. The Act is aimed primarily at securing air quality from the perspective of pollution prevention, and it contains no express reference to "climate change".[15] Climate change is not defined in the Act, but "greenhouse gas" is defined as meaning "gaseous constituents of the atmosphere, both natural and anthropogenic, that absorb and re-emit infrared radiation and includes carbon dioxide, methane and nitrous oxide".[16]

The Act does require the Minister of Environmental Affairs to "establish a national framework for achieving the object of the Act",[17] and this framework must include "mechanisms, systems and procedures to give effect to [South Africa's] obligations in terms of international agreements".[18] The same section requires that national norms and standards established[19] must be aimed at ensuring[20] "compliance with [South Africa's] obligations in terms of international agreements".[21] Arguably, this gives the Minister the opportunity – some might even argue, the obligation – to promulgate regulations under the Act which contain binding strictures to combat GHG emissions.[22] Realistically, however, the obligations which states assume in

[12] Act 107 of 1998. See Ch.1, s 2 for the principles. The s 2 principles are justiciable (see s 32(1)).

[13] Section 2(2).

[14] Act 39 of 2004. Hereafter referred to as the Air Quality Act.

[15] There is an oblique reference in the Preamble: "whereas atmospheric emissions of ozone-depleting substances, greenhouse gases and other substances have deleterious effects on the environment both locally and globally".

[16] Section 1.

[17] Section 7(1).

[18] Section 7(1)(c).

[19] Established under s 7(1).

[20] Section 7(2).

[21] Section 7(2)(h). In addition, s 8 provides that the national framework must establish national standards for collecting and managing data necessary to assess compliance with South Africa's obligations in terms of international agreements (s 8(c)(v)); and s 16 requires that air quality management plans, which must (per s 15) be included by national or provincial departments in environmental implementation plans or environmental management plans (per s 15(1)) or by municipalities in integrated development plans (s 15(2)), must "seek" (s 16(1)(a)) to "implement [South Africa's] obligations in respect of international agreements" (s 16(1)(a)(vii)). Similar obligations apply in respect of "controlled emitters" (per s 23(2)(c)); and "controlled fuels" (s 26(2)(c)).

[22] The Minister is empowered, although not compelled, to make regulations "that are not in conflict with this Act" regarding "any matter necessary to give effect to the Republic's obligations in terms of an international agreement relating to air quality" (s 53(a)).

international law tend not to be highly specific and so their usefulness for enhancing national law is necessarily limited.

Despite its general lack of specificity on climate change, the Act does provide that an atmospheric emission licence issued in terms of the Act must contain greenhouse gas (GHG) emission measurement and reporting requirements.[23] Such a licence is required for activities which result in atmospheric emissions and that are listed pursuant to a decision by the Minister, the decision being made if he or she reasonably believes the activity has or may have "a significant detrimental effect on the environment, including health, social conditions, economic conditions, ecological conditions or cultural heritage".[24] In addition, s 29(1) of the Act provides for the declaration of a "priority air pollutant",[25] and this mechanism could be used to declare GHGs as priority pollutants. Priority pollutants would then be specially regulated by means of pollution prevention plans provided for in respect of the specific pollutants.[26] The White Paper envisages that s 29(1) will be used to "manage GHG emissions from all significant industrial sources (those responsible for more than 0.1% of total emissions for the sector) in line with approved mitigation plans that conform to the Act's requirements for pollution prevention plans prepared by identified industries and sectors".[27] At the time of writing, the power in s 29 has not yet been used for this purpose, and it should be noted that it is a directory provision, not a mandatory one, so that there is no compulsion on the Minister to make such a declaration.

Further, the Act does provide for licensing authorities to take into account, when considering applications for atmospheric emission licences, "any relevant tradable emission scheme[s]"[28]; and provides that "greenhouse gas emission measurement and reporting requirements" must be specified in atmospheric emission licenses.[29]

Finally, the Act contains a section[30] which provides that the Minister "may investigate any situation which creates, or may reasonably be anticipated to contribute to":

(a) air pollution across the Republic's boundaries; or
(b) air pollution that violates, or is likely to violate, an international agreement binding on the Republic in relation to the prevention, control or correction of pollution[31]

---

[23] Section 43(1)(l).

[24] Section 21(1).

[25] This subsection reads:

> The Minister or MEC may, by notice in the *Gazette*—

(a) declare any substance contributing to air pollution as a priority air pollutant; and
(b) require persons falling within a category specified in the notice to prepare, submit to the Minister or MEC for approval, and implement pollution prevention plans in respect of a substance declared as a priority air pollutant in terms of paragraph (a).

Note that the "MEC" is the Member of the Executive Council for environmental affairs, essentially the provincial equivalent of a national Minister.

[26] Section 29.

[27] White Paper, supra note 4, at §10.6.

[28] Section 39(e).

[29] Section 43(1)(l).

[30] Section 50, which is headed "Transboundary air pollution".

[31] Section 50(1).

and to "prescribe measures to prevent, control or correct the releases within" South Africa, if the investigation "reveals that the release of a substance into the air from a source in [South Africa] may have a significant detrimental impact on air quality, the environment or health in a country other than [South Africa]".[32]

Cumulatively, these sections certainly provide the Minister with the legal wherewithal to take measures to curb GHG emissions within South Africa. However, there is still a marked lack of firm direction to the Minister, and much will rely on the Minister to elect to deal firmly with climate change issues.

Other current legislation that is relevant is the Electricity Regulation Act,[33] under which regulations were promulgated requiring the periodic production of the Integrated Resource Plans.[34] The Act contains amongst its objectives the achievement of efficient, effective, sustainable and orderly development and operation of electricity supply infrastructure in South Africa; ensuring that the interests and needs of present and future electricity customers and end users are safeguarded and met, having regard to the governance, efficiency, effectiveness and long-term sustainability of the electricity supply industry within the broader context of economic energy regulation in the Republic; and the promotion of the use of diverse energy sources and energy efficiency.[35] The Integrated Resource Plans set out how electricity is to be produced in future, which undoubtedly has an important role to play in climate change mitigation, given South Africa's circumstances. There is, however, no explicit reference to climate change considerations in the Act.

Finally, in 2010, the Minister of Finance announced in his budget speech a flat rate CO_2 emissions tax on new motor vehicles, with effect from 1 September 2010.[36] A carbon tax is envisaged by the White Paper,[37] but is not yet in place.

Although South Africa's law relating to climate change is currently sparse, legislative innovation seems certain in the light of the White Paper's recommendations. The mitigation targets set out in the White Paper will certainly require legislative implementation of mitigation efforts and possibilities for new legislation are discussed in the analysis of the White Paper below.

[32] Section 50(2).

[33] Act 4 of 2006.

[34] Electricity Regulations on New Generation Capacity: GN R721 *GG* 32378 of 5 August 2009.

[35] Section 2.

[36] See National Treasury, "Press Release Regarding CO_2 Vehicle Emissions Tax", 26 August 2010, available at: http://www.treasury.gov.za/comm_media/press/2010/2010082601.pdf (last accessed 7 February 2011).

[37] White Paper, supra, note 4, at §10.7. See also National Treasury, *Reducing Greenhouse Gas Emissions: The Carbon Tax Option*, December 2010, Discussion Paper for Public Comment.

28.4 Climate Change Policy in South Africa

This section considers the policy that directly addresses climate change, and which culminated in the 2011 White Paper,[38] timeously released shortly before South Africa hosted COP17. It necessarily examines also what could broadly be called energy policy, since there seem to be worrying elements of the latter that could serve to undermine the objectives of the climate change policy. The White Paper will be examined first.

28.4.1 White Paper on the National Climate Change Response[39]

The White Paper is a culmination of a policy-development process that began with a 2004 policy statement,[40] although there relevant energy policies (which are discussed later) were also published before 2004. In March 2006, the Cabinet commissioned a process aimed at examining greenhouse gas mitigation options. This process had its outcome in the Long Term Mitigation Scenarios (LTMS) document,[41] which had as its purpose "to outline different scenarios of mitigation action by South Africa, to inform long-term national policy and to provide a solid basis for our position in multi-lateral climate negotiations on a post-2012 climate regime".[42] The scenarios were sketched between two limits: the "growth without constraints" (GWC) limit; and the "required by science" (RBS) limit, the latter being based on a reduction of emissions of between 30 and 40% from 2003 levels by 2050. These scenarios form the basis for the thinking in the White Paper.[43] Other important policy documents preceding the White Paper were the Carbon Tax discussion paper,[44] and the Green Paper which was the immediate predecessor of the White Paper.[45]

---

[38] In South Africa, a White Paper is not itself legislation (although it is a step toward possible legislation) and provides no binding obligations. However, a White Paper has value in that it provides an important guide for organs of state formulating and implementing policy; and for organs of state and the judiciary in interpreting legislation.

[39] Some of the discussion of the White Paper is based on Michael Kidd, "Environmental Law", *Juta's Quarterly Review of South African Law* (October to December 2011).

[40] Department of Environmental Affairs and Tourism, *A National Climate Change Response Strategy for South Africa* (September 2004).

[41] Energy Research Centre, Long Term Mitigation Scenarios: Technical Summary (October 2007).

[42] Ibid at 2.

[43] White Paper, supra, note 4, at §6.4.

[44] National Treasury, supra, note 36.

[45] GenN 1083 in *GG* 33801 of 11 November 2010. For discussion of the Green Paper, see Kidd, supra, note 1, at 318–323.

28.4.2 The Aims of the White Paper

The White Paper aims to:

- Effectively manage inevitable climate change impacts through interventions that build and sustain South Africa's social, economic and environmental resilience and emergency response capacity.
- Make a fair contribution to the global effort to stabilise GHG concentrations in the atmosphere at a level that avoids dangerous anthropogenic interference with the climate system within a timeframe that enables economic, social and environmental development to proceed in a sustainable manner.[46]

The White Paper's "strategic priorities" are risk reduction and management (aimed essentially at adaptation strategies); mitigation actions with significant outcomes (consistent with the targets mentioned above); sectoral responses, requiring measurable implementation measures; policy and regulatory alignment (expanded upon below); integrated planning, involving prioritization of the mainstreaming of climate change considerations; informed decision-making and planning; technology research, development and innovation; facilitated behaviour change (involving use of incentives and disincentives); behaviour change through choice; and resource mobilization.

The White Paper deals with adaptation in a way that highlights those sectors and the necessary interventions in each sector subject to particular risk of adverse impacts: water, agriculture and commercial forestry, health, biodiversity and ecosystems, urban, rural and coastal settlements, disaster risk reduction and management. Under mitigation, the key elements are setting the performance benchmark, identifying desired sectoral mitigation contributions, defining carbon budgets for significant GHG emitting sectors and/or sub-sectors, mitigation plans, the use of different types of mitigation approaches, policies, measures and actions, use of the market, and monitoring and evaluation.

The White Paper identifies eight "Near-Term Priority Flagship Programmes", some of which could be regarded as nationally appropriate mitigation actions. These programmes include mitigation in specific sectors (e.g. water, transport), renewable energy promotion and energy efficiency and energy demand management. One looks at carbon capture and sequestration and another is aimed at adaptation research. The major aspects in the rest of the White Paper are instruments and mechanisms aimed at achieving the strategies, including policy instruments; the mainstreaming of climate-change development; mobilization of resources and monitoring and evaluation. A section on job creation is also included.

[46] White Paper, supra, note 4, at §2.

28.4.3 Legal Aspects of the White Paper

Whereas many of the strategic interventions envisaged by the White Paper require investigation and prioritizing from a scientific/technical/economic perspective, there are several aspects that will require legal input. One of the strategic priorities is headed "policy and regulatory alignment" and this necessitates significant legal consequences in that it entails the prioritization of interventions that are already envisaged by existing legislation (etc.)[47] that will have climate change co-benefits. This suggests that the legislation contemplated is not aimed directly at climate change but may have relevant benefits therefor. The second aspect of this strategic priority is the review of existing legislation with a view to optimising and maximising climate change co-benefits. The third component is the integration into the relevant existing or new legislation of "those climate change response interventions that stimulate new economic activities as well as those that improve the efficiency and competitive advantage of existing activities".[48] This suggests that those climate change interventions that do not have what could be called a positive economic spin-off are not prioritized, even if they could have significant impacts on reducing GHG emissions. Unfortunately, in our view, the White Paper often emphasizes interventions with positive economic consequences, which is a flaw. While efforts to avoid economic detriment should be pursued wherever possible and economic benefits ought to be pursued where they arise, it must be recognized that climate change mitigation and adaptation will in all likelihood not always have positive economic spin-offs and will often be economically difficult. The White Paper glosses over this, which is disingenuous.

The mainstreaming of climate change considerations into all relevant sectors and spheres of government is absolutely critical if any significant progress is to be made in addressing climate change. It is no good, for example, for the Department of Environment to have first-class climate change policies in place if the Department of Energy continues to insist on coal as being the primary source of energy in the country, as discussed below. This mainstreaming (integrated planning) will probably have a legal consequence in that there will have to be legal requirements for such integrated planning. It is important, however, that the efforts not be left at the planning stage and that whatever integrated plans are decided upon be implemented.

Adaptation responses are required to be "mainstreamed" into sectoral plans, which will require the appropriate legislative duty to be enacted. For example, the White Paper explicitly recognizes that adaptation responses will have to be included in the National Water Resource Strategy, a second edition of which is currently being drafted. In the water sector, the White Paper indicates that a "key element" of climate change response is the provision of, inter alia, legal and regulatory resources

[47] The White Paper speaks of "national policies, legislation or strategies". For ease of reference, this discussion will refer simply to "legislation", but this may incorporate strategies and policies as well.

[48] White Paper, supra, note 4, at §4.2.

and capacity to deal with the long-term effects of climate change. As for the health aspect of adaptation, the White Paper requires the reduction of respiratory diseases and improvement of air quality through the reduction of ambient particulate matter and ozone and sulphur dioxide concentrations "by legislative and other measures" to ensure full compliance with National Ambient Air Quality Standards by 2020.

28.4.4 Biodiversity, Ecosystems and the White Paper

In the biodiversity and ecosystems section of the adaptation aspect, one of the proposed interventions is the expansion of the protected areas network "where it improves climate change resilience".[49] This is associated with the intention to explore a regulatory framework to support investment in conservation or land rehabilitation as a way of offsetting the environmental impacts of new property developments. It would be a good idea in this regard to consider the use of protected areas not just for the obvious purpose of conserving ecosystems and habitats and thereby species, but also to use them for conserving areas of high water yield, given that water is likely to suffer severe impacts as a result of climate change. It is disconcerting at present to see how much of the high-yield water catchment areas are subject to mining and prospecting rights, which does not seem to accord with sustainability thinking, particularly in the context of climate change.

As for urban human settlements, the White Paper recommends that land-use zoning regulations must be enforced and that urban land-use planning must consider the impacts of climate change and the need to sustain ecosystem services. This really does not need further explanation and is strongly supported.

28.4.5 Further Legal Aspects of the White Paper

There are several legal interventions required or at least suggested in the mitigation section of the White Paper. The first, overarching aspect that warrants some discussion is the "commitment" by South Africa to implement mitigation actions that will collectively result in a 34% deviation from "business as usual" by 2020 and 42% by 2025. This announcement (first made at the Copenhagen Conference of the Parties – COP 15 – to the UNFCCC in 2009) was made conditional on the provision of finance and technical assistance and on the implementation of a binding multilateral climate agreement. As observed earlier, this may be conditional on the international plane but now forms part of domestic policy that, even though it not strictly binding, ought to set a target that is not so flexible as to be meaningless.

There are a number of key elements with legal significance in the White Paper's mitigation approach. First, the White Paper speaks of defining carbon budgets for

[49] Ibid at §5.5.6.

significant GHG emitting sectors and/or sub sectors. This will entail adoption of sectoral carbon budgets within 2 years. This will be followed by a "mechanism and process to translate the Carbon Budgets for each relevant sector and/or sub-sector into company level desired emission reduction outcomes [that] will be developed and implemented within 3 years of the publication of this policy for companies above a minimum emissions threshold".[50] Although the White Paper is not explicit about regulatory or legislative development in this regard, it will not be possible to implement a carbon budget process (at least, not in a way that involves securing compliance) unless it is cast in legal terms, so legislation in this regard will be necessary. The White Paper expands on the initial statement of carbon budgeting being a key element by spelling out that sectors particularly likely to be targeted by this mechanism are major energy supply (electricity and liquid fuels) and energy use (mining, industry and transport) sectors. This is likely to be controversial, as evidenced by several recent government hearings and media reports.[51]

Companies and economic sectors for which desired emission reduction outcomes have been established will also be required to prepare and submit mitigation plans that set out how they intend to achieve such reduction outcomes. These may form part of Pollution Prevention Plans already envisaged by the Air Quality Act,[52] or may require new legislative intervention.

28.4.6 *The White Paper and Alternate Instruments*

One set of mechanisms identified by the White Paper is the use of economic instruments, including the "appropriate pricing of carbon and economic incentives, as well as the possible use of emissions offset or emission reduction trading mechanisms" for the relevant companies and sectors.[53] Although the White Paper is silent as to the legal consequences where this "key element" is first raised, it later states that a mix of economic instruments "complemented by appropriate regulatory policy measures"[54] are an essential element of mitigation efforts. National Government will take the lead on this. At first glance, it is often mistakenly assumed that economic instruments are alternatives to legal control, but the law is necessary at least to set the parameters within which the economic instruments will operate. This prevents the problem of "free riders".

---

[50] Ibid at §6.1.3.

[51] See, for instance, Parliamentary Monitoring Group, "White Paper on Climate Change: Public Hearings", 1 November 2011, available at: http://www.pmg.org.za/report/20111102-public-hearings-national-climate-change-white-paper-2011-south-africa (last accessed 1 March 012); and Sue Blaine, "White Paper Sets Industry Carbon 'Budgets'", *Business Day*, 14 October 2011, available at: http://www.businessday.co.za/articles/Content.aspx?id=156085 (last accessed 1 March 2012).

[52] See supra, note 13.

[53] White Paper, supra, note 4, at §6.1.6.

[54] Ibid at §6.3.

"Carbon tax" is specifically mentioned as a fiscal measure that will require consultation with the National Treasury,[55] and the Departments of Trade and Industry[56] and Economic Development.[57] This is not the only official policy engagement with carbon tax. The National Treasury released a discussion paper entitled *Reducing Greenhouse Gas Emissions: The Carbon Tax Option* in December 2010.[58] In the White Paper's section on market-based instruments, the National Treasury is tasked with continuing to develop carbon tax policy and the White Paper sets out key considerations that will inform this process. These address issues of the rate of the tax; technical and administrative feasibility; distributional implications; effects on competitiveness; timing of the implementation of the tax; regressive impacts; and relief measures. In the February 2012 government budget speech, it was announced that 2012 would see a follow-up discussion document on the carbon tax. It is anticipated that the tax will be calculated on percentage-based emission thresholds rather than absolute thresholds and the rate will be R120.00[59] per tonne of CO_2e[60] above the suggested thresholds, which is proposed to take effect during 2013/2014, with annual increases of 10% until 2019/20.[61]

Legislation will also be necessary if reporting of emissions data is to become mandatory (as required by the White Paper) for entities that emit more than 0.1 Mt of GHGs annually, or that consume electricity which results in more than 0.2 Mt of emissions from the electricity sector.[62]

28.4.7 Programmes Under the White Paper

When dealing with the Near-Term Priority Flagship Programmes, the White Paper makes some general observations about the prerequisites for meeting these priorities, some of which entail regulatory or legal change. The White Paper states[63] that:

> [t]he Flagship Programmes also utilise, test and/or demonstrate a suite of policy interventions including regulatory measures, market-based instruments, tax incentives and fiscal subsidies, and information and awareness initiatives. Regulatory measures include renewable energy

---

[55] See, generally, http://www.treasury.gov.za

[56] See, generally, http://www.dti.gov.za

[57] See, generally, http://www.economic.gov.za

[58] See, infra, note 2; and discussion in Kidd, supra, note 1, at 317–318.

[59] Equivalent: 16.06 US$ at 1 March 2012.

[60] This symbol meaning "equivalent amounts of carbon dioxide", so as to include other greenhouse gases.

[61] Wendy Gardner, "Climate Change Taxes", *Moneywebtax*, 22 February 2012, available at: http://www.moneywebtax.co.za/moneywebtax/view/moneywebtax/en/page34677?oid=65818&sn=Detail&pid=34677 (last accessed 27 February 2012).

[62] White Paper, supra, note 4, at §6.7.

[63] Ibid at §8.

> and energy efficiency targets complemented by appropriate standards; market-based instruments including the electricity generation levy and taxes on motor vehicle emissions and incandescent light bulbs; tax incentives and fiscal subsidies are targeted at various programmes that support climate change mitigation and adaptation objectives; and information and awareness initiatives including the motor vehicle emissions labelling scheme.

These include specific measures, some of which are already in place.[64] Further measures to be investigated are set out in the carbon tax discussion paper discussed above.

When looking at the individual priority flagship programmes, the "Energy Efficiency and Energy Demand Management" programme explicitly requires "regulation" for the industry energy efficiency programme; the residential energy efficiency programme (particularly in regard to specifications for low-income housing); and in respect of commercial and residential building standards to enforce green building construction practices. In this regard, new energy efficiency standards in the National Building Regulations came into effect on 9 November 2011.[65] For the Waste Management Flagship Programme, a detailed Waste-Related GHG Emission Mitigation Action Plan will be established that will, inter alia, detail the development and implementation of any "policy, legislation and/or regulations required to facilitate the implementation of the plan".[66]

28.4.8 Coordination Under the White Paper

From an overarching macro perspective, one of the most important aspects is the mainstreaming of climate change actions and this requires coordination and alignment of government policies and actions. This is critical and, in our view, is one of the most serious deficiencies in current governmental arrangements. Unless changes are made, silo-thinking will seriously militate against successful implementation of the White Paper. In order to achieve this, the White Paper envisages a "comprehensive review of all government legislation, policy, strategies, plans and regulatory frameworks"[67] as underpinning the successful implementation of the White Paper. This will be followed by review "on a regular basis" so as to ensure that regulations, etc., falling within the jurisdiction of all spheres of government, including state-owned enterprises, are fully aligned with the climate policy. In this review process, particular attention will be given to local government legislation.

The responsibility for implementation of the climate change policy involves both identification of roles and institutional arrangements. The White Paper envisages Parliament as overseeing the development and implementation of the policy through the portfolio committees, particularly those on water and environmental affairs; energy; agriculture, forestry and fisheries; trade and industry; mining; science and

[64] See the list of those measures already implemented in ibid, at §10.7.

[65] See GN R211 in GG 34586 of 9 September 2011.

[66] White Paper, supra, note 4, at §8.6.

[67] Ibid, at §10.1.

technology; and transport. The portfolio committees are tasked with reviewing legislation "to determine the legal requirements to support the institutional and regulatory arrangements proposed in this White Paper, and to ensure policy and legislative alignment".[68] The committees are intended to work together with the Department of Environmental Affairs and the Inter-Ministerial Committee on Climate Change to "draft any Bills, or an amendment to NEMA,[69] that may be required within 3 years of the publication of this policy". This should probably refer to amendment of any relevant legislation, not only NEMA. Potentially, changes could be made to the Air Quality Act,[70] for example, to serve some of the purposes of the White Paper. Provinces are expected to develop individual climate response strategies and these will probably require some kind of legislative authority as well.

As far as carbon markets are concerned, the White Paper envisages that these may include cap-and-trade mechanisms and offset schemes, both of which will need legislative infrastructure in order to operate. National Treasury is given the responsibility of investigating the feasibility of an emissions trading scheme "as a medium- to long-term response to climate change".[71] Another type of market-based instrument to be explored, which will also require a legislative source, is that of incentives. There are some existing incentives (lower fuel taxes on cleaner fuels, for example) and others will be explored as part of a suite of policy instruments that are aimed at influencing climate change response.

28.4.9 The White Paper: Conclusion

The White Paper recommends that South Africa must continue "proactively [to] contribute to the technical and institutional reform debates of the UNFCCC financing measures to ensure that developing and least-developed countries such as those in [SADC][72] can access the additional and necessary resources in a fair, transparent and timely manner".[73] Should parties agree on a new binding climate instrument in the future (as envisaged by the Durban Platform for Enhanced Action),[74] it is likely

[68] Ibid, at §10.2.1.

[69] National Environmental Management Act 107 of 1998.

[70] Act 39 of 2004.

[71] White Paper, supra, note 4, at §10.7.2.

[72] The Southern African Development Community; see, generally, http://www.sadc.int

[73] White Paper, supra, note 4, at §11.1.2.a.

[74] Meaning the programme agreed to at COP 17/CMP 7 of the UNFCCC/Kyoto Protocol process in Durban in December 2011, in terms of which the parties extended the Ad Hoc Working Groups on Long-term Cooperative Action; launched "a process to develop a protocol, another legal instrument or an agreed outcome with legal force"; set a deadline of 2015 for adopting this instrument, and of 2020 for its coming into force; and made certain other commitments. Available at: http://unfccc.int/files/meetings/durban_nov_2011/decisions/application/pdf/cop17_durbanplatform.pdf (last accessed 1 March 2012).

that countries will have to follow monitoring, reporting and verification (MRV) procedures in respect of emissions. One of the interventions suggested in the White Paper is that research institutions (including universities) be encouraged to develop national MRV guidelines for South Africa. These guidelines are intended to focus on mitigation and adaptation actions "such as land-use practises (sic) and change".[75]

South Africa's climate change response actions will have to be evaluated, and the White Paper provides that the Department of Environmental Affairs will be responsible for defining review mechanisms as well as developing the White Paper into a "suite of regulatory and legislative instruments where required".[76] More specifically, it is provided that South Africa will, within 2 years of the publication of the White Paper, "design and publish a draft Climate Change Response Monitoring and Evaluation System".[77] It is intended that this system will evolve with international MRV requirements.

28.5 South Africa's Foreign Policy

Whether, and how, countries ("states") incorporate their international obligations into their national legal systems is, obviously, of crucial importance to the success or failure of the international instruments in terms of which countries make those commitments. The stronger the commitments which countries make internationally, the more likely they are to attempt to translate these commitments into meaningful action at home. With South Africa clearly making efforts to position itself within the international community as a perceived leader on environmental matters, and as a committed driver of change in the climate change issue-area,[78] it might be expected that South Africa would ensure that its foreign policy was mirrored by domestic action. In May 2011, South Africa released its *White Paper on Foreign Policy* ("White Paper FP").[79] Being so recent a document, it can be expected with some confidence that South Africa will not depart from its tenets dramatically in the few years to come.

According to the White Paper FP, South Africa's "unique" approach to foreign policy is driven by "ubuntu", which is a concept or philosophy that "translates

[75] White Paper, supra, note 4, at §11.2.6.

[76] Ibid, at §12.

[77] Ibid, at §12.3.

[78] On this, *vide* South Africa's drive firstly to host COP17/CMP7 of the UNFCCC/Kyoto Protocol Process in 2011; and then to see a successful outcome from the Conference.

[79] "Building a Better World: The Diplomacy of Ubuntu", White Paper on South Africa's Foreign Policy, 13 May 2011, available at: http://www.info.gov.za/view/DownloadFileAction?id=149749 (last accessed 1 March 2012).

into an approach to international relations that respects all nations, peoples, and cultures [and which] recognises that it is in our national interest to promote and support the positive development of others".[80] It is then suggested that South Africa "accords central importance to our immediate African neighbourhood and continent; working with countries of the South to address shared challenges of underdevelopment; promoting global equity and social justice; working with countries of the North to develop a true and effective partnership for a better world"; and, additionally, "doing our part to strengthen the multilateral system, including its transformation, to reflect the diversity of our nations, and ensure its centrality in global governance".[81]

What is perhaps worrying for the environmental lawyer seeking increased action within the climate change issue-area, is that there is a strong theme running through the White Paper FP of South Africa's need to seize every opportunity to increase economic growth and meet the employment-related aspirations of the underprivileged.[82] While nobody would dispute the need for strong poverty-alleviation measures to be taken, it needs to be recognised that in the area of climate change mitigation efforts, great sacrifices are going to need to be made which might prove impossible to reconcile with strong economic growth. While innovative technologies will have an important role to play,[83] it is not going to be easy for South Africa to meet its lofty goals of lifting from poverty both its own people and people in the Southern African region[84] without making extensive use of present industry infrastructure and established industrial techniques.

The White Paper FP does acknowledge climate change, stating that:

> [s]ignposts of climate change include environmental degradation, desertification, melting of the icecaps, rising sea levels and more volatile and extreme weather patterns. Both natural and man-made environmental changes impact on all aspects of human development. These changes will increasingly hinder sustainable development and have a significant impact on the world's social and economic systems.[85]

However, many of the aims of the White Paper FP remain if not contradictory then at least internally difficult (and perhaps impossible) to reconcile. The conundrum of providing environmental protection and making lifestyle sacrifices whilst at the same time providing poverty alleviation to many, perhaps even the majority, of people, appears starkly from comparing South Africa's stated goals in the climate change area with its current policy on energy, as will appear from the next section.

[80] Ibid. at 4.

[81] Ibid. at 4–5.

[82] Ibid. at 8, 13–14, 18–19, 26–28, 29, 31–32.

[83] Ibid. at 14.

[84] Ibid. at 4–5, 8, 13, 19–23.

[85] Ibid. at 15.

28.6 Energy Policy in South Africa

Whereas the White Paper on climate change emphasizes the mainstreaming of climate change considerations in all government activities, this seems to be a clarion call that is not being heeded elsewhere in government. The White Paper, at first glance replete with planned interventions to address climate change mitigation and adaptation, has to be considered in the light of South Africa's energy policy (amongst other plans and objectives). During the time that South Africa's climate policy was being developed, culminating in the White Paper of 2011, South Africa was also planning for its ongoing energy needs.

In 1998, the White Paper on the Energy Policy of the Republic of South Africa made passing reference to climate change, providing it as one of many factors to take into account in the development of energy policy; but perhaps the main message to emerge from the energy White Paper is that coal will "dominate other energy sources in South Africa for many years to come".[86] This was followed by the integrated energy plan of 2003, which was aimed at ensuring that supply meets projected demand. Various scenarios are considered, which do take into account climate change considerations.[87] The dominant consideration, however, is clearly cost. In its conclusions, the plan states that – coal remains the dominant primary energy source over the planning horizon. In all circumstances where cost is the major driver, coal generally emerges as the least expensive option. The use of such coal energy presupposes the increased use of clean coal technologies. Moreover, coal remains the largest indigenous energy resource currently available.[88]

Clearly, the "cost" referred to does not take into account the myriad externalities relating to the mining and use of coal. To be fair, the plan does recognise a role to be played by renewable energy and indicates that the "current target for renewable energy is 10,000 GWh by the year 2012".[89] For the year 2000, South Africa's primary energy supply was approximately 4,782 PJ[90] and final energy demand was 2,363 PJ for the same year.[91] The target for renewable energy, therefore, is 0.75% of the supply and 1.5% of demand at the 2000 levels. This indicates the relatively peripheral role to be played by renewable energy, especially if the levels of supply and demand do not remain at 2000 levels, but instead increase significantly.

---

[86] Department of Minerals and Energy, White Paper on the Energy Policy of the Republic of South Africa (December 1998) at 92.

[87] Department of Minerals and Energy, *Integrated Energy Plan for the Republic of South Africa* (March 2003) at 20.

[88] Ibid, at 25.

[89] Ibid. By way of comparison, according to International Energy Agency data, in 2008 Denmark produced 58,426 GWh from renewable sources and Mexico 47,303 GWh, available at: http://www.iea.org/stats/renewdata.asp?COUNTRY_CODE=DK and http://www.iea.org/stats/renewdata.asp?COUNTRY_CODE=MX respectively (last accessed 27 February 2012).

[90] Department of Minerals and Energy, supra, note 87, at 6. PJ denotes Peta Joules, which is 10^{15} J.

[91] Ibid at 7.

On the subject of renewable energy, the White Paper on the Renewable Energy Policy for the Republic of South Africa of 2003 sets out "Government's vision, policy principles, strategic goals and objectives for promoting and implementing renewable energy in South Africa".[92] The overall vision of the White Paper is to increase the contribution of renewable energy to the energy mix, "thus contributing to sustainable development and environmental conservation".[93] This is an admirable objective, but the vision is somewhat limited. The White Paper sets a rather conservative target:

10,000 GWh (0.8 Mtoe) renewable energy contribution to final energy consumption by 2013, to be produced mainly from biomass, wind, solar and small-scale hydro. The renewable energy is to be utilised for power generation and non-electric technologies such as solar water heating and bio-fuels. This is approximately 4% (1,667 MW) of the estimated electricity demand (41,539 MW) by 2013. This is equivalent to replacing two (2×660 MW) units of Eskom's combined coal fired power stations.[94]

One of the shortcomings of the White Paper, in our view, is that renewable energy is seen in the "big picture" largely as an energy source to feed "into the grid", rather than as a source that can power individual consumers' needs. This is despite the fact that solar power, for example, is considered in the document to be appropriate for use by individuals. It is said that "South Africa experiences some of the highest levels of solar radiation in the world"[95] and that average daily solar radiation in South Africa varies between 4.5 and 6.5 kWh/m^2 (16 and 23 MJ/m^2), compared to about 3.6 kWh/m^2 for parts of the United States and about 2.5 kWh/m^2 for Europe and the United Kingdom.[96] These facts suggest that far greater emphasis should be given to use of solar energy for domestic power generation (not just water heating), given that solar power is used extensively in Western Europe (with far lower solar radiation) for domestic generation, to such an extent that many users sell power back to the grid.

Closely related to overarching energy policy is electricity planning, since such a large proportion of the country's energy usage is for electricity generation. The Integrated Resource Plan (IRP) for Electricity is required by electricity regulations on new generation capacity in terms of the Electricity Regulation Act.[97] The IRP "gives effect to the following policy objectives"[98]:

1. 10,000 GWh (approximately 4% of the energy mix) of renewable energy usage,

---

[92] Department of Minerals and Energy, White Paper on the Renewable Energy Policy of the Republic of South Africa (November 2003), at 1.

[93] Ibid.

[94] Ibid, at 25. Eskom ("Electricity Supply Commission", translated from the Afrikaans) is South Africa's parastatal electricity supply company; on which see, generally, http://www.eskom.co.za

[95] Ibid at 20.

[96] Ibid, citing Gideon Stassen, *Towards a Renewable Energy Strategy for South Africa*, Ph.D. Thesis on file at the University of Pretoria (1996).

[97] Act 4 of 2006.

[98] GN 1243 in *GG* 32837 of 31 December 2009 at 10.

2. the implementation of Energy Efficiency and Demand Side Management through financial incentives scheme (*sic*), and
3. installation of one million solar water heaters.

The IRP contains a schedule of power-generation sources including two new coal-fired power stations (Medupi and Kusile) and makes it clear (although not using express words to this effect) that coal remains the primary energy source. From the policy objectives stated above it is also manifestly clear that renewable energy sources are regarded very much as peripheral. A second IRP is still awaited, although it is unlikely to differ significantly from the first plan in respect of the energy mix for electricity generation.

In the 2003 Integrated Energy Plan, untapped coal reserves in South Africa were estimated at 55 billion tones, and coal was regarded as "plentiful and inexpensive to exploit",[99] leading to the conclusion that it would remain the primary energy source into the future (as pointed out above). This thinking has not changed in the intervening period and renewable energy is still seen as a fringe source. In the 2009 Integrated Resource Plan, central sources of electricity for the short- to medium-term are seen to be the Medupi coal-fired power station (the first unit of which will be commissioned in 2012) and the Kusile coal-fired station (the first unit of which to be commissioned in 2013). When Medupi is fully operational, it will be producing 4,800 MW of power (more than a tenth of total current capacity), and also producing 30 Mt of CO_2.[100] It is due to commence operation in 2012, but to become fully operational only some time after that. Controversially, Eskom obtained a loan from the World Bank of US$3-billion in order to construct the project.[101] The power station will reportedly require 14.6 Mt of coal annually for the next 40 years[102] and will require enormous amounts of water in a region of the country which is already facing water stress.

28.7 Conclusion

South Africa, a major global GHG emitter, has – at the time of writing of this chapter – very little legislation aimed directly at responding to climate change. In the White Paper on the National Climate Change Response, however, the Department of Environmental Affairs has set out a blueprint for considerable regulatory innovation

[99] Department of Minerals and Energy, supra note 53, at para. 3.

[100] Carol Paton, "Hot Air v Action", *Financial Mail*, 29 July 2010, available at: http://www.fm.co.za/Article.aspx?id=116438 (last accessed 27 February 2012).

[101] Janice Roberts, "World Bank approves Eskom Loan" *Mail & Guardian*, 9 April 2010. For criticism, see, for example, Khadija Sharife, "South Africa: Country's Dirty Secret – Eskom and the Medupi Power Plant", 14 May 2010, available at: http://allafrica.com/stories/201005140838.html (last accessed 27 February 2012).

[102] Jonathan Faurie, "Medupi Project on Track for Scheduled Delivery", *Engineering News Online*, 5 December 2008, available at: http://www.engineeringnews.co.za/article/medupi-project-on-track-for-scheduled-delivery-2008-12-05 (last accessed 27 February 2012).

in addressing climate change, both from the perspective of adaptation and mitigation. If this were the only climate change-related message coming out of South Africa at the present time, there would be cause for considerable optimism. The voice of the government's environmental Department, however, has consistently been drowned out by those sectors of government that are seen as more oriented toward development and rapid poverty alleviation (and hence as pushing economic rather than environmental agendas). Consequently, it is possible to have energy policy that flies in the face of the climate change response policy. This contradiction could be seen in the President's 2012 "state of the nation" address,[103] delivered in February 2012 shortly before this chapter was written. In the address, the only reference to climate change was reference to the "successful" hosting of COP17; and the major thrust of the speech was on plans for major infrastructural development throughout the country, with no reference being made to the potential environmental (and climate) impacts of these developments.[104] This should arguably not be surprising. Although South Africa clearly has the potential to respond more appropriately to climate change than it has done so far, the country faces huge – even overwhelming – challenges. An important task for environmental lawyers and climate change issue-area activists is to convince South Africa's government that the challenge of mainstreaming climate change thinking, the clear need for which is highlighted by the government's own 2011 White Paper, is one of the most important of these challenges.

[103] "State of the Nation Address By His Excellency Jacob G. Zuma, President of the Republic of South Africa on the occasion of the Joint Sitting of Parliament, Cape Town", 9 February 2012, available at: http://www.info.gov.za/speech/DynamicAction?pageid=461&sid=24980&tid=55960 (last accessed 1 March 2012).

[104] See Patrick Bond, "South Africa: "Global Sustainability" Wilts in Hot Political Air", *Pambazuka News*, 16 February 2012, available at: http://www.pambazuka.org/en/category/features/80007 (last accessed 27 February 2012).

Chapter 29
Climate Change Policy and Legislation in Brazil

Haroldo Machado-Filho

Abstract The starting point of this chapter is an overview of the common obligations for all the Parties established under the United Nations Framework Convention on Climate Change, Article 4.1. These obligations encompass a set of measures to be taken by a responsible State, under normal circumstances, through the adoption of legislations and administrative control, as well as other available means, with a view to respecting the obligations formulated under international law, which is close to the idea of "due diligence". Moving from the abstract to the concrete, the chapter also focuses on recent policies and legislation on climate change adopted in Brazil, which are fundamental for the implementation of commitments under the UNFCCC. These new developments, including the voluntary quantified goal for reducing emissions announced by Brazil in 2009, encapsulated in its National Policy on Climate Change, demonstrates that the country has moved from "due diligence" measures, with a view to respecting the obligations formulated under international law, towards the goal for real contribution to the combat to climate change.

Haroldo Machado-Filho holds a Ph.D. in International Law from the Graduate Institute of International Studies, Geneva. He is also Lead Author of the Fifth Assessment Report of the Intergovernmental Panel on Climate Change (IPCC) in Working Group III; Member of the Global Climate Change and International Relations Network at the University of Brasília – UnB; and Brazilian negotiator under the multilateral climate change regime, since 1998. The views expressed by the author in this paper represent his personal opinion and do not necessarily reflect those of the Brazilian government, the UNDP or the IPCC.

H. Machado-Filho (✉)
Environment and Development Unit, SQSW 303, Bloco J, Ap 521, Brasilia DF, Brazil
e-mail: haroldoclima@gmail.com

E.J. Hollo et al. (eds.), *Climate Change and the Law*, Ius Gentium: Comparative Perspectives on Law and Justice 21, DOI 10.1007/978-94-007-5440-9_29,
© Springer Science+Business Media Dordrecht 2013

29.1 From Common Commitments Under the UNFCCC to National Policy and Legislation

Based on the principles set out in Article 3 of the United Nations Framework Convention on Climate Change (UNFCCC),[1] the Convention establishes common obligations for all the Parties.[2] The principle of common but differentiated responsibilities (CBDR) was especially taken into consideration when the general commitments were drafted. The "chapeau" of UNFCCC Article 4.1 makes it clear that all Parties shall implement such commitments "taking into account their common but differentiated responsibilities and their specific national and regional development priorities, objectives and specific circumstances". This provision was extremely important in ensuring wider participation of countries in the UNFCCC, since it left room for each Party "to determine its own level of implementation."[3]

The common commitments have a qualitative nature and do not directly establish timetables or deadlines, an indication that they should be implemented progressively, as a long-term strategy. An extensive list of these commitments was established, addressing both mitigation and adaptation issues, as well as reporting, public awareness and the scientific aspects of climate change.

The Parties' first common commitment under the UNFCCC is to develop and periodically update national inventories of anthropogenic emissions by sources and removals by sinks of all greenhouse gases not controlled by the Montreal Protocol.[4]

Despite not establishing quantitative targets, another important common commitment is the formulation, implementation and regular updating of national and, where appropriate, regional programmes with a view to mitigating and adapting to climate change.[5] The mention of CBDR in the "chapeau" of Article 4, as well as specific national and regional circumstances and development priorities, was important to reduce the fear of certain Parties that this formulation could interfere with their sovereign function of establishing national programmes. Parties will necessarily take these elements into account when establishing programmes and activities containing measures that contribute to addressing climate change and its adverse effects. Moreover, the Parties shall take into consideration, to the extent feasible, climate change issues in social, economic and environmental policies. In undertaking

[1] United Nations Framework Convention on Climate Change, United Nations Framework Convention on Climate Change, New York, 9 May 1992, in force 21 March 1994, 31 *International Legal Materials* (1992), 849.

[2] Article 4.1 of the UNFCCC.

[3] Farhana Yamin and Joanna Depledge, *The International Climate Change Regime: A Guide to Rules, Institutions and Procedures*, 1st edition (Cambridge: Cambridge University Press, 2004), at 93.

[4] Article 4.1 (a) of the UNFCCC. The expression "Montreal Protocol" refers to the 1987 Montreal Protocol on Substances that Deplete the Ozone Layer, as adjusted and amended on 29 June 1990.

[5] Article 4.1 (b) of the UNFCCC.

projects and measures to promote mitigation or adaptation to climate change, Parties must also employ adequate methods to minimise adverse effects on the economy, on public health and on the quality of the environment.[6]

The fundamental role of the development and application of technologies to mitigate greenhouse gas emissions in all the relevant sectors of economy is also recognised under the general commitments. All Parties shall, therefore, promote and cooperate in the development, application and diffusion, including transfer, of technologies, as well as of practices and process that promote mitigation of GHG emissions.[7] Since most Parties do not have access to such technologies, a specific commitment was established under the Convention to enable this access.

Although the most frequent concern relates to the processes and activities that release greenhouse gases into the atmosphere, those processes, activities and components that store or remove these gases from the atmosphere are equally important in stabilising concentrations. In this regard, the Parties have also committed themselves to conserving and enhancing sinks and reservoirs of greenhouse gases, including biomass, forests and oceans, as well as other ecosystems.[8]

The adaptation aspects related to climate change are also addressed in the general commitments, by means of the cooperation in the preparatory process for adaptation to the impacts of climate change. Such cooperation must be extended to the development and elaboration of integrated plans of coastal management, water resources and agriculture, as well for the protection and rehabilitation of vulnerable areas, particularly those affected by flooding, drought and desertification.[9] A special reference (another example of contextual norms) is made to Africa in this regard, which is the only region mentioned by name in the Convention.

Considering the remaining scientific uncertainties related to climate change, and the still limited understanding of many social and economically related issues, the gathering and dissemination of information will play a crucial role in achieving the Convention's ultimate objective. Parties shall therefore promote and cooperate in research, systematic observation and data collection related to the climate change system,[10] as well as in the exchange of scientific, technological, technical, socio-economic and legal information related to climate change and response strategies.[11]

The promotion of and cooperation in education and training, as well as in raising public awareness and participation related to climate change, is another vital feature

---

[6] Article 4.1 (f) of the UNFCCC.

[7] Article 4.1 (c) of the UNFCCC.

[8] Article 4.1 (d) of the UNFCCC.

[9] Article 4.1 (e) of the UNFCCC.

[10] Article 4.1 (g) of the UNFCCC. See also Article 5 that fleshes out this commitment. It is worth mentioning that international and intergovernmental programmes and organisations play a fundamental role in promoting and co-operating in research and systematic observation under the UNFCCC.

[11] Article 4.1 (h) of the UNFCCC.

in the implementation of the UNFCCC.[12] Climate change is a highly complex and technical issue, difficult for non-experts to understand, and one of the main problems in this regard is that there is very little written material available in non-UN languages. The Convention makes a special reference to the participation of non-governmental organisations in this process.[13]

As can be noticed – by means of the reference to CBDR and to "specific national and regional development priorities" in the "chapeau" of Article 4 – the text on common commitments in the UNFCCC reflects general formulations, which do not impose to the Parties specific standards of conduct. This is precisely the trend of modern environmental international law: to set forth particular standards of conduct so that they take the adequate preventive measures to protect the environment, as well as reduce and control activities that can damage it, without necessarily prohibiting these activities. In recent decades one has seen, especially through the adoption of multilateral environmental agreements (MEAs), the efforts of the international community in regulating such a conduct of States mainly through the establishment of co-operative measures, such as regulation and supervision by international institutions.

Thus, States should take the necessary measures applicable both to public and private conduct in order to fulfil their international responsibilities,[14] which also implies that States have to act with "due diligence". In this regard, "due diligence" encompasses a set of actions to be taken by a responsible State, under normal circumstances, through the adoption of legislations and administrative control, as well as other available means, with a view to respecting the obligations formulated under international law. In the context of international environmental law, these actions shall be taken under the jurisdiction of this State aiming to protect the environment, as well as to reduce and control activities (and substances resulting from such activities) that can damage other States or areas beyond the limits of its national jurisdiction. Most MEAs – as well as several instruments adopted by international conferences, international organisations and scientific institutions – are aimed mainly at establishing obligations limited by due diligence, which is considered to be a primary environmental obligation of States.[15]

Flexibility is an essential characteristic of "due diligence", taking into consideration characteristics that might vary from country to country, and from time to time. Thus,

---

[12] Article 4.1 (i) of the UNFCCC. See also Article 6 that fleshes out this commitment.

[13] This is one of the two references to non-governmental organisations in the UNFCCC. The other reference is contained in its Article 7.2 (l) and is related to cooperation and the provision of information to the Conference of the Parties.

[14] Xue Hanqin, *Transboundary Damage in International Law* (Cambridge: Cambridge University Press, 2003), at 163.

[15] For instance, Article 194 of the 1982 UNCLOS; Article 2 of the 1979 LRTAP Convention; Article 2 of the 1985 Vienna Convention on the Protection of the Ozone Layer; Article 1 of the 1996 Protocol to the 1972 London Dumping Convention; Principle 21 of the World Charter for Nature. Cf. Patricia Birnie and Alan Boyle, *International Law and the Environment*, 2nd edition (Oxford: Oxford University Press, 2002), at 113.

in this context, since it accommodates differentiated standards of conduct for different States, there is no absolute guarantee that all States will effectively prevent harm. After all, even if a State takes a diligent conduct, it may fail to fulfil the standard of conduct expected as good governance, given that the result could also depend on objective factors that might be outside its control.[16]

The complex discussion on obligation of States certainly goes beyond the scope of this chapter. Nevertheless, these elements are important to understand, from a more theoretical perspective, how the adoption of legislation, policies and other administrative measures taken by a State can be seen as significant steps towards the compliance with obligations under international law.

Moving from the abstract to the concrete, the following section will focus on recent policies and legislation on climate change adopted in Brazil, which are fundamental for the implementation of commitments under the UNFCCC, especially those contained in its Article 4, as analysed above.

29.2 Recent Policies and Legislation on Climate Change Adopted in Brazil

29.2.1 National Plan on Climate Change

Although Brazil does not have any quantified commitments on greenhouse gas emission limitation or reduction under the multilateral climate change regime, the country has not been idle and is playing a critical role in fighting against climate change. As reported in this National Communication, various government programs and initiatives in Brazil are bringing about major reductions in greenhouse gas emissions, some of which are responsible for the fact that Brazil has a clean energy mix compared to other countries, with low greenhouse gas emissions per unit of energy produced or consumed.[17]

Most of the programmes and actions implemented do not have the direct objective of reducing greenhouse gas emissions, although they do have significant impacts on emission reductions from different sources. A good example of this is the programme related to the use of ethanol (produced from sugar cane) as vehicle fuel. The National Alcohol Program – Proálcool – in Brazil was originally developed to avoid increasing dependence on foreign oil and foreign currency evasion during the oil price shocks.

Nevertheless, more recently, given the increasing awareness related to climate change, Brazil is deliberately moving towards undertaking voluntary commitments

---

[16] Ibid.

[17] Brazil, Second National Communication of Brazil to the UNFCCC (Brasília: MCT, 2010), volume 1, at 17.

that represent a significant reduction in the emission of greenhouse gases and protection to sinks.

Hence, in 2007, the President of the Republic included in the agenda of government activities the development of a plan, initially called "National Action Plan to Combat Climate Change," aimed at structuring and coordinating government actions concerning the effects of global warming arising from anthropogenic activities.

In 2007, the federal government created the Interministerial Committee on Climate Change (CIM),[18] coordinated by the Executive Office of the Presidency of the Republic ("Casa Civil da Presidência da República") and encompassing 17 ministries, with a mandate to develop the National Plan on Climate Change and the National Policy on Climate Change.

The Executive Group on Climate Change (GEx),[19] which is coordinated by the Ministry of Environment and reports to the CIM, is responsible for elaborating, implementing, monitoring and evaluating the National Plan on Climate Change. As a result of GEx's work, a bill for the National Policy on Climate Change was submitted to the Legislative Branch.[20]

Another practical result of GEx's work was the draft National Plan on Climate Change. In its initial phase of drafting, consultation questionnaires were forwarded to the ministries that comprise the CIM in order to bring together the actions already in place for each of them and their related bodies, such as programs and projects that contribute to preventing climate change.

This process included public consultations of the utmost importance: the III National Conference on the Environment[21] and the meetings held by the Brazilian Forum on Climate Change,[22] the so-called "Sector Dialogues."[23]

[18] Presidential Decree no. 6,263, of 21 November 2007.

[19] The GEx is a smaller group, composed of representatives from eight Ministries plus a representative from the Brazilian Forum on Climate Change.

[20] Bill no. 3,535, of 10 June 2008. This bill became the basis for negotiations in the National Congress that resulted in Law no. 12,187, which was sanctioned by the President of the Republic on 29 December 2009, as discussed in the next section.

[21] The National Conferences on the Environment are part of the Federal Government's policy for social mobilization in decision-making processes. They have been held since 2003, with the 1st National Conference on the Environment becoming a source of social legitimization and democratic stability.

[22] The Brazilian Climate Change Forum (FBMC), chaired by the President of the Republic, was created (Decree no. 3,515 of 20 June 2000) with the objective of including the organized civil society in discussions related to global climate change, as well as educating and mobilizing society to debate, and providing inputs for decision-making on problems resulting from global climate change and regarding the CDM. The FBMC should also assist the government to incorporate global climate change issues in the various levels of public policies. The Forum has the participation of the Ministers as well as civil society personalities and representatives, appointed by the President of the Republic due to their renowned expertise or relevant knowledge on climate change.

[23] In the sector dialogues held in this first phase of the Plan, several sectors of society were heard, such as industry, forestry, finance, agriculture, forest and changes in land use, municipal movements, civil society and NGOs.

The overall objective of the National Plan on Climate Change is to identify, plan and coordinate actions and measures that can be undertaken to mitigate greenhouse gas emissions generated in Brazil, as well as those necessary for the adaptation of society to the impacts of climate change.

The National Plan, which was launched on 1 December 2008, was a significant step towards a more structured and organized set of mitigation actions with the aim of collaborating in the international efforts to combat climate change. The National Plan must be guided by the National Policy on Climate Change, which came up afterwards. With the adoption of the National Policy, the National Plan has been reviewed and updated in the light of this more comprehensive legal instrument.

29.2.2 *National Policy on Climate Change (PNMC)*

The year of 2009 was characterised by intense debates on climate change related issues all around the globe, given that a comprehensive deal under the multilateral regime on climate change was expected at the 15th Conference of the Parties to the Convention (COP-15) and the 5th session of the Conference of the Parties serving as the Meeting of the Parties to the Kyoto Protocol (COP/MOP 5), held in Copenhagen. In Brazil, the situation was not different, and in October 2009, shortly before COP 15, there was an intensification of debates within the federal government for more ambitious actions to reduce emissions, considering that some segments within the government resisted the adoption of this proposal. On 13 November 2009, important mitigation actions by the government were announced.

The mitigation goals announced by Brazil have the following features:

1. They are voluntary, but involve political will of undertaking actions in an eventual international agreement;
2. Refer to the deviation of the growth curve of emissions relative to expectations of the future emissions based on a "business as usual" scenario and do not relate to a base year as do the commitments of the European Union, Japan, Korea South, Switzerland, and Norway presented at COP-15. It was a compromise between the progressive sectors and conservative members of the government, who argued that Brazil is not obliged to commit to legally binding and quantified mitigation targets.
3. Brazil has committed to reduce emissions between 36 and 39 % compared to expectations for 2020 projected emissions based on a "business as usual" scenario.

The announcement of the Brazilian mitigation actions was the outcome of a combination of business, civil society and political-election pressure, taking into consideration that the core of government (the Presidency, "Casa Civil" and Foreign Affairs) was not favourable to the adoption of quantified goal for reducing emissions. However, the establishment of such actions took place by means of a non-transparent decision-making process by the government. Thus, no systematic study to produce a consistent, appropriate and precise plan to reduce emissions has been produced and if it has been produced, it has not been presented to the general public.

Having made these caveats, the announcement of these measures by the President of the Republic during the High Level Segment of the COP-15 and COP/MOP-5 implied a fundamental change in the history of Brazilian climate foreign policy, given that it states that the country is willing to make mitigation action in a clear, quantifiable and verifiable manner, as a contribution to the international combat against climate change.

As a proof that it would not be only a political announcement, the measures were incorporated into the National Policy on Climate Change, enacted by law[24] of 29 December 2009.

The National Policy on Climate Change (PNMC, Portuguese for "Política Nacional de Mudança do Clima"), which established its own principles, objectives, guidelines, and instruments, aims, among other things, at the harmonisation of social and economic development while protecting the climate system; reduction of anthropogenic greenhouse gas emissions in relation to their various sources; strengthening of anthropogenic removals by sinks of greenhouse gases in the country; and implementation of measures to promote adaptation to climate change by the three levels of government (at federal, state and local level), with the participation and collaboration of the economic and social stakeholders, particularly those especially vulnerable to its adverse effects.

The objectives of the National Policy on Climate Change should be in line with sustainable development in order to pursue economic growth, poverty eradication and reduction of social inequalities.

The following are considered to be instruments of the National Policy on Climate Change: the National Plan on Climate Change, the National Fund for Climate Change[25]; Action Plans for the prevention and control of deforestation in the biomes; Brazil's National Communication to the United Nations Framework Convention on Climate Change, according to the criteria established by the Convention and by the Conference of the Parties; the resolutions of the Interministerial Commission on Global Climate Change[26]; fiscal and tax measures to encourage emission reductions and removal of greenhouse gases, including differentiated tax rates, exemptions, compensations and incentives, to be established by specific legislation; lines of credit and financing of specific public and private financial agents; the development

[24] Federal Law no. 12,187, 29 December 2009.

[25] Economic instruments are fundamental to implement the strategies contained in the Policy. Approved by the Brazilian Senate in November 2009, and signed by the President on 10 December 2009, Federal Law 1,204 created the National Fund on Climate Change, with the goal of securing resources (part of the revenues from the petroleum and natural gas industry) for supporting projects and studies that are directed toward mitigating climate change and adapting to its impacts.

[26] The Interministerial Commission on Global Climate Change is composed by 11 ministries and functions as the Designated National Authority (DNA) for the Clean Development Mechanism (CDM).

of research programs by funding agencies; specific allocations for actions on climate change in the federal budget; financial and economic mechanisms related to climate change mitigation and adaptation to the effects of climate change that exist under the United Nations Framework Convention on Climate Change and the Kyoto Protocol; and financial and economic mechanisms, at national level, pertaining to mitigation and adaptation to climate change.

Furthermore, instruments of the PNMC also include existing or future measures that encourage the development of processes and technologies that contribute to the reduction of greenhouse gas emissions and removals, as well as to adaptation, among which the establishment of eligibility criteria in tenders and bids, including public-private partnerships, and the authorizations, permits, grants and concessions of public services and natural resources; to proposals that provide greater savings of energy, water and other natural resources; and to reduction of greenhouse gas emissions and waste.

The official financial institutions will provide specific lines of credit and financing for the development of actions and activities that meet the objectives of the Law on Climate Change that are aimed at encouraging private players to act in compliance with and enforce the PNMC as part of their social responsibilities and actions.

The principles, objectives, guidelines and instruments of public policies and governmental programmes should be made compatible with the principles, objectives, guidelines and instruments of the National Policy on Climate Change.

As announced at COP 15 and COP/MOP-5, the text of the law provides that, in order to achieve the goals of the PNMC, the country will adopt, as a voluntary commitment at national level, actions to mitigate greenhouse gas emissions with a view to reducing its projected emissions by 36.1–38.9 % by 2020.

Accordingly, in January 2010 the Government of Brazil informed the Secretariat of the Framework Convention of the nationally appropriate mitigation actions that it intends to undertake,[27] for the information of the Parties to this international instrument. These actions are as follows:

- Reduction in Amazon deforestation (range of estimated reduction: 564 million tons of CO_2 eq. in 2020);
- Reduction in Cerrado deforestation (range of estimated reduction: 104 million tons of CO_2 eq. in 2020);
- Restoration of grazing land (range of estimated reduction: 83–104 million tons of CO_2 eq. in 2020);
- Integrated crop-livestock system (range of estimated reduction: 18–22 million tons of CO_2 eq. in 2020);
- No-till farming (range of estimated reduction: 16–20 million tons of CO_2 eq. in 2020);
- Biological N_2 fixation (range of estimated reduction: 16–20 million tons of CO_2 eq. in 2020);

[27] Available http://unfccc.int/files/meetings/application/pdf/brazilcphaccord_app2.pdf (last accessed on 2 February 2012).

- Energy efficiency (range of estimated reduction: 12–15 million tons of CO_2 eq. in 2020);
- Increase the use of bio-fuels (range of estimated reduction: 48–60 million tons of CO_2 eq. in 2020);
- Increase in energy supply by hydroelectric power plants (range of estimated reduction: 79–99 million tons of CO_2 eq. in 2020);
- Alternative energy sources (range of estimated reduction: 26–33 million tons of CO_2 eq. in 2020);
- Iron and steel (replace coal from deforestation with coal from planted forests) (range of estimated reduction: 8–10 million tons of CO_2 eq. in 2020).

It should be emphasized that these are voluntary actions, and that they will be implemented in accordance with the principles and provisions of the Framework Convention, particularly Article 4, paragraph 1; Article 4, paragraph 7; Article 12, paragraph 1(b); Article 12, paragraph 4; and Article 10, paragraph 2(a). In fact, the elements of the PNMC are completely in line with these provisions of the UNFCCC, especially with the commitments contained in Article 4.1, as described in the first section of this chapter.

The PNMC provides that the projected emissions for 2020, as well as the detailed actions to achieve the reduction goal above will be established by Decree, based on the Second Brazilian Inventory of Anthropogenic Emissions by Sources and Removals by Sinks of Greenhouse Gases not Controlled by the Montreal Protocol.

Moreover, the law of the PNMC anticipated that an Executive Decree shall, in accordance with the National Policy on Climate Change Plans, establish sectoral plans on mitigation and adaptation to climate change, with a view to consolidate a low-carbon economy related to the generation and distribution of electricity; public urban transport systems and modal inter-state transportation systems of cargo and passengers; in manufacturing and durable consumer goods; fine chemical industries based on pulp and paper industry; mining; construction industry; health services; and agriculture, in order to meet gradual goals related to the reduction of anthropogenic emissions in a measurable and verifiable manner, considering the specificities of each sector, including by means of the Clean Development Mechanism (CDM) and Nationally Appropriate Mitigation Actions (NAMAs).

In 2010, work on the measures to implement the PNMC started, with a view to establishing the following priority sectoral plans to achieve the goal expressed in the PNMC regarding mitigation actions:

- Action Plan to Prevent and Control Deforestation in the Legal Amazon (PPCDAM);
- Action Plan to Prevent and Control Deforestation in the Cerrado (Brazilian savannah) (PPCerrado);
- Plan to reduce emissions related to the production and consumption of Energy in Brazil by 2020;
- Mitigation and Adaptation Plan for a Low Carbon Agriculture and Livestock Sectors;

- Replace coal from deforestation with coal from planted forests in the iron and steel industry.

According to the Decree no. 7,390, of 9 December 2010, the other plans mentioned in Law 12,187/2009 should be completed by 15 December 2011. However, the elaboration of such plans was delayed, considering the difficulties to come up with measurable and verifiable indicators and the resistance of some stakeholders, especially from the private sector, in establishing mitigation actions. Thus, a new Decree (no. 7,643) postponed to 16 April 2012, the deadline for the conclusion of such plans, which shall contain the following minimum elements:

I – emissions reduction target by 2020, including incremental goals with a maximum interval of 3 years;
II – actions to be implemented;
III – definition of indicators for the monitoring and evaluation of their effectiveness;
IV – proposal for regulatory tools and incentives for implementation of their plan, and
V – sectoral competitiveness studies with estimated costs and impacts.

Also according to the Decree no. 7,390, the projection of national emissions of greenhouse gases for the year 2020 referred to in the Law no. 12,187/2009, is 3,236 million tonCO_2eq, following the methodology described in detail in the Annex to this Decree. In order to achieve the national voluntary goal announced by the PNMC, actions shall be implemented that aim to reduce between 1,168 million and 1,259 million tonCO_2eq of the total emissions estimated for the year 2020.

Although the practical results of most of these plans remain to be seen and verified, it is worth pointing out the remarkable results have already been achieved related to the combat against deforestation, particularly in the Amazon. Administrative, economic and legal measures have been adopted, according to a political action strategy, among which there is the Action Plan for the Prevention and Control of Deforestation in the Legal Amazon (PPCDAM). In fact, the PPCDAM is not a direct outcome of the PNMC, given that it was first released in 2004, but it has been integrated to the National Policy of Climate Change.

It is worth pointing out that Brazil's emissions profile is different from that of developed countries, where emissions from fossil fuel combustion are the most significant. According to the Brazilian Second National Communication to the UNFCCC, in 2005, CO_2 emissions were estimated at 1,638 Tg, with the Land-Use Change and Forestry sector as the main contributor, accounting for 77% of emissions, followed by the Energy sector, which was responsible for 19% of total emissions. Net emissions for this sector totaled 1,259 Tg CO_2, driven by the Amazon biome (67%) and Cerrado biome (22%).[28]

[28] Brazil, supra, note 17, volume 1, at 15.

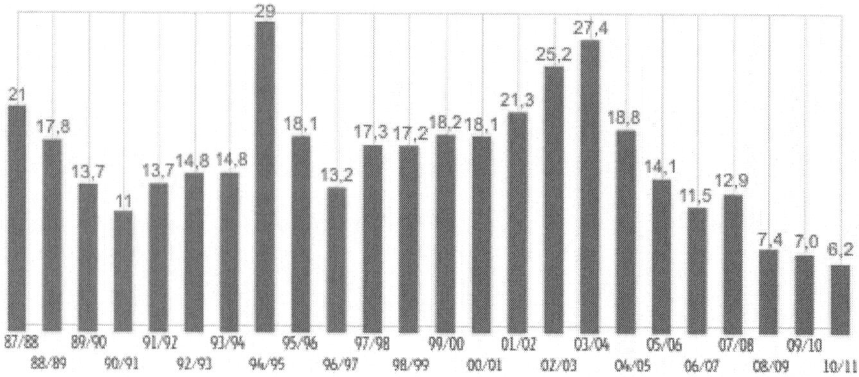

Fig. 29.1 Brazil: Deforestation in the Amazon Region between 1988 and 2011 (Source: INPE-PRODES, 2011)

From 1995 to 1997, deforestation in the Amazon region increased and then decreased. In 1995, deforestation reached its peak, with 29,059 km^2, compared to the lowest record for the decade, in 1991, of 11,130 km^2. In 1997, deforestation was reported at 13,227 km^2, confirming the tendency towards decreasing that began in 1996, when it fell nearly 40 %. However, the deforestation rate began to grow again in the period related to 1998, remaining more or less stable between 1998 and 2001. The deforestation rate saw considerable growth between 2002 and 2004, when it reached 27,772 km^2, near the peak of 1995. Since then, with a series of measures that have been adopted, the deforestation rate has been falling significantly, as proven by the figures for 2011 (estimate), of 6,200 km^2 (Fig. 29.1).

In fact, the reduction of emissions of CO_2 (in millions of tons per year) in the Amazon region verified in 2010 equals to the reduction equivalent to 67% of the emissions projected to 2020.

Nevertheless, Brazil still needs to face many challenges until 2020 to check if this reduction in deforestation rates is sustainable. For instance, the proposal that is to be adopted by the National Congress to change the Forest Code, which includes a provision for amnesty for those who have committed illegal deforestation, may undermine the legislation rather than improve it.[29] Another aspect that may influence on deforestation in the Amazon is the recently observed trend regarding the reduction of interest rates in Brazil, which can benefit loans for agricultural activities.

Moreover, the effectiveness of the implementation of the other sectoral mitigation plans is yet to be verified. The sustainable growth of the country's economy in the

---

[29] Center for Strategic Studies and Management (CGEE), Amazon Environmental Research Institute (IPAM), Secretariat for Strategic Affairs of the Presidency of Brazil (SAE/PR), *REDD in Brazil: A Focus on the Amazon. Principles, Criteria, and Institutional Structures for a National Program for Reducing Emissions from Deforestation and Forest Degradation – REDD* (Brasília: Center for Strategic Studies and Management, 2011), at 35.

last years, although positive from a economic and financial point of view, can also put some pressure on the demand side and result in more GHG emissions beyond what has been projected.

It is also worth noting that some policies have been adopted at state level, recalling that Brazil is a Federative government. One of the most advanced ones is the legislation related to climate change adopted in the state of São Paulo, which was the first state law containing quantified goals approved in the country. The state law on climate changes (state law 13.798) was approved on 13 October 2009 and requires that, by 2020, emissions GHG emissions are reduced by 20% over the base year 2005. However, the assessment of state level legislation goes beyond the scope of this chapter.

Despite the uncertainties of the successful implementations of the plans and policies adopted at state and national level, it should be recognised that in recent years there has been an increasing number of initiatives in various stages of implementation that contribute and/or will contribute to the inflection in the growth rate of the greenhouse gas emissions curve in the country, which reflects the commitment by many stakeholders to combating climate change.

Most importantly, the voluntary quantified goal for reducing emissions announced by Brazil in 2009, encapsulated in its National Policy on Climate Change, demonstrates that the country has moved from "due diligence" measures, with a view to respecting the obligations formulated under international law, towards the goal for real contribution to the combat to climate change.

Chapter 30
Climate Law in Latin American Countries

Soledad Aguilar and Eugenia Recio

Abstract Climate law in Latin America is in its infancy, although advancing at a steady pace. Most countries in the region have adopted soft law instruments, including climate change strategies, and, in some cases, climate change plans of action or sectoral action plans for adaptation or forestry. Brazil, Mexico, Colombia and Ecuador have more coherent legal frameworks for climate change, although at the time of writing, only Brazil had adopted a substantive climate change law. This chapter finds that frameworks related to climate change mitigation are more advanced than those dealing with adaptation, even though several countries in the region identify adaptation as a key priority for their future development. It argues that policy implementation remains challenging, with mainstreaming across sectors, allocation of budget and presidential support being identified as crucial elements and recurring challenges. The chapter also finds that subnational entities are increasingly involved in the development and implementation of climate change policy tools at the local level.

Soledad Aguilar holds an LL.M from the London School of Economics and Political Science. She specializes in international environmental law and policy, was a former negotiator for the Government of Argentina on climate change and biodiversity, and currently works for international organizations and academic institutions in the assessment and design of national and international climate change regulations. Eugenia Recio holds an M.Phil. from the Complutense University, Madrid, Spain. She specializes in international and Latin American environmental law and policy, and works for international organizations in the assessment and design of national and international climate change regulations. The authors would like to thank Carlos Salgado for his review and thoughtful comments to this chapter.

S. Aguilar (✉)
Ambiente y Comercio,
Carlos Pellegrini 1255, 1009 Buenos Aires, Argentina
e-mail: aguilar.envt@gmail.com

E. Recio
N.397 B, City of Knowledge, Panama City, Panama
e-mail: eugeniarecio@gmail.com

E.J. Hollo et al. (eds.), *Climate Change and the Law*, Ius Gentium: Comparative Perspectives on Law and Justice 21, DOI 10.1007/978-94-007-5440-9_30,

© Springer Science+Business Media Dordrecht 2013

30.1 Introduction: Climate Legislation in Latin America

Climate law in Latin America is in its infancy, although advancing at a steady pace. Most countries in the region have adopted soft-law instruments on climate change, including strategies, plans of action for implementing strategies or sectoral action plans for adaptation or forestry. At the time of writing, only one country in the region, Brazil, has passed a substantive law on climate change. Several countries, including Colombia, Dominican Republic, Mexico and Ecuador, have incorporated mitigation and adaptation objectives into their national development plans and are implementing related targets, providing budget support to climate change activities and mainstreaming the issue across sectors.

This chapter reviews the salient characteristics of climate change legislation in Latin America, focusing first on laws and decrees regulating climate change and then looking at the broader normative context, also including soft-law instruments, such as strategies, policies and action plans that guide policy-makers with respect to climate change in Latin America. This chapter does not attempt to provide an exhaustive analysis of climate change law in the region, but represents key examples and identifies best practices and normative trends that guide the development of climate law in Latin America.

Interviews by the authors with climate change decision-makers in 17 countries in Latin America[1] on climate change regulation and, in particular, on adaptation revealed that when dealing with the environment and climate change, it is important to take into consideration political will at the highest level of government, budgetary allocations and mainstreaming of climate change objectives within a broad range of sectors. It is certainly not adequate to infer, without looking at these additional variables, that the mere approval of a climate change policy entails its implementation and/or enforcement.

Our analysis of climate legislation in Latin America will thus complement the description of main trends in climate change law in the region with examples of national climate change strategies and policies at subregional, national and subnational levels, including considerations about the allocation of competences and examples on enforcement and budgetary allocations.

30.2 Constitutional Law and Climate Change in Latin America

Countries in Latin America have relatively similar Constitutions and codified civil law systems. Originally inspired by the 1787 United States Constitution, the 1812 Spanish Constitution from Cadiz and the French civil codes, legal systems in Latin

[1] The authors conducted some of the interviews as part of a consultancy for the Regatta project carried out by UNEP/Regional Office for Latin America and the Caribbean on adaptation law in Latin America. Interviewees included experts from Argentina, Bolivia, Brazil, Chile, Colombia, Costa Rica, Cuba, Dominican Republic, Ecuador, El Salvador, Guatemala, Honduras, Mexico, Panama, Paraguay, Peru and Uruguay.

America were designed as presidential democracies with a balanced distribution of powers. In practice, they often have, however, strong presidencies with a wide degree of discretion and control by the Executive Power over Congress.[2] In line with this, we found that engagement by presidents and presidential support were identified by policy makers interviewed as key to the advancement, or slow-down, of climate-related policies in Latin America, and are thus a relevant aspect to consider when evaluating climate-related legislation. Brazil presents the most visible example, with a flurry of climate-related activity during the presidency of Luiz Inácio Lula da Silva (2003–2011) and a considerable slow-down during the current presidency of Dilma Rousseff, a former mines and energy minister.

In terms of environmental rights, most countries in Latin America have undertaken constitutional reforms since the mid-1990s, incorporating environmental rights into their constitutions.[3] Notably, constitutions both in Ecuador and Dominican Republic make an explicit reference to addressing climate change as a responsibility of the State. The Constitution of Ecuador tasks the national government with addressing climate change mitigation by adopting measures to limit greenhouse gas emissions and deforestation, and protecting the population at risk.[4] The Constitution of Dominican Republic establishes land planning taking into account the climate change adaptation needs as a national priority.[5] Both of these examples highlight some of the key priorities for the region in addressing climate change, which involve building the resilience of vulnerable populations through adaptation to extreme weather events and the protection of forests for climate change mitigation and adaptation purposes.

30.3 National Climate Change Laws in Latin America

Addressing climate change through hard law instruments, congressional laws or presidential decrees that regulate climate change from a substantive perspective is not abounding in Latin America. However, all of the 17 countries evaluated have developed regulations and institutions for the implementation of projects under the Kyoto Protocol's Clean Development Mechanism (CDM), with each of them having

[2] Jorge Carpizo, "Características esenciales del sistema presidencial e influencias para su instauración en América Latina", 8 *Anuario de Derechos Humanos* (2007), at 49, 62 and 74.

[3] To consult the Constitutions of countries in Latin America and their reforms, see: Georgetown University Center for Latin American Studies, "Political Database of the Americas", 2012, available at: http://pdba.georgetown.edu/ (last accessed on 15 February 2012).

[4] Constitution of Ecuador (*Constitución de la República del Ecuador*), 2008, Art. 414. Art. 413 of the Constitution of Ecuador also indicates that the State will promote energy efficiency, the development and use of environmentally clean and healthy technologies and practices, as well as diversified, low impact renewable energy, that do not place food sovereignty, the environmental balance of ecosystems, or the right to water, at risk.

[5] Constitution of the Dominican Republic (*Constitución de la República Dominicana, 26 January 2010*), 2010, Art. 194.

registered at least one CDM project.[6] Apart from this, at the time of writing, Brazil is the only country in Latin America with a substantive climate change law approved by its Congress. Several countries in the region, like Guatemala, are in the process of developing, or have a draft climate change law under consideration by the Congress.[7] Other countries have adopted national strategies or policies on climate change through presidential decrees or ministerial resolutions. Given the programmatic approach of these instruments, we have chosen to analyse them below in the Sect. 4.

30.3.1 The Brazilian Climate Change Law

Brazil adopted Law N. 12.187 in December 2009, immediately following the 2009 UN Climate Change Conference in Copenhagen. The Law includes a voluntary national target to reduce greenhouse gas emissions compared to business as usual by 2020. The National Policy on Climate Change Law,[8] and its regulatory Decree,[9] provide a leading example of climate change law in Latin America given that they design a complete architecture for climate change regulation for Brazil. They establish institutional arrangements and legal mechanisms to fulfil mitigation and adaptation objectives, as well as to adopt an economy-wide mitigation target and sectoral targets and objectives.

Law N. 12.187 first focuses on the principles applicable to all climate change – related activities, such as the consideration of future generations and sustainable development in Article 3. It then outlines objectives and directives for climate change policy in Articles 4 and 5, including the development of a low-carbon economy and the promotion of scientific and technical research to identify vulnerabilities and adopt adequate adaptation measures.

---

[6] CDM projects in Latin America and the Caribbean represent 15 % of the overall number of CDM projects, with Brazil issuing more than 7 % of Certified Emission Reductions globally, and Mexico around 1.5 %. Source: UNFCCC, "CDM Database", 2012, available at: http://cdm.unfccc.int (last accessed on 23 February 2012).

[7] The draft climate change law currently under debate in Guatemala is law N. 4139, named "Framework law for the regulation of vulnerability reduction, mandatory adaptation to the effects of climate change and mitigation of greenhouse gases" (*Ley Marco para Regular la Reducción de la Vulnerabilidad, la Adaptación Obligatoria ante los efectos del Cambio Climático y la Mitigación de Gases Efecto Invernadero*) and it includes main components of Guatemala's National Policy on Climate Change. The project can be accessed at: http://www.infoiarna.org.gt/red%20iarna/2010/Red%20IARNA%2010(30)/adjuntos/dictamen-iniciativa-ley-cambio-climatico.pdf (last accessed on 19 March 2012).

[8] Law N. 12 187 (2009) on the National Policy on Climate Change, *Diario Oficial da União (D.O.U.)* of 29 December 2009, at 109 (extra edition).

[9] Decree N. 7390 (2010) on the Regulation of articles 6, 11 and 12 of Law 12.187, *Diario Oficial da União (D.O.U.)* of 10 December 2010, at 4.

Article 7 of the Law defines the institutional structure for dealing with climate change at the national level, including: a Climate Change Forum led by the president of Brazil and involving all major stakeholders and governments agencies; three inter-ministerial commissions; and a research network. Article 6 defines instruments for climate change policy in Brazil, including strategies, plans, monitoring mechanisms and financial schemes to support climate change action, including the creation of the Brazilian emission reductions market.

It is also interesting to note that the draft law approved by Congress included several sections related to the gradual phase-off from fossil fuel dependence, subsequently vetoed by the president.

The Regulatory Decree N. 7390 establishes specific mitigation targets for different economic sectors in Brazil concerning land-use change, energy, industrial processes and wastes, as well as agriculture. It calls for the adoption of sectoral mitigation and adaptation plans that include targets, actions to be implemented, policy instruments and incentives, as well as sectoral competitiveness studies to estimate the costs and impacts of such policies.[10]

In addition, Brazil is the only country in Latin America to have created a specific structure to compile advances in research related to climate change in Brazil, including on vulnerability and adaptation. The Brazilian Panel on Climate Change, designed as a mirror to the Intergovernmental Panel on Climate Change, is expected to become a model in the region on how regulations can be used to enhance the compilation and analysis of scientific evidence on climate change, and to provide regular scientific updates for policy makers.[11]

The result of Brazil's climate legal architecture, compared to most countries in the region, is a high degree of institutional engagement and the spread of climate change-related activities at different levels of government. The legal architecture has also lead to the allocation of significant resources, including through the channelling of oil revenues, to enforce the country's climate change policy (this topic will be addressed in Sect. 8).

30.4 National Climate Strategies and Policies

Most countries in Latin America rely on policy instruments, such as strategies and action plans, to guide policy makers on climate change issues. These soft-law instruments thus currently constitute the primary source for climate change law in most Latin American countries. The process leading to the adoption of climate

[10] Ibid.

[11] Information on PBMC – Brazilian Panel on Climate Change (*Painel Brasileiro de Mudancas Climaticas*) is available at: http://www.pbmc.coppe.ufrj.br/en/ (last accessed on 24 February 2012).

strategies as well as their legal status differs between countries. Some countries, like Colombia, Dominican Republic, Mexico and Ecuador, have included climate change into their national planning laws or strategies; others have simply adopted climate change strategies within ministries or as action plans for state programs and activities.[12] A few countries, like Peru[13] and Guatemala,[14] sought to give legal force and permanence to their climate change strategies by adopting them as Decrees or equivalent legal norms, even if their elaboration was led by one particular ministry or body.

In terms of civil society's involvement, only a few countries, like Colombia, have regulated the involvement of civil society in the design of their climate change strategies.[15] Several other countries have integrated civil society's contributions through *ad hoc* participatory processes. In some cases this has been done for the purposes of adopting a climate change specific strategy, including in Argentina, Brazil, Mexico, Uruguay, Costa Rica and Chile. In other cases this has been done for the adoption of national development plans or strategies, like in the Dominican Republic.[16]

The instruments to define national priorities for climate change mitigation and adaptation vary in Latin America. From broader to more specific instruments, countries have incorporated climate change priorities in:

- national development plans and strategies (Ecuador, Colombia, El Salvador, the Dominican Republic, Mexico);
- environmental policies and strategies (Cuba, Peru); and/or

[12] Uruguay, for example, created a national system for climate change response and variability, under which an action plan was published and endorsed by its former President. Uruguay, "*Plan Nacional de Respuesta al Cambio Climático*", January 2010, available at: http://www.cambioclimatico.gub.uy/index.php/plan-nacional (last accessed on 13 March 2012).

[13] Supreme Decree N. 086-2003-PCM on Peru's Climate Change Strategy, *Diario Oficial "El Peruano"* of 27 October 2003.

[14] *Acuerdo Gubernativo* N. 329–2009 on Guatemala's Climate Change Policy, of 9 December 2009.

[15] The principle of public participation in the environmental sector was incorporated in Colombia's Law N. 99 of 1993 as part of the duties of the Ministry of Environment to elaborate environmental policies with the participation of communities and within the functions of the regional autonomous corporations (Arts. 2 and 31). Concerning the diffusion of relevant information on climate change, Colombia also adopted a National Strategy for Environmental education and awareness on climate change in 2010 (*Estrategia Nacional de educación, formación y sensibilización de públicos sobre cambio climático*) which was prepared by an intersectoral and interinstitutional body named "Table Article 6 of UNFCCC". The Strategy is available at: http://www.pnud.org.co//img_upload/36353463616361636163616361636163/Estrategia_de_Educacion_CC.pdf (accessed on 27 February 2012).

[16] Soledad Aguilar and Eugenia Recio, *Fichas de País: Integración de la Adaptación en la planificación y los marcos regulatorios nacionales en América Latina*, Report prepared for the REGATTA project, (UNEP/ROLAC, 2011), (official publication forthcoming).

- specific climate change policies or strategies (Brazil, Chile, Mexico, Guatemala, Costa Rica, Honduras,[17] Panama[18] and Peru).[19]

The countries that are still in the process of adopting climate change specific policies or strategies include Argentina and Ecuador.

A good combination, in this regard, seems to be that presented by some countries like Mexico,[20] which have a specific climate change policy stemming from their national development plan. In the case of Mexico, this allows climate-related actions to access budget funding provided by the government for national development priorities despite the fact that the strategy has not been approved by law.[21]

In terms of substance, we find that strategies for climate change present a broad variety in terms of scope and focus. In most cases, these strategies have been developed by the ministry of the environment, often in collaboration with other relevant ministries.[22] These policy instruments establish national objectives and priorities both for mitigation and adaptation. As planning tools, they are aimed at public authorities, providing orientation and mandates to be integrated within their activities. However, as their objective is to provide guidelines for action, they are, in most cases, not subject to specific controls concerning their implementation and enforcement. In some countries, however, the legal framework requires that institutions provide information on the implementation of their plans and policies, as for example in Peru[23]

---

[17] Honduras adopted a National Strategy on Climate Change in 2010 (*Estrategia Nacional de Cambio Climático Honduras – ENCC*) in line with its national development plan. The strategy is available at: http://cambioclimaticohn.org/uploaded/content/category/2129570286.pdf (last accessed on 28 February 2012).

[18] Panama adopted in 2007 the National Policy on Climate Change including principles, objectives and lines for action on the issue and is currently working on a more specific strategy for climate change. See Executive Decree N. 35 of 4 April 2007 on the national policy on climate change.

[19] Aguilar and Recio, Fichas de país: Integración de la Adaptación en la planificación y los marcos regulatorios nacionales en América Latina, supra, note 17.

[20] Mexico adopted a Special Programme for Climate Change 2009–2012, elaborated on the basis of the National Strategy for Climate Change (2007). Government of Mexico. See "Programa Especial de Cambio Climático 2008–2012 (PECC)", 2009, available at: http://www.cambioclimatico.gob.mx/images/stories/PDF/pecc.pdf (last accessed on 17 February 2012).

[21] For further information on the role of national development plans in integrating climate change priorities see Sect. 4.1.

[22] Soledad Aguilar and Eugenia Recio, Estudio Regional de la integración de la Adaptación en los marcos legales y la planificación en América Latina, Report prepared for the REGATTA project (UNEP/ROLAC, 2011).

[23] For further information on the process applied by the Office of the Comptroller General, please refer to "La Contraloría General de la República", 2012, available at: http://www.contraloria.gob.pe (last accessed on 17 February 2012).

and Paraguay,[24] where the National Office of the Comptroller General performs audits on the implementation of the United Nations Framework Convention on Climate Change (UNFCCC)[25] at the national level.

Based on such strategies, some countries are working to further specify actions through the adoption of climate change action plans: for example, Chile developed a climate change action plan[26] and Costa Rica is working on one.[27]

Other countries are focusing their actions on adopting sectoral plans on climate change. For example, Colombia is focused on developing a plan for adaptation, a low-carbon strategy focused on mitigation issues, a strategy for REDD+and a strategy for financial protection for disasters.[28] Along the same lines, numerous countries are working or already adopted agriculture or agroforestry sectoral plans, including Chile, Costa Rica and Brazil.[29]

Some countries have even adopted several climate change-related planning instruments, which could represent a challenge both in terms of avoiding overlaps and ensuring their implementation. This is, for example, the case of Peru, which has adopted climate change related objectives in its National Environmental Policy (2009),[30]

---

[24] A report presented in 2009 by the Office of the Comptroller General in Paraguay on the implementation of the UNFCCC for the period 2005–2008 highlighted the need to enhance implementation of institutions and provisions adopted for that purpose, and required the Secretary of Environment to present a Management plan for enhancing implementation. For further information see: Contraloría General de la República de Paraguay, Dirección General de Control de la Gestión Ambiental, "Informe Final", 2009, available at: http://www.contraloria.gov.py/index.php?option=com_docman&task=doc_view&gid=3586 (last accessed on 23 February 2012).

[25] United Nations Framework Convention on Climate Change, New York, 9 May 1992, in force 21 March 1994, 31 *International Legal Materials* (1992), 849.

[26] Chile's Plan of Action for Climate Change aims at refining future vulnerability scenarios in prioritized sectors to assess the environmental, socio-economic and health impacts of climate change and enable the adoption of national and sectoral measures. Government of Chile, "Plan of Action for Climate Change 2008–2012" (*Plan de Acción National de Cambio Climático*), available at: http://www.mma.gob.cl/1257/articles-49744_plan_01.pdf (last accessed on 18 September 2011).

[27] Despite having adopted a National Strategy on Climate Change in 2009, Costa Rica is currently working on a Climate Change Plan of Action to incorporate specific information on activities required to implement the Strategy. Ministry of Environment, Energy and Telecommunications of Costa Rica, *National Strategy for Climate Change* (San José: Calderón y Alvarado, 2009). The Strategy identifies five main areas for action: mitigation, vulnerability and adaptation, metrics, capacity development and technology transfer and education and public awareness.

[28] The mandate to develop these sectoral strategies emerges from the National Council for Economic and Social Policies (CONPES) "Institutional Strategy for the articulation of policies and actions on climate change in Colombia, Document N. 3700" (*Estrategia institucional para la articulación de políticas y acciones en materia de cambio climático*), available at: http://www.dnp.gov.co/LinkClick.aspx?fileticket=2yrDLdRTUKY%3D&tabid=1260 (last accessed on 20 February 2012.); and Law N. 1450 of 16 June 2011 on the National Development Plan 2010–2014 "Prosperity for all."

[29] Aguilar and Recio, Fichas de país: Integración de la Adaptación en la planificación y los marcos regulatorios nacionales en América Latina, supra, note 17.

[30] Supreme Decree N. 012-2009-MINAM on the National Environmental Policy, *Diario Oficial "El Peruano"*, of 23 May 2009.

National Strategy on Climate Change (ENCC, 2003),[31] Plan of Action for climate change mitigation and adaptation introduced in December 2010 by the Ministry of Environment (MINAM) and the National Environmental Action Plan 2010–2012[32] adopted in 2011. As long as these plans target different areas of public policy, or translate into concrete actions, overarching principles and goals, they are useful components of a legal framework. However, the proliferation of strategic instruments can lead to a more chaotic and possibly less effective implementation, in particular since climate change plans already need to be harmonized with existing sectoral instruments, such as disaster reduction and agriculture strategies to ensure coherent policy frameworks.

30.4.1 Mainstreaming Climate Change Priorities Within Development Plans and Sectors

After a period of abandoning the exercise of planning for development,[33] many Latin American countries have returned to a stronger involvement of governments in state planning. Despite difficulties and challenges, particularly in terms of further articulating general objectives with their implementation at the sectoral and institutional level and the allocation of budget, the return of planning for development in many countries in Latin America is seen as an indication of a "renewed political interest in defining each country's future."[34] The consideration of climate change in national development plans and strategies provides a stronger basis for mainstreaming climate change mitigation and adaptation priorities into relevant sectors. The examples of Colombia, Ecuador, Costa Rica, El Salvador,[35] Mexico and Dominican Republic, which have already included climate change in their highest policy instrument for development planning, also give an indication of those countries that already see this topic as pivotal for development.

---

[31] Supreme Decree N. 086-2003-PCM on Peru's Climate Change Strategy, supra, note 14. By Executive Decree PCM-046-2010 the Strategy is considered a government's policy.

[32] Supreme Decree N. 014-2011-MINAM on the National Environmental Action Plan 2010–2012, *Diario Oficial "El Peruano"*, of 9 July 2011.

[33] Jorge Leiva, *Instituciones e Instrumentos para el Planeamiento Gubernamental en América Latina* (Santiago: UN Economic Comission for Latin America and the Caribbean – ECLAC, 2009).

[34] Instituto Latinoamericano y del Caribe de Planificación Económica y Social (ILPES), UN Economic Comission for Latin America and the Caribbean – ECLAC, "Panorama de la gestión pública en América Latina. En la Hora de la Igualdad", 2011, available at: http://www.eclac.org/publicaciones/xml/9/42339/PANORAMA_GP_H_FINAL.pdf (last accessed on 15 February 2012), at 17–20.

[35] El Salvador incorporated climate change mitigation and adaptation as a priority in the National Development Plan 2010–2014 (*Plan Quinquenal de Desarrollo 2010–2014*), San Salvador, 2010.

Colombia is a good example of including climate change priorities into its national development plan and mainstreaming climate change from national planning into sectoral planning. Colombia's 2010–2014 National Development Plan, entitled "Prosperity for All," includes climate change as one of its strategic lines and provides a mandate for the implementation of the National Policy on Climate Change and the configuration of the National System for Climate Change to strengthen information management and the provision of financial resources for adaptation and mitigation projects. It also mandates the development of a national plan for climate change adaptation based on a vulnerability assessment and a low-carbon strategy, in collaboration with regions and sectors, which also considers reducing emissions from deforestation. The Plan incorporates specific targets to be developed up to 2014, including the adoption of four sectoral low-carbon strategies, three additional adaptation sectoral plans and 83 updated plans for river basin management considering climate risks.[36] It also seeks to increase the amount of CDM projects from 158 to 300, and ensure 200.000 hectares of avoided deforestation.

The Dominican Republic has recently adopted its National Development Strategy 2030 by law.[37] The Strategy includes among its four objectives in Article 10 the search for a society of environmentally sustainable production and consumption that manages with equity and efficacy risks, environmental and natural resources protection and promotes an adequate adaptation to climate change. The Strategy also includes indicators that must be subject to systematic assessment, including on: environmental protection and enhancing adaptation to climate change[38]; poverty and inequality reduction; and innovative and environmentally sustainable economy.

Mainstreaming climate change also takes place through the amendment or reform of existing planning and policy instruments. For example, Cuba's updated national environmental strategy, currently in the process of adoption, incorporates emerging concerns and priorities, including climate change.[39]

Costa Rica presents another example of efforts to mainstream climate change issues. It first incorporated the objective of achieving the carbon neutrality in its National Development Plan 2011–2014,[40] then established objectives and priorities in its Climate Change Strategy, and finally adopted a Government Agreement that requires all public institutions, local governments and autonomous institutions to

---

[36] Law N. 1450 of 16 June 2011 on Colombia's National Plan for Development 2010–2014 (*Prosperidad para Todos*), at 559–593.

[37] Law N. 1–12 on the Dominican Republic's National Development Strategy 2030 (*Estrategia Nacional de Desarrollo 2030), Gaceta Oficial* (G.O.) 26 January 2011.

[38] Objective 4.3.1, Dominican Republic's National Development Strategy 2030, supra, note 38, at 50–51.

[39] The Cuban Environmental Strategy 2011–2015 (Estrategia Ambiental Nacional) is reportedly in process of adoption. Aguilar and Recio, Fichas de país: Integración de la Adaptación en la planificación y los marcos regulatorios nacionales en América Latina, supra, note 17.

[40] Government of Costa Rica, "National Development Plan 2011–2014 (*Plan Nacional de Dessarollo 2011–2014. "María Teresa Obregón Zamora"*), 2010, available at: http://documentos.mideplan.go.cr/alfresco/d/d/workspace/SpacesStore/122fcd1c-53a7-47a7-a0ad-84cac6f1d7b9/PND-2011-2014-Maria-Teresa-Obregon-Zamora.pdf (last accessed on 20 February 2012).

elaborate and implement a plan of action for the short-, medium- and long-term, considering clear objectives on climate change mitigation, adaptation, capacity building and technology transfer, and education.[41]

Notably, some of the countries that included climate change priorities in their national development plans and are working to develop substantive and robust strategies, like Colombia, Ecuador and Mexico, reportedly show significant progress in mainstreaming climate change throughout sectors.[42]

30.4.2 Balance Between Mitigation and Adaptation

Circumstances in Latin American countries differ in terms of climate change impacts and mitigation needs, but there are several key traits shared by the region. For example, countries are shifting from a focus on mitigation, originally geared towards regulating the implementation of CDM projects, towards an increasing consideration of adaptation issues. There is a growing acknowledgement of the shared vulnerability to extreme weather events among those countries with coastlines in the Caribbean or subject to El Niño/La Niña-Southern Oscillation Phenomenon,[43] which, coupled with several incidences of weather-related disasters, are reported by policy-makers as reasons behind such shift.[44] Central American countries, in particular, have identified adaptation as their main concern related to climate change, having encountered considerable economic and human losses in the past years due to the occurrence of natural disasters. In most cases, their efforts focused on developing planning tools and systems to respond to disasters.[45]

---

[41] Acta N. 056 of Costa Rica's Government Council (*Consejo de Gobierno*) on the National Climate Change Strategy of 1 August 2007.

[42] Aguilar and Recio, Integración de la Adaptación en la planificación y los marcos regulatorios nacionales en América Latina, supra, note 20.

[43] UN Economic Commission for Latin America and the Caribbean – ECLAC, "Efectos del cambio climático en la costa de América Latina y el Caribe: Dinámicas, tendencias y variabilidad climática", 2011, available at: http://www.cepal.org/cgi-bin/getprod.asp?xml=/publicaciones/xml/2/45542/P45542.xml&xsl=/dmaah/tpl/p9f.xsl&base=/dmaah/tpl/top-bottom.xsl (last accessed on 17 February 2012).

[44] Aguilar and Recio, Estudio Regional sobre la Integración de la Adaptación en la planificación y los marcos regulatorios nacionales en América Latina, supra, note 23.

[45] Costa Rica is the only country that kept its focus on becoming a carbon neutral country, while also increasing the relevance of adaptation. Central American countries are of a relative small size, very vulnerable to natural disasters and highly dependent on the provision of external funds for climate change issues. For example, UN Economic Commission for Latin America and the Caribbean – ECLAC assessed that the tropical depression 12-E that occurred in Central America in 2011, caused in El Salvador losses for more than US$840 million (equivalent to almost 4 % of its GDP). For the reconstruction of the affected areas, ECLAC estimated that more than US$575 million will be needed up to 2015. Source: ECLAC, "CEPAL evalúa costos de daños ocasionados por lluvias en América Central", 3 November 2011, available at: http://www.eclac.cl/cgi-bin/getProd.asp?xml=/prensa/noticias/comunicados/4/44944/P44944.xml (last accessed on 16 February 2012).

This shift towards adaptation is reflected within National Development Plans, climate change strategies and in the development of sectoral plans for adaptation. For example, Colombia mandates in its 2010–2014 National Development Plan "Prosperity for All" the development of a national plan for adaptation based on a vulnerability assessment.[46] Ecuador's development strategy, the "National Plan for Good Living", emphasizes adaptation and considers, among other issues, a target to reduce the vulnerability of ecosystems by 2013 by 23%.[47] Being one of the first countries in the region to focus its national strategy on adaptation, Uruguay designed its National Plan of Response to Climate Change as a strategic framework identifying lines for action and measures for addressing climate change.[48] The Plan identifies areas for adaptation, including risk management, water resources, energy, ecosystems and biodiversity, production and consumption, quality of life and health.[49]

Notwithstanding the shift of emphasis noted at the policy level, mitigation efforts in the region are reported to continue to dominate the policy-making agenda. As we will see in Sect. 5, most targets adopted by Latin American countries indeed focus on mitigation. This trend is, in our view, related to various factors, including that financial flows from international cooperation have a stronger focus on mitigation activities and that countries in the region are increasingly interested in generating resources from mitigation, creating an enabling environment for sustainable investments and the development and transfer of technologies. In addition, issues relevant for adaptation are usually handled by several ministries dealing with agriculture, disasters, health, housing, and so on, and may be more scattered rather than unified in a single identifiable regulation.

30.4.3 Sectoral Approaches

Streamlining climate change objectives within different sectors in the public administration, such as infrastructure, transport, health, industry and agriculture, is also a necessary condition for implementing climate change measures. Promoting and requiring the adoption of sectoral climate change plans is a recent trend in a good number of Latin American countries, particularly with respect to the agriculture

---

[46] Colombia's National Development Plan, supra, note 37, at 559–593.

[47] Ecuador, "National Plan for Development: National Plan for the Good Living 2009–2013: Building a plurinational and intercultural state" (*Plan Nacional para El Buen Vivir: Construyendo un Estado Plurinacional y Intercultural*), available at: http://www.ambiente.gob.ec/sites/default/files/users/mponce/Versión%20Completa_4.pdf (last accessed on 15 February 2012).

[48] Government of Uruguay, "National Plan for Climate Change" (*Plan Nacional de Respuesta al Cambio Climático*), 2010, available at: http://www.cambioclimatico.gub.uy/index.php/plan-nacional (last accessed on 15 September 2011).

[49] Government of Uruguay, "Third National Communication to the UNFCCC", 2010, available at: http://unfccc.int/resource/docs/natc/urync3.pdf (last accessed on 15 September 2011).

sector. Countries that are undertaking efforts to develop sectoral climate change strategies for agriculture or agroforestry include Costa Rica, Colombia, Chile, Mexico, Brazil and Uruguay. For example, Chile is working on the development of an adaptation plan for the forestry and agricultural sector,[50] while Brazil has adopted a plan for low-carbon agriculture.[51]

Forestry, watersheds and coastal management are also relevant sectors for climate change regulation in the region. Some countries, like Colombia, have used the opportunity to update old water laws to incorporate climate change concerns. In its National Policy for the Integral Management of Water Resources, adopted in 2010, Colombia incorporated as a priority the adoption of measures for reducing and adapting to risks associated with water supply. In particular, this priority foresees the implementation of climate change adaptation and mitigation measures by water users exposed to natural disasters.[52] Peru also included provisions to adapt to climate change in its National Strategy for Water Resources.[53] Moreover, regarding risk management and response to disasters, many countries in the region have developed strategies that indirectly consider climate change consequences and contribute to responses to natural disasters, to minimize economic and human losses.[54] For example, Peru passed a law in 2011[55] creating a National System for Disaster Risk Management (SINAGERD), mainstreaming risk management for natural disasters in all public institutions and planning processes. Cuba combined the approach to risk prevention and reduction with coastal zone management through the implementation of a project on scenarios of danger and vulnerability in the Cuban coastline, associated with

[50] Chile is developing a Plan for Adaptation to Climate Change for agro forestry led by the Ministry of Environment and the Ministry of Agriculture that promotes a participatory programme to develop an Adaptation Plan for the agro forestry sector in the region of Magallanes and the Antarctic. See: Aguilar and Recio, *Estudio Regional sobre la integración de la Adaptación en los marcos legales y la planificación en América Latina*, supra, note 23.

[51] Brazil Ministry of Agriculture, "Programa ABC", 2010, available at: http://www.agricultura.gov.br/desenvolvimento-sustentavel/programa-abc (last accessed on 17 February 2012).

[52] Government of Colombia, "National Policy for the Integral Management of Water Resources" (*Política Nacional para la Gestión Integral del Recurso Hídrico*), 2010, available at: http://www.minambiente.gov.co/documentos/5774_240610_libro_pol_nal_rec_hidrico.pdf.pdf (last accessed on 16 February 2012), Strategy 4.3, at 102.

[53] Art. 119 of Peru's National Strategy for Water Resources (*Política y Estrategia Nacional de Recursos Hídricos del Perú*), 2009, available at: http://www.ana.gob.pe/media/290336/politicas_estrategias_rh.pdf (last accessed on 27 February 2012).

[54] For example, Mexico, Colombia, Cuba, El Salvador, Nicaragua, Honduras and Costa Rica have developed systems of response to natural disasters, including organizing the population as well as institutions for states of emergency. See: Aguilar and Recio, Estudio Regional sobre la *Integración de la Adaptación en la planificación y los marcos regulatorios nacionales en América Latina*, supra, note 23.

[55] Law N. 29664 on the National System for Disaster Risk Management, *Diario Oficial "El Peruano"*, 19 February 2011.

sea-level rise expected for 2050 and 2100.[56] This project is increasingly influencing the adoption of regulations to counter consequences of the sea-level rise.

30.5 Mitigation and Adaptation Targets

The countries in Latin America that have submitted nationally appropriate mitigation actions (known as NAMAs in UNFCCC jargon) or mitigation targets to the UNFCCC include Argentina, Brazil, Chile, Colombia, Costa Rica, Mexico and Peru.[57] Most NAMAs submitted to the UNFCCC are linked to climate change regulations and strategies adopted in each of these countries, although most have presented their NAMAs as succinct, less detailed versions of their domestic policy instruments.

30.5.1 Economy-Wide Targets

As mentioned previously, Brazil is the first country in the region to have adopted a national voluntary mitigation target by law, to reduce "projected emissions" in 36.1 to 38.9 % by 2020[58] based on projections established by Decree on the basis of the Second National Communication data.[59] Mexico submitted to the UNFCCC an aggregate target of a 30 % emission reductions from business-as-usual by 2020, a target that is based on the more ambitious aspirational long-term goal set out in its Special Programme on Climate Change (PECC),[60] to reduce emissions by 50 % by 2050 compared to 2000 levels.[61] Chile aims at 20 % reduction below business-as-usual emissions by 2020,[62] while Costa Rica has stated in its Climate

[56] Macroproyecto 11, Escenarios de peligro y vulnerabilidad de la zona costera cubana, asociados al ascenso del nivel medio del mar para los años 2050 y 2100 (Havana: Instituto de Planificación Física, 2007).

[57] UNFCCC, Compilation of information on nationally appropriate mitigation actions to be implemented by Parties not included in Annex I to the Convention, Note by the Secretariat, FCCC/AWGLCA/2011/INF.1, 18 March 2011.

[58] Law N. 12.187 (2009), supra, note 9.

[59] Decree N. 7390 (2010), supra, note 10.

[60] Government of Mexico. Programa Especial de Cambio Climático 2008–2012 (PECC), supra, note 21.

[61] Ibid.

[62] UNFCCC, Compilation of information on nationally appropriate mitigation actions to be implemented by Parties not included in Annex I to the Convention, supra, note 58. See also Government of Chile, "National Strategy for Climate Change" (*Estrategia Nacional de Cambio Climático*), 2006, available at: http://www.bcn.cl/carpeta_temas_profundidad/temas_profundidad.2007-04-11.5841476988/Estrategia%20nacional%20_2006.pdf (last accessed on 12 March 2012).

Change Strategy and National Development Plan that it aims to become a carbon neutral country by 2021.[63]

30.5.2 Sectoral Targets

In their NAMA submissions to the UNFCCC, Brazil and Colombia incorporated specific emission reduction targets that derive from their national climate change strategies. Other countries did not state specific sectoral targets, but identified the sectors they would be working on to achieve their overall mitigation objectives. The submissions are useful to identify the most relevant sectors for mitigation in the region.[64] Interestingly, in some cases, like Mexico, the national strategy contains a large number of specific targets that have not been reflected in NAMAs submitted to the UNFCCC.[65]

For example, in their NAMA submissions to the UNFCCC for the energy sector: Argentina, Peru, Chile, Colombia, Brazil and Costa Rica, include policies to increase the share of renewables in the energy matrix, while Colombia, Brazil, Argentina include a specific focus on biofuels. Brazil, Chile and Argentina also consider energy efficiency.[66]

Regarding forests, Chile, Brazil, Costa Rica and Argentina include a focus on reducing emissions from deforestation or dealing with land use, land-use change and forestry issues, while Brazil has a specific deforestation reduction target and Peru and Colombia have a target to achieve zero net deforestation rates for native forests or specific areas.[67] Peru, Costa Rica, Argentina also add efforts to reduce emissions from solid wastes, while Costa Rica mentions transport, as well.[68]

Clearly, energy and forests constitute the key share of mitigation efforts in Latin America. Our studies also found that both Ecuador and Mexico have adopted

[63] Ministry of Environment, Energy and Telecommunications (MEET) of Costa Rica, *National Strategy for Climate Change* (San José: MEET, 2009). The Strategy identifies five main areas for action: mitigation; vulnerability and adaptation; metrics; capacity development and technology transfer; and education and public awareness. See also Costa Rica, National Development Plan, supra, note 41.

[64] UNFCCC, "Appendix II – Nationally appropriate mitigation actions of developing country Parties", available at: http://unfccc.int/meetings/cop_15/copenhagen_accord/items/5265.php (last accessed on 24 February 2012).

[65] Government of Mexico, PECC, supra, note 21.

[66] UNFCCC, Compilation of information on nationally appropriate mitigation actions to be implemented by Parties not included in Annex I to the Convention, supra, note 58.

[67] Ibid. See also Supreme Decree N. 014-2011-MINAM on the National Environmental Action Plan 2010–2012, supra, note 33.

[68] UNFCCC, Compilation of information on nationally appropriate mitigation actions to be implemented by Parties not included in Annex I to the Convention, supra, note 58.

specific targets on adaptation. Ecuador included in its national development plan a target to reduce by 23% the "high level of threat" in the ecosystem vulnerability to climate change index, and by 69% the "medium level of threat" by 2013.[69]

30.6 Allocation of Competences for Climate Change in Latin America

While often lacking specific climate change regulations, Latin American countries tend to have relatively complex systems to distribute competencies and jurisdiction on climate-related matters. Most countries in the region, for example, have developed regulations for the implementation of CDM projects, including the creation of specific authorities to deal with the assessment and national approval of CDM projects. These authorities do not necessarily have jurisdiction over other climate change issues. In order to provide a more complete picture of climate law in Latin America, we have also briefly reviewed the regulations determining the distribution of tasks and responsibilities related to climate regulation in the region.

30.6.1 Federal vs. Centralized Governments

Constitutions in Latin America usually define the allocation of powers over environmental regulation among central and provincial or local governments.[70] The region varies concerning a centralized or descentralized approach to environmental regulation, although the former is more common. Some countries with a centralized approach, like Chile, Cuba, Panama, Paraguay, Uruguay and Bolivia, stipulate that environmental issues are eminently under central government jurisdiction, with variations on how they address the attribution of power to indigenous communities to regulate resources located in their territories. However, structures and degrees of descentralization among centralized countries differ. For example, in some Central American countries like Panama, the environmental national authority has created offices in the provinces or departments in charge of implementing environmental policies. These offices are granted some local or regional competences, such as powers to grant permits over certain resources and the enforcement of environmental law. In some other centralized countries, those functions are allocated to regional or provincial governments, as in the case of Peru, where the regions have competences to develop

[69] Ecuador, "National Plan for Development", supra, note 48.

[70] Aguilar and Recio, Estudio Regional sobre la Integración de la Adaptación en los marcos legales y la planificación en América Latina, supra, note 23.

regulations and policy instruments, in addition to other competences. In an effort to promote descentralization of environmental protection, Colombia, Peru and Ecuador distribute power over environmental regulation and enforcement between the central government and subnational regional entities.[71]

In the case of federal states, the allocation of competences tends to be descentralized, but the situation is very different among countries. Argentina allocates most regulatory powers to the provinces, and gives the national government solely the power of adopting minimum national principles for environmental protection, a fact that may have influenced the relatively low level of development of climate law in this country. Brazil, on the other hand, establishes a common jurisdiction by the central government, provinces (states) and municipalities, for environmental issues, a situation that leads to a more complex architecture, with several layers of regulations dealing with the same issues.[72]

During interviews by the authors throughout the region, the distribution of competencies among federal and state governments was not identified as a key trend affecting the effectiveness of climate change policy, but rather an element of added complexity that must be contemplated when studying each country's legislation.

30.6.2 Competence and Jurisdiction Over Climate-Related Issues

The distribution of competences regarding climate change among different ministries often offers a glimpse of the political importance awarded to climate change issues in a given country, and whether there is an engagement of the presidency in climate change policy-making. In particular, the design and implementation of climate change strategies at the country level requires a good level of engagement by sectors other than the environment, including relevant authorities for finance, agriculture, industry, transport, watersheds and coastal management.

The trend in most countries in the region is the allocation of competences related to climate change issues to the ministry of environment and, in most cases, the mentioned ministry led the process of adoption of strategies and other policy instruments on climate change priorities, for instance, in Costa Rica, Cuba and Guatemala,. Many countries additionally created a climate change directorate or unit to coordinate all the related policies within the ministry of environment, which were in many cases adopted by secondary regulations. Usually, competences on other related issues such as agriculture are allocated to other ministries. In the case of water protection, the allocation of competences is much more varied among countries and presents increasing challenges in terms of coordination.

[71] Aguilar and Recio, Fichas de país: Integración de la Adaptación en la planificación y los marcos regulatorios nacionales en América Latina, supra, note 17.

[72] Brazil's Constitution of 1998 (*Constituição da República Federativa do Brasil*, of 1988 with amendments until 13 July 2010), Art. 123.

Additionally, given the transectoral nature of climate change, most countries in the region have created different types of transectoral and interinstitutional coordinating entities or bodies for policy making on climate change issues. Creating a new coordinating body has the advantage that, as a general trend, all the relevant ministries participate in it and that, according to the political level of relevance provided to this entity, their influence can be substantive, particularly if the President is appointed as the head of the body. For example, Brazil's Forum on Climate Change includes more than a hundred representatives of state agencies, the private sector and non-governmental organizations, and is headed by the Brazilian president.[73] The Dominican Republic created the National Council for Climate Change and CDM, headed by the president and composed by all the ministries, which acquired a relevant role in the definition of climate change policies in the country.[74]

Inter-institutional bodies are also found to be relevant when leading the development of national strategies on climate change. A good example, from our perspective, is presented by Peru, which elaborated its strategy through an *ad hoc* inter-institutional body, the National Commission on Climate Change, which was composed by different technical groups.

However, interviews with decision makers also showed that the creation of new bodies usually entails difficulties in obtaining the necessary budget. While countries tend to give the ministry of environment a coordinating role in these inter-institutional bodies, many of the experts interviewed highlighted the need to provide an active engagement of the entity in charge of national planning in order to incorporate climate change as an integral aspect of a country's development strategy. In the case of Colombia, for example, the institutional strategy on climate change was adopted by the National Council for Economic and Social Policies (CONPES),[75] headed by the president of Colombia, led by the Secretary of National Development, and with the participation of all ministries.

30.6.3 Regional Legislation

Countries in Latin America also participate in regional and subregional entities such as the Caribbean Community (CARICOM),[76] the Andean Community

[73] Information on Brazil's Forum on Climate Change (*Forum Brasileiro de Mudancas Climaticas*), including on its composition, is available at: http://www.forumclima.org.br/ (last accessed on 24 February 2012).

[74] Adopted by Presidential Decree N. 601, of 20 September 2008. The Decree creates and integrates the National Council for Climatic development and CDM (*Consejo Nacional para el Desarrollo Climático y Mecanismo de Desarrollo Limpio*).

[75] The National Council for Economic and Social Policies (CONPES) was created by Law N. 19, D.O, 9 December 1958.

[76] CARICOM member States are: Antigua and Barbuda, The Bahamas, Barbados, Belize, Dominica, Grenada, Guyana, Haiti, Jamaica, Montserrat, Saint Lucia, St. Kitts and Nevis, St. Vincent and the Grenadines, Suriname, and Trinidad and Tobago.

(CAN),[77] MERCOSUR[78] and the System of Central American Integration (SICA),[79] which also contribute to the development of policies on climate change related issues. For example, countries that are part to SICA adopted regional strategies on climate change issues, including the Regional Strategy on Agriculture, Environment and Health (*Estrategia Regional Agroambiental y de Salud – ERAS*), based on five strategic axis: soils sustainable management; climate change and climatic variability; biodiversity; agricultural and environmental business; and healthy spaces and living styles. The strategic frameworks for vulnerability and disaster reduction in Central America to face the situation of food and nutritional insecurity associated with drought and climate change, includes a plan of action for agriculture and livestock for climate change. This region was very active in terms of adopting climate change related regional strategies,[80] including: a 1993 Regional Convention on Climate Change, *Convenio Regional sobre Cambios Climáticos*, a 1999 Strategic Framework for Reduction of Vulnerabilities and Disasters Impacts – *Marco Estratégico para la Reducción de las Vulnerabilidades y el Impacto de los Desastres* – and a subregional Strategy on Climate Change.[81] As a general trend in the region, subregional strategies proliferated in the past years, which can be understand as a sign of willingness of countries to collaborate and enhance coordination in facing climate change related challenges. However, we found that their implementation remains particularly challenging given the shortage in funds, overlaps among several strategies and institutional difficulties.[82]

30.6.4 Sub-National Climate Change Laws

Depending on each country's distribution of the relevant competences, some Latin American countries developed subnational legislation, plans and strategies relating to climate change. The development of subnational legislation is a recent trend that

[77] CAN members are: Ecuador, Bolivia, Colombia and Peru.

[78] MERCOSUR members are: Argentina, Brazil, Uruguay and Paraguay.

[79] SICA members are: Guatemala, El Salvador, Honduras, Nicaragua, Costa Rica, Panama and Belize.

[80] Among other strategies, they adopted: the Regional Convention on Climate Change (*Convenio Regional sobre Cambios Climáticos*, 1993), and the Strategic Framework for Reduction of Vulnerabilities and Disasters Impacts (*Marco Estratégico para la Reducción de las Vulnerabilidades y el Impacto de los Desastres*, 1999).

[81] Comisión Centroamericana de Ambiente y Desarrollo (CCAD), SICA, "Estrategia Regional de Cambio Climático, Executive Document", 2010, available at: http://www.sica.int/busqueda/busqueda_archivo.aspx?Archivo=info_55238_1_29112010.pdf (last accessed on 27 February 2012).

[82] Aguilar and Recio, Estudio regional sobre la integración de la Adaptación en los marcos legales y la planificación en América Latina, supra, note 23, at 11–13.

was identified particularly in federal countries such as Brazil, while the development of subnational strategies is also being promoted in some centralized countries that are going through descentralization processes, such as Peru. The advantage of adopting regional or local strategies on climate change is that measures and planning can be adjusted to local conditions and local institutions involved in the process of implementation. In the next subsections we will explore some particular examples of subnational legislation, without pretending to be exhaustive.

30.6.4.1 State-Level Legislation in Brazil and Peru

As previously mentioned, the Brazilian Constitution allocates shared responsibilities on environmental issues among the central government, provinces/states and municipalities.[83] States and local governments in Brazil provide a good example of the types of laws and policies that are likely to be increasingly adopted by subnational entities in Latin America to promote emission reductions, protect forests and support alternative energies.[84] Among the states that adopted extensive regulations, the state of Sao Paulo adopted the first mandatory mitigation target to reduce emissions in 20 % by 2020 based on 2005 emissions. The Sao Paulo State regulation addresses both mitigation and adaptation, and stipulates in Article 31 that the state will define real, measurable and verifiable measures to reduce emissions, using among other instruments: emission reduction or stabilization targets, sectoral efficiency targets, and additional mechanisms to trade carbon rights.[85] In addition, ten out of the 27 states in Brazil have already adopted policies on climate change, with five having draft regulations under consideration.[86] These regulations have a stronger focus on mitigation, but in some cases also incorporated adaptation measures adjusted to the vulnerabilities of each region.[87]

Peru is a centralized country but it is increasingly promoting descentralization of certain environmental competences, including the development of strategies on climate change at the subnational level. In addition to the national environmental strategy and a climate change strategy, the Law on Regional Governments[88] requests

[83] Brazil's Constitution of 1998, supra, note 73, Art. 123.

[84] Instituto de Pesquisa Economica Aplicada (IPEA), "Perspectivas sobre as negociações climáticas e seus impactos na política brasileira", in *Comunicado* N.45 (Brasilia: IPEA, 2010).

[85] Sao Paulo State's Law N.13, 798 of 2009 on the state's policy on climate change (PEMC), Diario Oficial del Estado (D.O.E.), 10 November 2009, Section I, at 1.

[86] Viviane Romeiro and Virginia Parente, "Regulacão Das Mudancas Climáticas No Brasil e o Papel dos Governos Subnacionais", in Ronaldo Seroa da Motta et al. (eds), *Mudança do clima no Brasil: aspectos econômicos, sociais e regulatórios* (Brasilia: IPEA, 2011), 43, at 45–46.

[87] Ibid.

[88] Organic Law of Regional Governments, Law N. 27867, Diario Oficial (D.O.), 18 November 2002, available at: http://biblioteca.unmsm.edu.pe/redlieds/Recursos/archivos/goblocales/ley_27867_ley_org_gob_reg.pdf (last accessed on 27 February 2012).

regional governments to generate regional strategies on climate change and on biodiversity protection. In order to guide regions, the Ministry of Environment adopted guidelines for the regional strategies on climate change.[89] At the time of writing, the regions that adopted regional strategies on climate change include: Junín (2007), Amazonas (2008), Ayacucho (2010), Lambayeque (2010), Loreto and Apurímac (2011). Other 11 regions were at different stages in the design and adoption of such strategies.[90] In cases such as in the Junín region, a technical group that included representatives from different sectors at the regional level carried out the elaboration of the strategy.[91]

30.6.4.2 City-Level Legislation

Local governments throughout Latin America are in the process of adopting regulations to reduce greenhouse gas emissions, enhance adaptation and resilience and monitor and report on achievements in these subjects. In many cases, such initiatives are reported to result from the participation by Latin American cities in broader initiatives that allow them to adopt climate-friendly policies and obtain financing for such efforts. For example, several local governments, including 38 cities in Mexico and 22 cities in Brazil, are participating in the ICLEI – Local Governments for Sustainability – initiative, and through this network have developed specific regulations and targets related to climate change mitigation and adaptation at the local level.

In a related initiative, the Global Cities Covenant on Climate or Mexico City Pact was signed by many local governments in Latin America, and both the city of Buenos Aires, and Mexico City, have already submitted their first reports on mitigation commitments, actions and greenhouse gas performance to a public registry.[92]

In the case of Brazil, the cities of Rio de Janeiro and Sao Paulo have developed climate change regulations and adopted mitigation targets. The city of Sao Paulo has adopted a 30 % emission reduction target by 2010 based on 2005 emissions,[93]

[89] Ministry of Environment of Peru, "Advances in Regional Planning" (*Avances en la Planificación Regional*), available at: http://cambioclimatico.minam.gob.pe/la-gestion-del-cc/avances-en-la-planificacion-regional/ (last accessed on 16 February 2012).

[90] Ibid.

[91] Enfrentando el Cambio. Regional Decree N. 002-2007-GRJunín/PR, of 4 December 2007, Regional Strategy on Climate Change. Also available at: http://www.regionjunin.gob.pe/documents/pdf/DR/DR2.pdf (last accessed on 27 February 2012).

[92] Carbonn Cities Climate Registry, "Reporting Cities", available at: http://citiesclimateregistry.org/cities/reporting-cities/ (last accessed on 18 February 2012). See also: Yunus Arikan et al., *Carbonn Cities Climate Registry 2011 Annual Report* (Bonn: Bonn Center for Local Climate Action and Reporting – carbonn, 2011).

[93] Sao Paulo Municipal Law N. 14933 (2009) on the policy of climatic change (PMMC) in the municipality of Sao Paulo, *Secretaria do Governo Municipal*, 5 June 2009.

and the city of Rio de Janeiro has a eight target to reduce emissions by 2012, a 16 % target by 2016 and a 20 % target by 2020 based on 2005 emissions.[94]

30.7 Implementation, Enforcement and Budgetary Allocations

Considering that the scope of this paper does not allow an evaluation of the effectiveness of climate change policy in Latin America, we nevertheless deem it key to incorporate implementation and enforcement in any analysis of legislation in the region. In this respect, we regard the allocation of funds for the implementation of specific climate change or laws as a relevant indicator concerning the enforcement of climate regulation. Thus, we will provide some examples of countries that have allocated such funds and other aspects that lead to a more effective enforcement of climate regulations.

Despite some initial arguments by climate experts in the region that the main benefit of having a climate change law – rather than relying on flexible strategies – is the possibility to ensure a budgetary allocation for climate change purposes, we found that this is not necessarily always the case. For example, in Mexico, the adoption of a national strategy by the Executive is sufficient to allocate funding, and all special programs derived from the national development strategy, including the climate change program, require a budgetary allocation by law.[95] Some countries, nevertheless, report having trouble finding funding sources for climate change strategies, while others receive multiple sources having to face challenges in harmonizing these sources. As stated in Costa Rica's national report to the UNFCCC in 2009, there is a disconnect between public spending planning at the national level and the development of strategies and action plans derived from international conventions, leading to a low level of integration of climate change objectives into the national public planning process.[96] Current initiatives supporting the development of climate change strategies in the region also point towards the need to include the climate change dimension into the public and private budget at the national, subnational and sectoral level.[97]

Expert voices from the region on the matter of enforcement and implementation effectiveness also point towards a combination of factors including budgetary

[94] Rio de Janeiro Municipal Law N. 5.248 on the municipal policy on climate change and sustainable development, Diario Oficial (D.O.) N. 210 28 January 2011, at 3.

[95] Aguilar and Recio, Fichas de país: Integración de la Adaptación en la planificación y los marcos regulatorios nacionales en América Latina, supra, note 17.

[96] Ministry of Environment, Energy and Telecommunications (MEET) of Costa Rica, Costa Rica 2009: Segunda comunicación nacional a la Convención Marco de las Naciones Unidas sobre Cambio Climático (San Jose: MEET, 2009), at 65–66.

[97] Martha Perdomo, *Proyecto Políticas Climáticas 2012: Preparando Estrategias Climáticas* (Panama: Programa para el Desarrollo de Naciones Unidas-PNUD, Centro Regional LAC, 2011), at 11.

allocations, political will and engagement by presidents with climate change, as well as generating sufficient scientific knowledge to back decision making. The following examples show some combinations that seem to have worked – or failed to succeed – for implementing climate-related legislation.

Argentina presents an example of a weak institutional and legal framework for climate change regulation. We find this country lacks budgetary allocations to climate change and has a low level of political engagement in this matter by the country's leadership. Some indicators of the lack of political leadership engagement are, for example, the absence of Argentina's current president from the 2009 UN Climate Change Conference in Copenhagen – the largest historical gathering of heads of state to discuss climate issues – the fact that Argentina's interagency coordination mechanism for climate change is led by a secretary of state – not a minister or president-, and that, in terms of normative development, the country is still in the process of adopting a climate change strategy and has no climate change law. This combination of legal and institutional factors, is possibly the reason why the country has no specific allocation for climate change in the national budget. We must mention, however, that Argentina does have an advanced regulation and assigned funding for the promotion of renewable energy sources, although motivated mainly by energy security issues, not by policies related to climate change.[98]

Stronger examples of implementation of climate change law are presented by Brazil and Mexico. In the case of Brazil, having a climate change law leads to the assignation of specific resources for the implementation of climate change regulations in the national budget. There are also budget lines for research and development of technologies for adaptation in agribusiness, as well as to implement the sectoral plan for a low-carbon emissions agriculture.[99] The strength of the framework may also be related to having a climate change forum led by the President and the creation, under Brazil's former presidency of specific funds to deal with climate change, such as the National Climate Change Fund and the Amazonian Fund. Even though the current president of Brazil seems to have lowered the priority of climate change issues within the national policy agenda, and as a result institutions are not showing the level of activity they had during President Lula's era, Brazil still has the largest budgetary allocation for climate issues in South America. As an example, the Brazilian national development bank (BNDES) announced in February 2012 a credit line to support mitigation and adaptation projects, with an estimated budget of US$ 570 million until 2014, which will be obtained from royalties from the oil industry.[100]

---

[98] Law 26.728, 2011 on Argentina's General Budget for the National Administration (*Presupuesto General De La Administracion Nacional*), *Boletín Oficial*, of 28 December 2011.

[99] Planning Ministry of Brazil, "Orçamento anual de 2012", 2012, available at: http://www.planejamento.gov.br/secretaria.asp?cat=50&sub=539&sec=8 (last accessed on 24 February 2012).

[100] O Banco Nacional do Desenvolvimento (BNDES), "MMA e BNDES lançam linha de crédito para projetos que reduzam emissões", 30 February 2012, available at: http://www.bndes.gov.br/SiteBNDES/bndes/bndes_pt/Institucional/Sala_de_Imprensa/Destaques_Primeira_Pagina/20120213_fundo_clima.html (last accessed on 18 February 2012).

Another example of a strong climate change framework is presented by Mexico, a country that with a special climate change program 2009–2012 as part of the country's national development plan, even if it has not approved a climate change law,[101] and a strong commitment by the president to climate change mitigation. Mexico's planning laws determine that all special programs deriving from the country's development plan, such as the special climate change program, must have a corresponding budget line, and thus receive specific budgetary allocations for the implementation of the different objectives of the plan.[102] In addition, the Mexican President Felipe Calderon has shown a personal commitment to climate change as an engaged host during the Cancun Climate Change Conference, and in the incorporation of climate change as a key topic to be addressed during Mexico's 2012 presidency of the Group of 20. The case of Mexico is interesting because, even in the absence of a specific climate law, it has developed one of the most advanced frameworks for dealing with mitigation and adaptation in Latin America, based on policy instruments (a national strategy and sectoral action plans), budgetary allocations and a strong presidential commitment.

Colombia presents another example of increasing efforts to mainstream climate change in national planning and across sectors. As previously mentioned, the National Development Plan included climate change priorities and targets. The Multiyear Plan for Investments defines budget allocations for the strategic areas identified in the National Development Plan. Then, the institutional strategy on climate change adopted by the CONPES set the basis for the development of required strategies on adaptation, low-carbon development and avoided deforestation in the country, as well as for the development of a financial strategy for risks. Furthermore, the Strategy establishes the creation of a Financial Committee to manage funds for climate change projects. The Committee is expected to have a relevant role in assessing possible sources of funding from international cooperation, multilateral agreements or the recently adopted National Adaptation Fund for adaptation and mitigation projects.[103]

Finally, Ecuador, despite being a small country in terms of economic size and emissions, also has a well-developed climate change strategy that has been incorporated into the country's national development strategy or *National Plan for Good Living*. As part of the national development strategy, many climate change-related projects and initiatives have been able to access national budget funding through the general secretary of planning, the organ tasked with assigning funds for the implementation of the country's national development plan.[104]

[101] Government of Mexico, PECC, supra, note 21.

[102] Aguilar and Recio, Fichas de país: Integración de la Adaptación en la planificación y los marcos regulatorios nacionales en América Latina, supra, note 17.

[103] Ibid.

[104] Ibid.

30.8 Conclusion

This Chapter has reviewed the main developments regarding climate change law in Latin America, drawing attention to the many layers of regulation currently in place, the balance between mitigation and adaptation objectives, and the diversity in the ways in which countries regulate climate challenge in this region.

Among the main trends, we found that most countries in Latin America have included climate change priorities in their national development plans and/or specific climate change strategies and are working to implement these through the development of sectoral programs and specific action plans. Such strategies vary in scope and in terms of their adoption process. However, they provide guidelines for policy makers and, depending on the national legal framework, can have a budget allocated to their implementation. They generally lack, however, mechanisms for enforcement and monitoring. Regarding climate change laws, only Brazil has already adopted comprehensive climate change legislation, and some other countries, like Guatemala, have projects under debate by Congress.

Our research also showed that some countries in the region are undergoing processes to enhance mainstreaming of climate change priorities into relevant sectors of the public arena through, *inter alia*, the creation of inter-institutional bodies, the adoption of sectoral strategies and the incorporation of climate change among national development plan's priorities. We found countries vary in their institutional designs, with most interesting approaches presented by those that implement president-led forums or bodies to allow for interdisciplinary and inter-agency consideration of the climate change challenge. Work is also underway at the sectoral level, with numerous countries working on agriculture and adaptation plans. Finally, the incorporation of adaptation and mitigation priorities into national development plans is increasingly taking place in the region, inscribed in a process of rebirth of national planning at the mid – and long-term. Some of the countries that included climate change priorities in their national development plans seem to have stronger basis and be making further progress in mainstreaming climate change throughout sectors.

In general, a shift from a focus on mitigation in the past towards adaptation in most recent legal developments was also identified, with many countries signalling adaptation as their main priority owing to the frequent occurrence of disasters that caused human and economic losses. Nevertheless, mitigation, originally concentrated on regulations for the implementation of CDM projects, maintains the largest share of climate related regulations, with many countries adopting economy-wide mitigation targets or aspirational goals.

Finally, the layers formed by strategies and action plans at national level are intersected and overlapped – and often enriched – by subnational legislation, which shows increased levels of activity regarding climate change regulation. At the subnational level, a myriad of initiatives are taking place, with some cities in the region (with populations surpassing that of many other countries as a whole, like Buenos Aires, Sao Paulo and Mexico City) adopting specific targets for climate change mitigation. The colourful picture is complemented at the international level, with many countries participating in subregional organizations that have adopted strategies for subregional cooperation on climate issues.

In terms of implementation, the picture is less intense with most policy makers reporting serious shortages of funds from national budgets for climate-related strategies, although interesting examples of frameworks that intend to provide adequate funds to climate change are provided by Brazil Mexico and Colombia.

In essence, we find climate change legislation in Latin America, having been developed during the past decade, is still in a programmatic stage, but advancing at a steady pace at both national and sub national levels. While originally mainly focused on mitigation, the emphasis on the prevention of disasters and adaptation is being strengthened. Operative clauses, or binding obligations that could trigger enforcement measures for mitigation or adaptation, however, do not abound.

Index

A

Aarhus Convention, 157, 312–313
Adaptation
  adaptation committee, 107–109, 270–271
  adaptation fund, 50, 109, 192, 199, 202–206, 217, 218, 272–274
  Adaptation Private Sector Initiative, 57
  agricultural sector, 665
  and biodiversity, 261–263, 269, 275, 282–283
  Birds Directive (EU), 283
  Buenos Aires Programme of Work on Adaptation and Response Measures, 269–270
  building law, 275, 281
  Cancun Adaptation Framework, 89, 90, 102, 107, 270–271
  coastal adaptation, 4, 275–278
  commitments, 268
  costs, 77, 194, 212, 262, 265
  desertification, 89, 267
  draught(s), 279, 284
  energy and telecommunications, 4, 275, 281–282
  floods, 281
  Floods Directive (EU), 277
  guidance on water and adaptation to climate change, 4
  Habitats Directive (EU), , 276, 283
  land use and spatial planning law, 280–281
  marine adaptation, 275–278
  Marine Strategy Framework Directive (EU), 276, 277
  measures, 4, 84, 89, 107, 109, 110, 213, 258–263, 267–269, 273–279, 284, 319, 362, 364, 365, 379, 381, 382, 556, 562, 656, 672
  and migration, 4, 261, 271, 275, 282
  Nairobi Work Programme on Impacts, Vulnerability and Adaptation, 57, 269–270
  national adaptation plans (NAPs), 108
  national adaptation programmes of action (NAPAs), 203, 268–270, 273, 274
  New York City Panel on Climate Change (NYPCC), 260
  objectives, 89–91, 277, 629, 654, 656, 677
  planning law, 261, 262, 280–281
  protected areas, 261, 262, 282, 283, 379, 380
  risks, 263–264
  transport infrastructure, 280
  uncertainties, 263–264
  Water Framework Directive (EU), 277
  water law, 261, 278
  water scarcity, 279
Adaptation Fund Board (AFB)
  Accreditation Panel, 205
  CER Monetization Guidelines, 205
Adaptive management, 369, 376, 379, 380, 382, 385
Addis Ababa Principles and Guidelines for the Sustainable Use of Biodiversity, 371, 379, 387
Ad Hoc Working Group for Further Commitments by Annex I Parties under the Kyoto Protocol (AWG-KP), 87, 88, 206, 253, 254, 581
Ad Hoc Working Group on Long-term Cooperative Action under the Convention (AWG-LCA), 35, 84, 87, 112, 160, 194, 206–209, 254, 270, 319, 360, 394–396, 428, 560, 666

E.J. Hollo et al. (eds.), *Climate Change and the Law*, Ius Gentium: Comparative Perspectives on Law and Justice 21, DOI 10.1007/978-94-007-5440-9,
© Springer Science+Business Media Dordrecht 2013

Ad Hoc Working Group on the Durban Platform for Enhanced Action (ADP), 36, 84, 89, 102, 114, 117, 159, 254, 517, 581, 630
Aerosols, 448, 450
Afforestation, 37, 40, 262, 263, 377, 398, 402, 467, 558
Africa, 35, 54, 89, 109, 126, 192, 199, 205, 218, 229, 240, 242, 245, 250, 267, 274, 284, 351, 403, 406, 618, 639
Agreement on Subsidies and Countervailing Measures (SCM) (WTO), 22, 425
Akwé: Kon Guidelines for the conduct of cultural, environmental and social impact assessment on sacred sites and on lands and waters traditionally occupied or used by indigenous and local communities, 373
Andean Community (CAN), 670
Appellate Body (WTO), 340, 342, 423, 432, 435, 440, 443, 445
AR4. *See* Fourth Assessment Report (AR4)
Architecture, 3, 13, 19, 24, 25, 33, 35, 44, 46, 47, 128, 130–133, 139, 140, 146, 158–160, 198, 210, 221, 223, 334, 336, 338, 395, 577, 656, 657, 669
Arctic Climate Impact Assessment (ACIA), 299–303
Argentina, 126, 238, 459, 460, 653, 654, 658, 659, 666, 667, 669, 671, 675
Assigned amount unit (AAU), 49, 104, 105, 504, 552, 561, 562
Atmosphere, 12, 19, 23, 77, 84, 118, 121, 228, 232, 264, 266, 288, 320, 330, 381, 447, 448, 450, 452, 455, 456, 490, 524, 542, 620, 624, 639
Australia, 6, 85, 88, 120, 121, 126, 211, 258, 276, 277, 355, 362, 367, 404, 407, 434, 466, 495, 565–582, 590, 618
Aviation emissions, 45, 53, 436, 437, 517, 518, 598
Avoided deforestation, 394, 662, 676
AWG-KP. *See* Ad Hoc Working Group for Further Commitments by Annex I Parties under the Kyoto Protocol (AWG-KP)
AWG-LCA. *See* Ad Hoc Working Group on Long-term Cooperative Action under the Convention (AWG-LCA)

B

Bali Action Plan, 90, 160, 161, 206, 207, 221, 394
Benefit-sharing, 364, 367, 370, 371, 374, 375, 379, 381, 383, 385–387
Bern Convention on the Conservation of European Wildlife and Natural Habitats, 283
Biocultural protocol, 387
Biodiversity
  conservation, 4, 262, 365, 370, 375, 376, 381, 382, 386, 392, 397, 402
  ecosystem approach, 283, 360, 368–383
  ecosystem services, 372, 375, 380
  forest biodiversity management, 377
  genetic resources, 367, 386
  invasive alien species, 364, 377
  law, 4, 261, 285, 331, 359–390
  mainstreaming of, 370
  marine biodiversity, 275, 276, 361, 382, 453
  marine ecosystems, 382
  mountain ecosystems, 380
  peatlands, 377
  protected areas, 261, 262, 276, 283, 382, 626
  restoration of ecosystems, 375
  terrestrial ecosystems, 418
  valuation of, 370
  wetlands, 376
Biodiversity-related conventions, 360, 367, 389
Biofuels
  custom classification of, 433
  sustainability criteria (EU), 52, 53, 431–433
  sustainable production, 378
Bolivia, 116, 133, 238, 421, 466, 654, 668, 671
Bottom up approach, 35, 36, 46, 135, 138, 140, 159, 164, 384, 386
Brazil, 7, 8, 85, 96, 105, 120–122, 125, 126, 129–131, 211, 229, 236, 237, 240, 241, 278, 293, 396, 428, 637–649, 654–660, 665–667, 669–673, 675, 677, 678
Buenos Aires, 197, 269–270, 673, 677
Bulgaria, 171, 173, 182–183, 186

C

California, 45, 50, 385, 393, 435, 481–482, 485, 486
Canada-Certain Measures Affecting the Renewable Energy Generation Sector (WTO), 437

Cancun Adaptation Framework, 84, 89, 90, 98, 102, 107, 270–271
Cancun agreements, 47, 84, 160, 194, 208, 209, 215, 216, 219, 220, 270, 319, 321, 360, 395–397, 401, 404, 408, 410–412, 414, 415, 560, 591
Capacity building, 2, 90, 116, 167, 210, 215, 217, 220, 245, 249–251, 270, 272, 277, 313, 351, 370, 380, 389, 406, 663
Cap-and-trade scheme, 54, 435
Capitalism, 12, 420–422
Carbon capture and storage (CCS)
  under the CDM, 452, 469
  in the EU, 524
Carbon crediting, 395
Carbon credits, 40, 49, 192, 395, 496, 552, 580, 600
Carbon dioxide (CO_2), 15, 21, 44, 45, 78, 85, 88, 104, 117, 119, 181, 228, 232, 233, 243, 245, 276, 279, 427, 430, 447, 452–456, 466, 476–478, 482, 484, 485, 504, 511, 513, 517, 518, 541, 550, 556, 561, 572, 574, 585, 589, 605, 611, 618, 620, 622, 628, 635, 645–648
Carbon dioxide removal (CDR), 450, 452–456, 459, 466, 467
Carbon leakage, 5, 54, 434–436, 575, 576
Carbon market(s)
  new market mechanism, 254, 550
  REDD+ mechanism, 395, 397, 418
Carbon price, 6, 434, 517, 541, 567, 575, 577–580
Carbon tax, 99, 433, 434, 492, 495, 498, 507, 573, 622, 623, 628, 629
Caribbean Community (CARICOM), 670
Cartagena Dialogue, 135, 140
Cartagena Protocol on Biosafety, 342, 344, 378, 442, 445
CDM. *See* Clean development mechanism (CDM)
Certified Emission Reduction (CER) Monetization Guidelines, 204–205
Charter of Fundamental Rights (ECFR), 65, 67–70, 72, 515
Chicago Climate Exchange (CCX), 57, 58
Chicago Convention on International Civil Aviation, 22
Chile, 195, 239, 240, 335, 337, 654, 658–660, 665–668
China-Measures concerning Wind Farm Equipment (WTO), 437, 438
CITES. *See* Convention on International Trade in Endangered Species (CITES)
Cities
  Buenos Aires, 197, 269–270, 673, 677
  Cities for Climate Protection (CCP), 51
  Mexico City, 673, 677
  New York, 12, 32, 50, 84, 86, 88, 112, 155, 192, 193, 227, 230, 231, 255, 258, 260, 261, 263, 265, 278, 288, 292, 295–298, 300, 331, 333, 335, 337, 338, 343, 354, 363, 393, 400, 409, 419, 421, 424, 429, 432, 450, 484, 550, 579, 584, 640, 660
  Rio de Janeiro, 96, 102, 199, 200, 331, 360, 393, 402, 449, 457, 517, 673, 674
  Sao Paulo, 672, 673, 677
Clean development mechanism (CDM)
  access to a review process, 317–318
  appeals procedure, 38
  Bajo Acuan palm oil project, 318
  barriers, 240–249
  CDM Executive Board, 36–39, 56, 242, 250, 315–319
  CDM Gold Standard, 36, 41
  CDM potential, 232–240, 243, 249, 252
  CDM Rulebook, 37
  Certified Emission Reductions (CERs), 36, 106, 204, 227, 273, 398, 580, 656
  contribution to sustainable development, 41, 319, 398
  Designated Operational Entity (DOE), 56, 315–317
  distributive justice, 4, 227, 229–231, 240–249, 251, 255
  equitable distribution, 229–231, 233, 234, 240, 242–246, 248–251, 255, 256
  equity, 227, 398
  levy, 204
  market-based nature, 56, 246–248, 250, 251, 253, 255
  project design document (PDD), 56, 247
  public-private partnership, 37, 55
  transaction costs, 115, 243–245, 250
  unilateral structure, 242, 248–249
Clean energy, 120, 211, 422, 437–439, 444, 476, 507, 565–582, 599, 600, 609, 612, 613, 641
Clean Energy Ministerial (CEM), 121
Climate and Energy Package (EU), 45, 46, 50, 421, 431, 516, 517
Climate capitalism, 422
Climate Change Act
  Scotland, 6, 537, 541
  UK, 6, 537, 541, 544

Climate change impacts
adverse health impacts, 300
draughts, 279, 284
extreme weather events, 15, 259, 261
floods, 279
sea level rise, 206, 258, 604
Climate change regime, 3–5, 32, 44, 46, 48, 83–92, 98, 100, 102–109, 118, 131, 139, 145, 155, 165–188, 227, 228, 289, 290, 304–325, 334, 344, 353, 360, 361, 363, 365, 366, 368, 372, 383–390, 399, 411, 422, 423, 427, 440, 444, 551–552, 638, 641
Climate finance
adaptation fund, 192, 199, 202–206, 217, 218
Adaptation Fund Board (AFB), 199, 204–206
Article 4.3 of the UNFCCC, 194, 197
estimates, 192, 196, 212, 213
fast-start finance, 210–212
fast-track finance, 208, 211
financial mechanism, 193, 198–206
funding sources, 194, 207, 208, 213
GEF Trust Fund, 199–202
Global Environment Facility (GEF), 195–197, 199–204, 215, 217, 218, 222, 223
Green Climate Fund, 216–218
Green Climate Fund Board, 199, 217, 218, 222
High-Level Advisory Group on Climate Change Financing, 208, 213, 214
incremental costs, 194, 196–199, 212, 217, 218
LDC Fund, 197, 199, 203, 211, 222, 223
long-term finance, 211–214
new and additional funding, 196–198
official development assistance (ODA), 196, 197, 209, 222
sources, 192, 194, 196, 202, 204, 206–211, 213, 214, 217, 218, 222
Special Climate Change Fund (SCCF), 197, 199, 202–203, 222, 223
Standing Committee, 208–210, 213, 216, 219–220
Transitional Committee, 208, 216, 217
Colombia, 8, 238, 654, 658, 660–667, 669–671, 676, 678
Common but differentiated responsibilities (CBDR), 3, 54, 65, 66, 92, 101–104, 195, 228, 385, 386, 436, 516, 568, 569, 596, 605, 638, 640
Compliance, 14, 27, 29, 30, 36, 38, 48, 55, 56, 73, 88, 107, 114, 115, 119, 132, 138, 139, 144, 146, 147, 149, 150, 152, 154, 155, 160, 164–188, 215, 247, 248, 297, 302, 311, 317, 337, 344, 346, 361, 371, 385, 387, 390, 405, 416, 431, 440, 474, 482, 484–486, 496, 499, 502–504, 580, 585, 586, 591, 599, 600, 607, 611, 620, 626, 627, 641, 645
Compliance Committee
Bulgaria, 171, 173, 182–183, 186
Canada's national registry, 180
enforcement branch (EB), 3, 166–169, 172–183
facilitative branch (FB), 166, 167, 169–172
Greece, 172–180
triggering, 184
Conference of the Parties (COP)
COP 13, Bali (2007), 144, 160, 161, 592
COP 16, Cancun (2010), 193, 208, 428
COP 15, Copenhagen (2009), 161, 193, 207, 560
COP 17, Durban (2011), 159, 193, 581, 593
COP 11, Montreal (2005), 270
COP 12, Nairobi (2006), 270
Conference of the Parties serving as the Meeting of the Parties to the Kyoto Protocol (COP/MOP)
COP/MOP 5, Copenhagen (2009), 253, 643
COP/MOP 7, Durban (2011), 87, 254, 593
COP/MOP 1, Montreal (2005), 37, 592
COP/MOP 2, Nairobi (2006), 49
Conflict clauses, 343–346, 348, 400, 401. *See also* Treaty interpretation
Conflicts
avoidance, 399, 412, 417
definition of, 346–348
Consensus, 23, 35, 100, 103, 116, 133, 137, 144, 146, 156, 163, 166, 200, 207, 216, 288, 318, 331, 349, 351, 361, 365, 374, 384, 393, 423, 440, 462, 463, 467, 495, 505, 522, 569, 573
Conservation, 4–6, 66, 94, 261, 262, 277, 278, 283, 319, 337, 341, 359, 363–365, 367, 368, 370, 372, 375, 376, 378–383, 385, 386, 392, 393, 395, 397, 398, 401, 402, 404, 418, 432, 437, 453, 477, 493, 529, 548, 558, 559, 564, 585, 586, 588, 590, 592, 606–607, 610, 611, 619, 626, 634
Control theory, 64

Convention of Biological Diversity (CBD)
  CBD COP, 352, 361, 365, 366, 368–372, 375, 376, 378, 382, 388, 389, 406
  CBD Expert Group on Climate Change, 372
  normative activity of CBD COP, 362, 365, 371
  secretariat, 37, 38, 351, 365, 370, 372, 406, 407, 463
Convention on International Trade in Endangered Species (CITES), 367
Convention on Migratory Species (CMS), 283, 367, 377, 380, 382
Coordination, 5, 13, 25, 42, 46, 54, 129, 192, 212, 219, 220, 222, 270, 271, 333, 335, 351, 355, 357, 359, 399, 403, 406, 408, 416–418, 427, 448, 466, 533, 629–630, 669, 671, 675
COP decisions
  Bali Action Plan, 160, 206
  Bonn Agreements, 197
  Buenos Aires Plan of Action, 197
  Cancun Agreements, 84, 208, 210, 219, 270, 319, 408
  Durban Platform, 89
  Marrakesh Agreements, 425
  use of, 48
Copenhagen, 13, 35, 44, 47, 48, 88, 95, 100, 101, 112, 114, 118, 121–123, 126, 129, 131, 134, 135, 138, 143, 145, 170, 193, 198, 206, 207, 210, 274, 279, 312, 317, 392, 395, 405, 428, 490, 505, 510, 513, 516, 517, 556, 560, 573, 596, 608, 626, 643, 656, 675
Copenhagen Accord, 13, 47, 114, 128, 139, 143, 160, 207–210, 213, 214, 216, 365, 395, 474, 505, 588
Corfu Channel case (ICJ), 457
Corporate environmental accountability, 388
Costa Rica, 213, 239, 654, 658–663, 665–667, 669, 671, 674
Cultural heritage, 381, 621
Cultural impact assessment, 462

D

Deforestation, 5, 44, 50, 78, 285, 319, 361, 381, 392, 394–397, 399, 402, 404, 406, 408, 412, 416, 431, 644–648, 655, 662, 667, 676. *See also* REDD+
  drivers of, 392, 396
  international trade in timber, 392
Deformalization, 32–34, 36–43, 55–61
Development cooperation, 52, 385, 396, 414
Dispute, 5, 40, 71, 93, 148, 166, 171, 181, 291, 304, 305, 317, 335–337, 340–342, 420, 423, 426, 432, 437–444, 459, 476, 519, 632
Dispute settlement system (WTO), 5, 44, 337, 340, 420, 423, 424, 426, 437, 439–445
Dispute Settlement Understanding (DSU) (WTO), 435, 440, 441
Distributive justice
  egalitarianism, 230
  Rawls's difference principle, 230
  utilitarianism, 230
Drivers of deforestation, 382, 396
Durban Platform, 36, 87, 89, 117, 581, 630

E

Economies in transitions (EITs), 181, 185, 195, 200
Ecosystem
  approach, 5, 283, 360, 368–384, 387
  services, 74, 285, 369, 372, 375, 377, 378, 380, 382, 383, 392, 404, 406, 418, 626
Emission Reductions Purchase Agreement (ERPA), 36, 40, 41
Emission Reduction Units (ERUs), 105, 580
Emissions trading, 2, 6, 15, 20–23, 25, 34, 45, 48, 49, 58, 64, 77, 78, 104, 106, 115, 167, 254, 433–436, 476, 482, 486, 492, 501, 503, 512, 513, 517, 518, 525–527, 529, 530, 535, 539, 552, 561, 566, 567, 570, 571, 573, 575–578, 584, 585, 586, 587, 597, 598, 630
Emissions trading scheme (ETS), 6, 22, 25, 39, 40, 45, 46, 48, 50, 53, 54, 58, 78, 233, 252, 254, 434–436, 517–519, 528, 533, 536–538, 544, 545, 552, 570, 571, 573, 577, 578, 586, 587, 597–599, 630
Energy
  biofuels, 52, 362, 430, 431, 513, 667
  renewable energy, 20–22, 26, 78, 125, 211, 280, 430, 481, 482, 485, 487, 512, 526, 546, 589, 599, 609, 633–635
  smart grids, 281
Enforcement, 3, 13, 21, 30, 68, 74, 139, 144, 145, 162, 164–188, 269, 305, 317, 336, 386, 390, 404, 405, 416, 480, 498, 527, 528, 534, 535, 585, 599, 654, 659, 668, 669, 674–678
Enforcement branch (Compliance Committee), 3, 166–169, 172–183, 317
ENMOD Convention, 461–462

Environmental effectiveness, 16, 28, 30, 577
Environmental impact assessment, 39, 99, 260, 264, 284, 364, 372, 373, 375, 456, 457, 460–461
Environmental integrity, 48, 56, 115, 255, 298, 318, 352, 397, 405, 435
Equal subsistence, 75
Equitable use, 96, 97
Equity, intergenerational, 91, 96, 97, 227, 457
ERPA. *See* Emission Reductions Purchase Agreement (ERPA)
ERUs. *See* Emission Reduction Units (ERUs)
Ethics, 32, 64, 65, 101, 151, 152, 265, 397, 421, 568
EU Emissions trading scheme (ETS)
  aviation emissions, 53, 436, 516, 517, 598
  carbon equalization system, 435
  EU-wide emissions cap, 434
  Phase III, 252, 599
  second phase, 434
European Charter of fundamental rights, 65
European convention on human rights (ECHR), 65, 68–70, 72, 292, 299
European Court of Human Rights (ECtHR), 68, 294, 299
European Court of Justice (ECJ), 67, 68, 283, 597, 598
European Union
  biofuels sustainability criteria, 53
  Birds Directive, 283
  climate and energy package, 45, 46, 50, 421, 431, 516, 517, 525
  Directive on the promotion of the use of energy from renewable sources, 431
  emissions trading scheme (ETS), 6, 22, 25, 39, 40, 45, 46, 48, 50, 53, 54, 58, 78, 233, 252, 254, 434–436, 517, 518, 528, 533, 536–538, 544, 545, 552, 570, 571, 573, 577, 578, 586, 587, 597–599, 630
  European Commission, 149, 153, 252, 254, 404, 421, 512, 597
  external relations, 45, 52, 515–519
  Floods Directive, 277
  Habitats Directive, 261, 276, 283
  Marine Strategy Framework Directive, 276, 277
  Natura 2000, 283
  Treaty on the Functioning of the European Union, 52, 515
  unilateralism, 2, 45, 46, 54, 436, 517
  Water Framework Directive (WFD), 277
Expert review team (ERT), 168, 169, 171–178, 180–184, 186, 187

F
Facilitative branch (FB) (Compliance Committee), 3, 166, 167, 169–173, 183–185, 187, 188
Fast-start finance, 210–211
FCPF. *See* World Bank
Flexibility mechanisms, 115, 315–318, 561–562
Forest carbon partnership facility (FCPF), 396, 397, 401, 405, 406, 413, 415, 416
Forests
  afforestation, 37, 40, 377, 398, 402, 558
  avoided deforestation, 394, 662, 676
  carbon sequestration, 397, 401
  deforestation, 5, 285, 319, 392, 394–396, 404, 406, 647, 648
  forest biodiversity management, 377
  Forest Carbon Partnership Facility (FCPF), 396, 397, 401, 405, 406, 413, 415, 416
  forest loss, 392
  management, 285, 371, 377, 378, 393, 402, 403, 558
  national forest reference emission levels, 396
  REDD, 319, 362, 394, 395, 397, 407–408
  REDD+, 5, 285, 392, 395–397, 399–403
  REDD-readiness, 417
  reforestation, 37, 377
  sustainable forest management, 378, 393
  timber, 369, 392
  UN-REDD Programme, 396, 402
Fourth Assessment Report (AR4), 27, 86, 87, 97, 258, 263, 288, 299, 300, 392, 424, 566, 592
Fourth Assessment Report (IPCC), 86, 299
Fragmentation
  conflict avoidance, 339, 399, 417
  institutional fragmentation, 333, 335, 337
  of international law, 5, 33, 42, 44, 153, 329–357, 361, 399–401, 420, 423, 442, 444
  International Law Commission (ILC), 33, 332–337, 339, 341, 342, 345, 348, 356, 399–401, 423, 443, 461
  managing fragmentation, 333, 334, 339–356
  negative impacts, 336
  Nicaragua case, 336
  positive impacts, 336
  proliferation of international instruments, 333
  regime boundaries, 354–356

structural bias, 153–154
Swordfish dispute, 335, 337
Tadić case, 336
treaty congestion, 333

G
General Agreement on Tariffs and Trade (GATT)
Article III, 432, 436
Article XX, 341, 428, 432, 436, 438
non-discrimination requirement, 432
Shrimps dispute, 445
General Agreement on Trade in Services (GATS), 425, 430, 437
Geoengineering
definition, 447–449
governance, 464–466
Oxford principles, 468
research, 467–468
Geoengineering techniques
biomass storage in the ocean, 454
carbon dioxide removal (CDR), 452–456
cloud brightening, 451
injecting of aerosols into the stratosphere, 450
ocean fertilization, 453, 454, 460, 462
ocean liming, 454, 455
solar radiation management (SRM), 450–452
Germany, German Constitution, 65, 67, 73
Global administrative law, 14, 33, 38, 39, 41, 42, 348
Global administrative space, 38
Global Cities Covenant on Climate, 673
Global climate governance, 3, 13, 77–79, 145, 157–159, 164, 332, 333, 336, 399
Global Environmental Facility (GEF)
GEF Trust Fund, 199–202
Memorandum of Understanding with the UNFCCC COP, 196, 198, 201, 218
Resource Allocation Framework, 198
Strategic Priority, 199, 273
Globalization, 33, 41–43, 50, 59, 144, 147
Gold Standard for the CDM, 58
Good faith, 94, 366, 403, 408, 418
Governance, hybrid governance initatives, 55
Greece
compliance plan, 174, 175, 179–180, 182
proceedings against, 173–180
Green Climate Fund (GCF), 4, 109, 193, 199, 202, 207–210, 216–222, 274
Green economy, 96, 127–129, 422, 439, 444, 507
Greenhouse Gas Management Act, 27, 28
Green Investment Scheme (GIS), 34, 48, 49, 562
Gross domestic product (GDP), 76, 198, 201, 209, 222, 234, 321, 421, 507, 550, 557–560, 563, 604, 663
Group of 77 and China (G-77/China), 119, 171, 195
Group of Eight (G-8), 114, 122–123, 125–126, 587
Group of Twenty (G-20), 126–127

H
Hard law, 48, 143–164, 357, 655
Harmonization, 351, 400, 587
Heiligendamm L’ Aquila Process (HAP), 125
Heiligendamm Process (HDP), 125
Historical responsibility, 98, 193, 209, 226, 605
Human Development Index (HDI), 234, 235
Human rights
American Declaration of the Rights and Duties of Man, 292–294, 296, 301–303
approach, 290, 291, 300–304, 321, 324, 362, 385, 389
compensation, 321, 324
European Court of Human Rights (ECtHR), 68, 294, 299
European Human Rights Convention, 294, 299
Inter-American Commission on Human Rights, 292–294, 299, 301, 304, 305, 409
Inter-American Court on Human Rights, 292, 294
International Covenant on Civil and Political Rights (CCPR), 292, 293, 314, 321, 409
International Covenant on Economic, Social and Cultural Rights (CESCER), 70, 296, 297, 385, 409
Inuit, 289, 301, 303, 322, 323
Inuit Human Rights Petition, 289
law, 5, 154, 288, 291, 293, 302, 324, 332, 364, 372, 385, 387–390
Office of the High Commissioner for Human Rights (OHCHR), 289, 295, 300, 314, 320, 323, 408–411
right to health, 296–300
right to life, 292–296

Human rights (*cont.*)
right to property, 294
social safeguards, 408, 409
UN Committee on the Rights of the Child, 297, 298
UN Convention on the Rights of the Child, 297
UN Human Rights Council (HRC), 288, 319–320
Universal Declaration of Human Rights, 292, 296, 414
violations, 282, 291, 321, 409–411
Human rights-based approach, 5, 359–390, 414

I

ICA. *See* International Consultation and Analysis (ICA)
ICC. *See* Inuit Circumpolar Council (ICC)
ICLEI. *See* International Council for Local Environmental Initiatives (ICLEI)
IEA. *See* International Energy Agency (IEA)
ILC. *See* International Law Commission (ILC)
IMF. *See* International Monetary Fund (IMF)
IMO. *See* International Maritime Organization (IMO)
India, 7, 53, 77, 85, 120–122, 125, 126, 128, 130, 131, 157, 159–161, 229, 236, 237, 240, 428, 430, 437, 516, 518, 519, 560, 566, 570, 603–616
Indigenous peoples
Inuit, 301–304, 362
traditional knowledge, 302, 361, 364, 370–372, 374
UN Permanent Forum on Indigenous Issues, 415
Informal networking, 61
Institutional cooperation, 334, 349–357, 399, 407, 411, 417, 420, 444. *See also* Fragmentation
Joint Liaison Group, 351
Proposed Terms of Reference and Modus Operandi for the Joint Liaison Group, 407
Instrument choice, 11–30
Instrument mix, 17, 18, 28, 78
Integrated Greenhouse Gas Management, 25–28
Inter-American Commission on Human Rights, 292–294, 299, 301, 304, 305, 409
Intergenerational equity, 91, 96, 97, 227, 457
Intergovernmental cooperation, 13
Intergovernmental Panel on Climate Change (IPCC)
Fifth Assessment Report (AR5), 449, 469, 637
Fourth Assessment Report (AR4), 27, 86, 87, 97, 258, 263, 288, 299, 300, 392, 424, 425, 429, 433, 439, 566, 592
Third Assessment Report (TAR), 84, 86, 226, 295
International Analysis and Review (IAR), 160
International aviation, 45, 52, 53, 344, 436, 437, 517, 518, 597
International Consultation and Analysis (ICA), 161
International Council for Local Environmental Initiatives (ICLEI), 51, 673
International Court of Justice (ICJ)
advisory opinion, 457, 458
Corfu Channel case, 457
Nicaragua case, 336
Pulp Mills case, 460
International Covenant on Civil and Political Rights (CCPR), 292, 293, 304, 314, 321, 409
International Criminal Tribunal for the Former Yugoslavia, 336
International economic law, 332, 337, 423, 426, 427
International Energy Agency (IEA), 127, 128, 136, 212, 605, 633
International Law Commission (ILC), 33, 332–337, 339, 341, 342, 345, 348, 356, 399–401, 423, 443, 461
International Maritime Organization (IMO), 276, 344, 452, 453
International Monetary Fund (IMF), 144, 604
International trade law, 357, 423, 425, 431, 432, 436, 439, 598
International trade regime, 422
International Tribunal for the Law of the Sea (ITLOS), 335, 337, 390, 459
Inuit Circumpolar Council (ICC), 301, 304, 323
Inuit human rights petition, 289
Invasive alien species, 364, 377, 378, 382
IPCC. *See* Intergovernmental Panel on Climate Change (IPCC)
Italy, 92, 120–123, 126, 171, 299, 391, 510

ITLOS. *See* International Tribunal for the Law of the Sea (ITLOS)

J

Japan, 6, 85, 88, 120–123, 126, 170, 211, 392, 404, 428, 434, 437, 438, 462, 495, 583–593, 643
Joint Implementation (JI), 104, 105, 115, 116, 167, 310, 397, 550, 552, 561, 580
Joint Liaison Group, 351, 354, 355, 403, 407, 408
  Proposed Terms of Reference and Modus Operandi for the Joint Liaison Group, 407
Justice, 3, 4, 17, 24, 35, 56, 57, 63–79, 84, 90, 97, 98, 101, 155, 225–256, 265, 283, 312, 321, 322, 324, 337, 362, 386, 387, 389, 421, 440, 458, 513, 518, 535, 575, 597. 606, 607, 632
  distributive, 4, 227–231, 240–251, 255, 533

K

Kyoto Protocol
  adaptation fund, 50, 109, 203
  adaptation fund board, 272
  Clean Development Mechanism (CDM), 4, 34, 37–39, 104, 109, 115, 192, 203, 227–229, 232, 242, 243, 245–247, 253, 254, 271, 319, 394, 452
  compliance committee, 3, 166, 167, 169, 171
  COP/MOP, 36, 37, 49, 87, 146, 166, 181, 182, 199, 203, 229, 242, 253, 254, 272, 307, 587, 643
  enforcement branch, 166
  facilitative branch, 3, 166, 167, 169–172
  first commitment period, 37, 87, 88, 94, 105, 114, 169, 228, 253, 428, 581
  Marrakesh Accords, 36, 37, 48, 146, 562, 587
  second commitment period, 7, 87, 88, 139, 160, 170, 185, 253, 254, 560, 581, 591

L

Land use, land-use change and forestry (LULUCF), 233, 392, 455, 556, 580, 605, 667
Latvia, 48, 171
Least developed countries (LDCs), 35, 108, 109, 192, 203, 204, 217, 218, 233, 243, 244, 249, 250, 252, 254, 268, 273, 274, 466, 599, 630
Least developed country fund (LDCF), 109, 197, 199, 203, 211, 222, 223, 268, 269, 273, 274
Legalization, 60, 144–156, 158–162, 164
Legal pluralism, 33, 42, 59–62
Legitimacy, 14, 21, 28, 30, 35, 39, 44, 48, 60, 113, 116, 124, 127, 136–138, 140, 143, 144, 146, 148, 149, 151, 152, 154–156, 158, 162–164, 310, 315, 318, 334, 355, 357, 420, 440, 463, 465
Linkages, 4, 23, 36, 54, 108, 157, 290, 343, 345, 349–351, 353, 354, 368, 420, 423–445, 587
Local communities, 41, 285, 290, 305, 306, 315, 318, 319, 361, 364, 367, 368, 370–382, 384, 385, 387–389, 397, 398, 408, 412, 416
Long-term negotiations
  AWG-KP, 253
  AWG-LCA, 206
  Durban Platform, 87, 254, 630
Low-carbon economy, 7, 420, 433, 439, 445, 470, 514, 559, 560, 570, 596, 597, 600, 601, 614, 646, 656
Low-carbon strategies, 614, 660, 662

M

Major Economies Forum on Energy and Climate Change (MEF), 119–122, 135, 146, 593
Major Economies Meeting on Energy Security and Climate Change (MEM), 119, 120, 122, 133, 134, 592
Market-based approach, 56, 503
Marrakesh Accords, 36, 37, 48, 146, 344, 562, 585, 587, 592
Massachusetts *v.* EPA decision, 477
Measurement, reporting and verification (MRV), 135, 140, 172, 194, 210, 211, 214–216, 219–222, 631
  of financial support, 194
  partnership, 140
MERCOSUR, 671
Mexico, 8, 120–122, 125, 126, 128, 138, 195, 207, 229, 237, 240, 241, 396, 485, 633, 654, 656, 658, 659, 661, 663, 665–667, 673–678
Mexico City, 673, 677

Migration, 2–4, 45, 51, 79, 261, 271, 275, 282, 283, 382, 524
Migratory species, 283, 367, 373, 376, 377, 380, 382
Millennium Development Goals, 300
Mitigation
  anchoring pledges, 47
  bottom-up, 36, 46, 135, 139, 140
  1.5°C, 89, 114
  2°C, 89
  historical emissions, 66
  historical responsibility, 98, 193
  nationally appropriate mitigation actions (NAMAs), 47, 268, 394, 428, 646, 666, 667
  objectives, 24, 87–89, 132, 567, 572, 667
  obligations under the Kyoto Protocol, 166, 455, 499
  per capita emissions, 77, 321, 474, 524, 566
  pledge-and-review, 46, 47, 138
  pledges, 47, 48, 88, 89, 100, 114, 139, 427, 428
  QELROs, 88, 105, 115
  regime, 103–107
  subsistence/survival emissions, 321
  targets-and-timetables, 162
  'top-down', 35, 36, 46, 135, 138, 140
Mitigation regime, 103–107, 110
Monitoring, 109, 157, 169, 272, 291, 295, 351, 359, 369, 373, 377, 380–382, 390, 396, 398, 402, 460, 511, 543, 553–556, 611, 624, 627, 631, 642, 647, 657, 677
Montreal, 22, 166, 170, 270, 272, 331, 343, 359, 370, 371, 373, 402, 451, 453, 489, 492, 498, 499, 592
Montreal protocol, 13, 113, 117–119, 134, 141, 157, 170, 331, 336, 343, 353, 451, 490, 638, 646
Multilateral Environmental Agreement (MEA), 117, 118, 165, 170, 172, 199, 200, 298, 310, 334, 337, 342, 345, 349–351, 353, 354, 363, 366, 400, 413, 429, 445, 640
Multilateralism, 61, 113–119, 516, 517, 520
Mutual supportiveness, 44, 49340, 343, 353, 356, 360, 363–368, 424, 442, 444, 445

N

Nairobi Work Programme on Impacts, Vulnerability and Adaptation, 57, 269–270
NAMAs. *See* nationally appropriate mitigation actions
NAPAs. *See* Mitigation
National Adaptation Plans, 108
National Adaptation Programmes of Action (NAPAs), 203, 268–270, 273, 274
Nationally appropriate mitigation actions (NAMAs), 47, 207, 210, 220–221, 268, 394, 397, 428, 624, 645, 646, 666, 667
Neoliberalism, 60, 61
Non-state actors
  access to information, 315–316
  access to negotiations, 307–309
  observer organizations, 307
  private sector, 34, 55–58, 61, 386, 388
  public participation, 310–313

O

Obligation, 4, 5, 7, 12–14, 24, 26, 27, 36, 40, 41, 49, 50, 53, 65, 68, 71, 72, 74, 78, 87, 89, 91, 92, 95, 96, 101–105, 114, 115, 117, 129, 146, 148, 156–160, 162, 163, 166–169, 172, 181, 184, 185, 188, 194, 195, 259, 265, 267, 283, 289, 292–294, 298, 303, 305, 306, 312–314, 316, 319, 323, 344–346, 348, 349, 355, 362, 364, 365, 367, 376, 384, 390–418, 427, 433, 441, 450, 452, 455–461, 465, 483, 485, 498–501, 504, 506, 519, 531–533, 541–545, 548, 568, 572, 577, 579–581, 591, 597, 601, 605, 613, 620, 623, 631, 638, 640, 641, 649, 678
Official Development Assistance (ODA), 196, 197, 209, 222
Organisation for Economic Cooperation and Development (OECD), 14, 15, 17, 76–79, 125, 127, 130–131, 136, 195, 198, 222, 388, 528, 539, 558, 566, 589, 605
Overlaps
  of instruments, 417
  of international legal regimes, 4, 434, 443
Ozone regime, 119, 145, 155, 157

P

Pacta sunt servanda-principle, 146, 156, 402
Palau, 458
Panama, 239, 404, 653, 654, 659, 668, 671, 674
PDD. *See* Project Design Document; Project Design Document (PDD)

Per capita emissions, 77, 321, 474, 524, 566, 605
Peru, 238, 299, 654, 658–660, 661, 665–673
Pledge-and-review, 46, 48, 138, 159, 162. *See also* Bottom up approach
Polluter pays principle, 20, 75–77, 92, 100–101
PPMs. *See* Processes and production methods (PPMs)
Precautionary approach, 292, 347, 368–370, 378, 382, 389, 459, 462
Precautionary principle, 65, 89, 90, 92, 99–100, 110, 363, 368, 369, 456–460, 464, 555
Precision, 148, 158, 160, 162, 163, 233
Preventative action, 3, 94, 103
Primary objective. *See* Ultimate objective
Principles
 Article 3 of the UNFCCC, 91, 94, 436
 common but differentiated responsibility, 65, 66, 92, 101–103, 385, 386, 596, 605
 cooperation, 91, 94–95
 equitable use, 96, 97
 equity, 91, 96, 97, 226
 pacta sunt servanda, 146, 156, 402
 polluter pays, 20, 75–77, 92, 100–101
 precautionary, 65, 89, 90, 92, 99–100, 110, 363, 368, 369, 456–460, 464, 555
 preventative action, 3, 94, 103
 Principle 21 of the Stockholm Declaration, 91, 93, 393
 systemic integration, 340–342, 356, 442
Private sector, 3, 33, 34, 38, 55–58, 61, 116, 192, 193, 205, 209, 211, 222, 223, 269, 357, 364, 386–388, 390, 479, 500, 599, 601, 607, 612, 614–616, 647, 670
Processes and production methods (PPMs), 5, 431, 433, 435
Project Design Document (PDD), 56, 242, 247, 249, 316
Protected areas, 261, 262, 276, 282, 283, 370, 371, 373, 375, 376, 378–380, 382, 626
Protection of the Ozone Layer, 117, 145, 331, 350, 352, 450, 490, 640
 Montreal Protocol, 117–119
Public-private partnerships, 33, 34, 37, 55–58, 60, 61, 211, 357, 645
Pulp Mills case (ICJ), 460

Q
Quantified Emission Limitation and Reduction Objectives (QELROs), 88, 104, 105, 115

R
Ramsar Convention on Wetlands of International Importance, 367
RECs. *See* United States of America
REDD+
 co-benefits, 97, 398
 conservation of forest carbon stocks, 401
 Forest Carbon Partnership Facility (FCPF), 396, 397, 405
 forest reference levels, 396
 Guidelines for Seeking the Free, Prior, and Informed Consent (FPIC) of Indigenous Peoples and other Forest Dependent Communities, 414
 market mechanism, 163, 399
 national forest monitoring systems, 396
 national forest reference emission levels, 396
 REDD-readiness, 396, 405, 406, 413, 415, 417
 Safeguards, 399–403, 406–409, 412, 416–418
 social safeguards, 408, 409, 411–416
 UN-REDD Programme, 396, 397, 402, 412, 413, 416
 World Bank Operational Policy 4.10 on indigenous peoples, 415
Reforestation, 37, 40, 262, 377, 398, 402, 467, 558, 572
Regime(s), conflicts, 153
Regional Greenhouse Gas Initiative (RGGI), 51, 483–485
Renewable energy
 Directive on the promotion of the use of energy from renewable sources, 431
 feed-in tariff, 437, 438, 507, 599, 613
 renewable energy certificates (RECs), 482, 483, 613
 wind power equipment subsidies, 438
Reporting
 biennial reporting, 161, 215
 format, 210, 211, 214–216
 guidelines, 214, 215, 222
 MRV, 172, 194, 211, 214, 216, 222, 631
 obligations, 105, 115, 167, 172, 194, 413, 450, 506, 543
 period, 197, 206, 210, 215
 requirements, 106, 110, 161, 544, 546, 549, 621
 and review, 170
 system, 163, 186, 216
 and transparency, 160

Response measures, 86, 98, 268–270, 325, 360, 362, 365, 372–375, 380, 389, 390, 429, 440
Forum on, 429, 440
RGGI. *See* Regional Greenhouse Gas Initiative (RGGI)
Rio Conventions, 351, 352, 354, 355, 402, 403, 406–408
Rio Declaration, 91, 94, 96, 99, 100, 102, 335, 340, 393, 457
Russian Federations, 88, 301, 304, 550–555, 557–560, 563

S
Safeguards, 26, 66, 117, 185, 294, 303, 309, 318, 319, 368, 388, 396, 399–409, 411–418, 459–461, 468, 622
SCCF. *See* Special Climate Change Fund (SCCF)
Scientific uncertainty, 93, 99, 320, 331, 459, 460, 555, 639
Scotland, 6, 359, 537–539, 541–549
Sectoral
action plans, 8, 654, 676
approaches, 530, 588, 591, 664–666
carbon budgets, 627
planning, 109, 662
plans, 625, 646, 660, 662, 664, 675
targets, 656, 667–668, 672
SIDS. *See* Small Island Developing States (SIDS)
Slovenia, 171
Small Island Developing States (SIDS), 35, 108, 109, 192, 204, 217, 218, 274
Smart grid(s), 281
Social impact assessment, 373, 382
Soft law, 3, 8, 32, 34, 47, 48, 54, 58, 59, 61, 143–164, 302, 357, 393, 654, 657
Solar radiation management (SRM), 448, 450–452, 459, 468
South Africa, 7, 114, 115, 120, 121, 125, 126, 130, 131, 138, 161, 171, 172, 187, 237, 241, 297–298, 300, 395, 428, 617–636
Special Climate Change Fund (SCCF), 109, 197, 199, 202–203, 222, 223, 273, 274
SRM. *See* Solar radiation management (SRM)
Stabilization, 15, 74, 84, 86, 266, 672
Standards, 2, 12, 14–16, 20, 27, 36, 41, 43–45, 52, 53, 58, 64, 77, 85, 97, 110, 135, 144, 149, 156, 163, 172, 205, 234, 289, 291,297-299, 317, 346, 361, 377, 378, 383, 384, 386–388, 398–399, 401, 405, 410, 415–418, 422, 425, 426, 429–433, 444, 457, 458, 476–480, 482–483, 502–504, 525, 529, 530, 536, 539, 558, 574, 590, 591, 600, 604, 607, 615, 620, 626, 629, 640, 641
Standing Committee (climate finance), 208–210, 213, 216, 219–222, 499, 575, 576
State responsibility, 61, 87, 93, 105, 321, 322, 336, 457, 458, 461
ILC Articles on Responsibility of States for Internationally Wrongful Acts, 461
State sovereignty, 3, 92–93, 355
Stern review, 15, 35, 99, 300, 394, 421, 538, 570
Sub-national
action, 50
governments, 6
initiatives, 50–51
plans, 614
Subsidiary Body for Implementation, 213, 268, 306, 307, 309, 311
Subsidiary Body for Scientific and Technological Advice (SBSTA), 106, 215, 268, 270, 306, 351, 354, 355, 375, 401, 403, 404, 406–409, 413, 466
Subsidies, 14–15, 21, 22, 27, 124, 127–129, 137, 425–427, 433, 435–439, 526, 528, 581, 586, 598, 600, 628, 629
Sustainable development
indicators, 398, 512
potential, 232–235, 237–239, 248, 251, 256
Sustainable forest management, 378, 393, 403, 558
Synergies, 4, 5, 29, 32, 36, 43, 61, 117, 334, 343, 344, 351, 352, 357, 364, 365, 401–405, 408–411, 413, 417, 418, 420, 430
System of Central American Integration (SICA), 671

T

Technology, 2, 3, 15–17, 20, 27, 33, 55, 56, 70, 77, 79, 83, 90, 95, 102, 106, 107, 109, 110, 114, 116, 117, 120–122, 128, 163, 167, 191, 194, 197, 198, 202, 206, 207, 210, 211, 215, 217, 220, 226, 245, 249, 267, 268, 270, 271, 273, 277, 306, 351, 374, 375, 376, 378, 381, 386, 397, 398, 401–403, 422, 425, 426, 437, 439, 444, 448, 451, 452, 455, 463, 465, 467, 468, 478, 483, 497, 502, 503, 521, 525, 526, 532, 541,555, 557, 558, 559, 562, 577, 587, 590, 607, 609, 610, 613, 614, 624, 630, 632, 633, 634, 639, 645, 655, 660, 663, 664, 667, 675
Theory of justice, 64, 65, 230
Third Assessment Report of the IPCC (TAR), 86
Top-down approach, 35, 36, 44, 46, 47, 48, 138, 140, 159, 164, 614
Trade
  regime, 5, 348, 422, 444
  sanctions, 423, 426
Traditional knowledge, 302, 361, 364, 370–372, 374, 376, 378–384, 386
Trail Smelter arbitration, 457
Transaction costs, 16, 28, 115–116, 150, 243–245, 250
Transitional Committee, 208, 216, 217, 274
Transnational
  associations, 59
  bodies, 59
  cooperation, 45, 51
  environmental law, 33, 43, 51, 56, 61, 277
  governance, 14
  human rights protection, 69
  networks, 51
  regulation, 60
  standard-setting, 144
Transparency
  access to a review process, 317
  access to information, 309–310
  access to negotiations, 307–309
  of actions, 3
  community participation, 371
  compliance committee procedure, 166, 168
  public participation, 306, 310–313
  Task Force on Public Participation in International Forums, 310, 312–313
Treaty interpretation, 339, 340, 342, 356, 400, 442
  Article 31.3(c) of the Vienna Convention on Law of Treaties, 339–342, 356, 442
  conflict clauses, 343–346, 400
  evolutionary approach, 342
  lex posterior, 339, 347, 348
  lex specialis, 339, 348, 357
  relevant rules of international law, 340, 342, 442, 443, 451
  systemic integration, 340–342, 356, 401, 442
  teleological interpretation, 342
Triple-win outcomes, 418

U

Ultimate objective, 3, 35, 84, 87, 89, 91, 104, 106, 110, 114, 140, 228, 232, 236, 363, 424, 426, 500, 639
  Article 2 of the UNFCCC, 86, 89
UNCCD. *See* UN Convention to Combat Desertification(UNCCD)
UN Climate Regime, 35, 46, 55, 57, 113, 114, 116, 118, 133–141, 144, 146, 159, 161, 332, 341
  post-2012 regime, 332
UN Convention on the Law of the Sea (UNCLOS), 275–277, 335, 345, 451, 452, 454, 459, 460, 640
UN Convention to Combat Desertification (UNCCD), 351, 355, 403, 404
UNECE Convention on Access to Information, Public Participation in Decision-Making and Access to Justice in Environmental Matters (Aarhus Convention), 312
UN Framework Convention on Climate Change (UNFCCC)
  adaptation objectives, 89–91, 654
  Article 2, 86, 89
  Article 3, 91, 94, 99, 436
  mitigation objectives, 87–89, 667
  post-2012 regime, 87
  principles, 91–103
  Secretariat, 38, 47, 57, 115, 122, 131, 137, 177, 191, 210, 312, 313, 319–320, 350, 354
  ultimate objective, 35, 84, 89, 91, 104, 106, 114, 228, 232, 363, 424, 426, 500, 639
UN General Assembly, 34–35, 113, 277
UN Human Rights Council, 98, 288, 319, 320, 361, 383, 387, 410, 411

Unilateralism
  action, 54
  and uncoordinated geoengineering, 465
  CDM structure, 242, 248–249
  emissions target, 517
  measures, 6, 54, 516, 518, 519
  trade measures, 46, 429, 444, 516
United Kingdom (UK), 6, 32, 40, 93, 97, 99, 100, 120–123, 126, 197, 277, 299, 330, 332, 337, 359, 388, 397, 421, 422, 424, 442, 448, 457, 516–518, 537–549, 634, 660
United States of America
  Bush Administration, 88, 155, 480
  CAFE standards, 477
  Clean Air Act, 477, 479, 480, 501, 504
  Clean Energy and Security Act of 2009, 476
  Congress, 53, 451, 486, 487, 518
  Environmental Protection Agency (EPA), 476–480, 605
  Federal Government, 45, 482, 487
  Judicial Branch, 476, 479–480
  Massachusetts *v.* EPA decision, 477, 479
  Obama Administration, 120, 435, 480
  Regional Greenhouse Gas Initiative, 51, 483–485
  renewable energy certificates (RECs), 482–483, 613
  U.S. Constitution, 475, 480, 483, 654
  U.S. Supreme Court, 476, 479
  Western Climate Initiative (WCI), 51, 483, 485–486, 507
UN Permanent Forum on Indigenous Issues, 415
UN-REDD Programme, 396, 397, 402, 405, 406, 412–416
Uruguay, 232, 239, 241, 341, 422, 423, 425, 440, 457, 459, 460, 654, 658, 664, 665, 668, 671

V

Vienna Convention on the Law of Treaties (VCLT)
  Article 31.3(c), 339–342, 356, 442
  relevant rules of international law, 340, 342, 442, 443, 451
Voluntary
  action(s), 13, 425, 584, 585, 646
  agreements, 27, 425, 430
  Climate, Community and Biodiversity Standard, 58
  declaration(s), 21
  intiatives, 3
  Japanese Voluntary Emissions Trading Scheme (JVETS), 6, 584, 586, 587, 592
  Keidanren's Voluntary Action Plan, 6, 584-586
  Pledges, 13
  Standards, 36, 149, 429
  Target, 586, 591
  Voluntary Carbon Standard (VCS), 58

W

Western Climate Initiative (WCI), 50–51, 483, 485–486, 507
Wicked problem, 330, 331
World Bank
  Forest Carbon Partnership Facility (FCPF), 396, 397, 401, 405, 406, 413, 415, 416
  Operational Policy 4.10 on indigenous peoples, 415
World Business Council for Sustainable Development, 58
World Heritage Convention, 367, 380
World People's Conference on Climate Change and the Rights of Mother Earth, 421
World Trade Organization (WTO)
  Agreement on Subsidies and Countervailing Measures, 22, 425
  agreements, 335, 337, 342, 344, 432, 441, 444
  appellate body, 340, 342, 423, 432, 440, 443, 445
  applicable law, 441, 442
  biotech, 342, 442, 443, 445
  Canada-Certain Measures Affecting the Renewable Energy Generation Sector, 437
  China-Measures concerning Wind Farm Equipment, 437, 438
  dispute settlement body, 336, 341, 342
  dispute settlement system, 5, 44, 337, 340, 420, 423, 424, 426, 430, 437, 439–445
  Dispute Settlement Understanding (DSU), 435, 440, 441
  factual evidence, 441, 443, 445
  General Agreement on Tariffs and Trade, 341, 423, 431
  General Agreement on Trade in Services (GATS), 425, 430, 437
  general council, 352

law, 339, 346, 420, 423, 425–433, 435, 436, 439–443, 445, 533
non-discrimination requirement, 432
processes and production methods, 431, 433, 435–436
secretariat, 332, 351, 355
shrimps, 340, 435–436, 443, 445
TBT Agreement, 425, 429, 432, 444
TBT Committee, 430